Physics of Cancer, Volume 6 (Second Edition)

Cellular mechanisms to foster or fight cancer

Online at: https://doi.org/10.1088/978-0-7503-4007-6

About the Series

The Biophysical Society and IOP Publishing have forged a new publishing partnership in biophysics, bringing the world-leading expertise and domain knowledge of the Biophysical Society into the rapidly developing IOP ebooks program.

The program publishes textbooks, monographs, reviews, and handbooks covering all areas of biophysics research, applications, education, methods, computational tools, and techniques. Subjects of the collection will include: bioenergetics; bioengineering; biological fluorescence; biopolymers *in vivo*; cryo-electron microscopy; exocytosis and endocytosis; intrinsically disordered proteins; mechanobiology; membrane biophysics; membrane structure and assembly; molecular biophysics; motility and cytoskeleton; nanoscale biophysics; and permeation and transport.

A list of recently published and forthcoming titles in this series can be found here: https://iopscience.iop.org/bookListInfo/iop-series-in-biophysical-society.

Physics of Cancer, Volume 6 (Second Edition)

Cellular mechanisms to foster or fight cancer

Claudia Tanja Mierke
Department of Biological Physics, Peter Debye Institute for Soft Matter Physics, Faculty of Physics and Earth System Sciences, Leipzig University, Leipzig, Germany

IOP Publishing, Bristol, UK

ISBN 978-0-7503-4007-6 (ebook)
ISBN 978-0-7503-4005-2 (print)
ISBN 978-0-7503-4008-3 (myPrint)
ISBN 978-0-7503-4006-9 (mobi)

DOI 10.1088/978-0-7503-4007-6

Version: 20250501

IOP ebooks

British Library Cataloguing-in-Publication Data: A catalogue record for this book is available from the British Library.

Published by IOP Publishing, wholly owned by The Institute of Physics, London

IOP Publishing, No.2 The Distillery, Glassfields, Avon Street, Bristol, BS2 0GR, UK

US Office: IOP Publishing, Inc., 190 North Independence Mall West, Suite 601, Philadelphia, PA 19106, USA

The book would never come into existence without the tireless and extraordinary support and everlasting encouragements by Thomas M L Mierke.

Contents

Preface

Dear Reader,

In 2013, on the fringes of a small, unimportant congress in Leipzig that I co-organized, I was approached by an employee of IOP and asked if I would be interested in writing a scientific textbook on my research area, *Physics of Cancer*, in which I had been working for over a decade. I agreed, and in 2015 my first book '*Physics of Cancer*' was published by IOP as the first book in a new book series in collaboration with the American Biophysical Society (BPS). Since then, five more books by me on this topic have followed. A decade has passed and for now my last book in this series *Physics of Cancer, Volume 6* is being published.

This book focuses on the topics of extracellular vesicular communication of cancer cells with their environment, such as matrix networks and neighboring cells like immune and stromal cells, the role of autophagy in cancer, the dual function of mast cells in solid tumors in relation to immune regulation and mechanobiology, the effect of the mechanical characteristics of cancer cells and tumors on cancer progression and their possible potential for mechanical reprogramming of cancer cells, and the combination of mechanical and molecular analyses for dynamic and multiple analyses.

It has always been a great pleasure for me to write another book in my field of expertise, especially since I have always received great recognition for the *Physics of Cancer* books worldwide outside of Leipzig and my books can be found in the libraries of many important universities. I am very grateful to IOP and the BPS for giving me the opportunity to publish these books for over a decade and I hope that this book will also attract a lot of attention and a large readership.

Werdau, January 2025
Yours sincerely,
Claudia Tanja Mierke

Acknowledgements

I thank Thomas M L Mierke for inspiring me, for the ongoing discussions, for formatting, reviewing and proofreading all chapters and for helping me to keep the schedule for writing this book and to decide on the topics.

Author biography

Claudia Tanja Mierke

Claudia Tanja Mierke studied biology at the TU Braunschweig in Germany and received her doctoral degree from the Hannover Medical School in March 2001; her dissertation focused on human endothelial and mast cell–cell interactions. As a postdoctoral fellow, she conducted research at various institutes of the University of Erlangen-Nuremberg in the fields of cancer research, endothelial cell inflammation and molecular cancer research, and biophysical cancer research. In 2012, she habilitated in biophysics.

Since 2010, she has been the female professor of Physics at the Leipzig University.

Based on education in biology and molecular oncology, the main emphasis of her research was biophysical cancer research with a focus on cell mechanics, cell motility and transmigration through endothelial vascular linings, where she had intensively considered the biological principles.

In teaching, she gives lectures and seminars on molecular and cell biology, methods of biophysics, soft matter physics and physics of cancer.

She has published numerous review articles and book chapters, mainly on the topic of physics-driven cancer research, in addition to the experimental findings of her research group.

Along the way, she has authored the first five volumes of this *Physics of Cancer* book series and published in 2020 the book *Cellular Mechanics and Biophysics*.

She is also a Chief/Senior Editor and reviewer for several scientific journals. Most recently, she has served as a reviewer and panel member for research granting societies in several European and Asian countries.

Introduction

Cancer cells can dynamically alter their own mechanical properties through specific structural and mechanical modification mechanisms. In addition, these cells can change the structural and mechanical properties of their environment by altering the composition of extracellular matrix (ECM) components and their location, breaking down the ECM with enzymes, or cross-linking the structures with molecules. All of this can influence the degree of homogeneity of the matrix, the structural architecture and the mechanics of the matrix. Cancer cells use various mechanisms such as exosome absorption, endosome formation and the process of autophagy. Along with cancer cells, immune cells with exosomes can also alter the tumor environment. Cancer stem cells and their elimination from primary tumors can also influence the characteristics of the tumor environment. The functions of endosomes and lysosomes in cancer are discussed. The dual role of mast cells (MCs) in cancer is presented and discussed with a focus on immune regulation and mechanobiological effects. The paradoxical phenomenon of autophagy is discussed in the context of carcinogenesis and cancer development. The effects of passive and dynamic mechanical characteristics of cancer cells and cancerous diseases are discussed in terms of cancer development and malignant progression, such as metastases. The multi-omics analysis known as ELASTomics, which combines mechanical and molecular analysis of cancer cells, is presented and the hypothesis of mechanical adaptability is discussed. Finally, shortcomings in the scientific field of cancer combating are highlighted, with emphasis on cell and matrix mechanics.

IOP Publishing

Physics of Cancer, Volume 6 (Second Edition)

Cellular mechanisms to foster or fight cancer

Claudia Tanja Mierke

Chapter 1

Matrix-modification by cells: exosomes and cancer

1.1 Summary

Cancer research focuses currently on the remodeling of the surrounding ECM of the primary solid tumor, with emphasis on a new type of structural unit referred to as extracellular vesicles (EVs) in cancer. Originally, these EVs have been considered waste products or products of cell damage rather than elements that perform biological functions. This view has changed dramatically. By releasing exosomes, cancer cells can even change their environment, like the ECM. Apart from cancer cells, even other cell types, such as endothelial cells, can secrete exosomes that are also capable of altering the tumor environment. The liberated EVs by cancer cells and other neighboring cells and their enclosed cargo can modify not only the structure of the ECM, but also the mechanical characteristics of the ECM scaffold. This represents another relevant way to modify the environment of cancer cells. It can therefore be hypothesized that exosomes can act in five ways. Firstly, they can support the malignant progression of cancer by altering neighboring cells and/or the ECM environment. Secondly, they can be used as unique or general biomarkers for cancer progression of specific types of cancers. Thirdly, they can act as mechanosensors and facilitate mechanotransduction due to altered mechanical properties of their environment, while at the same time containing other mechanosensors in the membrane region whose mechanosensory activity can support and even enhance the mechanosensing process in the receiving cells. Fourthly, EVs can serve as a diagnostic tool as they can protect their contents from the nucleases and proteases present in biological fluids and reflect the original cell phenotype of the cell of origin. Fifthly, the EVs can carry cargos in their membrane, such as cell-matrix adhesion receptors, such as integrins and carry in their lumen, such as several non-coding RNAs (ncRNAs), cytokines, chemokines and proteins. Moreover, EVs are present in almost every biological fluid, so they can be isolated easily and in a non-invasive

doi:10.1088/978-0-7503-4007-6ch1

manner. A comprehensive overview of the discovery of exosomes, their assembly and the path of their biogenesis is provided in this chapter, which are relevant in gaining an in-depth knowledge of these mechanisms. The chapter additionally considers the function of exosomes in cancer and highlights their role in the malignant progression of cancers from a biophysical viewpoint. The key focus of the chapter will be on the specific contribution of exosomes to cancer, with particular emphasis on exosomes in remodeling the tumor microenvironment and how they influence various cancer hallmarks, including proliferation, adhesion, migration, intercellular communication and responsiveness to cancer therapy. Thereafter, exosomal cargos like ncRNAs including long intergenic non-coding RNA (linkRNA), circRNA, and miRNAs, are introduced as well as how they can influence cancer cell initiation and advancement. Subsequently, the most crucial molecular signaling routes of exosomes that govern cancer progression are unveiled and discussed. It can be hypothesized that the epigenetic phenotype of cancer cells and neighboring cells is altered permanently, which can dramatically impact of cancer treatments. Ultimately, a short overview of tumor-derived exosomes and the therapeutic uses of exosomes in the clinic that are relevant for the treatment of cancer patients is discussed. A comprehensive new hypothesis of how these exosomes cause both direct and indirect epigenetic alterations and how this knowledge can finally be utilized to improve cancer research and consequently also personalized medicine.

1.2 Introduction to the role of exosomes in cancer

EVs comprise small, non-replicative, lipid bilayer-surrounded entities that are liberated in nearly every cell type in every organism (figure 1.1) (Zaborowski *et al* 2015, Doyle and Wang 2019, Patel *et al* 2023, Xu *et al* 2023). EVs incorporate a variety of bioactive cargos, ranging from lipids, metabolites, nucleic acids to proteins, and can be classified into different categories according to their biogenesis (Shurtleff *et al* 2018, Jeppesen *et al* 2019). In general, the size of exosomes falls between 30 and 150 nm (figure 1.1), and they are generated by the endosomal route as they mature from early endosomes toward late endosomes/multivesicular bodies (MVBs), during which inward budding of the MVB membrane generates intraluminal vesicles (ILVs) (Doyle and Wang 2019, Gurung *et al* 2021). In the next step, ILVs are liberated as exosomes during the process of fusing MVBs with the cell's plasma membrane (Hessvik and Llorente 2018, Xu *et al* 2023). Microvesicles with a diameter of about 50–1000 nm are formed when the plasma membrane is knotted outwards, which means they display outwards budding. Other EV subtypes comprise apoptotic bodies, which serve an integral part in apoptosis (figure 1.1) (Gregory and Rimmer 2023), and oncosomes, which are sequestered from cancer cells to support the growth of tumors and the evolution of the tumor microenvironment (TME) (Da Costa *et al* 2021, Kumar *et al* 2024).

In recent years, emphasis in cancer research has been placed on a new structural element of cells, referred to as EVs (Ciardiello *et al* 2016, Zhang *et al* 2021c). EVs

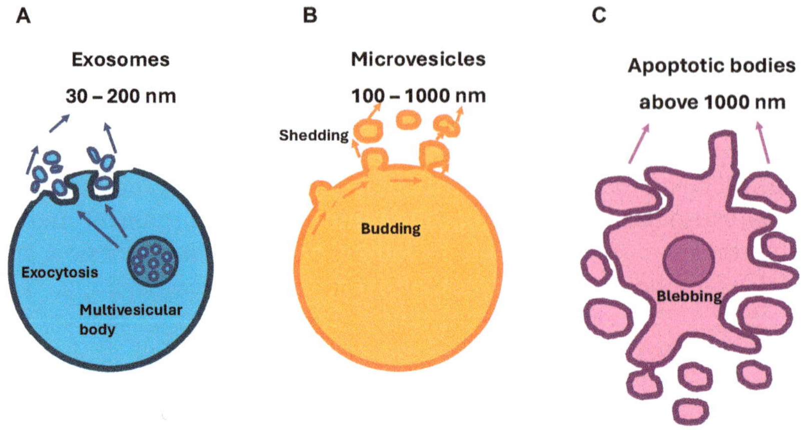

Figure 1.1. Classification of EVs. There exist three major classes of EVs. (A) Exosomes are produced via the endocytic route and liberated through exocytosis. Their shape is spherical, and they have a typical diameter of 30–200 nm. (B) Microvesicles (MVs) are liberated via budding out from the cell membrane. Their form is non-regular and they possess a diameter of 100–1000 nm. (C) Apoptotic bodies are secreted via blebbing through cells that perform apoptosis and are above 1000 nm in diameter.

can change the fate of surrounding cells, which is especially relevant in cancer, as cancer cells exploit EVs to alter the phenotype of neighboring cells in the TME, thereby enhancing growth of the primary tumor, malignant progression of cancer, such as metastasis and resistance to therapy. EVs offer a diagnostic value as they protect their cargo from nucleases and proteases in biological liquids and can mirror the original cell phenotype. The key point is that EVs can be detected in virtually all biological liquids, enabling easy and non-invasive harvesting (Colombo *et al* 2014).

EVs arise in the membrane of cells and are referred to as micro- or nanovesicles. These types of structures can be released by all prokaryotic and eukaryotic cells in a way that is evolutionarily conserved (Kim *et al* 2015, Liebana-Jordan *et al* 2021, Mobarak *et al* 2024). The EVs enable the exchange between hosts and viruses (Zhu *et al* 2020), fungi and plants (Rutter *et al* 2022) or *E. coli* bacteria and the gastrointestinal tract in diseases (Carrière *et al* 2016). Originally, EVs have been recognized as cellular debris or as formations resulting from cellular injury (Yáñez-Mó *et al* 2015). Additional studies on EVs, nonetheless, have demonstrated that they fulfill vital biological purposes and are essential cellular elements (Al-Jipouri *et al* 2023, Liu and Wang 2023). There are various kinds of EVs that are classified due to their size, origin, and cellular localization (Doyle and Wang 2019, Petroni *et al* 2023, Lee *et al* 2024). Among the most well-known EVs are apoptotic bodies, exosomes, microparticles, prominosomes, proteasomes, shedding vesicles, and tolerosomes (Doyle and Wang 2019, Battistelli and Falcieri 2020). Biophysical, biochemical and functional heterogeneity of EVs has been revealed in a way never seen before since their first characterization (figure 1.2) (Théry *et al* 2018, Verweij *et al* 2021, Van Niel *et al* 2022).

There are two basic kinds of EVs according to their biogenesis (Buzas 2023). Exosomes originate from endosomes and are liberated when the boundary membrane of MVBs or amphisomes merges with the plasma membrane (Jeppesen *et al* 2019, Mathieu *et al* 2021, Van Niel *et al* 2022). Newer evidence indicates that additional endomembranes like the endoplasmic reticulum (Barman *et al* 2022) and the nuclear envelope (Arya *et al* 2022) are implicated in exosome biogenesis. The alternative basic pathway of evolutionary EV biogenesis is the liberation of plasma membrane-derived EVs that are commonly referred to as ectosomes. Certain

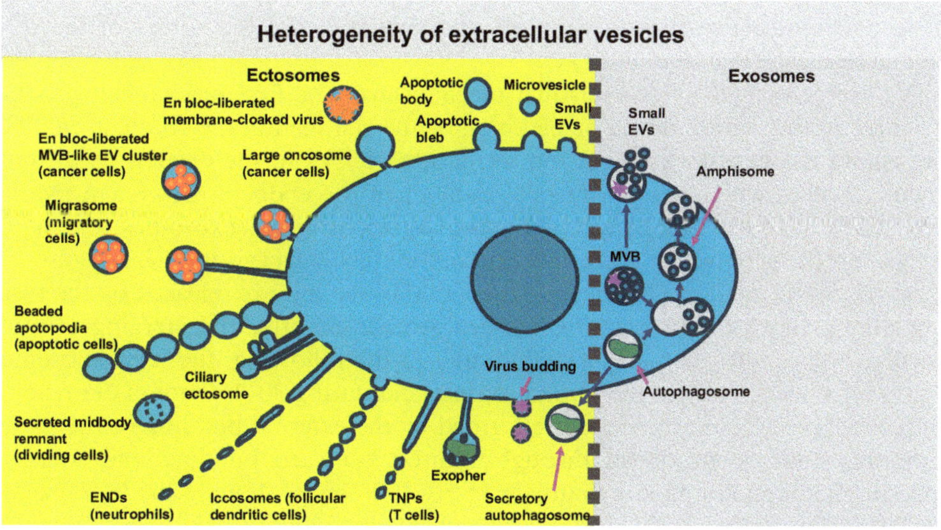

Figure 1.2. EVs are a heterogenous group of structures surrounded by a phospholipid membrane. The two major types of EVs are characterized based on their distinct biogenesis and are referred to as exosomes and ectosomes. Exosomes represent small EVs that originate from endosomes and are liberated through exocytosis of MVBs and amphisomes. Amphisomes are created by the merging of MVBs with autophagosomes. In contrast, ectosomes are produced from the plasma membrane through budding or blebbing. It is important to note that ectosomes contain endosome cargo elements. Ectosomes harbor small-sized EVs like small ectosomes and arrestin domain-containing protein1-faciltitated microvesicles, medium-sized microvesicles and larger-sized apoptotic bodies. Viruses can bud from the plasma membrane or be liberated by MVBs. En bloc released virus clusters are a new type of large EV that resemble en bloc released MVB-like EV clusters generated by cancer cells. Oncosomes represent large EVs generated by cancer cells. Long protrusions of motile cells cause the formation of migrasomes, which break away from the tip of long retraction fibers of migrating cells. Secreted midbody remnants are liberated due to the end of cytokinesis of dividing cells. A specific type of ectosomes are ciliary ectosomes that shed the plasma membrane of cilia. Beaded apoptopodia secrete apoptotic vesicles when apoptosis occurs. Neutrophils crawl on the vascular endothelium and deposit elongated neutrophil-derived structures (ENDs). Follicular dendritic cells form long thread-like processes from which iccosomes are formed by a beading process. In the immune synapse, T cell microvilli undergo fragmentation like the beading mechanism to liberate EVs referred to as T cell microvilli particles (TMPs). Exophers comprise large vesicles located at the tip of a stalk carrying injured organelles and proteins aggregates. Secretory autophagosomes are liberated by cells.

molecular biomarkers for the various biogenetic pathways are, however, not yet identified and functional concepts have been proposed to discriminate EV types on the basis of their biophysical or biochemical characteristics (Théry *et al* 2018). The EVs that occur in highest abundance in biological liquids are small EVs that range from 50–150 nm in estimated diameter, large oncosomes and en bloc released MVB-like small EV clusters (table 1.1) (Valcz *et al* 2019) constitute the lowest abundant community of EVs. The diversity of EVs is a result of the various kinds and functional conditions of the liberating cells and the diverse biogenetic routes. EVs comprise also vesicles that are produced through various cell death mechanisms like apoptosis, necroptosis or pyroptosis. There is also growing appreciation that EV biogenesis may be related to viral escape (Kerviel *et al* 2021), secretory autophagy, the cellular senescence-elicited secretory characteristic and the nature of the DNA damage reaction (Wu *et al* 2021).

EVs are generated by two distinct mechanisms. The first mechanism produces EVs directly through budding or blebbing of the cell membrane (Fang *et al* 2022). The second mechanism involves the formation of EVs in the exocytosis of multi-vesicular bodies which are an integral element of the endocytosis system (Moulin *et al* 2023). EVs are categorized into two principal types: exosomes and micro-vesicles, on the grounds of their differing nature of biogenesis and size (figure 1.1). Exosomes are formed as part of the endosomal route and are released by the fusion of MVBs, a type of endosome, which are exocytosed after the merging of multi-vesicular bodies into the plasma membrane. In the following, the term exosome is used to refer specifically to membrane vesicles measuring 30–100 nm in diameter. In complete opposite, microvesicles are formed by the external budding of the plasma membrane and are produced through a direct outward budding and squeezing process. Microvesicles have a diameter of 50–1000 nm (Liu *et al* 2017a, Bebelman *et al* 2018, Van Niel *et al* 2018). EVs are implicated in biological roles in cells and are involved in pathological processes. They can exchange different molecules among cells and are a means of intercellular communication (Meldolesi, 2018, Kumar *et al* 2024). Therefore, the emphasis is placed on the specific role of exosomes in diseases with the focus on cancer (Keller *et al* 2006, Dai *et al* 2020, Zhou *et al* 2021, Wandrey *et al* 2023, Lyu *et al* 2024). EVs perform two contrasting functions: they can be used as waste collection vehicles or as delivery vehicles. Finally, the EVs are in the focus of therapeutic approaches in cancer (Kumar *et al* 2024).

1.3 Biogenesis of EVs

EVs are classified roughly into three categories depending on their biogenesis track (figure 1.1): exosomes, ectosomes, and apoptotic bodies (Busatto *et al* 2021). Mammalian EVs are largely heterogenous, as they possess various proteins in their lipid membranes, multiple proteins, different types of RNAs, and DNAs (Kowal *et al* 2016). Exosomes represent intraluminal vesicles, usually with a diameter of fewer than 150 nm, which result from the merging of MVBs and the plasma membrane. In contrast, ectosomes or microvesicles with a size of about 100–1000

Table 1.1. Definition and functions of several kinds of EVs.

	Definition	Function	References
Amphisomes	They are intermediate/hybrid organelles formed after the outer autophagosome membrane has fused with the MVBs.	The generation of an amphisome is a required step during a sequential maturation process of autophagosomes prior to their ultimate merging with lysosomes for breakdown of cargos. Crosstalk between autophagy and endolysosomal routes. CA-IKKβ promotes the generation and trafficking of amphisomes and thus controls the homeostasis of cancer cells, which may shed light on a particular survival mechanism in cancer cells under stress.	Ganesan and Cai (2021), Peng *et al* (2021)
Apoptotic bodies (ApoBDs)	They are a special type of EV with a diameter of 500 nm to 2 μm and are the leftovers of cells that have suffered apoptosis, or programmed cell death.	Crosstalk between cells. Involved in removing cellular contents, such as waste. Carry biomolecules comprising microRNA and DNA to govern exchange between cells.	Zernecke *et al* (2009), Battistelli and Falcieri (2020)
Ectosomes	They are an extracellular vesicle with a diameter of 100–500 nm, which is formed through budding of the plasma membrane to the outside and subsequent constriction and liberation into the extracellular environment.	Broaden the boundaries of the cells beyond the plasma membrane. Create communication systems through which specific characteristics and pieces of data can be exchanged between the cells.	Cocucci and Meldolesi (2015), Surman *et al* (2017), Meldolesi (2018)

	Display biomarkers, such TyA and C1q.	Can have functional involvement in oncogenic transformation, progression of the cancer, invasion, metastasis, promotion of angiogenesis, immune evasion and resistance to medication, thereby promoting disease advancement.	
En bloc released MVB-like small EV clusters	These are MVB-like aggregates of ALIX/CD63-positive EV clusters with a diameter of 0.6 nm to 2 μm, which are liberated en bloc from migrating cancer cells. Released as amphisomes (or amphiectosomes).	Contribute to cancer malignant progression. Bleeding into the extracellular environment via a 'tear-sack mechanism'.	Valcz *et al* (2019)
Endosomes	They are intracellular membrane-bound organelles that are formed by endocytosis and are categorized as early or late in their generation according to the time they remain in the cytoplasm after their emergence. They are compartments lying along the endocytic pathway which starts at the plasma membrane and is finalized at the lytic organelle, such as lysosome or vacuole. Endosome-like structures have been identified at the presynaptic compartment.	Regulate numerous critical physiological processes, comprising nutrient absorption, hormone-mediated signaling, immune surveillance and presentation of antigens. Function in the abscission process in endocytosis, possibly by delivering new membrane and by signaling. Provide multipurpose platform for signaling.	Barr and Gruneberg (2007), Gould and Lippincott-Schwartz (2009), Jähne *et al* (2015)

(*Continued*)

Table 1.1. (*Continued*)

	Definition	Function	References
Exomers	They are extracellular vesicles with a diameter of 40–120 nm. They are liberated when multivesicular bodies and the plasma membrane merge, eventually liberating smaller vesicles. They develop as intraluminal vesicles during the coalescence of the multivesicular bodies (MVBs). Display the biomarkers CD63 and CD61	Provide new mechanism of cell–cell crosstalk. Transfer of mRNA and miRNA via exosomes. Communication between cells to both send and receive messages.	Théry *et al* (2002), Valadi *et al* (2007), Simons and Raposo (2009), Guescini *et al* (2010), Cocucci and Meldolesi (2015)
Exophers	They are extruded by neurons and are large membrane-surrounded vesicles with a diameter of about 4 μm that can contain protein aggregates and organelles.	The production of exophers provides cells with an alternative route to eliminate aggregated/toxic proteins or dysfunctional organelles, such as mitochondria in conditions of proteostasis. Chaperone expression, autophagy, proteasome and mitochondrial quality impairments increase exopher generation. Exophers are part of a conserved mechanism that represents a fundamental arm of neuronal proteostasis and mitochondrial quality regulation Functions as transporters for muscle-generated yolk proteins in aiding offspring development.	Melentijevic *et al* (2017), Turek *et al* (2021)

Exosomes	They are membrane vesicles because they emerge directly from the plasma membrane, have a size of 100–500 nm and contain cytoplasmic material. Their biogenesis takes place by directly breaking out and pinching the plasma membrane to the outside, whereby the resulting microvesicle is liberated into the extracellular cavity.	Comprise different amounts of adhesion molecules, like integrins, which, when liberated by stem cells at different steps of the cell cycle, could influence the transport and capture of vesicles. With distinct, localized alterations in the protein and lipid constituents of the plasma membrane they evoke changes in the curvature and rigidity of the membrane. Transport proteins, nucleic acids and bioactive lipids from the cell of origin. After liberation in the extracellular environment and entry into the bloodstream, these vesicles can pass their cargo to neighboring or remote cells, causing phenotypic and functional alterations that are pertinent in various physio-pathological states. Facilitators of cell–cell communication involved in the preservation of function.	Kalra *et al* (2012), El Andaloussi *et al* (2013), Zaborowski *et al* (2015), Tricarico *et al* (2017), Nieuwland *et al* (2018)
ILVs	They are vesicles of a typical diameter of 50 nm produced by budding of the boundary membrane of late endosomes that develop into MVBs. Upon exocytosis of MVBs, ILVs are liberated as exosomes.	Perform a key function in the mitigation of growth factor receptor signaling pathways.	Wenzel *et al* (2024)

(*Continued*)

Table 1.1. (*Continued*)

	Definition	Function	References
Large oncosomes	These are gigantic EVs with a diameter of 1 μm nm to 10 μm, which emerge from large protrusions of the plasma membrane of cells. They are membrane-bound microvesicles that are released by cancer cells and transmit oncogenic messages and protein complexes beyond cell limits.	Promote the amoeboid migration mode of human metastatic prostate cancer cells. The generation of these EVs depended on cell transformation, involving activation of the AKT1 and EGFR signaling cascades, and entailed aberrant accumulation of molecular cargos, comprising proteins and nucleic acids. This process mirrored both oncogenic transformation and the shift to a fast-migrating and severely metastatic amoeboid phenotype of cancer cells.	Di Vizio *et al* (2009), Minciacchi *et al* (2015)
MVs	They are characterized as intact, submicroscopic, phospholipid-rich vesicles with a diameter of 200 nm to over 10 μm, which are liberated from the plasma membrane of a multitude of cells on activation.	They harbor a wide range of molecular cargo, reflecting both the cell type from which the MV was liberated and the intracellular trafficking routes that convey the cargo toward the cell surface. The identified cargos comprise various types of proteins, ranging from integrin receptors, active proteases, multiple small GTPases to multidrug resistance proteins and miRNA processing modules.	Clancy *et al* (2021)
Migrasomes	They are new types of cellular organelles that arise during cell migration and their biogenesis relies on the migration mechanism. They are formed in a broad range	They play a role in communication between cells, in preserving homeostasis, in the development of embryos and in numerous diseases.	Ma *et al* (2015), Zhang *et al* (2020, 2023)

	of cells like immune cells, metastatic cancer cells, other specialized functional cells like podocytes and cells in developing organisms. They are a newly identified extracellular vesicle whose formation relies on cell migration. The formation of migrasomes is tightly linked to TSPANs and integrins and is controlled by different elements.	They serve critical purposes in several fields, ranging from cell–cell communication, maintenance of homeostasis, embryonic development, and the appearance, progression and diagnosis of various diseases. Cytosolic constituents can be carried and liberated from the cell by migrasomes, with the migration-dependent liberation mechanism referred to as migracytosis.	
MVBs	MVBs are spherical organelles (late endosomes with a diameter of about 250 nm) characterized by a single outer confining membrane that surrounds a variable amount of charge-containing ILVs.	MVBs act to sort proteins between early and late endocytic compartments and can redistribute proteins to the plasma membrane or break down their components by merging with lysosomes. MVB-like organelles are generated in melanocytes throughout melanosome biogenesis, and von Willebrand factor is accumulated in ILVs secreted by platelets and endothelial cells as exosomes.	Maurya *et al* (2022)
Multivesicular endosomes (MVEs)	They are often synonymously used with MVBs. They generate EVs by budding intraluminally and fusing with the plasma membrane to liberate EVs.	Production of EVs. The recycling of plasma membrane proteins usually takes place in early and recycling endosomes; their destructive sorting is carried out in intermediate/late endosomes, also referred to as MVBs or multivesicular endosomes.	Gruenberg (2020)

(Continued)

Table 1.1. (*Continued*)

	Definition	Function	References
Oncosomes	They are EVs with a diameter of 10 μm and facilitate the extracellular exit of structurally and functionally abnormal, mutant and potentially transforming macromolecules, which are referred to as oncogenes. This characteristic differentiates these oncogene-containing EVs profoundly and in qualitative terms from any other EVs that exist. The word 'onco-' in the root part indicates the oncogenic molecular load of cancer-derived EVs.	Contain oncogenic cargo. Function in remodeling innate immunity at metastatic cancer sites and are produced during generation of invadopodia and in the step of cancer cell extravasation of the metastatic cascade.	Leong *et al* (2014), Headley *et al* (2016), Meehan *et al* (2016)
Supermeres	They compromise a type of small non-membranous extracellular nanoparticle of a diameter below 50 nm, termed exomere, The protein and RNA content of supermeres is distinct from small EVs and exomers. Supermeres are strongly loaded with cargo involved in several cancers comprising glycolytic enzymes, such as (ENO1), TGFBI, miR-1246, MET, GPC1 and AGO2, Alzheimer's disease, such as APP, and cardiovascular diseases, like ACE2, ACE and PCSK9.	The bulk of extracellular RNA is linked to supermeres rather than small extracellular vesicles and exomers. Cancer-derived supermeres augment lactate release, transmit cetuximab resistance, and reduce hepatic lipids and glycogen *in vivo*.	Zhang *et al* (2021b)

nm are formed through budding and detachment from the plasma membrane (Li 2021, Lim *et al* 2021, Han *et al* 2022). In a similar way, exosomes and ectosomes facilitate communication between cells. The tetraspanin protein family has been recognized as an important key player in the vesicle biogenesis, although their specific functions in the formation of exosomes and ectosomes are distinct (Perez-Hernandez *et al* 2013). Apoptotic bodies, in contrast, are liberated in the shape of cell vesicles when cells undergo programmed cell death, in the form of apoptosis (Nagata 2018). The question arises as to how exosomes can be distinguished from ectosomes. Although they play a major function in intercellular interactions, the various types of EVs and their mechanisms of secretion are not completely understood: it remains elusive how and to what extent EVs are produced as intraluminal vesicles from endocytic entities, such as exosomes, or generated at the plasma membrane, such as ectosomes. Tracing the intracellular transportation of the EV markers CD9 and CD63 from the endoplasmic reticulum to their home compartments, namely the plasma membrane and late endosomes, is feasible. Temporary co-localization occurs at both locations before they ultimately diverge. CD9 and a CD63 mutant, which is stabilized at the plasma membrane, are secreted in EVs in larger quantities compared to CD63. In HeLa cells, ectosomes appear to be predominant over exosomes. A comparative proteomic study and a differential reaction to endosomal pH neutralization has revealed a few surface proteins that are probably typical for either exosomes, such as LAMP1, PDCD6IP (Alix), TSG101 and syntenin-1 (Kugeratski *et al* 2021), or ectosomes, such as brainstem glioma (BSG) and SLC3A2. This research paves the way for the molecular and functional differentiation of exosomes and small ectosomes in each cell type (Mathieu *et al* 2021).

1.3.1 Exosomes

Exosomes were first identified in 1967 when it was found that red blood cells can liberate vesicles and that these extracellular vesicles exhibit an anticoagulant effect (Wolf 1967). In later investigations, these vesicles are named exosomes (Johnstone *et al* 1987). The biology of exosomes is crucial for a large spectrum of human diseases like cancer. At present, some remarkable research results have been obtained on the mechanism of exosome generation, although numerous mysterious events persist. The generation of exosomes proceeds with the following steps. The plasma membrane first endocytoses extracellular compounds to create early endosomes (Van Niel *et al* 2018). Exosomal membrane loads undergo a process of internalization from the plasma membrane or Golgi apparatus into the endosome and are afterwards categorized as ILVs in the endosomal maturation phase (Klumperman and Raposo 2014). Exosomes are formed by the endosomal membrane budding inwards, which results in the creation of MVBs (Zhang and Yu 2019) (figure 1.1). These MVBs merge either with lysosomes for breakdown or coalesce with the plasma membrane to liberate exosomes (Han *et al* 2022). Exosomes possess a characteristic phospholipid bilayer membrane architecture and exhibit minimal immunogenicity and can penetrate the blood–brain barrier and the barrier of the placenta with high efficiency. By transferring biological messages like proteins and genetic material from the hosting cells to the receiving cells, altering their physiological and pathological condition or triggering the onward progression of signaling cascades, exosomes hold tremendous power as natural carriers of vaccines (Van Niel *et al* 2018, McAndrews and Kalluri 2019). Exosomes serve as the principal vehicles of intercellular communication and therefore have a key function in the intricate signaling system connecting cancer cells and immune cells. For one thing, exosomes assist the immune cells to talk to one another and therefore trigger the activation of subsequent effector cells. Alternatively, they provide the immune system with cancer cell-specific antigens and thus prevent the cancer cells from evading the immune system (Kahlert and Kalluri 2013, Kalluri 2016).

Early endosomal membranes continue to fold to build MVEs harboring ILVs (Catoni *et al* 2021), whose cargos incorporate in a selective or passive fashion a multitude of intracellular elements including RNAs, such as mRNAs, miRNAs and other ncRNAs, DNAs and lipids into vesicles (Colombo *et al* 2014). The accumulation of cargos and the trafficking of MVEs within ILVs can be accomplished through endosomal sorting complexes required for transport (ESCRT) or by ESCRT-unrelated pathways (Hurley 2008). The ESCRT-dependent mechanism encompasses four complexes. ESCRT-0 consists of two subunits and enlists exosomal cargos by using ubiquitination. ESCRT-I becomes enlisted by ESCRT-0 and is subsequently carried to the membrane. ESCRT-I interferes with ESCRT-II to generate buds (Henne *et al* 2011, Vietri *et al* 2020). Finally, ESCRT-III dissociates the buds to generate ILVs and facilitate the detachment of MVEs, which is referred to as shedding. The polymerized ESCRT-III compound can be released from the MVE membrane using the energy supplied via the sorting protein Vps4 (Larios *et al* 2020). Several investigations, nevertheless, have identified some ESCRT-unrelated processes for sorting exosomal loads, suggesting that the underlying mechanisms are broader and more intricate. Exosome biogenesis consists of the orchestration of several specific routes and can be divided into two main paths, such as the ESCRT-induced route and the ESCRT-independent route (figure 1.3). The ESCRT-elicited route entails the generation of ILVs through the sequential effects of ESCRT-0, -I, -II and -III complexes, comprising ALIX, TSG101 and VPS4 (Baietti *et al* 2012). In contrast, the non-ESCRT pathway entails the generation of ILVs through tetraspanin-enriched microdomains (TEMs) and lipid rafts.

Thereafter, MVEs may merge with lysosomes for breakdown (Fader and Colombo 2009) or with plasma membranes to liberate ILVs, termed exosomes, into the extracellular compartment to fulfill their physiological roles (Bebelman *et al* 2018). This process is controlled via Rabs and different Ras-GTP enzymes, soluble N-ethylmaleimide-sensitive factor attachment protein receptors (SNARE proteins) and its regulatory factors (Pfeffer 2007), as well as intracellular Ca^{2+} (Tucker and Chapman 2002). Rab GTPases constitute the broadest class of small GTPases participating in the coordination of MVE ripening and its homing toward the plasma membrane (Stenmark 2009). For instance, Rab27a and Rab27b have been

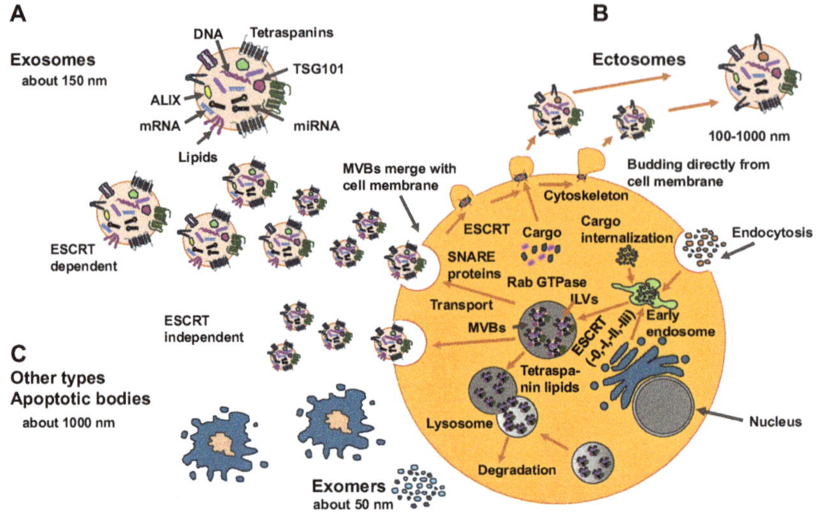

Figure 1.3. Biogenesis of EVs. EVs can be classified into (A) exosomes, (B) ectosomes and (C) other types like apoptotic bodies and exomeres. Exosome biogenesis takes place via two routes, namely an ESCRT-dependent and an ESCRT pathway. Multivesicular bodies = MVBs, Intraluminal vesicles = ILVs.

found to be involved in the tethering of MVEs to plasma membranes (Ostrowski *et al* 2010), and Rab27a can be blocked using KIBRA to regulate the liberation of exosomes (Song *et al* 2019). Rab GTPases are situated in exosomal membranes, and tetraspanins (CD9, CD63, CD81 and CD82) (Escola *et al* 1998, Huang *et al* 2004) and major histocompatibility complex (MHC) class I and II molecules (Théry *et al* 2002) are also shown to be accumulated and serve an essential purpose in the membranes of exosomes. SNARE proteins support the membrane coalescence of vesicles with plasma membranes or certain organelles (Bonifacino and Glick 2004). In addition, elevated intracellular Ca^{2+} concentration in K562 human erythrocytic leukemia cells exposed to monensin, which is a Na^+/H^+ exchanger causing intracellular modifications, enhanced exosome liberation (Savina *et al* 2003).

Tetraspanins, consisting of CD9, CD81 and CD63, are plentiful in TEMs and function to coordinate the assortment and wrapping of cargo in exosomes (van Niel *et al* 2011). A new purpose for the endoplasmic reticulum has been proposed on the basis of models for the liberation of exosomes. Located at the interface between the endoplasmic reticulum and the membranes of late endosomes, it has a decisive function in regulating the motility and ripening of late endosomes, which is connected to the small GTPases' activity. These processes are key for the merging of MVBs with the plasma membrane, which results in the subsequent liberation of exosomes (Gurung *et al* 2021). Small Rab GTPases, including RAB27a/b, RAB11, RAB7 and RAB35, regulate the budding and motility of vesicles and thus ease the trafficking of MVBs toward exosome liberation. Other participants comprise members of the SNARE proteins like Vamp7 and YKT6, which are essential for the process of fusing MVBs to the plasma membrane (Xie *et al* 2022), ADP ribosylation factor 6 (ARF6), syndecan heparan sulfate proteoglycans, phospholipase D2 (PLD2), and syntenin (Ghossoub *et al* 2014).

1.3.2 Ectosomes

Ectosomes represent a different type of extracellular vesicle that is produced by the external budding of the plasma membrane (figure 1.1(A)). In contrast to exosomes, the biogenesis of ectosomes is not strongly dependent on endosomal transport routes (Lim *et al* 2021). The determinants implicated in ectosome biogenesis continue to be enigmatic; nevertheless, the generation of ectosomes at the plasma membrane necessitates a stringent reorganization of membrane and cytoskeletal components (Kalra *et al* 2016, Meldolesi 2021). During the nucleation process, proteins containing lipid anchoring changes, such as myristoylation and palmitoylation, concentrate in the lumen to induce the development of membrane bending for budding (Cocucci and Meldolesi 2015). Transmembrane proteins and lipids also accumulate in certain membrane domains, and ESCRT-I subunits are attracted to the plasma membrane (Nabhan *et al* 2012). It is also recognized that a small GTPase, ARF, controls the assembly of load and release of ectosomes (Sedgwick *et al* 2015). During the final budding procedure, the ectosomes need to be disconnected from the plasma membrane. In this procedure, the dissociation of the cell's cytoskeleton is combined with the sorting of cytosolic proteins and RNA molecules into ectosomes (Cocucci and Meldolesi 2015). Ca^{2+} triggers membrane rearrangement and cytoskeletal breakdown, and ESCRT-III complexes are necessary for the delamination and liberation of ectosomes (Gurunathan *et al* 2021, Sun *et al* 2021).

1.3.3 Other types, such as apoptotic bodies, and exomers

Among the different types of EVs are apoptotic bodies and non-vesicular nanoparticles like exomers, which also merit consideration here (figure 1.1). Apoptotic bodies arise from the cells in the end phase of apoptosis and have so far been viewed as simply the remnants of deceased cells (Yu *et al* 2023). They are between 50 nm and 3 μm in size and include DNA fragments, histones and/or premature glycoepitopes. As with certain other EV types, their make-up is a mirror image of their cellular origin. When absorbed by other cells, the apoptotic bodies are able to elicit an anti-inflammatory or tolerogenic effect (Mohan *et al* 2020). These findings have caused some controversy about their classifications. While certain researchers favor the categorization of apoptotic bodies as a subclass of EVs due to their resemblance in size and composition, other researchers consider apoptotic bodies exclusively as the leftovers of dead cells (Caruso and Poon 2018). The exact mechanisms and genes implicated in the generation of apoptotic bodies are still not fully elucidated. Nevertheless, there is evidence that apoptotic bodies are shaped through a process referred to as apoptotic cell breakdown, which is typified by a number of strictly regulated morphological stages (Santavanond *et al* 2021). Exomers and supermers, which are commonly smaller than 50 nm in diameter, have recently been characterized as complementary non-vesicular nanoparticles (table 1.1). Exomers display differing proteomic characteristics and biodistribution profiles relative to small EVs. In a similar way, supermeres are fortified with RNA and exhibit increased enrichment in tissues in comparison to exomers and small EVs (Zhang *et al* 2021b).

1.4 Cargo and biological make-up of EVs, such as exosomes

Exosomes represent transport vehicles that are tethered to the membrane. Their cargos span a broad spectrum, such as proteins, nucleic acids and metabolites (figure 1.4) (Lee *et al* 2024), that mirror the characteristics of the donor cell and its underlying physiological condition (Dixson *et al* 2023).

Since the early 2000s, the emergence of sophisticated analytical methods like mass spectrometry and next-generation sequencing has highlighted that EVs harbor a multitude of biomolecules, such as proteins, lipids, metabolites and nucleic acids, as well as different kinds of DNA and RNA (Shurtleff *et al* 2018, Jeppesen *et al* 2019). This unveiling revolutionized the concept of exosomes and underscored the importance of EVs in cell–cell communication over and above the mere elimination of cellular trash. In the following the different types of cargo are outlined and the make-up of exosomes is presented.

1.4.1 Exosomal cargo

mRNAs consist of three basic regions, the 5′-untranslated region (5′-UTR), the coding region (CR, encoding the protein) and the 3′-UTR. Over the past decades, research has been conducted to unravel the mechanisms of EVs in the control of tumorigenesis and development, as well as to identify new and challenging biomarkers and promising therapeutic targets. Some findings indicate that mRNA located in EVs from the parent cell can be transferred to the recipient cell to carry out relevant tasks, which is referred to as horizontal transfer (Skog *et al* 2008, Kim *et al* 2017). MicroRNAs (miRNAs) comprise the body's own small RNAs with a size of about 19–22 nucleotides, which fulfill a number of critical regulatory functions in cells. The intricate regulatory framework of miRNAs is not just controlling the expression of numerous genes by a singular miRNA, but the combined expression of multiple miRNAs can also fine-tune the expression of a singular targeted gene (Bianchi *et al* 2017, O'Brien *et al* 2018). Long non-coding RNAs (lncRNAs) exceed 200 nucleotides in total number and are generally classified into five different groups comprising sense, antisense, bidirectional, intronic and intergenic nucleotides. LncRNAs participate in the inactivation of the X chromosome, genomic imprinting, the modification of chromatin, the activation of transcription, the interference of transcription, the trafficking in and out of the nucleus and numerous further major processes of regulation (Statello *et al* 2021, Mattick *et al* 2023). There is bidirectional interaction involving the trafficking

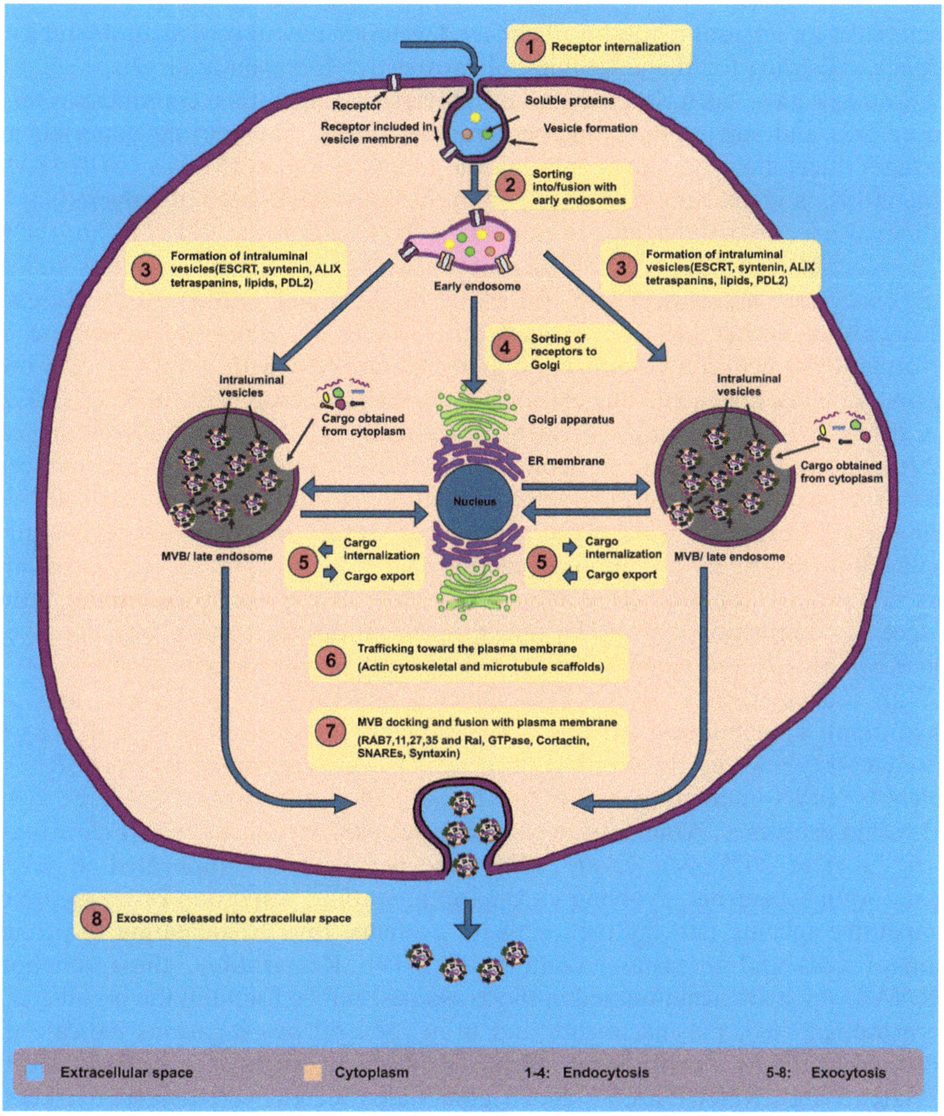

Figure 1.4. The biogenesis of exosomes. Within the endosomal system, (1) internalized loads such as receptors or soluble proteins are (2) sorted toward early endosomes, (3) which subsequently ripen into late endosomes or MVBs. Late endosomes/MVBs are distinct endosomal compartments that are rich in ILVs, which sequester proteins, lipids and cytosolic compartment and potential exosome loads. (4) Sorting of receptors to Golgi apparatus. (5) Cargos are transported from trans-Golgi network and possibly from the cytosol. It is also possible to export cargo via the Golgi apparatus. (6) MVBs containing exosome loads are translocated toward the plasma membrane, (7) merge with the plasma membrane and (8) the ILVs then get liberated as exosomes. Endoplasmatic reticulum = ER.

of membrane-coated particles carrying functional molecules from cancer cells and/or stromal cells among the cancer cells and the TME (Yuan *et al* 2016, Mierke *et al* 2018, Mierke 2019 2021a). By providing restraining or stimulating cues, the TME functions as an integral regulator of cancer development and propagation and as an efficient wellspring for the identifying of prospective therapeutic targets.

Exosomes have been studied as key players in facilitating communication of cancer cells and surrounding microenvironment cells, transporting materials and messages intercellularly, and regulating the immune system (Maia *et al* 2018, Xie *et al* 2019), and exosomes secreted by host cells have been discovered as an additional avenue of intercellular signaling (Quail and Joyce 2013, Li and Nabet 2019, Wortzel *et al* 2019). The formation process of exosomes comprises endocytosis, exocytosis, assortment and trafficking (Kalluri and LeBleu 2020). Research indicates that cancer cells or stromal cells secrete exosomes to orchestrate cell performance in cell–cell communication by filling them with a variety of growth factors, proteins, lipids, and nucleic acids, comprising microRNAs (miRNAs) (Fong *et al* 2015), lncRNAs (Zheng *et al* 2018), and circular RNAs (circRNAs) (Wang *et al* 2020), that are transferred horizontally from the donor to the recipient cells. Moreover, exosomes perform a similar function on antigen-presenting cells, and their surface harbors a wide range of molecules that are associated with antigen display and can trigger an immune reaction (Xu *et al* 2018). Circulating exosomes thus hold promise for clinical use as accurate vehicles for the targeted administration of specific compounds that aid in tumor propagation, metastasis, immunoregulation, angiogenesis and regeneration of tissues, and as anti-cancer vaccines.

With the emergence of RNA sequencing technologies (RNA-seq) and new bioinformatics techniques, various ncRNAs, comprising circRNAs, miRNAs (or synonymously referred to as miRs), transfer RNA-derived small RNAs (tsRNAs), lncRNAs, PIWI-interacting RNAs (piRNAs), and pseudogenes, have been identified within exosomes. Abundant in exosomes released by cancer cells (Felicetti *et al* 2016, Li *et al* 2018, Ma *et al* 2020) these ncRNAs are implicated in several regulatory mechanisms involving chromatin alteration, activation of transcription, competitive splicing and interactions with proteins, thus participating in proliferation of cells and metastasis (Zhu *et al* 2019). Remarkably, these functional ncRNAs, which are accumulated in the exosomes, can be found in the bloodstream, cerebrospinal fluid, plasma, serum and urine. Several investigations indicate that exosomal tsRNAs, comprising tRNA-ValTAC-3, tRNA-GlyTCC-5, tRNA-ValAAC-5 and tRNA-GluCTC-5 can potentially act as predictive biomarkers in liver cancer (Zhu *et al* 2019). Exosomal piRNAs, like piR-004800 (Ma *et al* 2020), piR-10506469 (Gu *et al* 2020), and piR-20548188 (Gu *et al* 2020), play a critical role in tumor development and in the progression of multiple myeloma, cholangiocarcinoma or gallbladder carcinoma. In addition, exosomes trigger the onset of carcinogenesis and subsequent development by contributing to the modification of the TME. Circulating exosomes encircling the TME control energy metabolism, deliver stimulatory or inhibitory inputs, and stimulate epithelial-mesenchymal transition (EMT), restructuring of the ECM, and promote vascular endothelial tubulogenesis, leading to metastatic cancer invasion, development of angiogenesis, and development of chemoresistance.

1.4.2 Biological make-up of exosomes

Exosomes have a spherical morphology when in solution, but become biconcave or beaker-shaped when prepared using artificial drying technology at the time of dissection (Yellon and Davidson 2014). The most relevant membrane-bound and cytosolic proteins contained in exosomes comprise tetraspanin family members such as CD9, CD63 and CD81, ESCRT proteins, such as Alix and TSG101, integrins, heat shock proteins (Hsp), actin and flotillins (Garcia *et al* 2015, Zhang *et al* 2019, Soe *et al* 2021). Some proteins such as heat shock proteins, CD63, ESCRT and elements of the cytoskeleton are shared across all exosomes, while other proteins like MHC class I and II are type-specific to the donor cell (Mashouri *et al* 2019). Exosomes' stiff bilayer membrane also contains lipid components including sphingomyelin, cholesterol and ceramides, which affect load assortment, the secretion of exosomes, their architecture and signal transduction (Skotland *et al* 2019, Skryabin *et al* 2020). A group of nucleic acids consisting of DNA, mRNA and non-coding RNA types can also be found in the exosome make-up (Doyle and Wang 2019, Mashouri *et al* 2019). MicroRNAs (miRs) constitute one of the most frequently occurring types of RNA within exosomes (Huang *et al* 2013, Zhang *et al* 2019). MiRs, involved in numerous biological mechanisms like exocytosis, hematopoiesis and angiogenesis, are implicated in exosome-regulated communication between cells (Zhang *et al* 2019). Some other exosomal RNA species comprise, lncRNA, p-element-induced wimpy testis (piwi)-interacting RNA, ribosomal RNA (rRNA), small nuclear RNA (snRNA), small nuclear RNA, and transfer RNA (tRNA), which all interfere with biological mechanisms, especially the development of cancers (Zhang *et al* 2019, Ge *et al* 2020). For this reason, their potential for application as a non-invasive instrument for the diagnosis and prediction of diseases has been investigated in various projects.

Originally considered an indispensable waste removal route (Johnstone *et al* 1991) or a special circumstance only process (Harding *et al* 1984, Pan *et al* 1985), the release of EVs is now regarded as a genuine mechanism for the transfer of molecules and the transmission of signals across cells. EVs can travel through the circulation to impact remote tissues or persist in the vicinity of the location of secretion to facilitate autocrine or paracrine signal transduction (Van Niel *et al* 2022). Besides exosomes, ectosomes and apoptotic bodies, other EV subtypes like migrasomes (table 1.1) (figure 1.5) have been characterized, which are less easy to classify into these two major groups and are yet to be defined (Ma *et al* 2015).

Cells transmit certain information to other cells so that they can synchronize, adjust and react to biological cues. The release of molecules into the extracellular compartment is an essential element of cell–cell communication. For decades, the way in which these molecules are chosen for release has been a key issue in the field of membrane trafficking (Dixson *et al* 2023). The key role of EVs in cell–cell interactions has recently been identified. They transport not just membrane proteins and lipids, but also RNAs, proteins in the cytosol and other signaling molecules to the target cells. To deliver the correct message, cargos in EVs must be organized and

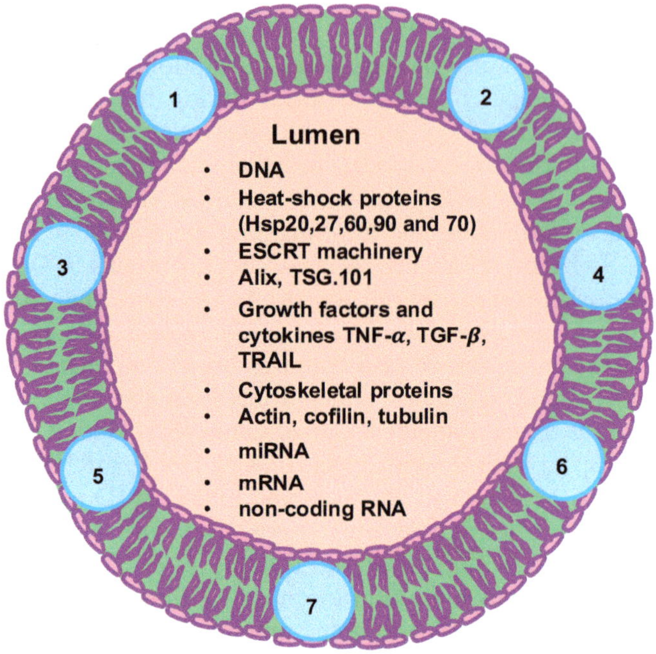

Figure 1.5. The composition of exosomes. Exosomes consist of a phospholipid bilayer membrane that contains membrane receptors, membrane-bound proteins and incorporated lipids. The membrane receptors are (1) adhesion molecules, (2) antigen-presenting molecules, (3) glycoproteins, (4) lipids, (5) membrane transport and fusion proteins, (6) tetraspanins and (7) other signaling receptors. (1) Among the adhesion molecules are integrins, P-selectins, ICAM-1. (2) The antigen-presenting molecules comprise MHC class I and II. (3) The glycoproteins include O-linked glycans, N-linked glycans and β-galactosidase. (4) Exosomes also incorporate multiple lipids, including cholesterol, ceramides, sphingomyelin, phosphatidylinositol (PI), phosphatidylserine (PS), phosphatidylcholine (PC), phosphatidylethanolamine (PE) and gangliosides (GM). (5) The membrane transport and fusion proteins are Annexins, Dynamin GTPases, RABs, and Syntaxin. (6) The tetraspanins on exosomes are CD9, CD37, CD53, CD63, CD81 and CD82. (7) The other signaling receptors contain Fas ligand (FasL), tumor necrosis factor (TNF) and receptor, transferrin receptor (TR). The phospholipid bilayer membrane of the exosome encloses a lumen. The lumen contains, proteins, lipids and nucleic acids. Transforming growth factor (TGF), TNF-related apoptosis-inducing ligand (TRAIL), tumor susceptibility gene (TSG).

contextually ordered. Over the past few years, a plethora of lipidome, proteome and RNA sequencing strategies have demonstrated that the make-up of EV cargo varies depending on donor cell type, metabolic factors and pathological conditions. Analyses of the different cargo 'fingerprints' have identified mechanistic connections between the activation of certain molecular signal transduction routes and the sorting of cargo. Moreover, cell biological investigations are starting to uncover new mechanisms of biogenesis that are controlled by the cellular background.

Exosomes are formed through the dual inward curvature of the plasma membrane, which also generates precocious endosomes that progress to ILVs prior to gradual maturation and release into MVBs (table 1.1) by inward budding, which avoids cytoplasmic lysosomes from breaking them down (Piper and Katzmann 2007, Crenshaw *et al* 2018) (figure 1.5). ESCRT, a discrete mechanism consisting of approximately 30 proteins that cluster into four complexes known as ESCRT-0, -I, -II and -III, regulates the formation of ILVs. Moreover, two important proteins, namely TSG101 and ALIX, are implicated in the production of exosomes, and syndecan and syntenin are two characteristic biomarkers that are specific for cancer cell-generated exosomes. Exosomes may also be released by a mechanism which is not governed by ESCRT. The Rab GTPase family, the cytoskeleton, molecular motors and membrane fusion assemblies (SNARE complex) are all implicated in the regulation of the transport of MVBs (Li and Marlin 2015, Langemeyer *et al* 2018). The secretion of exosomes is affected through various physiological determinants and cellular circumstances like lipopolysaccharide, tumor necrosis factor-α (TNF-α), interferon and hypoxia. Some secreted from damaged organs include healing RNAs and proteinaceous exosomes and are able to induce stem cells, helping to sustain tissue homeostasis, like the liver (Yang *et al* 2020a, Yuan *et al* 2023).

1.5 Cargo-centered view on the biogenesis of the exosomes and ectosomes

In traditional membrane transport routes, the cargos intended for a specific organelle enlist the necessary equipment for their self-sorting and conveyance (Henne *et al* 2011). In broad terms, it can be said that the 'trafficking machinery' attaches the cargos to membrane segments, reshapes the enclosing membrane into the shape of a vesicle and finally separates the vesicle away from the membrane site (Van Niel *et al* 2018). EV cargos seem to share the same fundamental route, tethering to trafficking effectors that accumulate cargos in endosomal and plasma membrane sites, and enlisting membrane curvature and cleavage mechanisms to produce an EV. One result of the cargo-specific biogenesis routes may be that several different subpopulations of EVs arise, which are each separately controlled. Cargo-generated biogenesis provides a candidate mechanism to account for the growing evidence of EV heterogeneity. In the following section, the present scientific knowledge on the fundamental mechanisms of EV cargo assortment is discussed. The subsequent section explains how the cell, tissue and pathological context governs these mechanisms of sorting. Most sorting principles for cargo have been uncovered for EV cargo within exosomes. Starting with a discussion of these principles, it is apparent that some of the sorting principles also hold for sorting cargo into specific types of ectosomes. In fact, inward budding of the endosome is equivalent in topological terms to outward budding of the plasma membrane, so that in theory analogous biogenesis mechanisms may be activated at various locations in the cell to produce exosomes and ectosomes (Jackson *et al* 2017). For example, the ESCRT apparatus is implicated in the biogenesis of both exosomes and ectosomes (Jackson *et al* 2017). The following is a discussion of situations in which the sorting mechanisms for ectosomes and exosomes are analogous, and other occasions where the sorting mechanisms seem to be unique.

1.5.1 Exosome biogenesis

Many EV cargo transport effectors have been recognized that are engaged in cargo tethering to the early endosomal membrane. The endosomal membrane then nodulates into the lumen of the endosome, and an ILV (table 1.1) is released through bud neck cleavage. The formation of ILVs, in conjunction with the clearance of other cargos to vesicles dedicated to different cellular sites, results in the maturation of endosomes to become late endosomal MVBs (Van Niel *et al* 2018). Therefore, the biogenesis of exosomes is characteristically characterized in terms of ILV generation, and exosomes per definition represent the subset of ILVs that are sequestered after MVB fusion to the plasma membrane.

1.5.1.1 ESCRT-facilitated biogenesis of exosomes

The first pathway of exosome biogenesis identified in mammalian cells became the ESCRT pathway, where all four ESCRT complexes such as ESCRT-0, -I, -II and -III, function in conjunction with enzymes for degradation and deubiquitination at the endosomal membrane. ESCRT-0, ESCRT-I, ESCRT-II and ESCRT-III are consecutively attracted to maturing endosomes in mammalian cells. The 'early' ESCRT machinery, consisting of ESCRT-0, -I and -II, enlists and sequesters transmembrane proteins like the epidermal growth factor receptor (EGFR) to ILVs through multivalent attachment to lysine-63-linked polyubiquitinated remnants or numerous monoubiquitinated residues (Raiborg *et al* 2002, Raiborg and Stenmark 2009). Because transmembrane receptors are often ubiquitinated and endocytosed when a ligand triggers signal transduction (Katzmann *et al* 2002), this mechanism of ESCRT enrollment results particularly in the trapping of ligand-receptor signaling assemblies and their integration into ILVs. In particular, the Hrs subunit of the ESCRT-0 dimer, such as Hrs and STAM1/2, orchestrates early steps in patterning ILV biogenesis through attachment to ubiquitinated cargos and enrollment of clathrin into the early endosome (Raiborg 2001). Hrs and clathrin interaction is critical for the clustered localization of Hrs in a specific type of endosomal microdomain and for the organization of cargos in ILVs (Raiborg *et al* 2002). The involvement of ESCRT proteins in ILV biogenesis is highly conserved, and ESCRT protein deficiency in yeast leads to prevacuolar endosomal complexes devoid of ILVs (Raiborg and Stenmark 2009, Coonrod and Stevens 2010). A shRNA screening of ESCRT proteins in HeLa cells revealed that removal of the ESCRT-0 proteins Hrs or STAM1 or the ESCRT-I subunit TSG101 impaired the release of small EVs. (Colombo *et al* 2013). In contrast, knocking down ESCRT-II and -III proteins barely had any consequences, which indicates that human cells express multiple redundant isoforms of ESCRT-II and -III proteins. Besides their cargo sequestering functions, the ESCRT-I and -II complexes are involved in the distortion of the endosomal membrane to form an ILV (Hurley and Hanson 2010,

Henne *et al* 2011). In reconstituted systems, the conjoining of ESCRT-II with cholesterol-rich membranes encourages the formation of liquid-order domains (L_o domains) (Boura *et al* 2012) and ILVs augmented with L_o lipids (Booth *et al* 2021), which suggests that the ESCRT pathway to facilitate EV biogenesis is reliant on the microenvironment of the L_o domains, and possibly specific lipid interactions in these domains. L_o domains themselves can promote the bending of membranes when bound through ESCRT-II. The selection of cargo may be facilitated by L_o domains may because of the enrichment with specific membrane proteins therein (Levental *et al* 2020). When the generation of L_o domains involves the ESCRT route in general, it can explain why L_o domains of small EVs contain elevated levels of lipids like sphingomyelin and cholesterol compared to cells (Llorente *et al* 2013). Progress in lipid imaging has permitted visualization of L_o domain in a near-natural environment (Levental *et al* 2020), thereby rendering the long-held hypothesis accessible for future studies. To finalize the process of exosome biogenesis, ESCRT-II triggers the generation of ESCRT-III filaments that cut off the neck of the forming exosome from the endosomal membrane (Hurley and Hanson 2010, Henne *et al* 2011). ESCRT-III complex can also be enlisted for this particular mechanism through ALIX, which is an ESCRT 'accessory' protein that attaches lysobisphosphatidic acid (LBPA) on the MVB membrane (Larios *et al* 2020). ESCRT-III appears to be targeted to the vesicle bud neck via detection of negative membrane curvature (Lee *et al* 2015) or by encouraging membrane bending to advance cleavage (Chiaruttini *et al* 2015); nevertheless, the exact mechanism is still being analyzed (Bertin *et al* 2020). Finally, the AAA-ATPase Vps4 strips the ESCRT-III filaments off the membrane thereby putting back the ESCRT system and potentially easing cleavage (Edgar *et al* 2014, Adell *et al* 2017).

1.5.1.2 Modulations of the ESCRT route

Exosomes that transport distinct cargos have distinct demands on ESCRT proteins throughout biogenesis, likely due to variations in how cargo can be recalled to the ESCRT route. These demands can be challenging to ascertain in mammalian cells because of the expression of multiple ESCRT protein isoforms with sometimes overlapping functionality and the pleotropic consequences of ESCRT depletion (Colombo *et al* 2013). Nevertheless, a variant of the ESCRT route has been found, such as the syndecan–syntenin ALIX route, which is typically affected via knockdown of the ESCRT I protein TSG101, the ESCRT II subunit VPS22 or the ESCRT III filament protein CHMP4 (Baietti *et al* 2012). A transmembrane heparan sulfate proteoglycan, syndecan-1 is oligomerized and facilitates signal transduction via multivalent attachment to growth factors and chemokines. In endosomes, the scaffold protein syntenin cooperates with the intracellular domain of syndecan-1 and connects it with CD63 and the ESCRT accessory protein Alix (Baietti *et al* 2012, Stepp *et al* 2015). Syndecans 2–4 are also able to regulate EV biogenesis (Baietti *et al* 2012, Ghossoub *et al* 2020). ALIX functions as a secondary organizing component on the surface of the MVB and engages with the lipid LBPA and the ESCRT proteins TSG101 and CHMP4B (Lee *et al* 2015, Elias *et al* 2020). Syntenin, syndecan-1 and

ALIX all seem to participate in cargo sorting throughout this pathway. Syndecan–syntenin–ALIX-induced exosome biogenesis is controlled through the triggering of activation of the oncogenic tyrosine kinase Src. Ultimately, the Src phosphorylates syndecan-1 at its intracellular domain, thereby triggering endocytosis of syndecan-1. Similarly, Src phosphorylates syntenin and Alix, that along with syndecan-1 phosphorylation promote the process of biogenesis of exosomes (Imjeti *et al* 2017). Syntenin-induced exosome biogenesis depends additionally on the GTPase Arf6, which regulates endocytosis and vesicle trafficking, and phospholipase D2 (PLD2) (Ghossoub *et al* 2014). Although their molecular inputs in this respect have not yet been delineated, Arf6 and PLD2 function downstream of Src (Imjeti *et al* 2017) and may possibly affect ILV budding through generating phosphatidic acid to attach syntenin (Muralidharan-Chari *et al* 2009, Ghossoub *et al* 2014). The syndecan–syntenin–ALIX route is also utilized by Epstein–Barr virus to charge exosomes with latent membrane protein 1 (LMP1), which is the key EBV oncogene (Rider *et al* 2018, Nkosi *et al* 2020). An additional modification of the ESCRT route comprises the ESCRT-III accessory proteins CHMP1, CHMP5 and IST1, which are essential for the generation of a specific class of exosomes in cells subjected to glutamine deficiency and/or mTOR/Akt suppression (Fan *et al* 2020, Marie *et al* 2020). Unexpectedly, these exosomes are produced in Rab11-positive recycling endosomes, in contrast to the traditional Rab5/Rab7-positive endosomes. These fascinating investigations disclose functionalities for the little-explored ESCRT-III accessory proteins and reveal their involvement in stress-activated EV biogenesis. In the future, there will be a great need for more investigations regarding ESCRT-III accessory proteins. Moreover, it needs to be explored whether the mechanical characteristics of the EVs are altered, and if so, how they impact the mechanical properties of the EVs receiving cells.

1.5.1.3 Role of lipids in the biogenesis of exosomes

EVs contain cholesterol, phosphatidylcholine, phosphatidylserine, sphingomyelin and ceramide, all of which are important in EV biogenesis, absorption and functional performance in receiving cells (Donoso-Quezada *et al* 2021). Lipids regulate multiple facets of endosome biology, among them cholesterol-regulated endosome localization (Rocha *et al* 2009) and endosome dimensions (Sobo *et al* 2007) endosome dimensions (Trajkovic *et al* 2008). Certain cargos have been found to be assorted into ILVs irrespective of ESCRT expression and ILVs harboring these cargos are produced through a lipid-dependent route. For example, proteolipid protein (PLP), a common membrane protein in the human central nervous system (Knapp 1996), can still be assorted into endosomes by oligodendrocytes even when Tsg101, ALIX or Hrs are knocked down using siRNA or when a dominant-negative VPS4 mutant is expressed (Trajkovic *et al* 2008). Endosomal PLP fails to localize to Hrs-containing membrane domains but rather to domains harboring flotillin-2, which is a Lo-domain hallmark, leading to the hypothesis that unique raft-associated lipids could function to promote the generation of ILVs harboring PLP (Trajkovic *et al* 2008, Otto and Nichols 2011). In fact, the conical shape of specific lipids like ceramide and phosphatidic acid can evoke a sudden negative membrane curvature resulting in an infolding into the endosomal membrane when cone-shaped lipids are generated on the external sheet (Matsuo *et al* 2004, Trajkovic *et al* 2008). This concept is underpinned by the fact that blocking cellular neutral sphingomyelinase-2 (nSMase2), a ceramide-producing enzyme, inhibits PLP from infiltrating ILVs (Trajkovic *et al* 2008). Besides PLP, nSMase2 regulates the EV-induced liberation of other proteins, such as TDP-43 (Iguchi *et al* 2016), interleukin-33 (Iguchi *et al* 2016, Katz-Kiriakos *et al* 2021), miRNAs (Kosaka *et al* 2010 2013, Mittelbrunn *et al* 2011, Tahara *et al* 2013, Cha *et al* 2015, Zhu *et al* 2017), prion proteins (Guo *et al* 2015), and vacuolar H+-ATPase (Choezom and Gross 2022), despite it not necessarily implying that the involved EVs are always exosomes. This discrimination is essential as blocking nSMase2 may also interfere with the liberation of ectosomes (Menck *et al* 2017) or the generation of apoptotic bodies (Shamseddine *et al* 2015). nSMase2-dependent and ALIX-dependent pathways can act simultaneously to generate unique subsets of EVs from the same cell (Shamseddine *et al* 2015). nSMase2-based and ALIX-dbased routes can operate concurrently to produce unique subsets of EVs derived from the exact same cell (Matsui *et al* 2021). For example, a polarized epithelial cell line has been identified to liberate nSMase-2-based EVs from the basolateral face alongside ALIX-based EVs originating from the apical face (Matsui *et al* 2021). These findings indicate that it is in principle possible to regulate the release of EVs in a side-specific manner in polarized cells. Subsequently, the direction of vesicle release may also contribute to maintain the polarity of the EV-liberating cell. These EV subpopulations harbored unique molecular biomarkers, providing a basis for the hypothesis that distinct EV cargos pursue distinct biogenesis pathways, resulting in EV heterogeneity (Matsui *et al* 2021), although certain of these cargos may sustain these routes merely because of their enrichment in apical or basal membranes. Exosome biogenesis is also affected by ceramide, in tandem with nSMase2, through the trafficking of existing ceramide from the endoplasmic reticulum (ER) to the endosomal membrane via the ceramide transporter (CERT) (Fukushima *et al* 2018, Barman *et al* 2022, Crivelli *et al* 2022) and via receptor-driven signaling to MVBs via the ceramide metabolite S1P73. CERT-induced transmission from the ER is also expected to affect ectosome biogenesis within ER plasma membrane locations (Barman *et al* 2022).

1.5.1.4 Function of tetraspanins in the biogenesis of exosomes

Tetraspanins constitute integral plasma membrane surface proteins that are commonly accumulated in small EVs. Across the cell membrane, tetraspanins can directly engage with integrins and related proteins to generate highly organized TEMs (Yáñez-Mó *et al* 2009). TEMs encompass homophilic and heterophilic types of engagement between tetraspanins, engagement of tetraspanins with other membrane proteins, such integrins, and engagement of tetraspanins with proteins at the membrane-cytoplasmic interface, such as focal adhesion proteins (Hemler 2005, Charrin *et al* 2009, 2014, Stipp 2010). In addition, these protein connections can arise through direct attachment of tetraspanins to other proteins or via tetraspanin interactions involving a joint binding cognate. Interactions of tetraspanin with signaling molecules have been identified for various kinds of proteins, including adhesion proteins like integrins and cadherins and signaling receptors like G protein-coupled receptors (GPCR), EGFR, c-kit, c-MET, ADAMs, and cytosolic signaling molecules like focal adhesion kinase (FAK), Rho GTPase, and β-catenin (Termini and Gillette 2017). The subsequent cellular effects of these interactions are diverse and vary from the control of cellular adhesion, motility, contractility and cell shape. Thus, traditional tetraspanins existing as EV cargo, comprising CD63, CD81 and CD9, can self-promote EV biogenesis by recruiting other TEM-affiliated proteins into EVs and regulating their sorting. Moreover, the tetraspanin CD63 couples the VEGFR2 to β1 integrins (Kummer *et al* 2020). In quiescent human umbilical vein endothelial cells (HUVECs), scant CD63-β_3 integrin complexes have been identified compared to the more abundant VEGFR2-β_3 integrin and CD63-β_1 integrin complexes (Tugues *et al* 2013), which indicated that the integrin partner for CD63 may be cell type dependent.

Consequently, the convergence between integrin and growth factor signal transduction in endothelial cells is dependent on CD63. Cell surface-anchored CD63 is implicated in the generation of the VEGFR2-β_1 integrin complex, and silencing of CD63 abrogates complex generation and downstream signal transduction of β1 integrin and VEGFR2. CD63 connects with β_1-integrin in human melanoma and osteosarcoma cells (Radford *et al* 1996, Iizuka *et al* 2011), $\alpha_3\beta_1$ integrin in lymphocytes and melanoma cells (Berditchevski *et al* 1995, 1996, 1997b), $\alpha_4\beta_1$ in T lymphoblasts (Mannion *et al* 1996), and $\alpha_6\beta_1$ in HT 1080 fibrosarcoma cells (Sincock *et al* 1997) and other different cell lines (Berditchevski 2001). Even though the interaction of the majority of tetraspanins with integrins can only be seen in the vicinity of what are termed 'mild' detergents, the CD151-$\alpha_3\beta_1$, CD151-$\alpha_6\beta_1$ and CD81-$\alpha_4\beta_1$ complexes appear to be stronger, as they can resist conditions (such as in the presence of Triton X-100 and digitonin) that perturb all other integrin–tetraspanin and tetraspanin–tetraspanin crosstalk interactions (Berditchevski *et al* 1997a, Yauch *et al* 1998, Serru *et al* 1999). It is known that there is a close molecular interaction between receptor tyrosine kinases and integrins, frequently engaging kinases of the Src family (Sundberg and Rubin 1996, Wang *et al* 1996). For example, c-Src activity is decisive for the integrin-based phosphorylation of EGFR and VEGFR3 (Moro 1998, Zhang *et al* 2005). In a reciprocal response, receptor tyrosine kinases can influence integrin signal transduction. The activation of FAK mediated by the VEGF receptor through c-Src enables the generation of FAK-αvβ5 integrin signaling complexes (Eliceiri *et al* 2002). In addition, VEGF triggers the tyrosine phosphorylation of the α_3 integrin subunit through c-Src and thus controls endothelial cell adhesion and migration (Mahabeleshwar *et al* 2007).

Using tetraspanin pulldowns, it has been determined that up to 45% of the EV proteome of lymphoblasts engages with TEMs (Perez-Hernandez *et al* 2013). Considering the involvement of tetraspanins in membrane curvature and vesicle assembly (McMahon and Boucrot 2015), the crystal structures illustrate that the four transmembrane helices shared by all tetraspanins create an inverted conical taper with a pinch that tethers lipids (Zimmerman *et al* 2016 Umeda *et al* 2019). In view of the structural resemblance between tetraspanins and ceramide, it is appealing to hypothesize that both classes of molecules share a specific biophysical mechanism to drive membrane curvature. Efforts to delineate the precise process through which tetraspanins promote exosome or ectosome biogenesis, nevertheless, have been hampered by contradictory knockdown results in different experimental approaches. For example, while knocking down CD63 in melanoma cells reduces the number of ILVs for each endosome (van Niel *et al* 2011), inverse impacts on the overall small EVs secreted are seen in other types of cells (Petersen *et al* 2011). Some similar disparities have been found for depletion of CD9 and knockout experiments (Chairoungdua *et al* 2010, Brzozowski *et al* 2018). A possible explanation for such mismatches could be that more than one tetraspanin could be available in the same tetraspanin-enriched membrane domains (Andreu and Yáñez-Mó 2014, Gurung *et al* 2021), and possibly impact exosome biogenesis in the identical route. For instance, syntenin-1 interfres directly with CD63 and tetraspanin-6 throughout the process of exosome biogenesis (Latysheva *et al* 2006, Baietti *et al* 2012, Ghossoub *et al* 2014, Guix *et al* 2017). Alternatively, downregulation of exosome biogenesis may result in increased regulation of ectosome biogenesis, resulting in an overall rise in EV abundance. The tetraspanin CD63 is known as a traditional marker for small EVs resulting from exosome biogenesis, as it is mainly distributed in intracellular endolysosomal compartments and is inserted into exosome-terminating ILVs (Sung *et al* 2020a, Mathieu *et al* 2021). Nevertheless, plasma membrane localization of CD63 has been noted (Booth *et al* 2006). Likewise, CD9 and CD81, which are mainly localized to the plasma membrane, also accompany late endosomes and are frequently accompanied by CD63 in MVB-derived EVs (Crescitelli *et al* 2013, Mathieu *et al* 2021). Therefore, the existence or non-existence of a particular tetraspanin in a cocktail of small EVs should not be utilized to categorize the EVs either as exosomes or ectosomes. The presence of tetraspanins in EVs together with integrins leads to the hypothesis that the function of integrins in cancer progression is enhanced via close proximity to tetraspanins that can elevate their mechanosensing and mechanotransduction mechanism and subsequently foster the promotion of cancer propagation.

1.5.2 Biogenesis of ectosomes

In contrast to exosomes, ectosomes arise as outer buds of the plasma membrane that cleave to liberate the vesicles (Ratajczak *et al* 2006, Antonyak and Cerione 2014). Compared to exosomes, greater size diversity has been found in ectosomes, and this could also mirror variations in biogenesis mechanisms.

1.5.2.1 Generation of small ectosomes

The generation of small ectosomes shares many characteristics with the biogenesis of exosomes, as many of the identical mechanisms are employed and vesicles of comparable size to exosomes with a diameter of about 100 nm are generated. These mechanisms comprise ESCRT proteins and tetraspanins. For example, the ESCRT-I protein TSG101 is attracted to the plasma membrane through engagement with the Hrs-mimicking PSAP motif of arrestin domain-containing protein 1 (ARRDC1), and participates in the VPS4-driven generation of small ectosomes (Nabhan *et al* 2012). Ectosomes produced via this pathway are referred to as 'ARMMs' (arrestin domain-containing protein 1-mediated microvesicles) (Nabhan *et al* 2012). Overexpression of ARRDC1 overexpression elevates levels of ESCRT and Notch signal transduction proteins in EVs, indicating that they constitute loads of ARMMs (Wang and Lu 2017). ARRDC1 and Notch2 entrance into EVs relies on the expressions of their corresponding E3 ubiquitin ligases (Nabhan *et al* 2012, Wang and Lu 2017), akin to the ubiquitin-based ESCRT sorting carried out on the endosome. Tetraspanins also participate in budding of ectosomes and the sorting of cargos. CD9 and CD81 are connected to the actin cytoskeleton via cooperating ERM and EWI proteins and are thus implicated in the organization of the plasma membrane. This interactome is hypothesized to affect signaling, cargo sorting and vesicle generation (Sala-Valdés *et al* 2006, Umeda *et al* 2019), such as the CD82 enrollment of the ERM protein ezrin into membrane vesicles for liberation into ectosomes (Huang *et al* 2020).

Several different types of membrane protrusions have now been demonstrated to release small ectosomes, including filopodia (Ando *et al* 2012, Scott 2012, Rilla *et al* 2013, Deen *et al* 2016, Noble *et al* 2020, Nishimura *et al* 2021), cilia (Dubreuil *et al* 2007, Kesimer *et al* 2009, Cao *et al* 2015, Wang *et al* 2021), and microvilli (Marzesco *et al* 2005 2009, McConnell *et al* 2009, Hara *et al* 2010). For instance, filopodia-like protrusions are detached at locations where the plasma membrane is compromised in a Vps4B-induced manner to form ectosomes (Mageswaran *et al* 2021). The ESCRT-III component CHMP4B becomes distributed at sites of damage and may play a similar function at the plasma membrane to that of exosome generation, certainly when overexpressed (Mageswaran *et al* 2021). The accumulation of glycocalyx complexes on the outer plasma membrane of cells also facilitates the detachment of ectosomes from filopodia following DNA damage, forming semi-linked strings of small 'beaded' EVs (Shurer *et al* 2019). Biogenesis of small ectosomes from intact cellular protrusions has also been characterized, through mechanisms engaging the actin cytoskeleton, cholesterol, hyaluronan production, and inverse BAR domain-containing proteins (Marzesco *et al* 2005 2009, Rilla *et al* 2013, Han *et al* 2016, Nishimura *et al* 2021). HIV-1 particles can also cluster at the apices of filopodia, which implies that filopodia assembly may participate in the biogenesis of ectosomes/retroviruses in a manner that is not yet completely elucidated (Aggarwal *et al* 2012, Bracq *et al* 2018). Similar to filopodia, cilia also secrete small ectosomes (Dubreuil *et al* 2007, Kesimer *et al* 2009, Cao *et al* 2015, Wang *et al* 2021). In numerous instances, these cilia-produced EVs contain cargos associated with diseases commonly referred to as 'ciliopathies', which encompass polycystic kidney disease and Bardet-Biedl syndrome (Reiter and Leroux 2017). In *Caenorhabditis elegans*, delivery of ectosomes harboring GFP-tagged polycystins, which is linked to polycystic kidney disease, has been seen at the basal end of ciliary sensory neurons, and these ectosomes control nematode mating activity (Wang *et al* 2014). Support for ciliary liberation of EVs also arises from interfering with cilia development by knocking out cilia-related genes, thereby diminishing EV-associated secretion of specific cargos, comprising Hedgehog and Wnt signal transduction proteins (Mohieldin *et al* 2021, Volz *et al* 2021, Cruz *et al* 2022). Impaired ectosome liberation is also linked to the retinal pathogenesis of ciliopathies. The specialized cilia of rod photoreceptor cells are competent to generate large amounts of ectosomes, but are inhibited from producing them due to the tetraspanin peripherin-2 (Salinas *et al* 2017). This suppression facilitates the ciliary outer segment expansion; subsequently, knockout of the peripherin-2 gene results in retinal degeneration (Salinas *et al* 2017).

1.5.2.2 Generation of large ectosomes

The generation of large ectosomes is not as well-known as that of exosomes and small ectosomes. Whether early ESCRT machinery or tetraspanins are implicated in this mechanism, or whether there are other commonalities with the biogenesis of exosomes or small ectosomes, is uncertain. On the contrary, restructuring of the actin cytoskeleton has been found to be the basis of the process of rupture and cleavage of the plasma membrane to liberate large EVs (Di Vizio *et al* 2009). Molecular restructuring within the plasma membrane is also involved in this process, encompassing alterations in proteins, lipids and electrolyte levels. For instance, altered Ca^{2+} concentrations trigger the activation of a family of lipid scramblases that perturb the lipid asymmetry of the membrane (Suzuki *et al* 2010, Bricogne *et al* 2019). It is hypothesized that this mechanism leads to a larger exposure of phosphatidylserine on the outer leaflet, which is a major characteristic of ectosomes (Lima *et al* 2009, Muralidharan-Chari *et al* 2009). Exposure of phosphatidylethanolamine on the external leaflet could also enhance ectosome production, as in early *C. elegans* embryos the production of ectosomes with a size of $\sim$200 nm is blocked through the phosphatidylethanolamine flippase TAT-5 (Beer *et al* 2018). Losing TAT-5 activity at the plasma membrane results in the aggregation of large ectosomes in the intercellular spaces, and depletion of TAT-5 interferes with gastrulation (Wehman *et al* 2011, Beer *et al* 2018). EVs, both large and small, incorporate sphingomyelin and ceramide (Llorente *et al* 2013, Haraszti *et al* 2016) as well as the lipid raft biomarker caveolin-1 (Mariscal *et al* 2020), which suggests that EVs possess or are derived from lipid raft-associated membrane domains. Caveolin has also been found to control the biogenesis of small EVs by tethering to cholesterol (Albacete-Albacete *et al* 2020); this mechanism may also operate in ectosomes. Nevertheless, there are no studies that consistently investigate the involvement of cholesterol and lipid rafts in EV biogenesis, and it appears feasible that the ectosomal membrane originates solely from the plasma membrane and the corresponding lipid rafts.

In addition to changes in membrane constitution, local dissolution of the cortical actin cytoskeleton in conjunction with actomyosin contractility may facilitate plasma membrane blebbing and consequent generation of large ectosomes (Crespin *et al* 2009, Muralidharan-Chari *et al* 2009, Bergert *et al* 2012, Ciardiello *et al* 2019). This mechanism is utilized from abnormally large blisters on the plasma membranes of non-apoptotic cells, which transform themselves from a highly metastatic, mesenchymal condition to an enhanced migratory and metastatic 'amoeboid' phenotype. Nevertheless, it is still under discussion whether the amoeboid phenotype of cancer cells displays an increased state of cell migration and not just merely another mode of cell migration employed due to mechanical and biochemical cues of the surrounding microenvironment (Mierke *et al* 2018, 2018, Mierke 2020, 2021a). Similarly, in both prostate and breast cancer cells, depletion of the actin nucleator diaphanous-related formin 3 (DIAPH3) leads to the generation of large oncosomes (Di Vizio *et al* 2009). The small GTPase RhoA, in conjunction with its subsequent target molecules ROCK and LIM kinase (LIMK), also controls the liberation of large ectosomes from breast cancer cells by raising contractility and reducing the activity of the actin-separating protein cofilin, respectively (Li *et al* 2012).

Besides their involvement in the syndecan–syntenin–ALIX route of exosome biogenesis, Arf6 and PLD2 control ectosome generation. At the plasma membrane, ARF6 activates phospholipase D, which causes the enrollment of Erk and the subsequent phosphorylation of myosin light chain kinase and myosin light chain, thereby enhancing actomyosin contractility and splitting of the membrane blister (Muralidharan-Chari *et al* 2009). Arf6 and PLD2 likewise promote the selective integration of pre-miRNAs within ectosomes. Subsequent to the export of pre-miRNAs to the nucleus through exportin-5, Arf6 in concert with GRP1 facilitates the transport of pre-miRNAs and exportin-5 to the cell perimeter and into ectosomes (Clancy *et al* 2019). Collectively, it appears that ectosome biogenesis is highly reliant on a mechanism that couples phospholipid repositioning in the plasma membrane with the activity of the actomyosin contraction apparatus.

1.6 Regulation of cargo based on the cellular state

As EV studies have progressed, it has emerged that there exist no definitive absolute guidelines for defining which cargo constitutes an EV and which does not. In many instances, regulation studies of EV cargos through cell signaling or other alterations in cell fate have yielded this insight, pinpointing instances where cargos that are rarely present in EVs are becoming upregulated in a particular circumstance. A few joint topics are emphasized in the following. The EV communication hypothesis continues to focus on the ability of cells to control the selection of EV cargo and thereby transmit certain signals in a controlled and specific mode.

1.6.1 Growth factor and oncogene signal transduction

An effect of growth factor signal transduction, which can pass on its impact to the recipient cells is the modification of the loading of EV cargo. For instance, fibroblast growth factor signaling in cultured hippocampal neurons, signaling to govern neuronal plasticity and wound healing *in vivo*, strongly changes the charge of sequestered EVs and promotes MVB-plasma membrane fusion (Kumar *et al* 2020a, 2020b). EV loads are also modified through growth factor signal transduction in lung tissue, in which EVs facilitate endothelial and smooth muscle cell communication. Excessive pacing of pulmonary artery smooth muscle cells with transforming growth factor-β1 (TGF-β1) triggers both liberation and ingestion of EVs spiked with

RNAs that code for cytoskeletal factors and the transcriptional regulator bHLHE40 (De La Cuesta *et al* 2019). In a similar manner, TGF-β2 signaling affects the amount of proteasome subunits, fibronectin and histones transported within EVs of cultured colon cancer cell lines (Fricke *et al* 2019), and triggers the entry of mRNAs codified for nuclear factor of activated T-cells 5 (NFAT5) and histone deacetylase 5 (HDAC5) into cardiomyocyte-derived EVs (Gennebäck *et al* 2013). The ability of cancer cells to transform is strongly enhanced through the impact of oncogenic signal transduction on the rearrangement of the EV cargo. A perfect illustration of such a process is glioblastoma cancers, where heterogeneity of the cancer is sustained through different oncogenic cell subtypes by the replacement of EVs with different pro-tumorigenic protein cargos (Ricklefs *et al* 2016). Among these subtypes, EGFRvIII oncogene expression drives deep modifications in both the expression of constituents of EV biogenesis and the sorting of EV cargos, including invasion-inducing factors and adhesion molecules (Al-Nedawi *et al* 2008, Skog *et al* 2008, Choi *et al* 2018). In a similar manner, the oncogenic type of guanine nucleotide exchange factor Dbl promotes the generation of ectosomes that transmit FAK to receiving cells, thereby fostering apoptotic circumvention and anchorage-independent cell proliferation (Kreger *et al* 2016). The oncogenic non-receptor protein kinase Src promotes the incorporation of the syndecan binding partners β_1 integrin and fibronectin into small EVs in breast cancer cells, encouraging the migration of receiving cells (Imjeti *et al* 2017). This type of cell–cell communication via EVs could also be an explanation for the orchestration of cell clusters or collective cell migration during malignant progression of cancer. At the molecular scale, phosphorylation of syndecan and syntenin through Src enhances endocytosis and exosomal sorting of these cargos through ARF6-PLD2 route (Imjeti *et al* 2017). Signaling by oncogenes can also influence the sorting of RNA cargos in EVs. Oncogenic KRAS mutations occur in multiple cancer types, among them in over 30% of adenocarcinomas of the colon and lung and in over 80% of pancreatic cancers (Timar and Kashofer 2020). Isogenic colorectal cancer cell lines that express wild-type or mutant KRAS generate small EVs with very variable cargos. Cells which express mutant KRAS yield small EVs that are fortified with cell signal transduction proteins and metabolic enzymes, while cells expressing wild-type KRAS yield small EVs that are fortified with RBPs (Demory Beckler *et al* 2013). The effects of KRAS mutation on EV sorting have also been found to be comparable to that in non-small cell lung cancer cells (Clark *et al* 2016). Specific miRNAs like miR-100 are also accumulated in small EVs derived from mutant KRAS-expressing cells and are able to be transmitted in a functional manner to receiving cells (Cha *et al* 2015). In addition, EVs originating from mutant KRAS enhance the proliferation of cancer cells in soft agar; the specific cargos driving this pro-tumorigenic response, yet, are still unverified (Hinger *et al* 2020). Ultimately, in mutant KRAS-expressing cells, RBP and RNA sorting are regulated through downstream MEK/Erk signal transduction, which impedes the sorting of Ago2 into EVs via displacing its subcellular partitioning at a location distant from endosomes and into RNA processing elements (McKenzie *et al* 2016). The functional implications of these alterations in RNA-RBP transport are still to be identified but are probably related to changes in the RNA expression pattern of both the donor and acceptor cells.

1.6.2 Metabolic control of the cargo sorting process

The metabolic regulation of the cargo sorting can be performed in four different ways. The first way is through the fatty acid metabolism route. The second way is the autophagy route. The third path is via oxidative stress. The fourth route involves senescence. All these different kinds of regulation of the cargo sorting event are outlined in the following.

1.6.2.1 Fatty acid metabolism route

As outlined in the section on EV biogenesis, lipids act to control EV generation through their ability to promote membrane bending and create orderly lipid domains that serve as a sorting stage for proteins. An overabundance of fatty acid metabolism, as seen in non-alcoholic steatotic hepatitis (NASH), also leads to profound alterations in EV sorting and induces the selective liberation of integrin β_1 as EV cargo from hepatocytes (Guo *et al* 2019). Comparable changes in EV cargos are induced by hepatic and hepatocellular carcinoma cell therapy with fatty acids, which comprise myristic acid (Speziali *et al* 2018) and palmitic acid (Fukushima *et al* 2018), where the latter promotes EV biogenesis at ER membrane attachment points. Palmitic acid can also facilitate EV biogenesis and the charging of EV cargos through palmitoylation of EV cargos, possibly directing them to the membrane at EV biogenesis locations (Mariscal *et al* 2020).

1.6.2.2 Autophagy route

Autophagy refers to the mechanism whereby cells secrete cytoplasmic constituents and transfer them to lysosomes for metabolic recycling. Autophagy includes both macroautophagy, whereby cargos are transferred to lysosomes through an intermediate organelle referred to as an autophagosome, and microautophagy, whereby cargos are taken up directly from endosomes or lysosomes (figure 1.6) (Schuck 2020).

Part of the autophagy mechanism also operates in secretory processes, among them EV biogenesis (Murrow *et al* 2015, Leidal *et al* 2020). A landmark work recently revealed that RNA sorting in exosomes is strongly associated with secretory autophagy (Leidal *et al* 2020). By this newly suggested mechanism, the autophagy adaptor protein LC3B and cognate proteins bind to the endosomal membrane and sense RBPs comprising a four-amino-acid-long LC3-interacting region (LIR) (Leidal *et al* 2020). To link ILV nodulation to cargo sorting, LC3B also recruits a factor associated with neutral sphingomyelinase activation (FAN) via its LIR to enable the activation of local ceramide synthesis (Leidal *et al* 2020). LC3-based sorting in EVs has also verified for the nuclear protein HMGB1 (Kim *et al* 2021). This emerging regulatory function of LC3 differs from its traditional function in substrate uptake into the autophagosome, which is a double-membrane organelle that directly merges with the lysosome, as silencing of genes essential for the

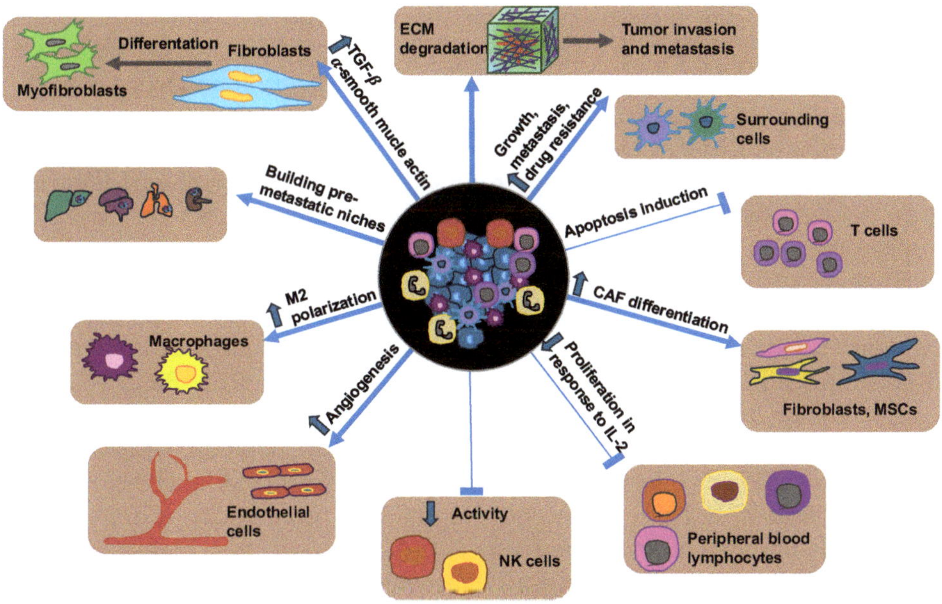

Figure 1.6. The impact of exosomes on the advancement of cancer. Exosomes act in the formation of pre-metastatic niches, angiogenesis, the breakdown of ECM scaffolds, growth of tumors, promote the differentiation of cancer-associated fibroblasts (CAFs), regulate the functionality of immune cells, such as polarization of macrophages from M1 to M2 phenotypes, trigger apoptosis in T cells, inhibit the proliferation of peripheral blood lymphocytes in reaction to IL-2 or blockage of NK cells, induce the differentiation of fibroblasts into myofibroblasts, promote the metastatic cascade and therapy resistance.

autophagosome located downstream of LC3B lipidation fails to repress LC3-driven exosome biogenesis (Leidal *et al* 2020). Nevertheless, the two functions of LC3 may be linked in instances where the outer autophagosome membrane merges with MVBs to create an 'amphisome', that in turn can merge with the plasma membrane to liberate traditional exosomes along with the internal autophagosome cargo (Fader *et al* 2008, Hessvik *et al* 2016). In a different type of endosomal micro-autophagy, the chaperone HSC70 catches protein cargos that carry a KFERQ peptide and supplies them not solely to lysosomes, but also to endosomes for release into exosomes (Sahu *et al* 2011, Ferreira *et al* 2022).

1.6.2.3 Oxidative stress

In reaction to multiple environmental stresses such as hypoxia, hyperoxia and chemotherapeutic substances, the generation of excess reactive oxygen species (ROS) induces an oxidative stress reaction that changes metabolic signal transduction routes and EV load. For instance, chemotherapy leads to a huge accumulation of oxidized and ubiquitylated proteins, which are excreted as EV cargo (Yarana and St. Clair 2017). ROS, which are produced as by-products of metabolism or immune response, also alter the EV load and can have either cytoprotective or damaging consequences for the receiving cells. In hypoxia environments like microenvironments or tumors, fibrotic lungs or ischemic myocardial tissue, the protein content of the EV cargo is significantly modified. Glioblastoma cells subjected to hypoxic environments, for instance, liberate EVs that are fortified with anti-apoptotic and pro-metastatic proteins (Kore *et al* 2018). In additional hypoxic cancer models, EVs can carry specific adhesion proteins and factors promoting EMT or partial EMT (He *et al* 2021, Kumar *et al* 2024). Hypoxia probably utilizes secretory autophagy routes to control the loading of EVs, as the hypoxia-based rise in EV loading of pro-angiogenic factors relies on the LC3-like autophagy factor GABARAPL1 (Keulers *et al* 2021). The type of cargo of EVs can also be regulated by hypoxia. Moreover, the expressions of multiple miRNAs are transcriptionally increased through the hypoxia inducible factor 1a (HIF-1a) and thus more elevated within hypoxic EVs during hypoxic settings. Among them are miR-210 and miR-21 (Tadokoro *et al* 2011, King *et al* 2012, Bang *et al* 2014, Makiguchi *et al* 2016, Zhu *et al* 2017, Chen *et al* 2018). MiR-210 and miR-21 have shown anti-apoptotic, cardioprotective activity in myocardial hypoxia models (Zhang *et al* 2017, Namazi *et al* 2018). In multiple cancer and heart models, hypoxia-induced EVs harbor elevated amounts of pro-angiogenic miRNAs and stimulate angiogenesis *in vivo* (Umezu *et al* 2014, Hsu *et al* 2017, Namazi *et al* 2018, Bister *et al* 2020). These types of miRNA roles can intriguingly be abolished by circular RNAs, and hypoxic signaling seemingly can fine-tune the downstream actions of EVs through either miRNA or circular RNA cargo alterations. For example, in hypoxic pancreatic cancer cells, circular RNA circZNF91 is highly induced through HIF-1α and delivered as EV cargo to receiving cells, in which it functions as an inhibitory shield to block miR-23b-3p and enhance HIF-1α-regulated transcriptional reprogramming (Zeng *et al* 2021). Hypoxic dysregulation of circular RNA cargo, such as upregulation, in EVs is linked to ischemic heart disease, colorectal cancer and (Wang *et al* 2019, Yang *et al* 2020b, 2021b, Ye *et al* 2021, Zhang *et al* 2021a). Hypoxic stress can cause the induction of HIF1-promoted enhanced expression of mRNAs (and their respective cognate proteins) in glioblastoma EVs (Kucharzewska *et al* 2013) and changed cardiac fibrosis-driving long non-coding RNA concentrations in cardiomyocyte EVs (Kenneweg *et al* 2019). Intriguingly, HIF-1α also enhances the expression of neutral sphingomyelinase-2, thereby driving a proprietary selection mechanism for the generation of EVs containing hypoxia-elicited cargos (Zhu *et al* 2017). Hyperoxia refers to an iatrogenic type of oxidative stress to which lung cells may be exposed as a result of the production of free oxygen radicals after the intake of a surfeit of oxygen. Using models of hyperoxia-induced acute lung damage, vigorous release of EVs augmented with caspase-3, miR-221 and miR-320a stimulates pro-inflammatory reactions of macrophages that induce tissue regeneration (Moon *et al* 2015, Lee *et al* 2016). Reciprocally, under hyperoxia stress in a rat model of bronchopulmonary dysplasia, EVs harboring the pyroptotic p30 fragment of gas dermin D are liberated and trigger pervasive inflammation, lung and brain damage (Ali *et al* 2021). Hyperoxia exposure experiments have also provided important mechanistic evidence for the enrollment of RNA in EVs. Hyperoxia or H_2O_2 stress induces phosphorylation of the lipid raft protein caveolin-1 on tyrosine-14, subsequently tethering hnRNPA2B1. This specific engagement prevents hnRNPA2B1 from being broken down and enhances its O-GlcNAcylation in the RRM1 domain (Lee *et al* 2019). There is also sumoylation of hnRNPA2B1 at its RRM2 domain. Sumoylation and O-GlcNAcylation increase the engagement of hnRNPA2B1 with certain miRNAs, thereby permitting hnRNPA2B1 to assort miRNAs into EVs (Villarroya-Beltri *et al* 2013, Lee *et al* 2019).

1.6.2.4 Senescence

Uncontrolled ROS can also cause damage to DNA, the escape of DNA into the cytoplasm and the triggering of cell senescence or apoptosis. EV-induced elimination of harmful cytoplasmic DNA offers a possibility to avoid this dangerous cell destiny (Takahashi *et al* 2017, Hitomi *et al* 2020). Senescent macrophages and fibroblasts release increased amounts of small EVs and various microRNAs in comparison to non-senescent peers, with macrophages having reduced miR-223 sorting and fibroblasts increasing miR-15b-5p and miR-30a-3p sorting toward EVs (Terlecki-Zaniewicz *et al* 2018, Alibhai *et al* 2020). Senescence in primary fibroblasts also enhances the liberation of EVs bearing the interferon-induced transmembrane protein 3, resulting in an increased senescent phenotype of receiving cells (Borghesan *et al* 2019).

1.7 Cellular functions of exosomes in immune response

Exosomes' main function is to facilitate two-way communication between cells. In relation to cancer, exosomes have an important function as facilitators of inter-cellular interactions that contribute to tumor propagation. They can also function as biomarkers for the identification of cancer and its advancement. Hence, their application in cancer treatments has been proposed, as a vehicle for medication or as a diagnostic instrument. Exosomes have also been widely recognized to be implicated in anti-cancer drug resistance by passing on drug resistance messages to susceptible cells.

1.7.1 General immunomodulatory function of exosomes

In the case of immunomodulatory or apoptotic cues, exosomal surface ligands attach themselves directly to the surface receptor of the receiving cell and trigger a downstream signaling pathway (Gurung *et al* 2021). For instance, exosomes released from dendritic cells contain MHC-peptide complexes that induce the activation of T cells (Tkach *et al* 2017). These cells also express surface ligands like tumor necrosis factor (TNF), Fas ligand (FasL) and TNF-related apoptosis-inducing ligand (TRAIL), which implies that they are able to tether to TNF receptors on cancer cells, inducing the activation of caspase and triggering apoptosis in cancer cells. In addition, exosomes originating from dendritic cells stimulate the activation of natural killer (NK) cells by engaging with their TNF receptor, which results in the release of interferon γ (IFNγ) (Munich *et al* 2012). Exosomes thus activate innate immune reactions.

1.7.2 Transfer of cargo

Exosomes can also merge with the plasma membrane of the target cells, leading to their cargo being transported into the cytoplasm. Internalization of exosomes can proceed through a distinct, receptor-dependent route or through non-specific micro- or macropinocytosis. Integrins, lectins, proteoglycans and T cell immunoglobin, mucin domain-containing protein 4 (Tim4) and synctin-1 and syncytin-2 are probably involved in exosome incorporation (Prada and Meldolesi 2016, Mathieu *et al* 2021). The endocytosis that takes place when exosomes are made internal can be either driven by or unrelated to clathrin-induced endocytosis, which entails the creation of clathrin-coated pits. Afterwards, the clathrin-coated vesicles undergo a process of uncoating and merge with the endosomes (Mettlen *et al* 2018). Mechanisms of clathrin-unrelated exosome endocytosis involve lipid force-driven endocytosis, caveolin-driven endocytosis and phagocytosis. Crucially, these routes can simultaneously persist alongside each other throughout internalization of the exosome and the liberated cargo circumvents direct lysosomal breakdown through multiple ways (Gurung *et al* 2021).

1.8 Cellular function of exosomes in cancer

The cellular function of exosomes in the progression of cancer, such as metastasis, may be driven by the selective capture of integrins in the plasma membrane of exosomes. In contrast, it can also be hypothesized that there is no relation between the integrin localization in exosomes, organotrophic metastasis and subsequently the advancement of cancer.

1.8.1 Exosomes promote cancer advancement via intercellular communication

In cancer, exosomes facilitate the intercellular crosstalk, resulting in cancer propagation (Dai *et al* 2020). In addition, they can act as biomarkers for cancer identification and monitoring of its malignant progression (Wang *et al* 2022). Cancer cells can release exosomes that impact the advancement of cancer through the generation of tumor-supporting stroma and the induction of the differentiation of fibroblasts toward tumor-supporting stromal myofibroblasts (Webber *et al* 2015). This kind of differentiation relies on the increased expression of α-smooth muscle actin (α-SMA) and activation of TGF-β signal transduction (Webber *et al* 2010). Cancer-associated fibroblasts (CAFs) function in elevating the aggressiveness of various cancers, as they liberate exosomes that contribute to the invasiveness of cancer cells (Ye *et al* 2023). Exosomes liberated from CAFs trigger the protrusive activity and the migratory capacity of breast cancer cells via Wnt–planar cell polarity (Wnt-PCP) signal transduction (Luga *et al* 2012). CAFs can be identified mostly through the expression of so-called CAF markers, comprised as fibroblast activation protein alpha (FAP) and α-SMA that enable the distinction of CAFs from normal fibroblasts (Nurmik *et al* 2020). CAFs differ from normal fibroblasts in that they engage with tumorigenic cells in the TME and remain in a hyperactivated mode that promotes cancer propagation via multiple routes. The expression of CAF markers is regrettably extremely heterogeneous and strongly differs among various CAF subpopulations, as outlined below. A few prevalent CAF biomarkers that have been pinpointed comprise α-SMA, vimentin, FAP, fibroblast-specific protein 1 (FSP1) and platelet-derived growth factor receptor alpha/beta (PDGFR-α/β) (Liu *et al* 2020, Nurmik *et al* 2020, Ortiz-Otero *et al* 2020, Shen *et al* 2020). To circumvent the absence of an omnipresent label, continuing research is concentrating on finding new ways to screen CAFs on the basis of their cellular functioning, as well as efforts to find new CAF labels, and contraction screens have been utilized to distinguish between CAFs and normal fibroblasts owing to their superior contractility (Su *et al* 2018). Three-dimensional hydrogels, comprising collagen or Matrigel, are an additional way to functionally characterize CAFs and their subtypes (Tilghman *et al* 2012, Nebuloni *et al* 2016, Zoetemelk *et al* 2019). Exosomes derived from ovarian cancer cell lines such as SK-OV-3 and OVCAR-3 have been isolated and delivered into adipose-derived mesenchymal stem cells (ADSCs), leading to both myofibroblastic functionality and ADSC phenotypes (Cho *et al* 2011). Exosomes released by cancer cells directly affect the adjacent stroma and other cells, altering their functions and encouraging cancer advancement (Qin *et al* 2018, Landskron *et al* 2019, Tan *et al* 2020, Song *et al* 2021). In addition, exosomes from CAFs can modify the tumor stroma by secreting molecules (Peng *et al* 2023). Various solute factors released by CAF like IL-6, IL-33, TGF-β, stromal cell-derived factor-1 (SDF-1) and CAF-derived cardiotrophin-like cytokine factor 1 (CLCF1) produce tumor-promoting signaling that facilitates signal transduction from cancer cells or other cells in the TME (Qin *et al* 2018, Wei *et al* 2018, Landskron *et al* 2019, Song *et al* 2021). SDF-1 liberated by CAFs fosters the development and progression of pancreatic cancer and results in gemcitabine resistance through elevated special AT-rich binding protein-1 (SATB-1) expression in cancer cells (Wei *et al* 2018). Overexpression of SATB-1 in pancreatic cancer cells favors a positive feedback circuit that perpetuates the cancer-inducing phenotype of CAFs. Overexpression of SATB-1 in pancreatic cancer cells favors a positive feedback circuit that perpetuates the cancer-inducing phenotype of CAFs. (Tan *et al* 2020, Wu *et al* 2022). The exosomal circTBPL1 of CAFs possibly could be delivered to breast cancer cells and stimulate their proliferation, migration and invasion (Ye *et al* 2023). Thus, exosomes liberated by CAFs impact in a direct manner via released exosomal circRNA and molecules their surrounding stromal environment and other adjacent cells, such as cancer cells, whereby cancer cells functionality and encouraging cancer advancement.

1.8.2 Cancer-cell-derived exosomes induce immunoregulation

Cancer cell-derived exosomes also perform immunoregulatory roles that result in immune suppression or evasion. It has been shown that the proliferation of lymphocytes in the peripheral blood in reaction to IL-2 is blocked by exosomes originating from cancer cells and NK cell activity is reduced (Clayton *et al* 2007). TS/A mouse mammary carcinoma cells release exosomes to deliver IL-6, which impairs the differentiation of myeloid tumor progenitor cells into dendritic cells

(Yu *et al* 2007). Exosomes obtained from human ovarian cancer specimens repressed the expression of signaling elements of T cell activation, including JAK3 and CD3-ζ; and induced apoptosis of T cells (Taylor and Gerçel-Taylor 2005). *In vitro* experiments have revealed that cancer cell-based exosomes impact T cell gene expression and upregulate inhibitory genes in $CD4^+$ T cells, causing depletion of CD69 and loss of function, even though exosomes are not interiorized through T cells (Muller *et al* 2016). It has been demonstrated that colorectal cancer cells release miR-145-containing exosomes that are taken up by macrophages, resulting in their phenotype shift towards the M2 macrophage phenotype (Shinohara *et al* 2017).

1.8.3 Cancer-cell-derived exosomes function in angiogenesis

Exosomes released by cancer cells may have an essential part to perform in angiogenesis, which is critical for cancer advancement. Melanoma cells were demonstrated to release exosomes containing factors regulating angiogenesis, such as the vascular endothelial growth factor (VEGF), IL-6, and matrix metalloproteinase 2 (MMP2), due to WNT5A signaling. Moreover, the depletion of WNT5A resulted in decreased endothelial cell branching (Ekström *et al* 2014). The human immortalized myelogenous leukemia cell line K562 -derived exosomes that harbor the miR-17–92 cluster, which can be delivered to human umbilical vein endothelial cells (HUVECs), resulting in diminished expression of the integrin α_5. Exosomes can thus contribute to the exchange between leukemia and endothelial cells because of their miRNA cargo (Umezu *et al* 2013). In a similar way, colorectal cancer cells stimulate the proliferation of endothelial cells through exosomes and their mRNA cargo (Hong *et al* 2009). Metastatic cancer cells release exosomes that carry miR-105 and aim at the tight junction protein zonula occludens-1 (ZO-1), which causes blood vessel permeability (Zhou *et al* 2014). Exosomes derived from kidney cancer cells enhance angiogenesis in HUVECs through the augmentation of VEGF expression (Zhang *et al* 2013a). These results indicate that exosomes released by cancer cells are implicated in tumor angiogenesis and metastasis as the disease evolves.

1.8.4 Cancer-cell-derived exosomes interact with the ECM

Cancer-cell-originated exosomes also modify the ECM and fulfill critical tasks in the movement of cancer cells through facilitating the assembly of focal adhesions and consequently cell adhesion. The exosomal fibronectin may play a crucial role in cell migration and invasion (Sung *et al* 2015). It has been shown that tumor-originated exosomes are capable of attaching to individual elements of the ECM, such as hyaluronic acid or laminin (Mu *et al* 2013). In addition, these exosomes are abundant in proteases that can breakdown collagens, laminins or fibronectin, resulting in pre-metastatic niche preconditioning (Mu *et al* 2013). Metastatic abilities can be passed through exosomes from metastatic to non-metastatic cancer cells (Le *et al* 2014). EVs harboring miR-200 released from metastatic breast cancer cell lines have been demonstrated to modify gene expression and enhance mesenchymal-to-epithelial transition (MET) in non-metastatic cells (Le *et al* 2014). Pancreatic ductal adenocarcinoma is an additional case in point, where exosomes have been proven to cause pre-metastatic niche development in the liver of naïve mice. Organ-seeking vesicles, such as brain-seeking exosomes (Busatto *et al* 2022), lung-seeking exosomes have been observed in breast cancer (Grigoryeva *et al* 2023) and liver-seeking exosomes have been seen associated with liver diseases (Kim *et al* 2020).

It has been shown that the liberation of TGF-β and the increased expression of fibronectin in the receiving liver cells favors the development of a fibrotic micro-environment, whereas the macrophage migration inhibitory factor (MIF) present in the exosomes opposes the bone marrow-derived macrophages and promotes metastasis (Costa-Silva *et al* 2015). Exosomal integrins appear to be especially relevant in defining organ-specific metastasis, as integrins $\alpha_6\beta_4$ and $\alpha_6\beta_1$ have been implicated in lung metastasis, whereas integrin $\alpha_V\beta5$ has been associated with liver metastasis (Hoshino *et al* 2015). The involvement of exosomes in the advancement of cancer is illustrated in figure 1.7. Cancer cell liberation of vesicles and the intercellular transfer of $\alpha v\beta 6$-containing exosomes fosters the aggressive disease progression (figure 1.7A).

In cancer research, the question of organ-specific metastasis and how this phenomenon, referred to as metastatic organotropism, is regulated has long been raised.

If the environment of the niche is not suitable for the metastasizing tumor cells, they fall into a dormant state and are only 'brought back to life' later when conditions are better (Mierke 2024). Metastatic organotropism has continued to be a mystery in cancer research dating back to Stephen Paget's hypothesis in 1889. It has

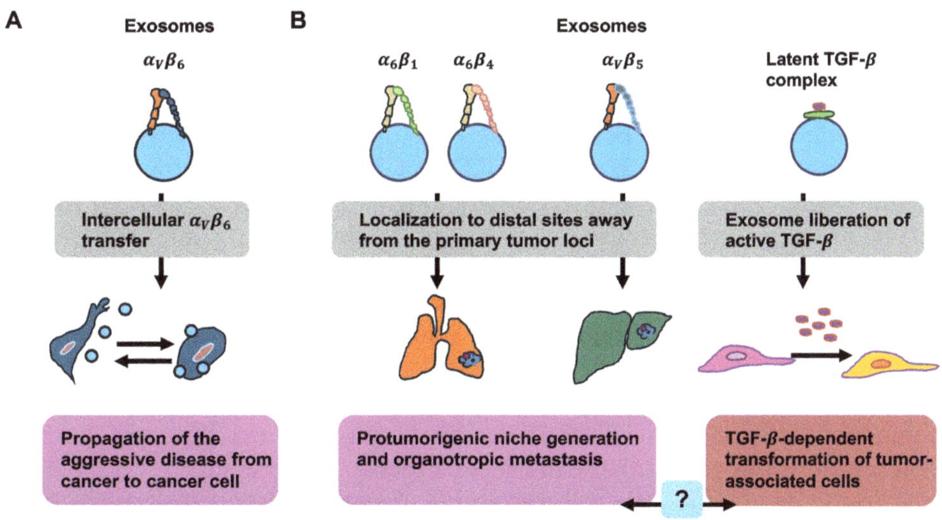

Figure 1.7. Integrins and integrin-related signal transduction pathways promote cancer evolution. (A) The secretory activity of cancer cells and intercellular delivery of $\alpha v\beta 6$-containing exosomes favor disease aggressiveness. (B) Tumor-secreted exosomes bearing specific integrin profiles determine metastatic seeding sites and help establish a pre-metastatic niche formation (left). TGF-β secretion by exosomes also participates in tumor microenvironment establishment (right).

been proven that exosomes from lung, liver and brain cancer cells of mice and humans prefer to merge with the cells that are located at their intended destination, specifically lung fibroblasts and epithelial cells, Kupffer cells of the liver and endothelial cells of the brain (Hoshino *et al* 2015). In addition, tumor-derived exosomes ingested by organ-specific cells have been identified to prime the pre-metastatic niche. Exosome treatment from lung-tropic models diverted metastasis of bone-tropic cancer cells. Exosome proteomics yielded differing integrin expression profiles, with the exosomal integrins $\alpha_6\beta_4$ and $\alpha_6\beta_1$ being linked to pulmonary metastases, whereas the exosomal integrin $\alpha_V\beta_5$ has been associated with hepatic metastases (figure 1.7B) (Hoshino *et al* 2015, Soe *et al* 2021). Targeting integrins $\alpha_6\beta_4$ and $\alpha_V\beta_5$ reduced exosome incorporation and metastasis to both lung and liver. Incorporation of exosome integrins by locoregional cells has been found to activate Src phosphorylation and gene expression of pro-inflammatory S100. Ultimately, the clinical evidence indicates that exosomal integrins can be utilized to forecast organ-specific metastasis.

In a mouse model, exosomal integrins can cause organotropic metastasis. The expression of integrins β_3, β_4 and $\alpha_V\beta_5$ on exosomes and cancer cells (circulating tumor cells and primary tumor) has been explored and their connection with the positioning of remote metastases has been determined (Grigoryeva *et al* 2023). The link between exosomal integrin β_4 and lung metastases in breast cancer patients has been verified. Nevertheless, it has not been feasible to evaluate the involvement of integrin β_3 in brain metastasis, since this localization is extremely seldom. In a cohort of breast cancer patients, no correlation has been identified of exosomal integrin $\alpha_V\beta_5$ with liver metastasis (Grigoryeva *et al* 2023). The additional evaluation of β_3, $\beta 4$, and $\alpha_V\beta_5$ integrin expression on CTCs demonstrated a correlation association of integrin $\beta 4$ and $\alpha_V\beta_5$ with liver metastases occurrences, whereas there is no correlation with lung metastases occurrences. The integrin β_4 in the primary tumor has been implicated in liver metastases. Additionally, an in-depth assessment of the phenotypic characteristics of $\beta 4^+$ cancer cells demonstrated a markedly elevated amount of E-cadherin$^+$ and $CD44^+CD24^-$ cells in patients with liver metastases when compared to patients with lung or no remote metastases. In a very smart *in vivo* experiment, it was possible to prove that the exosomal integrins may have a decisive part to play in the organotropism of remote metastasis in a breast cancer model. Using an animal model, it has been found that the exosomal integrin β_3 carries cancer cells to the brain, β_4 to the lungs and $\alpha_V\beta_5$ to the liver (Hoshino *et al* 2015). Various exosomal integrins enriched selectively in the lung and liver ECM. In particular, integrin $\alpha_6\beta_4$ in S100A4-positive pulmonary stromal cells fortified with laminin and integrin $\alpha_V\beta_5$ in F4/80$^+$ hepatic stromal macrophages harboring fibronectin. It is recognized that exosomes can serve as carriers for several bioactive molecules (Othman *et al* 2019). For example, the exosomes of cancer cells include molecules that increase vascular permeability in remote organs, lead to the formation of inflammation and the enrollment of progenitor cells derived from the bone marrow. All these regulatory mechanisms are implicated in the creation of pre-metastatic niches in a remote organ. Notwithstanding the existing experimental data, the bulk of the mechanisms governing the mechanism behind the involvement of exosomes with specified integrins in organotropic metastasis have yet to be elucidated. The question emerges whether it is feasible that the integrin-driven targeting mechanism is of universal origin, considering that exosomes are generated by cancer cells (at the primary site or in circulation). CTCs are the key drivers of cancer re-emergence and cancer metastasis. There exists practically not much knowledge about the expression of integrins on CTCs. For instance, the expression of integrins β_3, β_4, and $\alpha_V\beta_5$ has been explored on exosomes and cancer cells, comprising CTCs and primary cancer cells, and its relationship to the positioning of remote metastases (Grigoryeva *et al* 2023). CTCs of breast cancer patients have been seen to co-express β_3, β_4 and $a_V\beta_5$ integrins. In addition, a correlation between exosomal integrin β_4 and lung metastases in breast cancer patients has been identified (Grigoryeva *et al* 2023). The $\alpha_V\beta_3$ integrin is targeted to circulating melanoma cells in the lung using the melanoma model, as demonstrated by the impairment of $a_V\beta_3$ integrin by the selective inhibitor of $\alpha_V\beta_3$ integrin labeled MK-0429 (Pickarski *et al* 2015). CTCs with no integrin gene expression or with expression of non-complementary α and β subunits that are unable to produce heterodimers predominate. Approximately only 15% of CTCs showed expression of integrin subunits capable of generating heterodimers. The lowest potential activity has been seen in CTCs with no integrin expression, whereas the highest frequency of expression of tumor progression-related genes, specifically genes for stemness, EMT, invasion, pro-inflammatory chemokines and cytokines, and laminin subunits, has been found in CTCs that co-express ITGA6 and ITGB4 (Grigoryeva *et al* 2024). Protein-level validation suggested that the median of integrin-β4+ CTCs comprised higher concentrations in patients exhibiting more aggressive molecular subtypes and in metastatic breast cancer patients. CTCs with ITGA6 and ITGB4 expression are anticipated to have marked metastatic potency, evident in the expression of EMT and stemness-related genes, and the ability to potentially generate chemokines/pro-inflammatory cytokines and laminins.

Integrin expression in cancer cells has been widely analyzed, but this mostly concerns cell cultures rather than the primary tumor. It is thus recognized that the expression of β_3 in breast cancer cells affects both the lymph nodes (Desgrosellier *et al* 2014) and bone metastases (Sloan *et al* 2006, Weilbaecher *et al* 2011). In addition, $\alpha_V\beta_3$ can promote the adhesion of cancer cells to the pulmonary vasculature in the 66cl4 mammary carcinoma line via interaction with the endothelial counter receptor PECAM-1 (Sloan *et al* 2006). In the MDA-MB-435 human cancer cell model, integrin $\alpha_V\beta_3$ crucially promotes tumor cell growth within the brain, even though it is not necessary for growth in the fat pads of the breast (Yoshioka *et al* 2010). Integrin $\alpha_V\beta$5 increases cell migration and hepatic metastasis of colon carcinomas (Leng *et al* 2016). The upregulation of integrin β_4 is linked to enhanced tumor growth and pulmonary metastasis of the hepatocellular carcinoma cell line (Pickarski *et al* 2015).

Paget concluded that specific organs like the liver seemed to be particularly susceptible to metastasis and that this could not be explained by blood flow on its own. It has also been proposed that the 'soil' or local microenvironment of these organs is more hospitable to the spread of cancer cells than that of other organs like the spleen. Four decades later, Paget's theory has been questioned by James Ewing, who suggested that metastasis is governed by the vascular and lymphatic channel anatomy draining the primary tumor (Ewing 2005). Ewing's view then gained acceptance until the pioneering research of Isaiah Josh Fidler proved convincingly that while cancer cells could access the vasculature of all organs, metastases developed exclusively in specific organs and not elsewhere (Fidler and Kripke 1977, Hart and Fidler 1980). Metastatic ground awareness has been re-awakened, and a flurry of subsequent studies have examined the pathophysiology of the local tissue microenvironment or 'niche' of primary cancer cells and cancer cells at metastatic locations (Psaila and Lyden 2009). Therefore, integrin expression patterns are rather dispersed and there are no insights into how the expression of the identical integrins is linked in primary cancer cells, in CTCs and in exosomes. In addition, further questions emerge. Can multiple distinct integrins be expressed within the same exosome? What specific factor dictates the existence of metastases in a particular organ and the lack of metastases in others when different integrins are expressed on the exosomes and cancer cells? These questions have been looked at in breast cancer patients. The variants most frequently found included exosomes lacking integrin expression and those with mono-expression of integrin β_4, which is present in nearly all instances. Mono-expression of $\alpha_V\beta_5$ and β_3 integrins appeared in about half of the instances. Co-expression of two and three integrins has been identified in 37.5%–79.2% of the screened instances (Grigoryeva *et al* 2023). The analysis of β_3, β_4 and $\alpha_V\beta_5$ co-expression on CTCs indicated no significant variation in patients with differing sites of metastasis. Thus, on CTCs, the mono-expression of β_3-, β_4- and $\alpha_V\beta_5$-integrins has been analyzed with respect to the positioning of remote metastases. Mono-expression of exosomal β4 integrin alone has been linked to the occurrence of lung metastases, which is in agreement with the existing experimental findings (Hoshino *et al* 2015). The occurrence of exosomes with the expression of integrin $\alpha_V\beta_5$ has been detected to be unrelated to liver metastases. The role of integrin β_3 in brain metastasis could not be assessed as instances with this site are very scarce. It is established that β_4-integrin is expressed in several tumors and has a function in tumor invasion, proliferation, EMT and angiogenesis. The integrin β_4 extracellular domain attaches to the basement membrane compound laminin. In breast cancer, integrin β_4 leads to invasive mode, increases the viability of cancer cells and stimulates angiogenesis (Yang *et al* 2021a). Expression of $\alpha_6\beta_4$ integrin on human breast cancer cells has been reported to enhance metastasis to the lungs through engagement with hCLCA2, which is expressed on the endothelial cells of the luminal surface of pulmonary arteries, arterioles and venules (Abdel-Ghany *et al* 2001). Cancer cells in the primary focus that express integrin β4 have been linked to hepatic metastases. In addition, this correlation has been noted for both primary cancer cells and CTCs. For instance, in a different investigation, β_4 integrin has been linked to metastasis, with exosomal expression in the lung and expression on cancer cells, including CTCs and primary tumor, in the liver. These results could be accounted for due to the circumstance that β_4^+ cancer cells involved in liver metastases also express EpCam, cytokeratins, E-cadherin and N-cadherin, which means that these cells may be in two modes—devoid of EMT manifestations or with a hybrid EMT phenotype, which is the result of co-expression of E-cadherin and N-cadherin. In addition, the hybrid phenotype in this instance could also develop at the MET. Indications are emerging as to whether it is cells with EMT hybrid phenotypes displaying epithelial characteristics that trigger carcinoma metastasis (Jolly 2015). Moreover, β4+ cancer cells that metastasize in the liver showed evidence of stemness ($EpCam^{+}CD44^{+}CD24^{-}ALDH1^{-}\beta4^{+}$). It has also been revealed that the expression of exosomal integrins may not reflect the expression of cellular integrins (Grigoryeva *et al* 2023). The selective expression of integrins on exosomes seems to account for this. Thus, an attempt has been made to identify not only the function of exosomal integrins, but also that of integrins in the potential sources of exosomes, such as cancer cells of the primary tumor. The hypothesis has been put forward that the targeting mechanism can be a universal one and is not restricted to exosomes as promoters of pre-metastatic sites. Correlation analysis in instances where integrin co-expression data have been available in exosomes, CTCs and cancer cells yielded zero correlations (Grigoryeva *et al* 2023). The correlation between the exosomal integrins and metastasis on the one side and the integrins of the primary tumor and the CTC on the other side has therefore been divergent. Exosomes, which are directed to specific sites by integrins, can serve as transporters for several bioactive molecules. They can transport phlogogenic substances like pro-caspase-1, IL-1, IL-6, IL-18 and elements of inflammosomes and thus induce the creation of pre-metastatic niches. Integrins on cancer cells are presumably critical for the selective binding of CTCs toward the endothelium.

1.8.4.1 Exosomes degrade the ECM via enzymes liberated from cancer cells, myeloid cells, platelets or CAFs

The interplay between cancer cell-derived exosomes can be through enzymes that can modify the ECM vis degradation. The rearrangement of the local tissue is crucial for the invasion of cancer cells and the metastatic spread of the tumor. The function of ECM-degrading enzymes such as matrix metalloproteinases (MMPs) is of decisive importance in this context. The expression of MMPs is also strongly upregulated in the pre-metastatic niche (Hiratsuka *et al* 2002, Kaplan *et al* 2005), and hence, it can be hypothesized that restructuring of this niche is a prerequisite for formation of secondary tumors at this targeted site. MMPs are involved in the breakdown of ECM components in inflammatory reactions and in tissue damage as well as in the growth of primary tumors (López-Otín and Matrisian 2007, Page-McCaw *et al* 2007). MMP9 expression is elevated in a specific, VEGFA-dependent manner in endothelial cells and $MAC1^{+}$ and $VEGFR1^{+}$ myeloid cells from pre-metastatic lungs (Hiratsuka *et al* 2002, Kaplan *et al* 2005). Expression of MMP9 at pre-metastatic locations is capable of enhancing invasion of cancer cells as well as liberating growth factors and chemokines, among them the soluble KIT ligand, which in turn attracts bone marrow-derived progenitor cells and cancer cells expressing the KIT receptor (Kaplan *et al* 2005).

It can be hypothesized that a critical role of tumor-associated myeloid cells at the primary tumor locus is to direct other cells of the immune response to encourage an immunosuppressive, anti-inflammatory phenotype and enable the tumor to evade immune recognition (Yang and Moses 2008). For instance, TGFβ production by GR1$^+$CD11b$^+$ myelocytes interferes directly with the functionality of CD8$^+$ cytotoxic T lymphocytes, and these cells also hamper natural killer cells, B cells and dendritic cell functional maturation (Yang and Moses 2008). Myeloid cells recruited to pre-metastatic areas may have a comparable purpose: they provide immunological refuges where malignant cells manage to survive and proliferate unrecognized. The expression of osteopontin through myeloid cells, a protein involved in the adhesion and survival of cancer cells and in the control of MMP activity, also impedes the immune defense of the host (Bellahcène *et al* 2008, Wai and Kuo 2008).

The molecular and functional phenotype of myeloid cells recruited to pre-metastatic areas remains to be completely analyzed; inter-laboratory differences in the surface markers utilized to distinguish the cells reinforce this difficulty. In primary cancer trials, a separation has been drawn between GR1$^+$MAC1$^+$ immature myeloid cells, which are synonymously referred to as myeloid-derived suppressor cells (MDSCs), and end-differentiated, MAC1$^+$F4/80$^+$ tumor-associated macrophages (Pollard 2004, Yang *et al* 2008). MAC1 and VEGFR1 are expressed on a broad spectrum of myeloid cells, such as progenitor cells, and it is probable that both completely mature cells and immature cells of the pre-metastatic and metastatic niches are affected. While there is considerable crossover between VEGFR1$^+$ and CD11b$^+$ cell subpopulations, the exact lineage connection is uncertain. It is hypothesized that VEGFR1$^+$ cells are the first cells to be committed to the niche and that these cells subsequently secrete factors that either encourage the recruitment of other myeloid cells or promote their proliferation (Lin and Pollard 2007). Besides myeloid cells, additional cell types also fullfill a role in generation of the pre-metastatic niche. For instance, the infiltration of VEGFR1+ hematopoietic progenitor cells, which also express the C–X–C chemokine receptor 4 (CXCR4), into locations of neovascularization in ischemic tissues and expanding tumors is reliant on SDF1, which is liberated from platelet granules (Jin *et al* 2006). While the involvement of platelets in the pre-metastatic niche generation process is still under scrutiny, it is possible that they also release chemokines and angiogenic regulatory agents (Italiano *et al* 2008, Rafii *et al* 2008). A number of cancer cell types are also expressing CXCR4 and may thus be influenced by platelet-derived SDF1 gradients, and platelet surface glycoprotein Ib-IX also seems to have a key involvement in lung colonization by metastatic melanoma cells using murine models (Jain *et al* 2007). Additional host cells residing in the pre-metastatic niche, like fibroblasts and endothelial cells, may also express chemokines and intercellular adhesive molecules that entice CTCs to anchor to these specific spots (Kucia *et al* 2005, Orimo *et al* 2005).

The transition of local fibroblasts is pathologically significant for the advancement of cancer. CAFs are constantly activated, proliferate more rapidly and accumulate larger quantities of ECM constituents than quiescent fibroblasts present in benign tissue (Kalluri 2016). There are processes that can impair the transition of local fibroblasts into CAFs, such as the neddylation mechanism. Neddylation represents a reversible post-translational alteration comparable to ubiquitination and has been proposed to act in reorganization of the ECM, such as the TME (Liu *et al* 2024). The process of neddylation is typified by the reversible covalent conjugation of the neural precursor cell expressed developmentally downregulated protein 8 (NEDD8) with certain protein substrates (Kamitani *et al* 1997, Xirodimas 2008). NEDD8, which is a highly conserved protein in eukaryotes, shows a prevalent expression in the nucleus and a relatively weak occurrence in the cytoplasm (Mergner and Schwechheimer 2014). NEDD8 is 60% identical to ubiquitin and 80% related to it, thus rendering it the molecule with the greatest similarity to ubiquitin of all ubiquitin-like proteins (Kumar *et al* 1993, Enchev *et al* 2015). Both the cullin protein and NEDD8 are highly expressed in different cancer cell types, like colon cancer and leukemia cells, supporting the evidence for the link between neddylation and cancer advancement (Du *et al* 1998, Hori *et al* 1999). MLN4924 has been identified as a powerful inhibitor of NAE that interferes with cullin-RING ligase-facilitated protein metabolism and triggers apoptosis in cancer cells (Soucy *et al* 2009). MLN4924 blocks neddylation as it attaches to the NAE enzyme, resulting in its breakdown. This inhibitory effect hinders the NEDD8 protein's activation, blocking the neddylation mechanism in its entirety. This inhibition leads to an abundance of non-modified cullin proteins, which ultimately suppress the activity of the ubiquitin-proteasome system (UPS). This sequence of steps culminates in the build-up of ubiquitinated proteins and induces DNA injury reactions in cancer cells, resulting in cell cycle arrest, apoptosis, senescence, autophagy and alterations in mitochondrial function (figure 1.8) (Jia *et al* 2011, Zhao *et al* 2012, Lin *et al* 2017, Zhou *et al* 2019).

The etiology of cancer is inextricably linked to the nuances of the TME. The environment, which is characterized by the fusion of cellular units including immune cells, fibroblasts and endothelial cells surrounded by a comprehensive ECM, has fundamental consequences for neoplastic development (Roma-Rodrigues *et al* 2019). Extensive intercellular and matrix-associated collaborations within the TME underlie tumorigenesis and contribute to the considerable difficulties of therapeutic compliance, like drug resistance (Quail and Joyce 2013). CAFs play an essential regulatory role in tumorigenesis initiation and malignancy advancement by promoting proliferation, invasion and motility of malignant cells and representing a resource of MMPs for matrix breakdown (Kaplan *et al* 2005, Knops *et al* 2020, Mao *et al* 2021). Evidence also suggests that fibroblasts have an essential function in the development of pre-metastatic niches. Activated fibroblasts can trigger the stromal reorganization necessary for the progression of liver metastases in a melanoma mouse model (Olaso *et al* 1997). Proliferation of stellate cells, the fibroblasts encircling the liver sinusoids, has been seen in conjunction with early melanoma micrometastases. These cells became hyperactivated and released MMPs and chemotactic factors that promoted a favorable early metastatic milieu (Piquet *et al* 2019). Thereafter, hypoxic stimulation of angiogenic growth factors (mainly VEGFA) in stellate cells enrolled endothelial progenitor cells in the metastatic cavity, which eased the transformation from micrometastases to angiogenic macrometastases (Markiewski *et al* 2020).

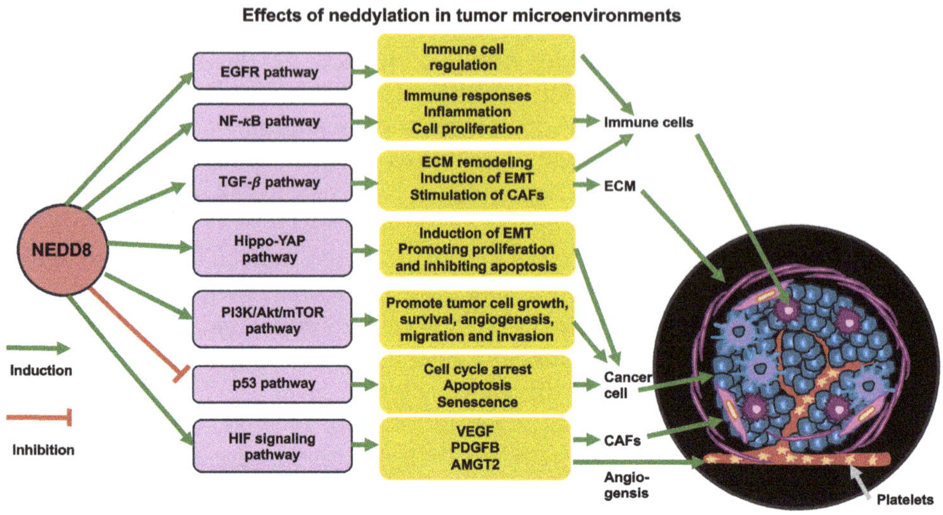

Figure 1.8. The intricate multi-step process of neddylation in cancer. Neddylation represents the post-translational binding of the neural precursor cell expressed developmentally downregulated protein 8 (NEDD8) protein to target proteins, fulfill different cellular functions and breakdown proteins. This mechanism comprises the maturation, activation, ligation and deneddylation steps and is carried out through specific enzymes like NEDP1, UBA3 and UBE2M. Moreover, the neddylation modification pronouncedly impacts the microenvironments of tumors. This modification can alter various factors, among them are vascular endothelial growth factor (VEGF), platelet-derived growth factor B (PDGFB), angiopoietin 2 (ANGPT2), the epithelial to mesenchymal transition (EMT) mechanism, cancer-associated fibroblasts (CAFs) and the ECM.

1.8.4.2 Cell-induced cross-linking of the ECM

EVs engage with the ECM, which serves as a physical framework for cell structure and functioning. The most recent research has demonstrated that EVs regulate the ECM and can function as a controller. This subsection begins by discussing the biogenesis of EVs and the mechanisms through which EVs are trafficked across the ECM. It also addresses how EVs serve as structural constituents of the matrix and as constituents that assist in ECM breakdown. Finally, the regulatory impact of EVs in modulating receiving cells to restructure the ECM in both pathological and therapeutic settings will be explored. The main constituents of the ECM comprise collagens, laminins, fibronectin and proteoglycans, which are arranged in a physically interconnected structure based on non-covalent cross-linking. The ECM engages with cells through cell adhesion molecules like integrins, cadherins and other transmembrane proteoglycans, which allow cells to travel through the matrix. Most importantly, cells can alter the matrix by laying down ECM constituents or by breaking down the ECM by releasing matrix-degrading enzymes like MMPs. The mechanical and physical characteristics of the ECM have been demonstrated to be of critical importance for the differentiation, movement and persistence of cells inside the ECM. These characteristics comprise: (1) mesh size, i.e. the spacing between two crosslinks inside a matrix; (2) stiffness, i.e. the degree to withstand deformation when the matrix is subjected to stress; and (3) viscoelastic properties, which is the capacity of the matrix to display both viscous and elastic characteristics when deformed (Mierke 2021b). Collectively, the constituents of the ECM and the mechanical properties of the matrix are crucial for the health and functioning of residing cells (Bonnans *et al* 2014, Humphrey *et al* 2014, Mouw *et al* 2014, Yue 2014). Therefore, EVs secreted by both healthy and pathological cells may function as facilitators of the ECM, either through direct EV-ECM mutual interactions or by affecting cell-ECM interference. For the accepting cell, EVs also travel through the ECM of the target tissue to arrive at the accepting cells and release their cargo to cause a cellular reaction. The transportation of EVs through the ECM could be a passive, diffusive operation that relies on the stiffness and viscoelastic characteristics of the ECM and the malleability of the EV itself (Chaudhuri *et al* 2015, Lenzini *et al* 2020). EVs may act as structural constituents of the matrix, either as triggers of calcification or as bioactive signaling cues embedded in the matrix, or they may mediate ECM modification indirectly by triggering receiving cells to synthesize or decompose ECM (Peinado *et al* 2012 2017, Costa-Silva *et al* 2015, Fong *et al* 2015, Nogués *et al* 2018, Siveen *et al* 2019). In view of their unexplored but important function, this subsection focuses on both the direct and indirect effects of EVs on the ECM. First, the biogenesis and characteristics of EVs and their trafficking through the ECM are outlined. In addition, the contribution of EVs in regulating the ECM in cancer propagation, as structural elements involved in bone/endochondral and vascular calcification and as ECM-connected bioactive signaling substances will be discussed. In addition, how EVs indirectly regulate the ECM in pathological as well as therapeutic situations will be examined.

1.8.4.3 Transportation of EVs through the ECM

EVs must cross the ECM from the mother cell to the receiving cell to partake in cell–cell signaling. Recent ECM stress relaxation experiments (Chaudhuri *et al* 2015, Lenzini *et al* 2020) have demonstrated that the mechanical characteristics of the ECM and EVs facilitate the diffusion of EVs through the extracellular space, notwithstanding the wider diameter of exosomes and microvesicles of 50–200 nm in comparison to the ECM mesh size of about 50 nm (Lenzini *et al* 2020). Thereby about 50% of the EVs carrying decellularized ECM of lung tissue have been liberated within 24 h. Using cross-linked alginate hydrogels with adjustable viscoelastic characteristics, EVs have been observed to have significantly higher diffusion coefficients in stiff, physically cross-linked alginate hydrogels when compared to soft viscoelastic and elastic hydrogels, implying that the ECM, which is subject to stress relaxation, results in enhanced diffusion of EVs across the matrix. These results have been valid for EVs from a wide range of cell sources, indicating that the mechanical characteristics of the matrix are important for EV transportation. In addition, the removal of the membrane water channel protein aquaporin-1 from the EV surface enhanced the stiffness of the EV and lowered the diffusion coefficient by a factor of about three. This points to the fact that the EVs are made more malleable by the water permeation and can therefore diffuse more effectively through the matrix (Clarke-Bland *et al* 2022).

1.8.4.4 Direct control of EVs of the ECM

It has been proposed that EVs can interact directly with the ECM and serve as both physical and bioactive structural constituents of the matrix. Moreover, EVs can bind and associate directly with the ECM and actively break down the matrix using surface-associated enzymes (Dolo *et al* 1999, Chaudhuri *et al* 2015, Sung *et al* 2015, Reiner *et al* 2017, Nawaz *et al* 2018, Sanderson *et al* 2019). In any event, EVs play an essential part in the development of the ECM in either physiological or pathological conditions (table 1.2). An example for the regulation of the ECM via EVs is the EV-driven calcification that is critical for bone formation and maintenance. Bone building and endochondral calcification mineralization of the ECM is a normal physiological occurrence in bone building and endochondral calcification, whereas it is a pathological occurrence in vascular calcification (Murshed *et al* 2004). The architecture of the ECM evolves rapidly during calcification from a non-crystalline lattice composed primarily of type 1 collagen fibrils to a crystalline matrix capable of withstanding high loads and stresses (Lin *et al* 2020). MVs with a diameter of 100–300 nm represent a subset of MVs that are critical for the ECM mineralization. The MVs secreted into the ECM by osteoblasts and chondrocytes trigger the generation of hydroxyapatite ($Ca_{10}(PO_4)_6(OH)_2$) crystals and the transition of the ECM from a purely organic organization to a composite of an organic and inorganic, such as mineralized, architecture. During the initial stage of mineralization, MVs bind Ca^{2+} ions through different calcium-binding molecules accumulated on the MV surface, among them the calcium-binding proteins annexin II, annexin V and annexin IV and the membrane phospholipid phosphatidylserine (PS) (Rojas *et al* 1992, Golub 2009, Veschi *et al* 2020). At the same time, the intra- and extravesicular levels of PO_4^{3-} ions are elevated through: firstly, MV membrane-bound tissue non-specific alkaline phosphate (TNAP), secondly, the turnover of ATP to ADP through ATPases, and thirdly Pit-1, a sodium-dependent phosphate transporter (Anderson 1995, Suzuki *et al* 2006, Golub 2009). The elevated intra-vesicular Ca^{2+} and PO_4^{3-} levels arising from this cause the precipitation of hydroxyapatite crystals inside the MV. In the second stage of mineralization, the hydroxyapatite crystals grow sufficiently large to penetrate through the MV into the ECM and proceed to mineralize the ECM alongside the type 1 collagen fibrils. It is noteworthy that MV-induced calcification appears not solely in endochondral calcification in development, but also in the continuous cycle of bone remodeling and bone formation in adults. Therefore, the MVs are significantly involved in the development of the ECM from a viscoelastic meshwork (Mierke 2021b) to a fully crystalline network under physiological calcification pathways (Patel *et al* 2019, Villa-Bellosta 2024). Vascular calcification refers to the mineralization of ECM in blood vessels and occurs in diseases like atherosclerosis, diabetes mellitus and chronic kidney disease (Wu *et al* 2013). Vascular calcification is triggered via transdifferentiation of VSMCs into an osteoblast-like phenotype. Under physiological circumstances, VSMCs liberate EVs comprising calcification blockers like Matrix Gla protein (MGP) and fetuin-A to sustain tissue homeostasis. Under pathogenic circumstances induced by chronic inflammation or disturbed mineral metabolism, nevertheless, osteoblast-like VSMCs liberate EVs that closely match

Table 1.2. EVs can directly alter the ECM.

Type of interaction	EV origin	Ev-driven ECM effects	Active EV constituents	References
EV-driven calcification	Osteoblasts	Foster calcification of the ECM in the process of bone and endochondral calcification	Annexin II, Annexin IV, Annexin V, Pit-1, Phosphatidylserine, TNAP	Rojas *et al* (1992), Golub (2009), Drabek *et al* (2011), Veschi *et al* (2020)
	Osteochondrogenic vascular smooth muscle cells	Facilitate calcification of the vasculature in the process of vascular calcification	Annexin I, Annexin II, Annexin IV, Annexin V, Annexin VI	Kapustin *et al* (2011), Rogers *et al* (2020)
	Macrophages	Facilitate procalcifying vesicles	S100A9 and Annexin V	New *et al* (2013), Kawakami *et al* (2020)
Bioactive signaling substances	ECM-attached vesicles	Induce anti-inflammatory signal transduction and cell growth in retinol ganglion cells and macrophages	—	Zhang *et al* (2013b), Huleihel *et al* (2016), Van Der Merwe *et al* (2019), Hussey *et al* (2020), Crum *et al* (2022), Turner *et al* (2022)
Direct alteration of the ECM	G361 melanoma cells	Break down collagen, fibronectin, and laminin; foster migration of cells	MMP-14	Page-McCaw *et al* (2007)
	HT-1080 fibro carcinoma cells	Break down collagen, fibronectin, and laminin; facilitate migration of cells	MMP-14	Hakulinen *et al* (2008)
	Nasal pharyngeal carcinoma cells	Break down collagen; induce migration of cells	MMP-13	You *et al* (2016)
	Vascular endothelial cells	Break down collagen, fibronectin, and laminin; encourage migration pf cells and angiogenesis	MMP-2, MMP-9, MMP-13	Taraboletti *et al* (2002)
	Myeloma cells	Break down heparin sulfate proteoglycans	Heparinase	Thompson *et al* (2013), Purushothaman *et al* (2016), Bandari *et al* (2018)
	Neutrophils	Break down collagen and elastase	Collagenase and elastase	De Jong *et al* (2016)
	Endothelial cells	Induce collagen cross-linking	Lysyl oxidase like-2	Genschmer *et al* (2019)
	Cancer cells	Contribute to matrix restructuring and breakdown; induce growth factor mobilization from ECM	Membrane-type 1 matrix metalloproteinase (MT1-MMP), insulin-degrading enzyme (IDE), sialidase, heparinases and glycosidases	Sanderson *et al* (2019)
	Cancer cells	Break down collagens as well as cleave and may activate soluble proteins that are shed from the cell surface	ADAMs and ADAMTS proteases	Shimoda and Khokha (2013)

MVs during the process of bone generation (Krohn *et al* 2016). These EVs are able to destroy the ECM of the blood vessel by calcifying certain areas of the vessel wall, which can result in thrombosis or tearing of the vessel (Ruiz *et al* 2016). The mechanism of EV-driven vascular calcification distinguishes itself from that of MV-driven mineralization of bone or cartilage. Under calcifying environmental parameters, VSMC-based EVs are fortified with phosphatidylserine and incorporate annexin I, annexin II, annexin V and annexin VI in a similar manner to MVs. Contrary to MVs, however, elevated extracellular Ca^{2+} concentration does not enhance TNAP activity within VSMC-originated EVs, implying that nucleation of calcifications is unrelated to TNAP (Kapustin *et al* 2011). Annexin I is concentrated in VSMC-generated EVs liberated through ectopic vascular calcification and facilitates the generation of EV aggregates in collagen fibrils of the ECM (Rogers *et al* 2020). These EV aggregates create mineralization nuclei on the EV surface, which subsequently expand in the face of elevated extracellular Ca^{2+} and PO_4^{3-} ion levels. Importantly, silencing of annexin-I in osteoblast-like VSMCs suppressed calcification development, suggesting that annexin-I-driven EV aggregation is required for calcification of vessels (Rogers *et al* 2020). In contrast to bone or cartilage, where mineralization takes place within the MV, vascular calcification is facilitated by the surface phospholipids and annexins of the EV assemblies distributed in the ECM.

1.9 Interplay between integrins and exosomes

Integrins fulfill a pivotal task in the function of exosomes and are especially important in targeting of exosomes to other recipient cells or the immune cell niche or pre-metastatic niche. As the role of integrins is so prominent, it is outlined and discussed in the following in more detail.

1.9.1 Integrins comprise specific functions on exosomes

The inheritance of integrins from cells to exosomes influences the biological mechanisms and processes of cancer and immune cells (figure 1.9).

They enhance cell adhesion and migration, escort cancer exosomes to certain metastatic locations, which is referred to as metastatic organotropism, pave the way for the development of pre-metastatic niches, regulate the angiogenic capacity of endothelial cells and enable the expression of specific genes such as Src and the pro-inflammatory and pro-migration S100 gene (Hoshino *et al* 2015, Gaballa *et al* 2020, Krishn *et al* 2020). The integrins expressed on the exosomes associate with their dedicated ligands on the receiving cells, and this crosstalk is an essential step in the interiorization of exosomes. Integrin $\alpha_4\beta_7$-expressing exosomes of T cells invade endothelial cells via binding to the ligand MAdCAM-1, that is expressed on the endothelial cells (Park *et al* 2019a, Soe *et al* 2019) (figure 1.9). Exosomes act as pivotal actors in cell–cell interaction by carrying integrins and corresponding kinases effectively between similar or dissimilar cells to exercise regulatory actions on receiving cells. For instance, the integrins α_2, $\alpha_V\beta_3$ and $\alpha_V\beta_6$ can be transferred

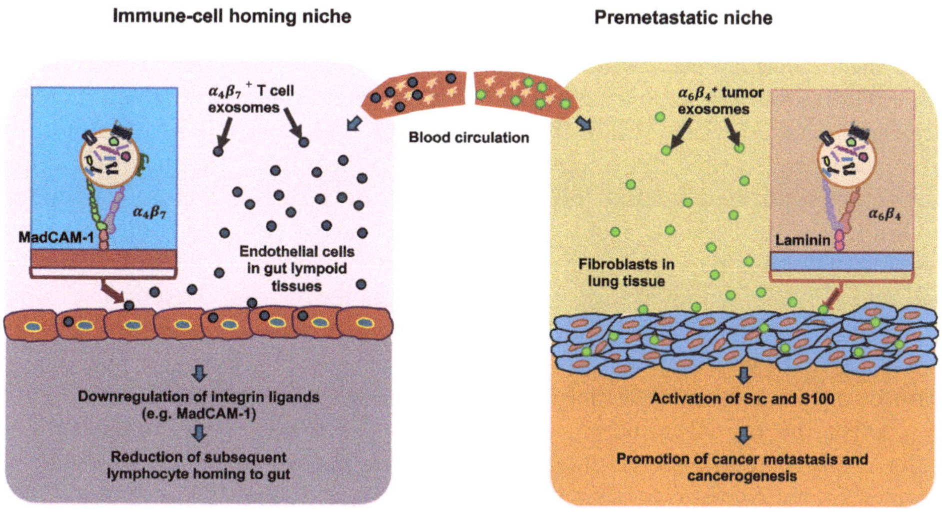

Figure 1.9. The homing niche is restructured by exosomal integrins of immune cells (left), whereas the pre-metastatic niche is altered by exosomal integrins of cancer cells (right).

between various prostate cancer cells (PrCa). Following the uptake of integrins, the receiving cells gain a more malignant cancer phenotype (Fedele *et al* 2015, Singh *et al* 2016, Lu *et al* 2018, Gaballa *et al* 2020, Krishn *et al* 2020). Apparently, exosomes can act as chemoattractive cues for the extravasation of neutrophils toward inflamed tissue areas. Exosomes secreted by neutrophils stuck to the endothelium in an integrin-β_2-driven fashion and locally liberated leukotriene B4 (LTB4), which is essential for neutrophil enrollment and extravasation in reaction to inflammatory responses (Subramanian *et al* 2020). The possible involvement of activated neutrophil exosomes in lung diseases like chronic obstructive pulmonary disease and bronchopulmonary dysplasia has also been identified. Homeostasis of ECM in lung tissues is disrupted through exosome elastase, which confers resistance to α1-antitrypsin, and exosome integrin $\alpha_M\beta_2$, which facilitates ECM protein degradation (Genschmer *et al* 2019). Moreover, the possible role of exosomal integrins in the pathogenesis of sepsis and systemic inflammatory syndrome has been observed (Kawamoto *et al* 2019). This study demonstrated that the expression of exosomal integrin β2 in the plasma of septic patients is considerably elevated compared to healthy subjects and appears to relate to hypotension and impaired renal performance (Kawamoto *et al* 2019). Exosomal integrins have also been found to confer resistance to medications and interfere with drug effectiveness. It has been revealed that integrin $\alpha_4\beta_7$-carrying circulating exosomes obtained from ulcerative colitis sufferers can impair the therapeutic effectiveness of the humanized monoclonal antibody vedolizumab, commonly employed for the therapy of inflammatory bowel disease. This is achieved by the sequestration of vedolizumab, whereby its attachment to $CD4^+$ T cells is compromised (Domenis *et al* 2020).

1.9.2 Governess of the function of integrins on exosomes

While the regulatory mechanisms underpinning integrin functioning in cells have been investigated for some decades, the precise mechanisms by which exosomal integrins are governed are still in their early stages of elucidation. For the first time, it has been proven that the integrin functions in exosomes, like in other cellular systems, are controlled by talin. TK-1 cells as a model (that is, a mouse T lymphoma cell line) were utilized to investigate the function of talin-2 within exosomes. Deletion of talin-2 had no effect on the level of vesicular expression of integrin $\alpha 4\beta 7$ and LFA-1 (synonymously referred to as integrin $\alpha_L\beta_2$). Deletion of talin-2, nevertheless, not exclusively decreased the engagement of exosomes with the integrin ligands MAdCAM-1 and ICAM-1, but also considerably decreased the incorporation of exosomes into endothelial cells. Remarkably, the existence of talin-1 did not abolish the talin-2-induced signaling effects seen in T cell exosomes. These results imply that talin-2 modulates integrin-dependent roles in exosomes via a comparable mechanism to that in cells (Soe *et al* 2019). These results justify additional investigations, especially of the roles of other integrin-regulating molecules in exosomes. Integrin-binding companion proteins like kindlins (as activator of integrins) and Shank-associated RH domain interactor (SHARPIN) (as inactivator of integrins) are also expressed in exosomes of cancer and immune cells. Therefore, they may be able to control integrin-driven biological processes in exosomes. The activities of integrins are driven alternatively through homotypic lateral engagement with themselves to promote integrin clustering and/or via heterotypic lateral commitment with tetraspanin and syndecan in the same cells to regulate Integrin-facilitated adhesiveness. Since there appears to be a lateral connection between integrins and tetraspanins, it is evident that tetraspanin proteins do not merely modulate integrin activity and clustering in cells, but also regulate integrin-driven cell adhesion and migration. Tetraspanins, like CD9, CD63 and CD81, are fortified compounds of the exosome membrane (Kowal *et al* 2016). There is growing indication that integrins and tetraspanins are co-expressed in the exosome partition. For example, tetraspanin 8 (Tspan8) selectively combines with integrin $\alpha 4\beta 1$ within pancreatic cancer exosomes, and the Tspan8-$\alpha 4\beta 1$ complex is necessary for exosomal engagement with endothelial cells and interiorization to drive stimulation of cancer angiogenesis (Nazarenko *et al* 2010). These findings raise the prospect that the functioning of exosomal integrins is regulated by tetraspanins, although additional scientific proof is required to confirm this hypothesis.

1.9.3 Integrin-controlled exosome biodistribution

The *in vivo* biodistribution scheme of exosomes is affected by many variables: origin of donor cells, delivery path, dosage of injected exosomes, identification of ligands and destination organs (Wiklander *et al* 2015). It has been shown that exosomes derived from a human breast cancer cell line (MDA-MB-231) and two pancreatic adenocarcinoma cell lines (BxPC-3 and HPAF-II) were preferentially dispersed in the lung and liver, respectively, after retro-orbital injection into nude mice (Hoshino *et al* 2015). MDA-MB-231-originated exosomes are uptaken threefold more efficiently in the lungs in comparison to BxPC-3 and HPAF-II exosomes. In complete opposition, the hepatic intake of BxPC-3 and HPAF-II exosomes is four to five times higher compared to MDA-MB-231 exosomes. It is hypothesized that this exosomal organotropism is facilitated through the interplay of exosomal integrins and cell-encircling ECM. Exosome proteomics, exosome capture experiments and integrin knockdown strategies have verified that the expression of integrin $\alpha_6\beta_4$ and $\alpha_6\beta_1$ in breast cancer exosomes is linked to metastasis in laminin-rich lung microenvironments. Moreover, the expression of integrin $\alpha v\beta 5$ in exosomes of pancreatic cancer has been related to metastasis in a fibronectin-rich hepatic microenvironment (Hoshino *et al* 2015). In this way, integrins bound to exosomes appear to have a crucial part to play in directing the spread of exosomes from tumors to their specific targets. In particular, the role of the exosomal integrin $\alpha_4\beta_7$ in controlling the tissue spreading of T cell exosomes has been investigated. It has been found that integrin $\alpha_4\beta_7$-expressing T cell exosomes are well dispersed in the small intestine, Peyer's patches, liver, spleen and mesenteric lymph node. Nevertheless, β_7 knockdown exosomes have not been well spread in the small intestinal mucosa when compared to control exosomes. This result indicates that $\alpha_4\beta_7$ is indispensable for the settlement of T cell exosomes within the small intestine (Park *et al* 2019a). Consequently, the integrin on the exosomes of T cells directs their destination and hence, the integrins serves as a label for exosomes trafficking and movement.

1.9.4 Integrin-driven exosome internalization

Exosomes utilize several mechanisms to enter the receiving cells including mechanisms of fusion, clathrin-induced endocytosis, caveolae-induced endocytosis, lipid raft-induced endocytosis, macropinocytosis and phagocytosis (Mulcahy *et al* 2014). Several research efforts have demonstrated the role of exosomal integrins in the process of signaling between receiving cells and exosomes. The integrins found on exosomes assist them in attaching to the surface of the receiving cells. The integrins $\alpha_V\beta_3$ and $\alpha_5\beta_1$ bound to exosomes are implicated in orchestrating the adhesion of exosomes to hepatic stellate cells (Chen and Brigstock 2016). B cell exosomes exhibit expression of integrins β_1 and β_2 on their external surface, and these integrins assist exosomes to attach to fibroblasts to elicit intracellular signaling pathways, like Ca^{2+} signal transduction (Clayton *et al* 2004). Even though no exosome incorporation tests have been conducted in these experiments, the attachment of exosomes to the cell surface is the first and crucial step for the following interiorization of exosomes into receiving cells. Integrin-induced incorporation of exosomes is thought to be essential for cancer cell to cancer cell interaction and metastasis of cancers. Integrin β_3 (through engagement with HSPG) is critical for exosome endocytosis in breast cancer cells and exosome-imposed colony development (Fuentes *et al* 2020). The interplay of exosomal integrins and cellular ligands determines the specific incorporation of cancer exosomes. Integrin $\alpha_6\beta_4$-expressing breast cancer exosomes tend to be taken up by pulmonary fibroblasts, while integrin $\alpha_V\beta_5$-expressing pancreatic cancer exosomes tend to be absorbed by liver Kupffer cells. These exosomes ultimately stimulate the expression of the pro-inflammatory S100 gene in the receiving cells and thus pave the way for the development of pre-metastatic niches (Hoshino *et al* 2015). Exosomal integrins are also essential for the incorporation of exosomes into the endothelial cells. T cell exosomes utilize their surface-attached LFA-1 and $\alpha_4\beta_7$ to invade mouse bEnd.3 brain endothelial cells, that display constitutive expression of integrin ligands ICAM-1 and MAdCAM-1 (Soe *et al* 2019). Moreover, T cell exosomes deliver their cargo, like miRNAs, into receiving cells to inhibit gene expression and regulate receiving cell functionality (Park *et al* 2019a). In a similar manner, macrophage-originated exosomes can pass into ICAM-1-expressing human brain endothelial cells through LFA-1. Neuroinflammation enhances the incorporation of these exosomes by 3.1-fold through triggering the expression of ICAM-1 on brain endothelial cells. In addition, it could be demonstrated *in vivo* that naive macrophage exosomes can overcome the blood–brain interface and transport a cargo protein, the brain-derived neurotrophic factor (BDNF), into the brain parenchyma, specifically in inflammatory situations (Yuan *et al* 2017). Tethering of rat adenocarcinoma cell-derived exosomes to endothelial cells and consequent interiorization has been found to be contingent on vesicular expression of the Tspan8 integrin-α_4 complex (Nazarenko *et al* 2010). Collectively, these investigations have demonstrated that integrin-ligand interactions are an essential driver of cellular incorporation of exosomes. The incorporation of exosomes is a requirement for proteins and nucleic acids to enter receiving cells where they can modify target cells' functional activities. An improved knowledge of the interplay between exosomal integrins and their target cells could provide an excellent opportunity for the administration of specific molecules for the therapy of different diseases.

1.9.5 Integrins and associated kinases can be transmitted via exosomes

The transfer of bioactive molecules conveyed by exosomes is a way of influencing the biological processes of receiving cells. The outcome of receiving cells can be modified through exosomal integrins and cognate kinases. There is no conclusive explanation of the principle of selective parceling of integrins into exosomes through the parent cells. Specific proteins, such as integrin α4, to be sequestered into the exosome partition could be conditionally integrated into tetraspanin-fortified, detergent-resistant MVB microdomains. Based on these experiments, it can be concluded that tetraspanins are probably implicated in the transfer of integrins into exosomes (Wubbolts *et al* 2003). Exosomal integrin trafficking of prostate cancer has been well established in multiple investigations. Androgen receptor-positive prostate cancer cells (PrCa) display aggressive characteristics, including enhanced proliferation, migration and invasion upon transmission of exosomal integrin α_2 from castration-resistant cells (Gaballa *et al* 2020). Integrin $\alpha_V\beta_3$, which is involved in aggressive cancers, is expressed in exosomes liberated from metastatic PrCa cells such as PC3 and CWR22Pc and is transmitted horizontally to non-cancerous BPH-1 cells or to other cancerous C4–2B cells. The two receiving cells then gain an $\alpha_V\beta_3$-derived migration phenotype (Singh *et al* 2016). A different research group also found that

$\alpha_V\beta_3$ integrin is enriched in exosomes derived from the human blood of PrCa patients and that integrin $\alpha_V\beta_3$ can be delivered onto $\alpha_V\beta_3$-negative PrCa cells through exosomes to elicit a migratory phenotype (Krishn *et al* 2020). Similarly, the integrin αVβ6, undetectable in healthy human prostate, is strongly expressed in PC-3 PrCa cells and can be delivered to $\alpha_V\beta_6$-negative DU-145 PrCa cells through exosomes. After exosome incorporation, DU-145 receiving cells exhibit enhanced adhesion and migration on $\alpha_V\beta_6$-specific latency-associated peptides (LAP) and transforming growth factor β (TGFβ).(Fedele *et al* 2015). In addition, $\alpha_V\beta_6$-expressing PC3 exosomes could be transmitted to β6-negative monocytes to trigger M2 polarization in the vicinity of the prostate cancers (Lu *et al* 2018). Through the exosome route, PrCa cells, such as PC-3, also transduce integrin $\alpha_V\beta_6$ to β_6-negative human microvascular endothelial cells (HMEC1) to enhance their proliferation, migration and tube development (Krishn *et al* 2020). These findings reveal that exosomes can be transferred to various types of cells. The exosomes carry specific receptor molecules or luminal cargo to modify the function and phenotype of receiving cells, whereby the tumor environment is altered and adapted to the cancers' requirements or the malignant progression. The general mechanism of exosome transformation of receiving cells is that targeting occurs via specific integrins located on the exosome membrane. Functional transmission of exosomal integrins is also commonly found in breast cancer. Current efforts have revealed that integrin $\alpha_V\beta_3$ of malignant MDA-MB-231 breast cancer cells is rapidly transduced to non-malignant MCF10A breast epithelial cells through exosomes (Altei *et al* 2020). In a similar manner, exosome-derived integrin β_4 from MDA-MB-231 breast cancer cells is relevant to breast cancer advancement and metabolic remodeling of CAFs, where integrin β_4 stimulates lactate formation and mitophagy (Sung *et al* 2020b). The presence of exosomal integrin $\alpha_M\beta_2$ has also been identified in hepatocellular carcinoma (HCC) metastasis, where the ability of HCC cells to migrate is enhanced by the presence of exosomal integrin secreted by M2 macrophages (Lu *et al* 2023). The participation of exosomal integrin β_1 in the promotion of non-alcoholic steatohepatitis is remarkable. Lipotoxin-treated hepatocytes liberate abundant integrin β1-fortified exosomes that stimulate β1-based enrollment of pro-inflammatory monocytes in the liver (Guo *et al* 2019).

ILK is contained in exosomes that are released by bone marrow-derived endothelial progenitor cells (EPCs). When ILK is delivered to mouse cardiac endothelial cells (MCECs), ILK-fortified EPC exosomes trigger the NF-κB signaling program and NF-κB-induced inflammatory gene expression in receiving cells. Lack of the anti-inflammatory cytokine IL-10 compromises the myocardial regeneration process via upregulating the accumulation of ILK within EPC exosomes (Yue *et al* 2020). In a similar way, the exosomal transport of ILK by breast cancer cells and normal breast epithelial cells is relevant for the development of mammary tumors (Bertolini *et al* 2020). ILK expression in cells, such as murine fibroblasts, can increase their invasiveness into 3D collagen matrices and contribute to a stiffer cell phenotype (Kunschmann *et al* 2017). Exosomes transmit also a downstream-effectors of integrins, which is FAK. Suppression of integrin β1 within PC3 cells decreases FAK concentration levels inside their exosomes, which has been associated with attachment-independent growth of PC3 PrCa cells (DeRita *et al* 2019). Similarly, receptor tyrosine kinases (EGFR, HER2) transferred from breast cancer exosomes encourage monocyte survival in inflammatory settings through activation of the mitogen-activated protein kinase (MAPK) signaling program within monocytes (Song *et al* 2016).

1.10 Exosomes can serve as biomarkers for cancer, regenerative reservoir and vaccine cargo-systems

EVs hold great promise as diagnostic and prognostic markers for a range of diseases, among them chronic cardiovascular diseases and cancer. In chronic cardiovascular pathologies like atherosclerosis, aortic stenosis and aortic aneurysms, the predictive value of EVs is linked to the identification of miRNAs like miR-222 and miR-143 as well as proteins like HSP60, and VCAM1 (Martin-Ventura *et al* 2022). In addition, specific molecules including proteins, proteoglycans and miRNAs in tumor-originated EVs have demonstrated diagnostic capability for specific cancer types (Melo *et al* 2015, Giallombardo *et al* 2016, Hu *et al* 2022). Interestingly, a subgroup analysis revealed that the utilization of small EVs as biomarkers in serum specimens of nervous system cancer is more precise compared to other types of cancer, suggesting that small EVs in serum are a powerful predictive instrument for the diagnosis of nervous system cancer (Liu *et al* 2021). In addition to their diagnostic value, EVs can also be used as therapeutic targets or instruments. One possible and promising way is to administer medication. EVs charged with therapeutic compounds can be released to the intended target cells. This delivery approach provides multiple benefits over conventional methods of delivering medications, such as protection of therapeutic agents from breakdown and the opportunity for accurate cell type targeting. This guarantees higher effectiveness and fewer adverse reactions (Sun *et al* 2023). In addition, EVs derived from available stem cells have become increasingly widespread in the regenerative medicine field. For example, EVs obtained from mesenchymal stem cells (MSCs) have therapeutic promise in a variety of diseases, among them cardiovascular, liver and renal diseases, encouraging tissue regeneration and healing (Park *et al* 2019b).

In terms of cancer, EVs can subtly adjust the tumor microenvironment and can potentially have a critical impact on cancer metastasis. Therefore, the incorporation of cancer-associated EVs may provide access to new therapeutic fields. Previous investigations have revealed phagocytosis as the mechanism responsible for the most efficient absorption of EVs by cancer and leukemia cells (Feng *et al* 2010, Emam *et al* 2018). This efficiency can be explained by the multiple surface receptors of phagocytic cells, which enable them to attach multiple ligands to EV surfaces, thus providing phagocytes with highly appropriate receiving cells (Gonda *et al* 2020). Therefore, phagocyte targeting of EVs can be beneficial for the investigation of new therapeutic approaches. This approach is not restricted to oncology, as immune cells have a substantial influence on many other diseases (Liu *et al* 2017b), like Parkinson's disease (Ren *et al* 2023) and neuroinflammation (Verweij *et al* 2021), where EV-induced macrophage polarization or phenotype shift in microglia has great implications. For this reason, much focus has been placed on the development of masking tactics to inhibit the detection of EVs by phagocytes. The introduction of molecules like CD47 or CD24 on EVs originating from cancer cells or fibroblasts protects them against phagocytosis (Kamerkar *et al* 2017, Karnas *et al* 2023). Additional molecules such as CD31, CD44 or β2-microglobulin are also being tested to enhance the evasion of phagocytosis and thus prolong the half-life of EVs in circulation (Parada *et al* 2021). Finally, EVs are investigated for application as cancer vaccines (Santos and Almeida 2021). More precisely, dendritic cell-originated exosomes (DCexos) and cancer cell-originated exosomes (Texos) are being investigated as cancer vaccine prospects. DCexos, which are carriers of major histocompatibility complex (MHC) class I/II and co-stimulatory molecules, have previously proven superior anti-tumor potency in preclinical models compared to dendritic cell vaccines; inversely, while Texos are alluring prospects for cancer vaccines, their inherent immunosuppressive capabilities represent a hurdle (Yao *et al* 2021).

1.11 Conclusions and future directions

The role of EVs, such as exosomes, has long been underestimated in the physics of cancer and also in oncology, as EVs are not only a product of the carcinogenesis process and malignant progression, but also an actor. The potential of EVs seems to be just emerging and the importance of the transported material, such as non-coding RNAs, is being recognized and the influence of epigenetic factors is becoming more and more important and is moving to the forefront of research. The usability of EVs in the treatment of diseases such as tumors should actually be sufficiently well known and established. However, this is not the case, as the mechanical effect of the vesicles and the uptake of the vesicles by non-cancer cells has not yet been worked out well. In particular, the effect of vesicles on tumor-associated cells such as fibroblasts and immune cells such as macrophages and T cells, which is often associated with a switch in their phenotype and mechanotype, should not be neglected. The integration of exosomes into the clinical field is limited by the absence of standardization of methods of exosome isolation and analysis (Doyle and Wang 2019). In addition, the proteomic profiling of EVs derived from the same origin has been shown to be sensitive to the isolation procedure (Yáñez-Mó *et al* 2015). In addition to the standardization of the isolation of EVs, their impact on mechanical and biological cues needs to be explored in more depth. Standardization of biophysical techniques and biological analysis is urgently needed. In particular, the vesicles seem to pass on the information of their lineage cells like the mechanical information to other cells of the same or a different type, for example to adapt the mechanical properties of the tumor metastasis niche. One can therefore also speak of a type of mechanical memory that is stored in the vesicles and perhaps in particular in the vesicle membrane and the membrane receptors. The membrane receptors of the vesicles may also have a completely different function.

The receptors in the membrane of EVs, such as aquaporin-1, can change the mechanical properties of the entire EVs. Therefore, I hypothesize that the mechanical properties of the receiving cell may be also altered in the receiving cell after the EV fuses with the receiving cell. Moreover, I suggest that this phenomenon can occur either locally at the site of EV integration into the plasma membrane of the receiving cell or the mechanical change spreads throughout the receiving cell. As the aquaporin-1 is associated with a specific migration and invasion mode of cancer cells, it seems to be likely that this specific migration mode can be induced through the aquaporin-1 carrying molecule directly in the receiving cells, such as cancer cells. I also hypothesize that this mechanism is a more universal one, as the EVs can target different kinds of cells. It can therefore be concluded that not only the cargo transported in the lumen of the vesicles drives the pathogenesis of diseases, but in particular, also the receptors and receptor-associated molecules anchored in the membrane, which can both themselves be passed on as information to other cells and/or determine the target cells, as these have counter-receptors for them. In addition, the composition of the vesicle membrane determines the mechanical phenotype of the vesicles and thus this mechanical information can also be transmitted to the target cells, which can be transferred locally at one site of the cell and overall, to the entire receiving cell. It should be investigated whether the receptor composition of the EVs or the mechanical characteristics of the EVs are decisive for the specific targeting to cells. Furthermore, it could be hypothesized that the mechanical properties of EVs are transferred to the receiving cells, thus leveling the mechanical floor of metastatic niches and determining the organotrophy of cancer metastasis. In the treatment of tumors, EVs could offer an escape mechanism for cancer cells to evade treatment with drugs. Research into the function of EVs is therefore a priority, especially as they are also to be used for drug delivery. There may be new hallmarks of cancer, such as the information of EVs transferred to cancer cells, cells embedded in the TME, such as fibroblasts and endothelial cells and immune cells, in the future, and it is being discussed whether there will be more and more hallmarks of cancer. The future gaps in EV research related to cancer are: what is the most important feature of EVs, e.g. charge, membrane receptors or mechanical properties? How to determine which components an EV contains. Can an EV be untethered from the target cell after some of its contents and membrane have been transferred to the target cell and fuse with other new target cells?

In summary, the role of vehicles in many different diseases can no longer be ignored. They also offer a very good and promising transport target for the delivery of therapeutic substances to a specific cell type in diseases such as cancer. The EVs enable the direct transfer of specific receptors to target cells such as cancer cells, which could change their function.

References and further reading

Abdel-Ghany M, Cheng H-C, Elble R C and Pauli B U 2001 The breast cancer β4 integrin and endothelial human CLCA2 mediate lung metastasis *J. Biol. Chem.* **276** 25438–46

Adell M A Y, Migliano S M, Upadhyayula S, Bykov Y S, Sprenger S, Pakdel M *et al* 2017 Recruitment dynamics of ESCRT-III and Vps4 to endosomes and implications for reverse membrane budding *eLife* **6** e31652

Aggarwal A, Iemma T L, Shih I, Newsome T P, McAllery S, Cunningham A L *et al* 2012 Mobilization of HIV spread by diaphanous 2 dependent filopodia in infected dendritic cells *PLoS Pathog.* **8** e1002762
Albacete-Albacete L, Navarro-Lérida I, López J A, Martín-Padura I, Astudillo A M, Ferrarini A *et al* 2020 ECM deposition is driven by caveolin-1–dependent regulation of exosomal biogenesis and cargo sorting *J. Cell Biol.* **219** e202006178
Ali A, Zambrano R, Duncan M R, Chen S, Luo S, Yuan H *et al* 2021 Hyperoxia-activated circulating extracellular vesicles induce lung and brain injury in neonatal rats *Sci. Rep.* **11** 8791
Alibhai F J, Lim F, Yeganeh A, DiStefano P V, Binesh-Marvasti T, Belfiore A *et al* 2020 Cellular senescence contributes to age-dependent changes in circulating extracellular vesicle cargo and function *Aging Cell* **19** e13103
Al-Jipouri A, Eritja À and Bozic M 2023 Unraveling the multifaceted roles of extracellular vesicles: insights into biology, pharmacology, and pharmaceutical applications for drug delivery *Int. J. Mol. Sci.* **25** 485
Al-Nedawi K, Meehan B, Micallef J, Lhotak V, May L, Guha A *et al* 2008 Intercellular transfer of the oncogenic receptor EGFRvIII by microvesicles derived from tumour cells *Nat. Cell Biol.* **10** 619–24
Altei W F, Pachane B C, Dos Santos P K, Ribeiro L N M, Sung B H, Weaver A M *et al* 2020 Inhibition of $\alpha v\beta 3$ integrin impairs adhesion and uptake of tumor-derived small extracellular vesicles *Cell Commun. Signal.* **18** 158
Anderson H C 1995 Molecular biology of matrix vesicles *Clin. Orthop. Rel. Res.* **314** 266–80
Ando H, Niki Y, Ito M, Akiyama K, Matsui M S, Yarosh D B *et al* 2012 Melanosomes are transferred from melanocytes to keratinocytes through the processes of packaging, release, uptake, and dispersion *J. Invest. Dermatol.* **132** 1222–9
Andreu Z and Yáñez-Mó M 2014 Tetraspanins in extracellular vesicle formation and function *Front. Immunol.* **5** 442
Antonyak M A and Cerione R A 2014 Microvesicles as mediators of intercellular communication in cancer *Cancer Cell Signaling* ed M Robles-Flores (New York: Springer) pp 147–73
Arya S B, Chen S, Jordan-Javed F and Parent C A 2022 Ceramide-rich microdomains facilitate nuclear envelope budding for non-conventional exosome formation *Nat. Cell Biol.* **24** 1019–28
Baietti M F, Zhang Z, Mortier E, Melchior A, Degeest G, Geeraerts A *et al* 2012 Syndecan–syntenin–ALIX regulates the biogenesis of exosomes *Nat. Cell Biol.* **14** 677–85
Bandari S K, Purushothaman A, Ramani V C, Brinkley G J, Chandrashekar D S, Varambally S *et al* 2018 Chemotherapy induces secretion of exosomes loaded with heparanase that degrades extracellular matrix and impacts tumor and host cell behavior *Matrix Biol.* **65** 104–18
Bang C, Batkai S, Dangwal S, Gupta S K, Foinquinos A, Holzmann A *et al* 2014 Cardiac fibroblast–derived microRNA passenger strand-enriched exosomes mediate cardiomyocyte hypertrophy *J. Clin. Invest.* **124** 2136–46
Barman B, Sung B H, Krystofiak E, Ping J, Ramirez M, Millis B *et al* 2022 VAP-A and its binding partner CERT drive biogenesis of RNA-containing extracellular vesicles at ER membrane contact sites *Dev. Cell* **57** 974–94.e8
Barr F A and Gruneberg U 2007 Cytokinesis: placing and making the final cut *Cell* **131** 847–60
Battistelli M and Falcieri E 2020 Apoptotic bodies: particular extracellular vesicles involved in intercellular communication *Biology* **9** 21

Bebelman M P, Smit M J, Pegtel D M and Baglio S R 2018 Biogenesis and function of extracellular vesicles in cancer *Pharmacol. Ther.* **188** 1–11

Beer K B, Rivas-Castillo J, Kuhn K, Fazeli G, Karmann B, Nance J F *et al* 2018 Extracellular vesicle budding is inhibited by redundant regulators of TAT-5 flippase localization and phospholipid asymmetry *Proc. Natl Acad. Sci.* **115** E1127–36

Bellahcène A, Castronovo V, Ogbureke K U E, Fisher L W and Fedarko N S 2008 Small integrin-binding ligand N-linked glycoproteins (SIBLINGs): multifunctional proteins in cancer *Nat. Rev. Cancer* **8** 212–26

Berditchevski F 2001 Complexes of tetraspanins with integrins: more than meets the eye *J. Cell Sci.* **114** 4143–51

Berditchevski F, Bazzoni G and Hemler M E 1995 Specific association Of CD63 with the VLA-3 and VLA-6 integrins *J. Biol. Chem.* **270** 17784–90

Berditchevski F, Chang S, Bodorova J and Hemler M E 1997a Generation of monoclonal antibodies to integrin-associated proteins *J. Biol. Chem.* **272** 29174–80

Berditchevski F, Tolias K F, Wong K, Carpenter C L and Hemler M E 1997b A novel link between integrins, transmembrane-4 superfamily proteins (CD63 and CD81), and phosphatidylinositol 4-kinase *J. Biol. Chem.* **272** 2595–8

Berditchevski F, Zutter M M and Hemler M E 1996 Characterization of novel complexes on the cell surface between integrins and proteins with 4 transmembrane domains (TM4 proteins) *Mol. Biol. Cell* **7** 193–207

Bergert M, Chandradoss S D, Desai R A and Paluch E 2012 Cell mechanics control rapid transitions between blebs and lamellipodia during migration *Proc. Natl Acad. Sci.* **109** 14434–9

Bertin A, De Franceschi N, De La Mora E, Maity S, Alqabandi M, Miguet N *et al* 2020 Human ESCRT-III polymers assemble on positively curved membranes and induce helical membrane tube formation *Nat. Commun.* **11** 2663

Bertolini I, Ghosh J C, Kossenkov A V, Mulugu S, Krishn S R, Vaira V *et al* 2020 Small extracellular vesicle regulation of mitochondrial dynamics reprograms a hypoxic tumor microenvironment *Dev. Cell* **55** 163–177.e6

Bianchi M, Renzini A, Adamo S and Moresi V 2017 Coordinated actions of MicroRNAs with other epigenetic factors regulate skeletal muscle development and adaptation *Int. J. Mol. Sci.* **18** 840

Bister N, Pistono C, Huremagic B, Jolkkonen J, Giugno R and Malm T 2020 Hypoxia and extracellular vesicles: a review on methods, vesicular cargo and functions *J. Extracell. Vesicles.* **10** e12002

Bonifacino J S and Glick B S 2004 The mechanisms of vesicle budding and fusion *Cell* **116** 153–66

Bonnans C, Chou J and Werb Z 2014 Remodelling the extracellular matrix in development and disease *Nat. Rev. Mol. Cell Biol.* **15** 786–801

Booth A M, Fang Y, Fallon J K, Yang J-M, Hildreth J E K and Gould S J 2006 Exosomes and HIV Gag bud from endosome-like domains of the T cell plasma membrane *J. Cell Biol.* **172** 923–35

Booth A, Marklew C J, Ciani B and Beales P A 2021 The influence of phosphatidylserine localisation and lipid phase on membrane remodelling by the ESCRT-II/ESCRT-III complex *Faraday Discuss.* **232** 188–202

Borghesan M, Fafián-Labora J, Eleftheriadou O, Carpintero-Fernández P, Paez-Ribes M, Vizcay-Barrena G *et al* 2019 Small extracellular vesicles are key regulators of non-cell autonomous intercellular communication in senescence via the interferon protein IFITM3 *Cell Rep.* **27** 3956–71.e6

Boura E, Ivanov V, Carlson L-A, Mizuuchi K and Hurley J H 2012 Endosomal Sorting Complex Required for Transport (ESCRT) complexes induce phase-separated microdomains in supported lipid bilayers *J. Biol. Chem.* **287** 28144–51

Bracq L, Xie M, Benichou S and Bouchet J 2018 Mechanisms for cell-to-cell transmission of HIV-1 *Front. Immunol.* **9** 260

Bricogne C, Fine M, Pereira P M, Sung J, Tijani M, Wang Y *et al* 2019 TMEM16F activation by Ca^{2+} triggers plasma membrane expansion and directs PD-1 trafficking *Sci. Rep.* **9** 619

Brzozowski J S, Bond D R, Jankowski H, Goldie B J, Burchell R, Naudin C *et al* 2018 Extracellular vesicles with altered tetraspanin CD9 and CD151 levels confer increased prostate cell motility and invasion *Sci. Rep.* **8** 8822

Busatto S, Morad G, Guo P and Moses M A 2021 The role of extracellular vesicles in the physiological and pathological regulation of the blood–brain barrier *FASEB BioAdv.* **3** 665–75

Busatto S, Morad G and Moses M 2022 Abstract 2190: brain-seeking extracellular vesicles derived from metastatic breast cancer cells modulate brain endothelial cell metabolism *Cancer Res.* **82** 2190–0

Buzas E I 2023 The roles of extracellular vesicles in the immune system *Nat. Rev. Immunol.* **23** 236–50

Cao M, Ning J, Hernandez-Lara C I, Belzile O, Wang Q, Dutcher S K *et al* 2015 Uni-directional ciliary membrane protein trafficking by a cytoplasmic retrograde IFT motor and ciliary ectosome shedding *eLife* **4** e05242

Carrière J, Bretin A, Darfeuille-Michaud A, Barnich N and Nguyen H T T 2016 Exosomes released from cells infected with Crohn's disease–associated adherent-invasive *Escherichia coli* activate host innate immune responses and enhance bacterial intracellular replication *Inflamm. Bowel Dis.* **22** 516–28

Caruso S and Poon I K H 2018 Apoptotic cell-derived extracellular vesicles: more than just debris *Front. Immunol.* **9** 1486

Catoni C, Di Paolo V, Rossi E, Quintieri L and Zamarchi R 2021 Cell-secreted vesicles: novel opportunities in cancer diagnosis, monitoring and treatment *Diagnostics* **11** 1118

Cha D J, Franklin J L, Dou Y, Liu Q, Higginbotham J N, Demory Beckler M *et al* 2015 KRAS-dependent sorting of miRNA to exosomes *eLife* **4** e07197

Charrin S, Jouannet S, Boucheix C and Rubinstein E 2014 Tetraspanins at a glance *J. Cell Sci.* **127** 3641–8

Charrin S, le Naour F, Silvie O, Milhiet P-E, Boucheix C and Rubinstein E 2009 Lateral organization of membrane proteins: tetraspanins spin their web *Biochem. J.* **420** 133–54

Chairoungdua A, Smith D L, Pochard P, Hull M and Caplan M J 2010 Exosome release of β-catenin: a novel mechanism that antagonizes WNT signaling *J. Cell Biol.* **190** 1079–91

Chaudhuri O, Gu L, Darnell M, Klumpers D, Bencherif S A, Weaver J C *et al* 2015 Substrate stress relaxation regulates cell spreading *Nat. Commun.* **6** 6365

Chen L and Brigstock D R 2016 Integrins and heparan sulfate proteoglycans on hepatic stellate cells (HSC) are novel receptors for HSC-derived exosomes *FEBS Lett.* **590** 4263–74

Chen X, Zhou J, Li X, Wang X, Lin Y and Wang X 2018 Exosomes derived from hypoxic epithelial ovarian cancer cells deliver microRNAs to macrophages and elicit a tumor-promoted phenotype *Cancer Lett.* **435** 80–91

Chiaruttini N, Redondo-Morata L, Colom A, Humbert F, Lenz M, Scheuring S *et al* 2015 Relaxation of loaded ESCRT-III spiral springs drives membrane deformation *Cell* **163** 866–79

Cho J A, Park H, Lim E H, Kim K H, Choi J S, Lee J H *et al* 2011 Exosomes from ovarian cancer cells induce adipose tissue-derived mesenchymal stem cells to acquire the physical and functional characteristics of tumor-supporting myofibroblasts *Gynecol. Oncol.* **123** 379–86

Choezom D and Gross J C 2022 Neutral sphingomyelinase 2 controls exosome secretion by counteracting V-ATPase-mediated endosome acidification *J. Cell Sci.* **135** jcs259324

Choi D, Montermini L, Kim D-K, Meehan B, Roth F P and Rak J 2018 The impact of oncogenic EGFRvIII on the proteome of extracellular vesicles released from glioblastoma cells *Mol. Cell. Proteomics* **17** 1948–64

Ciardiello C, Cavallini L, Spinelli C, Yang J, Reis-Sobreiro M, De Candia P *et al* 2016 Focus on extracellular vesicles: new frontiers of cell-to-cell communication in cancer *Int. J. Mol. Sci.* **17** 175

Ciardiello C, Leone A, Lanuti P, Roca M S, Moccia T, Minciacchi V R *et al* 2019 Large oncosomes overexpressing integrin alpha-V promote prostate cancer adhesion and invasion via AKT activation *J. Exp. Clin. Cancer Res.* **38** 317

Clancy J W, Schmidtmann M and D'Souza-Schorey C 2021 The ins and outs of microvesicles *FASEB BioAdv.* **3** 399–406

Clancy J W, Zhang Y, Sheehan C and D'Souza-Schorey C 2019 An ARF6–Exportin-5 axis delivers pre-miRNA cargo to tumour microvesicles *Nat. Cell Biol.* **21** 856–66

Clark D J, Fondrie W E, Yang A and Mao L 2016 Triple SILAC quantitative proteomic analysis reveals differential abundance of cell signaling proteins between normal and lung cancer-derived exosomes *J. Proteom.* **133** 161–9

Clarke-Bland C E, Bill R M and Devitt A 2022 Emerging roles for AQP in mammalian extracellular vesicles *Biochim. Biophys. Acta—Biomembr.* **1864** 183826

Clayton A, Mitchell J P, Court J, Mason M D and Tabi Z 2007 Human tumordDerived exosomes selectively impair lymphocyte responses to interleukin-2 *Cancer Res.* **67** 7458–66

Clayton A, Turkes A, Dewitt S, Steadman R, Mason M D and Hallett M B 2004 Adhesion and signaling by B cell-derived exosomes: the role of integrins *FASEB J.* **18** 977–9

Cocucci E and Meldolesi J 2015 Ectosomes and exosomes: shedding the confusion between extracellular vesicles *Trends Cell Biol.* **25** 364–72

Colombo M, Moita C, Van Niel G, Kowal J, Vigneron J, Benaroch P *et al* 2013 Analysis of ESCRT functions in exosome biogenesis, composition and secretion highlights the heterogeneity of extracellular vesicles *J. Cell Sci.* **126** 5553–65

Colombo M, Raposo G and Théry C 2014 Biogenesis, secretion, and intercellular interactions of exosomes and other extracellular vesicles *Annu. Rev. Cell Dev. Biol.* **30** 255–89

Coonrod E M and Stevens T H 2010 The yeast *vps* class *E mutants*: the beginning of the molecular genetic analysis of multivesicular body biogenesis *Mol. Biol. Cell* **21** 4057–60

Costa-Silva B, Aiello N M, Ocean A J, Singh S, Zhang H, Thakur B K *et al* 2015 Pancreatic cancer exosomes initiate pre-metastatic niche formation in the liver *Nat. Cell Biol.* **17** 816–26

Crenshaw B J, Gu L, Sims B and Matthews Q L 2018 Exosome biogenesis and biological function in response to viral infections *Open Virol. J.* **12** 134–48

Crescitelli R, Lässer C, Szabó T G, Kittel A, Eldh M, Dianzani I *et al* 2013 Distinct RNA profiles in subpopulations of extracellular vesicles: apoptotic bodies, microvesicles and exosomes *J. Extracell. Vesicles.* **2** 20677

Crespin M, Vidal C, Picard F, Lacombe C and Fontenay M 2009 Activation of PAK1/2 during the shedding of platelet microvesicles *Blood Coagul. Fibrin.* **20** 63–70

Crivelli S M, Giovagnoni C, Zhu Z, Tripathi P, Elsherbini A, Quadri Z *et al* 2022 Function of ceramide transfer protein for biogenesis and sphingolipid composition of extracellular vesicles *J. Extracell. Vesicles* **11** e12233

Crum R J, Hall K, Molina C P, Hussey G S, Graham E, Li H *et al* 2022 Immunomodulatory matrix-bound nanovesicles mitigate acute and chronic pristane-induced rheumatoid arthritis *Npj Regen. Med.* **7** 13

Cruz N M, Reddy R, McFaline-Figueroa J L, Tran C, Fu H and Freedman B S 2022 Modelling ciliopathy phenotypes in human tissues derived from pluripotent stem cells with genetically ablated cilia *Nat. Biomed. Eng.* **6** 463–75

Da Costa V R, Araldi R P, Vigerelli H, D'Ámelio F, Mendes T B, Gonzaga V *et al* 2021 Exosomes in the tumor microenvironment: from biology to clinical applications *Cells* **10** 2617

Dai J, Su Y, Zhong S, Cong L, Liu B, Yang J *et al* 2020 Exosomes: key players in cancer and potential therapeutic strategy *Signal Transduct. Target. Ther.* **5** 145

De Jong O G, Van Balkom B W M, Gremmels H and Verhaar M C 2016 Exosomes from hypoxic endothelial cells have increased collagen crosslinking activity through up-regulation of lysyl oxidase-like 2 *J. Cell. Mol. Med.* **20** 342–50

De La Cuesta F, Passalacqua I, Rodor J, Bhushan R, Denby L and Baker A H 2019 Extracellular vesicle cross-talk between pulmonary artery smooth muscle cells and endothelium during excessive TGF-β signalling: implications for PAH vascular remodelling *Cell Commun. Signal* **17** 143

Deen A J, Arasu U T, Pasonen-Seppänen S, Hassinen A, Takabe P, Wojciechowski S *et al* 2016 UDP-sugar substrates of HAS3 regulate its O-GlcNAcylation, intracellular traffic, extracellular shedding and correlate with melanoma progression *Cell. Mol. Life Sci.* **73** 3183–204

Demory Beckler M, Higginbotham J N, Franklin J L, Ham A-J, Halvey P J, Imasuen I E *et al* 2013 Proteomic analysis of exosomes from mutant KRAS colon cancer cells identifies intercellular transfer of mutant KRAS *Mol. Cell. Proteom.* **12** 343–55

DeRita R M, Sayeed A, Garcia V, Krishn S R, Shields C D, Sarker S *et al* 2019 Tumor-derived extracellular vesicles require β1 integrins to promote anchorage-independent growth *iScience* **14** 199–209

Desgrosellier J S, Lesperance J, Seguin L, Gozo M, Kato S, Franovic A *et al* 2014 Integrin αvβ3 drives slug activation and stemness in the pregnant and neoplastic mammary gland *Dev. Cell* **30** 295–308

Di Vizio D, Kim J, Hager M H, Morello M, Yang W, Lafargue C J *et al* 2009 Oncosome formation in prostate cancer: association with a region of frequent chromosomal deletion in metastatic disease *Cancer Res.* **69** 5601–9

Dixson A C, Dawson T R, Di Vizio D and Weaver A M 2023 Context-specific regulation of extracellular vesicle biogenesis and cargo selection *Nat. Rev. Mol. Cell Biol.* **24** 454–76

Dolo V, D'Ascenzo S, Violini S, Pompucci L, Festuccia C, Ginestra A *et al* 1999 Matrix-degrading proteinases are shed in membrane vesicles by ovarian cancer cells *in vivo* and *in vitro Clin. Exp. Metastasis* **17** 131–40

Domenis R, Marino M, Cifù A, Scardino G, Curcio F and Fabris M 2020 Circulating exosomes express α4β7 integrin and compete with CD4+ T cells for the binding to Vedolizumab *PLoS One* **15** e0242342

Donoso-Quezada J, Ayala-Mar S and González-Valdez J 2021 The role of lipids in exosome biology and intercellular communication: function, analytics and applications *Traffic* **22** 204–20

Doyle L and Wang M 2019 Overview of extracellular vesicles, their origin, composition, purpose, and methods for exosome isolation and analysis *Cells* **8** 727

Drabek K, Van De Peppel J, Eijken M and Van Leeuwen J P 2011 *GPM6B* regulates osteoblast function and induction of mineralization by controlling cytoskeleton and matrix vesicle release *J. Bone Miner. Res.* **26** 2045–51

Du M, Sansores-Garcia L, Zu Z and Wu K K 1998 Cloning and expression analysis of a novel salicylate suppressible gene, Hs-CUL-3, a member of Cullin/Cdc53 family *J. Biol. Chem.* **273** 24289–92

Dubreuil V, Marzesco A-M, Corbeil D, Huttner W B and Wilsch-Bräuninger M 2007 Midbody and primary cilium of neural progenitors release extracellular membrane particles enriched in the stem cell marker prominin-1 *J. Cell Biol.* **176** 483–95

Edgar J R, Eden E R and Futter C E 2014 Hrs- and CD63-dependent competing mechanisms make different sized endosomal intraluminal vesicles *Traffic* **15** 197–211

Ekström E J, Bergenfelz C, Von Bülow V, Serifler F, Carlemalm E, Jönsson G *et al* 2014 WNT5A induces release of exosomes containing pro-angiogenic and immunosuppressive factors from malignant melanoma cells *Mol. Cancer* **13** 88

El Andaloussi S, Mäger I, Breakefield X O and Wood M J A 2013 Extracellular vesicles: biology and emerging therapeutic opportunities *Nat. Rev. Drug Discov.* **12** 347–57

Elias R D, Ma W, Ghirlando R, Schwieters C D, Reddy V S and Deshmukh L 2020 Proline-rich domain of human ALIX contains multiple TSG101-UEV interaction sites and forms phosphorylation-mediated reversible amyloids *Proc. Natl Acad. Sci.* **117** 24274–84

Eliceiri B P, Puente X S, Hood J D, Stupack D G, Schlaepfer D D, Huang X Z *et al* 2002 Src-mediated coupling of focal adhesion kinase to integrin α v β 5 in vascular endothelial growth factor signaling *J. Cell Biol.* **157** 149–60

Emam S E, Ando H, Lila A S A, Shimizu T, Okuhira K, Ishima Y *et al* 2018 Liposome co-incubation with cancer cells secreted exosomes (extracellular vesicles) with different proteins expressions and different uptake pathways *Sci. Rep.* **8** 14493

Enchev R I, Schulman B A and Peter M 2015 Protein neddylation: beyond cullin–RING ligases *Nat. Rev. Mol. Cell Biol.* **16** 30–44

Escola J-M, Kleijmeer M J, Stoorvogel W, Griffith J M, Yoshie O and Geuze H J 1998 Selective enrichment of tetraspan proteins on the internal vesicles of multivesicular endosomes and on exosomes secreted by human B-lymphocytes *J. Biol. Chem.* **273** 20121–7

Ewing J 2005 Neoplastic diseases: a treatise on tumours *Br. J. Surg.* **16** 174–5

Fader C M and Colombo M I 2009 Autophagy and multivesicular bodies: two closely related partners *Cell Death Differ.* **16** 70–8

Fader C M, Sánchez D, Furlán M and Colombo M I 2008 Induction of autophagy promotes fusion of multivesicular bodies with autophagic vacuoles in K562 cells *Traffic* **9** 230–50

Fan S, Kroeger B, Marie P P, Bridges E M, Mason J D, McCormick K *et al* 2020 Glutamine deprivation alters the origin and function of cancer cell exosomes *EMBO J.* **39** e103009

Fang Y, Wang Z, Liu X and Tyler B M 2022 Biogenesis and biological functions of extracellular vesicles in cellular and organismal communication with microbes *Front. Microbiol.* **13** 817844

Fedele C, Singh A, Zerlanko B J, Iozzo R V and Languino L R 2015 The αvβ6 integrin is transferred intercellularly via exosomes *J. Biol. Chem.* **290** 4545–51

Felicetti F, De Feo A, Coscia C, Puglisi R, Pedini F, Pasquini L *et al* 2016 Exosome-mediated transfer of miR-222 is sufficient to increase tumor malignancy in melanoma *J. Transl. Med.* **14** 56

Feng D, Zhao W-L, Ye Y-Y, Bai X-C, Liu R-Q, Chang L-F *et al* 2010 Cellular internalization of exosomes occurs through phagocytosis *Traffic* **11** 675–87

Ferreira J V, Da Rosa Soares A, Ramalho J, Máximo Carvalho C, Cardoso M H, Pintado P *et al* 2022 LAMP2A regulates the loading of proteins into exosomes *Sci. Adv.* **8** eabm1140

Fidler I J and Kripke M L 1977 Metastasis results from preexisting variant cells within a malignant tumor *Science* **197** 893–5

Fong M Y, Zhou W, Liu L, Alontaga A Y, Chandra M, Ashby J *et al* 2015 Breast-cancer-secreted miR-122 reprograms glucose metabolism in premetastatic niche to promote metastasis *Nat. Cell Biol.* **17** 183–94

Fricke F, Michalak M, Warnken U, Hausser I, Schnölzer M, Kopitz J *et al* 2019 SILAC-based quantification of TGFBR2-regulated protein expression in extracellular vesicles of microsatellite unstable colorectal cancers *Int. J. Mol. Sci.* **20** 4162

Fuentes P, Sesé M, Guijarro P J, Emperador M, Sánchez-Redondo S, Peinado H *et al* 2020 ITGB3-mediated uptake of small extracellular vesicles facilitates intercellular communication in breast cancer cells *Nat. Commun.* **11** 4261

Fukushima M, Dasgupta D, Mauer A S, Kakazu E, Nakao K and Malhi H 2018 StAR-related lipid transfer domain 11 (STARD11)–mediated ceramide transport mediates extracellular vesicle biogenesis *J. Biol. Chem.* **293** 15277–89

Gaballa R, Ali H E A, Mahmoud M O, Rhim J S, Ali H I, Salem H F *et al* 2020 Exosomes-mediated transfer of itga2 promotes migration and invasion of prostate cancer cells by inducing epithelial-mesenchymal transition *Cancers* **12** 2300

Ganesan D and Cai Q 2021 Understanding amphisomes *Biochem.* J. **478** 1959–76

Garcia N A, Ontoria-Oviedo I, González-King H, Diez-Juan A and Sepúlveda P 2015 Glucose starvation in cardiomyocytes enhances exosome secretion and promotes angiogenesis in *Endothelial Cells PLoS One* **10** e0138849

Ge L, Zhang N, Li D, Wu Y, Wang H and Wang J 2020 Circulating exosomal small RNAs are promising non-invasive diagnostic biomarkers for gastric cancer *J. Cell. Mol. Med.* **24** 14502–13

Gennebäck N, Hellman U, Malm L, Larsson G, Ronquist G, Waldenström A *et al* 2013 Growth factor stimulation of cardiomyocytes induces changes in the transcriptional contents of secreted exosomes *J. Extracell. Vesicles.* **2** 20167

Genschmer K R, Russell D W, Lal C, Szul T, Bratcher P E, Noerager B D *et al* 2019 Activated PMN exosomes: pathogenic entities causing matrix destruction and disease in the lung *Cell* **176** 113–126.e15

Ghossoub R, Chéry M, Audebert S, Leblanc R, Egea-Jimenez A L, Lembo F *et al* 2020 Tetraspanin-6 negatively regulates exosome production *Proc. Natl Acad. Sci.* **117** 5913–22

Ghossoub R, Lembo F, Rubio A, Gaillard C B, Bouchet J, Vitale N *et al* 2014 Syntenin-ALIX exosome biogenesis and budding into multivesicular bodies are controlled by ARF6 and PLD2 *Nat. Commun.* **5** 3477

Giallombardo M, Chacártegui Borrás J, Castiglia M, Van Der Steen N, Mertens I, Pauwels P *et al* 2016 Exosomal miRNA analysis in non-small cell lung cancer (NSCLC) patients' plasma through qPCR: a feasible liquid biopsy tool *J. Vis. Exp.* 53900

Golub E E 2009 Role of matrix vesicles in biomineralization *Biochim. Biophys. Acta—Gen. Subj.* **1790** 1592–8

Gonda A, Moyron R, Kabagwira J, A. Vallejos P and R. Wall N 2020 Cellular-defined microenvironmental internalization of exosomes *Extracellular Vesicles and Their*

Importance in Human Health ed A Gil De Bona and J Antonio Reales Calderon (London: IntechOpen)

Gould G W and Lippincott-Schwartz J 2009 New roles for endosomes: from vesicular carriers to multi-purpose platforms *Nat. Rev. Mol. Cell Biol.* **10** 287–92

Gregory C D and Rimmer M P 2023 Extracellular vesicles arising from apoptosis: forms, functions, and applications *J. Pathol.* **260** 592–608

Grigoryeva E S, Tashireva L A, Savelieva O E, Zavyalova M V, Popova N O, Kuznetsov G A *et al* 2023 The association of integrins β3, β4, and αVβ5 on exosomes, CTCs and tumor cells with localization of distant metastasis in breast cancer patients *Int. J. Mol. Sci.* **24** 2929

Grigoryeva E, Tashireva L, Alifanov V, Savelieva O, Zavyalova M, Menyailo M *et al* 2024 Integrin-associated transcriptional characteristics of circulating tumor cells in breast cancer patients *PeerJ* **12** e16678

Gruenberg J 2020 Life in the lumen: the multivesicular endosome *Traffic* **21** 76–93

Gu X, Wang C, Deng H, Qing C, Liu R, Liu S *et al* 2020 Exosomal piRNA profiling revealed unique circulating piRNA signatures of cholangiocarcinoma and gallbladder carcinoma *Acta Biochim. Biophys. Sin.* **52** 475–84

Guescini M, Guidolin D, Vallorani L, Casadei L, Gioacchini A M, Tibollo P *et al* 2010 C2C12 myoblasts release micro-vesicles containing mtDNA and proteins involved in signal transduction *Exp. Cell. Res.* **316** 1977–84

Guix F X, Sannerud R, Berditchevski F, Arranz A M, Horré K, Snellinx A *et al* 2017 Tetraspanin 6: a pivotal protein of the multiple vesicular body determining exosome release and lysosomal degradation of amyloid precursor protein fragments *Mol. Neurodegener.* **12** 25

Guo B B, Bellingham S A and Hill A F 2015 The neutral sphingomyelinase pathway regulates packaging of the prion protein into exosomes *J. Biol. Chem.* **290** 3455–67

Guo Q, Furuta K, Lucien F, Gutierrez Sanchez L H, Hirsova P, Krishnan A *et al* 2019 Integrin β1-enriched extracellular vesicles mediate monocyte adhesion and promote liver inflammation in murine NASH *J. Hepatol.* **71** 1193–205

Gurunathan S, Kang M-H, Qasim M, Khan K and Kim J-H 2021 Biogenesis, membrane trafficking, functions, and next generation nanotherapeutics medicine of extracellular vesicles *Int. J. Nanomed.* **16** 3357–83

Gurung S, Perocheau D, Touramanidou L and Baruteau J 2021 The exosome journey: from biogenesis to uptake and intracellular signalling *Cell Commun. Signal.* **19** 47

Hakulinen J, Sankkila L, Sugiyama N, Lehti K and Keski-Oja J 2008 Secretion of active membrane type 1 matrix metalloproteinase (MMP-14) into extracellular space in microvesicular exosomes *J. Cell. Biochem.* **105** 1211–8

Han J, Yang B, Li Y, Li H, Zheng X, Yu L *et al* 2016 RhoB/ROCK mediates oxygen–glucose deprivation-stimulated syncytiotrophoblast microparticle shedding in preeclampsia *Cell Tissue Res.* **366** 411–25

Han Q-F, Li W-J, Hu K-S, Gao J, Zhai W-L, Yang J-H *et al* 2022 Exosome biogenesis: machinery, regulation, and therapeutic implications in cancer *Mol. Cancer* **21** 207

Hara M, Yanagihara T, Hirayama Y, Ogasawara S, Kurosawa H, Sekine S *et al* 2010 Podocyte membrane vesicles in urine originate from tip vesiculation of podocyte microvilli *Hum. Pathol.* **41** 1265–75

Haraszti R A, Didiot M, Sapp E, Leszyk J, Shaffer S A, Rockwell H E *et al* 2016 High-resolution proteomic and lipidomic analysis of exosomes and microvesicles from different cell sources *J. Extracell. Vesicles.* **5** 32570

Harding C, Heuser J and Stahl P 1984 Endocytosis and intracellular processing of transferrin and colloidal gold-transferrin in rat reticulocytes: demonstration of a pathway for receptor shedding *Eur. J. Cell Biol.* **35** 256–63

Hart I R and Fidler I J 1980 Role of organ selectivity in the determination of metastatic patterns of B16 melanoma *Cancer Res.* **40** 2281–7

He L, Wang G-P, Guo J-Y, Chen Z-R, Liu K and Gong S-S 2021 Epithelial–mesenchymal transition participates in the formation of vestibular flat epithelium *Front. Mol. Neurosci.* **14** 809878

Headley M B, Bins A, Nip A, Roberts E W, Looney M R, Gerard A *et al* 2016 Visualization of immediate immune responses to pioneer metastatic cells in the lung *Nature* **531** 513–7

Hemler M E 2005 Tetraspanin functions and associated microdomains *Nat. Rev. Mol. Cell Biol.* **6** 801–11

Henne W M, Buchkovich N J and Emr S D 2011 The ESCRT pathway *Dev. Cell* **21** 77–91

Hessvik N P and Llorente A 2018 Current knowledge on exosome biogenesis and release *Cell. Mol. Life Sci.* **75** 193–208

Hessvik N P, Øverbye A, Brech A, Torgersen M L, Jakobsen I S, Sandvig K *et al* 2016 PIKfyve inhibition increases exosome release and induces secretory autophagy *Cell. Mol. Life Sci.* **73** 4717–37

Hinger S A, Abner J J, Franklin J L, Jeppesen D K, Coffey R J and Patton J G 2020 Rab13 regulates sEV secretion in mutant KRAS colorectal cancer cells *Sci. Rep.* **10** 15804

Hiratsuka S, Nakamura K, Iwai S, Murakami M, Itoh T, Kijima H *et al* 2002 MMP9 induction by vascular endothelial growth factor receptor-1 is involved in lung-specific metastasis *Cancer Cell* **2** 289–300

Hitomi K, Okada R, Loo T M, Miyata K, Nakamura A J and Takahashi A 2020 DNA damage regulates senescence-associated extracellular vesicle release via the ceramide pathway to prevent excessive inflammatory responses *Int. J. Mol. Sci.* **21** 3720

Hong B S, Cho J-H, Kim H, Choi E-J, Rho S, Kim J *et al* 2009 Colorectal cancer cell-derived microvesicles are enriched in cell cycle-related mRNAs that promote proliferation of endothelial cells *BMC Genomics* **10** 556

Hori T, Osaka F, Chiba T, Miyamoto C, Okabayashi K, Shimbara N *et al* 1999 Covalent modification of all members of human cullin family proteins by NEDD8 *Oncogene* **18** 6829–34

Hoshino A, Costa-Silva B, Shen T-L, Rodrigues G, Hashimoto A, Tesic Mark M *et al* 2015 Tumour exosome integrins determine organotropic metastasis *Nature* **527** 329–35

Hsu Y-L, Hung J-Y, Chang W-A, Lin Y-S, Pan Y-C, Tsai P-H *et al* 2017 Hypoxic lung cancer-secreted exosomal miR-23a increased angiogenesis and vascular permeability by targeting prolyl hydroxylase and tight junction protein ZO-1 *Oncogene* **36** 4929–42

Hu Y, Tian Y, Di H, Xue C, Zheng Y, Hu B *et al* 2022 Noninvasive diagnosis of nasopharyngeal carcinoma based on phenotypic profiling of viral and tumor markers on plasma extracellular vesicles *Anal. Chem.* **94** 9740–9

Huang C, Hays F A, Tomasek J J, Benyajati S and Zhang X A 2020 Tetraspanin CD82 interaction with cholesterol promotes extracellular vesicle–mediated release of ezrin to inhibit tumour cell movement *J. Extracell. Vesicles.* **9** 1692417

Huang C, Liu D, Masuya D, Kameyama K, Nakashima T, Yokomise H *et al* 2004 MRP-1/CD9 gene transduction downregulates WNT signal pathways *Oncogene* **23** 7475–83

Huang X, Yuan T, Tschannen M, Sun Z, Jacob H, Du M *et al* 2013 Characterization of human plasma-derived exosomal RNAs by deep sequencing *BMC Genomics* **14** 319

Huleihel L, Hussey G S, Naranjo J D, Zhang L, Dziki J L, Turner N J *et al* 2016 Matrix-bound nanovesicles within ECM bioscaffolds *Sci. Adv.* **2** e1600502

Humphrey J D, Dufresne E R and Schwartz M A 2014 Mechanotransduction and extracellular matrix homeostasis *Nat. Rev. Mol. Cell Biol.* **15** 802–12

Hurley J H 2008 ESCRT complexes and the biogenesis of multivesicular bodies *Curr. Opin. Cell Biol.* **20** 4–11

Hurley J H and Hanson P I 2010 Membrane budding and scission by the ESCRT machinery: it's all in the neck *Nat. Rev. Mol. Cell Biol.* **11** 556–66

Hussey G S, Pineda Molina C, Cramer M C, Tyurina Y Y, Tyurin V A, Lee Y C *et al* 2020 Lipidomics and RNA sequencing reveal a novel subpopulation of nanovesicle within extracellular matrix biomaterials *Sci. Adv.* **6** eaay4361

Iguchi Y, Eid L, Parent M, Soucy G, Bareil C, Riku Y *et al* 2016 Exosome secretion is a key pathway for clearance of pathological TDP-43 *Brain* **139** 3187–201

Iizuka S, Kudo Y, Yoshida M, Tsunematsu T, Yoshiko Y, Uchida T *et al* 2011 Ameloblastin regulates osteogenic differentiation by inhibiting Src kinase via cross talk between integrin β1 and CD63 *Mol. Cell. Biol.* **31** 783–92

Imjeti N S, Menck K, Egea-Jimenez A L, Lecointre C, Lembo F, Bouguenina H *et al* 2017 Syntenin mediates SRC function in exosomal cell-to-cell communication *Proc. Natl Acad. Sci.* **114** 12495–500

Italiano J E, Richardson J L, Patel-Hett S, Battinelli E, Zaslavsky A, Short S *et al* 2008 Angiogenesis is regulated by a novel mechanism: pro- and antiangiogenic proteins are organized into separate platelet α granules and differentially released *Blood* **111** 1227–33

Jackson C E, Scruggs B S, Schaffer J E and Hanson P I 2017 Effects of inhibiting VPS4 support a general role for ESCRTs in extracellular vesicle biogenesis *Biophys. J.* **113** 1342–52

Jähne S, Rizzoli S O and Helm M S 2015 The structure and function of presynaptic endosomes *Exp. Cell. Res.* **335** 172–9

Jain S, Zuka M, Liu J, Russell S, Dent J, Guerrero J A *et al* 2007 Platelet glycoprotein Ibα supports experimental lung metastasis *Proc. Natl Acad. Sci.* **104** 9024–8

Jeppesen D K, Fenix A M, Franklin J L, Higginbotham J N, Zhang Q, Zimmerman L J *et al* 2019 Reassessment of exosome composition *Cell* **177** 428–45.e18

Jia L, Li H and Sun Y 2011 Induction of p21-dependent senescence by an NAE inhibitor, MLN4924, as a mechanism of growth suppression *Neoplasia* **13** 561–9

Jin D K, Shido K, Kopp H-G, Petit I, Shmelkov S V, Young L M *et al* 2006 Cytokine-mediated deployment of SDF-1 induces revascularization through recruitment of CXCR4+ hemangiocytes *Nat. Med.* **12** 557–67

Johnstone R M, Adam M, Hammond J R, Orr L and Turbide C 1987 Vesicle formation during reticulocyte maturation. Association of plasma membrane activities with released vesicles (exosomes) *J. Biol. Chem.* **262** 9412–20

Johnstone R M, Mathew A, Mason A B and Teng K 1991 Exosome formation during maturation of mammalian and avian reticulocytes: evidence that exosome release is a major route for externalization of obsolete membrane proteins *J. Cell. Physiol.* **147** 27–36

Jolly M K 2015 Implications of the hybrid epithelial/mesenchymal phenotype in metastasis *Front. Oncol.* **5** 155

Kahlert C and Kalluri R 2013 Exosomes in tumor microenvironment influence cancer progression and metastasis *J. Mol. Med.* **91** 431–7

Kalluri R 2016 The biology and function of fibroblasts in cancer *Nat. Rev. Cancer* **16** 582–98

Kalluri R and LeBleu V S 2020 The biology, function, and biomedical applications of exosomes *Science* **367** eaau6977

Kalra H, Drummen G and Mathivanan S 2016 Focus on extracellular vesicles: introducing the next small big thing *Int. J. Mol. Sci.* **17** 170

Kalra H, Simpson R J, Ji H, Aikawa E, Altevogt P, Askenase P *et al* 2012 Vesiclepedia: a compendium for extracellular vesicles with continuous community annotation *PLoS Biol.* **10** e1001450

Kamerkar S, LeBleu V S, Sugimoto H, Yang S, Ruivo C F, Melo S A *et al* 2017 Exosomes facilitate therapeutic targeting of oncogenic KRAS in pancreatic cancer *Nature* **546** 498–503

Kamitani T, Kito K, Nguyen H P and Yeh E T H 1997 Characterization of NEDD8, a developmentally down-regulated ubiquitin-like protein *J. Biol. Chem.* **272** 28557–62

Kaplan R N, Riba R D, Zacharoulis S, Bramley A H, Vincent L, Costa C *et al* 2005 VEGFR1-positive haematopoietic bone marrow progenitors initiate the pre-metastatic niche *Nature* **438** 820–7

Kapustin A N, Davies J D, Reynolds J L, McNair R, Jones G T, Sidibe A *et al* 2011 Calcium regulates key components of vascular smooth muscle cell–derived matrix vesicles to enhance mineralization *Circ. Res.* **109** e1–e12

Karnas E, Dudek P and Zuba-Surma E K 2023 Stem cell- derived extracellular vesicles as new tools in regenerative medicine—Immunomodulatory role and future perspectives *Front. Immunol.* **14** 1120175

Katz-Kiriakos E, Steinberg D F, Kluender C E, Osorio O A, Newsom-Stewart C, Baronia A *et al* 2021 Epithelial IL-33 appropriates exosome trafficking for secretion in chronic airway disease *JCI Insight* **6** e136166

Katzmann D J, Odorizzi G and Emr S D 2002 Receptor downregulation and multivesicular-body sorting *Nat. Rev. Mol. Cell Biol.* **3** 893–905

Kawakami R, Katsuki S, Travers R, Romero D C, Becker-Greene D, Passos L S A *et al* 2020 S100A9-RAGE axis accelerates formation of macrophage-mediated extracellular vesicle microcalcification in diabetes mellitus *Arterioscler. Thromb. Vasc. Biol.* **40** 1838–53

Kawamoto E, Masui-Ito A, Eguchi A, Soe Z Y, Prajuabjinda O, Darkwah S *et al* 2019 Integrin and PD-1 ligand expression on circulating extracellular vesicles in systemic inflammatory response syndrome and sepsis *Shock* **52** 13–22

Keller S, Sanderson M P, Stoeck A and Altevogt P 2006 Exosomes: from biogenesis and secretion to biological function *Immunol. Lett.* **107** 102–8

Kenneweg F, Bang C, Xiao K, Boulanger C M, Loyer X, Mazlan S *et al* 2019 Long noncoding RNA-enriched vesicles secreted by hypoxic cardiomyocytes drive cardiac fibrosis *Mol. Ther. —Nucleic Acids.* **18** 363–74

Kerviel A, Zhang M and Altan-Bonnet N 2021 A new infectious unit: extracellular vesicles carrying virus populations *Annu. Rev. Cell Dev. Biol.* **37** 171–97

Kesimer M, Scull M, Brighton B, DeMaria G, Burns K, O'Neal W *et al* 2009 Characterization of exosome-like vesicles released from human tracheobronchial ciliated epithelium: a possible role in innate defense *FASEB J.* **23** 1858–68

Keulers T G, Libregts S F, Beaumont J E J, Savelkouls K G, Bussink J, Duimel H *et al* 2021 Secretion of pro-angiogenic extracellular vesicles during hypoxia is dependent on the autophagy-related protein GABARAPL1 *J. Extracell. Vesicles* **10** e12166

Kim J H, Lee J, Park J and Gho Y S 2015 Gram-negative and Gram-positive bacterial extracellular vesicles *Semin. Cell Dev. Biol.* **40** 97–104

Kim J, Lee C, Kim I, Ro J, Kim J, Min Y *et al* 2020 Three-dimensional human liver-chip emulating premetastatic niche formation by breast cancer-derived extracellular vesicles *ACS Nano* **14** 14971–88

Kim K M, Abdelmohsen K, Mustapic M, Kapogiannis D and Gorospe M 2017 RNA in extracellular vesicles *WIREs RNA* **8** e1413

Kim Y H, Kwak M S, Lee B, Shin J M, Aum S, Park I H *et al* 2021 Secretory autophagy machinery and vesicular trafficking are involved in HMGB1 secretion *Autophagy* **17** 2345–62

King H W, Michael M Z and Gleadle J M 2012 Hypoxic enhancement of exosome release by breast cancer cells *BMC Cancer* **12** 421

Klumperman J and Raposo G 2014 The complex ultrastructure of the endolysosomal system *Cold Spring Harb. Perspect. Biol.* **6** a016857–a7

Knapp P E 1996 Proteolipid protein: is it more than just a structural component of myelin? *Dev. Neurosci.* **18** 297–308

Knops A M, South A, Rodeck U, Martinez-Outschoorn U, Harshyne L A, Johnson J *et al* 2020 Cancer-associated fibroblast density, prognostic characteristics, and recurrence in head and neck squamous cell carcinoma: a meta-analysis *Front. Oncol.* **10** 565306

Kore R A, Edmondson J L, Jenkins S V, Jamshidi-Parsian A, Dings R P M, Reyna N S *et al* 2018 Hypoxia-derived exosomes induce putative altered pathways in biosynthesis and ion regulatory channels in glioblastoma cells *Biochem. Biophys. Rep.* **14** 104–13

Kosaka N, Iguchi H, Hagiwara K, Yoshioka Y, Takeshita F and Ochiya T 2013 Neutral sphingomyelinase 2 (nSMase2)-dependent exosomal transfer of angiogenic MicroRNAs regulate cancer cell metastasis *J. Biol. Chem.* **288** 10849–59

Kosaka N, Iguchi H, Yoshioka Y, Takeshita F, Matsuki Y and Ochiya T 2010 Secretory mechanisms and intercellular transfer of microRNAs in living cells *J. Biol. Chem.* **285** 17442–52

Kowal J, Arras G, Colombo M, Jouve M, Morath J P, Primdal-Bengtson B *et al* 2016 Proteomic comparison defines novel markers to characterize heterogeneous populations of extracellular vesicle subtypes *Proc. Natl Acad. Sci.* **113** E968–77

Kreger B T, Dougherty A L, Greene K S, Cerione R A and Antonyak M A 2016 Microvesicle cargo and function changes upon induction of cellular transformation *J. Biol. Chem.* **291** 19774–85

Krishn S R, Salem I, Quaglia F, Naranjo N M, Agarwal E, Liu Q *et al* 2020 The $\alpha v\beta 6$ integrin in cancer cell-derived small extracellular vesicles enhances angiogenesis *J. Extracell. Vesicles* **9** 1763594

Krohn J B, Hutcheson J D, Martínez-Martínez E and Aikawa E 2016 Extracellular vesicles in cardiovascular calcification: expanding current paradigms *J. Physiol.* **594** 2895–903

Kucharzewska P, Christianson H C, Welch J E, Svensson K J, Fredlund E, Ringnér M *et al* 2013 Exosomes reflect the hypoxic status of glioma cells and mediate hypoxia-dependent activation of vascular cells during tumor development *Proc. Natl Acad. Sci.* **110** 7312–7

Kucia M, Reca R, Miekus K, Wanzeck J, Wojakowski W, Janowska-Wieczorek A *et al* 2005 Trafficking of normal stem cells and metastasis of cancer stem cells involve similar mechanisms: pivotal role of the SDF-1–CXCR4 axis *Stem Cells* **23** 879–94

Kugeratski F G, Hodge K, Lilla S, McAndrews K M, Zhou X, Hwang R F *et al* 2021 Quantitative proteomics identifies the core proteome of exosomes with syntenin-1 as the highest abundant protein and a putative universal biomarker *Nat. Cell Biol.* **23** 631–41

Kumar , Baba M A, Sadida S K, Marzooqi H Q, Al S, Jerobin J, Altemani F H *et al* 2024 Extracellular vesicles as tools and targets in therapy for diseases *Signal Transduct. Target. Ther.* **9** 27

Kumar R, Donakonda S, Müller S A, Lichtenthaler S F, Bötzel K, Höglinger G U *et al* 2020a Basic fibroblast growth factor 2-induced proteome changes endorse Lewy body pathology in hippocampal neurons *iScience* **23** 101349

Kumar R, Tang Q, Müller S A, Gao P, Mahlstedt D, Zampagni S *et al* 2020b Fibroblast growth factor 2-mediated regulation of neuronal exosome release depends on VAMP3/cellubrevin in hippocampal neurons *Adv. Sci.* **7** 1902372

Kumar S, Yoshida Y and Noda M 1993 Cloning of a cDNA which encodes a novel ubiquitin-like protein *Biochem. Biophys. Res. Commun.* **195** 393–9

Kummer D, Steinbacher T, Schwietzer M F, Thölmann S and Ebnet K 2020 Tetraspanins: integrating cell surface receptors to functional microdomains in homeostasis and disease *Med. Microbiol. Immunol. (Berl.)* **209** 397–405

Kunschmann T, Puder S, Fischer T, Perez J, Wilharm N and Mierke C T 2017 Integrin-linked kinase regulates cellular mechanics facilitating the motility in 3D extracellular matrices *Biochim. Biophys. Acta BBA—Mol. Cell Res.* **1864** 580–93

Landskron G, De La Fuente López M, Dubois-Camacho K, Díaz-Jiménez D, Orellana-Serradell O, Romero D *et al* 2019 Interleukin 33/ST2 axis components are associated to desmoplasia, a metastasis-related factor in colorectal cancer *Front. Immunol.* **10** 1394

Langemeyer L, Fröhlich F and Ungermann C 2018 Rab GTPase function in endosome and lysosome biogenesis *Trends Cell Biol.* **28** 957–70

Larios J, Mercier V, Roux A and Gruenberg J 2020 ALIX- and ESCRT-III–dependent sorting of tetraspanins to exosomes *J. Cell Biol.* **219** e201904113

Latysheva N, Muratov G, Rajesh S, Padgett M, Hotchin N A, Overduin M *et al* 2006 Syntenin-1 Is a new component of tetraspanin-enriched microdomains: mechanisms and consequences of the interaction of syntenin-1 with CD63 *Mol. Cell. Biol.* **26** 7707–18

Le M T N, Hamar P, Guo C, Basar E, Perdigão-Henriques R, Balaj L *et al* 2014 miR-200–containing extracellular vesicles promote breast cancer cell metastasis *J. Clin. Invest.* **124** 5109–28

Lee H, Li C, Zhang Y, Zhang D, Otterbein L E and Jin Y 2019 Caveolin-1 selectively regulates microRNA sorting into microvesicles after noxious stimuli *J. Exp. Med.* **216** 2202–20

Lee H, Zhang D, Zhu Z, Dela Cruz C S and Jin Y 2016 Epithelial cell-derived microvesicles activate macrophages and promote inflammation via microvesicle-containing microRNAs *Sci. Rep.* **6** 35250

Lee I-H, Kai H, Carlson L-A, Groves J T and Hurley J H 2015 Negative membrane curvature catalyzes nucleation of endosomal sorting complex required for transport (ESCRT)-III assembly *Proc. Natl Acad. Sci.* **112** 15892–7

Lee Y J, Shin K J and Chae Y C 2024 Regulation of cargo selection in exosome biogenesis and its biomedical applications in cancer *Exp. Mol. Med.* **56** 877–89

Leidal A M, Huang H H, Marsh T, Solvik T, Zhang D, Ye J *et al* 2020 The LC3-conjugation machinery specifies the loading of RNA-binding proteins into extracellular vesicles *Nat. Cell Biol.* **22** 187–99

Leng C, Zhang Z, Chen W, Luo H, Song J, Dong W *et al* 2016 An integrin beta4-EGFR unit promotes hepatocellular carcinoma lung metastases by enhancing anchorage independence through activation of FAK–AKT pathway *Cancer Lett.* **376** 188–96

Lenzini S, Bargi R, Chung G and Shin J-W 2020 Matrix mechanics and water permeation regulate extracellular vesicle transport *Nat. Nanotechnol.* **15** 217–23

Leong H S, Robertson A E, Stoletov K, Leith S J, Chin C A, Chien A E *et al* 2014 Invadopodia are required for cancer cell extravasation and are a therapeutic target for metastasis *Cell Rep.* **8** 1558–70

Levental I, Levental K R and Heberle F A 2020 Lipid rafts: controversies resolved, mysteries remain *Trends Cell Biol.* **30** 341–53

Li B, Antonyak M A, Zhang J and Cerione R A 2012 RhoA triggers a specific signaling pathway that generates transforming microvesicles in cancer cells *Oncogene* **31** 4740–9

Li B, Xu H, Han H, Song S, Zhang X, Ouyang L *et al* 2018 Exosome-mediated transfer of lncRUNX2-AS1 from multiple myeloma cells to MSCs contributes to osteogenesis *Oncogene* **37** 5508–19

Li G and Marlin M C 2015 Rab family of GTPases *Rab GTPases* ed G Li (New York: Springer) pp 1–15

Li I and Nabet B Y 2019 Exosomes in the tumor microenvironment as mediators of cancer therapy resistance *Mol. Cancer* **18** 32

Li S 2021 The basic characteristics of extracellular vesicles and their potential application in bone sarcomas *J. Nanobiotechnol.* **19** 277

Liebana-Jordan M, Brotons B, Falcon-Perez J M and Gonzalez E 2021 Extracellular vesicles in the fungi kingdom *Int. J. Mol. Sci.* **22** 7221

Lim H J, Yoon H, Kim H, Kang Y-W, Kim J-E, Kim O Y *et al* 2021 Extracellular vesicle proteomes shed light on the evolutionary, interactive, and functional divergence of their biogenesis mechanisms *Front. Cell Dev. Biol.* **9** 734950

Lima L G, Chammas R, Monteiro R Q, Moreira M E C and Barcinski M A 2009 Tumor-derived microvesicles modulate the establishment of metastatic melanoma in a phosphatidylserine-dependent manner *Cancer Lett.* **283** 168–75

Lin E Y and Pollard J W 2007 Tumor-associated macrophages press the angiogenic switch in breast cancer *Cancer Res.* **67** 5064–6

Lin S, Shang Z, Li S, Gao P, Zhang Y, Hou S *et al* 2017 Neddylation inhibitor MLN4924 induces G2 cell cycle arrest, DNA damage and sensitizes esophageal squamous cell carcinoma cells to cisplatin *Oncol. Lett.* **15** 2583–9

Lin X, Patil S, Gao Y-G and Qian A 2020 The bone extracellular matrix in bone formation and regeneration *Front. Pharmacol.* **11** 757

Liu D, Che X and Wu G 2024 Deciphering the role of neddylation in tumor microenvironment modulation: common outcome of multiple signaling pathways *Biomark. Res.* **12** 5

Liu L, Zhou Q, Xie Y, Zuo L, Zhu F and Lu J 2017a Extracellular vesicles: novel vehicles in herpesvirus infection *Virol. Sin.* **32** 349–56

Liu Q, Yu B, Tian Y, Dan J, Luo Y and Wu X 2020 P53 mutant $p53^{N236S}$ regulates cancer-associated fibroblasts properties through Stat3 pathway *OncoTargets Ther.* **13** 1355–63

Liu S, Liao Y, Hosseinifard H, Imani S and Wen Q 2021 Diagnostic role of extracellular vesicles in cancer: a comprehensive systematic review and meta-analysis *Front. Cell Dev. Biol.* **9** 705791

Liu Y, Wang Z, Liu Y, Zhu G, Jacobson O, Fu X *et al* 2017b Suppressing nanoparticle-mononuclear phagocyte system interactions of two-dimensional gold nanorings for improved tumor accumulation and photothermal ablation of tumors *ACS Nano* **11** 10539–48

Liu Y-J and Wang C 2023 A review of the regulatory mechanisms of extracellular vesicles-mediated intercellular communication *Cell Commun. Signal.* **21** 77

Llorente A, Skotland T, Sylvänne T, Kauhanen D, Róg T, Orłowski A *et al* 2013 Molecular lipidomics of exosomes released by PC-3 prostate cancer cells *Biochim. Biophys. Acta BBA—Mol. Cell Biol. Lipids* **1831** 1302–9

López-Otín C and Matrisian L M 2007 Emerging roles of proteases in tumour suppression *Nat. Rev. Cancer* **7** 800–8

Lu H, Bowler N, Harshyne L A, Craig Hooper D, Krishn S R, Kurtoglu S *et al* 2018 Exosomal $\alpha v\beta 6$ integrin is required for monocyte M2 polarization in prostate cancer *Matrix Biol.* **70** 20–35

Lu Y, Han G, Zhang Y, Zhang L, Li Z, Wang Q *et al* 2023 M2 macrophage-secreted exosomes promote metastasis and increase vascular permeability in hepatocellular carcinoma *Cell Commun. Signal.* **21** 299

Luga V, Zhang L, Viloria-Petit A M, Ogunjimi A A, Inanlou M R, Chiu E *et al* 2012 Exosomes mediate stromal mobilization of autocrine Wnt-PCP signaling in breast cancer cell migration *Cell* **151** 1542–56

Lyu C, Sun H, Sun Z, Liu Y and Wang Q 2024 Roles of exosomes in immunotherapy for solid cancers *Cell Death Dis* **15** 106

Ma H, Wang H, Tian F, Zhong Y, Liu Z and Liao A 2020 PIWI-interacting RNA-004800 is regulated by S1P receptor signaling pathway to keep myeloma cell survival *Front. Oncol.* **10** 438

Ma L, Li Y, Peng J, Wu D, Zhao X, Cui Y *et al* 2015 Discovery of the migrasome, an organelle mediating release of cytoplasmic contents during cell migration *Cell Res.* **25** 24–38

Mageswaran S K, Yang W Y, Chakrabarty Y, Oikonomou C M and Jensen G J 2021 A cryo–electron tomography workflow reveals protrusion-mediated shedding on injured plasma membrane *Sci. Adv.* **7** eabc6345

Mahabeleshwar G H, Feng W, Reddy K, Plow E F and Byzova T V 2007 Mechanisms of integrin–vascular ndothelial growth factor receptor cross-activation in angiogenesis *Circ. Res.* **101** 570–80

Maia J, Caja S, Strano Moraes M C, Couto N and Costa-Silva B 2018 Exosome-based cell-cell communication in the tumor microenvironment *Front. Cell Dev. Biol.* **6** 18

Makiguchi T, Yamada M, Yoshioka Y, Sugiura H, Koarai A, Chiba S *et al* 2016 Serum extracellular vesicular miR-21-5p is a predictor of the prognosis in idiopathic pulmonary fibrosis *Respir. Res.* **17** 110

Mannion B A, Berditchevski F, Kraeft S K, Chen L B and Hemler M E 1996 Transmembrane-4 superfamily proteins CD81 (TAPA-1), CD82, CD63, and CD53 specifically associated with integrin alpha 4 beta 1 (CD49d/CD29) *J. Immun., Balt.* **157** 2039–47

Mao X, Xu J, Wang W, Liang C, Hua J, Liu J *et al* 2021 Crosstalk between cancer-associated fibroblasts and immune cells in the tumor microenvironment: new findings and future perspectives *Mol. Cancer* **20** 131

Marie P P, Fan S-J, Mendes C C, Wainwright S M, Harris A L, Goberdhan D C I *et al* 2020 Accessory ESCRT-III proteins selectively regulate Rab11-exosome biogenesis in Drosophila secondary cells bioRxiv:2020.06.18.158725 https://www.biorxiv.org/content/10.1101/2020.06.18.158725v1

Mariscal J, Vagner T, Kim M, Zhou B, Chin A, Zandian M *et al* 2020 Comprehensive palmitoyl-proteomic analysis identifies distinct protein signatures for large and small cancer-derived extracellular vesicles *J. Extracell. Vesicles.* **9** 1764192

Markiewski M M, Daugherity E, Reese B and Karbowniczek M 2020 The role of complement in angiogenesis *Antibodies* **9** 67

Martin-Ventura J L, Roncal C, Orbe J and Blanco-Colio L M 2022 Role of extracellular vesicles as potential diagnostic and/or therapeutic biomarkers in chronic cardiovascular diseases *Front. Cell Dev. Biol.* **10** 813885

Marzesco A-M, Janich P, Wilsch-Bräuninger M, Dubreuil V, Langenfeld K, Corbeil D *et al* 2005 Release of extracellular membrane particles carrying the stem cell marker prominin-1 (CD133) from neural progenitors and other epithelial cells *J. Cell Sci.* **118** 2849–58

Marzesco A-M, Wilsch-Bräuninger M, Dubreuil V, Janich P, Langenfeld K, Thiele C *et al* 2009 Release of extracellular membrane vesicles from microvilli of epithelial cells is enhanced by depleting membrane cholesterol *FEBS Lett.* **583** 897–902

Mashouri L, Yousefi H, Aref A R, Ahadi A M, Molaei F and Alahari S K 2019 Exosomes: composition, biogenesis, and mechanisms in cancer metastasis and drug resistance *Mol. Cancer* **18** 75

Mathieu M, Névo N, Jouve M, Valenzuela J I, Maurin M, Verweij F J *et al* 2021 Specificities of exosome versus small ectosome secretion revealed by live intracellular tracking of CD63 and CD9 *Nat. Commun.* **12** 4389

Matsui T, Osaki F, Hiragi S, Sakamaki Y and Fukuda M 2021 ALIX and ceramide differentially control polarized small extracellular vesicle release from epithelial cells *EMBO Rep.* **22** e51475

Matsuo H, Chevallier J, Mayran N, Le Blanc I, Ferguson C, Fauré J *et al* 2004 Role of LBPA and alix in multivesicular liposome formation and endosome organization *Science* **303** 531–4

Mattick J S, Amaral P P, Carninci P, Carpenter S, Chang H Y, Chen L-L *et al* 2023 Long non-coding RNAs: definitions, functions, challenges and recommendations *Nat. Rev. Mol. Cell Biol.* **24** 430–47

Maurya D K, Berghard A and Bohm S 2022 A multivesicular body-like organelle mediates stimulus-regulated trafficking of olfactory ciliary transduction proteins *Nat. Commun.* **13** 6889

McAndrews K M and Kalluri R 2019 Mechanisms associated with biogenesis of exosomes in cancer *Mol. Cancer* **18** 52

McConnell R E, Higginbotham J N, Shifrin D A, Tabb D L, Coffey R J and Tyska M J 2009 The enterocyte microvillus is a vesicle-generating organelle *J. Cell Biol.* **185** 1285–98

McKenzie A J, Hoshino D, Hong N H, Cha D J, Franklin J L, Coffey R J *et al* 2016 KRAS-MEK signaling controls Ago2 sorting into exosomes *Cell Rep.* **15** 978–87

McMahon H T and Boucrot E 2015 Membrane curvature at a glance *J. Cell Sci.* **128** 1065–70

Meehan B, Rak J and Di Vizio D 2016 Oncosomes—large and small: what are they, where they came from? *J. Extracell. Vesicles.* **5** 33109

Meldolesi J 2018 Exosomes and ectosomes in intercellular communication *Curr. Biol.* **28** R435–44

Meldolesi J 2021 Extracellular vesicles (exosomes and ectosomes) play key roles in the pathology of brain diseases *Mol. Biomed.* **2** 18

Melentijevic I, Toth M L, Arnold M L, Guasp R J, Harinath G, Nguyen K C *et al* 2017 *C. elegans* neurons jettison protein aggregates and mitochondria under neurotoxic stress *Nature* **542** 367–71

Melo S A, Luecke L B, Kahlert C, Fernandez A F, Gammon S T, Kaye J *et al* 2015 Glypican-1 identifies cancer exosomes and detects early pancreatic cancer *Nature* **523** 177–82

Menck K, Sönmezer C, Worst T S, Schulz M, Dihazi G H, Streit F *et al* 2017 Neutral sphingomyelinases control extracellular vesicles budding from the plasma membrane *J. Extracell. Vesicles* **6** 1378056

Mergner J and Schwechheimer C 2014 The NEDD8 modification pathway in plants *Front. Plant Sci.* **5** 103

Mettlen M, Chen P-H, Srinivasan S, Danuser G and Schmid S L 2018 Regulation of clathrin-mediated endocytosis *Annu. Rev. Biochem.* **87** 871–96

Mierke C T 2019 The matrix environmental and cell mechanical properties regulate cell migration and contribute to the invasive phenotype of cancer cells *Rep. Prog. Phys.* **82** 064602

Mierke C T 2020 Mechanical cues affect migration and invasion of cells from three different directions *Front. Cell Dev. Biol.* **8** 583226

Mierke C T 2021a Bidirectional mechanical response between cells and their microenvironment *Front. Phys.* **9** 749830

Mierke C T 2021b Viscoelasticity acts as a marker for tumor extracellular matrix characteristics *Front. Cell Dev. Biol.* **9** 785138

Mierke C T 2024 *Intricate Synergy of Mechanical and Biochemical Cues in the Transmigration of Cancer Cells Across the Endothelium* (Cham: Springer International Publishing)

Mierke C T, Sauer F, Grosser S, Puder S, Fischer T and Käs J A 2018 The two faces of enhanced stroma: stroma acts as a tumor promoter and a steric obstacle *NMR Biomed.* **31** e3831

Minciacchi V R, You S, Spinelli C, Morley S, Zandian M, Aspuria P-J *et al* 2015 Large oncosomes contain distinct protein cargo and represent a separate functional class of tumor-derived extracellular vesicles *Oncotarget* **6** 11327–41

Mittelbrunn M, Gutiérrez-Vázquez C, Villarroya-Beltri C, González S, Sánchez-Cabo F, González M Á *et al* 2011 Unidirectional transfer of microRNA-loaded exosomes from T cells to antigen-presenting cells *Nat. Commun.* **2** 282

Mobarak H, Javid F, Narmi M T, Mardi N, Sadeghsoltani F, Khanicheragh P *et al* 2024 Prokaryotic microvesicles Ortholog of eukaryotic extracellular vesicles in biomedical fields *Cell Commun. Signal.* **22** 80

Mohan A, Agarwal S, Clauss M, Britt N S and Dhillon N K 2020 Extracellular vesicles: novel communicators in lung diseases *Respir. Res.* **21** 175

Mohieldin A M, Pala R, Beuttler R, Moresco J J, Yates J R and Nauli S M 2021 Ciliary extracellular vesicles are distinct from the cytosolic extracellular vesicles *J. Extracell. Vesicles.* **10** e12086

Moon H-G, Cao Y, Yang J, Lee J H, Choi H S and Jin Y 2015 Lung epithelial cell-derived extracellular vesicles activate macrophage-mediated inflammatory responses via ROCK1 pathway *Cell Death Dis* **6** e2016–6

Moro L 1998 Integrins induce activation of EGF receptor: role in MAP kinase induction and adhesion-dependent cell survival *EMBO J.* **17** 6622–32

Moulin C, Crupi M J F, Ilkow C S, Bell J C and Boulton S 2023 Extracellular vesicles and viruses: two intertwined entities *Int. J. Mol. Sci.* **24** 1036

Mouw J K, Ou G and Weaver V M 2014 Extracellular matrix assembly: a multiscale deconstruction *Nat. Rev. Mol. Cell Biol.* **15** 771–85

Mu W, Rana S and Zöller M 2013 Host matrix modulation by tumor exosomes promotes motility and invasiveness *Neoplasia* **15** 875–IN4

Mulcahy L A, Pink R C and Carter D R F 2014 Routes and mechanisms of extracellular vesicle uptake *J. Extracell. Vesicles.* **3** 24641

Muller L, Mitsuhashi M, Simms P, Gooding W E and Whiteside T L 2016 Tumor-derived exosomes regulate expression of immune function-related genes in human T cell subsets *Sci. Rep.* **6** 20254

Munich S, Sobo-Vujanovic A, Buchser W J, Beer-Stolz D and Vujanovic N L 2012 Dendritic cell exosomes directly kill tumor cells and activate natural killer cells via TNF superfamily ligands *OncoImmunology* **1** 1074–83

Muralidharan-Chari V, Clancy J, Plou C, Romao M, Chavrier P, Raposo G *et al* 2009 ARF6-regulated shedding of tumor cell-derived plasma membrane microvesicles *Curr. Biol.* **19** 1875–85

Murrow L, Malhotra R and Debnath J 2015 ATG12–ATG3 interacts with Alix to promote basal autophagic flux and late endosome function *Nat. Cell Biol.* **17** 300–10

Murshed M, Schinke T, McKee M D and Karsenty G 2004 Extracellular matrix mineralization is regulated locally; different roles of two gla-containing proteins *J. Cell Biol.* **165** 625–30

Nabhan J F, Hu R, Oh R S, Cohen S N and Lu Q 2012 Formation and release of arrestin domain-containing protein 1-mediated microvesicles (ARMMs) at plasma membrane by recruitment of TSG101 protein *Proc. Natl Acad. Sci.* **109** 4146–51

Nagata S 2018 Apoptosis and clearance of apoptotic cells *Annu. Rev. Immunol.* **36** 489–517

Namazi H, Mohit E, Namazi I, Rajabi S, Samadian A, Hajizadeh-Saffar E *et al* 2018 Exosomes secreted by hypoxic cardiosphere-derived cells enhance tube formation and increase pro-angiogenic miRNA *J. Cell. Biochem.* **119** 4150–60

Nawaz M, Shah N, Zanetti B R, Maugeri M, Silvestre R N, Fatima F *et al* 2018 Extracellular vesicles and matrix remodeling enzymes: the emerging roles in extracellular matrix remodeling, progression of diseases and tissue repair *Cells* **7** 167

Nazarenko I, Rana S, Baumann A, McAlear J, Hellwig A, Trendelenburg M *et al* 2010 Cell surface tetraspanin Tspan8 contributes to molecular pathways of exosome-induced endothelial Cell activation *Cancer Res.* **70** 1668–78

Nebuloni M, Albarello L, Andolfo A, Magagnotti C, Genovese L, Locatelli I *et al* 2016 Insight on colorectal carcinoma infiltration by studying perilesional extracellular matrix *Sci. Rep.* **6** 22522

New S E P, Goettsch C, Aikawa M, Marchini J F, Shibasaki M, Yabusaki K *et al* 2013 Macrophage-derived matrix vesicles: an alternative novel mechanism for microcalcification in atherosclerotic plaques *Circ. Res.* **113** 72–7

Nieuwland R, Falcon-Perez J M, Soekmadji C, Boilard E, Carter D and Buzas E I 2018 Essentials of extracellular vesicles: posters on basic and clinical aspects of extracellular vesicles *J. Extracell. Vesicles.* **7** 1548234

Nishimura T, Oyama T, Hu H T, Fujioka T, Hanawa-Suetsugu K, Ikeda K *et al* 2021 Filopodium-derived vesicles produced by MIM enhance the migration of recipient cells *Dev. Cell* **56** 842–859.e8

Nkosi D, Sun L, Duke L C, Patel N, Surapaneni S K, Singh M *et al* 2020 Epstein-barr virus LMP1 promotes syntenin-1- and Hrs-induced extracellular vesicle formation for its own secretion to increase cell proliferation and migration *mBio* **11** e00589-20

Noble J M, Roberts L M, Vidavsky N, Chiou A E, Fischbach C, Paszek M J *et al* 2020 Direct comparison of optical and electron microscopy methods for structural characterization of extracellular vesicles *J. Struct. Biol.* **210** 107474

Nogués L, Benito-Martin A, Hergueta-Redondo M and Peinado H 2018 The influence of tumour-derived extracellular vesicles on local and distal metastatic dissemination *Mol. Aspects Med.* **60** 15–26

Nurmik M, Ullmann P, Rodriguez F, Haan S and Letellier E 2020 In search of definitions: cancer-associated fibroblasts and their markers *Int. J. Cancer* **146** 895–905

O'Brien J, Hayder H, Zayed Y and Peng C 2018 Overview of MicroRNA biogenesis, mechanisms of actions, and circulation *Front. Endocrinol.* **9** 402

Olaso E, Santisteban A, Bidaurrazaga J, Gressner A M, Rosenbaum J and Vidal-Vanaclocha F 1997 Tumor-dependent activation of rodent hepatic stellate cells during experimental melanoma metastasis *Hepatology* **26** 634–42

Orimo A, Gupta P B, Sgroi D C, Arenzana-Seisdedos F, Delaunay T, Naeem R *et al* 2005 Stromal fibroblasts present in invasive human breast carcinomas promote tumor growth and angiogenesis through elevated SDF-1/CXCL12 secretion *Cell* **121** 335–48

Ortiz-Otero N, Clinch A B, Hope J, Wang W, Reinhart-King C A and King M R 2020 Cancer associated fibroblasts confer shear resistance to circulating tumor cells during prostate cancer metastatic progression *Oncotarget* **11** 1037–50

Ostrowski M, Carmo N B, Krumeich S, Fanget I, Raposo G, Savina A *et al* 2010 Rab27a and Rab27b control different steps of the exosome secretion pathway *Nat. Cell Biol.* **12** 19–30

Othman N, Jamal R and Abu N 2019 Cancer-derived exosomes as effectors of key inflammation-related players *Front. Immunol.* **10** 2103

Otto G P and Nichols B J 2011 The roles of flotillin microdomains—endocytosis and beyond *J. Cell Sci.* **124** 3933–40

Page-McCaw A, Ewald A J and Werb Z 2007 Matrix metalloproteinases and the regulation of tissue remodelling *Nat. Rev. Mol. Cell Biol.* **8** 221–33

Pan B T, Teng K, Wu C, Adam M and Johnstone R M 1985 Electron microscopic evidence for externalization of the transferrin receptor in vesicular form in sheep reticulocytes *J. Cell Biol.* **101** 942–8

Parada N, Romero-Trujillo A, Georges N and Alcayaga-Miranda F 2021 Camouflage strategies for therapeutic exosomes evasion from phagocytosis *J. Adv. Res.* **31** 61–74

Park E J, Prajuabjinda O, Soe Z Y, Darkwah S, Appiah M G, Kawamoto E *et al* 2019a Exosomal regulation of lymphocyte homing to the gut *Blood Adv.* **3** 1–11

Park K-S, Bandeira E, Shelke G V, Lässer C and Lötvall J 2019b Enhancement of therapeutic potential of mesenchymal stem cell-derived extracellular vesicles *Stem Cell Res. Ther.* **10** 288

Patel J J, Bourne L E, Davies B K, Arnett T R, MacRae V E, Wheeler-Jones C P *et al* 2019 Differing calcification processes in cultured vascular smooth muscle cells and osteoblasts *Exp. Cell. Res.* **380** 100–13

Patel N J, Ashraf A and Chung E J 2023 Extracellular vesicles as regulators of the extracellular matrix *Bioengineering* **10** 136

Peinado H, Alečković M, Lavotshkin S, Matei I, Costa-Silva B, Moreno-Bueno G *et al* 2012 Melanoma exosomes educate bone marrow progenitor cells toward a pro-metastatic phenotype through MET *Nat. Med.* **18** 883–91

Peinado H, Zhang H, Matei I R, Costa-Silva B, Hoshino A, Rodrigues G *et al* 2017 Pre-metastatic niches: organ-specific homes for metastases *Nat. Rev. Cancer* **17** 302–17

Peng X, Yang L, Ma Y, Li X, Yang S, Li Y *et al* 2021 IKKβ activation promotes amphisome formation and extracellular vesicle secretion in tumor cells *Biochim. Biophys. Acta BBA—Mol. Cell Res.* **1868** 118857

Peng Z, Tong Z, Ren Z, Ye M and Hu K 2023 Cancer-associated fibroblasts and its derived exosomes: a new perspective for reshaping the tumor microenvironment *Mol. Med.* **29** 66

Perez-Hernandez D, Gutiérrez-Vázquez C, Jorge I, López-Martín S, Ursa A, Sánchez-Madrid F *et al* 2013 The intracellular interactome of tetraspanin-enriched microdomains reveals their function as sorting machineries toward exosomes *J. Biol. Chem.* **288** 11649–61

Petersen S H, Odintsova E, Haigh T A, Rickinson A B, Taylor G S and Berditchevski F 2011 The role of tetraspanin CD63 in antigen presentation via MHC class II *Eur. J. Immunol.* **41** 2556–61

Petroni D, Fabbri C, Babboni S, Menichetti L, Basta G and Del Turco S 2023 Extracellular vesicles and intercellular communication: challenges for *in vivo* molecular imaging and tracking *Pharmaceutics* **15** 1639

Pfeffer S R 2007 Unsolved mysteries in membrane traffic *Annu. Rev. Biochem.* **76** 629–45

Pickarski M, Gleason A, Bednar B and Duong L T 2015 Orally active αvβ3 integrin inhibitor MK-0429 reduces melanoma metastasis *Oncol. Rep.* **33** 2737–45

Piper R C and Katzmann D J 2007 Biogenesis and function of multivesicular bodies *Annu. Rev. Cell Dev. Biol.* **23** 519–47

Piquet L, Dewit L, Schoonjans N, Millet M, Bérubé J, Gerges P R A *et al* 2019 Synergic interactions between hepatic stellate cells and uveal melanoma in metastatic growth *Cancers* **11** 1043

Pollard J W 2004 Tumour-educated macrophages promote tumour progression and metastasis *Nat. Rev. Cancer* **4** 71–8

Prada I and Meldolesi J 2016 Binding and fusion of extracellular vesicles to the plasma membrane of their cell targets *Int. J. Mol. Sci.* **17** 1296

Psaila B and Lyden D 2009 The metastatic niche: adapting the foreign soil *Nat. Rev. Cancer* **9** 285–93

Purushothaman A, Bandari S K, Liu J, Mobley J A, Brown E E and Sanderson R D 2016 Fibronectin on the surface of myeloma cell-derived exosomes mediates exosome-cell interactions *J. Biol. Chem.* **291** 1652–63

Qin X, Yan M, Wang X, Xu Q, Wang X, Zhu X *et al* 2018 Cancer-associated fibroblast-derived IL-6 promotes head and neck cancer progression via the osteopontin-NF-kappa B signaling pathway *Theranostics* **8** 921–40

Quail D F and Joyce J A 2013 Microenvironmental regulation of tumor progression and metastasis *Nat. Med.* **19** 1423–37

Radford K J, Thorne R F and Hersey P 1996 CD63 associates with transmembrane 4 superfamily members, CD9 and CD81, and with beta 1 integrins in human melanoma *Biochem. Biophys. Res. Commun.* **222** 13–8

Rafii D C, Psaila B, Butler J, Jin D K and Lyden D 2008 Regulation of vasculogenesis by platelet-mediated recruitment of bone marrow–derived cells *Arterioscler. Thromb. Vasc. Biol.* **28** 217–22

Raiborg C 2001 Hrs recruits clathrin to early endosomes *EMBO J.* **20** 5008–21

Raiborg C, Bache K G, Gillooly D J, Madshus I H, Stang E and Stenmark H 2002 Hrs sorts ubiquitinated proteins into clathrin-coated microdomains of early endosomes *Nat. Cell Biol.* **4** 394–8

Raiborg C and Stenmark H 2009 The ESCRT machinery in endosomal sorting of ubiquitylated membrane proteins *Nature* **458** 445–52

Ratajczak J, Wysoczynski M, Hayek F, Janowska-Wieczorek A and Ratajczak M Z 2006 Membrane-derived microvesicles: important and underappreciated mediators of cell-to-cell communication *Leukemia* **20** 1487–95

Reiner A T, Tan S, Agreiter C, Auer K, Bachmayr-Heyda A, Aust S *et al* 2017 EV-associated MMP9 in high-grade serous ovarian cancer is preferentially localized to annexin V-binding EVs *Dis. Markers* **2017** 1–9

Reiter J F and Leroux M R 2017 Genes and molecular pathways underpinning ciliopathies *Nat. Rev. Mol. Cell Biol.* **18** 533–47

Ren J, Zhu B, Gu G, Zhang W, Li J, Wang H *et al* 2023 Schwann cell-derived exosomes containing MFG-E8 modify macrophage/microglial polarization for attenuating inflammation via the SOCS3/STAT3 pathway after spinal cord injury *Cell Death Dis* **14** 70

Ricklefs F, Mineo M, Rooj A K, Nakano I, Charest A, Weissleder R *et al* 2016 Extracellular vesicles from high-grade glioma exchange diverse pro-oncogenic signals that maintain intratumoral heterogeneity *Cancer Res.* **76** 2876–81

Rider M A, Cheerathodi M R, Hurwitz S N, Nkosi D, Howell L A, Tremblay D C *et al* 2018 The interactome of EBV LMP1 evaluated by proximity-based BioID approach *Virology* **516** 55–70

Rilla K, Pasonen-Seppänen S, Deen A J, Koistinen V V T, Wojciechowski S, Oikari S *et al* 2013 Hyaluronan production enhances shedding of plasma membrane-derived microvesicles *Exp. Cell. Res.* **319** 2006–18

Rocha N, Kuijl C, Van Der Kant R, Janssen L, Houben D, Janssen H *et al* 2009 Cholesterol sensor ORP1L contacts the ER protein VAP to control Rab7–RILP–p150Glued and late endosome positioning *J. Cell Biol.* **185** 1209–25

Rogers M A, Buffolo F, Schlotter F, Atkins S K, Lee L H, Halu A *et al* 2020 Annexin A1–dependent tethering promotes extracellular vesicle aggregation revealed with single–extracellular vesicle analysis *Sci. Adv.* **6** eabb1244

Rojas E, Arispe N, Haigler H T, Burns A L and Pollard H B 1992 Identification of annexins as calcium channels in biological membranes *Bone Miner* **17** 214–8

Roma-Rodrigues C, Mendes R, Baptista P V and Fernandes A R 2019 Targeting tumor microenvironment for cancer therapy *Int. J. Mol. Sci.* **20** 840

Ruiz J L, Weinbaum S, Aikawa E and Hutcheson J D 2016 Zooming in on the genesis of atherosclerotic plaque microcalcifications *J. Physiol.* **594** 2915–27

Rutter B D, Chu T, Dallery J, Zajt K K, O'Connell R J and Innes R W 2022 The development of extracellular vesicle markers for the fungal phytopathogen *Colletotrichum higginsianum J. Extracell. Vesicles.* **11** e12216

Sahu R, Kaushik S, Clement C C, Cannizzo E S, Scharf B, Follenzi A *et al* 2011 Microautophagy of cytosolic proteins by late endosomes *Dev. Cell* **20** 131–9

Sala-Valdés M, Ursa , Charrin Á, Rubinstein S, Hemler E, M E, Sánchez-Madrid F *et al* 2006 EWI-2 and EWI-F link the tetraspanin web to the actin cytoskeleton through their direct association with ezrin-radixin-moesin proteins *J. Biol. Chem.* **281** 19665–75

Salinas R Y, Pearring J N, Ding J-D, Spencer W J, Hao Y and Arshavsky V Y 2017 Photoreceptor discs form through peripherin-dependent suppression of ciliary ectosome release *J. Cell Biol.* **216** 1489–99

Sanderson R D, Bandari S K and Vlodavsky I 2019 Proteases and glycosidases on the surface of exosomes: newly discovered mechanisms for extracellular remodeling *Matrix Biol.* 75–6 160–9

Santavanond J P, Rutter S F, Atkin-Smith G K and Poon I K H 2021 Apoptotic bodies: mechanism of formation, isolation and functional relevance *New Frontiers: Extracellular Vesicles* ed S Mathivanan, P Fonseka, C Nedeva and I Atukorala (Cham: Springer International Publishing) pp 61–88

Santos P and Almeida F 2021 Exosome-based vaccines: history, current state, and clinical trials *Front. Immunol.* **12** 711565

Savina A, Furlán M, Vidal M and Colombo M I 2003 Exosome release is regulated by a calcium-dependent mechanism in K562 cells *J. Biol. Chem.* **278** 20083–90

Schuck S 2020 Microautophagy—distinct molecular mechanisms handle cargoes of many sizes *J. Cell Sci.* **133** jcs246322

Scott G 2012 Demonstration of melanosome transfer by a shedding microvesicle mechanism *J. Invest. Dermatol.* **132** 1073–4

Sedgwick A E, Clancy J W, Olivia Balmert M and D'Souza-Schorey C 2015 Extracellular microvesicles and invadopodia mediate non-overlapping modes of tumor cell invasion *Sci. Rep.* **5** 14748

Serru V, Le Naour F, Billard M, Azorsa D O, Lanza F, Boucheix C *et al* 1999 Selective tetraspan-integrin complexes (CD81/alpha4beta1, CD151/alpha3beta1, CD151/alpha6beta1) under conditions disrupting tetraspan interactions *Biochem.* J% **340** 103–11

Shamseddine A A, Airola M V and Hannun Y A 2015 Roles and regulation of neutral sphingomyelinase-2 in cellular and pathological processes *Adv. Biol. Regul.* **57** 24–41

Shen T, Li Y, Zhu S, Yu J, Zhang B, Chen X *et al* 2020 YAP1 plays a key role of the conversion of normal fibroblasts into cancer-associated fibroblasts that contribute to prostate cancer progression *J. Exp. Clin. Cancer Res.* **39** 36

Shimoda M and Khokha R 2013 Proteolytic factors in exosomes *Proteomics* **13** 1624–36

Shinohara H, Kuranaga Y, Kumazaki M, Sugito N, Yoshikawa Y, Takai T *et al* 2017 Regulated polarization of tumor-associated macrophages by miR-145 via colorectal cancer–derived extracellular vesicles *J. Immunol.* **199** 1505–15

Shurer C R, Kuo J C-H, Roberts L M, Gandhi J G, Colville M J, Enoki T A *et al* 2019 Physical principles of membrane shape regulation by the glycocalyx *Cell* **177** 1757–70.e21

Shurtleff M J, Temoche-Diaz M M and Schekman R 2018 Extracellular vesicles and cancer: caveat lector *Annu. Rev. Cancer Biol.* **2** 395–411

Simons M and Raposo G 2009 Exosomes—vesicular carriers for intercellular communication *Curr. Opin. Cell Biol.* **21** 575–81

Sincock P M, Mayrhofer G and Ashman L K 1997 Localization of the transmembrane 4 superfamily (TM4SF) member PETA-3 (CD151) in normal human tissues: comparison with CD9, CD63, and $\alpha 5\beta 1$ integrin *J. Histochem. Cytochem.* **45** 515–25

Singh A, Fedele C, Lu H, Nevalainen M T, Keen J H and Languino L R 2016 Exosome-mediated transfer of $\alpha v\beta 3$ integrin from tumorigenic to nontumorigenic cells promotes a migratory phenotype *Mol. Cancer Res.* **14** 1136–46

Siveen K S, Raza A, Ahmed E I, Khan A Q, Prabhu K S, Kuttikrishnan S *et al* 2019 The role of extracellular vesicles as modulators of the tumor microenvironment, metastasis and drug resistance in colorectal cancer *Cancers* **11** 746

Skog J, Würdinger T, Van Rijn S, Meijer D H, Gainche L, Curry W T *et al* 2008 Glioblastoma microvesicles transport RNA and proteins that promote tumour growth and provide diagnostic biomarkers *Nat. Cell Biol.* **10** 1470–6

Skotland T, Hessvik N P, Sandvig K and Llorente A 2019 Exosomal lipid composition and the role of ether lipids and phosphoinositides in exosome biology *J. Lipid Res.* **60** 9–18

Skryabin G O, Komelkov A V, Savelyeva E E and Tchevkina E M 2020 Lipid rafts in exosome biogenesis *Biochem. Mosc.* **85** 177–91

Sloan E K, Pouliot N, Stanley K L, Chia J, Moseley J M, Hards D K *et al* 2006 Tumor-specific expression of $\alpha v\beta 3$ integrin promotes spontaneous metastasis of breast cancer to bone *Breast Cancer Res.* **8** R20

Sobo K, Le Blanc I, Luyet P-P, Fivaz M, Ferguson C, Parton R G *et al* 2007 Late endosomal cholesterol accumulation leads to impaired intra-endosomal trafficking *PLoS One* **2** e851

Soe Z Y, Park E J and Shimaoka M 2021 Integrin regulation in immunological and cancerous cells and exosomes *Int. J. Mol. Sci.* **22** 2193

Soe Z Y, Prajuabjinda O, Myint P K, Gaowa A, Kawamoto E, Park E J *et al* 2019 Talin-2 regulates integrin functions in exosomes *Biochem. Biophys. Res. Commun.* **512** 429–34

Song L, Tang S, Han X, Jiang Z, Dong L, Liu C *et al* 2019 KIBRA controls exosome secretion via inhibiting the proteasomal degradation of Rab27a *Nat. Commun.* **10** 1639

Song M, He J, Pan Q, Yang J, Zhao J, Zhang Y *et al* 2021 Cancer-associated fibroblast-mediated cellular crosstalk supports hepatocellular carcinoma progression *Hepatology* **73** 1717–35

Song X, Ding Y, Liu G, Yang X, Zhao R, Zhang Y *et al* 2016 Cancer cell-derived exosomes induce mitogen-activated protein kinase-dependent monocyte survival by transport of functional receptor tyrosine kinases *J. Biol. Chem.* **291** 8453–64

Soucy T A, Smith P G, Milhollen M A, Berger A J, Gavin J M, Adhikari S *et al* 2009 An inhibitor of NEDD8-activating enzyme as a new approach to treat cancer *Nature* **458** 732–6

Speziali G, Liesinger L, Gindlhuber J, Leopold C, Pucher B, Brandi J *et al* 2018 Myristic acid induces proteomic and secretomic changes associated with steatosis, cytoskeleton remodeling, endoplasmic reticulum stress, protein turnover and exosome release in HepG2 cells *J. Proteomics* **181** 118–30

Statello L, Guo C-J, Chen L-L and Huarte M 2021 Gene regulation by long non-coding RNAs and its biological functions *Nat. Rev. Mol. Cell Biol.* **22** 96–118

Stenmark H 2009 Rab GTPases as coordinators of vesicle traffic *Nat. Rev. Mol. Cell Biol.* **10** 513–25

Stepp M A, Pal-Ghosh S, Tadvalkar G and Pajoohesh-Ganji A 2015 Syndecan-1 and its expanding list of contacts *Adv. Wound Care* **4** 235–49

Stipp C S 2010 Laminin-binding integrins and their tetraspanin partners as potential antimetastatic targets *Expert Rev. Mol. Med.* **12** e3

Su S, Chen J, Yao H, Liu J, Yu S, Lao L *et al* 2018 CD10+GPR77+ cancer-associated fibroblasts promote cancer formation and chemoresistance by sustaining cancer stemness *Cell* **172** 841–56.e16

Subramanian B C, Melis N, Chen D, Wang W, Gallardo D, Weigert R *et al* 2020 The LTB4–BLT1 axis regulates actomyosin and $\beta 2$-integrin dynamics during neutrophil extravasation *J. Cell Biol.* **219** e201910215

Sun M, Xue X, Li L, Xu D, Li S, Li S C *et al* 2021 Ectosome biogenesis and release processes observed by using live-cell dynamic imaging in mammalian glial cells *Quant. Imaging Med. Surg.* **11** 4604–16

Sun Y, Sun F, Xu W and Qian H 2023 Engineered extracellular vesicles as a targeted delivery platform for precision therapy *Tissue Eng. Regen. Med.* **20** 157–75

Sundberg C and Rubin K 1996 Stimulation of beta1 integrins on fibroblasts induces PDGF independent tyrosine phosphorylation of PDGF beta-receptors *J. Cell Biol.* **132** 741–52

Sung B H, Ketova T, Hoshino D, Zijlstra A and Weaver A M 2015 Directional cell movement through tissues is controlled by exosome secretion *Nat. Commun.* **6** 7164

Sung B H, Von Lersner A, Guerrero J, Krystofiak E S, Inman D, Pelletier R *et al* 2020a A live cell reporter of exosome secretion and uptake reveals pathfinding behavior of migrating cells *Nat. Commun.* **11** 2092

Sung J S, Kang C W, Kang S, Jang Y, Chae Y C, Kim B G *et al* 2020b ITGB4-mediated metabolic reprogramming of cancer-associated fibroblasts *Oncogene* **39** 664–76

Surman M, Stępień E, Hoja-Łukowicz D and Przybyło M 2017 Deciphering the role of ectosomes in cancer development and progression: focus on the proteome *Clin. Exp. Metastasis* **34** 273–89

Suzuki A, Ghayor C, Guicheux J, Magne D, Quillard S, Kakita A *et al* 2006 Enhanced expression of the inorganic phosphate transporter Pit-1 is involved in BMP-2–induced matrix mineralization in osteoblast-like cells *J. Bone Miner. Res.* **21** 674–83

Suzuki J, Umeda M, Sims P J and Nagata S 2010 Calcium-dependent phospholipid scrambling by TMEM16F *Nature* **468** 834–8

Tadokoro S, Nakazawa T, Kamae T, Kiyomizu K, Kashiwagi H, Honda S *et al* 2011 A potential role for α-actinin in inside-out αIIbβ3 signaling *Blood* **117** 250–8

Tahara H, Kay M A, Yasui W and Tahara E 2013 MicroRNAs in cancer: the 22nd Hiroshima Cancer Seminar/the 4th Japanese Association for RNA Interference Joint International Symposium, 30 August 2012, Grand Prince Hotel Hiroshima *Jpn. J. Clin. Oncol.* **43** 579–82

Takahashi A, Okada R, Nagao K, Kawamata Y, Hanyu A, Yoshimoto S *et al* 2017 Exosomes maintain cellular homeostasis by excreting harmful DNA from cells *Nat. Commun.* **8** 15287

Tan H-X, Gong W-Z, Zhou K, Xiao Z-G, Hou F-T, Huang T *et al* 2020 CXCR4/TGF-β1 mediated hepatic stellate cells differentiation into carcinoma-associated fibroblasts and promoted liver metastasis of colon cancer *Cancer Biol. Ther.* **21** 258–68

Taraboletti G, D'Ascenzo S, Borsotti P, Giavazzi R, Pavan A and Dolo V 2002 Shedding of the matrix metalloproteinases MMP-2, MMP-9, and MT1-MMP as membrane vesicle-associated components by endothelial cells *Am. J. Pathol.* **160** 673–80

Taylor D D and Gerçel-Taylor C 2005 Tumour-derived exosomes and their role in cancer-associated T-cell signalling defects *Br. J. Cancer* **92** 305–11

Terlecki-Zaniewicz L, Lämmermann I, Latreille J, Bobbili M R, Pils V, Schosserer M *et al* 2018 Small extracellular vesicles and their miRNA cargo are anti-apoptotic members of the senescence-associated secretory phenotype *Aging* **10** 1103–32

Termini C M and Gillette J M 2017 Tetraspanins function as regulators of cellular signaling *Front. Cell Dev. Biol.* **5** 34

Théry C, Witwer K W, Aikawa E, Alcaraz M J, Anderson J D, Andriantsitohaina R *et al* 2018 Minimal information for studies of extracellular vesicles 2018 (MISEV2018): a position statement of the International Society for Extracellular Vesicles and update of the MISEV2014 guidelines *J. Extracell. Vesicles* **7** 1535750

Théry C, Zitvogel L and Amigorena S 2002 Exosomes: composition, biogenesis and function *Nat. Rev. Immunol.* **2** 569–79

Thompson C A, Purushothaman A, Ramani V C, Vlodavsky I and Sanderson R D 2013 Heparanase regulates secretion, composition, and function of tumor cell-derived exosomes *J. Biol. Chem.* **288** 10093–9

Tilghman R W, Blais E M, Cowan C R, Sherman N E, Grigera P R, Jeffery E D *et al* 2012 Matrix rigidity regulates cancer cell growth by modulating cellular metabolism and protein synthesis *PLoS One* **7** e37231

Timar J and Kashofer K 2020 Molecular epidemiology and diagnostics of KRAS mutations in human cancer *Cancer Metastasis Rev.* **39** 1029–38

Tkach M, Kowal J, Zucchetti A E, Enserink L, Jouve M, Lankar D *et al* 2017 Qualitative differences in T-cell activation by dendritic cell-derived extracellular vesicle subtypes *EMBO J.* **36** 3012–28

Trajkovic K, Hsu C, Chiantia S, Rajendran L, Wenzel D, Wieland F *et al* 2008 Ceramide triggers budding of exosome vesicles into multivesicular endosomes *Science* **319** 1244–7

Tricarico C, Clancy J and D'Souza-Schorey C 2017 Biology and biogenesis of shed microvesicles *Small GTPases* **8** 220–32

Tucker W C and Chapman E R 2002 Role of synaptotagmin in Ca^{2+}-triggered exocytosis *Biochem.* J. **366** 1–13

Tugues S, Honjo S, König C, Padhan N, Kroon J, Gualandi L *et al* 2013 Tetraspanin CD63 promotes vascular endothelial growth factor receptor 2-β1 integrin complex formation, thereby regulating activation and downstream signaling in endothelial cells *in vitro* and *in vivo* *J. Biol. Chem.* **288** 19060–71

Turek M, Banasiak K, Piechota M, Shanmugam N, Macias M, Śliwińska M A *et al* 2021 Muscle-derived exophers promote reproductive fitness *EMBO Rep.* **22** e52071

Turner N J, Quijano L M, Hussey G S, Jiang P and Badylak S F 2022 Matrix bound nanovesicles have tissue-specific characteristics that suggest a regulatory role *Tissue Eng. Part* A **28** 879–92

Umeda R, Nishizawa T and Nureki O 2019 Crystallization of the human tetraspanin protein CD9 *Acta Crystallogr. Sect. F: Struct. Biol. Commun.* **75** 254–9

Umezu T, Ohyashiki K, Kuroda M and Ohyashiki J H 2013 Leukemia cell to endothelial cell communication via exosomal miRNAs *Oncogene* **32** 2747–55

Umezu T, Tadokoro H, Azuma K, Yoshizawa S, Ohyashiki K and Ohyashiki J H 2014 Exosomal miR-135b shed from hypoxic multiple myeloma cells enhances angiogenesis by targeting factor-inhibiting HIF-1 *Blood* **124** 3748–57

Valadi H, Ekström K, Bossios A, Sjöstrand M, Lee J J and Lötvall J O 2007 Exosome-mediated transfer of mRNAs and microRNAs is a novel mechanism of genetic exchange between cells *Nat. Cell Biol.* **9** 654–9

Valcz G, Buzás E I, Kittel Á, Krenács T, Visnovitz T, Spisák S *et al* 2019 *En bloc* release of MVB-like small extracellular vesicle clusters by colorectal carcinoma cells *J. Extracell. Vesicles.* **8** 1596668

Van Der Merwe Y, Faust A E, Sakalli E T, Westrick C C, Hussey G, Chan K C *et al* 2019 Matrix-bound nanovesicles prevent ischemia-induced retinal ganglion cell axon degeneration and death and preserve visual function *Sci. Rep.* **9** 3482

Van Niel G, Carter D R F, Clayton A, Lambert D W, Raposo G and Vader P 2022 Challenges and directions in studying cell–cell communication by extracellular vesicles *Nat. Rev. Mol. Cell Biol.* **23** 369–82

Van Niel G, D'Angelo G and Raposo G 2018 Shedding light on the cell biology of extracellular vesicles *Nat. Rev. Mol. Cell Biol.* **19** 213–28 van Niel G, Charrin S, Simoes S, Romao M, Rochin L, Saftig P *et al* 2011 The tetraspanin CD63 regulates ESCRT-independent and -dependent endosomal sorting during melanogenesis *Dev. Cell* **21** 708–21

Verweij F J, Balaj L, Boulanger C M, Carter D R F, Compeer E B, D'Angelo G *et al* 2021 The power of imaging to understand extracellular vesicle biology *in vivo Nat. Methods* **18** 1013–26

Veschi E A, Bolean M, Strzelecka-Kiliszek A, Bandorowicz-Pikula J, Pikula S, Granjon T *et al* 2020 Localization of annexin A6 in matrix vesicles during physiological mineralization *Int. J. Mol. Sci.* **21** 1367

Vietri M, Radulovic M and Stenmark H 2020 The many functions of ESCRTs *Nat. Rev. Mol. Cell Biol.* **21** 25–42

Villa-Bellosta R 2024 Vascular calcification: a passive process that requires active inhibition *Biology* **13** 111

Villarroya-Beltri C, Gutiérrez-Vázquez C, Sánchez-Cabo F, Pérez-Hernández D, Vázquez J, Martin-Cofreces N *et al* 2013 Sumoylated hnRNPA2B1 controls the sorting of miRNAs into exosomes through binding to specific motifs *Nat. Commun.* **4** 2980

Volz A-K, Frei A, Kretschmer V, De Jesus Domingues A M, Ketting R F, Ueffing M *et al* 2021 Bardet-Biedl syndrome proteins modulate the release of bioactive extracellular vesicles *Nat. Commun.* **12** 5671

Wai P Y and Kuo P C 2008 Osteopontin: regulation in tumor metastasis *Cancer Metastasis Rev.* **27** 103–18

Wandrey M, Jablonska J, Stauber R H and Gül D 2023 Exosomes in cancer progression and therapy resistance: molecular insights and therapeutic opportunities *Life* **13** 2033

Wang J, Nikonorova I A, Silva M, Walsh J D, Tilton P E, Gu A *et al* 2021 Sensory cilia act as a specialized venue for regulated extracellular vesicle biogenesis and signaling *Curr. Biol.* **31** 3943–51.e3

Wang J, Silva M, Haas L A, Morsci N S, Nguyen K C Q, Hall D H *et al* 2014 *C. elegans* ciliated sensory neurons release extracellular vesicles that function in animal communication *Curr. Biol.* **24** 519–25

Wang Q and Lu Q 2017 Plasma membrane-derived extracellular microvesicles mediate non-canonical intercellular NOTCH signaling *Nat. Commun.* **8** 709

Wang R, Kobayashi R and Bishop J M 1996 Cellular adherence elicits ligand-independent activation of the met cell-surface receptor *Proc. Natl Acad. Sci.* **93** 8425–30

Wang X, Tian L, Lu J and Ng I O-L 2022 Exosomes and cancer—diagnostic and prognostic biomarkers and therapeutic vehicle *Oncogenesis* **11** 54

Wang X, Zhang H, Yang H, Bai M, Ning T, Deng T *et al* 2020 Exosome-delivered circRNA promotes glycolysis to induce chemoresistance through the miR-122-PKM2 axis in colorectal cancer *Mol. Oncol.* **14** 539–55

Wang Y, Zhao R, Liu W, Wang Z, Rong J, Long X *et al* 2019 Exosomal circHIPK3 released from hypoxia-pretreated cardiomyocytes regulates oxidative damage in cardiac microvascular endothelial cells via the miR-29a/IGF-1 pathway *Oxid. Med. Cell. Longev.* **2019** 1–28

Webber J P, Spary L K, Sanders A J, Chowdhury R, Jiang W G, Steadman R *et al* 2015 Differentiation of tumour-promoting stromal myofibroblasts by cancer exosomes *Oncogene* **34** 290–302

Webber J, Steadman R, Mason M D, Tabi Z and Clayton A 2010 Cancer exosomes trigger fibroblast to myofibroblast differentiation *Cancer Res.* **70** 9621–30

Wehman A M, Poggioli C, Schweinsberg P, Grant B D and Nance J 2011 The P4-ATPase TAT-5 inhibits the budding of extracellular vesicles in *C. elegans* embryos *Curr. Biol.* **21** 1951–9

Wei L, Ye H, Li G, Lu Y, Zhou Q, Zheng S *et al* 2018 Cancer-associated fibroblasts promote progression and gemcitabine resistance via the SDF-1/SATB-1 pathway in pancreatic cancer *Cell Death Dis* **9** 1065

Weilbaecher K N, Guise T A and McCauley L K 2011 Cancer to bone: a fatal attraction *Nat. Rev. Cancer* **11** 411–25

Wenzel E M, Pedersen N M, Elfmark L A, Wang L, Kjos I, Stang E *et al* 2024 Intercellular transfer of cancer cell invasiveness via endosome-mediated protease shedding *Nat. Commun.* **15** 1277

Wiklander O P B, Nordin J Z, O'Loughlin A, Gustafsson Y, Corso G, Mäger I *et al* 2015 Extracellular vesicle *in vivo* biodistribution is determined by cell source, route of administration and targeting *J. Extracell. Vesicles.* **4** 26316

Wolf P 1967 The nature and significance of platelet products in human plasma *Br. J. Haematol.* **13** 269–88

Wortzel I, Dror S, Kenific C M and Lyden D 2019 Exosome-mediated metastasis: communication from a distance *Dev. Cell* **49** 347–60

Wu M, Rementer C and Giachelli C M 2013 Vascular calcification: an update on mechanisms and challenges in treatment *Calcif. Tissue Int.* **93** 365–73

Wu Q, Zhang H, Sun S, Wang L and Sun S 2021 Extracellular vesicles and immunogenic stress in cancer *Cell Death Dis* **12** 894

Wu T, Wang W, Shi G, Hao M, Wang Y, Yao M *et al* 2022 Targeting HIC1/TGF-β axis-shaped prostate cancer microenvironment restrains its progression *Cell Death Dis* **13** 624

Wubbolts R, Leckie R S, Veenhuizen P T M, Schwarzmann G, Möbius W, Hoernschemeyer J *et al* 2003 Proteomic and biochemical analyses of human B cell-derived exosomes *J. Biol. Chem.* **278** 10963–72

Xie C, Ji N, Tang Z, Li J and Chen Q 2019 The role of extracellular vesicles from different origin in the microenvironment of head and neck cancers *Mol. Cancer* **18** 83

Xie S, Zhang Q and Jiang L 2022 Current knowledge on exosome biogenesis, cargo-sorting mechanism and therapeutic implications *Membranes* **12** 498

Xirodimas D P 2008 Novel substrates and functions for the ubiquitin-like molecule NEDD8 *Biochem. Soc. Trans.* **36** 802–6

Xu M, Ji J, Jin D, Wu Y, Wu T, Lin R *et al* 2023 The biogenesis and secretion of exosomes and multivesicular bodies (MVBs): intercellular shuttles and implications in human diseases *Genes Dis* **10** 1894–907

Xu R, Rai A, Chen M, Suwakulsiri W, Greening D W and Simpson R J 2018 Extracellular vesicles in cancer—implications for future improvements in cancer care *Nat. Rev. Clin. Oncol.* **15** 617–38

Yáñez-Mó M, Barreiro O, Gordon-Alonso M, Sala-Valdés M and Sánchez-Madrid F 2009 Tetraspanin-enriched microdomains: a functional unit in cell plasma membranes *Trends Cell Biol.* **19** 434–46

Yáñez-Mó M, Siljander P R, M, Andreu Z, Bedina Zavec A, Borràs F E, Buzas E I *et al* 2015 Biological properties of extracellular vesicles and their physiological functions *J. Extracell. Vesicles.* **4** 27066

Yang B, Duan W, Wei L, Zhao Y, Han Z, Wang J *et al* 2020a Bone marrow mesenchymal stem cell-derived hepatocyte-like cell exosomes reduce hepatic ischemia/reperfusion injury by enhancing autophagy *Stem. Cells Dev.* **29** 372–9

Yang H, Xu Z, Peng Y, Wang J and Xiang Y 2021a Integrin β4 as a potential diagnostic and therapeutic tumor marker *Biomolecules* **11** 1197

Yang H, Zhang H, Yang Y, Wang X, Deng T, Liu R *et al* 2020b Hypoxia induced exosomal circRNA promotes metastasis of colorectal cancer via targeting GEF-H1/RhoA axis *Theranostics* **10** 8211–26

Yang K, Zhang J and Bao C 2021b Exosomal circEIF3K from cancer-associated fibroblast promotes colorectal cancer (CRC) progression via miR-214/PD-L1 axis *BMC Cancer* **21** 933

Yang L, Huang J, Ren X, Gorska A E, Chytil A, Aakre M *et al* 2008 Abrogation of TGFβ signaling in ammary carcinomas recruits Gr-1+CD11b+ myeloid cells that promote metastasis *Cancer Cell* **13** 23–35

Yang L and Moses H L 2008 Transforming growth factor β: tumor suppressor or promoter? are host immune cells the nswer? *Cancer Res.* **68** 9107–11

Yao Y, Fu C, Zhou L, Mi Q-S and Jiang A 2021 DC-derived exosomes for cancer Immunotherapy *Cancers* **13** 3667

Yarana C and St. Clair D 2017 Chemotherapy-induced tissue injury: an insight into the role of extracellular vesicles-mediated oxidative stress responses *Antioxidants* **6** 75

Yauch R L, Berditchevski F, Harler M B, Reichner J and Hemler M E 1998 Highly stoichiometric, stable, and specific association of integrin α3β1 with CD151 provides a major link to phosphatidylinositol 4-kinase, and may regulate cell migration *Mol. Biol. Cell* **9** 2751–65

Ye F, Liang Y, Wang Y, Le Yang R, Luo D, Li Y *et al* 2023 Cancer-associated fibroblasts facilitate breast cancer progression through exosomal circTBPL1-mediated intercellular communication *Cell Death Dis* **14** 471

Ye L, Guo H, Wang Y, Peng Y, Zhang Y, Li S *et al* 2021 Exosomal circEhmt1 released from hypoxia-pretreated pericytes regulates high glucose-induced microvascular dysfunction via the NFIA/NLRP3 pathway *Oxid. Med. Cell. Longev.* **2021** 1–13

Yellon D M and Davidson S M 2014 Exosomes: nanoparticles involved in cardioprotection? *Circ. Res.* **114** 325–32

Yoshioka T, Nishikawa Y, Ito R, Kawamata M, Doi Y, Yamamoto Y *et al* 2010 Significance of integrin αvβ5 and erbB3 in enhanced cell migration and liver metastasis of colon carcinomas stimulated by hepatocyte-derived heregulin *Cancer Sci.* **101** 2011–8

You Y, Zheng Q, Dong Y, Xie X, Wang Y, Wu S *et al* 2016 Matrix stiffness-mediated effects on stemness characteristics occurring in HCC cells *Oncotarget* **7** 32221–31

Yu J *et al* 2024 Biogenesis and delivery of extracellular vesicles: harnessing the power of EVs for diagnostics and therapeutics *Front. Mol. Biosci.* **10** 1330400

Yu L, Zhu G, Zhang Z, Yu Y, Zeng L, Xu Z *et al* 2023 Apoptotic bodies: bioactive treasure left behind by the dying cells with robust diagnostic and therapeutic application potentials *J. Nanobiotechnol.* **21** 218

Yu S, Liu C, Su K, Wang J, Liu Y, Zhang L *et al* 2007 Tumor exosomes inhibit differentiation of bone marrow dendritic cells *J. Immunol.* **178** 6867–75

Yuan D, Zhao Y, Banks W A, Bullock K M, Haney M, Batrakova E *et al* 2017 macrophage exosomes as natural nanocarriers for protein delivery to inflamed brain *Biomaterials* **142** 1–12

Yuan Y, Jiang Y-C, Sun C-K and Chen Q-M 2016 Role of the tumor microenvironment in tumor progression and the clinical applications (review) *Oncol. Rep.* **35** 2499–515

Yuan Y-G, Wang J-L, Zhang Y-X, Li L, Reza A M M T and Gurunathan S 2023 Biogenesis, composition and potential therapeutic applications of mesenchymal stem cells derived exosomes in various diseases *Int. J. Nanomed.* **18** 3177–210

Yue B 2014 Biology of the extracellular matrix: an overview *J. Glaucoma* **23** S20–3

Yue Y, Wang C, Benedict C, Huang G, Truongcao M, Roy R *et al* 2020 Interleukin-10 deficiency alters endothelial progenitor cell–derived exosome reparative effect on myocardial repair via integrin-linked kinase enrichment *Circ. Res.* **126** 315–29

Zaborowski M P, Balaj L, Breakefield X O and Lai C P 2015 Extracellular vesicles: composition, biological relevance, and methods of study *BioScience* **65** 783–97

Zeng Z, Zhao Y, Chen Q, Zhu S, Niu Y, Ye Z *et al* 2021 Hypoxic exosomal HIF-1α-stabilizing circZNF91 promotes chemoresistance of normoxic pancreatic cancer cells via enhancing glycolysis *Oncogene* **40** 5505–17

Zernecke A, Bidzhekov K, Noels H, Shagdarsuren E, Gan L, Denecke B *et al* 2009 Delivery of MicroRNA-126 by apoptotic bodies induces CXCL12-dependent vascular protection *Sci. Signal.* **2**

Zhang C, Wang H, Li J and Ma L 2021a Circular RNA involvement in the protective effect of human umbilical cord mesenchymal stromal cell-derived extracellular vesicles against hypoxia/reoxygenation injury in cardiac cells *Front. Cardiovasc. Med.* **8** 626878

Zhang J, Ma J, Long K, Qiu W, Wang Y, Hu Z *et al* 2017 Overexpression of exosomal cardioprotective miRNAs mitigates hypoxia-induced H9c2 cells apoptosis *Int. J. Mol. Sci.* **18** 711

Zhang L, Wu X, Luo C, Chen X, Yang L, Tao J *et al* 2013a The 786-0 renal cancer cell-derived exosomes promote angiogenesis by downregulating the expression of hepatocyte cell adhesion molecule *Mol. Med. Rep.* **8** 272–6

Zhang L and Yu D 2019 Exosomes in cancer development, metastasis, and immunity *Biochim. Biophys. Acta BBA—Rev. Cancer.* **1871** 455–68

Zhang L, Zhang F, Weng Z, Brown B N, Yan H, Ma X M *et al* 2013b Effect of an inductive hydrogel composed of urinary bladder matrix upon functional recovery following traumatic brain injury *Tissue Eng. Part* A **19** 1909–18

Zhang Q, Jeppesen D K, Higginbotham J N, Graves-Deal R, Trinh V Q, Ramirez M A *et al* 2021b Supermeres are functional extracellular nanoparticles replete with disease biomarkers and therapeutic targets *Nat. Cell Biol.* **23** 1240–54

Zhang X, Groopman J E and Wang J F 2005 Extracellular matrix regulates endothelial functions through interaction of VEGFR-3 and integrin $\alpha_5\,\beta_1$ *J. Cell. Physiol.* **202** 205–14

Zhang X, Liu D, Gao Y, Lin C, An Q, Feng Y *et al* 2021c The biology and function of extracellular vesicles in cancer development *Front. Cell Dev. Biol.* **9** 777441

Zhang X, Yao L, Meng Y, Li B, Yang Y and Gao F 2023 Migrasome: a new functional extracellular vesicle *Cell Death Discov.* **9** 381

Zhang Y, Liu Y, Liu H and Tang W H 2019 Exosomes: biogenesis, biologic function and clinical potential *Cell Biosci.* **9** 19

Zhang Y, Wang J, Ding Y, Zhang J, Xu Y, Xu J *et al* 2020 Migrasome and tetraspanins in vascular homeostasis: concept, present, and future *Front. Cell Dev. Biol.* **8** 438

Zhao Y, Xiong X, Jia L and Sun Y 2012 Targeting Cullin-RING ligases by MLN4924 induces autophagy via modulating the HIF1-REDD1-TSC1-mTORC1-DEPTOR axis *Cell Death Dis.* **3** e386–6

Zheng R, Du M, Wang X, Xu W, Liang J, Wang W *et al* 2018 Exosome–transmitted long non-coding RNA PTENP1 suppresses bladder cancer progression *Mol. Cancer* **17** 143

Zhou Q, Li H, Li Y, Tan M, Fan S, Cao C *et al* 2019 Inhibiting neddylation modification alters mitochondrial morphology and reprograms energy metabolism in cancer cells *JCI Insight* **4** e121582

Zhou W, Fong M Y, Min Y, Somlo G, Liu L, Palomares M R *et al* 2014 Cancer-secreted miR-105 destroys vascular endothelial barriers to promote metastasis *Cancer Cell* **25** 501–15

Zhou Y, Zhang Y, Gong H, Luo S and Cui Y 2021 The role of exosomes and their applications in cancer *Int. J. Mol. Sci.* **22** 12204

Zhu J, Lu K, Zhang N, Zhao Y, Ma Q, Shen J *et al* 2017 Myocardial reparative functions of exosomes from mesenchymal stem cells are enhanced by hypoxia treatment of the cells via transferring microRNA-210 in an nSMase2-dependent way *Artif. Cells Nanomed. Biotechnol.* **46** 1659–70

Zhu L, Li J, Gong Y, Wu Q, Tan S, Sun D *et al* 2019 Exosomal tRNA-derived small RNA as a promising biomarker for cancer diagnosis *Mol. Cancer* **18** 74

Zhu Y, Yu S, Qiu H-J and Wang C 2020 Exosomes: another arena for the game between viruses and hosts *Sheng Wu Gong Cheng Xue Bao Chin. J. Biotechnol.* **36** 1732–40

Zimmerman B, Kelly B, McMillan B J, Seegar T C M, Dror R O, Kruse A C *et al* 2016 Crystal structure of a full-length human tetraspanin reveals a cholesterol-binding pocket *Cell* **167** 1041–51.e11

Zoetemelk M, Rausch M, Colin D J, Dormond O and Nowak-Sliwinska P 2019 Short-term 3D culture systems of various complexity for treatment optimization of colorectal carcinoma *Sci. Rep.* **9** 7103

IOP Publishing

Physics of Cancer, Volume 6 (Second Edition)
Cellular mechanisms to foster or fight cancer
Claudia Tanja Mierke

Chapter 2

Endosomes and receptor endocytic routes in cancer

2.1 Summary

The two membrane-anchored organelles endosomes and lysosomes are decisive for the proper operation of the eukaryotic cell. The main purpose of endosomes is to convey extracellular material intracellularly. Lysosomes, in contrast, are principally responsible for the breakdown of macromolecules. The interaction between endosomes and lysosomes takes place via two different mechanisms: kiss-and-run and direct fusion. Besides the process of internalizing particles, endosomes take an integral part in cell signal transduction and autophagy. Disturbances in any one of these processes can lead to cancer initiation and development. Lysosomal proteins such as cathepsins may be involved in both tumorigenesis and apoptosis of cancer cells. Because endosomal and lysosomal biogenesis and signal transduction are key elements of normal cell growth and proliferation, the proteins participating in these processes represent compelling candidates for anti-cancer drug development. Moreover, both organelles are hypothesized to contribute to mechanical characteristics of cells, such as cell stiffness and viscoelasticity. The process of endocytosis seems to be also critical for controlling the metastasis of cancers. The process of endocytosis compromises the internalization of cell-surface receptors through pinocytosis, phagocytosis or receptor-based endocytosis. The receptor-based endocytosis can be carried out via clathrin-, caveolae- and non-clathrin or caveolae-driven mechanisms. Endocytosis then passes through several intracellular compartments for cargo sorting and transfer and terminates with lysosomal breakdown, and the recycling mechanism returning to the cell surface or extracellular release. In cancer, several endocytic proteins are deregulated and are implicated in regulating tumor metastasis, specifically migration and invasion. Four metastasis suppressor genes may operate partially through the regulation of endocytosis, which appears to be based on NME, KAI, MTSS1 and KISS1 signaling routes. Metastasis suppressors may be altered in cancer metastasis, such as in signal transduction through endocytosis. This chapter

doi:10.1088/978-0-7503-4007-6ch2

concentrates on the multicomponent mechanism of endocytosis, including endocytic pathway, endolysosome assembly, and the disruption of endosomal/lysosomal biogenesis, which affects different steps of metastasis, and how metastasis suppressors such as genes utilize endocytosis to suppress metastasis.

2.2 Introduction to endosomes in cancer

Cancer is the main contributor to worldwide death rates (Bray *et al* 2024). It is estimated that in 2022, around one in five men and women will be diagnosed with cancer during their lifetime, while approximately one in nine men and one in twelve women will die from the disease (Bray *et al* 2024). The dissemination of cancer cells originating from the primary tumor to remote organs and their consequent progressive colonization is defined as metastasis (Liao *et al* 2019, Fares *et al* 2020, Nikkilä *et al* 2024). It is assumed that 90% of cancer-attributable deaths are caused due to metastases and not the actual growth of the primary tumor (Chaffer and Weinberg 2011). Is the number of deaths from metastases really that high? In solid tumors, 66.7% of cancer deaths and 60.1% of all cancer deaths have been recorded with metastases as a causative factor (Dillekås *et al* 2019). It can therefore be assumed that the percentage of deaths due to metastases is less than 90% but remains nevertheless very high and therefore extremely important. The usual approach to treating metastatic cancer is a systemic approach with chemotherapy or molecular therapeutics, hormonal medications, immune checkpoint medications, radiotherapy or surgery. Although advances have been made in prolonging survival times (Jemal *et al* 2017), there has been still very limited advancement in the development of treatments for metastatic cancer because of its inherent intricacy and insufficient comprehension of the molecular and biochemical mechanisms implicated.

Metastasis constitutes a process consisting of several steps, including invasion of cancer cells into adjacent areas, intravasation into the circulation, anchoring in the capillary embedding of a secondary organ, extravasation from the vascular network and settlement at the target location (Chambers *et al* 2002). Each of the stages listed above involves intricate reciprocal relationships between cancer cells and their microenvironment. Although the diversity and heterogeneity of the metastatic development process has been recognized, it has been shown that mutations or alterations in the expression of individual genes can alter the capacity to metastasize. Genes implicated in facilitating metastasis to remote locations are designated metastasis-inducing genes. Expression of these genes assists cancer cells in building appropriate interfaces with evolving microenvironments to support ongoing survival and propagation at secondary locations. In a similar way, genes that impede the process of metastasis but do not interfere with the growth of the primary tumor are known as metastasis suppressor genes.

This chapter emphasizes a frequently overseen facet of metastasis: the endocytic receptor routes. The spread of various cell surface receptors on cancer and microenvironment cells accounts for each stage of metastasis. Receptor signal transduction is itself regulated through endocytosis comprising internalization, recycling or breakdown. In the last few years, considerable advances have been achieved in gaining an appreciation of the mechanisms of the endocytosis route and its modifications throughout metastasis. A mounting evidence base points to the fact that endocytosis of receptors influences metastasis and may be a driver for the functioning of metastasis suppressors or metastasis enhancers (Jeger 2020). The hypothesis that the endocytosis process can alter cellular mechanical characteristics that consequently alter their migratory and invasive capacities is discussed. This chapter deals with the involvement of endocytosis in cancer metastasis and how these signaling routes are exploited via metastasis suppressors. Before discussing the involvement of mechanical factors in the process of endocytosis and cancer development and metastasis, the current knowledge about endosomes and their interaction with lysosomes is presented.

2.3 Endosomes and the process of endocytosis

The word ‘endocytosis’ comes from the Greek word ‘endon’, which means inside, ‘kytos’, which denotes the Greek word for a cell and ‘-osis’, which is referring to a process. Consequently, the process of endocytosis is an event through which cells actively internalize molecules and surface proteins utilizing an endocytic vesicle. Endosomes function as temporary trafficking vesicles for the inward and outward transfer of substances into and out of eukaryotic cells (Helenius *et al* 1983). In general, endocytosis represents a mechanism for internalization of plasma membrane together with its associated membrane proteins and lipids. Cells utilize endocytosis to modulate signal transduction routes and to assess the extracellular microenvironment for suitable reactions. It influences nearly all steps of metastasis and is utilized as a readout for metastasis suppressor functions. According to the literature, endocytosis governs the internalization, recycling, and turnover of receptors or can influence cytoskeletal dynamics to modify cancer cell invasiveness or metastatic potential (Khan and Steeg 2021). The global cellular mechanism of intracellular vesicle trafficking is the fundamental event required to sustain the homeostasis of membrane-enclosed organelles within eukaryotic cells. These organelles carry cargos from the donation membrane to the target membrane via the cargo-laden vesicles (Cohen *et al* 2018). The vesicle transportation mechanism comprises vesicle generation from the donor membrane, vesicle trafficking and vesicle incorporation into the target membrane. Vesicle assembly facilitated by the coat protein is a sensitive procedure that involves the choice of cargo molecules and their packaging into vesicle supports (Cai *et al* 2007). Vesicle transportation is in fact a dynamic and dedicated process for the displacement of cargo-carrying vesicles from the donor membrane to the destination membrane (Bonifacino and Glick 2004). This process involves a panel of conserved proteins like Rab GTPases, motor adaptors and motor proteins to assure vesicle trafficking across the cytoskeletal guidance pathway. Vesicle fusion facilitated via the soluble N-ethyl maleimide factor (NSF) provides the ultimate process whereby the vesicles offload the cargo molecules at the destination membrane (Jahn and Scheller 2006, Cai *et al* 2007). As many as one third of eukaryotic cell proteins are engaged in these vesicle trafficking routes. These processes are coordinated by a variety of proteins and protein assemblies, comprising envelope proteins, like envelope protein complex II (COPII), COPI and clathrin, small GTP-binding proteins, tethering proteins and fusogenic proteins (Jahn and Scheller 2006, McMahon and Boucrot 2011, Robinson and Pimpl 2014, Wang *et al* 2017c, Béthune and Wieland 2018, Langemeyer *et al* 2018, Sun and Brodsky 2019, Zhang and Hughson 2021). To guarantee that vesicle fusion is carried out at a specific location and at a specific time in the eukaryotic cell, several fusogenic proteins like synaptotagmin (Syt), complexin (Cpx), Munc13, Munc18 and other tethering determinants interact to tightly orchestrate the vesicle fusion event. Malfunction of fusogenic proteins in SNARE-based vesicle fusion is strongly correlated with numerous diseases. The stimulated membrane incorporation can be pharmacologically modulated by interfering with the junction of the SNARE complex and the Ca^{2+} sensor protein (Rizo 2018).

Endocytic transportation is crucial for the absorption of extracellular compounds, the internalization of receptors and the control of cell signaling events (Doherty and McMahon 2009). In the endocytic route, clathrin-coated vesicles with internalized molecules are trafficked to early endosomes for load sorting (figure 2.1). Certain molecules, like recycling receptors and lipid membranes, are trafficked back to the plasma membrane through recycling endosomes, whereas other molecules, like ubiquitylated proteins and downregulated receptors, are trafficked toward late endosomes and lysosomes for breakdown (Gruenberg 2001). Endocytic vesicle trafficking between multiple intracellular organelles, comprising early endosomes, late endosomes and lysosomes, also shares a common molecular pattern, even though the regulatory modules are distinct (Cantalupo 2001, Jordens *et al* 2001, Johansson *et al* 2007, Wieffer *et al* 2009, Hammer and Sellers 2012, Takahashi *et al* 2012, Xiang *et al* 2015). The endocytic transport of the clathrin-coated endocytic vesicle from plasma membrane to endosome is frequently chosen as an example for elucidation of the molecular mechanism of the principal controllers acting in this pathway.

The plasma membrane of the cells is characterized by a dynamic architecture that impairs the non-controlled passage of biomolecules at the interface of intracellular and extracellular compartments. Smaller molecules, like sugars and ions, can employ integral membrane proteins, including ion channels and ion pumps, to transit the plasma membrane.

In contrast, macromolecules are required to be fully internalized through primary endocytic vesicles (PEVs) that are generated through invaginations of the plasma membrane of the cell, which is termed phagocytosis (Conte and Sigismund 2016). These vesicles hand over their cargo to early endosomes (EEs) within the peripheral cytoplasm (Huotari and Helenius 2011). The engulfment and trafficking of extracellular cargo through the formation of membrane-bound vesicles can take place by a multitude of distinct routes, collectively referred to as endocytosis (Mukherjee *et al* 1997, Conte and Sigismund 2016). The different routes through which macromolecules cross the plasma membrane can be determined based on the size of the endocytosed cargo and consequently, the size of the engulfment they generate.

Micropinocytosis is a term used to describe invaginations below 200 nm and comprises both clathrin-mediated endocytosis (CME) and non-clathrin-mediated

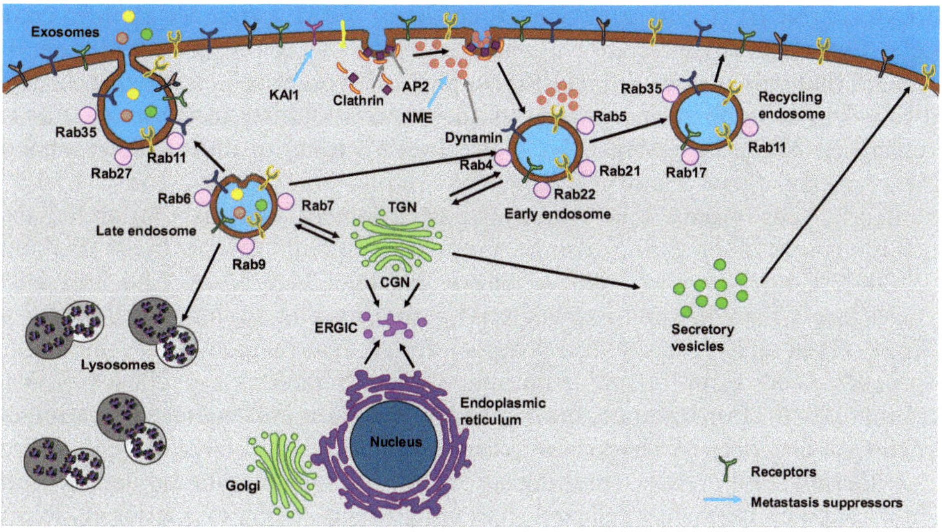

Figure 2.1. A multitude of receptors and their ligands are transported intracellularly through endocytosis. Clathrin-mediated endocytosis starts with the initialization and maturation of clathrin-coated pits through AP2 complexes that are attracted to the plasma membrane and serve as the main cargo recognizing molecule. As the nascent invagination expands, AP2 and additional cargo-specific adaptor proteins assemble and concentrate the cargo. AP2 combines with other adapter proteins to create complexes that attract clathrin. Clathrin recruitment establishes and stabilizes the curvature of the growing pit with the assistance of other BAR-domain-containing proteins. BAR-domain-containing proteins additionally attract dynamin to the throat of the budding vesicle, which then draws in the entire region and forms a closed vesicle. Dynamin forms a large GTPase helical oligomer that furls around the pinched neck of the vesicle and cleaves to disseminate the vesicle into the cytoplasm following GTP hydrolysis. After vesicle departure from the plasma membrane, the clathrin coat is broken down. The liberated vesicle undergoes an initial fusion phase that results in the generation of early endosomes, in which initial sorting occurs, and the destiny of the internalized sorting proteins and lipids is determined. RAB proteins, which are mainly associated with the early endosome, comprise RAB5 and RAB4 as well as the less prominent RAB21 and RAB22. These pathways regulate the motility of the early endosome on actin and microtubule routes, the homotypic fusion of endosomes with late endosomes, and the differentiated sorting and transport functions of the late endosome. The internalized receptors were sorted into recycling routes by extensive invagination of the early endosome membranes, with receptors sorted into the invaginated tubular membranes being trafficked back toward the plasma membrane via recycling endosomes. As an alternative, early growth and ripening of the endosomes into the trans-Golgi network (TGN) or late endosomes could arise. Mature late endosomes typically reach a diameter of about 250–1000 nm and are identified as such by the presence of a RAB7 domain. Late endosomes experience homotypic fusion, increase in size, and form increased numbers of intraluminal vesicles (ILVs). ILVs comprising late endosomes accumulate RAB35 and RAB27 and their effector proteins, which favor their merging with the plasma membrane to liberate exosomes, for instance, vesicles with a diameter of 40–100 nm. Mainly late endosomes travel to the perinuclear area of the cell, where they temporarily merge with one another and finally merge with lysosomes to break down their cargos. Cellular proteins produced in the rough endoplasmic reticulum (ER) are steadily transferred from the ER toward the Golgi complex through an ER–Golgi intermediate compartment (ERGIC) for all proteins in mammals. The points at which metastasis suppressors interfere with the endocytic pathway are marked in light blue.

endocytosis (NCE) (Conte and Sigismund 2016). CME produces clathrin-coated vesicles that endocytose receptor-ligand complexes of the plasma membrane. The molecules that gain access to the cell through CME comprise hormones, transferrin and low-density lipoproteins as well as their corresponding receptors (Kural and Kirchhausen 2012). In contrast, NCE involves a variety of diverse heterogeneous routes, among them clathrin-independent liquid-phase endocytosis, which is exploited by cells to sense the extracellular environment, and caveolar endocytosis, which is engaged in the endocytosis of multiple viruses and sphingolipids (Pelkmans *et al* 2001, Fittipaldi *et al* 2003, Cheng *et al* 2006, Hansen and Nichols 2009). Particles larger than 500 nm, like bacterial pathogens and apoptotic cell debris, are generally taken up by a specialized process referred to as phagocytosis, which occurs in specialized cells from the innate immune system (Schmitter *et al* 2004, Conte and Sigismund 2016). For instance, macrophages recognize and internalize pathogenic particles via a variety of phagocytic receptors (Ravetch and Clynes 1998, Aderem and Underhill 1999). After internalization, the plasma membrane-derived intracellular vacuoles, frequently referred to as phagosomes, perform a series of cleavage and merging events that endow the phagosome with decomposing characteristics that are decisive for its bactericidal capacity (Berón *et al* 1995, Tjelle *et al* 2000, Vieira *et al* 2002). Finally, particles with a size of 200–500 nm are endocytosed by a mechanism termed macropinocytosis. In both phagocytosis and macropinocytosis, the particles undergo internalization through major restructuring of the plasma membrane and actin cytoskeleton controlled by the Rho-family GTPase (Qualmann and Mellor 2003, Swanson 2008, Mooren *et al* 2012, Conte and Sigismund 2016). When the molecules are internalized, they are passed on to the EEs, which function as the primary sorting step of the endocytic route (Scott *et al* 2014). Endosomes offer an acidic microenvironment that is important for the dissociation of the interiorized receptor-ligand components (Helenius *et al* 1983). Receptors and specific other proteins can be returned to the plasma membrane, whereas other molecules are carried onwards into the cell through several membrane-connected intracellular organelles. This route is illustrated in greater detail in figure 2.1.

2.4 Generation of endosomes and their maturation

The EE is produced from PEVs that merge with one another and release both the membrane and the cargo to the nascent organelle (Huotari and Helenius 2011). EEs aggregate and then recycle cargo from these PEVs and interact with the trans-Golgi network (TGN) through bidirectional vesicle interchange. While the arrangement of EEs depends on the cell type, most EEs are generally distributed alongside the plasma membrane. The EEs screen patrol the peripheral cytoplasm whereby they move along the microtubules, which is controlled through Rab5 (Nielsen *et al* 1999, Huotari and Helenius 2011). From a structural viewpoint, EEs can be subdivided in two domains. Firstly, tubular domains, in which most of the membrane surface area is comprised, and secondly, vacuolar domains that exhibit most of the volume. A subset of the cargo that builds up in the peripheral EEs is returned to the plasma membrane either through a direct route or via recycled endosomes (Huotari and Helenius 2011).

In the course of time, the vacuolar domains ripen and develop late endosomes (LEs) by a process governed by Rab7 (Rink *et al* 2005, Vonderheit and Helenius 2005). The maturation is characterized primarily through acidification of the lumen, the replacement of membrane constituents and the shift to the perinuclear region (Huotari and Helenius 2011). The developing LEs bear a particular subgroup of the endocytosed cargo and travel alongside the microtubules. When the LEs travel towards the perinuclear region, homotypic merging events arise, and they increase in volume (Huotari and Helenius 2011). Moreover, the LEs can also accommodate intralumenal vesicles (ILVs), which are charge-laden vesicles that arise from luminal protuberances of the EE membrane (Scott *et al* 2014). The biogenesis of ILVs relies on a molecular mechanism termed endosomal sorting complex for transport (ESCRT), which targets clathrin-containing plaques to the cytosolic surface of the EE membrane (Raiborg 2001, Sachse *et al* 2002, Williams and Urbé 2007, Huotari and Helenius 2011). LEs act as a secondary sorting site where cargo can be transported to different targets. LEs ripen continuously and finally merge with lysosomes to transfer cargo and membrane constituents (Scott *et al* 2014). A considerable proportion of the internalized molecules are possibly transferred to the lysosomes for decomposition (Lim and Gleeson 2011, Conte and Sigismund 2016).

2.5 Lysosomes and creation of an endolysosome

Lysosomes represent dynamic, acid hydrolase-containing membrane-tethered organelles that function as the last compartment in the endocytosis route and perform a crucial task in the breakdown and proteolysis of internalized macromolecules (De Duve *et al* 1955, Luzio *et al* 2007). In addition, they are also necessary for the breakdown of intracellular debris in the autophagy reaction (Luzio *et al* 2007, Rajawat *et al* 2009). Most lysosomes are positioned perinuclearly, whereas the residual lysosomes are located on the periphery of the plasma membrane (Matteoni and Kreis 1987). Lysosomes exist in two distinct groups: endolysosomes and classical compacted lysosomes. Endolysosomes are temporary hybrid organelles that are produced by the merging of an LE and a lysosome and where active breakdown of the endocytosed cargo takes place (Huotari and Helenius 2011). Endolysosomes mature into classical dense lysosomes, serving as reservoir organelles for membrane constituents and hydrolases (Huotari and Helenius 2011).

Lysosomes exploit proton-pumping vacuolar ATPases to sustain an acidic microenvironment, such as a pH between 4.5 and 5.0, which eases digestive and proteolytic events (Mellman *et al* 1986, Boya 2012). Lysosomes comprise two main protein types that are critical for their functioning: integral membrane proteins and soluble hydrolases (Saftig and Klumperman 2009). In addition to their function in the breakdown of extra- and intracellular substances, lysosomes can also liberate their components into the extracellular cavity through merging with the plasma membrane (Repnik *et al* 2013). This fusion enables the lysosomes to liberate built-up, non-degradable matter and regenerate the plasma membrane when damage arises (Medina *et al* 2011).

The endosome and lysosome cargos are commingled through two distinct pathways: kiss-and-run and direct fusion. The kiss-and-run route comprises transient association and dissociation occurrences in which cargo is swapped over between the two membrane-bound organelles. In contrast, in the direct fusion route, the lysosome and endosome fuse completely to create a hybrid organelle (Bright *et al* 2005, Luzio *et al* 2007). In the lysosome, the components are broken down through hydrolytic processes catalyzed through proteases, lipases, nucleases and similar enzymes (Boya 2012). The joint action of these hydrolytic enzymes is accountable for the overall catabolic activity of the lysosome (Conus and Simon 2008, Saftig and Klumperman 2009).

2.6 Biogenesis of lysosomes

Lysosomal biogenesis involves transcriptional control via the transcription factor EB (TFEB) and is based on the integration of endocytic and biosynthetic cellular routes (Settembre *et al* 2011). As the correct sequencing of lysosomal activity is critical for the normal cellular functionality, it is important to strictly monitor lysosomal biogenesis. When the cell's nutritional status is elevated and a high number of amino acids are abundant, a key controller of cell growth, the mTORC1 kinase complex, is triggered across the lysosomal membrane (Shimobayashi and Hall 2014). Activated mTORC1 blocks TFEB and keeps it from migrating into the nucleus, thereby hindering lysosomal biogenesis (Sardiello *et al* 2009). In contrast, when the nutritional level of the cell is low, mTORC1 is blocked and TFEB is liberated (Settembre *et al* 2013, Hämälistö and Jäättelä 2016). Dephosphorylated TFEB migrates to the nuclei to trigger the transcription of genes implicated in lysosomal biogenesis and autophagy (Martina *et al* 2012).

Lysosomal proteins are produced in the endoplasmic reticulum and enter the nascent lysosome through two distinct routes. The first route is based on direct targeting of the TGN to the endosomal structure and the second route is based on indirect targeting, where proteins are delivered along the constitutive secretory pathway from the TGN to the plasma membrane, typically accompanied by endocytosis (Saftig and Klumperman 2009). When cargo is located in an endosome, it can be transferred to the lysosome through the kiss-and-run and fusion mechanisms. The principal constituent of the lysosomal membrane is a family of highly glycosylated, acidic membrane proteins, which comprise LAMP1 and LAMP2 (Howe *et al* 1988, Kornfeld and Mellman 1989). The lysosomal membrane proteins (LMPs) are stable toward lysosomal degradation and have an integral part in lysosomal homeostasis, motility, partitioning and integrity of the membrane (Saftig and Klumperman 2009, Boya 2012). LAMP1 and LAMP2 also take an instrumental place in lysosomal merging events in the autophagy route (Eskelinen 2006). AMPs can access the lysosome through either the direct or indirect routes outlined earlier during lysosomal biogenesis (Janvier and Bonifacino 2005, Saftig and Klumperman 2009). The route by which a particular LMP is trafficked differs according to where it leaves the TGN, which itself varies in relation to the type of LMP and a range of intracellular circumstances (Janvier and Bonifacino 2005, Saftig and Klumperman 2009, Carlson *et al* 2019).

The hydrolases take both direct and indirect routes to the lysosomes through EEs. The majority of lysosomal hydrolases that are trafficked via the direct route are linked to mannose-6-phosphate receptors (M6PRs). These receptors dissolve from the enzymes in the slightly acidic surrounding of the EEs and travel back to the TGN through specialized transporters to be subsequently recycled (Mari *et al* 2008, Saftig and Klumperman 2009). Most lysosomal hydrolases receive an M6P label when they cross the Golgi complex. M6PRs are found in the TGN and transport the attached hydrolases directly toward the endosomal matrix (Saftig and Klumperman 2009).

2.7 Biogenesis of endosomes during cancer development

Endosomes and the endocytic routes are essential for a variety of key cellular processes, among them cell proliferation, differentiation and metabolism (Doherty and McMahon 2009, McMahon and Boucrot 2011). The deregulation of endosomal biogenesis and functioning is linked to different pathophysiological mechanisms, even cancer formation and progression (McMahon and Boucrot 2011, Hu *et al* 2013). Changes in the expression of endosomal proteins have been identified in several cancer entities. LIMP-2, which is a transmembrane glycoprotein and an effective regulatory factor in endosome biogenesis, has been demonstrated to be overexpressed in prostate cancer and oral squamous cell carcinoma (Kuronita *et al* 2002, Pasini *et al* 2012, Johnson *et al* 2014b). Overexpression of LIMP-2 is linked to enhanced endosomal biogenesis and expansion of EEs and LEs (Kuronita *et al* 2002, Johnson *et al* 2014a). It has also been determined that endosomal subpopulations in prostate cancer cells exhibit aberrant intracellular localization schemes. These alterations in subcellular endosome positioning can promote cancer advancement through changes in signal transduction processes (Johnson *et al* 2014a). Overexpression and anomalous peripheral spreading of EEs could lead to enhanced nutrient absorption, enhanced membrane release and abnormal intracellular signal transductions (Glunde *et al* 2003, O'Sullivan and Lindsay 2020). The processes are all in accordance with cancer evolution and emphasize the link between modified endosomal biogenesis and carcinogenesis (Johnson *et al* 2014a).

In addition, certain endosomal gene expression profiles in cancer can be prognostic of successful therapy. For instance, elevated expression of endosomal acid ceramidase (ASAH1) in estrogen receptor-positive breast cancer is linked to improved survival rates (Ruckhäberle *et al* 2009). Cathepsin B, which is a protease primarily expressed by endosomes, is commonly found to be overexpressed in cancer as well as in premalignant sites (Podgorski and Sloane 2003). Heightened cathepsin B expression in gliomas and colorectal carcinomas relates to poorer prognosis and reduced survival of patients (Campo *et al* 1994, Koblinski *et al* 2000, Podgorski and Sloane 2003). In prostate cancer, EE-induced overexpression of APPL1, EEA1 and RAB5A genes is accompanied by more aggressive disease and worse therapeutic results (Johnson *et al* 2015). This shows that endosome-connected proteins can be utilized as cancer and disease propagation biomarkers.

2.8 Function of endosomes in signal transduction and cell fate regulation

As endocytosis facilitates the internalization and possible subsequent lysosomal breakdown of signaling receptors and their ligands, endosomes are also of crucial relevance for prolonged signal mitigation (Lanzetti and Di Fiore 2008). In tandem with sorting internalized receptors, endosomes may also play an active part in signal transduction across the endocytic pathway. Endosomal membranes are receptor-initiated signal transduction locations that facilitate spatiotemporal monitoring and transportation of biological cues, among them receptor tyrosine kinase (RTK) and G protein-coupled receptor (GPCR) signal transduction (Von Zastrow and Sorkin 2007). Endosomal malfunctions can therefore cause irregular signaling cascades and favor the development of tumors.

Endocytosis serves an integral part in establishing cell fate in asymmetric division of stem cells (Lanzetti and Di Fiore 2008). Interrupting this mechanism may lead to an exponential proliferation of cells, which favors the formation of cancer (Coumailleau and González-Gaitán 2008). Notch signal transduction governs the self-renewal in stem cells and cancer cells (Coumailleau and González-Gaitán 2008, Bian *et al* 2016). The Notch receptor Is a heterodimeric cell surface receptor that is being activated through DSL-family ligands present on the surface of adjacent cells. Endocytosis is necessary for internalization and the endosomal sorting of the ligand-bound Notch receptor, that facilitates the Notch signaling pathway (Seugnet *et al* 1997, Bian *et al* 2016). Asymmetric Notch activation in a Drosophila neuroblast experiencing cell division encourages self-renewal and represses the differentiation of one daughter cell. This leads to a daughter cell staying a neuroblast and a daughter cell continuing to differentiate (Wang *et al* 2006, 2007, Coumailleau and González-Gaitán 2008). The asymmetric endocytosis of signaling molecules in the two daughter cells is controlled through molecules that govern cell fate, like Numb (Betschinger *et al* 2003). In Drosophila with the Numb mutation, the neuroblast splits into two neuroblasts, resulting in exponential enlargement and the development of a neuroblastoma (Coumailleau and González-Gaitán 2008). For instance, a disruption of endocytosis-driven Notch signaling in stem cells can cause the development of tumors.

2.9 Relationship between endocytosis and autophagy

Endocytosis is also necessary for the repression of cancer progression through autophagy, a mechanism by which cytoplasmic material is transported to the lysosomes for breakdown (Lanzetti and Di Fiore 2008, Levine and Kroemer 2008). A malfunction of autophagy inhibits the degradation of intracellular substances and is linked to the formation of cancer (Mizushima *et al* 2008). Immature autophagic vacuoles merge with endosomes and lysosomes in the process of autophagy to generate hybrid organelles (Razi *et al* 2009). Evidence has demonstrated that endosomal malfunction caused by depletion of coatomer subunits impedes merging, leading to a build-up of immature autosomal vacuoles and impairment of autophagy (Razi *et al* 2009). Endosomes and endocytosis therefore have a key function in signal transduction pathways, cell fate specification and cancer mitigation by autophagy.

In addition, autophagy is instrumental in modulating intracellular reactive oxygen species (ROS) through the elimination of protein assemblies and defective mitochondria out of the intracellular milieu (Panda *et al* 2015, Kulikov *et al* 2017). The partial breakdown of mitochondria via the autophagic route, a mechanism also referred to as mitophagy, is an integral physiological process necessary for the maintenance of normal cell functionality (Kulikov *et al* 2017). In cells with autophagy deficiencies, defective mitochondria and ROS build-up, resulting in DNA defects and chromosomal fragility (Panda *et al* 2015). A lack of autophagy also leads to an excess build-up of p62, which is a selective autophagy protein and a polyfunctional protein that is implicated in different signal transduction routes (Komatsu *et al* 2010, Liu *et al* 2016b). It has been demonstrated that excessive production of p62 leads to an overstabilization of Nrf2, which is associated with the development of hepatocellular carcinomas in mice with autophagy deficiency (Inami *et al* 2011). In addition, the build-up of p62 has been found to further increase ROS concentration, leading to the triggering of NF-κB activation and antioxidant defense mechanisms, enhancing the survival of autophagy-deficient cells and facilitating tumorigenesis (White 2012, Panda *et al* 2015).

The induction of mitophagy metabolism comprises the activation of PTEN-induced putative kinase 1 (PINK1) and the E3 ubiquitin ligase Parkin (PARK2) (Kulikov *et al* 2017). PARK2 plays an essential part in mitochondrial coordination and acts as a tumor suppressor gene (Narendra *et al* 2008). PARK2 loss-of-function mutations are implicated in several diseases, among them a familial form of early-onset Parkinson's disease, several malignancies in humans, like glioblastoma, lung cancer and colon cancer, and liver cancer in a mouse model system (Fujiwara *et al* 2008, Veeriah *et al* 2010, Kulikov *et al* 2017). PINK1 has been demonstrated to be involved in cell survival and cell cycle progression, and PINK1 overexpression has the potential to enhance the proliferation of cancer cells and increase chemoresistance (Zhang *et al* 2013, O'Flanagan *et al* 2015).

2.10 Biogenesis of lysosomes during the development of cancer

The number of lysosomal subpopulations in the periphery of a cell considerably enlarges throughout the course of malignant transformation and performs an essential function in carcinogenesis through controlling cell adhesion and invasion (Kallunki *et al* 2013, Schiefermeier *et al* 2014, Bian *et al* 2016, Hämälistö and Jäättelä 2016). The mechanism whereby juxtanuclear lysosomes migrate to the peripheral reservoir through microtubule trails is termed anterograde trafficking and is controlled through multiple microtubule-associated kinesin proteins (Matteoni and Kreis 1987, Rosa-Ferreira and Munro 2011). Enhanced anterograde lysosomal transportation in malignant transformation appears to be associated with elevated expression of cathepsin proteases, which belong to a subgroup of lysosomal hydrolases (Rafn *et al* 2012, Hämälistö and Jäättelä 2016).

Besides the enhanced anterograde transport, the malignant transformation is simultaneously accompanied with an augmentation of biogenesis of the lysosomes (Kirkegaard and Jäättelä 2009, Giatromanolaki *et al* 2015). TFEB, which is the transcription factor involved in lysosomal biogenesis, has been demonstrated to be highly expressed in non-small cell lung cancer cell lines. Elevated cytoplasmic TFEB expression is related to an elevation of LAMP2a as well as cathepsin D expression (Giatromanolaki *et al* 2015). In addition, TFEB suppression using specific siRNAs decreases metastatic capability in several cell lines, indicating that decreased lysosomal functioning and biogenesis relates to a diminished capacity of cancers to metastasize and grow (Giatromanolaki *et al* 2015).

Yin Yang 1 (YY1), which is a co-transcriptional factor of TFEB, appears to be involved in the autophagy and lysosomal biogenesis pathways (Du *et al* 2019). YY1 interacts with DNA segments to modify gene expression through the assembly of an intranuclear complex with TFEB. Elevated YY1 expression is linked to enhanced LAMP1 levels, accompanied by an elevated generation of lysosomal membranes. Overexpression of YY1 is linked to neoplastic pathogenesis and malignant transformation within human BRAF-mutated melanoma cell lines (Du *et al* 2019). In addition, suppression of YY1 by lentiviral vector knockdown in human A375 melanoma cells has been demonstrated to have a synergistic impact on the anti-tumor effectiveness of vemurafenib, which is a BRAF V600E kinase blocker (Larkin and Fisher 2012, Motzer *et al* 2015, Du *et al* 2019). The emergence of TFEB blockers and YY1 limiting agents could therefore provide appealing candidates for future cancer therapeutics.

2.11 Lysosomal cathepsins and their action during carcinogenesis

According to their composition and position, cathepsin proteases may have either carcinogenic or anti- carcinogenic effects (Repnik *et al* 2013, Piao and Amaravadi 2016). For instance, specific cytosolic cathepsins can trigger the intrinsic apoptotic signaling route to prevent tumor growth, whereas extracellular cathepsins stimulate the growth of tumors through the activation of pro-tumorigenic proteins and breakdown of the basement membrane (Repnik *et al* 2012, Piao and Amaravadi 2016). Small-molecule blockers of cysteine cathepsins have been found to be a promising cancer therapeutic strategy for breast and pancreatic cancer using animal models (Bell-McGuinn *et al* 2007, Mikhaylov *et al* 2011). While overexpression of cathepsin D in breast and prostate cancer is linked to worse outcome, cathepsins B and L have been documented to dismantle the ECM as the tumor proliferates (Denhardt *et al* 1987, Sloane *et al* 1987, Cherry *et al* 1998, Foekens *et al* 1999, Nomura and Katunuma 2005). In addition, downregulation of cathepsin D has been found to impede lung metastasis of human breast cancer cells, and elevated levels of cathepsin B have been linked to enhanced metastatic capacity of melanoma in mouse models (Qian *et al* 1989, Glondu *et al* 2002). Cathepsin D is also responsible for breaking down chemokines like SCL and MIP-1α/β, which are involved in the activation of dendritic cells (Wolf *et al* 2003). As such, cathepsin D may attenuate the body's anti-tumoral immune response, resulting in increased tumor growth (Nomura and Katunuma 2005).

Lysosomal cathepsins likewise contribute to the invasion of cancer. While invading, cancer cells build invadopodia, which are specific adhesions that are particularly abundant in lysosomes to attach the expanding tumor to its immediate environment (Hämälistö and Jäättelä 2016, Du *et al* 2019). For the tumor to penetrate the neighboring tissues, the ECM requires disintegration, a process that can be achieved by both intracellular and extracellular mechanisms (Sevenich and Joyce 2014). Cathepsins, that are either transported to the plasma membrane or released from peripheral lysosomes into the extracellular compartment, ease the decomposition process of the ECM (Bian *et al* 2016, Hämälistö and Jäättelä 2016). ECM compounds can alternatively be endocytosed through certain cell surface receptors and consequently eliminated through lysosomal breakdown (Melander *et al* 2015). The lysosomal decomposed ECM proteins supply the invading cancer cells with nutrients and energy (Commisso *et al* 2013). Lysosomes thus promote the invasion of cancer as well as the growth and advancement of tumors.

2.12 Lysosomal cathepsins function in apoptosis of cancer cells

Cysteine cathepsins have been demonstrated to enhance apoptosis of cancer cells through death ligand-induced apoptosis routes in both human and murine cancer cell lines (Vasiljeva and Turk 2008). Cysteine cathepsins are involved in the TNF-α-triggered apoptosis pathway in human cervical cancer, fibrosarcoma, ovarian and prostate cancer in the mouse (Foghsgaard *et al* 2001, Liu *et al* 2006, Vasiljeva and Turk 2008). In addition, through activation of BID, which is a pro-apoptotic protein of the Bcl-2 family, cathepsins can initiate an apoptotic pathway induced via the TNF-related apoptosis-inducing ligand (TRAIL). This cathepsin-activated TRAIL signal transduction pathway has been identified in a number of different cancer cell lines, among them oral cavity cancer cells and osteosarcoma cells (Li *et al* 1998, Nagaraj *et al* 2006, Garnett *et al* 2007).

Cathepsins are also able to activate apoptosis through the degradation of anti-apoptotic proteins of the Bcl-2 family, including Bcl-2, Bcl-xL and Mcl-1, in the process of LMP-induced apoptosis (Droga-Mazovec *et al* 2008). It has been found that these anti-apoptotic proteins are highly expressed in different types of cancer, among them breast and colon cancer (Konishi *et al* 2006, Wang *et al* 2009, Repnik *et al* 2012). Knockdown of cathepsin B, which is a serine protease, using short hairpin RNA (shRNA) has been demonstrated to be linked to enhanced breakdown of the *X*-linked inhibitor of apoptosis protein (XIAP) in invasive meningioma cells, confirming the positive impact of cathepsin B on the proliferation of cancer cells (Gogineni *et al* 2012, Taniguchi *et al* 2015). Moreover, elevated cathepsin B amounts are critical for prostate cancer invasion and their metastasis in bone tissue (Miyake *et al* 2004, Podgorski *et al* 2005, Kumar *et al* 2018). Other investigations, nevertheless, have revealed that doxorubicin triggers cathepsin B-facilitated caspase-dependent apoptosis of HeLa cells, pointing to a pro-apoptotic function (Bien *et al* 2010). A correlation has been observed between a substantial reduction in cathepsin B concentrations and serum deprivation-activated apoptosis of PC12 cells (Shibata *et al* 1998, Taniguchi *et al* 2015). Cathepsin B blockade using CA-074 Me has been demonstrated to prevent both XIAP breakdown and apoptosis of lymphoma cells lacking IL-2, which implies that cathepsin B controls XIAP breakdown and serves a pro-apoptotic purpose (Taniguchi *et al* 2015). IL-2 withdrawal activates acid spingomyelinase-driven ceramide generation and aggregation in lysosomes, which induces apoptosis in natural killer (NK)/T-cell lymphoma cells (Taguchi *et al* 2005, Taniguchi *et al* 2015). Ceramide improves cellular susceptibility to apoptosis by modulating the expression of apoptosis-regulating genes like Bax, Bak and Bcl-*x* (Liu *et al* 2016a). Ceramide accumulation as a result of treatment with an acid ceramidase blocker has been demonstrated to sensitize human metastatic colon and breast cancer cell lines toward apoptosis (Paschall *et al* 2014). Ceramide is able to link directly to cathepsins and increase their enzymatic function (Heinrich 1999, Taniguchi *et al* 2015, Liu *et al* 2016a). Build-up of ceramide inside NK/T-cell lymphoma cells leads to cathepsin B activation and caspase-derived apoptosis (Taniguchi *et al* 2015). In myeloid-derived suppressor cells (MDSCs), immunosuppressive cells that are triggered by cancer, the build-up of ceramide induces the activation of cathepsins B and D, which initiate cell death through a cathepsin-dependent apoptotic route (Liu *et al* 2016a). MDSCs encourage the progression of the cancer and evasion of the immune response through blocking the effector roles of cytotoxic T lymphocytes and NK cells (Liu *et al* 2014, 2016a; OuYang *et al* 2015, Dufait *et al* 2016). The activation of cathepsins as a result of ceramide aggregation therefore fulfills a negative regulatory function in cancer development promoting apoptosis of cancer cells and apoptosis of oncogenic MDSCs. Cathepsins therefore have a dual function in carcinogenesis: they can stimulate both the formation of cancers and the apoptosis of cancer cells.

2.13 Endocytic routes and cancer metastasis

In this section, the emphasis is on the various routes of the endocytic pathway and its involvement in the metastatic progression of cancer. There are three general types of vesicular endocytic trafficking that coexist and proceed concurrently in the cell, according to the type of cargo, the route of endosomal capture and the cleavage mechanism, such as phagocytosis, pinocytosis and receptor-mediated endocytosis. In phagocytosis, the plasma membrane of the cell encloses a macromolecule, which are large solid particles above 0.5 μm, or even a whole cell from the extracellular surroundings and creates intracellular vesicles termed phagosomes (Underhill and Ozinsky 2002). Cellular pinocytosis is a mechanism by which liquids and nutrients are absorbed by the cell via entrapment, creating vesicles that are smaller than the phagosomes of a diameter of 0.5–5 μm (Haigler *et al* 1979). Phagocytosis and pinocytosis are both non-selective ways of capturing molecules. There are cases, nevertheless, in which the cells need specific molecules that are absorbed in a more efficient fashion through the mechanism of receptor-mediated endocytosis (RME). Endocytosis of distinct cargos through selective receptors can occur via clathrin-mediated (CME), caveolae-mediated (CavME), clathrin- and caveolae-independent endocytic pathways (CLIC/GEEC). The endocytic routes are shortly outlined hereafter. Table 2.1 connects specific endocytic proteins with *in vitro* participants of the metastatic cascade and *in vivo* metastasis in cancer. Moreover, associations of these endocytic proteins with mechanical characteristics are provided in table 2.1 and below.

Table 2.1. Endocytic proteins play a role in cancer metastasis and are influenced by mechanical stimuli.

Endocytic protein	Functions	Phenotypic impact (*in vitro*, or indicated)	Cancer types analyzed	Sensitive to mechanical cues or regulation of mechanics
Clathrin-medicated endocytosis (CME)				
AP2	Enrolls cargo and clathrin toward growing clathrin-coated pits.	Impacts cancer cell migration, invasion and chemotaxis via CXCR-2.	Ovarian, pancreatic and melanoma cancer (Azarnia Tehran *et al* 2019)	Membrane tension level and dynamics (Willy *et al* 2017)
Clathrin	Element of the envelope protein for membrane engraftment during endocytosis.	Clathrin light chain isoform (CLCb) is highly expressed and linked to bad outcome. NSCLC cells expressing CLCb exhibit enhanced cell migration and *in vivo* metastasis.	NSCLC (Chen *et al* 2017)	Membrane tension level and dynamics (Willy *et al* 2017)
Dynamin	Dynamins constitute a large GTPase that is encoded by three genes in mammals and is necessary for the cleavage of newly generated vesicles off the membrane.	Dynamin 1 and 2 are generally recognized to stimulate cancer cell proliferation, cancer invasion and metastasis, while dynamin 3 acts as a tumor suppressor. Overexpression of dynamin 2 is associated with a bad outcome.	NSCLC cells (Reis *et al* 2015) Prostate cancer (Xu *et al* 2014)	Each dynamin helix bundles 12–16 actin filaments and generates mechanically stiff actin super-bundles (Roux *et al* 2006). Dynamin regulates the dynamics and mechanical strength of the actin cytoskeletal matrix (Zhang *et al* 2020a). Longitudinal tension is needed for dynamin-driven constricting activity that leads to fission of the vesicle from the plasma membrane (Roux *et al* 2006).

(*Continued*)

Table 2.1. (*Continued*)

Endocytic protein	Functions	Phenotypic impact (*in vitro*, or indicated)	Cancer types analyzed	Sensitive to mechanical cues or regulation of mechanics
Caveolin-mediated endocytosis (CavME)				
Caveolin	Important envelope protein of the caveolae and participates in the lipid raft domain invagination.	In early phases of the disease, caveolin acts primarily as a tumor suppressor, while its expression in later stages is linked to tumor progression and metastasis. The effect of CAV-1 in late-stage tumor propagation and metastasis is hypothesized to be due to tyrosine phosphorylation (Tyr14) of its protein product through Src kinases. CAV-1 is frequently knocked down in human cancers and blocks the signaling of cytokine receptors. Knockdown of CAV-1 decreases the speed, direction and consistency of cellular migration. Positive expression of CAV-1 indicates histopathologic grade and bad prognosis (pancreatic cancer). Low expression of CAV-1 is linked to a worse prognosis (liver cancer).	Hepatomas, ovarian cancers, prostate cancer and breast cancer (Engelman *et al* 1998, Joshi *et al* 2008). Breast cancer (Joshi *et al* 2008, Urra *et al* 2012). Breast and prostate cancers (Joshi *et al* 2008, Urra *et al* 2012). Pancreatic adenocarcinoma and lung cancer; hepatocellular carcinoma (Chen and Che 2014).	Cav-1 can interact with the plasma membrane to cause membrane curvature and clustering of specific lipids (Prakash *et al* 2022, Tang *et al* 2023).

Clathrin-independent endocytosis (CIE)	Endocytic vesicles implicated in CIE possess no specific coating.	The CIE signaling route inhibits blistering and invasion of cancer cells via the GTPase-activating protein GRAF1.	Colon cancer (Holst *et al* 2017).	CIEs maintain the membrane tension (Hemalatha and Mayor 2019).
Endosomal trafficking proteins				
ARF subfamily: Small GTPase family				
ARF1	Controls the formation of various types of coat complexes on budding vesicles, participates in the secretory route and activates lipid-modifying enzymes.	Regulates cellular migration and proliferation through controlling the interaction between β_1-integrin and pivotal proteins of focal adhesions like paxillin, talin and FAK.	Breast cancer (Boulay *et al* 2011, Schlienger *et al* 2015).	Small G protein ARF1 and phosphatidate phosphatase lipin 1, which regulates the transportation of SREBPs between ER and Golgi, facilitated the impact of actomyosin contractility on the activity of SREBPs. Reduced actomyosin contractility impedes lipin 1 association with cytoplasmic membranes and impairs ARF1 activity and its enrollment to the Golgi (Otto 2019).
ARF4	Controls together with ARF1 retrograde trafficking from endosomes toward the TGN.	Tensin-driven cellular invasion and migration is modified through ARF4-based internalization of $\alpha5\beta1$ integrins into late endosomes/lysosomes and subsequent breakdown.	Ovarian cancer (Rainero *et al* 2015).	Internalization and recycling of the $\alpha5\beta1$ integrin is regulated by ARF4 (Rainero *et al* 2015), indicating that it impacts indirectly the generation of contractile forces.

(*Continued*)

Table 2.1. (*Continued*)

Endocytic protein	Functions	Phenotypic impact (*in vitro*, or indicated)	Cancer types analyzed	Sensitive to mechanical cues or regulation of mechanics
ARF6	Controls endocytic membrane transportation, polarized morphology and actin rearrangement.	ARF6 fosters the internalization of E-cadherin and promotes breakdown of adherens junction to encourage cellular movement and invasion. ARF6 inhibitory substance mitigates melanoma pulmonary metastasis.	MDCK cells (Palacios 2001, Palacios *et al* 2002), glioma, breast cancer (Schlienger *et al* 2016). Melanoma (Miao *et al* 2012).	ARF6 seems to control cellular mechanical properties, such as stiffness.
Ras-homolog (RHO) subfamily:				
RHOA	Promotes the generation of contractile actomyosin filaments within focal adhesion complexes and vesicle transportation.	RHOA depletion impedes endocytosis of several receptors and promotes breast cancer metastasis *in vivo* with a simultaneous rise in CCR5 and CXCR4 chemokine signal transduction. N-WASP controls endosomal recycling of LPAR1, which enhances RhoA-induced contractile reactions, cell guidance and spontaneous metastasis. Mice bearing N-WASP-depleted tumors experienced significantly longer survival.	Breast cancer (Kalpana *et al* 2019) Pancreatic ductal adenocarcinoma (Juin *et al* 2019).	Mechanical stress elevates the activation of RHOA (Smith *et al* 2003). RHOA is important for mechanotrans-duction (Burridge *et al* 2019).

RAC1	Controls macropinocytosis, membrane transportation and cellular shape.	Activation of RAC1 results in proliferation/survival of cancer cells, actin restructuring/migration with EMT transition phenotype, metastasis *in vivo* and angiogenesis.	Breast cancer (Ridley *et al* 1992, Morrison Joly *et al* 2017), gastric adenocarcinoma (Yoon *et al* 2017).	The Young's modulus (rigidity) of glioblastoma multiforme (GBM) cells is increased and their viscosity also enhanced upon inhibition of RAC1 (Xu *et al* 2020a). In contract the stiffness of Rac1 knockout cells is decreased in mouse embryonic fibroblasts (Kunschmann *et al* 2017, Mierke *et al* 2020).
CDC42	Acts in intracellular transportation, ER–Golgi interface anterograde and retrograde transport, post-Golgi trafficking and exocytosis.	Activation/overexpression enhanced tumor propagation and metastasis in several cancer types *in vivo*.	NSCLC, gastric cancer, breast cancer (Xiao *et al* 2018).	Matrix stiffness is a strong inducer of CDC42 activity (Kim *et al* 2014).
RAB subfamily:				
RAB1 (RAB1A and RAB1B)	Regulates ER–Golgi traffic.	In triple negative breast cancer, the lack of RAB1B expression relates to increased metastasis.	Breast cancer (Jiang *et al* 2015), NSCLC, gastric cancer, and esophageal squamous cell carcinoma (Yang *et al* 2016).	RAB1 seems to be involved in force transmission.
RAB2	Retrograde transfer of vesicles.	Overexpression of RAB2A leads to elevated cell invasiveness and the development of EMT characteristics.	Breast cancer (Reis *et al* 2015).	RAB2 seems to be affected by membrane tension.

(*Continued*)

Table 2.1. (*Continued*)

Endocytic protein	Functions	Phenotypic impact (*in vitro*, or indicated)	Cancer types analyzed	Sensitive to mechanical cues or regulation of mechanics
RAB3	Alters liberation of vesicles resulting in exocytosis.	Overexpression of RAB3C enhances migration, invasion and metastasis *in vivo*. RAB3D enhances breast cancer cell invasion and lung metastasis *in vivo* by activating the AKT/GSK-3β/Snail signaling route.	Colorectal cancer (Chang *et al* 2017) Breast cancer (Yang *et al* 2015).	Rab3 acts as a promotor of microdomain recycling and plasma membrane composition, indicating that it interferes with membrane tension (Diaz-Rohrer *et al* 2023).
RAB4	Controls vesicle recycling	The RAB4 recycling pathway plays a key role in driving the invasive characteristics of cancer cells mediated by integrin β3.	Breast cancer (Do *et al* 2017).	RAB4 indirectly seems to be alter force transmission.
RAB5, RAB21 and RAB22	Controls early endosome transportation.	RAB5 enhances integrin transportation, focal adhesion conversion, activation of RAC1 and migration and invasion of cancer cells. RAB5 modulates hypoxia-induced migration, invasion and metastasis of cancer cells *in vivo*. RAB21 modulates integrin-induced cell adhesion and motility. RAB22 and RAB163, whcih is a C-terminal protein of BRCA2, interfere selectively with the RAD51 protein.	Breast cancer (Mendoza *et al* 2013), colon adenocarcinoma and melanoma (Díaz *et al* 2014) Breast cancer and melanoma (Silva *et al* 2016). Cervical cancer cells (Pellinen *et al* 2006). Affects breast cancer susceptibility gene (BRCA2) (Mizuta *et al* 1997). NSCLC (Manshouri *et al* 2019).	They seem to be involved in cellular mechanical properties.

		TBC1D2b, which denotes a RAB22 GTPase-activating protein, is suppressed with the ZEB1/NuRD complex to enhance the internalization of E-cadherin and support lung cancer metastasis in a subcutaneous syngeneic mouse model.		
RAB6	Governs anterograde and retrograde transport ways between the Golgi apparatus, endoplasmic reticulum, plasma membrane, and endosomes.	Increased expression of RAB6A is associated with bad or good outcome. Suppression of RAB6A enhances cell migration through blocking myosin II phosphorylation and increasing Cdc42 activity.	Ovarian cancer (Ji *et al* 2022), cholangiocarcinoma (Yang *et al* 2023). Bone osteosarcoma, NSCLC cells (Vestre *et al* 2019).	RAB6 interferes with stiffness and membrane tension (Guet *et al* 2014)
RAB7	RAB7 governs the late endocytic route, which involves endosome maturation, the transition from early endosomes to late endosomes, aggregation and merging into lysosomes	The suppression of RAB7 is critical for the gain of invasive characteristics in melanoma cells and relates to an elevated metastatic potential.	Melanoma (Alonso-Curbelo *et al* 2014)	RAB7 is associated with tubular microdomains and hence it is likely to be involved in membrane tension (Markworth *et al* 2021).
RAB11	Controls the recycling of membrane proteins and protein transportation of the TGN toward the plasma membrane (slow recycling).	RAB11 controls RAC activity and polarization throughout collective cell migration and hypoxia-induced cell invasion in cancer cells.	Breast cancer (Yoon *et al* 2005). Pancreatic adenocarcinoma (Gundry *et al* 2017).	RAB11-positive EVs control the supracellular actomyosin scaffold to foster apical constriction (Chen and He 2022).

(*Continued*)

Table 2.1. (*Continued*)

Endocytic protein	Functions	Phenotypic impact (*in vitro*, or indicated)	Cancer types analyzed	Sensitive to mechanical cues or regulation of mechanics
		RCP, which is a RAB11 effector, drives cell–cell rejection and metastasis in an autochthonous mouse model of pancreatic adenocarcinoma through Eph receptor transport.		
RAB27	Controls secretory route/ exocytosis and melanosomes.	The two isoforms RAB27A and RAB27B are recognized to stimulate cell invasion and metastasis *in vivo*.	Bladder cancer, melanoma and breast cancer cells (Hendrix *et al* 2010).	RAB27 seems to be affected by mechanical cues.
RAB35	Controls rapid recycling of proteins toward the plasma membrane and in sorting endosomes.	It is assumed that constitutively active RAB35 is oncogenic because of the activation of PI3K/Akt signal transduction. It controls the migration and invasion of cancer cells. Mutant p53 promotes metastasis in autochthonous mouse models of pancreatic cancer through regulation of the output of sialomucin, podocalyxin and the GTPase RAB35, which interferes with podocalyxin sorting into exosomes. These exosomes affect the integrin transport within normal fibroblasts and encourage the accumulation of a highly invasion-inducing ECM.	Gastric cancer, cervical cancer cells (Ye *et al* 2018). Pancreatic cancer and NSCLC (Novo *et al* 2018).	RAB35 control apicobasal polarity and controls actin polymerization via restriction of Rac1 and RhoA activity (Francis *et al* 2022).

2.13.1 Clathrin-mediated endocytosis (CME)

The best-established endocytic mechanism examined is the clathrin-mediated endocytosis (CME). It has been established for the first time that it is a key factor in the absorption of low-density lipoproteins (Carpentier *et al* 1982) and transferrin (Neutra *et al* 1985). CME is implicated in the interiorization and recycling of several receptors involved in signal transduction, like G-protein and tyrosine kinase receptors, nutrient absorption and synaptic vesicle regeneration (Takei and Haucke 2001). Clathrin-coated pits (CCP) are clusters of cytosolic coat proteins initiated through assembly polypeptide 2 (AP2) complexes recruited to a plasma membrane site that is fortified with phosphatidylinositol (4,5)-bisphosphate lipid (Traub and Bonifacino 2013). AP2 functions as the main cargo sensing molecule and detects internalized receptors via a short sequence motif located in their cytoplasmic domains (Sorkin 2004). While the nascent invagination expands, AP2 and other cargo-specific adaptor proteins enlist and focus the cargo, which is now directed towards the interior of the vesicle. After sensing/accumulating cargo, AP2 forms a compound with other adaptor proteins to enlist clathrin. The recruitment of clathrin leads to stabilization of the bending of the growing CCP with the assistance of Bin-Amphiphysin-Rvs (BAR)-domain-containing proteins up to the point where the entire area involutes into a cohesive vesicle (Qualmann *et al* 2011).

Liberation of mature clathrin-coated vesicles of the plasma membrane is carried out via the large multi-domain GTPase, dynamin. Dynamin is entrapped around the necks of budding vesicles from proteins like amphiphysin, endophilin and sorting nexin 9 (proteins containing a BAR domain) (Ferguson *et al* 2009). Other dynamin partners such as Grb2 also attach to dynamin and enhance its oligomerization, which leads to increased GTPase activity (Barylko *et al* 1998). Oligomerized dynamin accumulates into collar-like structures encircling the necks of deeply depressed pits and is subject to GTP hydrolysis to promote membrane cleavage (Schmid and Frolov 2011). After separation of a vesicle from the plasma membrane, the clathrin envelope is broken down through the joint activity of the ATPase HSC70 and the envelope component auxilin (Ungewickell *et al* 1995, Newmyer and Schmid 2001). The liberated, non-coated vesicle is prepared to move and merges with its destination endosome.

Signaling via CME is crucial in the context of cancer and metastasis. The clathrin light chain (CLCb) isoform is typically upregulated in non-small cell lung cancer (NSCLC) cells and is related to bad outcome. NSCLC cells expressing CLCb show elevated CME levels by dynamin 1. This results in the activation of a positive feedback cycle that incorporates increased epidermal growth factor receptor (EGFR)-dependent Akt/glycogen synthase kinase 3β (GSK-3β) phosphorylation, leading to amplified cell migration and progression to metastasis (Chen *et al* 2017). Dynamin 2 is critical for the endocytosis of various proteins that are associated with cancer motility and invasiveness, such as β_1 integrin and focal adhesion kinase (FAK). Overexpression of dynamin 2 is associated with a bad outcome (Xu *et al* 2014).

Control of specific receptors implicated in cancer and metastasis such as EGFR and transforming growth factor-β receptor (TGFβR) through clathrin- and non-clathrin-mediated internalization routes primarily aims at the distinct fates of the receptors including recycling or breakdown (Di Guglielmo *et al* 2003, Sigismund *et al* 2008). Varying receptor fates dictate the overall net signal transduction in a cell and determine cancer advancement. Interestingly, CME is reported to direct the fate of EGFR in the direction of recycling rather than break down, resulting in a lengthened period of signaling (Sigismund *et al* 2008). In a similar way, the internalized EGF-EGFR complex may retain the capability to produce cell responses from endosomes that influence numerous downstream signal transduction routes (Wang *et al* 2002). It is established that this active endosomal EGFR acts to modulate the activity of oncogenic Ras through the co-internalization of its modulators such as Grb2, SHC, GAP and Cbl (Levkowitz *et al* 1998, Wang *et al* 2002).

Reacts sensitive to mechanical cues, such as tension gradients or tension alterations. The dynamics of endocytic clathrin-coated patterns can be strikingly diverse among various cell types, cells within the same culture or cells on various surfaces of the exact same cell. The underlying mechanism of this remarkable heterogeneity is still to be unraveled. Cellular events accompanied by alterations in effective plasma membrane tension can lead to substantial spatiotemporal variations in the dynamics of endocytic clathrin coating. Spatiotemporal heterogeneity in clathrin envelope dynamics is also seen in morphological alterations in evolving multicellular organisms. These results indicate that gradients in tension can result in tissue organization in patterns and diversification via mechanoregulation of clathrin-mediated endocytosis. Cellular events linked to membrane tension gradients, such as spreading and migration, cause enhanced spatiotemporal heterogeneity of endocytic clathrin envelope dynamics. The fluctuations in clathrin envelope dynamics correlate with plasma membrane tension gradients, which are a powerful controller of endocytic events (Dai and Sheetz 1995). Spatiotemporal changes in clathrin envelope dynamics can occur during developmental stages to shape embryos of *Drosophila melanogaster* (Willy *et al* 2017). Both endocytosis and exocytosis prevent cells experiencing strong fluctuations in membrane tension (Mao *et al* 2021). A mechanism needs to be established whereby the decrease in membrane tension triggers membrane budding and tubulation by endocytic proteins like endophilin A1. Membrane shape instability has been shown to arise at clearly defined membrane tensions and surface densities of endophilin A1 (Shi and Baumgart 2015). A diagram of membrane shape robustness based on the experimental data demonstrates an extraordinary match with a quantitative model (Shi and Baumgart 2015). The model is valid for all laterally diffusive curvature-coupling proteins and thus for a broad palette of endocytic proteins.

The question now arises as to whether there is a relationship between stiffness sensing and cellular tension. The rigidity of the ECM leads to varying tensions within the integrin-generated adhesions and thus induces various mechanoreactions. It is uncertain, whether the stiffness-dependent variation in tension is exclusively triggered by myosin activity. In the lack of myosin contractility, 3T3 fibroblasts remain capable of transmitting a differential degree of traction as a function of stiffness (Mittal *et al* 2024). Thereby, this myosin-independent selective traction is governed by polymerizing actin facilitated through the actin nucleators Arp2/3 and formin, with formin contributing more than Arp2/3 to both traction and actin flux. Interestingly, compared to cells with only myosin blocking, cells with combined restriction of Arp2/3 and myosin exhibit four times less traction than cells with only myosin blocking, although the F-actin flow velocity varies only slightly. The analyses show that conventional models focusing on stiff F-actin are not sufficient to deal with such a massive decrease in force at comparable actin flow. The incorporation of the viscoelastic characteristics of the F-actin structure is critical in this context. The emerging model, integrating the viscoelasticity of F-actin, demonstrates that Arp2/3 and formin increase sensitivity to stiffness via mechanically strengthening the F-actin meshwork, facilitating more efficacious propagation of flow-triggered forces. This model is confirmed through the determination of the cell stiffness using atomic force microscopy and the experimental monitoring of the stiffness-induced fluctuation of the actin flux predicted by the model.

Physical determinants that raise the energy cost of bending creation at the plasma membrane decelerate the rate of clathrin-coated vesicle generation (figure 2.2). This phenomenon can be visualized in living cells through quantitative imaging of fluorescently labeled clathrin mantle constituents such as clathrin or AP2 as an extended mantle lifespan (Boulant *et al* 2011). Actin dynamics oppose membrane tension in clathrin-mediated endocytosis. As an alternative, the mechanoregulation of CME dynamics at various surfaces of a cell can be observed through growth rate distributions compiled by quantifying the variations in the fluorescence signatures of individual clathrin envelopes over short time periods Thus, the standard deviation (SD) of the growth rate distributions is narrowed when the actual membrane tension is elevated due to cholesterol starvation or a hypotonic swelling (Dai *et al* 1998, p 19, Sun *et al* 2007, Khatibzadeh *et al* 2012, Diz-Muñoz *et al* 2016). Inversely, the SD of

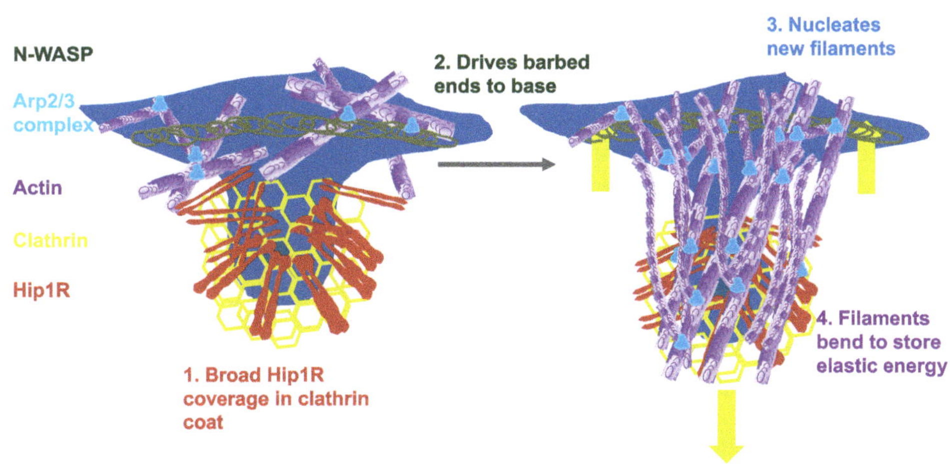

Figure 2.2. Schematic representation of the load-dependent adaptation of the self-organizing endocytic actin scaffold because of the spatial partitioning of the active Arp2/3 complex at the basal site and Hip1R in a broad distribution throughout the clathrin layer.

the growth rate distributions rises when the tension is lowered upon application of deoxycholate (Raucher and Sheetz 1999, Batchelder and Yarar 2010).

The membrane tension drops progressively as the cell spreads (Gauthier *et al* 2009). Rounded cells display about three times higher membrane tension in comparison to properly spread cells (Gauthier *et al* 2012). It has been found that the drivers that influence the dynamics of the CME vary considerably over the period of spreading. To assess the progression of spreading over time, the ventral (adherent) surfaces of BSC1 cells, which are kidney cells from an African green monkey, are visualized at multiple time points during spreading. The AP2-EGFP traces can be color-coded according to lifespan to illustrate the variations in clathrin shell density and dynamics during propagation. Declining membrane tension leads to a progressive decrease in the average lifespan of the clathrin envelope. Alterations in the SD of growth rates at both the ventral and dorsal surfaces reveal that clathrin coat dynamics intensify throughout the cell. Moreover, the initiation and disintegration densities of clathrin-coated patterns were observed to rise markedly with the termination of propagation. The increase in cell surface area is linked to rising membrane tension (Gauthier *et al* 2011, Houk *et al* 2012, Masters *et al* 2013). In accordance with this, a pronounced correlation has been found in spreading cells between the rate of surface expansion and the average lifetime of the clathrin coat. This effect is particularly noticeable in cells that experience multiple cycles of expansion. When the spreading is momentarily discontinued, the clathrin sheath lifetimes tend to convert to the levels seen during low tension periods. The lifetimes extend back to the levels found under high tension when the cells restart spreading. These results demonstrate that temporal fluctuations in tension have a direct impact on the dynamics and distribution of endocytic clathrin envelopes in the cells. The polarization of cells leads to spatial heterogeneity of the effective membrane tension (Dai and Sheetz 1999, Lieber *et al* 2015). Theoretical studies forecast a steep front-to-back stress gradient at the ventral surface of protrusive cells (Fogelson and Mogilner 2014). The spatial heterogeneity of clathrin coat dynamics has been shown to delineate the postulated stress gradient at the ventral surface of asymmetrically spreading cells. Long-lived clathrin-coated patterns are predominantly concentrated close to the leading edge. To improve the quantification of this tendency, the lifetime dipole moment, a vector pointing in the direction of the growing lifetime of the clathrin coating, is determined for each time point of the propagation. A clear correlation between the orientation of the lifetime dipole and the shift of the center of mass of the cells can be determined, even if the cells alter their orientation. As a control, the lifetime levels between the clathrin coats are randomly interchanged and the dipoles are recalculated. The rose diagrams generated using the angular distance between the simulated lifetime dipoles and the original displacement directions of the cells were omnidirectional, suggesting that the control plots had no directional predilection. It has been found that the distribution of clathrin coats is extremely heterogeneous, even if the net shift of cells is attributable to a slight asymmetry of spreading. The initiation and disintegration densities are at their minimum in the cell areas with the highest expansion speed.

Measurements of the tethering force indicated a marked tension gradient from anterior to posterior at the lamellipodial segments of the migrating keratocytes (Lieber *et al* 2015). Such segments cannot be separated from migrating astrocytes for tension analysis. Nevertheless, at the dorsal surface of these cells, a significant spatial heterogeneity in CME dynamics is noted, which parallels the anticipated tension gradient. Clathrin coats arising near the leading edge experienced a pronouncedly longer lifespan, such as 69 ± 51 s for the leading edge and 58 ± 39 s for the lamina and a tighter growth rate distribution of 0.035 ± 0.003 for the leading edge versus 0.041 ± 0.005 for the lamina. To visualize the spatial distribution of clathrin dynamics, growth rate charts were created where each pixel is assigned the value of the SD of the growth rates determined in a circular adjacency. In this representation, areas of the cell that exhibit slower clathrin dynamics display smaller SD values. A comparison of the clathrin coat initiation and disintegration densities in the two areas is not possible because of the complex three-dimensional (3D) geometries of the membrane ruffles occurring at the leading edge (Kural *et al* 2015). Overall, these findings indicate that the cellular mechanisms involved in spatial differences in plasma membrane tension enhance the heterogeneity of CME in cells (Willy *et al* 2017). The present investigation demonstrated that spatial and temporal fluctuations of cell membrane tension govern the dynamics of clathrin-coated structures. The endocytic apparatus needs to overwhelm the key components of effective membrane tension, which are in-plane tension and adhesion that occurs between the membrane and the cytoskeleton, to distort the plasma membrane (Sheetz 2001). It is hypothesized that the in-plane tension is in balance due to the rapid flux of membrane lipids throughout the plasma membrane. The adhesion between membrane and cytoskeleton can, nevertheless, be heterogeneous and lead to strong variations in clathrin dynamics across various surfaces of a cell (Dai and Sheetz 1999, Boulant *et al* 2011). In a similar way, adhesion to the substrate can restrain curvature generation and locally slow down the dynamics of the clathrin coat. Thus, the non-uniform adhesion of a cell to the substrate adds a further degree of heterogeneity to the clathrin dynamics (Batchelder and Yarar 2010, Ferguson *et al* 2016).

The spatiotemporal heterogeneity of clathrin coat dynamics appears to be important in key cellular events. Mechanical restraint of endocytosis at early phases of cell spreading could increase the velocity of expansion of the plasma membrane area (Gauthier *et al* 2009, 2011). Endocytosis restraint at the leading edge of migrating cells could ease cell protrusion through facilitating net membrane settlement in this zone (Bretscher 2014). In a similar way, elevated tension in the amnioserosa tissue of developing embryos could be responsible for the progressive decrease in cell volume due to the retardation of endocytosis in the late phases of dorsal closure. Prospective investigations should focus on the mechanoregulation of endocytosis *in situ* and the clarification of its function at the organismal scale.

2.13.2 Caveolae-mediated endocytosis (CavME)

CavME is the second most frequently investigated pathway of endocytosis and has been found to be essential for transcytotic transport through cells and mechanosensing (Parton and Simons 2007). The CavME mechanism entails the creation of an onion-shaped, 50–60 nm invagination of the plasma membrane termed caveolae (small cavities), which is fueled by both integral membrane proteins, the caveolins, and peripheral membrane proteins, the cavins, which are cytosolic envelope proteins. Caveolins, encoded by paralogs CAV-1, 2 and 3, are small integral membrane proteins that become incorporated into the internal face of the membrane bilayer through their cytosolic N-terminal moiety, which attaches to cholesterol. Approximately 50 cavin molecules bind to every caveolae and are present in homo- or hetero-oligomeric versions, which utilize four cavin family members (Gambin *et al* 2014). CavME is induced through the attachment of ligands to cargo receptors that are highly localized in caveolae. The detachment of caveolae from the plasma membrane is controlled by kinases and phosphatases like the Src tyrosine kinases and the serine/threonine protein phosphatases PP1 and PP2A (Kiss 2012). Dynamin is needed to detach caveolae vesicles away from the plasma membrane, as is the situation with CME (Henley *et al* 1998). The constituents of CavME play an important part in cell migration, invasion and cancer metastasis. CAV-1 is thought to play a twofold function in carcinogenesis and metastasis. In the initial phases of the disease, it acts primarily as a tumor suppressor, while its expression in later phases is linked to tumor advancement and metastasis (Quest *et al* 2004, 2008, Goetz *et al* 2008). Like a tumor suppressor, CAV-1 is frequently deleted in human cancers and is mechanistically recognized to function through the caveolin scaffold domain (CSD) through blocking cytokine receptor signal transduction (Engelman *et al* 1998, Goetz *et al* 2008). The effect of CAV-1 on advanced phase of tumor progression and metastasis is considered to be due to tyrosine phosphorylation (Tyr14) of its protein product through Src kinases, which results in enhanced Rho/ROCK signaling and consequently to the turnover of focal adhesions (Joshi *et al* 2008). Elimination of CAV-1 in breast and prostate cancer cells decreases the speed, orientation and consistency of cell migration (Joshi *et al* 2008, Urra *et al* 2012). In a similar way, the expression of CAV-1 has been utilized as a prognostic and overall survival biomarker in different types of human cancer. In pancreatic adenocarcinoma, positive expression of CAV-1 has been correlated with tumor size, histopathological grading and bad prognosis. In lung cancer, CAV-1 expression levels correspond statistically with low differentiation, disease stage, metastases to lymph nodes and bad prognosis. In hepatocellular carcinoma, nevertheless, low expression of CAV-1 is linked to a bad prognosis (Chen and Che 2014).

2.13.3 Clathrin-independent endocytosis (CIE)

As the name suggests, the endocytic vesicles implicated in clathrin-independent endocytosis (CIE) lack a clear envelope and they were first identified due to their opposition to blocking agents that prevent CME and CavME (Moya *et al* 1985). CIE involves multiple signaling pathways. First, an endophilin-, dynamin- and RhoA-driven signaling route for endocytosis of the interleukin-2 receptor (Lamaze *et al* 2001). Second, a clathrin- and dynamin-independent (CLIC/GEEC) route, where the GTPases RAC1 and CDC42 result in the actin-dependent generation of clathrin-independent carriers (CLICs). These in return generate the glycosylphosphatidylinositol (GPI)-AP-enriched endosomal compartments (GEECs) (Kirkham *et al* 2005, Huotari and Helenius 2011). Third, an ARF6-based route that comprises the small GTPase ARF6 to activate phosphatidylinositol 4-phosphate 5-kinase, which generates phosphatidylinositol (4,5)-bisphosphate, resulting in a stimulation of actin formation and endocytosis (Donaldson and Jackson 2011). The CIE route can impede the blebbing formation of cancer cells and invasion through GTPase-activating protein GTPase regulator associated with focal adhesion kinase-1 (GRAF1) (table 2.1) (Holst *et al* 2017). Several receptors undergo endocytosis through the CIE route, which encompasses the interleukin-2 receptor (IL-2R), the T-cell receptor (TCR) and GPI-linked proteins (Mayor *et al* 2014).

2.14 Downstream endosomal transport

Internalized receptor-ligand loads can fuse in a shared endosomal structure by passing through multiple cycles of merging. The first round of fusion results in the generation of early endosomes, in which initial sorting events take place and the fate of the internalized receptors is ultimately determined. Early endosomes are characterized through the combination of multiple proteins on their cytosolic interface, among them RAB5 and its effector VPS34/p150, which is a phosphatidylinositol 3-kinase complex. VPS34/p150 produces phosphatidylinositol-3-phosphate, that controls the spatiotemporal effects and partitioning of endosomal processes. (Christoforidis *et al* 1999, Zerial and McBride 2001). From a structural point of view, early endosomes consist of tubular (membrane) and vacuolar (vacuoles) regions. The majority of the membrane area is situated in the tubules, whereas a major portion of the volume is concentrated in the vacuoles. The membrane domains are fortified with proteins, like RAB5, RAB4, RAB11, ARF1/COPI, retromer and caveolin (Rojas *et al* 2008, Hayer *et al* 2010). These proteins are implicated in several activities, involving the molecular sorting of early endosomes to specific organelles, their recycling and their subsequent differentiation to late endosomes or the trans-Golgi network (TGN). The function of these endocytic proteins in metastasis *in vivo* and their prognostic value, if present, are presented in table 2.1.

A recycling route takes endosomes back to the cell surface either via a rapid recycling route through RAB4-positive endosomes or via a retarded recycling route through RAB11-positive endosomes (Stenmark 2009). Internalized receptors within early endosomes can be oriented into the recycling route through extensive tubulation of early endosomal membranes in a mechanism referred to as geometry-based sorting, in which receptors oriented into the newly generated tubular membranes of the early endosome are recruited back to the plasma membrane. Intralumenal vesicles (ILVs) develop naturally in early endosomes, powered by clathrin and elements of the endosomal sorting complex required for transport (ESCRT) (Raiborg 2001). ESCRT-driven sorting of receptors in ILVs represents an evolutionarily conserved mechanism that is necessary for multivesicular body (MVB) assembly. ESCRT utilizes its different complexes for receptor detection

(ESCRT-0), the inward budding (ESCRT-I and II) and the terminal ESCRT-III-based ablation (Raiborg and Stenmark 2009). This disconnects the cytoplasmic part of the receptors from the remainder of the cell, resulting in disruption of signal transduction. Importantly, depletion of ESCRT-0 and ESCRT-I subunits blocks EGFR breakdown and leads to increased recycling and persistent activation of extracellular signal-regulated kinase (ERK) signaling (Malerød *et al* 2007, Raiborg *et al* 2008).

A contributing function to endosomal acidification and ligand dissociation has also been identified. The recycling of receptors to the plasma membrane occurs when the ligands are liberated in the early endosome for example transferrin receptor, where the pH is kept at approximately 6.5 (Fuchs *et al* 1989). In contrast, certain signaling receptors like EGFR frequently preserve ligand binding and stay active at low pH values of around 4.5, resulting in continuous signaling from endosomal regions before they are processed into ILVs and broken down in the lysosome (Mellman 1996). Several internalized receptors present in early endosomes can be assorted to the TGN in a specific transport process termed retrograde transport, for instance, mannose-6-phosphate receptors and various toxins such as Shiga, cholera and ricin. The TGN is a meshwork of interlinked tubules and vesicles at the transfront of the Golgi apparatus. It is crucial for the preservation of cellular homeostasis and is recognized to be involved in protein sorting or the redirection of proteins and lipids prior to lysosomal breakdown.

Mature late endosomes have a diameter of approximately 250–1000 nm and are round/oval in appearance. The presence of RAB7 GTPase, which is essential for the maturation of early to late endosomes and for biogenesis of lysosomes, is a characteristic feature. The maturation of early to late endosomes relies on the generation of a hybrid RAB5/RAB7 endosome, where RAB7 is attracted to the early endosome through RAB5-GTP (Novick and Zerial, 1997). Late endosomes experience homotypic merging reactions, increase in their size and receive further intraluminal vesicles. When ILVs harboring late endosomes are augmented with RAB35, RAB27A, RAB27B and their effectors Slp4 and Slac2b, these vesicles merge with the plasma membrane and liberate exosomes (Huotari and Helenius 2011). The liberated exosomes consist of small 40–100 nm in diameter, single membrane-bound vesicles carrying proteins, DNA and RNA. In most cases, endosomes migrate by dynein-driven transportation into the perinuclear zone of the cell in the proximity of lysosomes. This is where the late endosomes temporarily merge with one another and finally merge with the lysosomes to produce a temporary hybrid organelle, namely the endolysosome. Most of the hydrolysis of the endocytosed load proceeds in the endolysosomes (Huotari and Helenius 2011). After a subsequent maturation phase, the endolysosome is transformed into a conventional dense lysosome. Cell constituents and organelles can also be transferred to lysosomes through a discrete mechanism termed autophagy. Autophagy, or self-consumption, is a one-of-a-kind membrane trafficking mechanism in which a newly generated isolation membrane can expand and engulf a portion of the cytoplasm or organelles to create autophagosomes that are passed to the lysosome for breakdown. There are an increasing number of reports pointing to a mechanistic role for autophagy in the process of tumor metastasis, detailed in a recent review (Mowers *et al* 2017).

A staggering number of proteins of the endosomal transport route are found to play an important role in cancer advancement and metastasis (table 2.1). Many of them have been evaluated in terms of cancer cell migratory activity and invasion, but a substantial amount of them have been demonstrated to influence metastasis *in vivo*. Changes seen include up- or downregulation of expression or mutations, usually causing abnormal receptor trafficking/recycling/destruction/signaling persistence, which has pervasive implications for cancer cell migration, invasion and/or proliferation. Most of these studies emphasize a single signal transduction route, but it is likely that multiple routes are implicated. These mechanistic approaches span a wide range of cancer types. More information on various members of the endosomal trafficking system and their involvement(s) in cancer and metastasis can be seen in the following reviews (Lanzetti and Di Fiore 2008, Mellman and Yarden 2013, Schmid 2017).

2.15 Integrin and ECM transportation during cancer metastasis

Cancer cells partially penetrate through the ECM via the production of matrix metalloproteinases (MMPs) and other proteinases that breakdown the ECM, thus providing routes for migration. Cells also bind to the ECM through integrins, which regulate cell adhesion, migration and proliferation. The interaction between integrins and ECM-digesting proteases is an effective controller of cancer invasion. In oral squamous cell carcinoma (SCC), elevated expression of $\alpha V\beta 6$ integrin results in activation of MMP-3 and drives proliferation and metastasis of oral SCC cells *in vivo* (Li *et al* 2003). MMP-14 (membrane-type 1 metalloprotease MT1-MMP) is localized at the protrusive terminals of invadopodia along with integrin $\alpha V\beta 3$, and its elevated local levels at the cell membrane enhance metastasis (Nakahara *et al* 1997). It is interesting to note that WDFY2 (a cytosolic protein) regulates the recycling of MT1-MMP to the plasma membrane, and depletion of WDFY2 results in enhanced release of MT1-MMP, resulting in active cell invasion (Sneeggen *et al* 2019). Latest investigations emphasize the relevance of integrin trafficking, such as endocytosis and recycling, as a regulator of fate of cancer cells. For instance, fast recycling of integrins from the anterior border of individual cells contributes to effective cell motility through supplying a stock of new receptors that are endogenously internalized at the posterior border. Further information on the transport of MMPs and integrins and their involvement in metastasis can be seen in these reviews (Ramsay *et al* 2007, Shay *et al* 2015).

2.16 The sorting process of endosomes requires molecular motors and specific adaptor proteins

For cellular motility, such as in cancer cells during the metastatic cascade, it is important that the migrating cell can recycle its cell surface receptors and release ECM remodeling enzymes or ECM crosslinking or storage proteins. After the decision for the recycling of internalized receptors or lipids, the sorting out of endosomes is facilitated by molecular motor proteins. Motor proteins are essential for the spatial and temporal regulation of intracellular localization and transport of membrane-enveloped cargo carriers such as endosomes. This section reviews how different motors and their associated cargo adaptors control the placement of cargos, such as endosomes, from the earliest phases of endocytosis (Khan and Steeg 2021). Two different mechanisms are involved: the first is the process of 'compartmentalization', which is responsible for creating isolated chambers and driving specific biochemical processes or avoiding the diffusion of certain concentrated substances (Diekmann and Pereira-Leal 2013, Gabaldón and Pittis 2015). The second mechanism concerns the process of 'active transport', which is responsible for bringing substances to the right place at the right time (Kamal *et al* 2001, Suh *et al* 2003, Qin *et al* 2024). Cells accomplish the second task by moving vesicles harboring motor proteins alongside an intricate microtubule framework. Kinesin and dynein, which are two oppositely polarized microtubule motors, perform this movement, respectively, towards the plus ends and minus ends of the microtubules, respectively. Intracellular substances are moved in bidirectional fashion, in particular, as two types of motor proteins engage with a single cargo at the same time, a process referred to as a tug-of-war; this is also characterized as a mechanical competition model. This mechanism eases the forward, backward or standstill movement of the cargo (Kural *et al* 2005, Hancock 2014). While this motion appears disordered, it can be hypothesized that the accurate and persistent delivery of cargo to its intended cellular destination is controlled by an intrinsic cellular organization (Jin *et al* 2024). The first mechanism can take place in the lysosomes and subsequently leads to the enzymatic degradation of the cargo within the lysosomes. The second mechanism involves the recycling of the cargo and delivers them toward their original location at the plasma membrane. Traditionally, most emphasis has been placed on two things, the *in vitro* separate analysis on the individual components, such as motor proteins and cargo adaptor proteins involved in cargo trafficking with and the investigation of traffic through the membranes. Recently, these experiments have also taken into account the effects of the environment and have therefore been carried out *in vivo* in living cell models that range. Thus, these experiments range from single molecules to whole organoids or tumoroids.

The endocytosis represents a key cellular process that regulates the absorption of cargoes such as different type of molecules like receptors, nutrients and lipids (Sigismund *et al* 2012, Thottacherry *et al* 2019). During this process, the substances to be transported are engulfed by the plasma membrane, where they either already reside or are internalized from the extracellular environment into the cell's interior. The process of endocytosis serves to remove deposits from the tissue, to regulate signal transduction and to ensure the homeostasis of the cell membrane. The original evolutionary driver for the emergence of endocytosis is likely to be signaling, but the system has been co-opted to actively manage various types of communication inside the cell and from the cell to its surrounding environment. For instance, after attachment with cognate ligands, signaling receptors are often internalized for subsequent destruction in lysosomes, leading to sustained signal dampening (Goh and Sorkin 2013, Irannejad *et al* 2015, Redpath *et al* 2020). Moreover, multiple surface molecules, among them are mainly proteins, are taken up independently of their interaction with extracellular components and either broken down or recirculated to the plasma membrane, offering a mechanism for the cell to tailor and adapt the pool of plasma membrane molecules for diverse functional demands.

An up-to-date perspective on endocytic transport sees it as an extensive program deeply rooted in the cellular blueprint and tightly linked to signaling, which is the principal communication system in the cell (Sigismund *et al* 2012). At the single-cell level, endosomes, for example, are important signaling hubs. This is exemplified by the concept of the signaling endosome as a framework that is able to uphold signaling through multiple mechanisms including endosome-specific signaling complex formation, interference, compartmental regulation of signaling persistence, and temporal and spatial signaling processing and resolution (Villaseñor *et al* 2016). Endocytosis also governs the performance of polarized cellular activities through the relocalization of surface molecules to areas in which polarized activities take place. The rapid and localized relocalization of membrane proteins is not accomplished through rather short-range planar diffusion across the plasma membrane, instead it is achieved by repeated cycles of endocytosis and targeted recycling (Disanza *et al* 2009, Eaton and Martin-Belmonte 2014, Lanzetti and Di Fiore 2017, Wilson *et al* 2018).

Independent of the importance of endocytosis and endosomal trafficking, its linkages of to the actin filament network and microtubule scaffold are critical for the control of cargo sorting for such recycling or deterioration events (Hinze and Boucrot 2018, Naslavsky and Caplan 2018). First, actin ramification aids extrusion endosome pinching off from the plasma membrane, succeeded by cleavage to release the endosome and its enclosed cargo from the plasma membrane. These endosomes can follow down the dense filamentous cortical actin underneath plasma membrane to enter deeper regions within the cytoplasm (Chakrabarti *et al* 2021). Within the cytoplasm, endosomes and their cargo merge with sorting endosomes that display the regulatory marker proteins, such as the master regulator GTPase Rab5 containing the isoforms Rab5a, Rab5b and Rab5c on their surface in mammals (Naslavsky and Caplan 2018, Nagano *et al* 2019). Rab5 GTPases function in the maturation of early endosomes to the late endosomes, whereby they part a prominent part (Rink *et al* 2005, Poteryaev *et al* 2010, Russell *et al* 2012). As part of the endosomal maturation a sequential relocation of the activity away from the early endosomal Rab5 toward the late endosomal Rab7 is necessitated. The process is referred to Rab conversion (Rink *et al* 2005, Poteryaev *et al* 2010). Most commonly, Rab converting is accomplished by guanine nucleotide exchange factors (GEFs), whereby an upstream Rab acquires a GEF for a downstream Rab (Markgraf *et al* 2007, Hutagalung and Novick 2011). Rab5 engages the Mon1–Ccz1 complex, which functions as a GEF for Rab7, in the process of early to late endosome ripening and stimulates the Rab5-to-Rab7 transformation (Nordmann *et al* 2010, Poteryaev *et al* 2010, Gerondopoulos *et al* 2012). Thus, a Rab5-positive sorting endosome ripens into a Rab7 (Rab7a and Rab7b)-positive late endosome that sequesters cargo not intended for recycling through Rab4 and Rab11. Loads earmarked for recycling by Rab4-directed, like Rab4a and Rab4b, or

Rab11-directed, like Rab11a and Rab11b, recycling routings are accumulated by the sorting endosome (Yudowski *et al* 2009, Campa *et al* 2018). Finally, the late endosome ultimately can ripen into or merge with the lysosome, which leaves the residual endosomal cargo exposed to breakdown enzymes and completes its endocytic processing (Guerra and Bucci 2016). Irrespective of destiny of the endosomes, recycling or breakdown involves long-range trafficking of endosomal cargo that is guided by microtubules toward the cell's interior, such as the perinuclear cavity. Rab11-positive recycling endosomes cluster in the endocytic recycling moiety in close proximity to the perinuclear area of the cell (Xie *et al* 2016), and the lysosomes similarly aggregate nearby the perinuclear area during cargo breakdown (Johnson *et al* 2016). In fact, depolymerization of microtubules has been demonstrated to impede the trafficking and maturation of Rab5-positive sorting endosomes and subsequently also the breakdown of cargos (Nielsen *et al* 1999, Mesaki *et al* 2011) and thereby underlines its essential part in these events. The depolymerization of microtubules also inhibits the recapture of recycling cargo from sorting endosomes (Delevoye *et al* 2014), while the retardation of stabilized detyrosinated microtubules blocks the transfer of cargo from the perinuclear recycling endosome partition toward their final destination back at the plasma membrane (Lin *et al* 2002, Delevoye *et al* 2014).

The endocytic transport and the involvement of motor proteins have been extensively studied in the past three decades. These events have been consistently characterized, but it is not yet clearly comprehended how these different mechanisms are orchestrated. Moreover, it is even not fully elucidated how stochastic processes, like motor-cargo engagement, result in distinct cargo transport within cells and how these processes controlled in time and space. In this section, tracing the passage of a cargo that is enclosed in the lumen of an endosome and travels through the tight actin cortex and thereafter through cytoplasmic the scaffold of microtubules, will help to elucidate how divergent sorting, either toward the lysosome in the vicinity of the nucleus or going back toward the plasma membrane via recycling mechanisms, is influenced by specific motor proteins. The results of *in vitro* reconstitution experimentation will be harmonized with those of *in vivo* and cellular models to reveal unifying underlying principles of motor-driven cargo transportation and to integrate the impact of a natural environment of these biochemical and cell biological processes. This section will also establish a general outline for future research in this research area and enable a comparative analysis of findings across various cell types and different cargo types (Redpath and Ananthanarayanan 2023).

2.16.1 Interplay between the cytoskeletal architecture and motor proteins in the trafficking of cargo relies on the length scale

On a global cellular length scale, rich and intricate dynamic events such as directional motions, punctuated pauses, reversals, and switching roads at the junction of cytoskeletal freeways occur during this transportation of endosomes (Watanabe *et al* 2007, Bálint *et al* 2013). Before the development of a new technique, it was unclear how these cargos could solve common traffic issues such as road closures and traffic jams in the highly crowded environment (Leduc *et al* 2012, Ross 2012). Today's insights into cellular trafficking have been made possible by fluorescence-based imaging approaches. The specificity, sensitivity, versatility and super-resolution potential of fluorescence microscopy are decisive advantages as a high-performance instrument for cell biology research. A number of questions related to cargo trafficking were tackled, ranging from motor protein dynamics within cytoplasmic milieus (Conway *et al* 2012, Sozański *et al* 2015), to 3D vesicle trafficking (Watanabe *et al* 2007), the cargo dynamics occurring at cytoskeletal junctions (Bálint *et al* 2013, Bergman *et al* 2018), the crosstalk of the cytoskeletal framework and organelles in trafficking at high spatiotemporal resolution (Guo *et al* 2018, Lu *et al* 2020). Fluorescence-based techniques are inherently constrained for purposes of prolonged live-cell applications because of photobleaching. In addition, unlabeled cellular components remain hidden from view, so all data about the immediate environment is lacking, excluding the labeled targets, although the non-labeled majority probably determines the obvious action of the labeled targets. This problem has been addressed by Park and colleagues. They found that the universal characteristic of intracellular trafficking can be captured in a highly parallel manner and for an unlimited time by following multiple unmarked cargos being carried along cytoskeletal freeways with interferometric scattering microscopy (iSCAT) (Park *et al* 2023). The fast and non-marking imaging of iSCAT provides the quantitative and physical insights into cargo trafficking and a faithful representation of nanoscale logistic processes inside living cells. The huge volume of cargo tracking information (more than 10^8) collected at 50 Hz over the whole observation period (about 30 min) permits the detailed architecture of the cytoskeletal network to be deciphered and the evolution of trafficking events in time alongside the active roads of the cytoskeleton to be imaged. Remarkably, cells naturally have an elegant way of avoiding intracellular road congestion by organizing themselves into a freight train or a convoy. Leading-edge iSCAT technologies permit the monitoring of the Brownian motion of nanoparticles in a 3D space, while the precise measurement of the fluctuation of their vertical displacements in the extremely crowded and non-homogeneous cytoplasm is demanding because of the existence of multiple optically heterogeneous cytoplasmic entities. Therefore, the map of cargo localization density obtained with CL-iSCAT microscopy has to be viewed as a 2D projection of all cargo particles that travel in 3D throughout the focus depth of iSCAT. Confocal iSCAT microscopes have recently been constructed, which can be employed to examine nanoscopic structures and dynamics occurring in living cells (Hsiao *et al* 2022, Küppers *et al* 2023). High-resolution imaging has significant potential for decoding the complexities of 3D cargo movements alongside complex cytoskeletal tracks, as it can resolve nanoscale movements along the axial axis. The *in vivo* work demonstrated that intracellular cargo undergoes frequent encounters with the dense transport network of the cytoskeleton. For instance, different dynamic occurrences in cargo trafficking like stopping, reversing, and congestion are illustrated by hindrances that stall the cargo's progress, front-on collisions with counter-trafficked cargo, and sudden stoppages at the junctions of cytoskeletal freeways. A cell seems to possess intrinsic mechanisms for the efficient transportation of cargo to its designated site through a severely crowded cellular landscape. Over the short period of time, the cargo movements seem to be intertwined and at a standstill, but over the long period of time, they move jointly in the same direction as a group of cargos. Occasionally, two or more cargos travel collectively and are directly attached to each other. There is strong support for the hypothesis of intracellular transport hitchhiking as a mechanism to enhance the overall trafficking velocity (Salogiannis and Reck-Peterson 2017, Saurabh Mogre *et al* 2020). Ultimately, the iSCAT approach offers a novel way to study the temporal development of nanoscopic cellular components in a physiological cytoplasmic environment with high resolution in space and time (Park *et al* 2023).

At subcellular or molecular length scales, the transportation of endosomes has been explored to a great extent. It has been found that both microfilaments and microtubules organize themselves in energy-regulated active processes and are polar polymers. Both are dynamic, which means that they can, for instance, change from a polymerization to a depolymerization period and back again. Due to their intrinsic polarity, these filaments exhibit a plus and a minus end with different characteristics. While the plus end is highly dynamic due to quicker (de)polymerization events, the minus end is less dynamic due to slower (de)polymerization events. For more details, there exist several valuable reviews on the structure of the cytoskeleton and its dynamic reshaping (Rottner *et al* 2017, Brouhard and Rice 2018). In nonpolar cells like HeLa, HEK293 and other frequently utilized cell lines, the polarity of the cortex and microtubules is categorized as ‘plus-end out’, indicating that the plus-ends are directed toward the exterior of the cell. The microtubules in these cells extend in a radial pattern from a microtubule organization center (MTOC), the centrosome, which is located near the cell nucleus. There is a capping of the microtubule’s minus end at the centrosome, which impairs the dynamic remodeling. Apart from centrosomal microtubules, there are also microtubules originating from the Golgi apparatus.

Motor proteins are capable of translating the chemical energy of ATP into mechanical work and are therefore driving the locomotion of a multitude of products in living cells. They cooperate with the underlying filaments of the cytoskeleton, like F-actin and microtubules, to facilitate the conveyance of cargo over long distances. The two principal categories of motor proteins differ in their selection of cytoskeletal filaments. Myosin motors rely on microfilaments, while kinesins and dyneins rely on microtubules. When the emphasis is placed on motor proteins of endocytic transport system, there are myosin V and VI that drive cargo across the cortex, and kinesins and cytoplasmic dynein 1 (which is in the following referred to as dynein) that are in charge of plus- and minus-end oriented transport on microtubules. All these motor proteins possess one or more ATPase domains that interact with and hydrolyze ATP, and an additional domain, usually referred to as the tail domain, that directly or indirectly binds to the plasma membrane of the endosome containing the cargo. A spiral stalk attaches the head to the microtubule-binding domains (dynein) or to the tail (kinesin and myosin), and a connector joins the head to the stalk. All motor proteins bind and dissociate stochastically from the filaments of the cytoskeleton and the cargo, and their activity is controlled, for example, by autoinhibition, which is transmitted through the motor by self-dimerization of the motor domains (Torisawa *et al* 2014), or by refolding the tail domain over the motor (Belyy *et al* 2016). The attachment of motor proteins to the cargo usually requires an adaptor protein and, in some cases, other regulatory proteins, such as dynactin for dynein, which attenuate this restriction and facilitate active motility of the motor. Considering the alignment of F-actin with the cortex and microtubules in a characteristic nonpolar cell, minus-end-driven myosin VI acts as the first motor targeting endosomes and transporting them via the cortex. Thereafter, the endosome is transferred to the minus-end-directed microtubule motor dynein (Watanabe and Higuchi 2007). Early endosomes travel on microtubules in both directions, moving towards the plus is kinesin-1 dependent and when the cargo is destined for lysosomal degradation, dynein transports it far towards the center of the cell. By contrast, when the cargo is intended for recirculation to the plasma membrane, the distinct recycling endosomes are initially propelled by kinesin motors and then pass to myosin V, which traverses the compact cortex before entering the plasma membrane. There exist extensive and excellent reviews of these motors that present additional insights (Wang *et al* 2015, Batters and Veigel 2016, Reck-Peterson *et al* 2018, Magistrati and Polo 2021).

2.16.2 Role of myosin VI in the movement of endocytic vesicles

The actin cortex represents a tight obstacle that newly emerging endocytic vesicles have to overcome before they can connect with the microtubules for extensive movement within the cell. The arrangement of the actin cortex is such that the barbed or positive end of the short actin filaments face the plasma membrane and the pointed or negative end face the cytoplasm (Chakrabarti *et al* 2021). Despite the participation of several myosin motors in endosome detachment from the plasma membrane, myosin VI is somewhat exceptional in that it doesn't merely participate in the act of detachment (Buss 2001), but also promotes the invasion of endosomes from the cell edge across the actin cortex (Aschenbrenner *et al* 2003), since it is the only myosin motor that targets the minus end of microfilaments (Wells *et al* 1999). Myosin VI is solely implicated in clathrin-dependent endocytosis (Puri 2009, 2010), but there are several clathrin-independent endocytic processes operating in the cell. The way in which motors and their adaptors control these clathrin-independent mechanisms is only now being elucidated (Jiang *et al* 2010, Williamson and Donaldson 2019, Renard *et al* 2020, Feng and Yu 2021, Ferreira *et al* 2021, Schink *et al* 2021, Wayt *et al* 2021, Tyckaert *et al* 2022). A brief overview can be found in table 2.2. The myosin VI cargo adaptor proteins DAB adaptor protein 2 (DAB2) and GAIP-interacting protein C-terminus (GIPC or synonymously referred to as GIPC1) impart myosin VI with the function to cleave and move on microfilaments, respectively. DAB2 controls the interference of myosin VI and clathrin-coated pits nearby the plasma membrane (Dance *et al* 2004, Spudich *et al* 2007). In contrast, GIPC govern the interaction of myosin VI and endocytic vesicles after clathrin decoating and subsequent transportation alongside actin filaments (Naccache *et al* 2006, Rai *et al* 2022).

Table 2.2. Function of motors in clathrin-independent endocytosis. Latest progress has resulted in the discovery of motors and adaptors implicated in endosomal motility in clathrin-independent endocytosis (CIE). In the following, these are described in detail and the still unknown elements of CIE motor transport are emphasized.

Motors in clathrin-independent endocytosis (CIE)	Advancements and remaining unknown elements
Fast endophilin-mediated endocytosis	In fast endophilin-mediated endocytosis (FEME), Bin1/amphiphysin II engages with FEME mediators and then enlists dynein toward the FEME endosome. Knockout of Bin1 decreases the engagement of dynein with FEME endosomes without compromising endocytosis of FEME cargos, suggesting that an as yet unknown actin-based motor could be necessary for endocytic capture and the clearance of cortex (Ferreira *et al* 2021).
Macropinocytosis	Macropinocytosis, which is an endocytic process in the liquid phase, needs myosin IE, myosin IIA and myosin IIB for the closing of the macropinosome (Jiang *et al* 2010, Schink *et al* 2021). After occlusion, dynein and its adaptors JIP3 and JIP4 are necessary for macropinosome trafficking alongside microtubules, and blocking dynein and knocking out JIP3 or JIP4 inhibit macropinosome uptake, possibly pointing to tight coupling of macropinosome assembly and linkage to microtubules (Williamson and Donaldson 2019).
Invadopodium endocytosis	In the invadopodium, integrin β3 endocytosis is accomplished through membrane tubulation in the paucity of clathrin. The dynein-activating adaptor Hook1 and the dynactin subunits p150 and Arp1 are positioned in the cytoplasmic node of this membrane tubule in advance of cleavage from the plasma membrane, even though dynein itself remains to be recognized in this endocytic architecture (Feng and Yu 2021).
Endophilin A3-dependent endocytosis	The CIE of CD166 is controlled through endophilin A3 and galectin-8 and is different from FEME (Renard *et al* 2020). Notably, removal of myosin-II and minus-end-directed members of the kinesin-14 family, though not dynein, impedes CD166 endocytosis, suggesting that a tight orchestration between CD166 endosome generation, cortex tension and microtubule engagement governs this mechanism (Tyckaert *et al* 2022).
MHC-I and CD59 endocytosis	Myosin-II has also been found to modulate the capture of the CIE cargos, such as MHC-I and CD59. Ablation of myosin-II causes MHC-I and CD59 endosomes to be caged in the cortex, but only marginally interferes with the incorporation of the CME cargo transferrin. Myosin-II is hypothesized to control CIE through modification of cortex tension; although how MHC-I and CD59 traverse the cortex is not yet determined (Wayt *et al* 2021).

2.16.2.1 Function of myosin VI in trafficking of multiple GPCRs in endocytosis

A β-arrestin-independent signaling mechanism for agonist transport of various GPCRs has been characterized by the differential recruitment and activation of the cytoskeletal motor myosin VI (Patel *et al* 2024). Subsequently, a model emerged in which activation of the receptor by agonists triggers autoregulatory crosstalk that entails the C-tail and the third intracellular loop of myosin VI (Patel *et al* 2024). The PBM sequence on the GPCR C-tail then binds GIPC, which serves as a PDZ adapter protein. The interplay between PBM and GIPC exposes the myosin interaction region (MIR), which removes the autoinhibition of myosin VI to trigger receptor uptake into endosomes. Myosin VI-stimulated endosomal transport attenuates G-protein signal transduction at the plasma membrane, whereas the phosphorylation of ERK1/2 is triggered in endosomal compartments. A series of GPCR-PBMs varies the activation of myosin VI, offering an adjustable mechanism for fine-tuning endosomal trafficking. Myosin VI mediates the transportation of a variety of cell surface receptors, among them PlexinD1, Megalin, TrkA/B, NMDA and AMPA (Lou *et al* 2001, Osterweil *et al* 2005, Naccache *et al* 2006, Varsano *et al* 2006, Yi *et al* 2007, Wieman *et al* 2009, Shang *et al* 2017, Wagner *et al* 2019). The functional multiplicity of myosin VI coincides precisely with its involvement in a variety of disease processes. Deficiency of myosin VI function causes chronic physiological disorders such as deafness (Avraham *et al* 1995, Avraham 1997). The overexpression of myosin VI is linked to increased cell migration in various types of cancer (Buss 2001, Tumbarello *et al* 2013). Nevertheless, a causal relationship between myosin VI and the transportation of members of the GPCR superfamily has not been established, and this is the subject of a systematic investigation by Park and colleagues (Patel *et al* 2024).

Myosin VI constitutes the sole minus end directed actin-based motor (Spudich and Sivaramakrishnan 2010, Magistrati and Polo 2021). The polarity of the actin cortex, in which the minus ends face the cell interior, facilitates the directed movement of the vesicles by myosin VI in a direction facing away from the plasma membrane (Buss 2001, Hasson 2003, Krueger *et al* 2019). Therefore, myosin VI, as a motor of the cytoskeleton, is in a unique situation to govern receptor desensitization and endocytosis away from the actin-rich apical periplasm. Accordingly, the inhibition of myosin VI causes receptors to remain in endosomes in the vicinity of the plasma membrane (Patel *et al* 2024). Myosin VI accomplishes its multifunctionality by coupling with diverse adaptors like DAB2, GIPC, Tom1/2 and LMTK2 (Tumbarello *et al* 2013), whereby the motor can orchestrate various steps of the endocytic process. Focusing on GIPC, a myosin VI adaptor previously found in uncoated vesicles (Aschenbrenner *et al* 2003) it is proposed that GIPC-myosin VI tethering in actin-rich peripheral compartments facilitates punctate endocytosis in a timely fashion (Dance *et al* 2004). Whereas GIPC is focused on PDZ adapters in the early endocytic cascade in HEK293 cells by Patel and colleagues, previous work has also implicated other PDZ adapters functioning in specialized cellular compartments, such as PSD-95 and SAP97 in synaptosomes, as possible myosin VI adapters (Hasson 2003, Osterweil *et al* 2005, Nash *et al* 2010). Moreover, the myosin VI interactome overlaps with endocytic adaptors that have been linked to GPCR trafficking earlier (O'Loughlin *et al* 2018). Thus, whereas the blocking of myosin VI using the ATPase inhibitor TIP abolished receptor endocytosis, dominant negative GIPC mutants had a relatively moderate or negligible effect in concurrent assays (Patel *et al* 2024). Nevertheless, the myosin VI inhibitor does not impair endocytosis of V2R, lacking the canonical PDZ motif, which also fails to trigger myosin VI driven motion. Therefore, although focused on GIPC, the conceptual implications of myosin VI-dependent GPCR regulation may be extended on other receptors. The magnitude of the interaction between the PDZ domain and the C-tail of the GPCR linearly corresponds to the activity of myosin VI (Patel *et al* 2024). Moreover, the sequence-optimized high-affinity PBM (KKETAV) (Saro *et al* 2007) led to the most pronounced augmentation of myosin VI motion, by a factor of about 1.8. In addition to the PBM sequence, however, it seems likely that the PBM effects on myosin VI are related to the whole C-terminus of the GPCR. Although the introduction of a high-affinity PBM is enough to improve myosin VI motility (V2D2R vs. V2R), the sequence context upstream of the PBM further enhances the activity of myosin VI. In particular, V2D2R is as fast as D2R, but V2VIPR1 is significantly less efficient than VIPR1. Myosin VI-mediated internalization of D2R is uncoupled from β-arrestin. D2R contains a short C-terminus that is missing the extended phosphorylation sites usually linked to GRK phosphorylation and ensuing recruitment of β-arrestin (Yang *et al* 2017). In addition, β-arrestin1/2 deletion does not interfere with internalization of D2R. Nevertheless, myosin VI repression in a β-arrestin-null context is enough to prevent internalization. Therefore, a β-arrestin-independent mechanism that proceeds through the activation of cytoskeletal motors by PBMs has been emphasized (Patel *et al* 2024). Considering the distinct activation of myosin VI through various GPCR C-tails, GPCR-PBM is proposed to correspond to the previously detected β-arrestin C-terminal barcode (Yang *et al* 2017), which variably influences and modulates receptor transportation and signal transduction. Although internalization of D2R requires no β-arrestin, it is conceivable that GPCRs with PBMs and GRK phosphorylation motifs exploit cooperative interactions between β-arrestin- and myosin VI-dependent events. A structural mechanism is postulated that accounts for the efficacy of agonists to control GPCR transportation. Latest studies utilizing β2AR have provided independent evidence for autoregulatory interplay between receptor cytosolic cavity-effector engagement and either its ICL361 or C-tail domain (Heng *et al* 2023). The direct, specific interaction between the D2R ICL3 and the C-tail modulates GIPC association and subsequent myosin VI-driven transportation (Patel *et al* 2024). A number of different factors can influence this interplay, for example the effectiveness of agonists, the overexpression of GIPC and proteins that interact with ICL3, such as NCAM (Xiao *et al* 2009). These factors regulate the attachment of myosin VI and thus the level of D2R transportation. Myosin VI does not work as a binary 'on-off' mechanism, but instead the effectiveness of agonists and environmental factors are deciphered to optimize receptor transportation and signal transduction. Considering the widespread utilization of apz in the therapy of neuropsychiatric disorders (Shapiro *et al* 2003, Tuplin and Holahan 2017, Mallet *et al* 2019), it is possible that GPCR therapeutics exert their cellular effects through the regulation of agonist transportation in the cytoskeleton. The sequence diversification of GPCR-PBMs and ICL3, a family of myosin VI-binding PDZ adaptors, and the alteration of myosin VI and adaptor expression in disease conditions like cancer opens new avenues for the development of context-sensitive therapeutic agents.

2.16.2.2 Combined function of DAB2 and myosin VI in endocytosis and endosomal trafficking

Myosin VI attaches to clathrin-coated endosomes via DAB2 (Morris *et al* 2002). In contrast, myosin VI tethers to uncoated endosomes via GIPC (Naccache *et al* 2006). Notably, myosin VI exhibits a series of unique structural characteristics, involving a flexible, extendable leverage arm (Spudich and Sivaramakrishnan 2010), cargo-driven dimerization (Phichith *et al* 2009), disengagement of an autoregulatory interplay (Batters *et al* 2016), and cargo-sensitive alterations of its chemomechanical circuitry (Altman *et al* 2004) that are engaged in a selective manner to fine-tune its performance. It is still elusive whether DAB2 in complexity with myosin VI particularly controls endocytosis or is also implicated in actin-based endosomal trafficking. DAB2 plays a critical role in regulating CME of a variety of receptors by interfering with clathrin-binding motifs on a number of receptors, phosphoinositide (4,5) bisphosphate (PI(4,5)P2) that is concentrated on clathrin-coated structures, and the clathrin adaptor AP-2, as extensively discussed by (Finkielstein and Capelluto 2016). Overexpression of DAB2 leads to the incorporation of myosin VI into clathrin-coated vesicles (Dance *et al* 2004), while a mutation of this locus impairs the incorporation of myosin VI into clathrin-coated vesicles (Spudich *et al* 2007). Myosin VI monomers are incapable of processivity, with DAB2 interaction myosin VI dimerization is induced and consequently myosin VI activation (Spudich *et al* 2007, Phichith *et al* 2009, Dos Santos *et al* 2022). In addition, myosin VI's processive travel alongside actin filaments *in vitro* is induced (Rai *et al* 2021). Thus, the cargo adaptor DAB2 targets myosin VI toward clathrin-coated endosomes. Typically, the interactions between a motor and its cargo adaptor are regarded as being stable. Nevertheless, force produced by multiple motors in stable clusters has been found to perturb localized cytoskeletal structure and may interfere with trafficking. It has been demonstrated that dynamic multimerization of myosin VI-Dab2 complexes promotes the processive movement of cargo in the absence of substantial rearrangement of cortical actin frameworks (Rai *et al* 2021). Notably, DAB2 myosin interactor region (MIR) interacts with myosin VI at a moderate affinity of 184 nM, and single-molecule kinetic assays reveal a high conversion rate at 1 s^{-1} of DAB2-MIR-myosin VI engagement (Rai *et al* 2021). Single-molecule motility indicates that saturation of DAB2-MIR at 2 μM enhances homodimerization and processivity of myosin VI, with run lengths similar to those of fully constitutive myosin VI dimers. DNA origami cargo-mimicking templates imprinted with DAB2-MIR-Myosin VI complexes are marginally processive and display limited motility on isolated actin filaments and halt-and-go movement on a cellular actin framework (Rai *et al* 2021). A minimal actin cortex built on lipid bilayers is remodeled and foci formed through unregulated processive motility by either constitutive myosin V or VI dimers. In opposition, DAB2-MIR-Myosin-VI-interference maintains the structural organization of a minimal cortical actin framework (Rai *et al* 2021). In conclusion, the dynamic motor-cargo connection is key to facilitating cargo trafficking while leaving the cytoskeleton organization unperturbed. In contrast to these experiments, the findings from in cellular and *in vivo* animal models indicate that DAB2 and myosin primarily modulate the detachment of endosomes from the plasma membrane, thereby permitting cargo capture. DAB2 and myosin VI both bind to clathrin-coated pits in living cells (Bond *et al* 2012). In fact, clathrin-coated cavity generation and the endocytosis of receptors are compromised in cells derived from myosin VI-deficient mice, and endocytic capture of a broad spectrum of cargos is impaired when DAB2 is absent *in vivo*, as reported in the review by Tao and colleagues (Tao *et al* 2016).

2.16.2.3 Interaction of GIPC and myosin VI on newly formed endosomes

The architecture reveals that autoinhibited myosin VI assumes a tight monomeric conformation through extended interactions between the head and tail domains, which is coordinated through an extended single-α-helix region that is similar to a 'backbone' (Niu *et al* 2024). Such an autoinhibited structure is expected to effectively obstruct the cargo recognition sites and thereby abolish ATPase activity. Certain cargo adapters, such as GIPC, can break up much inhibitory interference and enhance the motor activity, which indicates that there exists a cargo-driven activation of the entire processive motor (Niu *et al* 2024). In contrast to DAB2, GIPC is not necessary to acquire myosin VI toward clathrin-coated cavities (Spudich *et al* 2007). It alternatively interferes with myosin VI on newly generated endosomes after decoating of the clathrin envelope and stays attached to them as long as they arrive at Rab5-positive sorting endosomes (Aschenbrenner *et al* 2003). Silencing of GIPC in PC12 cells leads to TrkA receptor-laden endosomes remaining in the cell perimeter, whereas in control cells they are transported toward the perinuclear space (Varsano *et al* 2006). Using TIRF microscopy in living HEK293 cells, GIPC is actively enrolled to endosomes after clathrin dissociation that harbor the luteinizing hormone receptor. Elimination of GIPC decreased cAMP generation when exposed to luteinizing hormone (Jean-Alphonse *et al* 2014). In *in vitro* assays, GIPC augmented the velocity of movement of myosin VI-conjugated DNA-origami cargo on microfilaments, although it failed to enhance the walking distance or the amount of moving myosin VI motors (Rai *et al* 2022). In the aforementioned reformed cortical actin framework of keratocytes, however, DNA origami cargos traveled quicker in the presence of GIPC and myosin VI, moving at longer distances and with shorter halt periods compared to the sole presence of myosin VI (Rai *et al* 2022). GIPC triggers dimerization of myosin VI after abolishing the autoinhibition due to receptor binding *in vitro* (Shang *et al* 2021), and myosin VI turns into a processive dimer after attachment to emerging endosomes within the cell (Altman *et al* 2007). Abrogation of myosin VI dimerization cancels its processing capacity *in vitro* and disrupts endocytosis in cells (Mukherjea *et al* 2014). In conclusion, cellular and *in vitro* evidence heavily points to GIPC-myosin VI engagement with newly emerging endosomesand leads to endosomal exit from the actin cortex. GIPC, endocytic cargos and myosin VI function together, which is vital for the coordination of endosomal signaling interactions with effector molecules. There is a distinct endosomal signaling compartment for multiple receptors within APPL1-positive endosomes (Jean-Alphonse *et al* 2014, Sposini *et al* 2017). An adaptor protein that plays a role in the control of signaling cascades involves the adaptor protein comprising a pleckstrin homology domain (PH), a phosphotyrosine-binding domain (PTB) and the leucine zipper motif 1 (APPL1) (Diggins and Webb 2017). APPL1, synonymously referred to as DIP13α, engages a subset of Rab5-positive endosomes via its Bin-Amphiphysin-Rvs (BAR) and PH domains and orchestrates signaling routes via its PTB domain by engaging with multiple signaling receptors and proteins. APPL1 is thought to mediate the overlap between the Wnt and insulin signaling routes by converging proteins participating in these paths in endosomes (Schenck *et al* 2008, Pálfy *et al* 2012). APPL1 incorporates signaling interactions across multiple domains that facilitate protein and lipid interference (Mitsuuchi *et al* 1999, Miaczynska *et al* 2004). The N-terminal BAR domain is involved in perceiving or causing membrane curvature (Takei *et al* 1999, Habermann 2004), whereas the central PH and C-terminal PTB domains can attach to phospholipids (Li *et al* 2007). In fact, several different studies have identified the binding capacity of APPL1 to several phosphoinositides, among them PtdIns(3)P, PtdIns(4)P, PtdIns(5)P, PtdIns(3,4)P2 and PtdIns(3,5)P2 and PtdIns(3,4,5)P3 (Li *et al* 2007, Chial *et al* 2008). APPL1 is able to oligomerize to homo- or heterodimers with APPL2 via its BAR domain (Chial *et al* 2008). APPL1 differs from other BAR domain-containing proteins because the BAR and PH domains of APPL1 combine to generate a functional domain capable of engaging the small GTPase Rab5. Mutagenesis studies have established that Rab5 is capable of tethering to both the BAR and PH domains of APPL1, but if either domain is missing, RAB5 cannot interfere with APPL1 (Zhu *et al* 2007). The BAR-PH domain of APPL1 can also engage Rab21 (Zhu *et al* 2007) that is structurally comparable to Rab5 (Pereira-Leal and Seabra 2001). The PTB domain facilitates the engagement of APPL1 with a variety of receptor proteins, among them tropomyosin receptor kinase A (TrkA) (Lin *et al* 2006) and epidermal growth factor receptor (EGFR) At the C-terminal, APPL1 attaches to the PSD-95/Discs-large/ZO-1 (PDZ) domain of the adaptor protein GIPC (Varsano *et al* 2012), which is associated with the loading of cargos onto vesicles via its engagement with myosin VI (Katoh 2013). Myosin VI knockout and knockdown (Masters *et al* 2017) or GIPC knockdown (Varsano *et al* 2006), amplified the perinuclear positioning of APPL1-positive endosomes and also the abolishment of signal transduction inside the cells models. These findings highlighted a new function of GIPC and myosin VI in the localization of signaling endosomes.

2.16.3 Entry of dynein in the endosomal pathway

Although most of the research effort has been spent on kinesin motors, the function of dynein in endocytic transport is generally well recognized (Aniento 1993). Nonetheless, it is still not clearly elucidated how and where exactly dynein is transferred to endosomal cargo. The microtubule adaptor protein (MAP) tau potently impairs dynein-mediated motility of early phagosomes over long distances. Tau decreases the forces generated by dynein motor complexes on early phagosomes and enhances dynein disengagement when under load. Consequently, cargo proteins show divergent reactions to tau, whereby dynein complexes on early phagosomes react more sensitively to tau impairment than their counterparts on late phagosomes (Beaudet *et al* 2024). Dynein from vertebrates shows little motility on its own and is not subject to processive minus-end-directed motility (Trokter *et al* 2012). Dynamin contains two heavy chain subunits harboring the ATPase motor domain and the N-terminal region of the heavy chain binds light intermediate chains (LICs), such as LIC1 and LIC2, that subsequently enable dynein to create processive dynein–dynactin motor complexes. The N-terminal part of LICs, the G protein-like domain, binds to the heavy chain of dynein and the C-terminal part attaches to cargo adaptor proteins (Schroeder *et al* 2014). In addition, these cargo adapters serve as multi-functional proteins that can attach themselves to proteins such as a Rab GTPase, located on a membranous cargo, such as endosomes (Cianfrocco *et al* 2015). Classic *in vitro* recombination assays using the purified recombinant mammalian dynein showed that dynein also requires its known regulatory protein dynactin and a cargo adaptor, such as bicaudal D2 (BICD2), Hook1 and Hook3. Hook3 can bind with its Hook domain directly to the C-terminal domain of LIC1 (Schroeder *et al* 2014). Moreover, Hook3 can associate with the dynein–dynactin complex and thereby activate it. In this way, the cargo adaptor couples dynein and its cargo, ensuring dynein activation and directed movement (McKenney *et al* 2014, Schlager *et al* 2014). Hook1 and BICD2 have similar capabilities (Olenick *et al* 2016). Single-molecule motility assessments utilizing total-reflection fluorescence microscopy demonstrate that both Hook1 and Hook3 potently activate cytoplasmic dynein, thereby eliciting increased run-length and velocity compared to the prior identified dynein activator BICD2. As a result, dynein adapters appear to be able to control dynein in a differential manner to provide organelle-specific fine-tuning of the motor for accurate intracellular cargo transport (Olenick *et al* 2016).

2.16.3.1 Cargo uptake requires the attachment of dynein and dynactin to the plus-ends of microtubules

Cytoplasmic dynein exists as a dimer and constitutes a multi-subunit complex of motor proteins that are located at the minus end of microtubules, and it transports a range of cargos like membrane-bound organelles, such as endosomes (Reck-Peterson *et al* 2018). The engagement of dynein with microtubules is essential for the trafficking of Rab5-positive early endosomes into the cell's the interior, which represents a retrograde transport mechanism (Driskell *et al* 2007). The two processes, such as the dynein engagement with microtubules and the dynein-driven endosomal transport, which is referred to as minus-end directed motility, both require dynactin (Cianfrocco *et al* 2015). Dynactin is a multi-subunit complex that controls as a key regulatory protein the function of dynein attached to the minus end of microtubules. Therefore, dynactin also needs to attach to the growing plus ends of microtubules that are usually located at the cell circumference (Vaughan *et al* 1999). As dynactin can only weakly interact with dynein, adaptor proteins are necessary to recruit dynactin to cargos and for the stabilization of this attachment (McKenney *et al* 2014, Schlager *et al* 2014). Thereby, a ternary complex is formed, whereby it is assumed that dynein becomes activated through the liberation and disengagement of the two motor proteins from its autoinhibited state (Urnavicius *et al* 2015). Dynactin activates dynein and promotes dynein-driven motility. A subunit of the dynactin complex is the dynein interacting subunit p150^{glued}, which is synonymously referred to as DCTN1. The p150^{glued} comprises a N-terminal cytoskeleton-associated protein-Gly-rich (CAP-Gly) microtubule-binding domain that is attached to the microtubule plus-end-tracking proteins (+TIPs) that accumulate at the growing plus ends of microtubules to govern microtubule growth and attachment (Fong *et al* 2017). The +TIP end-binding protein 1 (EB1) (synonymously referred to MAPRE1) and EB3 (synonymously referred to MAPRE3) can autonomously attach to plus ends of microtubules and interfere with other +TIPs proteins (Ligon *et al* 2003, Moughamian *et al* 2013, Fong *et al* 2017, Tirumala and Ananthanarayanan 2020). Since dynactin and dynein interact and both cluster at the plus ends of microtubules at the cell border, they can capture arriving endosomes due to their vicinity to the plasma membrane (Vaughan *et al* 2002). The precise localization of early and late endosomes is important for their distinctive roles. The two motor proteins dynein and kinesin KIFC1 are required to varying degrees for the retrograde trafficking of early endosomes towards the minus ends of microtubules then for the trafficking of late endosomes/lysosomes (Villari *et al* 2020). The analysis of the endoplasmic reticulum transmembrane protein STIM1 in primary human umbilical vein endothelial cells demonstrated that STIM1 promotes the association of the microtubule plus end protein EB1 and the p150^{glued} subunit of the dynactin/dynein complex through its the SxIP motif and the coiled-coil domains (Villari *et al* 2020). Thereafter, the STIM1/p150^{glued}/dynein complex causes the concentration of late endosomes at the perinuclear vicinity of the endothelial cells. The precise cytosolic localization of early endosomes requires their simultaneous attachment to the STIM1/p150^{glued}/dynein and SxIP motif-containing KIFC1/p150^{glued} complexes. The adaptors HOOK1 and HOOK3 facilitate the association of p150^{glued} with the early endosome-bound dynein or KIFC1. Thus, a mechanism was uncovered that attributes the strongly differing steady-state distributions of early endosomes and late endosomes/lysosomes to the use of two different retrograde motor types (Villari *et al* 2020).

2.16.3.2 Dynein–dynactin interplay during early endosomal trafficking

Since it has been reported that the removal of p150^{glued} in HeLa cells fails to disrupt the transportation of early endosomes (Watson and Stephens 2006), it has been suggested that accumulation of dynein at the microtubule plus ends may not be a universal principle for endosomal transportation. In a further experimental study on HeLa cells, where individual dynein motors were made visible, these findings were supported by the observation that endogenous p150 was not only limited to the growing microtubule plus ends, but was also detected throughout the entire microtubule meshwork (Tirumala *et al* 2021). In contrast to dynein, which attaches to microtubules temporarily and then detaches from them again, the dynactin complex persists on microtubules and early endosomes. The random binding of dynein to such a dynactin cargo leads to a short-term movement of the cargo towards the minus end, and a series of such short movements could result in an extended transport of the cargo (Tirumala *et al* 2021) that are named pause and run movements. Consistent with these results, analogous pause and run events have been identified for early endosomal trafficking in HeLa cancer cells (Flores-Rodriguez *et al* 2011) and in human retinal pigment epithelial Arpe-19 cells (Zajac *et al* 2013). All of which may be attributable to a crowded cellular micro-environment that pronouncedly affects the motor-driven transport of endosomes. Transport directed towards the minus end along microtubules is mainly conveyed by the cytoplasmic dynein and its cofactor dynactin (Verma *et al* 2024). Dynein–dynactin complexes are quick, of about 1.2 $\mu m\ s^{-1}$, and travel characteristically over about 2.7 μm. The adapter BICD2 attracted a solitary dynein unit to dynactin, whereas BICDR1 and HOOK3 fostered the build-up of a twin-dynein complex (Urnavicius *et al* 2015 2018, Chaaban and Carter 2022). Remarkably, dynein–dynactin complex processivity and speed varied with the amount of dyneins, so that the BICDR1- and HOOK3-assembled double dynein complexes displayed a higher rate of processive motility occurrences and traveled faster than the individual dynein complexes formed by BICD2 (Urnavicius *et al* 2018). The quantification of the fluorescence intensities of the moving dots indicates that the dynein–dynactin movements are conveyed through at least one dynein heavy chain (DHC) dimer, while the speed is in agreement with the speed of double dynein complexes, such as two DHC dimers, recorded *in vitro* (Verma *et al* 2024).

2.16.3.3 Role of cargo adapter proteins regulate the enrollment of dynein and possibly its activation

A single cytoplasmic dynein can fulfill various trafficking work similar to several different kinesins. Thus, the question arises how a single dynein can carry out so many different tasks. Are there different adapter proteins for dynein due to the specific cargos? The answer is that several cargo adapter proteins exist that tether dynein/dynactin to specific types of cargo, such as organelles or the cell cortex (Kardon and Vale 2009).

Apart from being essential for the active minus end transportation system, the cargo adaptors have a critical function in the proper attracting of dynein into distinct membrane-bound compartments. Among the first dynein adaptor proteins to be identified was the specific adaptor protein Bicaudal-D2 (BICD2). It was shown to increase the dynein–dynein interaction and to facilitate the dynein activation to become a processive motor (Splinter *et al* 2012). It has been verified that BICD2 acts as an activating adaptor for dynein and thereby even other adaptor proteins including Hook1, Hook3, BICD2, BICDL1, Ninein, Ninein-like protein, Rab11FIP3 and Spindly have been identified (McKenney *et al* 2014, Schlager *et al* 2014, Reck-Peterson *et al* 2018). The highly conserved proteins Hook1, Hook2 and Hook3 function as adaptors linking the endosomal cargo with dynein. Hook proteins are integral components of a three-part FTS (synonymously referred to as AKTIP)–Hook–FH (FHF) complex that connects dynein–dynactin with the cargo.

In human cells, three Hook paralogues, such as Hook1, Hook2 and Hook3, which bind to dynein, create combinatorial complexes with four FHIP paralogues, such as FHIP1A, FHIP1B, FHIP2A and FHIP2B, which in turn interfere with broad range of membrane cargos, such as early endosomes, ER-to-Golgi vesicles, and mitochondria (Christensen *et al* 2021, Carvalho *et al* 2025). In *C. elegans*, it has been revealed that ZYG-12/Hook builds two diverse complexes to attract dynein to early endosomes and the nuclear envelope (Carvalho *et al* 2025). The dual function of the ZYG-12/Hook adaptor is controlled in a tissue-specific manner through alternative splicing of its C-terminal domain that is responsible for cargo tethering (Carvalho *et al* 2025). The FTS-Hook1/3-FHIP1B complex is directly associated with early endosomes via a direct interplay between FHIP1B and GTP-bound Rab5 (Christensen *et al* 2021). The motor-adaptor couples Dynein–Hook1 and KIFC1–Hook3 were identified for early endosomal trafficking in primary venous endothelial cells (Villari *et al* 2020). Similarly, the transport of Rab5-positive early endosomes and Rab7-positive late endosomes harboring brain-derived neurotrophic factor (BDNF) is reliant on dynein–Hook1 interaction within primary rat hippocampal neurons (Olenick *et al* 2019). Rab7 acts predominantly in the degradation pathway and is therefore crucial for regulating the amounts of degradation-destined cargos, a large proportion of them being signaling receptors. Hence, Rab7 activity can regulate the amounts of intracellular cues and the length of time that cues are present. Lysosomes and frequently also autophagosomes are Rab7-positive and linked to Rab7 effector proteins. Finally, Rab7's unique subcellular actions in distinct compartments, spanning from early endosomes toward autophagosomes, rely on the distinct selection of Rab7 effector proteins recruited at every endocytosis step (Mulligan and Winckler 2023).

Cellular cargos travel bidirectionally on microtubules using the dynein and kinesin motors with opposite polarity. In many cellular experiments, a codependency of microtubule motors has been revealed, in which the impairment or elimination of one motor leads to decreased cargo trafficking in both directions (Ashkin *et al* 1990, Gross *et al* 2002, Hancock 2014, Jongsma *et al* 2023). However, the exact mechanism is still elusive. Using *in vitro* reconstitutions of human proteins, kinesin-3 KIF1C was found to function as both an activator and a processivity factor for dynein. In addition, Hook3 can concurrently engage both dynein and the kinesin-3 KIF1C, whereby the association of Hook3 with dynein overrides the autoinhibited status of KIF1C (Siddiqui *et al* 2019). Activation demands only a fragment of the non-motor stalk of KIF1C, which engages the cargo adaptor HOOK3 (Abid Ali *et al* 2025). The interaction site is distinct from the constitutive factors FTS and FHIP, that connect HOOK3 with small G-proteins on cargos. Three models have been postulated to account for this phenomenon (Hancock 2014). The first model is the mechanical activation, whereby countervailing forces cause the motors to open and become active. The second model is the steric disinhibition, whereby the interaction of the motors with one another overcomes their self-inhibition. The third model is the microtubule attachment, whereby the interaction of both motors enables the cargo to stick to their microtubule tracks (Hancock 2014). A structural model for the autoinhibited FTS-HOOK3-FHIP1B (an FHF complex) has been postulated and it is elucidated how KIF1C unloads it (Abid Ali *et al* 2025). It has been demonstrated that dynein and kinesin can uncouple to enhance the frequency of processive motility on microtubules, which confirms the model of steric uncoupling (Abid Ali *et al* 2025). Initially, this model foresaw both motors engaging directly (Hancock 2014). Instead, disinhibition by the regulation of the scaffolding adaptor HOOK3 is postulated. Together, the codependency is unraveled by demonstrating how the mutual activation of dynein and kinesin is achieved by their joint adapter (Abid Ali *et al* 2025). Many adapters link both dynein and kinesin, indicating that this mechanism can be transferred to similar bidirectional complexes.

Wang and colleagues characterize two notable and novel dynein adapters, CRACR2a and Rab45, which integrate dynein activation and cargo acquisition capabilities into a sole polypeptide (Wang *et al* 2019). Both proteins belong to the Rab family and exhibit N-terminal EF-hand motifs, which are typically followed by coiled-coil regions (like the dynein adaptors Rab11FIP3 and Ninein) and a C-terminal Rab GTPase domain. Due to the similarity to the well-known dynein adaptors Rab11FIP3 and Ninein and the subcellular visualization in a perinuclear compartment close to the minus end of microtubules (Srikanth *et al* 2016), it seems to be likely that Rab45 and CRACR2a may function as dynein adaptors. Rab45 as well as CRACR2a activated processive microtubule-based motility of the purified dynein–dynactin complex, as measured by a single-molecule assay using total-reflection fluorescence microscopy (Sharma and Dwivedi 2019, Wang *et al* 2019). Importantly, CRACR2a, in contrast to Rab45, displayed an affinity for dynein–dynactin complexes solely when calcium was available, and calcium-binding defective (EF-hand mutants) CRACR2a could not engage dynein (Sharma and Dwivedi 2019). The physiological increase in calcium concentration, such as from 10 nM to 2 μM, markedly enhances the processivity of dynein–dynein CRACR2a complexes on microtubules, providing further support for the idea that calcium-dependent CRACR2a association with dynein is important in physiological settings. Moreover, CRACR2a-facilitated, but not Rab45-facilitated, motility could be activated through calcium *in vitro* (Wang *et al* 2019). Cargo adaptors fulfill various functions due to the cell type and type of the process. For instance, in Jurkat T cells, CRACR2a is involved in the early endosomal incorporation of the cell surface protein CD47 (Wang *et al* 2019).

2.16.4 Two-way motility of the early endosomes

Early endosomes are extensively characterized to be transported in a bidirectional manner (Nielsen *et al* 1999, Loubéry *et al* 2008). In contrast to mitochondria, endosomes can move bidirectional relatively fast in stop-and-go fashion (Welte 2004). Bidirectional stop-and-go vesicular traffic is a common characteristic of endosomal and vesicular trafficking, where the time spent at each stop, the frequency of stops, the time during which vesicles move, the length of their movements, and their velocity are all dependent on the vesicle type. These characteristics may be necessary to enable vesicles and endosomes to negotiate barriers within the crowded cell. Thereby, the concurrent association of plus-end-directed kinesin motors and minus-end-directed dynein motors triggers the bidirectional movement (Soppina *et al* 2009). Moreover, the bidirectional movement of early endosomes may induce their ripening toward late endosomes. Late endosomes undergoing bidirectional trafficking lack interchange of the Rab7 GTPase effectors RILP and FYCO1 or their cognate dynein and KIF5B motor proteins, respectively, in a physiological model of transport using a crowded cellular microenvironment (Jongsma *et al* 2024). During the maturation process, it can be chosen whether the endosomal cargo is broken down. For instance, a single early endosome frequently contains cargo that needs to be broken down, like EGF, and cargo like transferrin that needs to be recycled. Transferrin-laden vesicles have actually been observed to spill out of early endosomes harboring other cargo and then to coalesce with lysosomes (Driskell *et al* 2007). The elimination of bidirectionality could therefore lead to a rapid transportation of endosomes to the minus or plus ends of the microtubules but could also cause these cargos to be sorted incorrectly.

2.16.4.1 Comparison between a kinesin and a dynein motor and beyond the tug-of-war model

The prevailing paradigm for bidirectional cargo transport is the 'tug-of-war' model, which states that when both kinesin and dynein are simultaneously in place, they tug in opposite directions, and the most powerful motor dictates the destination of the cargo (Gross 2004). One remarkable experimental finding is the lengthening of endosomes immediately preceding a directional motility change in *Dictyostelium* cells (Soppina *et al* 2009). Generally, a single kinesin motor is considered to produce more force compared to a sole dynein motor. The kinesin motor produces holding forces of 5–6 pN (Svoboda and Block 1994) in contrast to 1 pN for a solitary dynein motor (Mallik *et al* 2004). Therefore, a single kinesin motor can engage in a tug-of-war with dynein motor clusters attached to the identical cargo. When under high load, the kinesin motor on Rab7-positive late phagosomes with entangled latex beads was more easily disengaged due to the 'catch-bond' characteristic of the dynein group, wherein the uncoupling rates of these motors decline with growing force on them (Rai *et al* 2013). This 'catch-bond' reaction of dynein upon exposure of high stress could thus favor the net progression of endosomes toward the minus end of microtubules. Results from a number of experimental and computational investigations indicate, nevertheless, that the tug-of-war model is inadequate to fully explain the spectrum of bidirectional trafficking operations in cells. A 'paradox of codependency' has been noted in various research studies (Hancock 2014), in which attenuation of plus-end-directed motors abrogates, rather than augments, minus-end-directed movement, and the reverse (Martin *et al* 1999, Gross *et al* 2002, Kunwar *et al* 2011). This intricacy points to mechanisms that go deeper than a mere mechanical tug-of-war, as exemplified by the need for both motors for complete activation, the binding of cargo and regulatory mechanisms (Gicking *et al* 2022). Bidirectional transportation of cargo, for example in neurons, involves competitive activity of the motors from the kinesin-1, kinesin-2 and kinesin-3 superfamilies against the cytoplasmic dynein-1. Earlier studies have demonstrated that motors bound to the kinesin-1–dynein–dynactin–BicD2 complex (DDB) walk slowly with a slight plus-end tilt, indicating that kinesin-1 overcomes the DDB complex, but that DDB imposes a considerable inhibition burden. Relative to kinesin-1, motors from the kinesin-2 and -3 families show greater sensitivity to cargo in single-molecule assays and are therefore expected to be overloaded by dynein complexes during cargo transportation. To verify this hypothesis, a DNA scaffold was used that fuses single-stranded DNA to each motor, couples DDB via complementary base pairing to different members of the kinesin-1, -2, and -3 families, recapitulates bidirectional transport *in vitro*, and monitors motor pairs by two-channel internal reflection fluorescence (TIRF) microscopy (Belyy *et al* 2016). This approach has been widely exploited to study the mechanical rivalry between kinesin-1 and different activated dynein complexes tethered to BicD2 (DDB), BicDR1 (DDR) and Hook3 (DDH) (Belyy *et al* 2016, Elshenawy *et al* 2019, Feng *et al* 2020, Ferro *et al* 2020). In these studies, it was found that DDB can considerably reduce the stepping rate of kinesin-1, but that kinesin-1 still dominates kinesin-DDB transportation. Conversely, DDR and DDH, which comprise two dyneins and possibly activate dyneins with greater efficiency (Grotjahn *et al* 2018, Urnavicius *et al* 2018), recruit kinesin-1 to the minus end with higher frequency compared to DDB (Elshenawy *et al* 2019). Within neurons and some other cells, dynein assemblies also transport cargo in the opposite direction to members of the kinesin-2 (Loubéry *et al* 2008, Hendricks *et al* 2010, 2012) and kinesin-3 (Schuster *et al* 2011) families. Interestingly, the kinesin-2 and -3 families possess movement and force-generating capabilities that differ from kinesin-1 (Arpağ *et al* 2014, Andreasson *et al* 2015, Lessard *et al* 2019, Zaniewski *et al* 2020, Budaitis *et al* 2021). Remarkably, when both kinesin and dynein are active and stepping on the microtubule, the kinesin-1, -2, and -3 motors are capable of resisting the inhibitory loads imposed by DDB. As mentioned above, bidirectional DDB-kinesin transportation was carried out by combining a kinesin and a DDB motor with complementary single-stranded DNA and directly examining how the various motility characteristics of the members of the kinesin-1, -2, and -3 families translate into the resulting bidirectional movement of the DDB-kinesin complexes. A stochastic step simulation of the three motor pairs revealed a mechanism whereby the fast-rebinding kinetics of kinesin-2 and kinesin-3 oppose their fast detachment under load, thereby facilitating strong force generation opposing DDB motors. These findings corroborate the assumption that load-dependent disengagement and reassociation are the crucial determinants of motor activity when under load and suggest that family-specific mechanochemical principles are used to ensure effective cargo transportation. These experiments are accompanied with stochastic stepping simulations demonstrating that kinesin-2 and kinesin-3 motors balance their faster uncoupling rate with increased reattachment kinetics when under load (Ohashi *et al* 2019, Gicking *et al* 2022). The near-equivalence of the three kinesin motor families underscores the critical importance of motor kinetics for the kinesin/dynein force equilibrium and highlights the central role of motor modulation by cargo adapters, regulatory proteins, and the microtubule cytoskeleton in coordinating the velocity and orientation of cargo trafficking in cells. Kinesin-1 resists considerable inhibitory forces over long periods of time, i.e. it does not tend to come untethered under load (Visscher *et al* 1999,

Schnitzer *et al* 2000, Blehm *et al* 2013, Pyrpassopoulos *et al* 2020). Conversely, members of the kinesin-2 and -3 families were found to dissociate rapidly under load (Andreasson *et al* 2015, Arpağ *et al* 2019, Budaitis *et al* 2021). Remarkably, these kinesin-2 and -3 motors were also found to reattach to the microtubule and re-establish force production more rapidly compared to kinesin-1, possibly balancing the fast disengagement (Andreasson *et al* 2015, Feng *et al* 2018, Arpağ *et al* 2019, Budaitis *et al* 2021). Overall, these findings imply that the best predictor of bidirectional transportation behavior for a given set of motor pairs or teams is the force-dependent dissociation, in preference to the stall force.

2.16.4.2 Microtubule-associated proteins govern the activity and enrollment of dynein and various kinesins

It has been postulated that post-translational modifications of tubulin serve as a 'tubulin code' that can be scanned by activated motor proteins to guide their motility towards certain cellular compartments (Janke 2014, Yu *et al* 2015, Gadadhar *et al* 2017, Park and Roll-Mecak 2018). Despite detyrosination lowering the motility of the activated dynein–dynein light chain (DDB) complex by a factor of four (McKenney *et al* 2016), the observed biophysical impact of certain tubulin modification types on kinesin-1 is rather small. Acetylation enhances the speed of kinesin-1 by a factor of 1.2 (Reed *et al* 2006) and detyrosination reduces the processivity of kinesin-1 by a factor of 1.4 without affecting its speed (Sirajuddin *et al* 2014), which raises the question of how such effects could have a direct and significant impact on kinesin-1 transport *in vivo*. In the cells, cargo-transporting motors need to engage with a plethora of non-enzymatic microtubule-associated proteins (MAPs) that line the microtubule cytoskeleton (Ramkumar *et al* 2018). A perturbation of this bidirectional trafficking system resulting from mutations in motor complexes or MAPs underlies a large variety of neurodevelopmental and neurodegenerative pathologies (Gleeson *et al* 1998, Chevalier-Larsen and Holzbaur 2006, Harms *et al* 2012), which emphasizes the significance of the crosstalk between these protein classes.

The first MAPs to be identified were the 'structural' MAPs, which were purified together with polymerized brain tubulin (Borisy *et al* 1975). Since then, MAPs have been characterized as stabilizers, nucleation-promoting factors, and regulators of microtubule bundling (Bulinski and Borisy 1979, Chen *et al* 1992, Bulinski and Bossler 1994, Moores *et al* 2006, Bechstedt and Brouhard 2012, Wieczorek *et al* 2015). Nevertheless, more recent work indicates that these MAPs may also have the role of regulating motor-based transportation (Lipka *et al* 2016, Ecklund *et al* 2017, Monroy *et al* 2018, Tan *et al* 2019). MAPs like tau, MAP2, MAP7 and MAP9 favor motor activity or enrollment (Monroy *et al* 2020). For example, tau has been observed to impair the processive motion of kinesin-1 and kinesin-3, while it causes just a slight reduction in motion of activated dynein–dynactin complexes (Vershinin *et al* 2007, Dixit *et al* 2008, McVicker *et al* 2011, Chaudhary *et al* 2018, Monroy *et al* 2018, Tan *et al* 2019). In distinct opposition, MAP7 directly enlists kinesin-1 to microtubules while not altering dynein motion (Chaudhary *et al* 2019, Métivier *et al* 2019),

In contrast, MAP9 leaves the processivity of kinesin-3 to microtubules unaltered but blocks that of dynein (Monroy *et al* 2020). Inside cells, variable enrollment of specific MAPs on microtubules might confer distorted transport of endosomes, for example, when endosomes are targeted for breakdown, they might be assembled on tau-decorated microtubules. Finally, a so-called general 'MAP code' has been assumed that drives the directed motion alongside microtubules and aids in elucidating the intricate intracellular sorting witnessed in highly polarized cells like neurons (Monroy *et al* 2020).

2.16.5 Ripening of early endosomes to late endosomes with emphasis on dynein's role

The formation of phopshoinositide(3)phosphate (PI(3)P) on Rab5 endosomes results in two distinct events: Acquisition of Rab7 and displacement of Rab5 and thereby maturation of the early endosomes toward a late endosome (Rink *et al* 2005), or enrollment of the kinesin family member KIF16B, that sequesters cargos from Rab5-positive early endosomes for redirection toward Rab11-positive recycling endosomes (Hoepfner *et al* 2005). The intricate regulation of endosomal ripening and endosomal recycling by PI(3)P is discussed by Redpath and colleagues (Redpath *et al* 2020). After the maturation into a Rab7-positive late endosome, the minus-end driven dynein motor has been demonstrated to be integral for the merging of Rab7-positive late endosomes with lysosomes and for lysosomal localization (Jordens *et al* 2001, Johansson *et al* 2007, Cabukusta and Neefjes 2018, Cason *et al* 2021). Even though lysosomes can be distributed all over the cytoplasm, perinuclear lysosomes have been found to have a higher breakdown efficiency. The maturation from early to late endosomal organelles is accompanied by a shift towards a more monodirectional, minus-end-directed motile behavior. Macrophage-derived phagosomes comprising latex beads move in a highly monodirectional manner as a result of cholesterol-triggered clustering of dynein motors (Rai *et al* 2016).

Apart from RABs, another strategy to enlist dynein toward late endosomes and lysosomes is based on septins is discussed. A new mechanism for retrograde lysosome trafficking involving septin GTPases has been uncovered. Septins can generate polymers that engage with cell membranes and filaments of the cytoskeletal network (Mostowy and Cossart 2012). Septins are found to be associated in lysosomal transportation (Dolat and Spiliotis 2016), which also covers that of bacteria (Krokowski *et al* 2018). Septin 9 (SEPT9) can bind to lysosomes and enhance the perinuclear distribution of lysosomes in a way that is independent of Rab7. The localization of SEPT9 to mitochondria and peroxisomes is adequate to cause dynein recruitment and perinuclear cluster formation. SEPT9 has been shown to engage both dynein and dynactin via its GTPase domain and N-terminal extension, respectively. It is noteworthy that SEPT9 binds favorably to dynein intermediate chain (DIC) in its GDP-bound form, promoting dimerization and multimerization of SEPT9 (Kesisova *et al* 2021). This is in opposition to Rab7, which relies on GTP attachment for binding with lysosomes. This distinction in nucleotide condition has been considered crucial for SEPT9-based lysosomal repositioning under acute oxidative stress (Kesisova *et al* 2021). In reply to arsenite-induced oxidative cell stress, the association of SEPT9 with lysosomes is increased, thereby stimulating the perinuclear accumulation of lysosomes. Septins were assumed to act as GDP-activated frameworks for the collaborative formation of dynein–dynactin and to offer an alternative mechanism for retrograde lysosome transportation in the steady state and while the cell is adjusting to stress. The small G protein Arl8b fulfills a well-established function in lysosomal localization (Khatter *et al* 2015). Arl8b, which is a specific Arf-like GTPase family localized to lysosomes and is crucial for multiple lysosome-related cellular functions like autophagy (Hofmann and Munro 2006). Besides the traditional motor proteins, several effectors have been found to control lysosomal positioning and functionality through Arl8 in mammals, such as the homotypic fusion and protein sorting (HOPS) (Khatter *et al* 2015), Pleckstrin Homology and RUN Domain Containing M1 (PLEKHM1) (Marwaha *et al* 2017) and PLEKHM2, which is synonymously referred to as SKIP (Rosa-Ferreira and Munro 2011, Keren-Kaplan and Bonifacino 2021). The active Arl8b may be preferentially targeted by the RUN domains of the two Arl8b effectors PLEKHM1 and SKIP, whereby the autophagosome/lysosome membrane fusion and intracellular lysosome localization are controlled (Qiu *et al* 2023). Like other small GTPases, Arl8 can act as a molecular on/off switch, toggling between a GDP-bound inactive state and a GTP-bound active state, a process that is tightly controlled by dedicated GTPase-activating proteins (GAPs) and guanine nucleotide exchange factors (GEFs). In terms of its structure, Arl8 comprises two switch regions (switch 1 and switch 2) that experience major conformational transitions when switching from the GDP-bound inactive conformation to the GTP-bound active conformation (Okamura *et al* 2011, Khatter *et al* 2015, Xu *et al* 2020b). RUN and FYVE domain-containing protein 3 (RUFY3) (Keren-Kaplan *et al* 2022, Kumar *et al* 2022) and RUFY4 (Keren-Kaplan *et al* 2022) can cause dynein–dynactin to predominantly exert control over the perinuclear positioning of lysosomes. RUFY3 has also been identified as an ARL8b signaling effector (Keren-Kaplan *et al* 2022, Kumar *et al* 2022). Consequently, deficiency of RUFY3 leads to a reduction in the size and a characteristic amount of lysosomes as well as to a decrease in the breakdown ability of lysosomes (Kumar *et al* 2022). In addition, RUFY3 is required for the localization of ARL8b^{+} /LAMP1^{+} -endolysosomes within pericentriolar organelles (Char *et al* 2023).

2.16.6 Kinesins and myosin V function in the endosomal recycling pathway

Recycling of endosomes is essential for adjusting the signaling events of receptors, as it maintains receptor plasma membrane abundance, regulates the circulating pool of proteins, and sustains the plasma membrane in its entirety (Goldenring 2015). The process is mainly carried out via Rab11-positive recycling endosomes, the majority of them is found in the perinuclear zone of the MTOC (Naslavsky and Caplan 2018). The trafficking of Rab5-positive sorting endosomes to the perinuclear recycling endosomes is facilitated via dynein; these are subsequently transported by kinesin-like protein 13A (KIF13A) and KIF13B, first along the microtubules, then by myosin V and finally by the actin cortex to the plasma membrane, where they finally fuse with the plasma membrane.

2.16.6.1 Dynein interacting proteins enhance cargo delivery to the perinuclear area

Rab11 endosomes sprout from Rab5-positive sorting endosomes and thereby select cargo from the endolysosomal route for the recycling process (Campa *et al* 2018). The FERARI complex, which includes Sorting Nexin 4 (SNX4) and Rab11 Family-Interacting Protein 2 (Rab11FIP2), is responsible for its cleavage (Solinger *et al* 2020). It has been demonstrated that SNX4 interferes with dynein, and when it is knocked down in cells, the Rab11-recirculated cargo transferrin cannot be delivered to the perinuclear area, but is destroyed in lysosomes (Traer *et al* 2007). Rab11FIP3, which is another Rab11 family-interacting protein, has been identified as an activation adaptor of dynein *in vitro* (McKenney *et al* 2014), and overexpression of mutants, that cannot engage with the intermediate light chain 1 of dynein and hence inhibits transferrin trafficking toward the perinuclear site. Consequently, it is sequestered at the cell perimeter (Horgan *et al* 2010), thereby offering additional support for the regulatory function of Rab11FIP3 in dynein-driven recycling endosome trafficking.

2.16.6.2 Recycling of cargo from perinuclear area via kinesin and myosin V

After dynein-driven trafficking to the perinuclear compartment, KIF13A and KIF13B moderate cargo trafficking toward the plasma membrane (Thankachan and Setty 2022). The small GTPase Rab10 was found to be necessary to generate tubular recycling endosomes within HeLa cells together with KIF13A and KIF13B (Etoh and Fukuda 2019, Kawai *et al* 2021). These Rab10-positive tubules lacked endosome/lysosome compartment marker proteins like Rab5, Rab7 or LAMP1, indicating that the Rab10-positive tubules are not components of the lysosomal degradative route. Rab10 deficiency leads to the clustering of LDL receptor and transferrin cargo in perinuclear Rab11 and Rab4 recycling endosomes (Khan *et al* 2022). KIF13A has been identified participating in the tubulation of GTP-Rab11-positive recycling endosomes, which is necessary for the return of cargo back to the plasma membrane (Delevoye *et al* 2014). Overall, these results suggest that Rab10 and KIF13A and B are the main effectors for the export of cargo from the Rab11-positive perinuclear area to be recycled toward the plasma membrane, indicating that Rab10 can be viewed as a candidate kinesin cargo adapter. The ultimate recycling step for cargos is the travel through the actin cortex to merge the recycling endosomes with the plasma membrane. This is accomplished by myosin V, which transports the cargos from the minus ends of the actin filaments, which point towards the interior of the cell, to the plus end directly under the plasma membrane (Hammer and Sellers 2012). Although myosin V is an actin-based motor, it seems to be instrumental in propelling recycling endosomes as they move out of the perinuclear area, possibly synergizing with kinesin. Myosin Vb tail overexpression can impair myosin V trafficking and leads to a clustering of cargo and Rab11-positive endosomes in the perinuclear area, which occurs in non-polarized and polarized cells (Lapierre *et al* 2001, Hales *et al* 2002, Xie *et al* 2016). This finding leads to the hypothesis that in the absence of myosin V activity, recycling cargos fail to contact microtubules. In agreement to this result, when mouse myosin V and Drosophila kinesin are coincubated *in vitro*, myosin V carries out longer movement phases on actin and kinesin travels longer on microtubules compared to each motor on its own (Ali *et al* 2008). Nevertheless, the synergistic action of myosin V and kinesin has not yet been shown within living cells. Myosin V and kinesin, however, can interfere (Ali *et al* 2008), and disruption of either myosin V or kinesin leads to an aggregation of cargo in the perinuclear area (Lapierre *et al* 2001, Hales *et al* 2002, Xie *et al* 2016). Both motors seem to be able to work together with the identical recycling cargos to drive cargo trafficking from microtubules toward actin-based routes in an efficient fashion

2.17 Metastasis and endocytosis suppressors

Metastasis suppressors comprise a group of genes that repress the metastatic capacity of cancer cells but do not significantly influence the size of the primary tumor (Khan and Steeg 2018). More than 20 metastasis suppressor genes, which encompass miRNAs, have been reported in different cancer types with a broad spectrum of biochemical activities (Hurst and Welch 2011). Several of the metastasis suppressor genes that function through changes in endocytosis are outlined in the following.

2.17.1 Non-metastatic clone 23, isoform H1 (NM23/NM23-H1) is named NME1

NME1 represents a multifunctional protein which is highly conserved from yeast to humans. Its enhanced expression suppressed metastasis in a wide range of cancer cell lines with no change in the growth of the primary tumor (Salerno *et al* 2003, Mátyási *et al* 2020). In addition to serving as a metastasis suppressor, it is also recognized for its developmental role. The *Drosophila* homolog of NME is abnormal wing disc (awd) and are recognized to control the cell differentiation and motility of multiple organs in late embryogenesis through controlling growth factor receptor signaling via endocytosis. These investigations identified a genetic interference between awd and dynamin (shi) (Dammai *et al* 2003). Aberrant endocytosis has been linked with mutant awd phenotypes and complementary RAB5 or shi genes (Dammai *et al* 2003, Woolworth *et al* 2009, Ignesti *et al* 2014). It has been reported that awd controlled tracheal cell motility during development through the regulation of fibroblast growth factor receptor (FGFR) concentrations via dynamin-driven endocytosis (Dammai *et al* 2003, Nallamothu *et al* 2008). In addition, deletion of the awd gene also inhibited Notch signal transduction via modifying notch receptor processing resulting in Notch aggregation inside early endosomes (Ignesti *et al* 2014). The role of NME as an endocytosis-interacting partner of dynamin has also been emphasized in various mammalian cancer models (Boissan *et al* 2014, Khan *et al* 2019). NME transfectants of several cell lines showed enhanced endocytosis of EGFR and transferrin in conjunction with inhibition of motility. Enhanced endocytosis and suppression of motility were both inhibited using dynamin inhibitors. In a lung metastasis experiment, overexpression of NME1 could not achieve meaningful inhibition of metastasis in cells in which dynamin 2 had been switched off. Employing the EGF/EGFR signaling axis as an *in vitro* model, NME1 reduced phospho-EGFR and phospho-Akt levels in a dynamin-2-dependent fashion, underscoring the relevance of this interaction for subsequent signal transduction. Notably, NME has been hypothesized to function as a GTP delivery/oligomerization facilitator of dynamin 2, resulting in higher GTPase activity of dynamin 2 and elevated endocytosis (Boissan *et al* 2014, Khan *et al* 2019). These data revealed a further mechanism of NME-dynamin engagement: *In vitro*, NME facilitated the oligomerization of dynamin and its elevated levels of GTPase activity required for vesicle cleavage (Khan *et al* 2019).

2.17.2 KAI1, synonymously referred to as cluster of differentiation 82 (CD82)

KAI1/CD82 is a family member of the evolutionarily conserved tetraspanin family and has originally been characterized as a metastasis suppressor for prostate cancer (Ichikawa *et al* 1991). KAI1 is now widely recognized as a metastasis suppressor in several solid cancers. Increased expression of KAI1 is associated with a better prognosis (Dong *et al* 1996, 1997, Ow *et al* 2000), while decreased expression of KAI1 is associated with aggressive cancer in various cancers, among them pancreatic cancer, hepatocellular carcinoma, bladder cancer, breast cancer and non-small cell lung cancer (Dong *et al* 1996, Kawana *et al* 1997, Uchida *et al* 1999). The suppression of metastasis conferred through KAI1 is assumed to be accomplished mainly through blocking the migration and invasion of cancer cells (Yang *et al* 2001). This phenotype is the consequence of the creation of oligomeric compounds with binding partners like integrins, EGFR and intracellular signaling proteins including protein kinase C (PKC). This complex usually results in involvement of multiple receptors in either reallocation or enhancement of internalization. For instance, overexpression of KAI1 results in a rearrangement of the urokinase-type plasminogen activator receptor (uPAR) into a stabilized complex involving integrin $\alpha5\beta1$ within focal adhesions (Bass *et al* 2005). The focal adhesion linkage of uPAR decreases its capacity to engage the ligand uPA and subsequently split and activate plasminogen. Likewise, KAI1 also tethers to EGFR, ErbB2 and ErbB3, resulting in enhanced endocytosis and desensitization of EGFR ((Odintsova *et al* 2000, 2003). KAI1 also selectively blocks ligand-induced EGFR dimerization and changes the EGFR dispersion in the plasma membrane, which affects EGFR activation (Odintsova *et al* 2003).

2.17.3 Metastasis suppressor protein 1 (MTSS1)/missing in metastasis (MIM)

MTSS1/MIM, first identified in bladder cancer cell lines, has been demonstrated to be abundant in non-metastatic but not metastatic bladder cancer cells (Lee *et al* 2002). MTSS1 is assumed to repress metastasis through acting as a scaffold protein that interferes with actin-associated proteins to control cytoskeletal dynamics and lamellipodia development, thereby influencing cancer cell invasion and metastatic outcome (Woodings *et al* 2003). In squamous cell carcinoma of the head and neck,

MTSS1 enhances EGF signal transduction through antagonizing EGFR endocytosis at low cell densities and enhances cellular proliferation in initial phases of tumor growth. At elevated cell densities, MTSS1 has a negative influence on EGF signal transmission and prevents metastasis (Dawson *et al* 2012).

2.17.4 Kisspeptin-1 (KISS1)

The KISS1 gene generates a peptide product termed kisspeptin (KP), which functions as an endogenous ligand for a G-protein-coupled receptor, KISS1R, which is synonymously referred to as GPR54 (Cvetković *et al* 2013). KISS1 serves as a metastasis suppressor gene in numerous human cancers, such as melanoma, pancreatic cancer and gastric cancer, through its KP/KISS1R signal transduction via blocking cellular motility, proliferation, invasion, chemotaxis and metastasis (Beck and Welch 2010). In breast cancer, however, KP induces cancer cell invasion and high expression of KISS1; GPR54 mRNA levels have been positively associated with reduced relapse-free survival. Intriguingly, GPR54 forms a direct complex with EGFR, and stimulation of breast cancer cells through EGF or KP-10 modulated the endocytosis of GPR54 and EGFR (Zajac *et al* 2011). The signaling has the contrary impact on breast cancer cells, that is, it enhances migration and invasion in human breast cancer cells. While metastasis suppressor genes frequently exhibit statistically significant inverse tendencies in the expression of cancer and the survival of patients, they are not expected to be used as clinical prognostic determinants in preference to more complex gene signatures. Like tumor suppressors, clinical translation is difficult. Reestablishing expression of the metastasis suppressor in each metastatic cancer cell would be required for maximal efficacy, but this is infeasible. The transcriptional upregulation of NME using high-dose medroxyprogesterone acetate has been examined (Palmieri *et al* 2005). A phase 2 trial had a technical default, as serum levels of medroxyprogesterone acetate failed to increase adequately, despite some long-term stability of disease noted (Miller *et al* 2014). The manner in which endocytic pathways may promote metastasis-suppressive clinical-translational performance is presently poorly understood but of much debate. Further investigation to determine the intricate mechanisms underpinning these processes is merited.

2.18 Transfer of invasive capacities by highly to lowly invasive cancer cells

Overexpression of the transmembrane membrane-type 1 matrix metalloproteinase (MT1-MMP, synonymously referred to as MMP14) facilitates invasion of cancer cells. MT1-MMP-positive cancer cells can switch on the invasive capacities in MT1-MMP-negative cells through the transfer of a soluble catalytic ectodomain of MT1-MMP (Wenzel *et al* 2024). Unexpectedly, this outcome relies on the availability of TKS4 and TKS5 in the donor cell, which are adaptor proteins involved in the formation of invadopodia. In donor cell endosomes, TKS4/5 facilitate ADAM-induced cleavage of MT1-MMP by bypassing the two proteases, and scission is incentivized through the low intraluminal pH of endosomes. Bypassing relies on the PX domains of TKS4/5, which happen to engage with the cytosolic tail of MT1-

MMP and endosomal phosphatidylinositol-3-phosphate. MT1-MMP enlists TKS4/5 inside multivesicular endosomes to co-secrete it along with the enzymatically active ectodomain into EVs. The secreted ectodomain switches non-invasive receiving cells toward invasive cells. Hence, TKS4/5 facilitate the intercellular transfer of cancer cell invasiveness via ADAM-mediated liberation of MT1-MMP in acidic endosomes. A requirement for the spread and consecutive metastasis of a cancer is the breakdown of the nearby ECM (figure 2.3). MMPs make up soluble or membrane-bound enzymes that carry out the digestion of ECM fibers and permit cancer cells to break through the basement membrane and penetrate the adjacent cell tissue. A pivotal factor in the advancement of cancer is MT1-MMP/MMP14, the overexpression of which is linked to a negative outcome in several types of cancer (Wu *et al* 2014). MT1-MMP is regarded as pro-invasive and pro-tumorigenic factor, which renders it an attractive candidate for cancer therapy (Shay *et al* 2015). It shows enzymatic activity against type I, II and III collagen, gelatin, fibronectin, laminin 1 and 5 and vitronectin and also triggers the activation of other soluble MMPs such as proMMP2 and proMMP13 (Sato *et al* 1994, Knäuper *et al* 1996), thus participating in both, the direct and indirect breakdown of the ECM.

The most well characterized role of MT1-MMP is at invadopodia and podosomes, in which MT1-MMP is specifically delivered through tubulation of late

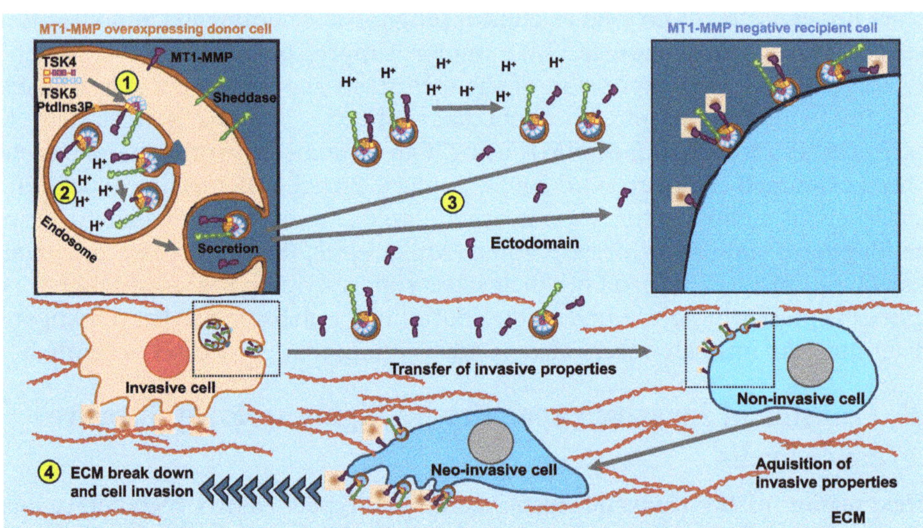

Figure 2.3. Proposed model for the intercellular transmission of invasiveness via endosomal protease delivery. TKS4/5 localizes to endosomes through interaction with the endosomal lipid PtdIns3P and the cytosolic tail of MT1-MMP and is then sequestered together with MT1-MMP in ILVs of multivesicular endosomes (1). TKS4/5 connects MT1-MMP to sheddaes of the ADAM family, and the low endosomal pH aids ADAM-based cleavage of a catalytically active ectodomain from MT1-MMP in the endosomal lumen (2). MT1-MMP-harboringg exosomes and their cleaved ectodomain are shed from donor cells and attach to MT1-MMP-negative, non-invasive receiving cells (3). This leads to breakdown of the ECM enveloping the receiving cell. As a result, the receiving cell turns invasive (4). The existence of TKS4/5 in EVs may promote the continued cleavage of MT1-MMP in the cancer's acidic microenvironment.

endosomes (Marchesin *et al* 2015, Pedersen *et al* 2020, Linder *et al* 2023). Podosomes contain small actin-rich cellular protrusions that are fortified with matrix metalloproteases, that break down the ECM and are a precondition for cell invasion by, for instance, macrophages and osteoclasts. Cancer cells can build corresponding structures, referred to as invadopodia, which facilitate cell invasion and metastasis (Linder *et al* 2023). The two scaffolding proteins, SH3 and PX domain-containing protein 2A and 2B (TKS5 and TKS4), are required for the generation of functional podosomes and invadopodia. Upon growth factor signal transduction, TKS4 and TKS5 are phosphorylated and activated, allowing them to detect the phospholipid phosphatidylinositol 3,4-bisphosphate (PtdIns(3,4)P2) at the invadopodial membrane via their amino-terminal phox homology (PX) domain. TKS4 and TKS5, which have four and five SH3 domains, respectively, function as adaptor proteins with multiple binding partners, among them actin regulators, cortactin and ADAM (a disintegrin and metalloprotease) sheddases (Abram *et al* 2003, Mao *et al* 2009, Kudlik *et al* 2020). In a similar way to MT1-MMP, overexpression of both, TSK4 and TSK5, proteins has been linked to the advancement of cancers including brain tumors, lung adenocarcinomas, prostate cancer, breast cancer and melanoma (Stylli *et al* 2012, Li *et al* 2013, Burger *et al* 2014, Blouw *et al* 2015, Iizuka *et al* 2016).

The transmembrane protein MT1-MMP is found on different cell membranes and is transported from the plasma membrane via the endocytic route (Castro-Castro *et al* 2016). The short C-terminal tail functions as a turntable for several cytosolic proteins and controls the functioning of MT1-MMP (Strouhalova *et al* 2023). Importantly, MT1-MMP has also been identified in conjunction with EVs, where it exhibits enzymatic activity and is able to promote pre-metastatic niche creation (Hakulinen *et al* 2008, Clancy *et al* 2015, Shimoda and Khokha 2017, Sanderson *et al* 2019). Moreover, MT1-MMP can split its catalytically active extracellular domain, which is facilitated through the ADAM family of sheddases (Toth *et al* 2005, 2006). How the ADAM-driven processing of MT1-MMP is regulated spatially and mechanistically is elusive. The operation and regulation of soluble or EV-bound MT1-MMPs remains poorly characterized. TKS4/5 has been shown to promote ADAM-induced splitting of MT1-MMP in acidic endosomes via forming a bridge between the two proteases, allowing spatiotemporal support of their activities. Consequent secretion of soluble and EV-associated MT1-MMP from highly invasive cancer cells renders MT1-MMP-negative cells subsequently to be able to break down ECM and invade the matrix. Therefore, overexpression of TKS4/5 and MT1-MMP promotes the intercellular transduction of the invasiveness to cancer cells. MT1-MMP and TKS4/5 overexpressions drive cancer advancement, which is attributed to their involvement in the development of invadopodia (Wu *et al* 2014, Kudlik *et al* 2020). An additional invadopodia-independent mechanism was uncovered that determines how regulators and cortactin interfere with their SH3 domains to promote actin filament polymerization and branching, which are critical for invadopodia growth and maturation (Saini and Courtneidge 2018, Kudlik *et al* 2020). Moreover, TKS4/5 were found to be attracted to endosomal membranes through their PX domain, as evidenced by the simultaneous presence of endosomal

PtdIns3P and the cytosolic tail of MT1-MMP. This offers an explanation for the hitherto undetected double specificity of the TKS4/5-PX domain for PtdIns3P and PtdIns(3,4)P (Buschman *et al* 2009, Lányi *et al* 2011). TKS5 has found to localized on endosomes (Daly *et al* 2020, Iizuka *et al* 2021), and TKS4 has been detected in cytoplasmic spots that build endosomes (Buschman *et al* 2009). Nevertheless, the function of endosomal TKS4/5 is to encourage ADAM-driven shedding of MT1-MMP within acidic endosomes. TKS4/5 proteins can engage with various members of the ADAM family of sheddases, including ADAM12, ADAM15 and ADAM19 through their C-terminal SH3 domains (Abram *et al* 2003, Kärkkäinen *et al* 2006, Zhong *et al* 2008, Mao *et al* 2009). Moreover, the co-expression of ADAM12-myc or ADAM15-myc-SNAP with mCh-MT1-MMP and TKS4/5-GFP demonstrated that these proteins are colocalized on endosomes (Wenzel *et al* 2024). In addition, co-immunoprecipitation detected ADAM15-myc-SNAP primarily in a complex with mCh-MT1-MMP in the presence of wt, but not PX-mutant TKS4 or TKS5, suggesting that TKS4/5 may couple ADAM sheddases to MT1-MMP through their N-terminal PX domains. This is confirmed by the fact that PX mutants of TKS4/5, which are incapable of engaging MT1-MMP, do not favor detachment. These findings suggest that TKS4/5 can drive ADAM-induced processing of MT1-MMPs via their capacity to engage with the cytosolic tail of MT1-MMPs. The pH-dependent splitting of MT1-MMP restricts this cleavage to acidic endosomes and offers some spatial and temporal regulatory constraint on the catalytically active MT1-MMP ectodomain. When the multivesicular endosomes merged with the plasma membrane, the ectodomain and EVs containing TKS4/5 and MT1-MMP became liberated. Notably, ADAM sheddases and MT1-MMP have previously been shown to be released into exosomes (Shimoda and Khokha 2013) and that sheddases can exit exosome membranes (Stoeck *et al* 2006, Shimoda and Khokha 2017). These findings indicate that the occurrence of TKS4/5 in EVs may mediate MT1-MMP processing, thus establishing a durable liberation mechanism of MT1-MMP out of EVs. This is reinforced with the concept that the TME is frequently acidic (Boedtkjer and Pedersen 2020). The possibility that TKS4/5-driven shedding also arises from the plasma membrane cannot be excluded. To substantiate this hypothesis, occasionally a recruitment of TKS4/5 to the MT1-MMP-positive plasma membrane was monitored, probably due to the simultaneous presence of MT1-MMP and PtdIns(3,4)P (Oikawa *et al* 2008, Lányi *et al* 2011). Acidification of the cell culture medium even mildly enhanced MT1-MMP cleavage. Deacidification of endosomes, nevertheless, had a much stronger impact on shedding, suggesting that significant MT1-MMP processing takes place in endosomes. It is important to note that the deposited ectodomain- and MT1-MMP-rich EVs were found to be linked to HeLa recipient cells after conditioned medium (CM) exposure of MT1-MMP-expressing donor cells. As a result, the formerly MT1-MMP-negative HeLa cells became ECM-attenuating and invasive cells. Intriguingly, CM by itself was not enough to trigger gelatin degradation, which absolutely demands the presence of cells. It can therefore be hypothesized that attachment of EVs and/or the ectodomain to the cell surface enables limited and extended engagement of the metalloprotease to its substrate (figure 2.1). The MT1-MMP-rich EVs are not only present at the cell surface of HeLa cells, but were also taken up by HeLa endosomes, thereby facilitating efficient transmission of the degradative features. These EVs could be broken down in lysosomes, or the endosomes could be recycled and merge with the plasma membrane, freeing their MT1-MMP-rich EVs elsewhere. Similarly, EVs derived from a donor cell have been demonstrated to elicit TLR9 signaling in endosomes of recipient cells, leading to enhanced endosome trafficking toward the cell periphery and enhanced ECM breakdown (Rabas *et al* 2021). The MT1-MMP-positive endosomes in CM-induced HeLa cells may undergo a related regulation that facilitates targeted release of internalized MT1-MMP, thus enhancing the ECM-attenuating action of CM.

Targeted transportation of MT1-MMP-positive endosomes toward invadopodia and merging with the invadopodial plasma membrane is necessary for invadopodia ripening and breakdown of the ECM (Castro-Castro *et al* 2016, Pedersen *et al* 2020). The observation that MT1-MMP enlists TKS4/5 to endosomes indicates that TKS4/5 could be transported along with MT1-MMP toward nascent invadopodia. This could offer an alternative and targeted pathway for the targeting of TKS proteins to nascent invadopodia in conjunction with MT1-MMP. TKS4/5 might ease the liberation of MT1-MMP and promote the development of invadopodia. In addition, exosome secretion through invadopodia is increased, thus reinforcing the invasive phenotype (Hoshino *et al* 2013). Hence, the directed trafficking of endosomes toward invadopodia is crucial for an efficient ECM breakdown, and the endosomal appearance of TKS4/5 and MT1-MMP seem to be support this functionality in several ways. Enlistment of TKS adaptor proteins toward MT1-MMP-positive endosomes enhances the capacity of MT1-MMP to breakdown ECM at remote sites. This can be partly attributed to the fact that TKS4/5 was shown to promote non-autocatalytic scission of its catalytically active ectodomain. It is possible that the TKS proteins may also affect MT1-MMP activity in different manners. These may modulate its activity towards soluble proteases like MMP2 and MMP9 and/or influence MT1-MMP transport and surface expression (Iizuka *et al* 2016). In addition, TKS proteins are involved in the generation of ROS (Diaz *et al* 2009), which are recognized to activate MMPs (Nelson and Melendez 2004). In fact, the enzymatic activity of MT1-MMP assessed by zymography was elevated in the presence of TKS4/5 compared to CM derived from TKS4/5-lacking donor cells, which may indicate a direct activation of enzyme activity mediated by TKS proteins. Nevertheless, as the enzyme activity was approximately equal to the level of secreted ectodomain, MT1-MMP splitting probably constitutes the primary mechanism for TKS-dependent transmission of invasiveness. MT1-MMP and TKS protein over-expression enhances cancer advancement in several cancer types (Stylli *et al* 2009, Wu *et al* 2014, Duan *et al* 2020). It is recognized that these proteins operate in concert and play independent yet similarly critical regulatory functions in supporting invadopodia development and functioning. The realization that TKS4/5 interfere with MT1-MMPs in endosomes and EVs offers a new insight into the close interplay between these proteins in ECM destruction, which will be relevant to the evolution of the knowledge of cancer advancement and metastasis.

2.19 Malfunction of recycling via endosomes in cancer

Impaired transport of growth factor receptors and adhesion proteins, such as integrins and cadherins, is a frequent feature of malignant cells. This may be due to a series of abnormal mutations, amplifications and deletions of genes that control the recycling process. In the following the Rab GTPases, p53 recycling, receptor tyrosine kinases (RTKs), including cMet and epidermal growth factor receptor (EGFR), are presented in contributing to the malfunction of endosomal recycling in cancer.

2.19.1 Rab GTPases function in cancer propagation

Rab25 is commonly being intensified in breast and ovarian cancer (Cheng *et al* 2004), and its overexpression has been shown in several cancer types, including prostate, lung, liver, glioblastoma, bladder, and gastric cancers (Wang *et al* 2017b). In these types of cancer, Rab25 functions as an oncogene and enhances epithelial-mesenchymal transition (EMT) and metastasis. It directly engages with β1-integrin and fuels invasive migration by guiding α5β1-integrin and EGFR to the anterior border of ovarian cancer cells (Caswell *et al* 2007). Nevertheless, the tumorigenic characteristics of Rab25 seem to be sensitive to specific contexts. Rab25 deficiency is widespread in colorectal cancer and squamous cell carcinoma of the head and neck, in which it seems to function as a tumor suppressor gene (Goldenring and Nam 2011, Seven *et al* 2015). Deficiency of Rab25 in these strongly polarized cell types is suspected to lead to a lack of targeted delivery of cell surface proteins, which may result in cell dedifferentiation. Rab25 deletion is also prevalent in basal subtypes of breast cancer. Thus, ectopic expression of Rab25 leads to apoptosis and represses the angiogenic characteristics of the triple negative breast cancer cell line MDA-MB-231 (Cheng *et al* 2010). The Rab25 gene is frequently hypermethylated and its expression is silenced in these types of cancer (Clausen *et al* 2016, Gu *et al* 2017). Like Rab25, Rab11a is also thought to have both tumor-fostering and tumor-suppressing functions in cancer. Decreased expression of Rab11a is frequent in colorectal cancer and is associated with advanced stages and reduced survival. Reduced Rab11a concentrations lead to hyperactivation of the Hippo signaling route and to an expansion of the intestinal stem cell reservoir (D'Agostino *et al* 2019). In contrast, overexpression of Rab11a positively relates to the stage and the outcome of non-small cell lung cancer and results in the activation of YAP, which is a key element of the Hippo signaling route (Dong *et al* 2017). Rab11a upregulation has also been seen in bladder and pancreatic cancer (Yu *et al* 2016, Gong *et al* 2018). It has been shown that Rab11b enhances brain metastases from breast cancer through regulating the surface expression of β1-integrin, which is necessary for effective engagement with the cerebral microenvironment (Howe *et al* 2020). A series of Rab11 effector proteins have also been linked to cancer (Tong *et al* 2017, Wang *et al* 2017a; Cho and Lee 2019). Rab14 and Rab35, both of which also have an important function in the recycling of endosomes, have been implicated in cancer advancement (Gundry *et al* 2017, Villagomez *et al* 2020). Rab34 is yet another Rab that associates endosomal recycling and cancer. Rab34 is primarily found in endosomes and the Golgi apparatus and has a role in secretion, lysosome placement, macropinosome generation and cilia formation. Recently, a high expression of Rab34 in aggressive breast cancer was demonstrated to link directly to the tail of β3 integrin and promote the recycling of internalized αvβ3 integrin back toward the plasma membrane. Rab34 is knocked down and restrains cell motility, while EGF receptor activation triggers Src kinase-induced phosphorylation of Rab34 at a tyrosine at its carboxy terminus, resulting in its consecutive relocation toward the plasma membrane. It has been proposed that αvβ3-integrin is internalized when activated within endosomes. Subsequently, β3 integrin induces the activation of Src, which is located within the same endosomes, causing Src to phosphorylate Rab34. This phosphorylation inhibits lysosomal breakdown of the integrin and promotes its recycling to the plasma membrane to generate new contact sites with the ECM to support migration and invasion (Sun *et al* 2018).

2.19.2 p53 and the recycling route via endosomes

p53 is a tumor suppressor protein frequently referred to as a molecule that acts as a custodian of the genome. In diseases, such as cancer, it is either lost or mutated in approximately half of the cancer types, which impacts the cell cycle checkpoint controls and the induction of apoptosis. In parallel to its anti-proliferative activity, p53 also participates in other phases of cancer development, including cell migration and invasion (Mak 2014, p 53). Mutations in the Tp53 gene occur mostly in the DNA-binding domain in the region between residues 102 and 292, with 90% of these being missense mutations that impair the DNA-binding ability of p53. Several of the most frequent cancer-associated p53 mutations lead to mutant proteins that are strongly expressed and, in addition to the reduction in transcriptional activity, display an increase in functionality (Oren and Rotter 2010, p 53). These acquired functions involve the capacity to drive tumor metastasis, in part through its capacity to antagonize the activity of p63, which is another member of the p53 family.

α5β1 integrin, fibronectin's major receptor in the ECM, is a critical factor for the migration and invasion of cancer cells. Integrins can be activated by 'inside-out' signals, which is basically an intracellular cue that stimulates the binding of proteins like talin to the β integrin tail. This enables the receptor to widen its conformation so that it exhibits a high affinity for its ligands, which are abundant in the ECM. When a ligand attaches to the receptor, it results in 'outside-in' signaling that engages protein complexes to control cellular functions (Moreno-Layseca *et al* 2019). Rab coupling protein (RCP) directly associates EGFR with β1 integrin and is essential for trafficking of α5β1 integrin- and EGFR-positive recycling endosomes to the plasma membrane to facilitate cancer cell invasion in fibronectin-rich 3D micro-environments (Caswell *et al* 2008). p63 impairs the linkage between RCP and α5β1 integrin; nevertheless, the expression of mutated p53 interferes with the activity of p63 and removes the inhibition. This results in an increase in α5β1 integrin and EGFR recycling, enhanced activation of Akt and a simultaneous rise in the incidental migration of cancer cells (Muller *et al* 2009). p53 also regulates the expression and release of many extracellular proteins that are either soluble or reside in extracellular vesicles, like exosomes. These substances are involved in cell–cell interaction and restructuring of the ECM. The participation of mutant p53 in the regulation of the 'secretome' of cancer has yielded a growing body of evidence on cancer invasion and metastasis (Pavlakis and Stiewe 2020). RCP is involved in the exosome-based crosstalk of mutant p53 gain-of-function between cancer cells and fibroblasts within the stroma and in the formation of pre-metastatic niches in remote organs (Novo *et al* 2018). HSP90 functions as a chaperone protein that promotes protein folding and is essential in carcinogenesis by modulating various functions, such as angiogenesis, cell proliferation, migration, invasion, and metastasis. The HSP90α isoform can be released from cancer cells, with extracellular HSP90α concentrations (eHSP90α) correlating positively with the malignancy of the tumor. Mutant p53 has been revealed to increase eHSP90α concentrations in an RCP-dependent fashion. RCP was found to interfere with HSP90α and functions as an adapter to package Rab11-positive recycling endosomes with HSP90α, which is later released into exosomes (Zhang *et al* 2020b, p 53). The mechanism through that mutant p53 enhances RCP activity is not yet understood, but characterizing the correlation between the impact of wild-type and mutant p53 on RCP expression will advance the knowledge of how mutant p53 facilitates endosomal recirculation.

2.19.3 Receptor tyrosine kinases underly recycling in endosomes

Receptor tyrosine kinases (RTKs) are central players in a plethora of signaling networks that are instrumental in mediating cell–cell communication and govern a wide array of complex biological processes, ranging from cell growth, motility, metabolism, and differentiation. They serve as high-affinity cell surface receptors for a variety of polypeptide growth factors, cytokines, and hormones, whose engagement triggers upregulated post-translational changes that govern intercellular and intracellular crosstalk. The human genome harbors 58 RTKs, among them EGFR, FGFRs, c-Met and vascular endothelial growth factor receptors (VEGFR) (Hubbard and Miller 2007). Disruption of RTK pathway signaling is often involved in the development of cancer, and this anomalous activation is conferred by four primary mechanisms, namely amplifications, gain-of-function mutations, reorganization of chromosomes and autocrine activation (Du and Lovly 2018). Intracellular membrane transport serves to determine the specificity and length of the activation of signaling cascades (Miaczynska 2013). Endocytic internalization and the equilibrium between breakdown and recirculation can profoundly influence the signaling characteristics of RTKs and is critical for revealing the net signaling outcome. The rate of ligand binding at the cell surface is a function of the rates of these three transport operations. After internalization, activated receptors are recycled back to the plasma membrane or broken down in lysosomes. The latter leads to the termination of the signal, while the former aids sustained signal transmission. In addition, many activated RTKs trigger various signaling paths according to whether they are localized at the plasma membrane or internalized in endosomes.

2.19.3.1 c-Met

When c-MET attaches to its ligand, hepatocyte growth factor (HGF), it triggers the wound repair and embryogenesis signaling paths in healthy cells. Abnormal c-Met activation, however, drives carcinogenesis through the stimulation of a multitude of signaling cascades, among them the Ras/MAPK, PI3K/Akt, Src, Wnt, and JAK/STAT signaling routes (Zhang *et al* 2018). c-Met has been overexpressed or mutated in multiple cancer types and represents an essential therapeutic target. The gene is overexpressed in 20%–30% of breast cancers and has been linked to the progression of the disease. High amounts of HGF are correlated with worse survival of patients with ductal breast carcinomas (Lengyel *et al* 2005). In HER2-positive breast cancer, the prognosis is frequently worse (Garcia *et al* 2007) and resistance to HER2-targeted therapies is often implicated. c-MET is recycled through endosomes, a step that is crucial for the regulation of signaling cascades activated through c-MET. For instance, c-MET activates the JAK/STAT route in endosomes. A series of oncogenic c-Met mutations display enhanced recycling and decreased breakdown activity, leading to an aggregation of activated receptors on endosomes and enhanced cell migration. Inhibition of endocytosis blocks the tumorigenic characteristics of the mutant c-Met (Joffre *et al* 2011). c-Met interferes with RCP and mutant p53 increases its recycling through being stimulated by RCP. This leads to the sustained activation of c-Met and a resulting pro-tumorigenic signal transduction. Consequently, elevated c-Met recycling is an integral driver of cancer advancement.

2.19.3.2 EGFR endosomal recycling is often deregulated in cancer cells

The ErbB family of RTKs harbors four members, namely EGFR, HER2 (ErbB2), HER3 (ErbB3), and HER4 (ErbB4). These receptors consist of single-chain transmembrane glycoproteins which are structurally similar. The attachment of ligands to their extracellular domain triggers the homo- and heterodimerization among receptors which is required for the activation of their intracellular tyrosine kinase domains and subsequently lead to phosphorylation of their C-termini (Wieduwilt and Moasser 2008). EGFR represents the prototypical ErbB family member and is critical for the perpetuation of multiple cellular functions such as proliferation, survival and migration. It has a range of ligands, among them EGF, TGF-α, HB-EGF and epiregulin. EGF receptors are fast internalized through CME and CIE following activation, however, various ligands may regulate their signal transduction and trafficking in different ways. The endosomal fate of EGFR, such as whether it is recycled back to the plasma membrane or degraded, is tightly determined after its internalization. Although signaling mainly relies on the plasma membrane, activated EGFR is also able to send messages from endosomes, suggesting that there are signaling routes that rely on its intracellular transportation. Moreover, EGFR can be trafficked to the nucleus or mitochondria through atypical transport routes (Tomas *et al* 2014). The misregulation of the intracellular trafficking of members of the EGFR family is a key event in oncogenesis, causing their misplacement and subsequent upregulation, resulting in amplified signal transduction (Sorkin and Von Zastrow 2009). Mutations that activate the EGFR kinase domain are commonly present in non-small cell lung cancers (NSCLC), and these mutants tend to undergo recycling, where they trigger the Src signal transduction route (Chung *et al* 2009).

To further elucidate the mechanisms underlying the differences in signaling responses to the same RTK and various ligands, a systematic proteomics strategy has been used to examine the cellular reaction to EGFR activation through EGF and TGF-α. (Francavilla *et al* 2016). Phosphorylation of Rab7 and subsequent recruitment of RCP have been identified as key modulators of EGFR signal transduction. EGF, unlike TGF-α, caused phosphorylation of Rab7 at tyrosine 183, which led to receptor turnover and cessation of signaling. In contrast, TGF-α triggered persistent activation of EGFR and enhanced RCP engagement with the receptor and its subsequent entering into the endosomal recycling route. The elimination of RCP via knockdown in cancer cell lines blocked TGF-α-triggered activation of MAPKs and suppressed cell proliferation and migration (Francavilla *et al* 2016). HER2, which belongs to the ErbB family, is found to be upregulated in 15%–20% of breast cancers and indicates a poor outcome. Antibodies and small-molecule HER2 antagonists have been effectively utilized in the clinical setting. HER2 is a uniquely orphan member of the ErbB family, as it has an unrecognized ligand and functions instead as a heterodimerization partner of ligand-activated EGFR or HER3. The intracellular transport of HER2 is not well characterized as compared to other RTKs, and it is controversial whether HER2 is capable of endosomal recirculation. A hypothesis proposes that HER2-containing dimers remain on the cell surface and are not subject to endocytosis, which explains the high surface density and resistance to EGF-driven downregulation. Another indication favors a model in which HER2 is quickly recycled, which explains its nearly exhaustive targeting to the cell surface (Bertelsen and Stang 2014). A recent study that confirms the latter hypothesis revealed that SORLA binds to HER2 and controls its return to the plasma membrane. SORLA depletion results in HER2 homing to lysosomes and leads to sensitization of HER2-overexpressing cells to lysosome-attacking medications (Pietilä *et al* 2019).

2.19.3.3 Immune checkpoint protein PD-L1 underlies endosomal recycling

Programmed death protein ligand 1 (PD-L1) is a member of the immune checkpoint protein family. PD-L1 is a type I transmembrane protein that can be found expressed on the membrane of cancer cells. It contains an extracellular N-terminal domain that engages with the PD-1 receptor located on the surface of T cells. This interaction hinders the activation, proliferation and effector capabilities of CD8+ T cells and represents a mechanism that cancer cells have developed to avoid being destroyed via the immune system (Boussiotis 2016). The relationship of PD-1 and PD-L1 establishes a delicate equilibrium between immune tolerance and autoimmunity. Disturbance of this equilibrium can result in a number of diseases and has been shown to promote the development of the autoimmune diseases arthritis and lupus (Zamani *et al* 2016). PD-L1 is widely expressed in 5%–40% of cancer cells, which resulted in massive research efforts to design therapies that interfere with its engagement with PD-1. Immune checkpoint blockers are monoclonal antibodies that attach to the extracellular domains of checkpoint proteins and hinder them from attaching to their companion proteins. As a result, the 'off' command never reaches the T cells, and they destroy the cancer cells. Immune checkpoint blockers have changed the oncology landscape by providing long-term sustained clinical benefit in a subgroup of patients and are perhaps the most transformational advance in cancer treatment. Seven immune checkpoint inhibitors have been authorized, including four that act on PD-L1 (Vaddepally *et al* 2020). In addition, there are hundreds of active clinical studies involving checkpoint inhibitors worldwide. The PD-L1 molecule has recently been characterized as being under the tight control of a so far non-characterized transmembrane protein complex named CMTM6. CMTM6 binds to PD-L1 on the cell surface and in recycling endosomes, thus inhibiting its ubiquitination and subsequent delivery to lysosomes for breakdown. Its exhaustion results in a decrease in overall PD-L1 protein levels by rerouting PD-L1 away from the recycling route and towards the catabolic pathway, where it is finally broken down in lysosomes. This results in a decrease in the capacity of cancer cells to suppress T cell activity *in vitro* and *in vivo* (Burr *et al* 2017, Mezzadra *et al* 2017). These results indicate that selective endosomal recycling blockade may be an alternative mechanism to inhibit PD-L1 and thus render cancer cells susceptible to immune suppression (O'Sullivan and Lindsay 2020).

Hypersecretory malignant cells exhibit therapeutic resistance, perform cancer metastasis, and lead to worse clinical outcomes. Nevertheless, the molecular foundation for malignant hypersecretion is still not yet clear. Epithelial-mesenchymal transition (EMT) has been demonstrated to initiate exocytic and endocytic vesicular transportation mechanisms in lung cancer. The EMT-inducing transcription factor zinc finger E-box-binding homeobox 1 (ZEB1) implemented a PI4KIIIβ-to-PI4KIIα (PI4K2A) reliance switch that governed PI4P synthesis within the Golgi and endosomes. EMT increased the susceptibility of lung cancer cells to PI4K2A antagonists using small molecules. PI4K2A assembled a MYOIIA-containing protein complex that promoted secretory vesicle generation in the Golgi apparatus, causing a hypersecretory condition in which osteopontin (SPP1) and other prometastatic ligands were secreted. In the endosomal recycling compartment, PI4K2A promoted the recycling of SPP1 receptors to enable execution of an SPP1-dependent autocrine circuit and cooperated with HSP90 to protect AXL receptor tyrosine kinase, a promoter of cell migration, from lysosomal breakdown. These findings demonstrate that EMT controls exocytic and endocytic vesicular transport to create a therapeutically effective hypersecretory condition that propels lung cancer advancement. Metastatic cancer cells show resistance to existing therapeutic strategies and are the leading source of cancer-related deaths (Weiss *et al* 2022). Since cancer cells gain metastatic characteristics and resistance to therapy due to transcription processes that are activated by EMT (Lu and Kang 2019), strategies for combating mesenchymal cancer cells should be envisaged. The results discussed here pinpoint PI4K2A as an EMT-based therapeutic weak spot. In a 'tumor-as-organizer' theory (Li and Stanger 2019), the main organizers of a TME that promotes metastases are actually the cancer cells themselves. Factors originating from hypersecretory cancer cells trigger autocrine circuits that sustain cancer cell viability and chemotactically attract a broad range of stromal cell types, generating an immunosuppressive, fibrotic, and vessel-rich TME

(Bajaj *et al* 2022). Hypersecretory conditions in cancer cells are triggered by frequently arising oncogenic mutations and chromosomal amplifications that advance neoplastic transformation (Dippold *et al* 2009, Halberg *et al* 2016, Capaci *et al* 2020, Tan *et al* 2020, Hafezi *et al* 2021, Tan *et al* 2021a), which forms the foundation for the prevailing hypothesis that hypersecretion originates from the cell itself (Baumann *et al* 2022). Nevertheless, the findings reported here imply that cancer cells acquire hypersecretory properties through EMT. This process is triggered by extracellular signals originating from the TME (Gibbons *et al* 2009). This implies that upon being activated by oncogenic mutations, the TME's capacity to organize is amplified by extracellular signals, accomplishing a feed-forward circuit with bidirectional cancer cell-TME communication.

miRs constitute a class of non-coding RNAs that precisely modulate the expression of target genes implicated in diverse prometastatic biological mechanisms, encompassing EMT, invasion and immunotherapy resistance (Zhang *et al* 2014). While miRs usually cause gene silencing by enlisting the miR-induced decay complex (Brodersen and Voinnet 2009), which causes translational attenuation or mRNA breakdown. MiRs can also enhance the mRNA abundance of target genes by interacting with 5′-UTR regulatory elements or by antagonizing RNA destabilizing proteins that associate with 3′-UTR elements (Ørom *et al* 2008, Murphy *et al* 2010, Lin *et al* 2011). It identified miR-34a as modulating PI4KB activity via this mechanism and ZEB1 as causing the silencing of miR-34a and miR-182/-183 to coordinately silence PI4KB and elevate PI4K2A abundance, thereby effecting a PI4KB-to-PI4K2A reliance switch that promotes tumor advancement. PI4P acts as a key modulator of vesicular transport and intracellular signal transduction and metabolic events (Waugh 2019). PI4P levels are maintained by four mammalian PI4K enzymes, comprising two type II kinases (PI4K2A and PI4K2B) and two type III kinases (PI4KA and PI4KB), which localize to discrete subcellular domains (Boura and Nencka 2015). High PI4KB, PI4KA and PI4K2A abundance is linked to unfavorable prognosis in various cancers and promotes malignant advancement via different mechanisms (Ilboudo *et al* 2014, Tan *et al* 2020, Adhikari *et al* 2021, Huang *et al* 2023). A chromosome 1q21.3 amplicon encompassing PI4KB sustains cancer cell survival and proliferation through activation of a PI4KB-reliant secretome (Tan *et al* 2020). The PI4K2B gene, in contrast, acts as a negative controller of cancer cell invasion and is often lost or downregulated in human cancers (Alli-Balogun *et al* 2016). PI4K2A ensures the survival of cancer cells and fosters tumor formation by increasing the stability of the EGFR protein (Li *et al* 2010, 2014) and assisting the transport of incorrectly folded proteins toward the lysosome (Pataer *et al* 2020). PI4K2A is able to stabilize AXL and orchestrates the exocytic and endocytic transportation of SPP1 and its receptors to enable a pro-tumorigenic autocrine circuit in human lung mesenchymal cancer cells (Tan *et al* 2023). Secretory impairment should be contemplated as a targeted therapeutic approach for cancer (Tan *et al* 2020, 2021b). Small-molecule PI4KB inhibitors elicit apoptosis in 1q21.3-amplified but not in diploid cancer cells (Tan *et al* 2020, Shi *et al* 2021). Likewise, the lack of p53 triggers a secretion mechanism that sustains the continued survival of p53-deficient cancer cells and can be blocked with a small-molecule antagonist (Tan *et al* 2021a, 2021b). It was determined that an enzymatic switch caused by EMT generates a susceptibility to PI4K2A antagonists. Frequently arising oncogenic mutations and epigenetic mechanisms fuel addictive secretory processes that can be therapeutically influenced.

2.20 Interplay between autophagy inhibitory agents and endosome-based secretory routes: a challenge for autophagy-directed therapy of solid cancers

Autophagy has become best recognized for its functions in organelle and protein degradation, quality assurance of the cell, and metabolism. The autophagic mechanism has also adjusted, however, to permit atypical protein transport and secretory routes, allowing organelles like autophagosomes and multivesicular bodies that deposit cargo for breakdown at lysosomes to modify their task from fusing with lysosomes to fusing with the plasma membrane, which is subsequently released from the cell. Certain signal factors bypass the classical secretory route and are released by autophagy. The positive clinical outcomes of certain autophagy inhibitor drugs are promising. Nonetheless, it is apparent that blocking autophagy can impact cancer disease progression in different ways, even in the exact same type of cancer. In addition, autophagy next-generation modulators may have important non-specific implications, such as effects on endosome-based secretory routes and EV release. Numerous investigations indicate that cancer cells secrete larger quantities of EVs in comparison to non-malignant cells, which is why the impact of autophagy blockers on EVs release is extremely relevant and appealing for cancer treatments (Raudenska *et al* 2021). Consequently, the regulation of autophagy has a significant influence not merely on the amount of EVs, but equally on their cargo, which may have a strong effect on the subsequent pro-tumorigenic or anticarcinogenic action of autophagy inhibitor drugs used in the antineoplastic therapies of solid tumors.

2.21 Conclusions and future directions

Most of the results on the function of three fundamental motor proteins, myosin, dynein and kinesin have been obtained from *in vitro* investigations. Thereby, the focus has been placed on studying the interactions with cargos and motor-adaptor proteins during the coupling to specific cargo and its transportation. Over the last two decades, however, with the increasing capability to image motor functioning under natural living cell conditions, it has emerged that the motor dynamics *in vivo* and in some instances the functionality of the adapters are not consistent with the motor properties measured *in vitro*. It is now time to set standards for the analysis of motor proteins, as many new methods and techniques are now available and the various findings are difficult to compare.

2.21.1 What are the perspectives and scientific frontiers or gaps in kownledge?

The process of endosomal trafficking in cells and the important role played by motor proteins are well characterized. The successful administration of therapeutic biomacromolecules into solid cancers is a challenging task as they can hardly infiltrate the intricate TME. Nanoparticles that are actively transported are used to efficiently carry biomacromolecular therapeutics into solid cancers by cell transcytosis (Qin *et al* 2024). In particular, a panel of molecularly-precise cyanine 5-core polylysine G5 dendrimers (Cy5 nanopods) with distinct peripheral amino acids (G5-AA) was constructed. The ability of these positively charged nanoparticles to trigger endocytosis, exocytosis and transcytosis in cells is assessed with the help of fluorescence-based high-throughput imaging readouts (Qin *et al* 2024). The optimized nanoparticles (G5-R) are fused with αPD-L1 (a therapeutic monoclonal antibody that binds to programmed cell death ligand 1) (αPD-L1-G5-R) to illustrate nanoparticle-driven tumor-targeting. The αPD-L1-G5-R can significantly improve the ability of tumor infiltration via adsorption-mediated transcytosis (AMT). The efficacy of αPD-L1-G5-R is evaluated in a mouse model with tumors that have been partly removed, thereby simulating local immunotherapy of remaining tumors in the clinic after surgical intervention. The αPD-L1-G5-R incorporated into fibrin gel is able to facilitate cancer cell transcytosis efficiently and distribute αPD-L1 throughout the entire tumor, thus increasing immune checkpoint blockade, decreasing tumor relapse and substantially increasing survival time (Qin *et al* 2024). Nanoparticles that actively transport therapeutic agents are an attractive approach for the efficient administration of therapeutic biomacromolecules to solid tumors. Understanding the differences between *in vitro* and *in vivo* performance of the motor proteins and adapters, and developing efficient drug delivery mechanisms to better combat tumors, is a major goal of research in this area. Biophysical aspects, such as stiffness or viscoelasticity, also play a crucial role here, since the mechanical conditions must also be taken into account during transportation.

2.21.2 Effect of the environment on the function of motors

Even though replication of motor functionality *in vitro* has emerged as fundamental to the field of motor trafficking, live-cell imaging and the ability to visualize individual motors in cells have revealed that the in-cell functionality of motors and functionality *in vitro* are sometimes divergent. For example, activated dynein and kinesin motors run on microtubules during *in vitro* testing over tens of micrometers (McKenney *et al* 2014, Schlager *et al* 2014), but none of the two motors has been observed to perform this task at the single-molecule level inside living cells (Cai *et al* 2009, Tirumala and Ananthanarayanan 2020, Tirumala *et al* 2021). Additionally, when endosomal trafficking is re-established using beads that couple to motors via biotin-streptavidin linkage, the coupling of the motor to the cargo is undone, which reflects an intrinsic property of the motor and may impact its overall function. In the same way, motors tugged by phagosomes attached to latex beads are not assessed for dynamic engagement and disengagement activities (Rai *et al* 2016). Consequently, the intrinsic complexity of the capture and release of cargo and its engagement during transportation is usually not probed in *in vitro* assays. These experiments need to be adapted in a way that they mimic the *in vivo* situation more closely. Another key characteristic is the intracellular milieu that is insufficiently characterized in *in vitro* experiments, especially the effect on crowding and consequently its impact on mechanical cues of the cytoplasmic environment. For instance, the cytoplasm contains proteins, RNAs, ribosomes, organelles and cytoskeletal filaments and exhibits a viscosity that is three times higher compared to water (Swaminathan *et al* 1997). This factor is seldom addressed in *in vitro* studies, since experiments are commonly conducted in buffers with a viscosity close to water, which is not mimicking the *in vivo* situation. Moreover, cytoskeletal filaments resembling microtubules are engaged by MAPs and membrane-bound compartments, and thus it is not unexpected that long-distance uninterrupted motions on microtubules are infrequent in a cell. These hurdles for the transport need to be included in future assays. In addition, the ionic strength of the buffer, the levels of the reactants, and the temperature at which *in vitro* tests are conducted differ from physiological parameters and may account for discrepancies in motor function and/or endosomal performance between *in vitro* and *in vivo* assays. Overall, the experiment should be conducted under more realistic conditions, taking into account the parameters mentioned. This applies in particular to biophysical measurements, such as optical or magnetic tweezer analyses.

2.21.3 Standardization for visualization and analysis of endosomal trafficking

Importantly, the experimental settings and analytical approaches selected to quantify endosomal motility differ from study to study. Experiments performed with the exact same cargo, in the identical cell line and under the same test parameters can lead to different results. Therefore, the scientific community needs to set standards for how endosome motion and motor functioning are characterized. It would be crucial to ensure reproducibility across laboratories by properly characterizing and comparing: firstly, ligand concentration used in visualizing ligand-mediated internalization of a receptor; secondly, imaging parameters employed for fluorescent visualization of endosomal motion, including time interval between successive images, 3D imaging or 2D, length of time of visualization; and thirdly, perhaps most critical, analysis tools utilized. This involves the analysis procedure for monitoring endosomes, the monitoring of all detectable endosomes and not just motile endosome subsets, the notification of momentary and net movements and speeds of all endosomes.

2.21.4 Does a unified mechanism for motor-based endosomal trafficking exist?

In conclusion, a possible mechanism is proposed that could be used by motor proteins to facilitate the long-range transportation of endosomal cargo in a multitude of cell types. Raising the cargo-motor ratio could possibly increase the length of the endosomes by assuring that there is at least one motor in touch with the microtubule and motile at any moment in time. While it is normally assumed that sets of motors are needed to move cargo across long distances (Hendricks *et al* 2010, Rai *et al* 2013, Gicking *et al* 2022), recent work (Tirumala *et al* 2021) showed that repeated tethering followed by short walks and subsequent untethering of the dynein motors is adequate to convey cargo transport over large length scales, on the several tens of micrometers, throughout cells (Shubeita *et al* 2008). Some evidence from endosome- and single-molecule-tracking experiments conducted in cells indicates that large quantities of motors are present on or attach to endosomes. First, when there are motors in large numbers, the image of fluorescently labeled motors should display clusters of motors (probably attached to endosomes) on microtubule paths. In live-cell imaging, these complexes can be visualized moving with the endosome, which seems to be not detected at all. Theoretical calculations made on the size and the association and dissociation kinetics of kinesin motors with assisted lipid bilayers imitating vesicles indicated that a 500 nm diameter vesicle would require 800 kinesin motors to travel a distance of 10 μm without interruption, so that each microtubule is effectively associated with three kinesin motors (Jiang *et al* 2019). It can be inferred that the creation of motor clusters is an impractical and inefficient mode of long-distance locomotion (Jiang *et al* 2019). Second, single-molecule imaging of motors would demonstrate the existence of multiple motors on the cargo. Even though dynein is seen to be localized in punctate clusters in a cholesterol-dependent fashion on late phagosomes, using a pulled down from cells (Rai *et al* 2016), it has not been possible to detect dynein (Shin *et al* 2019, Tirumala *et al* 2021) nor kinesin-1, -2, or -3 within these punctate clusters within cells (Cai *et al* 2009, Norris *et al* 2014, Guedes-Dias *et al* 2019, Siddiqui *et al* 2019). Similarly, the lengths and durations of individual dynein and kinesin motors within living cells are both similar and quite short, on the order of 1 s or less (Cai *et al* 2009, Tirumala *et al* 2021), and they coincide with the characteristic 'stop-and-go' movement seen for endosomal motility (Chaudhary *et al* 2018, Shin *et al* 2019, Villari *et al* 2020, Tirumala *et al* 2021). This suggests that the attachment and detachment of a single motor molecule to and from endosomes is required to facilitate their transportation into cells. Therefore, the engine attachment times are an important factor in regulating the overall velocity of the cargo transportation, with a quick reattachment of the motor to the cargo leading to accelerated cargo conveyance (Jiang *et al* 2019, Tirumala *et al* 2021). Ultimately, the cells could adjust the modulation of the motor binding speeds to ensure adequate cargo transportation. In fact, more recent *in vitro* reinstalls of kinesin-1 motion on supporting lipid bilayers and follow-up calculations have led to the conclusion that motor-linkage kinetics is the main factor for long-distance transportation (Jiang *et al* 2019).

2.21.5 Future perspective

Most of the above conclusions were drawn based on research conducted on cancer cell lines. These investigations could clearly be strengthened using patient tissues. Additional obstacles in this area include the limited high-resolution imaging knowledge of endosomal sorting complexes and their key regulatory proteins, as well as how signal transduction in cancer cells is modified at specific phases of endocytosis. These questions will doubtless be resolved during the ongoing research. Identifying these key regulatory molecules may provide drug trafficking hubs suitable for therapeutic intervention. A translational potential problem is the impact of an endocytic node inhibitor on multiple pathways that it turns on and how cumulative effects shape the metastatic phenotype. This problem is not only observed in endocytosis, but also in DNA methylation and many other oncogenic processes. Endolysosomal escapement is an extremely low-efficiency route that constitutes a hurdle for intracellular delivery of biologics, comprising proteins and nucleic acids. Construction has been shown of a lipid-based nanoscale molecular vehicle that delivers efficient cytosolic transportation of biologics through destabilization of endolysosomal machinery via nanomechanical effects induced by light irradiation. Lipid-based nanoscale molecular motors that are intended to carry out mechanical movements by absorbing photons were developed by co-assembling azobenzene lipidoids with auxiliary lipids. After penetrating cells, lipid-based nanoscale molecular machines can attach themselves to the endolysosome membrane. The sustained rotation-invasion motion of azo-lipidoids induced by UV/vis irradiation leads to membrane destabilization and thus delivers cargo like mRNAs and Cre proteins into the cytoplasm. Cytosolic transportation efficiency is 2.1 times better than traditional intracellular drug delivery systems. Finally, lipid-based nanoscale molecular vehicles were suitable for cytosolic transportation of tumor antigens into dendritic cells, which was able to elicit robust anti-tumor activity in a melanoma mouse model (Zhao *et al* 2023). Present endeavors to exert control over molecular-scale movements may appear cumbersome in comparison to the delicate functionalities exhibited in natural systems. Nevertheless, it is important to bear in mind that the molecular devices created by Nature are intricate systems that have developed gradually over time. The synthetic versions may not be as intricate as the natural systems. Notwithstanding, research continues to show that it is still possible to create artificial molecular devices with biological functionality through simple chemistry, which is of considerable importance for the design of soft robotics and the development of next-generation smart drug delivery vehicles. In summary, it can be said that the specific interference with the endocytic system could be a practical and effective therapeutic approach for combating cancer and metastases.

References and further reading

Abid Ali F, Zwetsloot A J, Stone C E, Morgan T E, Wademan R F, Carter A P *et al* 2025 KIF1C activates and extends dynein movement through the FHF cargo adapter *Nat. Struct. Mol. Biol.* https://doi.org/10.1038/s41594-024-01418-z

Abram C L, Seals D F, Pass I, Salinsky D, Maurer L, Roth T M *et al* 2003 The adaptor protein fish associates with members of the ADAMs family and localizes to podosomes of src-transformed cells *J. Biol. Chem.* **278** 16844–51

Aderem A and Underhill D M 1999 Mechanisms of phagocytosis in macrophages *Annu. Rev. Immunol.* **17** 593–623

Adhikari H, Kattan W E, Kumar S, Zhou P, Hancock J F and Counter C M 2021 Oncogenic KRAS is dependent upon an EFR3A-PI4KA signaling axis for potent tumorigenic activity *Nat. Commun.* **12** 5248

Ali M Y, Lu H, Bookwalter C S, Warshaw D M and Trybus K M 2008 Myosin V and Kinesin act as tethers to enhance each others' processivity *Proc. Natl Acad. Sci. USA* **105** 4691–6

Alli-Balogun G O, Gewinner C A, Jacobs R, Kriston-Vizi J, Waugh M G and Minogue S 2016 Phosphatidylinositol 4-kinase IIβ negatively regulates invadopodia formation and suppresses an invasive cellular phenotype *Mol. Biol. Cell* **27** 4033–42

Alonso-Curbelo D, Riveiro-Falkenbach E, Pérez-Guijarro E, Cifdaloz M, Karras P, Osterloh L *et al* 2014 RAB7 controls melanoma progression by exploiting a lineage-specific wiring of the endolysosomal pathway *Cancer Cell* **26** 61–76

Altman D, Goswami D, Hasson T, Spudich J A and Mayor S 2007 Precise Positioning of myosin VI on endocytic vesicles *in vivo PLoS Biol.* **5** e210

Altman D, Sweeney H L and Spudich J A 2004 The mechanism of myosin VI translocation and its load-induced anchoring *Cell* **116** 737–49

Andreasson J O, Milic B, Chen G-Y, Guydosh N R, Hancock W O and Block S M 2015 Examining kinesin processivity within a general gating framework *eLife* **4** e07403

Aniento F 1993 Cytoplasmic dynein-dependent vesicular transport from early to late endosomes [published erratum appears in 1994 *J. Cell Biol.* **124** 397] *J. Cell Biol.* **123** 1373–87

Arpağ G, Norris S R, Mousavi S I, Soppina V, Verhey K J, Hancock W O *et al* 2019 Motor dynamics underlying cargo transport by pairs of kinesin-1 and kinesin-3 motors *Biophys. J.* **116** 1115–26

Arpağ G, Shastry S, Hancock W O and Tüzel E 2014 Transport by populations of fast and slow kinesins uncovers novel family-dependent motor characteristics important for *in vivo* function *Biophys. J.* **107** 1896–904

Aschenbrenner L, Lee T and Hasson T 2003 Myo6 facilitates the translocation of endocytic vesicles from cell peripheries *Mol. Biol. Cell* **14** 2728–43

Ashkin A, Schütze K, Dziedzic J M, Euteneuer U and Schliwa M 1990 Force generation of organelle transport measured *in vivo* by an infrared laser trap *Nature* **348** 346–8

Avraham K 1997 Characterization of unconventional MYO6, the human homologue of the gene responsible for deafness in Snell's waltzer mice *Human Mol. Genet.* **6** 1225–31

Avraham K B, Hasson T, Steel K P, Kingsley D M, Russell L B, Mooseker M S *et al* 1995 The mouse Snell's waltzer deafness gene encodes an unconventional myosin required for structural integrity of inner ear hair cells *Nat. Genet.* **11** 369–75

Azarnia Tehran D, López-Hernández T and Maritzen T 2019 Endocytic adaptor proteins in health and disease: lessons from model organisms and human mutations *Cells* **8** 1345

Bajaj R, Warner A N, Fradette J F and Gibbons D L 2022 Dance of the golgi: understanding golgi dynamics in cancer metastasis *Cells* **11** 1484

Bálint Š, Verdeny Vilanova I, Sandoval Álvarez Á and Lakadamyali M 2013 Correlative live-cell and superresolution microscopy reveals cargo transport dynamics at microtubule intersections *Proc. Natl Acad. Sci. USA* **110** 3375–80

Barylko B, Binns D, Lin K-M, Atkinson M A L, Jameson D M, Yin H L *et al* 1998 Synergistic activation of dynamin GTPase by Grb2 and phosphoinositides *J. Biol. Chem.* **273** 3791–7

Bass R, Werner F, Odintsova E, Sugiura T, Berditchevski F and Ellis V 2005 Regulation of urokinase receptor proteolytic function by the tetraspanin CD82 *J. Biol. Chem.* **280** 14811–8

Batchelder E M and Yarar D 2010 Differential requirements for clathrin-dependent endocytosis at sites of cell–substrate adhesion *Mol. Biol. Cell* **21** 3070–9

Batters C, Brack D, Ellrich H, Averbeck B and Veigel C 2016 Calcium can mobilize and activate myosin-VI *Proc. Natl Acad. Sci. USA* **113**

Batters C and Veigel C 2016 Mechanics and activation of unconventional myosins *Traffic* **17** 860–71

Baumann Z, Auf Der Maur P and Bentires-Alj M 2022 Feed-forward loops between metastatic cancer cells and their microenvironment—the stage of escalation *EMBO Mol. Med.* **14** e14283

Beaudet D, Berger C L and Hendricks A G 2024 The types and numbers of kinesins and dyneins transporting endocytic cargoes modulate their motility and response to tau *J. Biol. Chem.* **300** 107323

Bechstedt S and Brouhard G J 2012 Doublecortin recognizes the 13-protofilament microtubule cooperatively and tracks microtubule ends *Dev. Cell* **23** 181–92

Beck B H and Welch D R 2010 The KISS1 metastasis suppressor: a good night kiss for disseminated cancer cells *Eur. J. Cancer* **46** 1283–9

Bell-McGuinn K M, Garfall A L, Bogyo M, Hanahan D and Joyce J A 2007 Inhibition of cysteine cathepsin protease activity enhances chemotherapy regimens by decreasing tumor growth and invasiveness in a mouse model of multistage cancer *Cancer Res.* **67** 7378–85

Belyy V, Schlager M A, Foster H, Reimer A E, Carter A P and Yildiz A 2016 The mammalian dynein–dynactin complex is a strong opponent to kinesin in a tug-of-war competition *Nat. Cell Biol.* **18** 1018–24

Bergman J P, Bovyn M J, Doval F F, Sharma A, Gudheti M V, Gross S P *et al* 2018 Cargo navigation across 3D microtubule intersections *Proc. Natl Acad. Sci. USA* **115** 537–42

Berón W, Alvarez-Dominguez C, Mayorga L and Stahl P D 1995 Membrane trafficking along the phagocytic pathway *Trends Cell Biol.* **5** 100–4

Bertelsen V and Stang E 2014 The mysterious ways of ErbB2/HER2 trafficking *Membranes* **4** 424–46

Béthune J and Wieland F T 2018 Assembly of COPI and COPII vesicular coat proteins on membranes *Annu. Rev. Biophys.* **47** 63–83

Betschinger J, Mechtler K and Knoblich J A 2003 The Par complex directs asymmetric cell division by phosphorylating the cytoskeletal protein Lgl *Nature* **422** 326–30

Bian B, Mongrain S, Cagnol S, Langlois M-J, Boulanger J, Bernatchez G *et al* 2016 Cathepsin B promotes colorectal tumorigenesis, cell invasion, and metastasis: CATHEPSIN B promotes colorectal cancer progression *Mol. Carcinog.* **55** 671–87

Bien S, Rimmbach C, Neumann H, Niessen J, Reimer E, Ritter C A *et al* 2010 Doxorubicin-induced cell death requires cathepsin B in HeLa cells *Biochem. Pharmacol.* **80** 1466–77

Blehm B H, Schroer T A, Trybus K M, Chemla Y R and Selvin P R 2013 *In vivo* optical trapping indicates kinesin's stall force is reduced by dynein during intracellular transport *Proc. Natl Acad. Sci. USA* **110** 3381–6

Blouw B, Patel M, Iizuka S, Abdullah C, You W K, Huang X *et al* 2015 The invadopodia scaffold protein Tks5 is required for the growth of human breast cancer cells *in vitro* and *in vivo PLoS One* **10** e0121003

Boedtkjer E and Pedersen S F 2020 The acidic tumor microenvironment as a driver of cancer *Annu. Rev. Physiol.* **82** 103–26

Boissan M, Montagnac G, Shen Q, Griparic L, Guitton J, Romao M *et al* 2014 Nucleoside diphosphate kinases fuel dynamin superfamily proteins with GTP for membrane remodeling *Science* **344** 1510–5

Bond L M, Arden S D, Kendrick-Jones J, Buss F and Sellers J R 2012 Dynamic Exchange of myosin VI on endocytic structures *J. Biol. Chem.* **287** 38637–46

Bonifacino J S and Glick B S 2004 The mechanisms of vesicle budding and fusion *Cell* **116** 153–66

Borisy G G, Marcum J M, Olmsted J B, Murphy D B and Johnson K A 1975 Purification of tubulin and associated high molecular weight proteins from porcine brain and characterization of microtubule assembly *in vitro Ann. N.Y. Acad. Sci.* **253** 107–32

Boulant S, Kural C, Zeeh J-C, Ubelmann F and Kirchhausen T 2011 Actin dynamics counteract membrane tension during clathrin-mediated endocytosis *Nat. Cell Biol.* **13** 1124–31

Boulay P-L, Schlienger S, Lewis-Saravalli S, Vitale N, Ferbeyre G and Claing A 2011 ARF1 controls proliferation of breast cancer cells by regulating the retinoblastoma protein *Oncogene* **30** 3846–61

Boura E and Nencka R 2015 Phosphatidylinositol 4-kinases: function, structure, and inhibition *Exp. Cell. Res.* **337** 136–45

Boussiotis V A 2016 Molecular and biochemical aspects of the PD-1 checkpoint pathway *N. Engl. J. Med.* **375** 1767–78

Boya P 2012 Lysosomal function and dysfunction: mechanism and disease *Antioxid. Redox Signal.* **17** 766–74

Bray F, Laversanne M, Sung H, Ferlay J, Siegel R L, Soerjomataram I *et al* 2024 Global cancer statistics 2022: GLOBOCAN estimates of incidence and mortality worldwide for 36 cancers in 185 countries *CA Cancer J Clin.* **74** 229–63

Bretscher M S 2014 Asymmetry of single cells and where that leads *Annu. Rev. Biochem.* **83** 275–89

Bright N A, Gratian M J and Luzio J P 2005 Endocytic delivery to lysosomes mediated by concurrent fusion and kissing events in living cells *Curr. Biol.* **15** 360–5

Brodersen P and Voinnet O 2009 Revisiting the principles of microRNA target recognition and mode of action *Nat. Rev. Mol. Cell Biol.* **10** 141–8

Brouhard G J and Rice L M 2018 Microtubule dynamics: an interplay of biochemistry and mechanics *Nat. Rev. Mol. Cell Biol.* **19** 451–63

Budaitis B G, Jariwala S, Rao L, Yue Y, Sept D, Verhey K J *et al* 2021 Pathogenic mutations in the kinesin-3 motor KIF1A diminish force generation and movement through allosteric mechanisms *J. Cell Biol.* **220** e202004227

Bulinski J C and Borisy G G 1979 Self-assembly of microtubules in extracts of cultured HeLa cells and the identification of HeLa microtubule-associated proteins *Proc. Natl Acad. Sci. USA* **76** 293–7

Bulinski J C and Bossler A 1994 Purification and characterization of ensconsin, a novel microtubule stabilizing protein *J. Cell Sci.* **107** 2839–49

Burger K L, Learman B S, Boucherle A K, Sirintrapun S J, Isom S, Díaz B *et al* 2014 Src-dependent Tks5 phosphorylation regulates invadopodia-associated invasion in prostate cancer cells *Prostate* **74** 134–48

Burr M L, Sparbier C E, Chan Y-C, Williamson J C, Woods K, Beavis P A *et al* 2017 CMTM6 maintains the expression of PD-L1 and regulates anti-tumour immunity *Nature* **549** 101–5

Burridge K, Monaghan-Benson E and Graham D M 2019 Mechanotransduction: from the cell surface to the nucleus via RhoA *Phil. Trans. R. Soc.* B **374** 20180229

Buschman M D, Bromann P A, Cejudo-Martin P, Wen F, Pass I and Courtneidge S A 2009 The novel adaptor protein Tks4 (SH3PXD2B) is required for functional podosome formation *Mol. Biol. Cell* **20** 1302–11

Buss F 2001 Myosin VI isoform localized to clathrin-coated vesicles with a role in clathrin-mediated endocytosis *EMBO J.* **20** 3676–84

Cabukusta B and Neefjes J 2018 Mechanisms of lysosomal positioning and movement *Traffic* **19** 761–9

Cai D, McEwen D P, Martens J R, Meyhofer E and Verhey K J 2009 Single molecule imaging reveals differences in microtubule track selection between kinesin motors *PLoS Biol.* **7** e1000216

Cai H, Reinisch K and Ferro-Novick S 2007 Coats, tethers, rabs, and SNAREs work together to mediate the intracellular destination of a transport vesicle *Dev. Cell* **12** 671–82

Campa C C, Margaria J P, Derle A, Del Giudice M, De Santis M C, Gozzelino L *et al* 2018 Rab11 activity and PtdIns(3)P turnover removes recycling cargo from endosomes *Nat. Chem. Biol.* **14** 801–10

Campo E, Muñoz J, Miquel R, Palacín A, Cardesa A, Sloane B F *et al* 1994 Cathepsin B expression in colorectal carcinomas correlates with tumor progression and shortened patient survival *Am. J. Pathol.* **145** 301–9

Cantalupo G 2001 Rab-interacting lysosomal protein (RILP): the Rab7 effector required for transport to lysosomes *EMBO J.* **20** 683–93

Capaci V, Bascetta L, Fantuz M, Beznoussenko G V, Sommaggio R, Cancila V *et al* 2020 Mutant p53 induces Golgi tubulo-vesiculation driving a prometastatic secretome *Nat. Commun.* **11** 3945

Carlson P, Dasgupta A, Grzelak C A, Kim J, Barrett A, Coleman I M *et al* 2019 Targeting the perivascular niche sensitizes disseminated tumour cells to chemotherapy *Nat. Cell Biol.* **21** 238–50

Carpentier J L, Gorden P, Anderson R G, Goldstein J L, Brown M S, Cohen S *et al* 1982 Co-localization of 125I-epidermal growth factor and ferritin-low density lipoprotein in coated pits: a quantitative electron microscopic study in normal and mutant human fibroblasts *J. Cell Biol.* **95** 73–7

Carvalho C, Moreira M, Barbosa D J, Chan F-Y, Koehnen C B, Teixeira V *et al* 2025 ZYG-12/Hook's dual role as a dynein adaptor for early endosomes and nuclei is regulated by alternative splicing of its cargo binding domain *Mol. Biol. Cell* **36** ar19

Cason S E, Carman P J, Van Duyne C, Goldsmith J, Dominguez R and Holzbaur E L F 2021 Sequential dynein effectors regulate axonal autophagosome motility in a maturation-dependent pathway *J. Cell Biol.* **220** e202010179

Castro-Castro A, Marchesin V, Monteiro P, Lodillinsky C, Rossé C and Chavrier P 2016 Cellular and molecular mechanisms of MT1-MMP-dependent cancer cell invasion *Annu. Rev. Cell Dev. Biol.* **32** 555–76

Caswell P T, Chan M, Lindsay A J, McCaffrey M W, Boettiger D and Norman J C 2008 Rab-coupling protein coordinates recycling of $\alpha 5\beta 1$ integrin and EGFR1 to promote cell migration in 3D microenvironments *J. Cell Biol.* **183** 143–55

Caswell P T, Spence H J, Parsons M, White D P, Clark K, Cheng K W *et al* 2007 Rab25 associates with $\alpha 5\beta 1$ integrin to promote invasive migration in 3D microenvironments *Dev. Cell* **13** 496–510

Chaaban S and Carter A P 2022 Structure of dynein–dynactin on microtubules shows tandem adaptor binding *Nature* **610** 212–6

Chaffer C L and Weinberg R A 2011 A perspective on cancer cell metastasis *Science* **331** 1559–64

Chakrabarti R, Lee M and Higgs H N 2021 Multiple roles for actin in secretory and endocytic pathways *Curr. Biol.* **31** R603–18

Chambers A F, Groom A C and MacDonald I C 2002 Dissemination and growth of cancer cells in metastatic sites *Nat. Rev. Cancer* **2** 563–72

Chang Y-C, Su C-Y, Chen M-H, Chen W-S, Chen C-L and Hsiao M 2017 Secretory RAB GTPase 3C modulates IL6-STAT3 pathway to promote colon cancer metastasis and is associated with poor prognosis *Mol. Cancer* **16** 135

Char R, Liu Z, Jacqueline C, Davieau M, Delgado M-G, Soufflet C *et al* 2023 RUFY3 regulates endolysosomes perinuclear positioning, antigen presentation and migration in activated phagocytes *Nat. Commun.* **14** 4290

Chaudhary A R, Berger F, Berger C L and Hendricks A G 2018 Tau directs intracellular trafficking by regulating the forces exerted by kinesin and dynein teams *Traffic* **19** 111–21

Chaudhary A R, Lu H, Krementsova E B, Bookwalter C S, Trybus K M and Hendricks A G 2019 MAP7 regulates organelle transport by recruiting kinesin-1 to microtubules *J. Biol. Chem.* **294** 10160–71

Chen D and Che G 2014 Value of caveolin-1 in cancer progression and prognosis: emphasis on cancer-associated fibroblasts, human cancer cells and mechanism of caveolin-1 expression (Review) *Oncol. Lett.* **8** 1409–21

Chen J, Kanai Y, Cowan N J and Hirokawa N 1992 Projection domains of MAP2 and tau determine spacings between microtubules in dendrites and axons *Nature* **360** 674–7

Chen P-H, Bendris N, Hsiao Y-J, Reis C R, Mettlen M, Chen H-Y *et al* 2017 Crosstalk between CLCb/Dyn1-mediated adaptive clathrin-mediated endocytosis and epidermal growth factor receptor signaling increases metastasis *Dev. Cell* **40** 278–88.e5

Chen W and He B 2022 Actomyosin activity-dependent apical targeting of Rab11 vesicles reinforces apical constriction *J. Cell Biol.* **221** e202103069

Cheng J, Volk L, Janaki D K M, Vyakaranam S, Ran S and Rao K A 2010 Tumor suppressor function of Rab25 in triple-negative breast cancer *Int. J. Cancer* **126** 2799–812

Cheng K W, Lahad J P, Kuo W, Lapuk A, Yamada K, Auersperg N *et al* 2004 The RAB25 small GTPase determines aggressiveness of ovarian and breast cancers *Nat. Med.* **10** 1251–6

Cheng Z-J, Deep Singh R, Marks D L and Pagano R E 2006 Membrane microdomains, caveolae, and caveolar endocytosis of sphingolipids (Review) *Mol. Membr. Biol.* **23** 101–10

Cherry J P, Mordente J A, Chapman J R, Choudhury M S, Tazaki H, Mallouh C *et al* 1998 Analysis of cathepsin D forms and their clinical implications in human prostate cancer *J. Urol.* **160** 2223–8

Chevalier-Larsen E and Holzbaur E L F 2006 Axonal transport and neurodegenerative disease *Biochim. Biophys. Acta (BBA)—Mol. Basis Dis.* **1762** 1094–108

Chial H J, Wu R, Ustach C V, McPhail L C, Mobley W C and Chen Y Q 2008 Membrane targeting by APPL1 and APPL2: dynamic scaffolds that oligomerize and bind phosphoinositides *Traffic* **9** 215–29

Cho K H and Lee H Y 2019 Rab25 and RCP in cancer progression *Arch. Pharm. Res.* **42** 101–12

Christensen J R, Kendrick A A, Truong J B, Aguilar-Maldonado A, Adani V, Dzieciatkowska M *et al* 2021 Cytoplasmic dynein-1 cargo diversity is mediated by the combinatorial assembly of FTS–Hook–FHIP complexes *eLife* **10** e74538

Christoforidis S, Miaczynska M, Ashman K, Wilm M, Zhao L, Yip S-C *et al* 1999 Phosphatidylinositol-3-OH kinases are Rab5 effectors *Nat. Cell Biol.* **1** 249–52

Chung B M, Raja S M, Clubb R J, Tu C, George M, Band V *et al* 2009 Aberrant trafficking of NSCLC-associated EGFR mutants through the endocytic recycling pathway promotes interaction with Src@ *BMC Cell Biol.* **10** 84

Cianfrocco M A, DeSantis M E, Leschziner A E and Reck-Peterson S L 2015 Mechanism and regulation of cytoplasmic dynein *Annu. Rev. Cell Dev. Biol.* **31** 83–108

Clancy J W, Sedgwick A, Rosse C, Muralidharan-Chari V, Raposo G, Method M *et al* 2015 Regulated delivery of molecular cargo to invasive tumour-derived microvesicles *Nat. Commun.* **6** 6919

Clausen M J A M, Melchers L J, Mastik M F, Slagter-Menkema L, Groen H J M, Laan B F A M V D *et al* 2016 RAB25 expression is epigenetically downregulated in oral and oropharyngeal squamous cell carcinoma with lymph node metastasis *Epigenetics* **11** 653–63

Cohen S, Valm A M and Lippincott-Schwartz J 2018 Interacting organelles *Curr. Opin. Cell Biol.* **53** 84–91

Commisso C, Davidson S M, Soydaner-Azeloglu R G, Parker S J, Kamphorst J J, Hackett S *et al* 2013 Macropinocytosis of protein is an amino acid supply route in Ras-transformed cells *Nature* **497** 633–7

Conte A and Sigismund S 2016 The ubiquitin network in the control of EGFR endocytosis and signaling *Progress in Molecular Biology and Translational Science* (Amsterdam: Elsevier) ch 6 pp 225–76

Conus S and Simon H-U 2008 Cathepsins: key modulators of cell death and inflammatory responses *Biochem. Pharmacol.* **76** 1374–82

Conway L, Wood D, Tüzel E and Ross J L 2012 Motor transport of self-assembled cargos in crowded environments *Proc. Natl Acad. Sci. USA* **109** 20814–9

Coumailleau F and González-Gaitán M 2008 From endocytosis to tumors through asymmetric cell division of stem cells *Curr. Opin. Cell Biol.* **20** 462–9

Cvetković D, Babwah A V and Bhattacharya M 2013 Kisspeptin/KISS1R system in breast cancer *J. Cancer* **4** 653–61

D'Agostino L, Nie Y, Goswami S, Tong K, Yu S, Bandyopadhyay S *et al* 2019 Recycling endosomes in mature epithelia restrain tumorigenic signaling *Cancer Res.* **79** 4099–112

Dai J and Sheetz M P 1995 Mechanical properties of neuronal growth cone membranes studied by tether formation with laser optical tweezers *Biophys. J.* **68** 988–96

Dai J and Sheetz M P 1999 Membrane tether formation from blebbing cells *Biophys. J.* **77** 3363–70

Dai J, Sheetz M P, Wan X and Morris C E 1998 Membrane tension in swelling and shrinking molluscan neurons *J. Neurosci.* **18** 6681–92

Daly C, Logan B, Breeyear J, Whitaker K, Ahmed M and Seals D F 2020 Tks5 SH3 domains exhibit differential effects on invadopodia development *PLoS One* **15** e0227855

Dammai V, Adryan B, Lavenburg K R and Hsu T 2003 *Drosophila awd*, the homolog of human *nm23*, regulates FGF receptor levels and functions synergistically with *shi/dynamin* during tracheal development *Genes Dev.* **17** 2812–24

Dance A L, Miller M, Seragaki S, Aryal P, White B, Aschenbrenner L *et al* 2004 Regulation of myosin-VI targeting to endocytic compartments *Traffic* **5** 798–813

Dawson J C, Timpson P, Kalna G and Machesky L M 2012 Mtss1 regulates epidermal growth factor signaling in head and neck squamous carcinoma cells *Oncogene* **31** 1781–93

De Duve C, Pressman B C, Gianetto R, Wattiaux R and Appelmans F 1955 Tissue fractionation studies. 6. Intracellular distribution patterns of enzymes in rat-liver tissue *Biochem.* J. **60** 604–17

Delevoye C, Miserey-Lenkei S, Montagnac G, Gilles-Marsens F, Paul-Gilloteaux P, Giordano F *et al* 2014 Recycling endosome tubule morphogenesis from sorting endosomes requires the kinesin motor KIF13A *Cell Rep.* **6** 445–54

Denhardt D T, Greenberg A H, Egan S E, Hamilton R T and Wright J A 1987 Cysteine proteinase cathepsin L expression correlates closely with the metastatic potential of H-ras-transformed murine fibroblasts *Oncogene* **2** 55–9

Di Guglielmo G M, Le Roy C, Goodfellow A F and Wrana J L 2003 Distinct endocytic pathways regulate TGF-β receptor signalling and turnover *Nat. Cell Biol.* **5** 410–21

Diaz B, Shani G, Pass I, Anderson D, Quintavalle M and Courtneidge S A 2009 Tks5-dependent, Nox-mediated generation of reactive oxygen species is necessary for invadopodia formation *Sci. Signal.* **2** ra53

Díaz J, Mendoza P, Ortiz R, Díaz N, Leyton L, Stupack D *et al* 2014 Rab5 is required for Caveolin-1-enhanced Rac1 activation, migration and invasion of metastatic cancer cells *J. Cell Sci.* **127** 2401-6

Diaz-Rohrer B, Castello-Serrano I, Chan S H, Wang H-Y, Shurer C R, Levental K R *et al* 2023 Rab3 mediates a pathway for endocytic sorting and plasma membrane recycling of ordered microdomains *Proc. Natl Acad. Sci. USA* **120** e2207461120

Diekmann Y and Pereira-Leal J B 2013 Evolution of intracellular compartmentalization *Biochem.* J. **449** 319–31

Diggins N L and Webb D J 2017 APPL1 is a multifunctional endosomal signaling adaptor protein *Biochem. Soc. Trans.* **45** 771–9

Dillekås H, Rogers M S and Straume O 2019 Are 90% of deaths from cancer caused by metastases? *Cancer Med.* **8** 5574–6

Dippold H C, Ng M M, Farber-Katz S E, Lee S-K, Kerr M L, Peterman M C *et al* 2009 GOLPH3 bridges phosphatidylinositol-4- phosphate and actomyosin to stretch and shape the golgi to promote budding *Cell* **139** 337–51

Disanza A, Frittoli E, Palamidessi A and Scita G 2009 Endocytosis and spatial restriction of cell signaling *Mol. Oncol.* **3** 280–96

Dixit R, Ross J L, Goldman Y E and Holzbaur E L F 2008 Differential regulation of dynein and kinesin motor proteins by Tau *Science* **319** 1086–9

Diz-Muñoz A, Thurley K, Chintamen S, Altschuler S J, Wu L F, Fletcher D A *et al* 2016 Membrane tension acts through PLD2 and mTORC2 to limit actin network assembly during neutrophil migration *PLoS Biol.* **14** e1002474

Do M T, Chai T F, Casey P J and Wang M 2017 Isoprenylcysteine carboxylmethyltransferase function is essential for RAB4A-mediated integrin β3 recycling, cell migration and cancer metastasis *Oncogene* **36** 5757–67

Doherty G J and McMahon H T 2009 Mechanisms of endocytosis *Annu. Rev. Biochem.* **78** 857–902

Dolat L and Spiliotis E T 2016 Septins promote macropinosome maturation and traffic to the lysosome by facilitating membrane fusion *J. Cell Biol.* **214** 517–27

Donaldson J G and Jackson C L 2011 ARF family G proteins and their regulators: roles in membrane transport, development and disease *Nat. Rev. Mol. Cell Biol.* **12** 362–75

Dong J T, Suzuki H, Pin S S, Bova G S, Schalken J A, Isaacs W B *et al* 1996 Down-regulation of the KAI1 metastasis suppressor gene during the progression of human prostatic cancer infrequently involves gene mutation or allelic loss *Cancer Res.* **56** 4387–90

Dong J-T, Isaacs W B and Isaacs J T 1997 Molecular advances in prostate cancer *Curr. Opin. Oncol.* **9** 101–7

Dong Q, Fu L, Zhao Y, Du Y, Li Q, Qiu X *et al* 2017 Rab11a promotes proliferation and invasion through regulation of YAP in non-small cell lung cancer *Oncotarget* **8** 27800–11

Dos Santos Á, Fili N, Hari-Gupta Y, Gough R E, Wang L, Martin-Fernandez M *et al* 2022 Binding partners regulate unfolding of myosin VI to activate the molecular motor *Biochem. J.* **479** 1409–28

Driskell O J, Mironov A, Allan V J and Woodman P G 2007 Dynein is required for receptor sorting and the morphogenesis of early endosomes *Nat. Cell Biol.* **9** 113–20
Droga-Mazovec G, Bojič L, Petelin A, Ivanova S, Romih R, Repnik U *et al* 2008 Cysteine cathepsins trigger caspase-dependent cell death through cleavage of bid and antiapoptotic Bcl-2 homologues *J. Biol. Chem.* **283** 19140–50
Du J, Ren W, Yao F, Wang H, Zhang K, Luo M *et al* 2019 YY1 cooperates with TFEB to regulate autophagy and lysosomal biogenesis in melanoma *Mol. Carcinog.* **58** 2149–60
Du Z and Lovly C M 2018 Mechanisms of receptor tyrosine kinase activation in cancer *Mol. Cancer* **17** 58
Duan F, Peng Z, Yin J, Yang Z and Shang J 2020 Expression of MMP-14 and prognosis in digestive system carcinoma: a meta-analysis and databases validation *J. Cancer* **11** 1141–50
Dufait I, Van Valckenborgh E, Menu E, Escors D, De Ridder M and Breckpot K 2016 Signal transducer and activator of transcription 3 in myeloid-derived suppressor cells: an opportunity for cancer therapy *Oncotarget* **7** 42698–715
Eaton S and Martin-Belmonte F 2014 Cargo sorting in the endocytic pathway: a key regulator of cell polarity and tissue dynamics *Cold Spring Harb. Perspect. Biol.* **6** a016899–a9
Ecklund K H, Morisaki T, Lammers L G, Marzo M G, Stasevich T J and Markus S M 2017 She1 affects dynein through direct interactions with the microtubule and the dynein microtubule-binding domain *Nat. Commun.* **8** 2151
Elshenawy M M, Canty J T, Oster L, Ferro L S, Zhou Z, Blanchard S C *et al* 2019 Cargo adaptors regulate stepping and force generation of mammalian dynein–dynactin *Nat. Chem. Biol.* **15** 1093–101
Engelman J A, Zhang X L, Galbiati F and Lisanti M P 1998 Chromosomal localization, genomic organization, and developmental expression of the murine caveolin gene family (Cav-1, -2, and -3): Cav-1 and Cav-2 genes map to a known tumor suppressor locus (6-A2/7q31) *FEBS Lett.* **429** 330–6
Eskelinen E-L 2006 Roles of LAMP-1 and LAMP-2 in lysosome biogenesis and autophagy *Mol. Aspects Med.* **27** 495–502
Etoh K and Fukuda M 2019 Rab10 regulates tubular endosome formation through KIF13A/B motors *J. Cell Sci.* **226977**
Fares J, Fares M Y, Khachfe H H, Salhab H A and Fares Y 2020 Molecular principles of metastasis: a hallmark of cancer revisited *Signal Transduct. Target Ther.* **5** 28
Feng Q, Gicking A M and Hancock W O 2020 Dynactin p150 promotes processive motility of DDB complexes by minimizing diffusional behavior of dynein *Mol. Biol. Cell* **31** 782–92
Feng Q, Mickolajczyk K J, Chen G-Y and Hancock W O 2018 Motor reattachment kinetics play a dominant role in multimotor-driven cargo transport *Biophys. J.* **114** 400–9
Feng Z and Yu C 2021 PI(3,4)P $_2$-mediated membrane tubulation promotes integrin trafficking and invasive cell migration *Proc. Natl Acad. Sci. USA* **118** e2017645118
Ferguson J P, Willy N M, Heidotting S P, Huber S D, Webber M J and Kural C 2016 Deciphering dynamics of clathrin-mediated endocytosis in a living organism *J. Cell Biol.* **214** 347–58
Ferguson S, Raimondi A, Paradise S, Shen H, Mesaki K, Ferguson A *et al* 2009 Coordinated actions of actin and BAR proteins upstream of dynamin at endocytic clathrin-coated pits *Dev. Cell* **17** 811–22
Ferreira A P A, Casamento A, Carrillo Roas S, Halff E F, Panambalana J, Subramaniam S *et al* 2021 Cdk5 and GSK3β inhibit fast endophilin-mediated endocytosis *Nat. Commun.* **12** 2424

Ferro L S, Eshun-Wilson L, Gölcük M, Fernandes J, Huijben T, Gerber E *et al* 2020 The mechanism of motor inhibition by microtubule-associated proteins bioRxiv:2020.10.22.351346 https://doi.org/10.1101/2020.10.22.351346

Finkielstein C V and Capelluto D G S 2016 Disabled-2: a modular scaffold protein with multifaceted functions in signaling *BioEssays* **38** s45–55

Fittipaldi A, Ferrari A, Zoppé M, Arcangeli C, Pellegrini V, Beltram F *et al* 2003 Cell membrane lipid rafts mediate caveolar endocytosis of HIV-1 Tat fusion proteins *J. Biol. Chem.* **278** 34141–9

Flores-Rodriguez N, Rogers S S, Kenwright D A, Waigh T A, Woodman P G and Allan V J 2011 Roles of dynein and dynactin in early endosome dynamics revealed using automated tracking and global analysis *PLoS One* **6** e24479

Foekens J A, Look M P, Vries J B, Gelder M E M, Putten W L J V and Klijn J G M 1999 Cathepsin-D in primary breast cancer: prognostic evaluation involving 2810 patients *Br. J. Cancer* **79** 300–7

Fogelson B and Mogilner A 2014 Computational estimates of membrane flow and tension gradient in motile cells *PLoS One* **9** e84524

Foghsgaard L, Wissing D, Mauch D, Lademann U, Bastholm L, Boes M *et al* 2001 Cathepsin B acts as a dominant execution protease in tumor cell apoptosis induced by tumor necrosis factor *J. Cell Biol.* **153** 999–1010

Fong K-W, Au F K C, Jia Y, Yang S, Zhou L and Qi R Z 2017 Microtubule plus-end tracking of end-binding protein 1 (EB1) is regulated by CDK5 regulatory subunit-associated protein 2 *J. Biol. Chem.* **292** 7675–87

Francavilla C, Papetti M, Rigbolt K T G, Pedersen A-K, Sigurdsson J O, Cazzamali G *et al* 2016 Multilayered proteomics reveals molecular switches dictating ligand-dependent EGFR trafficking *Nat. Struct. Mol. Biol.* **23** 608–18

Francis C R, Kincross H and Kushner E J 2022 Rab35 governs apicobasal polarity through regulation of actin dynamics during sprouting angiogenesis *Nat. Commun.* **13** 5276

Fuchs R, Mâle P and Mellman I 1989 Acidification and ion permeabilities of highly purified rat liver endosomes *J. Biol. Chem.* **264** 2212–20

Fujiwara M, Marusawa H, Wang H-Q, Iwai A, Ikeuchi K, Imai Y *et al* 2008 Parkin as a tumor suppressor gene for hepatocellular carcinoma *Oncogene* **27** 6002–11

Gabaldón T and Pittis A A 2015 Origin and evolution of metabolic sub-cellular compartmentalization in eukaryotes *Biochimie* **119** 262–8

Gadadhar S, Bodakuntla S, Natarajan K and Janke C 2017 The tubulin code at a glance *J. Cell Sci.* **130** 1347–53

Gambin Y, Ariotti N, McMahon K-A, Bastiani M, Sierecki E, Kovtun O *et al* 2014 Single-molecule analysis reveals self assembly and nanoscale segregation of two distinct cavin subcomplexes on caveolae *eLife* **3** e01434

Garcia S, Dalès J-P, Charafe-Jauffret E, Carpentier-Meunier S, Andrac-Meyer L, Jacquemier J *et al* 2007 Poor prognosis in breast carcinomas correlates with increased expression of targetable CD146 and c-Met and with proteomic basal-like phenotype *Human Pathol.* **38** 830–41

Garnett T O, Filippova M and Duerksen-Hughes P J 2007 Bid is cleaved upstream of caspase-8 activation during TRAIL-mediated apoptosis in human osteosarcoma cells *Apoptosis* **12** 1299–315

Gauthier N C, Fardin M A, Roca-Cusachs P and Sheetz M P 2011 Temporary increase in plasma membrane tension coordinates the activation of exocytosis and contraction during cell spreading *Proc. Natl Acad. Sci. USA* **108** 14467–72

Gauthier N C, Masters T A and Sheetz M P 2012 Mechanical feedback between membrane tension and dynamics *Trends Cell Biol.* **22** 527–35

Gauthier N C, Rossier O M, Mathur A, Hone J C and Sheetz M P 2009 Plasma membrane area increases with spread area by exocytosis of a GPI-anchored protein compartment *Mol. Biol. Cell* **20** 3261–72

Gerondopoulos A, Langemeyer L, Liang J-R, Linford A and Barr F A 2012 BLOC-3 mutated in Hermansky-Pudlak syndrome is a Rab32/38 guanine nucleotide exchange factor *Curr. Biol.* **22** 2135–9

Giatromanolaki A, Kalamida D, Sivridis E, Karagounis I V, Gatter K C, Harris A L *et al* 2015 Increased expression of transcription factor EB (TFEB) is associated with autophagy, migratory phenotype and poor prognosis in non-small cell lung cancer *Lung Cancer* **90** 98–105

Gibbons D L, Lin W, Creighton C J, Rizvi Z H, Gregory P A, Goodall G J *et al* 2009 Contextual extracellular cues promote tumor cell EMT and metastasis by regulating miR-200 family expression *Genes Dev* **23** 2140–51

Gicking A M, Ma T-C, Feng Q, Jiang R, Badieyan S, Cianfrocco M A *et al* 2022 Kinesin-1, -2, and -3 motors use family-specific mechanochemical strategies to effectively compete with dynein during bidirectional transport *eLife* **11** e82228

Gleeson J G, Allen K M, Fox J W, Lamperti E D, Berkovic S, Scheffer I *et al* 1998 Doublecortin, a brain-specific gene mutated in human *x*-linked lissencephaly and double cortex syndrome, encodes a putative signaling protein *Cell* **92** 63–72

Glondu M, Liaudet-Coopman E, Derocq D, Platet N, Rochefort H and Garcia M 2002 Down-regulation of cathepsin-D expression by antisense gene transfer inhibits tumor growth and experimental lung metastasis of human breast cancer cells *Oncogene* **21** 5127–34

Glunde K, Guggino S E, Solaiyappan M, Pathak A P, Ichikawa Y and Bhujwalla Z M 2003 Extracellular acidification alters lysosomal trafficking in human breast cancer cells *Neoplasia* **5** 533–45

Goetz J G, Lajoie P, Wiseman S M and Nabi I R 2008 Caveolin-1 in tumor progression: the good, the bad and the ugly *Cancer Metastasis Rev.* **27** 715–35

Gogineni V R, Gupta R, Nalla A K, Velpula K K and Rao J S 2012 uPAR and cathepsin B shRNA impedes TGF-β1-driven proliferation and invasion of meningioma cells in a XIAP-dependent pathway *Cell Death Dis.* **3** e439–9

Goh L K and Sorkin A 2013 Endocytosis of receptor tyrosine kinases *Cold Spring Harb. Perspect. Biol.* **5** a017459–a9

Goldenring J R 2015 Recycling endosomes *Curr. Opin. Cell Biol.* **35** 117–22

Goldenring J R and Nam K T 2011 Rab25 as a tumour suppressor in colon carcinogenesis *Br. J. Cancer* **104** 33–6

Gong X, Liu J, Zhang X, Dong F, Liu Y and Wang P 2018 Rab11 functions as an oncoprotein via nuclear factor kappa B (NF-κB) signaling pathway in human bladder carcinoma *Med. Sci. Monit.* **24** 5093–101

Gross S P 2004 Hither and yon: a review of bi-directional microtubule-based transport *Phys. Biol.* **1** R1–R11

Gross S P, Welte M A, Block S M and Wieschaus E F 2002 Coordination of opposite-polarity microtubule motors *J. Cell Biol.* **156** 715–24

Grotjahn D A, Chowdhury S, Xu Y, McKenney R J, Schroer T A and Lander G C 2018 Cryo-electron tomography reveals that dynactin recruits a team of dyneins for processive motility *Nat. Struct. Mol. Biol.* **25** 203–7

Gruenberg J 2001 The endocytic pathway: a mosaic of domains *Nat. Rev. Mol. Cell Biol.* **2** 721–30

Gu Y, Zou Y M, Lei D, Huang Y, Li W, Mo Z *et al* 2017 Promoter DNA methylation analysis reveals a novel diagnostic CpG-based biomarker and RAB25 hypermethylation in clear cell renel cell carcinoma *Sci. Rep.* **7** 14200

Guedes-Dias P, Nirschl J J, Abreu N, Tokito M K, Janke C, Magiera M M *et al* 2019 Kinesin-3 responds to local microtubule dynamics to target synaptic cargo delivery to the presynapse *Curr. Biol.* **29** 268–282.e8

Guerra F and Bucci C 2016 Multiple roles of the small GTPase Rab7 *Cells* **5** 34

Guet D, Mandal K, Pinot M, Hoffmann J, Abidine Y, Sigaut W *et al* 2014 Mechanical role of actin dynamics in the rheology of the golgi complex and in golgi-associated trafficking events *Curr. Biol.* **24** 1700–11

Gundry C, Marco S, Rainero E, Miller B, Dornier E, Mitchell L *et al* 2017 Phosphorylation of Rab-coupling protein by LMTK3 controls Rab14-dependent EphA2 trafficking to promote cell:cell repulsion *Nat. Commun.* **8** 14646

Guo Y, Li D, Zhang S, Yang Y, Liu J-J, Wang X *et al* 2018 Visualizing intracellular organelle and cytoskeletal interactions at nanoscale resolution on millisecond timescales *Cell* **175** 1430–1442.e17

Habermann B 2004 The BAR-domain family of proteins: a case of bending and binding?: the membrane bending and GTPase-binding functions of proteins from the BAR-domain family *EMBO Rep.* **5** 250–5

Hafezi S, Saber-Ayad M and Abdel-Rahman W M 2021 Highlights on the role of KRAS mutations in reshaping the microenvironment of pancreatic adenocarcinoma *IJMS* **22** 10219

Haigler H T, McKanna J A and Cohen S 1979 Rapid stimulation of pinocytosis in human carcinoma cells A-431 by epidermal growth factor *J. Cell Biol.* **83** 82–90

Hakulinen J, Sankkila L, Sugiyama N, Lehti K and Keski-Oja J 2008 Secretion of active membrane type 1 matrix metalloproteinase (MMP-14) into extracellular space in microvesicular exosomes *J. Cell. Biochem.* **105** 1211–8

Halberg N, Sengelaub C A, Navrazhina K, Molina H, Uryu K and Tavazoie S F 2016 PITPNC1 recruits RAB1B to the golgi network to drive malignant secretion *Cancer Cell* **29** 339–53

Hales C M, Vaerman J-P and Goldenring J R 2002 Rab11 family interacting protein 2 associates with myosin Vb and regulates plasma membrane recycling *J. Biol. Chem.* **277** 50415–21

Hämälistö S and Jäättelä M 2016 Lysosomes in cancer—living on the edge (of the cell) *Curr. Opin. Cell Biol.* **39** 69–76

Hammer J A and Sellers J R 2012 Walking to work: roles for class V myosins as cargo transporters *Nat. Rev. Mol. Cell Biol.* **13** 13–26

Hancock W O 2014 Bidirectional cargo transport: moving beyond tug of war *Nat. Rev. Mol. Cell Biol.* **15** 615–28

Hansen C G and Nichols B J 2009 Molecular mechanisms of clathrin-independent endocytosis *J. Cell Sci.* **122** 1713–21

Harms M B, Ori-McKenney K M, Scoto M, Tuck E P, Bell S, Ma D *et al* 2012 Mutations in the tail domain of *DYNC1H1* cause dominant spinal muscular atrophy *Neurology* **78** 1714–20

Hasson T 2003 Myosin VI: two distinct roles in endocytosis *J. Cell Sci.* **116** 3453–61

Hayer A, Stoeber M, Ritz D, Engel S, Meyer H H and Helenius A 2010 Caveolin-1 is ubiquitinated and targeted to intralumenal vesicles in endolysosomes for degradation *J. Cell Biol.* **191** 615–29

Heinrich M 1999 Cathepsin D targeted by acid sphingomyelinase-derived ceramide *EMBO J.* **18** 5252–63

Helenius A, Mellman I, Wall D and Hubbard A 1983 Endosomes *Trends Biochem. Sci.* **8** 245–50

Hemalatha A and Mayor S 2019 Recent advances in clathrin-independent endocytosis *F1000Res.* **8** 138

Hendricks A G, Holzbaur E L F and Goldman Y E 2012 Force measurements on cargoes in living cells reveal collective dynamics of microtubule motors *Proc. Natl Acad. Sci. USA* **109** 18447–52

Hendricks A G, Perlson E, Ross J L, Schroeder H W, Tokito M and Holzbaur E L F 2010 Motor coordination via a tug-of-war mechanism drives bidirectional vesicle transport *Curr. Biol.* **20** 697–702

Hendrix A, Maynard D, Pauwels P, Braems G, Denys H, Van Den Broecke R *et al* 2010 Effect of the secretory small GTPase Rab27B on breast cancer growth, invasion, and metastasis *J. Natl Cancer Inst.* **102** 866–80

Heng J, Hu Y, Pérez-Hernández G, Inoue A, Zhao J, Ma X *et al* 2023 Function and dynamics of the intrinsically disordered carboxyl terminus of β2 adrenergic receptor *Nat. Commun.* **14** 2005

Henley J R, Krueger E W A, Oswald B J and McNiven M A 1998 Dynamin-mediated internalization of caveolae *J. Cell Biol.* **141** 85–99

Hinze C and Boucrot E 2018 Local actin polymerization during endocytic carrier formation *Biochem. Soc. Trans.* **46** 565–76

Hoepfner S, Severin F, Cabezas A, Habermann B, Runge A, Gillooly D *et al* 2005 Modulation of receptor recycling and degradation by the endosomal kinesin KIF16B *Cell* **121** 437–50

Hofmann I and Munro S 2006 An N-terminally acetylated Arf-like GTPase is localised to lysosomes and affects their motility *J. Cell Sci.* **119** 1494–503

Holst M R, Vidal-Quadras M, Larsson E, Song J, Hubert M, Blomberg J *et al* 2017 Clathrin-independent endocytosis suppresses cancer cell blebbing and invasion *Cell Rep.* **20** 1893–905

Horgan C P, Hanscom S R, Jolly R S, Futter C E and McCaffrey M W 2010 Rab11-FIP3 links the Rab11 GTPase and cytoplasmic dynein to mediate transport to the endosomal-recycling compartment *J. Cell Sci.* **123** 181–91

Hoshino D, Kirkbride K C, Costello K, Clark E S, Sinha S, Grega-Larson N *et al* 2013 Exosome secretion is enhanced by invadopodia and drives invasive behavior *Cell Rep.* **5** 1159–68

Houk A R, Jilkine A, Mejean C O, Boltyanskiy R, Dufresne E R, Angenent S B *et al* 2012 Membrane tension maintains cell polarity by confining signals to the leading edge during neutrophil migration *Cell* **148** 175–88

Howe C L, Granger B L, Hull M, Green S A, Gabel C A, Helenius A *et al* 1988 Derived protein sequence, oligosaccharides, and membrane insertion of the 120-kDa lysosomal membrane glycoprotein (lgp120): identification of a highly conserved family of lysosomal membrane glycoproteins *Proc. Natl Acad. Sci. USA* **85** 7577–81

Howe E N, Burnette M D, Justice M E, Schnepp P M, Hedrick V, Clancy J W *et al* 2020 Rab11b-mediated integrin recycling promotes brain metastatic adaptation and outgrowth *Nat. Commun.* **11** 3017

Hsiao Y-T, Wu T-Y, Wu B-K, Chu S-W and Hsieh C-L 2022 Spinning disk interferometric scattering confocal microscopy captures millisecond timescale dynamics of living cells *Opt. Express* **30** 45233

Hu C-T, Wu J-R and Wu W-S 2013 The role of endosomal signaling triggered by metastatic growth factors in tumor progression *Cell. Signal.* **25** 1539–45

Huang X, Cao Y, Bao P, Zhu B and Cheng Z 2023 High expression of PI4K2A predicted poor prognosis of colon adenocarcinoma (COAD) and correlated with immunity *Cancer Med.* **12** 837–51

Hubbard S R and Miller W T 2007 Receptor tyrosine kinases: mechanisms of activation and signaling *Curr. Opin. Cell Biol.* **19** 117–23

Huotari J and Helenius A 2011 Endosome maturation: endosome maturation *EMBO J.* **30** 3481–500

Hurst D R and Welch D R 2011 Metastasis suppressor genes *International Review of Cell and Molecular Biology* (Elsevier) ch 3 pp 107–80

Hutagalung A H and Novick P J 2011 Role of Rab GTPases in membrane traffic and cell physiology *Physiol. Rev.* **91** 119–49

Ichikawa T, Ichikawa Y and Isaacs J T 1991 Genetic factors and suppression of metastatic ability of prostatic cancer *Cancer Res.* **51** 3788–92

Ignesti M, Barraco M, Nallamothu G, Woolworth J A, Duchi S, Gargiulo G *et al* 2014 Notch signaling during development requires the function of awd, the *Drosophila* homolog of human metastasis suppressor gene Nm23 *BMC Biol.* **12** 12

Iizuka S, Abdullah C, Buschman M D, Diaz B and Courtneidge S A 2016 The role of Tks adaptor proteins in invadopodia formation, growth and metastasis of melanoma *Oncotarget* **7** 78473–86

Iizuka S, Quintavalle M, Navarro J C, Gribbin K P, Ardecky R J, Abelman M M *et al* 2021 Serine-threonine kinase TAO3-mediated trafficking of endosomes containing the invadopodia scaffold TKS5A promotes cancer invasion and tumor growth *Cancer Res.* **81** 1472–85

Ilboudo A, Nault J-C, Dubois-Pot-Schneider H, Corlu A, Zucman-Rossi J, Samson M *et al* 2014 Overexpression of phosphatidylinositol 4-kinase type IIIα is associated with undifferentiated status and poor prognosis of human hepatocellular carcinoma *BMC Cancer* **14** 7

Inami Y, Waguri S, Sakamoto A, Kouno T, Nakada K, Hino O *et al* 2011 Persistent activation of Nrf2 through p62 in hepatocellular carcinoma cells *J. Cell Biol.* **193** 275–84

Irannejad R, Tsvetanova N G, Lobingier B T and Von Zastrow M 2015 Effects of endocytosis on receptor-mediated signaling *Curr. Opin. Cell Biol.* **35** 137–43

Jahn R and Scheller R H 2006 SNAREs—engines for membrane fusion *Nat. Rev. Mol. Cell Biol.* **7** 631–43

Janke C 2014 The tubulin code: molecular components, readout mechanisms, and functions *J. Cell Biol.* **206** 461–72

Janvier K and Bonifacino J S 2005 Role of the endocytic machinery in the sorting of lysosome-associated membrane proteins *Mol. Biol. Cell* **16** 4231–42

Jean-Alphonse F, Bowersox S, Chen S, Beard G, Puthenveedu M A and Hanyaloglu A C 2014 Spatially restricted G protein-coupled receptor activity via divergent endocytic compartments *J. Biol. Chem.* **289** 3960–77

Jeger J L 2020 Endosomes, lysosomes, and the role of endosomal and lysosomal biogenesis in cancer development *Mol. Biol. Rep.* **47** 9801–10

Jemal A, Ward E M, Johnson C J, Cronin K A, Ma J, Ryerson A B *et al* 2017 Annual report to the nation on the status of cancer, 1975–2014, featuring survival *J. Natl. Cancer Inst.* **109** djx030

Ji J, Li C, Wang J, Wang L, Huang H, Li Y *et al* 2022 Hsa_circ_0001756 promotes ovarian cancer progression through regulating IGF2BP2-mediated RAB5A expression and the EGFR/MAPK signaling pathway *Cell Cycle* **21** 685–96

Jiang H-L, Sun H-F, Gao S-P, Li L-D, Hu X, Wu J *et al* 2015 Loss of RAB1B promotes triple-negative breast cancer metastasis by activating TGF-β/SMAD signaling *Oncotarget* **6** 16352–65

Jiang R, Vandal S, Park S, Majd S, Tüzel E and Hancock W O 2019 Microtubule binding kinetics of membrane-bound kinesin-1 predicts high motor copy numbers on intracellular cargo *Proc. Natl Acad. Sci. USA* **116** 26564–70

Jiang X, Tsitsiou E, Herrick S E and Lindsay M A 2010 MicroRNAs and the regulation of fibrosis: miRNAs and fibrosis *FEBS J.* **277** 2015–21

Jin S, Ahn Y, Park J, Park M, Lee S, Lee W J *et al* 2024 Temporal patterns of angular displacement of endosomes: insights into motor protein exchange dynamics *Adv. Sci.* **11** 2306849

Joffre C, Barrow R, Ménard L, Calleja V, Hart I R and Kermorgant S 2011 A direct role for met endocytosis in tumorigenesis *Nat. Cell Biol.* **13** 827–37

Johansson M, Rocha N, Zwart W, Jordens I, Janssen L, Kuijl C *et al* 2007 Activation of endosomal dynein motors by stepwise assembly of Rab7–RILP–p150Glued, ORP1L, and the receptor βlll spectrin *J. Cell Biol.* **176** 459–71

Johnson D E, Ostrowski P, Jaumouillé V and Grinstein S 2016 The position of lysosomes within the cell determines their luminal pH *J. Cell Biol.* **212** 677–92

Johnson I R D, Parkinson-Lawrence E J, Keegan H, Spillane C D, Barry-O'Crowley J, Watson W R *et al* 2015 Endosomal gene expression: a new indicator for prostate cancer patient prognosis? *Oncotarget* **6** 37919–29

Johnson I R D, Parkinson-Lawrence E J, Shandala T, Weigert R, Butler L M and Brooks D A 2014a Altered endosome biogenesis in prostate cancer has biomarker potential *Mol. Cancer Res.* **12** 1851–62

Johnson I R, Parkinson-Lawrence E J, Butler L M and Brooks D A 2014b Prostate cell lines as models for biomarker discovery: performance of current markers and the search for new biomarkers: alternative biomarkers for prostate cancer *Prostate* **74** 547–60

Jongsma M L M, Bakker N and Neefjes J 2023 Choreographing the motor-driven endosomal dance *J. Cell Sci.* **136** jcs259689

Jongsma M L M, Bakker N, Voortman L M, Koning R I, Bos E, Akkermans J J L L *et al* 2024 Systems mapping of bidirectional endosomal transport through the crowded cell *Curr. Biol.* **34** 4476–94.e11

Jordens I, Fernandez-Borja M, Marsman M, Dusseljee S, Janssen L, Calafat J *et al* 2001 The Rab7 effector protein RILP controls lysosomal transport by inducing the recruitment of dynein-dynactin motors *Curr. Biol.* **11** 1680–5

Joshi B, Strugnell S S, Goetz J G, Kojic L D, Cox M E, Griffith O L *et al* 2008 Phosphorylated Caveolin-1 regulates Rho/ROCK-dependent focal adhesion dynamics and tumor cell migration and invasion *Cancer Res.* **68** 8210–20

Juin A, Spence H J, Martin K J, McGhee E, Neilson M, Cutiongco M F A *et al* 2019 N-WASP control of LPAR1 trafficking establishes response to self-generated lpa gradients to promote pancreatic cancer cell metastasis *Dev. Cell* **51** 431–445.e7

Kallunki T, Olsen O D and Jäättelä M 2013 Cancer-associated lysosomal changes: friends or foes? *Oncogene* **32** 1995–2004

Kalpana G, Figy C, Yeung M and Yeung K C 2019 Reduced RhoA expression enhances breast cancer metastasis with a concomitant increase in CCR5 and CXCR4 chemokines signaling *Sci. Rep.* **9** 16351

Kamal A, Almenar-Queralt A, LeBlanc J F, Roberts E A and Goldstein L S B 2001 Kinesin-mediated axonal transport of a membrane compartment containing β-secretase and presenilin-1 requires APP *Nature* **414** 643–8

Kardon J R and Vale R D 2009 Regulators of the cytoplasmic dynein motor *Nat. Rev. Mol. Cell Biol.* **10** 854–65

Kärkkäinen S, Hiipakka M, Wang J, Kleino I, Vähä-Jaakkola M, Renkema G H *et al* 2006 Identification of preferred protein interactions by phage-display of the human Src homology-3 proteome *EMBO Rep.* **7** 186–91

Katoh M 2013 Functional proteomics, human genetics and cancer biology of GIPC family members *Exp. Mol. Med.* **45** e26–6

Kawai K, Nishigaki A, Moriya S, Egami Y and Araki N 2021 Rab10-positive tubular structures represent a novel endocytic pathway that diverges from canonical macropinocytosis in RAW264 macrophages *Front. Immunol.* **12** 649600

Kawana Y, Komiya A, Ueda T, Nihei N, Kuramochi H, Suzuki H *et al* 1997 Location of KAI1 on the short arm of human chromosome 11 and frequency of allelic loss in advanced human prostate cancer *Prostate* **32** 205–13

Keren-Kaplan T and Bonifacino J S 2021 ARL8 relieves SKIP Autoinhibition to enable coupling of lysosomes to kinesin-1 *Curr. Biol.* **31** 540–54.e5

Keren-Kaplan T, Šarić A, Ghosh S, Williamson C D, Jia R, Li Y *et al* 2022 RUFY3 and RUFY4 are ARL8 effectors that promote coupling of endolysosomes to dynein-dynactin *Nat. Commun.* **13** 1506

Kesisova I A, Robinson B P and Spiliotis E T 2021 A septin GTPase scaffold of dynein–dynactin motors triggers retrograde lysosome transport *J. Cell Biol.* **220** e202005219

Khan I, Gril B and Steeg P S 2019 Metastasis suppressors NME1 and NME2 promote dynamin 2 oligomerization and regulate tumor cell endocytosis, motility, and metastasis *Cancer Res.* **79** 4689–702

Khan I and Steeg P S 2018 Metastasis suppressors: functional pathways *Lab. Invest.* **98** 198–210

Khan I and Steeg P S 2021 Endocytosis: a pivotal pathway for regulating metastasis *Br. J. Cancer* **124** 66–75

Khan T G, Ginsburg D and Emmer B T 2022 The small GTPase RAB10 regulates endosomal recycling of the LDL receptor and transferrin receptor in hepatocytes *J. Lipid Res.* **63** 100248

Khatibzadeh N, Gupta S, Farrell B, Brownell W E and Anvari B 2012 Effects of cholesterol on nano-mechanical properties of the living cell plasma membrane *Soft Matter* **8** 8350

Khatter D, Sindhwani A and Sharma M 2015 Arf-like GTPase Arl8: moving from the periphery to the center of lysosomal biology *Cell. Logist.* **5** e1086501

Kim S J, Wan Q, Cho E, Han B, Yoder M C, Voytik-Harbin S L *et al* 2014 Matrix rigidity regulates spatiotemporal dynamics of Cdc42 activity and vacuole formation kinetics of endothelial colony forming cells *Biochem. Biophys. Res. Commun.* **443** 1280–5

Kirkegaard T and Jäättelä M 2009 Lysosomal involvement in cell death and cancer *Biochim. Biophys. Acta (BBA)—Mol. Cell Res.* **1793** 746–54

Kirkham M, Fujita A, Chadda R, Nixon S J, Kurzchalia T V, Sharma D K *et al* 2005 Ultrastructural identification of uncoated caveolin-independent early endocytic vehicles *J. Cell Biol.* **168** 465–76

Kiss A L 2012 Caveolae and the regulation of endocytosis *Caveolins and Caveolae* ed J-F Jasmin, P G Frank and M P Lisanti (New York: Springer) pp 14–28

Knäuper V, Will H, López-Otin C, Smith B, Atkinson S J, Stanton H *et al* 1996 Cellular mechanisms for human procollagenase-3 (MMP-13) activation *J. Biol. Chem.* **271** 17124–31

Koblinski J E, Ahram M and Sloane B F 2000 Unraveling the role of proteases in cancer *Clin. Chim. Acta* **291** 113–35

Komatsu M, Kurokawa H, Waguri S, Taguchi K, Kobayashi A, Ichimura Y *et al* 2010 The selective autophagy substrate p62 activates the stress responsive transcription factor Nrf2 through inactivation of Keap1 *Nat. Cell Biol.* **12** 213–23

Konishi T, Sasaki S, Watanabe T, Kitayama J and Nagawa H 2006 Overexpression of hRFI inhibits 5-fluorouracil-induced apoptosis in colorectal cancer cells via activation of NF-κB and upregulation of BCL-2 and BCL-XL *Oncogene* **25** 3160–9

Kornfeld S and Mellman I 1989 The biogenesis of lysosomes *Annu. Rev. Cell. Biol.* **5** 483–525

Krokowski S, Lobato-Márquez D, Chastanet A, Pereira P M, Angelis D, Galea D *et al* 2018 Septins recognize and entrap dividing bacterial cells for delivery to lysosomes *Cell Host Microbe* **24** 866–874.e4

Krueger D, Izquierdo E, Viswanathan R, Hartmann J, Pallares Cartes C and De Renzis S 2019 Principles and applications of optogenetics in developmental biology *Development* **146** dev175067

Kudlik G, Takács T, Radnai L, Kurilla A, Szeder B, Koprivanacz K *et al* 2020 Advances in understanding TKS4 and TKS5: molecular scaffolds regulating cellular processes from podosome and invadopodium formation to differentiation and tissue homeostasis *IJMS* **21** 8117

Kulikov A V, Luchkina E A, Gogvadze V and Zhivotovsky B 2017 Mitophagy: link to cancer development and therapy *Biochem. Biophys. Res. Commun.* **482** 432–9

Kumar A, Dhar S, Campanelli G, Butt N A, Schallheim J M, Gomez C R *et al* 2018 MTA 1 drives malignant progression and bone metastasis in prostate cancer *Mol. Oncol.* **12** 1596–607

Kumar G, Chawla P, Dhiman N, Chadha S, Sharma S, Sethi K *et al* 2022 RUFY3 links Arl8b and JIP4-Dynein complex to regulate lysosome size and positioning *Nat. Commun.* **13** 1540

Kunschmann T, Puder S, Fischer T, Perez J, Wilharm N and Mierke C T 2017 Integrin-linked kinase regulates cellular mechanics facilitating the motility in 3D extracellular matrices *Biochim. Biophys. Acta (BBA)—Mol. Cell Res.* **1864** 580–93

Kunwar A, Tripathy S K, Xu J, Mattson M K, Anand P, Sigua R *et al* 2011 Mechanical stochastic tug-of-war models cannot explain bidirectional lipid-droplet transport *Proc. Natl Acad. Sci. USA* **108** 18960–5

Küppers M, Albrecht D, Kashkanova A D, Lühr J and Sandoghdar V 2023 Confocal interferometric scattering microscopy reveals 3D nanoscopic structure and dynamics in live cells *Nat. Commun.* **14** 1962

Kural C, Akatay A A, Gaudin R, Chen B-C, Legant W R, Betzig E *et al* 2015 Asymmetric formation of coated pits on dorsal and ventral surfaces at the leading edges of motile cells and on protrusions of immobile cells *Mol. Biol. Cell* **26** 2044–53

Kural C, Kim H, Syed S, Goshima G, Gelfand V I and Selvin P R 2005 Kinesin and dynein move a peroxisome *in vivo*: a tug-of-war or coordinated movement? *Science* **308** 1469–72

Kural C and Kirchhausen T 2012 Live-cell imaging of clathrin coats *Methods in Enzymology* (Amsterdam: Elsevier) pp 59–80

Kuronita T, Eskelinen E-L, Fujita H, Saftig P, Himeno M and Tanaka Y 2002 A role for the lysosomal membrane protein LGP85 in the biogenesis and maintenance of endosomal and lysosomal morphology *J. Cell Sci.* **115** 4117–31

Lamaze C, Dujeancourt A, Baba T, Lo C G, Benmerah A and Dautry-Varsat A 2001 Interleukin 2 receptors and detergent-resistant membrane domains define a clathrin-independent endocytic pathway *Mol. Cell* **7** 661–71

Langemeyer L, Fröhlich F and Ungermann C 2018 Rab GTPase function in endosome and lysosome biogenesis *Trends Cell Biol.* **28** 957–70

Lányi Á, Baráth M, Péterfi Z, Bőgel G, Orient A, Simon T *et al* 2011 The Homolog of the five SH3-domain protein (HOFI/SH3PXD2B) regulates lamellipodia formation and cell spreading *PLoS One* **6** e23653

Lanzetti L and Di Fiore P P 2008 Endocytosis and cancer: an 'Insider' network with dangerous liaisons *Traffic* **9** 2011–21

Lanzetti L and Di Fiore P P 2017 Behind the scenes: endo/exocytosis in the acquisition of metastatic traits *Cancer Res.* **77** 1813–7

Lapierre L A, Kumar R, Hales C M, Navarre J, Bhartur S G, Burnette J O *et al* 2001 Myosin Vb is associated with plasma membrane recycling systems *Mol. Biol. Cell* **12** 1843–57

Larkin J and Fisher 2012 Vemurafenib: a new treatment for BRAF-V600 mutated advanced melanoma *Cancer Manag Res.* **4** 243–52

Leduc C, Padberg-Gehle K, Varga V, Helbing D, Diez S and Howard J 2012 Molecular crowding creates traffic jams of kinesin motors on microtubules *Proc. Natl Acad. Sci. USA* **109** 6100–5

Lee Y-G, Macoska J A, Korenchuk S and Pienta K J 2002 MIM, a potential metastasis suppressor gene in bladder cancer *Neoplasia* **4** 291–4

Lengyel E, Prechtel D, Resau J H, Gauger K, Welk A, Lindemann K *et al* 2005 C-Met overexpression in node-positive breast cancer identifies patients with poor clinical outcome independent of Her2/neu *Int. J. Cancer* **113** 678–82

Lessard D V, Zinder O J, Hotta T, Verhey K J, Ohi R and Berger C L 2019 Polyglutamylation of tubulin's C-terminal tail controls pausing and motility of kinesin-3 family member KIF1A *J. Biol. Chem.* **294** 6353–63

Levine B and Kroemer G 2008 Autophagy in the pathogenesis of disease *Cell* **132** 27–42

Levkowitz G, Waterman H, Zamir E, Kam Z, Oved S, Langdon W Y *et al* 1998 c-Cbl/Sli-1 regulates endocytic sorting and ubiquitination of the epidermal growth factor receptor *Genes Dev.* **12** 3663–74

Li C M-C, Chen G, Dayton T L, Kim-Kiselak C, Hoersch S, Whittaker C A *et al* 2013 Differential *Tks5* isoform expression contributes to metastatic invasion of lung adenocarcinoma *Genes Dev.* **27** 1557–67

Li H, Zhu H, Xu C and Yuan J 1998 Cleavage of BID by Caspase 8 mediates the mitochondrial damage in the fas pathway of apoptosis *Cell* **94** 491–501

Li J, Lu Y, Zhang J, Kang H, Qin Z and Chen C 2010 PI4KIIα is a novel regulator of tumor growth by its action on angiogenesis and HIF-1α regulation *Oncogene* **29** 2550–9

Li J, Mao X, Dong L Q, Liu F and Tong L 2007 Crystal structures of the BAR-PH and PTB domains of human APPL1 *Structure* **15** 525–33

Li J and Stanger B Z 2019 The tumor as organizer model *Science* **363** 1038–9

Li J, Zhang L, Gao Z, Kang H, Rong G, Zhang X *et al* 2014 Dual inhibition of EGFR at protein and activity level via combinatorial blocking of PI4KIIα as anti-tumor strategy *Protein Cell* **5** 457–68

Li X, Yang Y, Hu Y, Dang D, Regezi J, Schmidt B L *et al* 2003 αvβ6-Fyn signaling promotes oral cancer progression *J. Biol. Chem.* **278** 41646–53

Liao Z, Chua D and Tan N S 2019 Reactive oxygen species: a volatile driver of field cancerization and metastasis *Mol. Cancer* **18** 65

Lieber A D, Schweitzer Y, Kozlov M M and Keren K 2015 Front-to-rear membrane tension gradient in rapidly moving cells *Biophys. J.* **108** 1599–603

Ligon L A, Shelly S S, Tokito M and Holzbaur E L F 2003 The microtubule plus-end proteins EB1 and dynactin have differential effects on microtubule polymerization *Mol. Biol. Cell* **14** 1405–17

Lim J P and Gleeson P A 2011 Macropinocytosis: an endocytic pathway for internalising large gulps *Immunol. Cell Biol.* **89** 836–43

Lin C-C, Liu L-Z, Addison J B, Wonderlin W F, Ivanov A V and Ruppert J M 2011 A KLF4–miRNA-206 autoregulatory feedback loop can promote or inhibit protein translation depending upon cell context *Mol. Cell. Biol.* **31** 2513–27

Lin D C, Quevedo C, Brewer N E, Bell A, Testa J R, Grimes M L *et al* 2006 APPL1 associates with TrkA and GIPC1 and is required for nerve growth factor-mediated signal transduction *Mol. Cell. Biol.* **26** 8928–41

Lin S X, Gundersen G G and Maxfield F R 2002 Export from pericentriolar endocytic recycling compartment to cell surface depends on stable, detyrosinated (Glu) microtubules and kinesin *Mol. Biol. Cell* **13** 96–109

Linder S, Cervero P, Eddy R and Condeelis J 2023 Mechanisms and roles of podosomes and invadopodia *Nat. Rev. Mol. Cell Biol.* **24** 86–106

Lipka J, Kapitein L C, Jaworski J and Hoogenraad C C 2016 Microtubule-binding protein doublecortin-like kinase 1 (DCLK1) guides kinesin-3-mediated cargo transport to dendrites *EMBO J.* **35** 302–18

Liu F, Li X, Lu C, Bai A, Bielawski J, Bielawska A *et al* 2016a Ceramide activates lysosomal cathepsin B and cathepsin D to attenuate autophagy and induces ER stress to suppress myeloid-derived suppressor cells *Oncotarget* **7** 83907–25

Liu J, Guo Q, Chen B, Yu Y, Lu H and Li Y-Y 2006 Cathepsin B and its interacting proteins, bikunin and TSRC1, correlate with TNF-induced apoptosis of ovarian cancer cells OV-90 *FEBS Lett.* **580** 245–50

Liu T, Xie C, Ma H, Zhang S, Liang Y, Shi L *et al* 2014 Gr-1 + CD11b + cells facilitate Lewis lung cancer recurrence by enhancing neovasculature after local irradiation *Sci. Rep.* **4** 4833

Liu W J, Ye L, Huang W F, Guo L J, Xu Z G, Wu H L *et al* 2016b p62 links the autophagy pathway and the ubiqutin–proteasome system upon ubiquitinated protein degradation *Cell. Mol. Biol. Lett.* **21** 29

Lou X, Yano H, Lee F, Chao M V and Farquhar M G 2001 GIPC and GAIP form a complex with TrkA: a putative link between G protein and receptor tyrosine kinase pathways *Mol. Biol. Cell* **12** 615–27

Loubéry S, Wilhelm C, Hurbain I, Neveu S, Louvard D and Coudrier E 2008 Different microtubule motors move early and late endocytic compartments *Traffic* **9** 492–509

Lu M, Van Tartwijk F W, Lin J Q, Nijenhuis W, Parutto P, Fantham M *et al* 2020 The structure and global distribution of the endoplasmic reticulum network are actively regulated by lysosomes *Sci. Adv.* **6** eabc7209

Lu W and Kang Y 2019 Epithelial-mesenchymal plasticity in cancer progression and metastasis *Dev. Cell* **49** 361–74

Luzio J P, Pryor P R and Bright N A 2007 Lysosomes: fusion and function *Nat. Rev. Mol. Cell Biol.* **8** 622–32

Magistrati E and Polo S 2021 Myomics: myosin VI structural and functional plasticity *Curr. Opin. Struct. Biol.* **67** 33–40

Mak A S 2014 p53 in cell invasion, podosomes, and invadopodia *Cell Adhes. Migr.* **8** 205–14

Malerød L, Stuffers S, Brech A and Stenmark H 2007 Vps22/EAP30 in ESCRT-II mediates endosomal sorting of growth factor and chemokine receptors destined for lysosomal degradation *Traffic* **8** 1617–29

Mallet J, Gorwood P, Le Strat Y and Dubertret C 2019 Major depressive disorder (MDD) and schizophrenia– addressing unmet needs with partial agonists at the D2 receptor: a review *Int. J. Neuropsychopharmacol.* **22** 651–64

Mallik R, Carter B C, Lex S A, King S J and Gross S P 2004 Cytoplasmic dynein functions as a gear in response to load *Nature* **427** 649–52

Manshouri R, Coyaud E, Kundu S T, Peng D H, Stratton S A, Alton K *et al* 2019 ZEB1/NuRD complex suppresses TBC1D2b to stimulate E-cadherin internalization and promote metastasis in lung cancer *Nat. Commun.* **10** 5125

Mao F, Yang Y and Jiang H 2021 Endocytosis and exocytosis protect cells against severe membrane tension variations *Biophys. J.* **120** 5521–9

Mao M, Thedens D R, Chang B, Harris B S, Zheng Q Y, Johnson K R *et al* 2009 The podosomal-adaptor protein SH3PXD2B is essential for normal postnatal development *Mamm Genome* **20** 462–75

Marchesin V, Castro-Castro A, Lodillinsky C, Castagnino A, Cyrta J, Bonsang-Kitzis H *et al* 2015 ARF6–JIP3/4 regulate endosomal tubules for MT1-MMP exocytosis in cancer invasion *J. Cell Biol.* **211** 339–58

Mari M, Bujny M V, Zeuschner D, Geerts W J C, Griffith J, Petersen C M *et al* 2008 SNX1 defines an early endosomal recycling exit for sortilin and mannose 6-phosphate receptors *Traffic* **9** 380–93

Markgraf D F, Peplowska K and Ungermann C 2007 Rab cascades and tethering factors in the endomembrane system *FEBS Lett.* **581** 2125–30

Markworth R, Dambeck V, Steinbeck L M, Koufali A, Bues B, Dankovich T M *et al* 2021 Tubular microdomains of Rab7-positive endosomes retrieve TrkA, a mechanism disrupted in Charcot–Marie–Tooth disease 2B *J. Cell Sci.* **134** jcs258559

Martin M, Iyadurai S J, Gassman A, Gindhart J G, Hays T S and Saxton W M 1999 Cytoplasmic dynein, the dynactin complex, and kinesin are interdependent and essential for fast axonal transport *Mol. Biol. Cell* **10** 3717–28

Martina J A, Chen Y, Gucek M and Puertollano R 2012 MTORC1 functions as a transcriptional regulator of autophagy by preventing nuclear transport of TFEB *Autophagy* **8** 903–14

Marwaha R, Arya S B, Jagga D, Kaur H, Tuli A and Sharma M 2017 The Rab7 effector PLEKHM1 binds Arl8b to promote cargo traffic to lysosomes *J. Cell Biol.* **216** 1051–70

Masters T A, Pontes B, Viasnoff V, Li Y and Gauthier N C 2013 Plasma membrane tension orchestrates membrane trafficking, cytoskeletal remodeling, and biochemical signaling during phagocytosis *Proc. Natl Acad. Sci. USA* **110** 11875–80

Masters T A, Tumbarello D A, Chibalina M V and Buss F 2017 MYO6 regulates spatial organization of signaling endosomes driving AKT activation and actin dynamics *Cell Rep.* **19** 2088–101

Matteoni R and Kreis T E 1987 Translocation and clustering of endosomes and lysosomes depends on microtubules *J. Cell Biol.* **105** 1253–65

Mátyási B, Farkas Z, Kopper L, Sebestyén A, Boissan M, Mehta A *et al* 2020 The function of NM23-H1/NME1 and its homologs in major processes linked to metastasis *Pathol. Oncol. Res.* **26** 49–61

Mayor S, Parton R G and Donaldson J G 2014 Clathrin-independent pathways of endocytosis *Cold Spring Harb. Perspect. Biol.* **6** a016758–a8

McKenney R J, Huynh W, Tanenbaum M E, Bhabha G and Vale R D 2014 Activation of cytoplasmic dynein motility by dynactin-cargo adapter complexes *Science* **345** 337–41

McKenney R J, Huynh W, Vale R D and Sirajuddin M 2016 Tyrosination of α-tubulin controls the initiation of processive dynein–dynactin motility *EMBO J.* **35** 1175–85

McMahon H T and Boucrot E 2011 Molecular mechanism and physiological functions of clathrin-mediated endocytosis *Nat. Rev. Mol. Cell Biol.* **12** 517–33

McVicker D P, Chrin L R and Berger C L 2011 The nucleotide-binding state of microtubules modulates kinesin processivity and the ability of Tau to inhibit kinesin-mediated transport *J. Biol. Chem.* **286** 42873–80

Medina D L, Fraldi A, Bouche V, Annunziata F, Mansueto G, Spampanato C *et al* 2011 Transcriptional activation of lysosomal exocytosis promotes cellular clearance *Dev. Cell* **21** 421–30

Melander M C, Jürgensen H J, Madsen D H, Engelholm L H and Behrendt N 2015 The collagen receptor uPARAP/Endo180 in tissue degradation and cancer (Review) *Int. J. Oncol.* **47** 1177–88

Mellman I 1996 Endocytosis and molecular sorting *Annu. Rev. Cell Dev. Biol.* **12** 575–625

Mellman I, Fuchs R and Helenius A 1986 Acidification of the endocytic and exocytic pathways *Annu. Rev. Biochem.* **55** 663–700

Mellman I and Yarden Y 2013 Endocytosis and cancer *Cold Spring Harb. Perspect. Biol.* **5** a016949–a9

Mendoza P, Ortiz R, Díaz J, Quest A F G, Leyton L, Stupack D *et al* 2013 Rab5 activation promotes focal adhesion disassembly, migration and invasiveness of tumor cells *J. Cell Sci.* **126** 3835–47

Mesaki K, Tanabe K, Obayashi M, Oe N and Takei K 2011 Fission of tubular endosomes triggers endosomal acidification and movement *PLoS One* **6** e19764

Métivier M, Monroy B Y, Gallaud E, Caous R, Pascal A, Richard-Parpaillon L *et al* 2019 Dual control of Kinesin-1 recruitment to microtubules by Ensconsin in *Drosophila* neuroblasts and oocytes *Development* **146** dev171579

Mezzadra R, Sun C, Jae L T, Gomez-Eerland R, De Vries E, Wu W *et al* 2017 Identification of CMTM6 and CMTM4 as PD-L1 protein regulators *Nature* **549** 106–10

Miaczynska M 2013 Effects of membrane trafficking on signaling by receptor tyrosine kinases *Cold Spring Harb. Perspect. Biol.* **5** a009035–a5

Miaczynska M, Christoforidis S, Giner A, Shevchenko A, Uttenweiler-Joseph S, Habermann B *et al* 2004 APPL proteins link Rab5 to nuclear signal transduction via an endosomal compartment *Cell* **116** 445–56

Miao B, Skidan I, Yang J, You Z, Fu X, Famulok M *et al* 2012 Inhibition of cell migration by PITENINs: the role of ARF6 *Oncogene* **31** 4317–32

Mierke C T, Puder S, Aermes C, Fischer T and Kunschmann T 2020 Effect of PAK inhibition on cell mechanics depends on Rac1 *Front. Cell Dev. Biol.* **8** 13

Mikhaylov G, Mikac U, Magaeva A A, Itin V I, Naiden E P, Psakhye I *et al* 2011 Ferri-liposomes as an MRI-visible drug-delivery system for targeting tumours and their microenvironment *Nat. Nanotech* **6** 594–602

Miller K D, Althouse S K, Nabell L, Rugo H, Carey L, Kimmick G *et al* 2014 A phase II study of medroxyprogesterone acetate in patients with hormone receptor negative metastatic breast cancer: translational breast cancer research consortium trial 007 *Breast Cancer Res Treat* **148** 99–106

Mitsuuchi Y, Johnson S W, Sonoda G, Tanno S, Golemis E A and Testa J R 1999 Identification of a chromosome 3p14.3-21.1 gene, APPL, encoding an adaptor molecule that interacts with the oncoprotein-serine/threonine kinase AKT2 *Oncogene* **18** 4891–8

Mittal N, Michels E B, Massey A E, Qiu Y, Royer-Weeden S P, Smith B R *et al* 2024 Myosin-independent stiffness sensing by fibroblasts is regulated by the viscoelasticity of flowing actin *Commun. Mater.* **5** 6

Miyake H, Hara I and Eto H 2004 Serum level of cathepsin B and its density in men with prostate cancer as novel markers of disease progression *Anticancer Res.* **24** 2573–7

Mizushima N, Levine B, Cuervo A M and Klionsky D J 2008 Autophagy fights disease through cellular self-digestion *Nature* **451** 1069–75

Mizuta R, LaSalle J M, Cheng H-L, Shinohara A, Ogawa H, Copeland N *et al* 1997 RAB22 and RAB163/mouse BRCA2: proteins that specifically interact with the RAD51 protein *Proc. Natl Acad. Sci. USA* **94** 6927–32

Monroy B Y, Sawyer D L, Ackermann B E, Borden M M, Tan T C and Ori-McKenney K M 2018 Competition between microtubule-associated proteins directs motor transport *Nat. Commun.* **9** 1487

Monroy B Y, Tan T C, Oclaman J M, Han J S, Simó S, Niwa S *et al* 2020 A combinatorial MAP code dictates polarized microtubule transport *Dev. Cell* **53** 60–72.e4

Mooren O L, Galletta B J and Cooper J A 2012 Roles for actin assembly in endocytosis *Annu. Rev. Biochem.* **81** 661–86

Moores C A, Perderiset M, Kappeler C, Kain S, Drummond D, Perkins S J *et al* 2006 Distinct roles of doublecortin modulating the microtubule cytoskeleton *EMBO J.* **25** 4448–57

Moreno-Layseca P, Icha J, Hamidi H and Ivaska J 2019 Integrin trafficking in cells and tissues *Nat. Cell Biol.* **21** 122–32

Morris S M, Arden S D, Roberts R C, Kendrick-Jones J, Cooper J A, Luzio J P *et al* 2002 Myosin VI binds to and localises with Dab2, potentially linking receptor-mediated endocytosis and the actin cytoskeleton *Traffic* **3** 331–41

Morrison Joly M, Williams M M, Hicks D J, Jones B, Sanchez V, Young C D *et al* 2017 Two distinct mTORC2-dependent pathways converge on Rac1 to drive breast cancer metastasis *Breast Cancer Res.* **19** 74

Mostowy S and Cossart P 2012 Septins: the fourth component of the cytoskeleton *Nat. Rev. Mol. Cell Biol.* **13** 183–94

Motzer R J, Escudier B, McDermott D F, George S, Hammers H J, Srinivas S *et al* 2015 Nivolumab versus everolimus in advanced renal-cell carcinoma *N. Engl. J. Med.* **373** 1803–13

Moughamian A J, Osborn G E, Lazarus J E, Maday S and Holzbaur E L F 2013 Ordered recruitment of dynactin to the microtubule plus-end is required for efficient initiation of retrograde axonal transport *J. Neurosci.* **33** 13190–203

Mowers E E, Sharifi M N and Macleod K F 2017 Autophagy in cancer metastasis *Oncogene* **36** 1619–30

Moya M, Dautry-Varsat A, Goud B, Louvard D and Boquet P 1985 Inhibition of coated pit formation in Hep2 cells blocks the cytotoxicity of diphtheria toxin but not that of ricin toxin *J. Cell Biol.* **101** 548–59

Mukherjea M, Ali M Y, Kikuti C, Safer D, Yang Z, Sirkia H *et al* 2014 Myosin VI must dimerize and deploy its unusual lever arm in order to perform its cellular roles *Cell Rep.* **8** 1522–32

Mukherjee S, Ghosh R N and Maxfield F R 1997 Endocytosis *Physiol. Rev.* **77** 759–803

Muller P A J, Caswell P T, Doyle B, Iwanicki M P, Tan E H, Karim S *et al* 2009 Mutant p53 drives invasion by promoting integrin recycling *Cell* **139** 1327–41

Mulligan R J and Winckler B 2023 Regulation of endosomal trafficking by rab7 and its effectors in neurons: clues from charcot–marie–tooth 2B disease *Biomolecules* **13** 1399

Murphy A J, Guyre P M and Pioli P A 2010 Estradiol suppresses NF-κB activation through coordinated regulation of let-7a and miR-125b in primary human macrophages *J. Immunol.* **184** 5029–37

Naccache S N, Hasson T and Horowitz A 2006 Binding of internalized receptors to the PDZ domain of GIPC/synectin recruits myosin VI to endocytic vesicles *Proc. Natl Acad. Sci. USA* **103** 12735–40

Nagano M, Toshima J Y, Siekhaus D E and Toshima J 2019 Rab5-mediated endosome formation is regulated at the trans-golgi network *Commun. Biol.* **2** 419

Nagaraj N S, Vigneswaran N and Zacharias W 2006 Cathepsin B mediates TRAIL-induced apoptosis in oral cancer cells *J. Cancer Res. Clin. Oncol.* **132** 171–83

Nakahara H, Howard L, Thompson E W, Sato H, Seiki M, Yeh Y *et al* 1997 Transmembrane/cytoplasmic domain-mediated membrane type 1-matrix metalloprotease docking to invadopodia is required for cell invasion *Proc. Natl Acad. Sci. USA* **94** 7959–64

Nallamothu G, Woolworth J A, Dammai V and Hsu T 2008 *awd*, the homolog of metastasis suppressor gene *Nm23*, regulates *Drosophila* epithelial cell invasion *Mol. Cell. Biol.* **28** 1964–73

Narendra D, Tanaka A, Suen D-F and Youle R J 2008 Parkin is recruited selectively to impaired mitochondria and promotes their autophagy *J. Cell Biol.* **183** 795–803

Nash J E, Appleby V J, Corrêa S A L, Wu H, Fitzjohn S M, Garner C C *et al* 2010 Disruption of the interaction between myosin VI and SAP97 is associated with a reduction in the number of AMPARs at hippocampal synapses *J. Neurochem.* **112** 677–90

Naslavsky N and Caplan S 2018 The enigmatic endosome—sorting the ins and outs of endocytic trafficking *J. Cell Sci.* **131** jcs216499

Nelson K K and Melendez J A 2004 Mitochondrial redox control of matrix metalloproteinases *Free Radical Biol. Med.* **37** 768–84

Neutra M R, Ciechanover A, Owen L S and Lodish H F 1985 Intracellular transport of transferrin- and asialoorosomucoid-colloidal gold conjugates to lysosomes after receptor-mediated endocytosis *J. Histochem. Cytochem.* **33** 1134–44

Newmyer S L and Schmid S L 2001 Dominant-interfering Hsc70 mutants disrupt multiple stages of the clathrin-coated vesicle cycle *in vivo J. Cell Biol.* **152** 607–20

Nielsen E, Severin F, Backer J M, Hyman A A and Zerial M 1999 Rab5 regulates motility of early endosomes on microtubules *Nat. Cell Biol.* **1** 376–82

Nikkilä R, Hirvonen E, Pitkäniemi J, Räsänen J, Malila N and Mäkitie A 2024 Risk of second primary cancer among patients with cardio-esophageal cancer in finland: a nationwide population-based study *Clin. Epidemiol.* **16** 475–85

Niu F, Li L, Wang L, Xiao J, Xu S, Liu Y *et al* 2024 Autoinhibition and activation of myosin VI revealed by its cryo-EM structure *Nat. Commun.* **15** 1187

Nomura T and Katunuma N 2005 Involvement of cathepsins in the invasion, metastasis and proliferation of cancer cells *J. Med. Invest.* **52** 1–9

Nordmann M, Cabrera M, Perz A, Bröcker C, Ostrowicz C, Engelbrecht-Vandré S *et al* 2010 The Mon1-Ccz1 complex is the GEF of the late endosomal Rab7 homolog Ypt7 *Curr. Biol.* **20** 1654–9

Norris S R, Soppina V, Dizaji A S, Schimert K I, Sept D, Cai D *et al* 2014 A method for multiprotein assembly in cells reveals independent action of kinesins in complex *J. Cell Biol.* **207** 393–406

Novick P and Zerial M 1997 The diversity of Rab proteins in vesicle transport *Curr. Opin. Cell Biol.* **9** 496–504

Novo D, Heath N, Mitchell L, Caligiuri G, MacFarlane A, Reijmer D *et al* 2018 Mutant p53s generate pro-invasive niches by influencing exosome podocalyxin levels *Nat. Commun.* **9** 5069

Odintsova E, Sugiura T and Berditchevski F 2000 Attenuation of EGF receptor signaling by a metastasis suppressor, the tetraspanin CD82/KAI-1 *Curr. Biol.* **10** 1009–12

Odintsova E, Voortman J, Gilbert E and Berditchevski F 2003 Tetraspanin CD82 regulates compartmentalisation and ligand-induced dimerization of EGFR *J. Cell Sci.* **116** 4557–66

O'Flanagan C H, Morais V A, Wurst W, De Strooper B and O'Neill C 2015 The Parkinson's gene PINK1 regulates cell cycle progression and promotes cancer-associated phenotypes *Oncogene* **34** 1363–74

Ohashi K G, Han L, Mentley B, Wang J, Fricks J and Hancock W O 2019 Load-dependent detachment kinetics plays a key role in bidirectional cargo transport by kinesin and dynein *Traffic* **20** 284–94

Oikawa T, Itoh T and Takenawa T 2008 Sequential signals toward podosome formation in NIH-src cells *J. Cell Biol.* **182** 157–69

Okamura H, Nishikiori M, Xiang H, Ishikawa M and Katoh E 2011 Interconversion of two GDP-bound conformations and their selection in an Arf-family Small G protein *Structure* **19** 988–98

Olenick M A, Dominguez R and Holzbaur E L F 2019 Dynein activator Hook1 is required for trafficking of BDNF-signaling endosomes in neurons *J. Cell Biol.* **218** 220–33

Olenick M A, Tokito M, Boczkowska M, Dominguez R and Holzbaur E L F 2016 Hook adaptors induce unidirectional processive motility by enhancing the dynein-dynactin interaction *J. Biol. Chem.* **291** 18239–51

O'Loughlin T, Masters T A and Buss F 2018 The MYO6 interactome reveals adaptor complexes coordinating early endosome and cytoskeletal dynamics *EMBO Rep.* **19** e44884

Oren M and Rotter V 2010 Mutant p53 gain-of-function in cancer *Cold Spring Harb. Perspect. Biol.* **2** a001107–a7

Ørom U A, Nielsen F C and Lund A H 2008 MicroRNA-10a binds the 5′UTR of ribosomal protein mRNAs and enhances their translation *Mol. Cell* **30** 460–71

Osterweil E, Wells D G and Mooseker M S 2005 A role for myosin VI in postsynaptic structure and glutamate receptor endocytosis *J. Cell Biol.* **168** 329–38

O'Sullivan M J and Lindsay A J 2020 The endosomal recycling pathway—at the crossroads of the cell *IJMS* **21** 6074

Otto G 2019 May the force be with your lipids *Nat. Rev. Mol. Cell Biol.* **20** 196–7

OuYang L-Y, Wu X-J, Ye S-B, Zhang R, Li Z-L, Liao W *et al* 2015 Tumor-induced myeloid-derived suppressor cells promote tumor progression through oxidative metabolism in human colorectal cancer *J. Transl. Med.* **13** 47

Ow K, Delprado W, Fisher R, Barrett J, Yu Y, Jackson P *et al* 2000 Relationship between expression of the KAI1 metastasis suppressor and other markers of advanced bladder cancer *J. Pathol.* **191** 39–47

Palacios F 2001 An essential role for ARF6-regulated membrane traffic in adherens junction turnover and epithelial cell migration *EMBO J.* **20** 4973–86

Palacios F, Schweitzer J K, Boshans R L and D'Souza-Schorey C 2002 ARF6-GTP recruits Nm23-H1 to facilitate dynamin-mediated endocytosis during adherens junctions disassembly *Nat. Cell Biol.* **4** 929–36

Pálfy M, Reményi A and Korcsmáros T 2012 Endosomal crosstalk: meeting points for signaling pathways *Trends Cell Biol.* **22** 447–56

Palmieri D, Halverson D O, Ouatas T, Horak C E, Salerno M, Johnson J *et al* 2005 Medroxyprogesterone acetate elevation of Nm23-H1 metastasis suppressor expression in hormone receptor–negative breast cancer *J. Natl Cancer Inst.* **97** 632–42

Panda P K, Mukhopadhyay S, Das D N, Sinha N, Naik P P and Bhutia S K 2015 Mechanism of autophagic regulation in carcinogenesis and cancer therapeutics *Semin. Cell Dev. Biol.* **39** 43–55

Park J H and Roll-Mecak A 2018 The tubulin code in neuronal polarity *Curr. Opin. Neurobiol.* **51** 95–102

Park J-S, Lee I-B, Moon H-M, Hong S-C and Cho M 2023 Long-term cargo tracking reveals intricate trafficking through active cytoskeletal networks in the crowded cellular environment *Nat. Commun.* **14** 7160

Parton R G and Simons K 2007 The multiple faces of caveolae *Nat. Rev. Mol. Cell Biol.* **8** 185–94

Paschall A V, Zimmerman M A, Torres C M, Yang D, Chen M R, Li X *et al* 2014 Ceramide targets xIAP and cIAP1 to sensitize metastatic colon and breast cancer cells to apoptosis induction to suppress tumor progression *BMC Cancer* **14** 24

Pasini F S, Maistro S, Snitcovsky I, Barbeta L P, Rotea Mangone F R, Lehn C N *et al* 2012 Four-gene expression model predictive of lymph node metastases in oral squamous cell carcinoma *Acta Oncol.* **51** 77–85

Pataer A, Ozpolat B, Shao R, Cashman N R, Plotkin S S, Samuel C E *et al* 2020 Therapeutic targeting of the PI4K2A/PKR lysosome network is critical for misfolded protein clearance and survival in cancer cells *Oncogene* **39** 801–13

Patel N M, Ripoll L, Peach C J, Ma N, Blythe E E, Vaidehi N *et al* 2024 Myosin VI drives arrestin-independent internalization and signaling of GPCRs *Nat. Commun.* **15** 10636

Pavlakis E and Stiewe T 2020 p53's extended reach: the mutant p53 secretome *Biomolecules* **10** 307

Pedersen N M, Wenzel E M, Wang L, Antoine S, Chavrier P, Stenmark H *et al* 2020 Protrudin-mediated ER–endosome contact sites promote MT1-MMP exocytosis and cell invasion *J. Cell Biol.* **219** e202003063

Pelkmans L, Kartenbeck J and Helenius A 2001 Caveolar endocytosis of simian virus 40 reveals a new two-step vesicular-transport pathway to the ER *Nat. Cell Biol.* **3** 473–83

Pellinen T, Arjonen A, Vuoriluoto K, Kallio K, Fransen J A M and Ivaska J 2006 Small GTPase Rab21 regulates cell adhesion and controls endosomal traffic of β1-integrins *J. Cell Biol.* **173** 767–80

Pereira-Leal J B and Seabra M C 2001 Evolution of the rab family of small GTP-binding proteins *J. Mol. Biol.* **313** 889–901

Phichith D, Travaglia M, Yang Z, Liu X, Zong A B, Safer D *et al* 2009 Cargo binding induces dimerization of myosin VI *Proc. Natl Acad. Sci. USA* **106** 17320–4

Piao S and Amaravadi R K 2016 Targeting the lysosome in cancer *Ann. N.Y. Acad. Sci.* **1371** 45–54

Pietilä M, Sahgal P, Peuhu E, Jäntti N Z, Paatero I, Närvä E *et al* 2019 SORLA regulates endosomal trafficking and oncogenic fitness of HER2 *Nat. Commun.* **10** 2340

Podgorski I, Linebaugh B E, Sameni M, Jedeszko C, Bhagat S, Cher M L *et al* 2005 Bone microenvironment modulates expression and activity of cathepsin B in prostate cancer *Neoplasia* **7** 207–23

Podgorski I and Sloane B F 2003 Cathepsin B and its role(s) in cancer progression *Biochem. Soc. Symp.* **70** 263–76

Poteryaev D, Datta S, Ackema K, Zerial M and Spang A 2010 Identification of the switch in early-to-late endosome transition *Cell* **141** 497–508

Prakash S, Malshikare H and Sengupta D 2022 Molecular mechanisms underlying Caveolin-1 mediated membrane curvature *J. Membr. Biol.* **255** 225 236

Puri C 2009 Loss of myosin VI No insert isoform (NoI) induces a defect in clathrin-mediated endocytosis and leads to caveolar endocytosis of transferrin receptor *J. Biol. Chem.* **284** 34998–5014

Puri C 2010 Effects of loss of myosin VI No-insert isoform on clathrin-mediated endocytosis of plasma-membrane receptors *Commun. Integr. Biol.* **3** 234–7

Pyrpassopoulos S, Shuman H and Ostap E M 2020 Modulation of kinesin's load-bearing capacity by force geometry and the microtubule track *Biophys. J.* **118** 243–53

Qian F, Bajkowski A S, Steiner D F, Chan S J and Frankfater A 1989 Expression of five cathepsins in murine melanomas of varying metastatic potential and normal tissues *Cancer Res.* **49** 4870–5

Qin Y, Wang G, Chen L, Sun Y, Yang J, Piao Y *et al* 2024 High-throughput screening of surface engineered cyanine nanodots for active transport of therapeutic antibodies into solid tumor *Adv. Mater.* **36** 2302292

Qiu X, Li Y, Wang Y, Gong X, Wang Y and Pan L 2023 Mechanistic insights into the interactions of Arl8b with the RUN domains of PLEKHM1 and SKIP *J. Mol. Biol.* **435** 168293

Qualmann B, Koch D and Kessels M M 2011 Let's go bananas: revisiting the endocytic BAR code: let's go bananas: revisiting the endocytic BAR code *EMBO J.* **30** 3501–15

Qualmann B and Mellor H 2003 Regulation of endocytic traffic by Rho GTPases *Biochem.* J. **371** 233–41

Quest A F G, Gutierrez-Pajares J L and Torres V A 2008 Caveolin-1: an ambiguous partner in cell signalling and cancer *J Cell. Mol. Med.* **12** 1130–50

Quest A F G, Leyton L and Párraga M 2004 Caveolins, caveolae, and lipid rafts in cellular transport, signaling, and disease *Biochem. Cell Biol.* **82** 129–44

Rabas N, Palmer S, Mitchell L, Ismail S, Gohlke A, Riley J S *et al* 2021 PINK1 drives production of mtDNA-containing extracellular vesicles to promote invasiveness *J. Cell Biol.* **220** e202006049

Rafn B, Nielsen C F, Andersen S H, Szyniarowski P, Corcelle-Termeau E, Valo E *et al* 2012 ErbB2-driven breast cancer cell invasion depends on a complex signaling network activating myeloid zinc finger-1-dependent cathepsin B expression *Mol. Cell* **45** 764–76

Rai A K, Rai A, Ramaiya A J, Jha R and Mallik R 2013 Molecular adaptations allow dynein to generate large collective forces inside cells *Cell* **152** 172–82

Rai A, Pathak D, Thakur S, Singh S, Dubey A K and Mallik R 2016 Dynein clusters into lipid microdomains on phagosomes to drive rapid transport toward lysosomes *Cell* **164** 722–34

Rai A, Shrivastava R, Vang D, Ritt M, Sadler F, Bhaban S *et al* 2022 Multimodal regulation of myosin VI ensemble transport by cargo adaptor protein GIPC *J. Biol. Chem.* **298** 101688

Rai A, Vang D, Ritt M and Sivaramakrishnan S 2021 Dynamic multimerization of Dab2–Myosin VI complexes regulates cargo processivity while minimizing cortical actin reorganization *J. Biol. Chem.* **296** 100232

Raiborg C 2001 Hrs recruits clathrin to early endosomes *EMBO J.* **20** 5008–21

Raiborg C, Malerød L, Pedersen N M and Stenmark H 2008 Differential functions of Hrs and ESCRT proteins in endocytic membrane trafficking *Exp. Cell. Res.* **314** 801–13

Raiborg C and Stenmark H 2009 The ESCRT machinery in endosomal sorting of ubiquitylated membrane proteins *Nature* **458** 445–52

Rainero E, Howe J D, Caswell P T, Jamieson N B, Anderson K, Critchley D R *et al* 2015 Ligand-occupied integrin internalization links nutrient signaling to invasive migration *Cell Rep.* **10** 398–413

Rajawat Y S, Hilioti Z and Bossis I 2009 Aging: central role for autophagy and the lysosomal degradative system *Age. Res. Rev.* **8** 199–213

Ramkumar A, Jong B Y and Ori-McKenney K M 2018 ReMAPping the microtubule landscape: how phosphorylation dictates the activities of microtubule-associated proteins *Dev. Dyn.* **247** 138–55

Ramsay A G, Marshall J F and Hart I R 2007 Integrin trafficking and its role in cancer metastasis *Cancer Metastasis Rev.* **26** 567

Raucher D and Sheetz M P 1999 Membrane expansion increases endocytosis rate during mitosis *J. Cell Biol.* **144** 497–506

Raudenska M, Balvan J and Masarik M 2021 Crosstalk between autophagy inhibitors and endosome-related secretory pathways: a challenge for autophagy-based treatment of solid cancers *Mol. Cancer* **20** 140

Ravetch J V and Clynes R A 1998 Divergent roles for fC receptors and complement *in vivo Annu. Rev. Immunol.* **16** 421–32

Razi M, Chan E Y W and Tooze S A 2009 Early endosomes and endosomal coatomer are required for autophagy *J. Cell Biol.* **185** 305–21

Reck-Peterson S L, Redwine W B, Vale R D and Carter A P 2018 The cytoplasmic dynein transport machinery and its many cargoes *Nat. Rev. Mol. Cell Biol.* **19** 382–98

Redpath G M I and Ananthanarayanan V 2023 Endosomal sorting sorted—motors, adaptors and lessons from *in vitro* and cellular studies *J. Cell Sci.* **136** jcs260749

Redpath G M I, Betzler V M, Rossatti P and Rossy J 2020 Membrane heterogeneity controls cellular endocytic trafficking *Front. Cell Dev. Biol.* **8** 757

Reed N A, Cai D, Blasius T L, Jih G T, Meyhofer E, Gaertig J *et al* 2006 Microtubule acetylation promotes kinesin-1 binding and transport *Curr. Biol.* **16** 2166–72

Reis C R, Chen P, Srinivasan S, Aguet F, Mettlen M and Schmid S L 2015 Crosstalk between Akt/GSK 3β signaling and dynamin-1 regulates clathrin-mediated endocytosis *EMBO J.* **34** 2132–46

Renard H-F, Tyckaert F, Lo Giudice C, Hirsch T, Valades-Cruz C A, Lemaigre C *et al* 2020 Endophilin-A3 and Galectin-8 control the clathrin-independent endocytosis of CD166 *Nat. Commun.* **11** 1457

Repnik U, Cesen M H and Turk B 2013 The endolysosomal system in cell death and survival *Cold Spring Harb. Perspect. Biol.* **5** a008755–a5

Repnik U, Stoka V, Turk V and Turk B 2012 Lysosomes and lysosomal cathepsins in cell death *Biochim. Biophys. Acta (BBA)—Proteins Proteom.* **1824** 22–33

Ridley A J, Paterson H F, Johnston C L, Diekmann D and Hall A 1992 The small GTP-binding protein rac regulates growth factor-induced membrane ruffling *Cell* **70** 401–10

Rink J, Ghigo E, Kalaidzidis Y and Zerial M 2005 Rab conversion as a mechanism of progression from early to late endosomes *Cell* **122** 735–49

Rizo J 2018 Mechanism of neurotransmitter release coming into focus *Protein Sci.* **27** 1364–91

Robinson D G and Pimpl P 2014 Clathrin and post-golgi trafficking: a very complicated issue *Trends Plant Sci.* **19** 134–9

Rojas R, Van Vlijmen T, Mardones G A, Prabhu Y, Rojas A L, Mohammed S *et al* 2008 Regulation of retromer recruitment to endosomes by sequential action of Rab5 and Rab7 *J. Cell Biol.* **183** 513–26

Rosa-Ferreira C and Munro S 2011 Arl8 and SKIP act together to link lysosomes to kinesin-1 *Dev. Cell* **21** 1171–8

Ross J L 2012 The impacts of molecular motor traffic jams *Proc. Natl Acad. Sci. USA* **109** 5911–2

Rottner K, Faix J, Bogdan S, Linder S and Kerkhoff E 2017 Actin assembly mechanisms at a glance *J. Cell Sci.* **130** 3427–35

Roux A, Uyhazi K, Frost A and De Camilli P 2006 GTP-dependent twisting of dynamin implicates constriction and tension in membrane fission *Nature* **441** 528–31

Ruckhäberle E, Holtrich U, Engels K, Hanker L, Gätje R, Metzler D *et al* 2009 Acid ceramidase 1 expression correlates with a better prognosis in ER-positive breast cancer *Climacteric* **12** 502–13

Russell M R G, Shideler T, Nickerson D P, West M and Odorizzi G 2012 Class E compartments form in response to ESCRT dysfunction in yeast due to hyperactivity of the Vps21 Rab GTPase *J. Cell Sci.* **125** 5208–20

Sachse M, Urbé S, Oorschot V, Strous G J and Klumperman J 2002 Bilayered clathrin coats on endosomal vacuoles are involved in protein sorting toward lysosomes *Mol. Biol. Cell* **13** 1313–28

Saftig P and Klumperman J 2009 Lysosome biogenesis and lysosomal membrane proteins: trafficking meets function *Nat. Rev. Mol. Cell Biol.* **10** 623–35

Saini P and Courtneidge S A 2018 Tks adaptor proteins at a glance *J. Cell Sci.* **131** jcs203661

Salerno M, Ouatas T, Palmieri D and Steeg P S 2003 Inhibition of signal transduction by the nm23 metastasis suppressor: possible mechanisms *Clin. Exp. Metastasis* **20** 3–10

Salogiannis J and Reck-Peterson S L 2017 Hitchhiking: a non-canonical mode of microtubule-based transport *Trends Cell Biol.* **27** 141–50

Sanderson R D, Bandari S K and Vlodavsky I 2019 Proteases and glycosidases on the surface of exosomes: newly discovered mechanisms for extracellular remodeling *Matrix Biol.* **75–76** 160–9

Sardiello M, Palmieri M, Di Ronza A, Medina D L, Valenza M, Gennarino V A *et al* 2009 A gene network regulating lysosomal biogenesis and function *Science* **325** 473–7

Saro D, Li T, Rupasinghe C, Paredes A, Caspers N and Spaller M R 2007 A Thermodynamic ligand binding study of the third PDZ domain (PDZ3) from the mammalian neuronal protein PSD-95 *Biochemistry* **46** 6340–52

Sato H, Takino T, Okada Y, Cao J, Shinagawa A, Yamamoto E *et al* 1994 A matrix metalloproteinase expressed on the surface of invasive tumour cells *Nature* **370** 61–5
Saurabh Mogre S, Brown A I and Koslover E F 2020 Getting around the cell: physical transport in the intracellular world *Phys. Biol.* **17** 061003
Schenck A, Goto-Silva L, Collinet C, Rhinn M, Giner A, Habermann B *et al* 2008 The endosomal protein Appl1 mediates Akt substrate specificity and cell survival in vertebrate development *Cell* **133** 486–97
Schiefermeier N, Scheffler J M, De Araujo M E G, Stasyk T, Yordanov T, Ebner H L *et al* 2014 The late endosomal p14–MP1 (LAMTOR2/3) complex regulates focal adhesion dynamics during cell migration *J. Cell Biol.* **205** 525–40
Schink K O, Tan K W, Spangenberg H, Martorana D, Sneeggen M, Stévenin V *et al* 2021 The phosphoinositide coincidence detector Phafin2 promotes macropinocytosis by coordinating actin organisation at forming macropinosomes *Nat. Commun.* **12** 6577
Schlager M A, Hoang H T, Urnavicius L, Bullock S L and Carter A P 2014 *In vitro* reconstitution of a highly processive recombinant human dynein complex *EMBO J.* **33** 1855–68
Schlienger S, Campbell S, Pasquin S, Gaboury L and Claing A 2016 ADP-ribosylation factor 1 expression regulates epithelial-mesenchymal transition and predicts poor clinical outcome in triple-negative breast cancer *Oncotarget* **7** 15811–27
Schlienger S, Ramirez R A M and Claing A 2015 ARF1 regulates adhesion of MDA-MB-231 invasive breast cancer cells through formation of focal adhesions *Cell. Signal.* **27** 403–15
Schmid S L 2017 Reciprocal regulation of signaling and endocytosis: implications for the evolving cancer cell *J. Cell Biol.* **216** 2623–32
Schmid S L and Frolov V A 2011 Dynamin: functional design of a membrane fission catalyst *Annu. Rev. Cell Dev. Biol.* **27** 79–105
Schmitter T, Agerer F, Peterson L, Münzner P and Hauck C R 2004 Granulocyte CEACAM3 is a phagocytic receptor of the innate immune system that mediates recognition and elimination of human-specific pathogens *J. Exp. Med.* **199** 35–46
Schnitzer M J, Visscher K and Block S M 2000 Force production by single kinesin motors *Nat. Cell Biol.* **2** 718–23
Schroeder C M, Ostrem J M, Hertz N T and Vale R D 2014 A Ras-like domain in the light intermediate chain bridges the dynein motor to a cargo-binding region *eLife* **3** e03351
Schuster M, Lipowsky R, Assmann M-A, Lenz P and Steinberg G 2011 Transient binding of dynein controls bidirectional long-range motility of early endosomes *Proc. Natl Acad. Sci. USA* **108** 3618–23
Scott C C, Vacca F and Gruenberg J 2014 Endosome maturation, transport and functions *Semin. Cell Dev. Biol.* **31** 2–10
Settembre C, De Cegli R, Mansueto G, Saha P K, Vetrini F, Visvikis O *et al* 2013 TFEB controls cellular lipid metabolism through a starvation-induced autoregulatory loop *Nat. Cell Biol.* **15** 647–58
Settembre C, Di Malta C, Polito V A, Arencibia M G, Vetrini F, Erdin S *et al* 2011 TFEB links autophagy to lysosomal biogenesis *Science* **332** 1429–33
Seugnet L, Simpson P and Haenlin M 1997 Requirement for dynamin during notch signaling indrosophilaneurogenesis *Dev. Biol.* **192** 585–98
Seven D, Dogan S, Kiliç E, Karaman E, Koseoglu H and Buyru N 2015 Downregulation of Rab25 activates Akt1 in head and neck squamous cell carcinoma *Oncol. Lett.* **10** 1927–31
Sevenich L and Joyce J A 2014 Pericellular proteolysis in cancer *Genes Dev* **28** 2331–47

Shang G, Brautigam C A, Chen R, Lu D, Torres-Vázquez J and Zhang X 2017 Structure analyses reveal a regulated oligomerization mechanism of the PlexinD1/GIPC/myosin VI complex *eLife* **6** e27322

Shang M, Lim S B, Jiang K, Yap Y S, Khoo B L, Han J *et al* 2021 Microfluidic studies of hydrostatic pressure-enhanced doxorubicin resistance in human breast cancer cells *Lab Chip* **21** 746–54

Shapiro D A, Renock S, Arrington E, Chiodo L A, Liu L-X, Sibley D R *et al* 2003 Aripiprazole, a novel atypical antipsychotic drug with a unique and robust pharmacology *Neuropsychopharmacology* **28** 1400–11

Sharma M and Dwivedi D 2019 A CRACker of an adaptor connects dynein-mediated transport to calcium signaling *J. Cell Biol.* **218** 1429–31

Shay G, Lynch C C and Fingleton B 2015 Moving targets: emerging roles for MMPs in cancer progression and metastasis *Matrix Biol.* **44–46** 200–6

Sheetz M P 2001 Cell control by membrane–cytoskeleton adhesion *Nat. Rev. Mol. Cell Biol.* **2** 392–6

Shi L, Tan X, Liu X, Yu J, Bota-Rabassedas N, Niu Y *et al* 2021 Addiction to Golgi-resident PI4P synthesis in chromosome 1q21.3–amplified lung adenocarcinoma cells *Proc. Natl Acad. Sci. USA* **118** e2023537118

Shi Z and Baumgart T 2015 Membrane tension and peripheral protein density mediate membrane shape transitions *Nat. Commun.* **6** 5974

Shibata M, Kanamori S, Isahara K, Ohsawa Y, Konishi A, Kametaka S *et al* 1998 Participation of cathepsins B and D in apoptosis of PC12 cells following serum deprivation *Biochem. Biophys. Res. Commun.* **251** 199–203

Shimobayashi M and Hall M N 2014 Making new contacts: the mTOR network in metabolism and signalling crosstalk *Nat. Rev. Mol. Cell Biol.* **15** 155–62

Shimoda M and Khokha R 2013 Proteolytic factors in exosomes *Proteomics* **13** 1624–36

Shimoda M and Khokha R 2017 Metalloproteinases in extracellular vesicles *Biochim. Biophys. Acta (BBA)—Mol. Cell Res.* **1864** 1989–2000

Shin K, Song S, Song Y H, Hahn S, Kim J-H, Lee G *et al* 2019 Anomalous dynamics of *in vivo* cargo delivery by motor protein multiplexes *J. Phys. Chem. Lett.* **10** 3071–9

Shubeita G T, Tran S L, Xu J, Vershinin M, Cermelli S, Cotton S L *et al* 2008 Consequences of motor copy number on the intracellular transport of kinesin-1-driven lipid droplets *Cell* **135** 1098–107

Siddiqui N, Zwetsloot A J, Bachmann A, Roth D, Hussain H, Brandt J *et al* 2019 PTPN21 and Hook3 relieve KIF1C autoinhibition and activate intracellular transport *Nat. Commun.* **10** 2693

Sigismund S, Argenzio E, Tosoni D, Cavallaro E, Polo S and Di Fiore P P 2008 Clathrin-mediated internalization is essential for sustained EGFR signaling but dispensable for degradation *Dev. Cell* **15** 209–19

Sigismund S, Confalonieri S, Ciliberto A, Polo S, Scita G and Di Fiore P P 2012 Endocytosis and signaling: cell logistics shape the eukaryotic cell plan *Physiol. Rev.* **92** 273–366

Silva P, Mendoza P, Rivas S, Díaz J, Moraga C, Quest A F G *et al* 2016 Hypoxia promotes Rab5 activation, leading to tumor cell migration, invasion and metastasis *Oncotarget* **7** 29548–62

Sirajuddin M, Rice L M and Vale R D 2014 Regulation of microtubule motors by tubulin isotypes and post-translational modifications *Nat. Cell Biol.* **16** 335–44

Sloane B F, Rozhin J, Hatfield J S, Crissman J D and Honn K V 1987 Plasma membrane-associated cysteine proteinases in human and animal tumors *Pathobiology* **55** 209–24

Smith P G, Roy C, Zhang Y N and Chauduri S 2003 Mechanical stress increases rhoa activation in airway smooth muscle cells *Am. J. Respir. Cell Mol. Biol.* **28** 436–42

Sneeggen M, Pedersen N M, Campsteijn C, Haugsten E M, Stenmark H and Schink K O 2019 WDFY2 restrains matrix metalloproteinase secretion and cell invasion by controlling VAMP3-dependent recycling *Nat. Commun.* **10** 2850

Solinger J A, Rashid H-O, Prescianotto-Baschong C and Spang A 2020 FERARI is required for Rab11-dependent endocytic recycling *Nat. Cell Biol.* **22** 213–24

Soppina V, Rai A K, Ramaiya A J, Barak P and Mallik R 2009 Tug-of-war between dissimilar teams of microtubule motors regulates transport and fission of endosomes *Proc. Natl Acad. Sci. USA* **106** 19381–6

Sorkin A 2004 Cargo recognition during clathrin-mediated endocytosis: a team effort *Curr. Opin. Cell Biol.* **16** 392–9

Sorkin A and Von Zastrow M 2009 Endocytosis and signalling: intertwining molecular networks *Nat. Rev. Mol. Cell Biol.* **10** 609–22

Sozański K, Ruhnow F, Wiśniewska A, Tabaka M, Diez S and Hołyst R 2015 Small crowders slow down kinesin-1 stepping by hindering motor domain diffusion *Phys. Rev. Lett.* **115** 218102

Splinter D, Razafsky D S, Schlager M A, Serra-Marques A, Grigoriev I, Demmers J *et al* 2012 BICD2, dynactin, and LIS1 cooperate in regulating dynein recruitment to cellular structures *Mol. Biol. Cell* **23** 4226–41

Sposini S, Jean-Alphonse F G, Ayoub M A, Oqua A, West C, Lavery S *et al* 2017 Integration of GPCR signaling and sorting from very early endosomes via opposing APPL1 mechanisms *Cell Rep.* **21** 2855–67

Spudich G, Chibalina M V, Au J S-Y, Arden S D, Buss F and Kendrick-Jones J 2007 Myosin VI targeting to clathrin-coated structures and dimerization is mediated by binding to Disabled-2 and PtdIns(4,5)P2 *Nat. Cell Biol.* **9** 176–83

Spudich J A and Sivaramakrishnan S 2010 Myosin VI: an innovative motor that challenged the swinging lever arm hypothesis *Nat. Rev. Mol. Cell Biol.* **11** 128–37

Srikanth S, Kim K-D, Gao Y, Woo J S, Ghosh S, Calmettes G *et al* 2016 A large Rab GTPase encoded by *CRACR2A* is a component of subsynaptic vesicles that transmit T cell activation signals *Sci. Signal.* **9** ra31

Stenmark H 2009 Rab GTPases as coordinators of vesicle traffic *Nat. Rev. Mol. Cell Biol.* **10** 513–25

Stoeck A, Keller S, Riedle S, Sanderson M P, Runz S, Le Naour F *et al* 2006 A role for exosomes in the constitutive and stimulus-induced ectodomain cleavage of L1 and CD44 *Biochem.* J **393** 609–18

Strouhalova K, Tolde O, Rosel D and Brábek J 2023 Cytoplasmic tail of MT1-MMP: a hub of MT1-MMP regulation and function *IJMS* **24** 5068

Stylli S S, I S T T, Kaye A H and Lock P 2012 Prognostic significance of Tks5 expression in gliomas *J. Clin. Neurosci.* **19** 436–42

Stylli S S, I, S T T, Verhagen A M, Xu S S, Pass I, Courtneidge S A *et al* 2009 Nck adaptor proteins link Tks5 to invadopodia actin regulation and ECM degradation *J. Cell Sci.* **122** 2727–40

Suh J, Wirtz D and Hanes J 2003 Efficient active transport of gene nanocarriers to the cell nucleus *Proc. Natl Acad. Sci. USA* **100** 3878–82

Sun L, Xu X, Chen Y, Zhou Y, Tan R, Qiu H *et al* 2018 Rab34 regulates adhesion, migration, and invasion of breast cancer cells *Oncogene* **37** 3698–714

Sun M, Northup N, Marga F, Huber T, Byfield F J, Levitan I *et al* 2007 The effect of cellular cholesterol on membrane-cytoskeleton adhesion *J. Cell Sci.* **120** 2223–31

Sun Z and Brodsky J L 2019 Protein quality control in the secretory pathway *J. Cell Biol.* **218** 3171–87

Svoboda K and Block S M 1994 Force and velocity measured for single kinesin molecules *Cell* **77** 773–84

Swaminathan R, Hoang C P and Verkman A S 1997 Photobleaching recovery and anisotropy decay of green fluorescent protein GFP-S65T in solution and cells: cytoplasmic viscosity probed by green fluorescent protein translational and rotational diffusion *Biophys. J.* **72** 1900–7

Swanson J A 2008 Shaping cups into phagosomes and macropinosomes *Nat. Rev. Mol. Cell Biol.* **9** 639–49

Taguchi T, Xu L, Kobayashi H, Taniguchi A, Kataoka K and Tanaka J 2005 Encapsulation of chondrocytes in injectable alkali-treated collagen gels prepared using poly(ethylene glycol)-based 4-armed star polymer *Biomaterials* **26** 1247–52

Takahashi S, Kubo K, Waguri S, Yabashi A, Shin H-W, Katoh Y *et al* 2012 Rab11 regulates exocytosis of recycling vesicles at the plasma membrane *J. Cell Sci.* **125** 4049–57

Takei K and Haucke V 2001 Clathrin-mediated endocytosis: membrane factors pull the trigger *Trends Cell Biol.* **11** 385–91

Takei K, Slepnev V I, Haucke V and De Camilli P 1999 Functional partnership between amphiphysin and dynamin in clathrin-mediated endocytosis *Nat. Cell Biol.* **1** 33–9

Tan R, Lam A J, Tan T, Han J, Nowakowski D W, Vershinin M *et al* 2019 Microtubules gate tau condensation to spatially regulate microtubule functions *Nat. Cell Biol.* **21** 1078–85

Tan X, Banerjee P, Pham E A, Rutaganira F U N, Basu K, Bota-Rabassedas N *et al* 2020 PI4KIIIβ is a therapeutic target in chromosome 1q–amplified lung adenocarcinoma *Sci. Transl. Med.* **12** eaax3772

Tan X, Banerjee P, Shi L, Xiao G-Y, Rodriguez B L, Grzeskowiak C L *et al* 2021a p53 loss activates prometastatic secretory vesicle biogenesis in the Golgi *Sci. Adv.* **7** eabf4885

Tan X, Shi L, Banerjee P, Liu X, Guo H-F, Yu J *et al* 2021b A protumorigenic secretory pathway activated by p53 deficiency in lung adenocarcinoma *J. Clin. Invest.* **131** e137186

Tan X, Xiao G-Y, Wang S, Shi L, Zhao Y, Liu X *et al* 2023 EMT-activated secretory and endocytic vesicular trafficking programs underlie a vulnerability to PI4K2A antagonism in lung cancer *J. Clin. Invest.* **133** e165863

Tang D, Zhang Y, Mei J, Zhao J, Miao C and Jiu Y 2023 Interactive mechanisms between caveolin-1 and actin filaments or vimentin intermediate filaments instruct cell mechanosensing and migration *J. Mol. Cell. Biol.* **14** mjac066

Taniguchi M, Ogiso H, Takeuchi T, Kitatani K, Umehara H and Okazaki T 2015 Lysosomal ceramide generated by acid sphingomyelinase triggers cytosolic cathepsin B-mediated degradation of *X*-linked inhibitor of apoptosis protein in natural killer/T lymphoma cell apoptosis *Cell Death Dis.* **6** e1717–7

Tao W, Moore R, Smith E R and Xu X-X 2016 Endocytosis and physiology: insights from disabled-2 deficient mice *Front. Cell Dev. Biol.* **4** 129

Thankachan J M and Setty S R G 2022 KIF13A—a key regulator of recycling endosome dynamics *Front. Cell Dev. Biol.* **10** 877532

Thottacherry J J, Sathe M, Prabhakara C and Mayor S 2019 Spoiled for choice: diverse endocytic pathways function at the cell surface *Annu. Rev. Cell Dev. Biol.* **35** 55–84

Tirumala N A and Ananthanarayanan V 2020 Role of dynactin in the intracellular localization and activation of cytoplasmic dynein *Biochemistry* **59** 156–62

Tirumala N A, Redpath G, Skerhut S V, Dolai P, Kapoor-Kaushik N, Ariotti N *et al* 2021 Single-molecule imaging of cytoplasmic dynein in cellulo reveals the mechanism of motor activation and cargo movement bioRxiv:2021.04.05.438428 https://doi.org/10.1101/2021.04.05.438428

Tjelle T E, Løvdal T and Berg T 2000 Phagosome dynamics and function *Bioessays* **22** 255–63

Tomas A, Futter C E and Eden E R 2014 EGF receptor trafficking: consequences for signaling and cancer *Trends Cell Biol.* **24** 26–34

Tong D, Liang Y-N, Stepanova A, Liu Y, Li X, Wang L *et al* 2017 Increased Eps15 homology domain 1 and RAB11FIP3 expression regulate breast cancer progression via promoting epithelial growth factor receptor recycling *Tumour Biol.* **39** 101042831769101

Torisawa T, Ichikawa M, Furuta A, Saito K, Oiwa K, Kojima H *et al* 2014 Autoinhibition and cooperative activation mechanisms of cytoplasmic dynein *Nat. Cell Biol.* **16** 1118–24

Toth M, Osenkowski P, Hesek D, Brown S, Meroueh S, Sakr W *et al* 2005 Cleavage at the stem region releases an active ectodomain of the membrane type 1 matrix metalloproteinase *Biochem. J.* **387** 497–506

Toth M, Sohail A, Mobashery S and Fridman R 2006 MT1-MMP shedding involves an ADAM and is independent of its localization in lipid rafts *Biochem. Biophys. Res. Commun.* **350** 377–84

Traer C J, Rutherford A C, Palmer K J, Wassmer T, Oakley J, Attar N *et al* 2007 SNX4 coordinates endosomal sorting of TfnR with dynein-mediated transport into the endocytic recycling compartment *Nat. Cell Biol.* **9** 1370–80

Traub L M and Bonifacino J S 2013 Cargo recognition in clathrin-mediated endocytosis *Cold Spring Harb. Perspect. Biol.* **5** a016790–a0

Trokter M, Mücke N and Surrey T 2012 Reconstitution of the human cytoplasmic dynein complex *Proc. Natl Acad. Sci. USA* **109** 20895–900

Tumbarello D A, Kendrick-Jones J and Buss F 2013 Myosin VI and its cargo adaptors—linking endocytosis and autophagy *J. Cell Sci.* **095554**

Tuplin E W and Holahan M R 2017 Aripiprazole, a drug that displays partial agonism and functional selectivity *Curr. Neuropharmacol.* **15** 1192–207

Tyckaert F, Zanin N, Morsomme P and Renard H-F 2022 Rac1, the actin cytoskeleton and microtubules are key players in clathrin-independent endophilin-A3-mediated endocytosis *J. Cell Sci.* **135** jcs259623

Uchida S, Shimada Y, Watanabe G, Li Z G, Hong T, Miyake M *et al* 1999 Motility-related protein (MRP-1/CD9) and KAI1/CD82 expression inversely correlate with lymph node metastasis in oesophageal squamous cell carcinoma *Br. J. Cancer* **79** 1168–73

Underhill D M and Ozinsky A 2002 Phagocytosis of microbes: complexity in action *Annu. Rev. Immunol.* **20** 825–52

Ungewickell E, Ungewickell H, Holstein S E H, Lindner R, Prasad K, Barouch W *et al* 1995 Role of auxilin in uncoating clathrin-coated vesicles *Nature* **378** 632–5

Urnavicius L, Lau C K, Elshenawy M M, Morales-Rios E, Motz C, Yildiz A *et al* 2018 Cryo-EM shows how dynactin recruits two dyneins for faster movement *Nature* **554** 202–6

Urnavicius L, Zhang K, Diamant A G, Motz C, Schlager M A, Yu M *et al* 2015 The structure of the dynactin complex and its interaction with dynein *Science* **347** 1441–6

Urra H, Torres V A, Ortiz R J, Lobos L, Díaz M I, Díaz N *et al* 2012 Caveolin-1-enhanced motility and focal adhesion turnover require tyrosine-14 but not accumulation to the rear in metastatic cancer cells *PLoS One* **7** e33085

Varsano T, Dong M-Q, Niesman I, Gacula H, Lou X, Ma T *et al* 2006 GIPC is recruited by APPL to peripheral TrkA endosomes and regulates TrkA trafficking and signaling *Mol. Cell. Biol.* **26** 8942–52

Varsano T, Taupin V, Guo L, Baterina O Y and Farquhar M G 2012 The PDZ Protein GIPC regulates trafficking of the LPA1 receptor from APPL signaling endosomes and attenuates the cell's response to LPA *PLoS One* **7** e49227

Vasiljeva O and Turk B 2008 Dual contrasting roles of cysteine cathepsins in cancer progression: apoptosis versus tumour invasion *Biochimie* **90** 380–6

Vaughan K T, Tynan S H, Faulkner N E, Echeverri C J and Vallee R B 1999 Colocalization of cytoplasmic dynein with dynactin and CLIP-170 at microtubule distal ends *J. Cell Sci.* **112** 1437–47

Vaughan P S, Miura P, Henderson M, Byrne B and Vaughan K T 2002 A role for regulated binding of p150 *Glued* to microtubule plus ends in organelle transport *J. Cell Biol.* **158** 305–19

Veeriah S, Taylor B S, Meng S, Fang F, Yilmaz E, Vivanco I *et al* 2010 Somatic mutations of the Parkinson's disease–associated gene PARK2 in glioblastoma and other human malignancies *Nat. Genet.* **42** 77–82

Verma V, Wadsworth P and Maresca T J 2024 Human dynein-dynactin is a fast processive motor in living cells *eLife* **13** RP94963

Vershinin M, Carter B C, Razafsky D S, King S J and Gross S P 2007 Multiple-motor based transport and its regulation by Tau *Proc. Natl Acad. Sci. USA* **104** 87–92

Vestre K, Kjos I, Guadagno N A, Borg Distefano M, Kohler F, Fenaroli F *et al* 2019 Rab6 regulates cell migration and invasion by recruiting Cdc42 and modulating its activity *Cell. Mol. Life Sci.* **76** 2593–614

Vieira O V, Botelho R J and Grinstein S 2002 Phagosome maturation: aging gracefully *Biochem. J.* **366** 689–704

Villagomez F R, Medina-Contreras O, Cerna-Cortes J F and Patino-Lopez G 2020 The role of the oncogenic Rab35 in cancer invasion, metastasis, and immune evasion, especially in leukemia *Small GTPases* **11** 334–45

Villari G, Enrico Bena C, Del Giudice M, Gioelli N, Sandri C, Camillo C *et al* 2020 Distinct retrograde microtubule motor sets drive early and late endosome transport *EMBO J.* **39** e103661

Villaseñor R, Kalaidzidis Y and Zerial M 2016 Signal processing by the endosomal system *Curr. Opin. Cell Biol.* **39** 53–60

Visscher K, Schnitzer M J and Block S M 1999 Single kinesin molecules studied with a molecular force clamp *Nature* **400** 184–9

Von Zastrow M and Sorkin A 2007 Signaling on the endocytic pathway *Curr. Opin. Cell Biol.* **19** 436–45

Vonderheit A and Helenius A 2005 Rab7 associates with early endosomes to mediate sorting and transport of semliki forest virus to late endosomes *PLoS Biol.* **3** e233

Wagner W, Lippmann K, Heisler F F, Gromova K V, Lombino F L, Roesler M K *et al* 2019 Myosin VI drives clathrin-mediated AMPA receptor endocytosis to facilitate cerebellar long-term depression *Cell Rep.* **28** 11–20.e9

Wang H, Ouyang Y, Somers W G, Chia W and Lu B 2007 Polo inhibits progenitor self-renewal and regulates Numb asymmetry by phosphorylating Pon *Nature* **449** 96–100

Wang H, Somers G W, Bashirullah A, Heberlein U, Yu F and Chia W 2006 Aurora-A acts as a tumor suppressor and regulates self-renewal of *Drosophila* neuroblasts *Genes Dev.* **20** 3453–63

Wang J, Yang S, Ye F, Xia X, Shao X, Xia S *et al* 2017a Hypoxia-induced Rab11-family interacting protein4 expression promotes migration and invasion of colon cancer and correlates with poor prognosis *Mol. Med. Rep.* **17** 3797–806

Wang M-Y, Chen P-S, Prakash E, Hsu H-C, Huang H-Y, Lin M-T *et al* 2009 Connective tissue growth factor confers drug resistance in breast cancer through concomitant up-regulation of Bcl-xL and cIAP1 *Cancer Res.* **69** 3482–91

Wang S, Hu C, Wu F and He S 2017b Rab25 GTPase: functional roles in cancer *Oncotarget* **8** 64591–9

Wang T, Li L and Hong W 2017c SNARE proteins in membrane trafficking *Traffic* **18** 767–75

Wang W, Cao L, Wang C, Gigant B and Knossow M 2015 Kinesin, 30 years later: recent insights from structural studies *Protein Sci.* **24** 1047–56

Wang Y, Huynh W, Skokan T D, Lu W, Weiss A and Vale R D 2019 CRACR2a is a calcium-activated dynein adaptor protein that regulates endocytic traffic *J. Cell Biol.* **218** 1619–33

Wang Y, Pennock S, Chen X and Wang Z 2002 Endosomal signaling of epidermal growth factor receptor stimulates signal transduction pathways leading to cell survival *Mol. Cell. Biol.* **22** 7279–90

Watanabe T M and Higuchi H 2007 Stepwise movements in vesicle transport of HER2 by motor proteins in living cells *Biophys. J.* **92** 4109–20

Watanabe T M, Sato T, Gonda K and Higuchi H 2007 Three-dimensional nanometry of vesicle transport in living cells using dual-focus imaging optics *Biochem. Biophys. Res. Commun.* **359** 1–7

Watson P and Stephens D J 2006 Microtubule plus-end loading of p150Glued is mediated by EB1 and CLIP-170 but is not required for intracellular membrane traffic in mammalian cells *J. Cell Sci.* **119** 2758–67

Waugh M G 2019 The great escape: how phosphatidylinositol 4-kinases and pi4p promote vesicle exit from the golgi (and drive cancer) *Biochem. J.* **476** 2321–46

Wayt J, Cartagena-Rivera A, Dutta D, Donaldson J G and Waterman C M 2021 Myosin II isoforms promote internalization of spatially distinct clathrin-independent endocytosis cargoes through modulation of cortical tension downstream of ROCK2 *Mol. Biol. Cell* **32** 226–36

Weiss F, Lauffenburger D and Friedl P 2022 Towards targeting of shared mechanisms of cancer metastasis and therapy resistance *Nat. Rev. Cancer* **22** 157–73

Wells A L, Lin A W, Chen L-Q, Safer D, Cain S M, Hasson T *et al* 1999 Myosin VI is an actin-based motor that moves backwards *Nature* **401** 505–8

Welte M A 2004 Bidirectional transport along microtubules *Curr. Biol.* **14** R525–37

Wenzel E M, Pedersen N M, Elfmark L A, Wang L, Kjos I, Stang E *et al* 2024 Intercellular transfer of cancer cell invasiveness via endosome-mediated protease shedding *Nat. Commun.* **15** 1277

White E 2012 Deconvoluting the context-dependent role for autophagy in cancer *Nat. Rev. Cancer* **12** 401–10

Wieczorek M, Bechstedt S, Chaaban S and Brouhard G J 2015 Microtubule-associated proteins control the kinetics of microtubule nucleation *Nat. Cell Biol.* **17** 907–16

Wieduwilt M J and Moasser M M 2008 The epidermal growth factor receptor family: biology driving targeted therapeutics *Cell. Mol. Life Sci.* **65** 1566–84

Wieffer M, Maritzen T and Haucke V 2009 SnapShot: endocytic trafficking *Cell* **137** 382.e1–3

Wieman H L, Horn S R, Jacobs S R, Altman B J, Kornbluth S and Rathmell J C 2009 An essential role for the Glut1 PDZ-binding motif in growth factor regulation of Glut1 degradation and trafficking *Biochem.* J. **418** 345–67

Williams R L and Urbé S 2007 The emerging shape of the ESCRT machinery *Nat. Rev. Mol. Cell Biol.* **8** 355–68

Williamson C D and Donaldson J G 2019 Arf6, JIP3, and dynein shape and mediate macropinocytosis *Mol. Biol. Cell* **30** 1477–89

Willy N M, Ferguson J P, Huber S D, Heidotting S P, Aygün E, Wurm S A *et al* 2017 Membrane mechanics govern spatiotemporal heterogeneity of endocytic clathrin coat dynamics *Mol. Biol. Cell* **28** 3480–8

Wilson B J, Allen J L and Caswell P T 2018 Vesicle trafficking pathways that direct cell migration in 3D matrices and *in vivo Traffic* **19** 899–909

Wolf M, Clark-Lewis I, Buri C, Langen H, Lis M and Mazzucchelli L 2003 Cathepsin D specifically cleaves the chemokines macrophage inflammatory protein-1α, macrophage inflammatory protein-1β, and SLC that are expressed in human breast cancer *Am. J. Pathol.* **162** 1183–90

Woodings J A, Sharp S J and Machesky L M 2003 MIM-B, a putative metastasis suppressor protein, binds to actin and to protein tyrosine phosphatase delta *Biochem.* J. **371** 463–71

Woolworth J A, Nallamothu G and Hsu T 2009 The *Drosophila* metastasis suppressor gene *Nm23* homolog, *awd*, regulates epithelial integrity during oogenesis *Mol. Cell. Biol.* **29** 4679–90

Wu K, Li Q, Lin F, Li J, Wu L, Li W *et al* 2014 MT1-MMP is not a good prognosticator of cancer survival: evidence from 11 studies *Tumor Biol.* **35** 12489–95

Xiang X, Qiu R, Yao X, Arst H N, Peñalva M A and Zhang J 2015 Cytoplasmic dynein and early endosome transport *Cell. Mol. Life Sci.* **72** 3267–80

Xiao M-F, Xu J-C, Tereshchenko Y, Novak D, Schachner M and Kleene R 2009 Neural cell adhesion molecule modulates dopaminergic signaling and behavior by regulating dopamine d_2 receptor internalization *J. Neurosci.* **29** 14752–63

Xiao X-H, Lv L-C, Duan J, Wu Y-M, He S-J, Hu Z-Z *et al* 2018 Regulating Cdc42 and its signaling pathways in cancer: small molecules and MicroRNA as new treatment candidates *Molecules* **23** 787

Xie S, Bahl K, Reinecke J B, Hammond G R V, Naslavsky N and Caplan S 2016 The endocytic recycling compartment maintains cargo segregation acquired upon exit from the sorting endosome *Mol. Biol. Cell* **27** 108–26

Xu B, Teng L H, Silva S D D, Bijian K, Al Bashir S, Jie S *et al* 2014 The significance of dynamin 2 expression for prostate cancer progression, prognostication, and therapeutic targeting *Cancer Med.* **3** 14–24

Xu J, Galvanetto N, Nie J, Yang Y and Torre V 2020a Rac1 promotes cell motility by controlling cell mechanics in human glioblastoma *Cancers* **12** 1667

Xu X, Liu J, Wang Y, Wang Y, Gong X and Pan L 2020b Mechanistic insights into the interactions of Ras subfamily GTPASES with the SPN domain of autism-associated SHANK3 *Chin. J. Chem.* **38** 1635–41

Yang J, Liu W, Lu X, Fu Y, Li L and Luo Y 2015 High expression of small GTPase Rab3D promotes cancer progression and metastasis *Oncotarget* **6** 11125–38

Yang L, Zhu Z, Zheng Y, Yang J, Liu Y, Shen T *et al* 2023 RAB6A functions as a critical modulator of the stem-like subsets in cholangiocarcinoma *Mol. Carcinog.* **62** 1460–73

Yang X, Wei L L, Tang C, Slack R, Mueller S and Lippman M E 2001 Overexpression of KAI1 suppresses *in vitro* invasiveness and *in vivo* metastasis in breast cancer cells *Cancer Res.* **61** 5284–8

Yang X-Z, Li X-X, Zhang Y-J, Rodriguez-Rodriguez L, Xiang M-Q, Wang H-Y *et al* 2016 Rab1 in cell signaling, cancer and other diseases *Oncogene* **35** 5699–704

Yang Z, Yang F, Zhang D, Liu Z, Lin A, Liu C *et al* 2017 Phosphorylation of G protein-coupled receptors: from the barcode hypothesis to the flute model *Mol. Pharmacol.* **92** 201–10

Ye B, Duan B, Deng W, Wang Y, Chen Y, Cui J *et al* 2018 EGF stimulates Rab35 activation and gastric cancer cell migration by regulating DENND1A-Grb2 complex formation *Front. Pharmacol.* **9** 1343

Yi Z, Petralia R S, Fu Z, Swanwick C C, Wang Y-X, Prybylowski K *et al* 2007 The role of the PDZ protein GIPC in regulating NMDA receptor trafficking *J. Neurosci.* **27** 11663–75

Yoon C, Cho S-J, Chang K K, Park D J, Ryeom S W and Yoon S S 2017 Role of Rac1 pathway in epithelial-to-mesenchymal transition and cancer stem-like cell phenotypes in gastric adenocarcinoma *Mol. Cancer Res.* **15** 1106–16

Yoon S-O, Shin S and Mercurio A M 2005 Hypoxia stimulates carcinoma invasion by stabilizing microtubules and promoting the rab11 trafficking of the α6β4 integrin *Cancer Res.* **65** 2761–9

Yu I, Garnham C P and Roll-Mecak A 2015 Writing and reading the tubulin code *J. Biol. Chem.* **290** 17163–72

Yu L, Li X, Li H, Chen H and Liu H 2016 Rab11a sustains GSK3β/Wnt/β-catenin signaling to enhance cancer progression in pancreatic cancer *Tumor Biol.* **37** 13821–9

Yudowski G A, Puthenveedu M A, Henry A G and Von Zastrow M 2009 Cargo-mediated regulation of a rapid Rab4-dependent recycling pathway *Mol. Biol. Cell* **20** 2774–84

Zajac A L, Goldman Y E, Holzbaur E L F and Ostap E M 2013 Local cytoskeletal and organelle interactions impact molecular-motor-driven early endosomal trafficking *Curr. Biol.* **23** 1173–80

Zajac M, Law J, Cvetkovic D D, Pampillo M, McColl L, Pape C *et al* 2011 GPR54 (KISS1R) transactivates EGFR to promote breast cancer cell invasiveness *PLoS One* **6** e21599

Zamani M R, Aslani S, Salmaninejad A, Javan M R and Rezaei N 2016 PD-1/PD-L and autoimmunity: a growing relationship *Cell. Immunol.* **310** 27–41

Zaniewski T M, Gicking A M, Fricks J and Hancock W O 2020 A kinetic dissection of the fast and superprocessive kinesin-3 KIF1A reveals a predominant one-head-bound state during its chemomechanical cycle *J. Biol. Chem.* **295** 17889–903

Zerial M and McBride H 2001 Rab proteins as membrane organizers *Nat. Rev. Mol. Cell Biol.* **2** 107–17

Zhang R, Lee D M, Jimah J R, Gerassimov N, Yang C, Kim S *et al* 2020a Dynamin regulates the dynamics and mechanical strength of the actin cytoskeleton as a multifilament actin-bundling protein *Nat. Cell Biol.* **22** 674–88

Zhang S, Wang C, Ma B, Xu M, Xu S, Liu J *et al* 2020b Mutant p53 Drives cancer metastasis via RCP-mediated Hsp90α secretion *Cell Rep.* **32** 107879

Zhang Y and Hughson F M 2021 Chaperoning SNARE folding and assembly *Annu. Rev. Biochem.* **90** 581–603

Zhang Y, Tan Y-F, Jiang C, Zhang K, Zha T-Z and Zhang M 2013 High ADAM8 expression is associated with poor prognosis in patients with hepatocellular carcinoma *Pathol. Oncol. Res.* **19** 79–88

Zhang Y, Xia M, Jin K, Wang S, Wei H, Fan C *et al* 2018 Function of the c-Met receptor tyrosine kinase in carcinogenesis and associated therapeutic opportunities *Mol. Cancer* **17** 45

Zhang Y, Yang P and Wang X-F 2014 Microenvironmental regulation of cancer metastasis by miRNAs *Trends Cell Biol.* **24** 153–60

Zhao Y, Ye Z, Song D, Wich D, Gao S, Khirallah J *et al* 2023 Nanomechanical action opens endo-lysosomal compartments *Nat. Commun.* **14** 6645

Zhong J L, Poghosyan Z, Pennington C J, Scott X, Handsley M M, Warn A *et al* 2008 Distinct functions of natural ADAM-15 cytoplasmic domain variants in human mammary carcinoma *Mol. Cancer Res.* **6** 383–94

Zhu G, Chen J, Liu J, Brunzelle J S, Huang B, Wakeham N *et al* 2007 Structure of the APPL1 BAR-PH domain and characterization of its interaction with Rab5 *EMBO J.* **26** 3484–93

IOP Publishing

Physics of Cancer, Volume 6 (Second Edition)
Cellular mechanisms to foster or fight cancer
Claudia Tanja Mierke

Chapter 3

The two faces of mast cells in cancer from a mechanobiological perspective

3.1 Summary

Mast cells (MCs) have figured out an essential part as critical regulators of both defensive immune response and immunomodulation. They also exert regulatory actions in modulating pathological processes in different allergic diseases and cancer. The effects of the mere presence of MCs in tumor tissues have attracted extensive scrutiny but continue to generate conflicting results. Some studies suggest that MCs in tumor tissues facilitate tumor initiation and progression, whereas others support the opposite view. It is highlighted that the dual role of mast cells in cancer, both as facilitators and suppressors, is essential for understanding the entire complexity of cancers and their malignant progression and is needed for refining cancer therapeutic approaches. These contradictory perspectives have led to significant debate and emphasize the requirement for a complete picture of the dual role of mast cells in tumor immune responses. Perhaps the effect of altered mechanical properties of the tumor and the tumor environment could enable the regulation between the two roles.

3.2 Introduction the interaction of cancer cells and MCs

MCs are a specific type of immune cell derived from $CD34^{+}/CD117^{+}/CD13^{+}$ multipotent hematopoietic progenitors of the bone marrow, with maturation occurring in peripheral tissues (Moon *et al* 2010, Falduto *et al* 2021, Shi *et al* 2023). Like basophils, MCs act as the principal effector cells that contribute to immunoglobulin E (IgE)-based inflammation. MCs were identified as granular cells in frog mesentery by Dr von Recklinghausen in 1863 and named 'Mastzellen' by Dr Paul Ehrlich in 1878, who was the first to describe them in tumors. Since then, a growing amount of scientific work has concentrated on the immunomodulatory roles of mast cells in the context of both innate and adaptive immune responses,

doi:10.1088/978-0-7503-4007-6ch3

rather than on their heterogeneity. Studies have also addressed their effects on the course and prognosis of immune-related diseases (Derakhshani *et al* 2019). As a component of the sentinel immune cell community, MCs have attracted increasing recognition due to their immunomodulatory and effector activities in the tumor microenvironment (TME) (Shefler *et al* 2021). Available evidence underscores the widespread prevalence of MCs across a range of tumors, where they may either drive or suppress tumorigenesis through different regulatory mechanisms unique to the specific cancer type (Derakhshani *et al* 2019). Nonetheless, the exact functionalities and regulatory mechanisms of MCs in different tumor microenvironments are still a matter of active controversy. This chapter presents a focused overview examining the effects of MCs on both pro-tumorigenic and anti-tumorigenic activities, together with their participations in reshaping the TME by infiltrative crosstalk. The interaction between MCs and cancer cells appears to be a general phenomenon that plays a role in various types of cancer at different stages of cancer, such as cancer metastasis. Therefore, it seems likely that there are fundamental mechanisms that MCs use to interact with cancer cells and the TME that can be identified in various kinds of tumors. It is possible that mechanical stimuli play a role, changing the MCs in such a way that they are either tumor-promoting or tumor-inhibiting. Therefore, this aspect is also considered, along with biological aspects.

3.3 Origin, location and function of MCs

In the beginning, mast cells were identified as the key effector cells in allergies and anaphylaxis and are considered faithful sentinels that protect the body from infections. They are supposed to differentiate from early and late erythro-myeloid progenitors (EMPs) in the yolk sac and hematopoietic stem cells (HSCs) in the bone marrow (Popescu *et al* 2008, Gentek *et al* 2018, Li *et al* 2018, Shi *et al* 2023). In murine models, yolk-sac-derived mesenchymal stem cells migrate into the peripheral tissues and mature while the embryo is developing (Gentek *et al* 2018, Li *et al* 2018). In contrast, bone marrow-derived MC progenitor cells enter the peripheral tissues after circulating in the bloodstream in response to specific chemotactic stimuli (Gurish *et al* 2001, Hallgren and Gurish 2007). Therefore, MCs originating from these two sources show a different predominant tissue infiltration. Early EMP-derived MCs are found only in the pleural and fat cavities; late EMP-derived MCs form the dominant population in the majority of connective tissues, whereas HSC-derived MCs are dispersed in almost all mucous membranes (Gentek *et al* 2018). Specifically, the yolk-sac-derived MCs become the dominant connective tissue MCs (CTMCs) after birth, whereas the HSC-derived MCs primarily become mucosal MCs (MMCs). Both maturation paths of the MC were supported by substantial evidence from mice, but it is not yet clear whether these paths also exist in humans. In addition, MC precursors and mature MCs were determined to express several of the identical chemokine receptors, like CXCR4 and CCR1, indicating that mature MCs can be relocated to certain tissues (Salomonsson *et al* 2020). This process normally takes place during the enlistment of MCs at foci of inflammation and is accomplished through actin reorganization. When MCs move under the control of chemokines, the actin scaffold aggregates pericentrally, stabilizes the secretory granules in the cell center and resists flattening of the cell. When MCs arrive at the site of inflammation and are activated by granule-releasing agents, actin is reorganized and actin network density is diminished to alleviate inhibition of secretory events and favor degranulation (Klein *et al* 2019).

MCs are equipped with many receptors to react to the many complex signals in their surroundings. The best-known receptor is the high-affinity immunoglobulin E (IgE) receptor (FcεRI), which causes the activation of MCs through an antigen-specific signaling cascade. There are other types of receptors, like immunoglobulin receptors, pattern recognition receptors, G-protein-coupled receptors, nuclear receptors, and alarm receptors, that also trigger the activation of MCs, though this time via a non-IgE-dependent pathway (Yu *et al* 2016). Interaction of one or more of these activated receptors with endogenous or exogenous agents results in the activation of MCs and the liberation of downstream effectors through one of three principal mechanisms. The pre-synthesized mediators present in the secretory granules of the MCs are liberated most quickly by degranulation (Wernersson and Pejler 2014, Ménasché *et al* 2021). This instant reaction can take place in minutes, enabling MCs to quickly induce an immune reaction to fight the invasion of pathogens. Via this mechanism several mediators, such as cytokines like tumor necrosis factor-α (TNF-α), transforming growth factor-β (TGF-β), fibroblast growth factor2/basic-FGF (FGF2/b-FGF), nerve growth factor (NGF), stem cell factor (SCF), and vascular endothelial growth factor (VEGF), proteases, including tryptases, chymases, granzyme B, and matrix-metaaloproteinase-9 (MMP-9), biological amines like histamine and serotonin, and chemokines encompassing CCL2, CCL3, and CCL5, are liberated. Several proteins including interleukin-4 (IL-4), VEGF, interferon-γ (IFN)-γ, and CCL2, and lipid substances like prostaglandin D2, leukotriene C4,and platelet activating factor, are *de novo* synthesized following activation of MCs and can be secreted via a degranulation-independent route (Theoharides *et al* 2007). MCs also release extracellular vesicles (EVs) of their own accord, which comprise exosomes, microvesicles and apoptotic bodies (Shefler *et al* 2021). Substances present in these EVs, among them proteins, enzymes, RNAs and miRNAs, are incorporated and transferred to other cells in the surrounding microenvironment. The number of identified MC-derived cytokines, chemokines, growth factors and other mediators is rather long (Mukai *et al* 2018). Their numerous cellular receptors and wide range of subsequent signaling mediators enable MCs to process environmental cues and to activate fast, sustained and specific immune reactions to triggers. Table 3.1 contains a list of some important receptors on MCs, the relevant common stimuli and the corresponding mediators secreted afterwards. MCs are important players in both innate and adaptive immune responses, and their health-protective function has been substantiated by extensive lines of evidence. When fighting off a pathogen infection, MCs have a protective function by direct antimicrobial, antiviral and antiparasitic activity and by stimulating the aggregation, activation and action of other inflammatory cells at the site of infection (Abraham and St. John 2010, Vukman *et al* 2016, Piliponsky and Romani 2018).

Table 3.1. Selected MC receptors and their stimuli including mechanoreceptors.

Receptor	Function	Cues or stimuli	References
TRPV2	Mechanoreceptor	Stiffness, stretching or trauma	Li *et al* (2024)
ADGRE2	Mechanoreceptor	Vibration	Li *et al* (2024)
Pattern recognition receptors (PRRs)	Mechanoreceptor	Specific molecular structures on the surface of pathogens, apoptotic host cells, and damaged senescent cells	St. John and Abraham (2013), Redegeld *et al* (201)8, Li and Wu (2021)
Toll-like receptors (TLRs)	Receptor releases pro-inflammatory cytokines, activation of cell death for clearance of pathogens	TLR ligands are bacterial and viral proteins	St. John and Abraham (2013), Redegeld *et al* (2018)
Retinoic acid-inducible gene-I (RIG-I)-like (Rig-I) family receptors		Increased after IL-33-based induction	St. John and Abraham (2013), Redegeld *et al* (2018)
Nod-like receptors (NLRs)	Sense molecules associated with intracellular infection and stress, such as reactive oxygen species (ROS)	Transcription of MHC molecules, activation of an intracellular complex referred to as inflammasomes	St. John and Abraham (2013), Redegeld *et al* (2018), Almeida *et al* (2023)
c-KIT	Type III cell surface tyrosine kinase receptor	Ligand is SCF	Tsai *et al* (2022)
FcγRI	Fc gamma receptor I complex	IgG-mediated activation	Chen *et al* (2008), Cianferoni (2021)
MRGPRX2	Mechanoreceptor?	Substance P, interactions with tendons (possible mechanical stimulation?)	Cianferoni (2021), Mousavizadeh *et al* (2024)
Adhesion G-protein coupled receptors (aGPRCs)	Mechanosensory receptor	Vibratory stimulation *in vitro* evokes degranulation of MCs	Naranjo *et al* (2020)

During the whole injury healing process, MCs are essential actors in every step. It is commonly assumed that tissue repair encompasses three phases: inflammation, proliferation and reorganization. In the early phase of inflammation, MCs infiltrate the lesion site and undergo degranulation, secreting multiple vasoactive and pro-inflammatory mediators like histamine and serotonin, which enhance vascular permeability and facilitate vasodilation (Kennedy *et al* 2021). In addition, products derived from MCs, including TNF-α, MIP-2, and IL-8, enhance neutrophil trafficking to areas of inflammation and liberate MCP-1 to promote monocyte conversion into phagocytes (Komi *et al* 2020). Events in which MCs are engaged at this phase assist in the removal of cellular debris and pathogens, along with the stimulation of angiogenesis and the activation of fibroblasts to synthesize extracellular matrix (ECM) proteins, thereby connecting the inflammatory phase to the proliferative phase. The major features of the proliferative phase comprise re-epithelialization, angiogenesis, and fibroplasia. In this phase, mediators derived from MCs, among them FGF-2, VEGF, TGF-β, IL-8, and MMP-9, facilitate neovascularization, fibrinogenesis, or re-epithelialization (Artuc *et al* 1999, Rao and Brown 2008). In the restructuring phase, granulation tissue converts into avascular scar tissue, which is abundant in collagen, and vascular degeneration outpaces angiogenesis (Nishikori *et al* 2014). At this stage, tryptase, FGF-2, VEGF and TGF-β released from MCs also promote fibroblast migration, proliferation and growth, and assist in the synthesis of ECM (Bagher *et al* 2018).

Even though health benefits of MCs are well established, including their involvement in host defense against infection, facilitation of tissue healing and upholding of the body's life-sustaining activities, the pathological alterations and diseases evoked by the mediators released by MCs needs to be considered. Typically, MCs are mostly recognized for their potential to trigger allergies and anaphylaxis in humans (Lieberman and Garvey 2016, Gülen and Akin 2022). The inadequate activation of MCs by IgE receptors is a key mechanism underpinning the etiopathogenesis of anaphylaxis and allergic diseases like food and drug allergies, allergic asthma, allergic rhinitis and allergic dermatitis (Li *et al* 2014, Méndez-Enríquez and Hallgren 2019, Kato *et al* 2021). IgE-driven anaphylaxis is subdivided into two phases: the first is sensitization, where IgE is liberated upon challenge with allergens into the bloodstream and attaches to FcεRI on the surface of MCs. The second phase consists of a re-exposure/activation step (Bagher *et al* 2018). As soon as the organism is re-exposed to allergens, MCs become activated via IgE-dependent mechanisms and immediately secrete pre- and newly synthesized effectors like histamine, prostaglandins, leukotrienes, serotonin, and several cytokines and chemokines, leading to an immediate-type hypersensitivity response (Banafea *et al* 2022, Fernandez-Bravo *et al* 2022). When confined to the respiratory tract, this hypersensitivity response can lead to contraction of the smooth muscles of the lungs and secretion of mucus, causing shortness of breath and wheezing. In case of a systemic response, serious hypotension, even shock and ultimately death can arise due to extensive vasodilation and vascular permeability. Besides IgE/FcεRI-driven anaphylaxis, non-IgE-driven MC stimulation and anaphylaxis have been equally recognized. For instance, IgG immune complexes, as well as various other ligands or cues, also activate MCs via FcγRI, MRGPRX2, complement receptors, and additional proteins capable of inducing anaphylaxis (Cianferoni 2021). In parallel to anaphylaxis and allergic diseases, an emerging body of evidence has recently linked

MCs to diseases beyond allergies, involving cardiovascular diseases like atherosclerosis, myocarditis and myocardial infarction, autoimmune diseases like rheumatoid arthritis and multiple sclerosis (Higuchi *et al* 2008, Ngkelo *et al* 2016, Rivellese *et al* 2017, Kovanen 2019, Ribatti *et al* 2020) and several kinds of cancer (Maciel *et al* 2015, Komi and Redegeld 2020). In mouse models an amount of MCs is elevated in various cancer types, encompassing pancreatic islet tumors, lung cancer, breast cancer, and colon cancer (Soucek *et al* 2007, Ko *et al* 2017). In humans, an elevated level of MCs or infiltration of MCs have been identified in the tumor and adjacent tissue of patients suffering from squamous cell carcinoma of the lung. Hodgkin lymphoma (Molin *et al* 2002), cholangiocarcinoma (Pham *et al* 2022), gastric cancer, prostate cancer, pancreatic adenocarcinoma (Strouch *et al* 2010, Molderings *et al* 2017), esophageal squamous cell carcinoma, hepatocellular carcinoma (Terada and Matsunaga 2000), breast cancer, renal carcinoma and melanoma (Yano *et al* 1999, Elpek *et al* 2001, Ribatti *et al* 2003, Tuna *et al* 2006, Rajput *et al* 2008, Johansson *et al* 2010, Strouch *et al* 2010, Stankovic *et al* 2019). These elevated MC numbers are mostly inked to worse prognosis and aggressive cancer progression. MC infiltration is correlated with disease progression in multiple cancer types. All these findings support evidence that suggests MCs seems to be implicated in tumor initiation and advancement.

3.4 Phenotype heterogeneity via polarization of MCs: tissue-specific phenotype of MCs

MCs are notable for their plasticity, which can lead to the emergence of phenotypically divergent subsets of MCs at distinct anatomical locations (Jiménez *et al* 2021). MCs possess multiple stimulatory and inhibitory ligands, which result in the incorporation of input cues and the release of a variety of stored mediators located in secretory granules, along with *de novo* synthesized mediators. Secretory granules enclose proteases like tryptase and chymase, histamines, heparin, lysosomal enzymes and pro-inflammatory cytokines like TNF-α, which are involved in serious allergic disorders including urticaria and anaphylaxis. MCs display considerable heterogeneity in their granuloma load, with a subtype categorization based on whether they comprise tryptase without chymase (MC_T), chymase without tryptase (MC_C), or both (MC_{TC}) (Moon *et al* 2014, Valent *et al* 2020). Murine studies suggest that MC phenotypes are versatile and can change based on the surrounding microenvironment including cytokine exposure as well as the stage of MC development (Levi-Schaffer *et al* 1986, Sonoda *et al* 1986, Otsu *et al* 1987, Kanakura *et al* 1988, Friend *et al* 1996, Valent *et al* 2020). Two forms of granule release have been identified: anaphylactic degranulation, where the whole granule load is suddenly liberated into the extracellular environment, and piecewise granule release, where only part of the preformed granule load is liberated in a more phased and specific fashion (Theoharides *et al* 2007, Moon *et al* 2014). Both have been characterized in humans and also in other species. Additionally or independently of degranulation, activation of MCs can lead to the liberation of several preformed, but mostly *de novo* produced, growth factors, eicosanoids including prostaglandins, leukotrienes, chemokines like CXCL10 and multiple cytokines (Turner and Kinet 1999, Mukai *et al* 2018, Valent *et al* 2020). Cytokines revealed to be liberated through MCs encompass inflammatory cytokines like TNF-α, IL-6, and IL-1, nevertheless, also anti-inflammatory cytokines like IL-10 and TGF-ß are secreted; MC-released growth factors, cytokines, and chemokines are described in more detail in this review article (Mukai *et al* 2018).

The primary mechanism by which MC degranulation arises is the antigen-specific IgE crosslinking of the high-affinity IgE-bearing surface receptor FcεRI after encountering a related antigen, resulting in fast MC degranulation (Turner and Kinet, 1999). MCs are also activated through alternative mechanisms, like damage- and pathogen-associated molecular patterns via toll-like receptors, complement proteins, cytokines and various other inducers. There is considerable heterogeneity between MCs in the expression of several surface receptors, among them complement receptors, and this leads to functional implications, although the elucidation of the mechanism responsible for this variability is insufficient (Füreder *et al* 1995, Kiener *et al* 1998, Oskeritzian *et al* 2005, Valent *et al* 2020).

The consequences of MC activation, degranulation and/or secretion of inflammatory molecules range from activation or attraction of other immune, stromal, neuronal and epithelial cells, resulting in alterations of the local tissue microenvironment like vasodilation and angiogenesis, to activation of systemic immune reactions (figure 3.1). MC activation and/or degranulation can follow the classical fast pathway, resulting in potent release of inflammatory molecules and causing severe clinical manifestations including anaphylaxis and angioedema. Alternatively, these processes can proceed gradually through the low-level secretion of certain signaling mediators, resulting in chronic inflammation and localized tissue remodeling. This particular mode of MC activation is especially pertinent in cancers, in which MCs

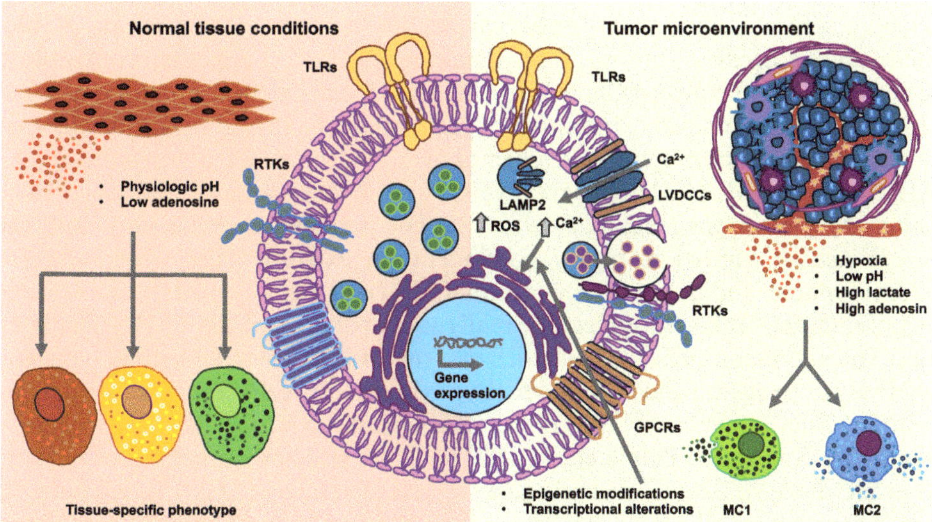

Figure 3.1. Heterogeneity of MCs in normal tissues and heterogeneity of MCs residing in the TME.

act as pivotal regulatory cells in tissue rearrangement and as sentinel immune cells orchestrating innate and adaptive immune responses (Oldford and Marshall 2015).

3.5 Crosstalk between MCs and the TME

There exists a special bidirectional crosstalk between the TME and the solid tumor bidirectional interaction and also a bidirectional interaction between MCs and the TME has been identified (Mierke 2019, 2021). While the MC–TME interaction MCs can polarize in two subtypes, such as MC1 and MC2 phenotypes. The polarization can be triggered via the classical or mechanical stimulation of MCs. In addition, the well-recognized function of MCs to restructure the local microenvironment is emphasized and its possible transfer to cancer is discussed.

3.5.1 Effect of the TME on the MC phenotype

Potential MC polarization toward phenotypes MC1 and MC2, is induced through the impact of TME. MCs found in normal organs provide crosstalk with neighboring cells through the perception of tissue-specific molecule production. This crosstalk facilitates the generation of a restricted set of mediators that are beneficial for tissue homeostasis. Exposed to the harsh extremes of TME like hypoxia, oxidizing environments and high concentrations of adenosine, MCs undergo alterations that comprise the accumulation of intracellular ROS, the relocation of L-type voltage-dependent calcium channels (LVDCC) from LAMP2-positive pools toward the plasma membrane, and potential epigenetic and transcriptional alterations. Present views on the role of MCs in tumor growth involve the differentiation of at least two distinct phenotypes, termed MC1 (anti-tumoral) and MC2 (pro-tumoral). The implications of the presence of MC in the TME or at the tumor edges are not yet clear, as they are linked to a favorable or dismal prognosis depending on the type and anatomical position of the tumor. Besides cancer cells, a plethora of cells (including stromal cells and fibroblasts), ECM, a complex meshwork of blood vessels, and molecules (among them signaling molecules) together constitute the TME of the tumor (Hui and Chen 2015). The TME can be conceptualized as a focus of low-grade inflammation, in which a vast array of infiltrating or resident cells synthesize and liberate cytokines, chemokines, and enzymes comprising TNF-α, MMP-9, Cox-2, IL-6, iNOS, and VEGF, that are potent mediators of the inflammatory reactions (Huang *et al* 2008). The persistence, expansion, metastasis or clearance of tumors is highly contingent on extrinsic cues emanating from adjacent immune and non-immune cells of TME (Hui and Chen 2015). The ultimate effect of immune orchestration can be malignant evolution of the TME (Huang *et al* 2008). The anomalous vascular network of a tumor is unable to adequately meet the cancer cells' oxygen needs. In turn, hypoxic cancer cells secrete angiogenesis-promoting factors, primarily VEGF-A, which binds to VEGFR2 that is expressed on endothelial cells (De Palma *et al* 2017). MCs are situated at the border of tumors and inside the TME, often encircling the vessels (Tamma *et al* 2017), whereby they interact with endothelial cells (Mierke *et al* 2000). Inside the TME, MCs exert both pro- and anti-tumorigenic effects (figure 3.2). After activation and degranulation,

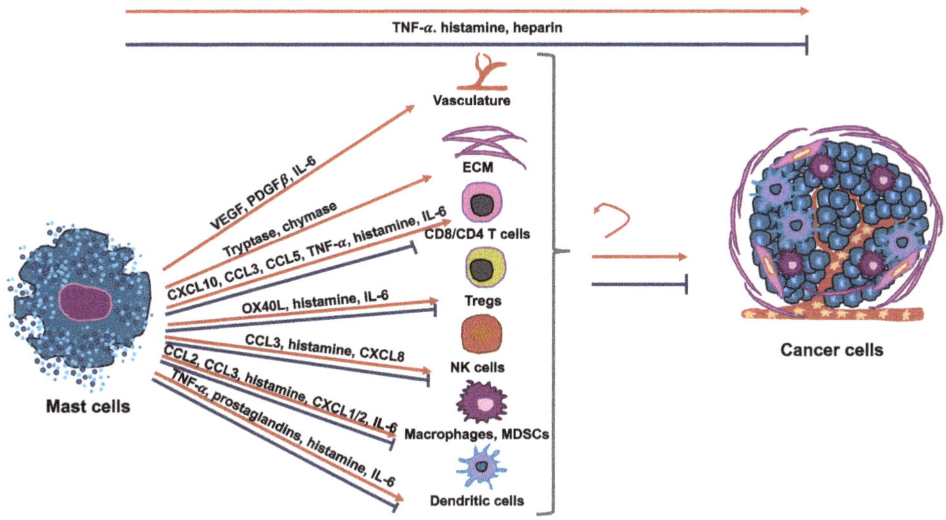

Figure 3.2. MCs drive anti-tumor immunity (blue color) and regulate the tumor development and progression (red color). MCs can directly affect cancer cells, immune cells and non-immune elements of the TME via the release of cytokines and other mediators. They lead to either cancer-promoting or cancer-suppressive functions. Moreover, the MC-released components of the TME can interfere with one another, which indirectly affects cancer cells via structural or mechanical cues. In addition, the components of the TME can alter cancer cells in a direct manner by binding to specific receptors. Both the direct and indirect interactions can even lead to cumulative signals on cancer cells.

MCs strongly promote inflammation and actively attract cells of the innate immune system, primarily neutrophils, macrophages and eosinophils, and cells of the adaptive immune system like B and T cells to coordinate the immune response against the tumor (Hempel *et al* 2017). By contrast, the result of MC presence may foster tumor advancement by secretion of VEGF to facilitate angiogenesis and MMP9 to break down the ECM and ease metastasis (Hempel *et al* 2017). The non-uniform and contradictory predictive power of the presence of MCs in the TME can be due to the heterogeneous character of the tumors investigated and the nature of the animal model (Fu *et al* 2017, Ghouse *et al* 2018).

Chemokine ligand 2, 3 or 5 (CCL2, CCL3 or CCL5), C-*X*-C motif chemokine ligand 1, 2, 8 or 10 (CXCL1, CXCL2, CXCL8 or CXCL10), Interleulin-6 (IL-6), myeloid-derived suppressor cells (MDSCs), natural killer cells (NK cells), OX40 ligand (OX40L), platelet-derived growth factor-β (PDGF-β), regulatory T cells (Tregs), tumor necrosis factor-α (TNF-α), VEGF.

MCs cluster in the tumor microenvironment with the help of chemoattractants secreted by cancer cells, like SCF or CCL15 (Yu *et al* 2018). In addition, MCs can inhibit T cells and natural killer (NK) cells through the secretion of adenosine into the cellular microenvironment (Huang *et al* 2008). After the infiltration of MCs into the tumor stroma, the spread and activation of Tregs is increased. Thus, Tregs drive immune tolerance, which results in cancer advancement (Ribatti *et al* 2020). Up-to-date information on the impact of MCs on the TME has been gained by *in vivo*

Features of human mast cell lines

Cell line	**HMC1.1**	**HMC1.2**	**LAD2**	**LUVA**
Functional FcεRI	**No**	**No**	**Yes**	**Yes**
Relies on SCF	**No**	**No**	**No**	**Yes**
Histamine (pg/cell)	**0.9**	**0.9**	**3.1**	**-**
Growth media	**Iscove's 20% FCS**	**Iscove's 20% FCS**	**StemPro Serum free**	**StemPro Serum free**
Doubling time	**2-3 days**	**2-3 days**	**2 weeks**	**2-3 days**
Mutations	**K509I**	**K509I**	**-**	
	-	**D816V**	**-**	**-**

Figure 3.3. Selected major features of human MC lines.

cytological analysis of animal and human cancers and by co-culturing cancer cell lines with primary and established MC lines, mostly Human Mast Cell leukemia-1 (HMC-1) and Laboratory of allergic diseases 2 (LAD2), *in vitro* (figure 3.3). Both MC lines resample weakly human MCs, whereby HMC-1 represents very immature MCs and lacks sufficient high-affinity IgE-bearing surface receptor FcεRI. LAD2 represents intermediate mature MCs and possesses consistent FceR1 mediated degranulation (Kirshenbaum *et al* 2014).

The LAD2 MC line comes closest to primary human MC cultures because of firstly the presence of functional FcεRI receptors and the capability of degranulation when stimulated immunologically and secondly the growth dependency on the availability of SCF (Kirshenbaum *et al* 2003). Work that addressed the functional crosslinking of FcεRI receptors in conjunction with certain G-protein coupled receptors expressed on LAD2 cells and not on HMC-1 cells has yielded findings that have advanced knowledge of human MC biology most significantly. These include the following, such as inhibiting phosphorylation of STAT3 serine727 in human MCs results in compromised FcεRI-driven proximity and downstream signal transduction, as well as diminished degranulation, which is in line with the lack of clinical signs of atopy in the autosomal dominant hyper-IgE syndrome (Siegel *et al* 2013). In addition, CD84 is strongly expressed in human MCs and attenuates FcεRI-based calcium release and degranulation with secretion of IL-8 and GM-CSF following its co-crosslinking with FcεRI (Álvarez-Errico *et al* 2011). Moreover, MrgX2 has been identified as a novel GPCR on human MCs that is activated upon binding of the antibacterial peptide cathelicidin (LL-37). MrgX2 is implicated in innate immunity and drives inflammation, angiogenesis, healing of injuries, and cancer metastasis (Subramanian *et al* 2011). Human beta-defensins and LL-37 trigger the release of a new pruritogenic factor, IL-31, from human MCs, indicating a novel mechanism through which skin-derived antimicrobial peptides/proteins may participate in cutaneous inflammation (Niyonsaba *et al* 2010). Leukotriene (LT)E (4), unlike LTD(4), is able to activate peroxisome proliferator-activated receptor gamma and drive prostaglandin D2 generation, thus potentially accounting for the ability of LTE(4) to exacerbate airway hyperresponsiveness and drive bronchial eosinophilia *in vivo* (Paruchuri *et al* 2008). Consequently, these findings lead to the conclusion that the LAD2 human MC line is suitable to resemble the human MC phenotype. A comparison of the advantages and disadvantages of the four existing lines of human MCs is illustrated in figure 3.3. A cell line that is very similar to primary cultured human MCs and has functional FcεRI receptors could be advantageous for research. LAD2 and LUVA cells express functional FcεRI receptors. LUVA cells double in size weekly, whereas LAD2 cells have the misfortune of growing more gradually, doubling in size every two weeks. HMC-1.1 and HMC-1.2 cells undergo weekly doublings; however, they are deficient in functional FcεRI. HMC-1.2 cells have the advantage of harboring the D816V mutation for the purpose of the study, nevertheless, the mutation is not present in LAD2 and LUVA cells. Many MC effectors can impact the TME through the induction of angiogenesis, ECM breakdown and tumor growth stimulation. Notably, MCs have been implicated to recruit and facilitate infiltration of MDSCs through CCL2/CCR2 axis in the TME wherein they generate IL-17 that attracts Tregs. The IL-9 released by Treg rounds off this positive feedback circuit by promoting the survival of MCs (Yang *et al* 2010). Neutrophils can be attracted into the TME via chemokines released by MCs, among them CCL1, CCL2, CCL3, CCL4, CCL5 and CXCL8 (Paolino *et al* 2017). Neutrophils can release VEGF-A, FGF2, and CXCL8 that contribute to angiogenesis (De Palma *et al* 2017).

MCs maintain a tightly regulated metabolism within the TME due to the crosstalk of locally generated chemokines and receptors that are expressed on MCs. Among the most critical chemoattractant factors generated by cancer cells is SCF, which acts also as the key survival factor for MCs. In addition, a multitude of other chemokine/receptor pairings play a crucial role in attracting MCs, among them LTB4 with BLT1 and BLT2 (Godot *et al* 2007), PGE2 with the EP2 receptor, VEGF via VEGFR-1 and VEGFR-2, angiopoietin 1 (Angl) which works on Tie2 receptor, and also CXCL8/IL-8 interplay with CXCR1 and CXCR2 are critical for the attraction of MCs to the locations of enhanced chronic inflammation, such as a TME. The localization of MCs in the TME is governed by the engagement of CCR2, CXCR2, and CXCR3 with their cognate ligands CCL2, CXCL1, and CXCL10, respectively (Varricchi *et al* 2017). The investigation of the recruitment pattern of MCs toward gliomas revealed that glioma-derived plasminogen activator inhibitor-1 (PAI-1) enhances MC infiltration, and the amount of PAI-1 corresponds to the amount of MC infiltration (Roy *et al* 2015). In addition, the macrophage migration inhibitory factor (MIF) secreted by glioma cells promotes the infiltration of MCs via the induction of STAT5 phosphorylation (Põlajeva *et al* 2014). In addition, glioma cells secrete CXCL12, which functions as an MC chemotaxin through activation of CXCR4 (Põlajeva *et al* 2011). It has been found that both anti-SCF and anti-c-Kit antibodies impaired the infiltration of injected bone marrow–cultured MCs into inoculated H22 tumors within mice (Huang *et al* 2008).

Regarding quantity, MCs are not a disregarded population in the TME immune landscape, but their cancer-related implications have long been debated. In the majority of cancers, MCs congregate in the tumor and surrounding tissue, suggesting a universal MC-specific hallmark of cancer (Cheng *et al* 2021). Following attraction by several chemokines, like soluble SCF, CCL5, CCL11, CXCL12, CCL15, IL-3 and IL-33, MCs are triggered to execute their tasks through EVs or activity modifiers secreted by cancer cells or by direct physical connection with cancer cells (Huang *et al* 2008, Visciano *et al* 2015a, Gorzalczany *et al* 2017, Yu *et al* 2018, Gorzalczany *et al* 2019, Salamon *et al* 2020). In the TME, MCs secrete a plethora of angiogenic factors, among them VEGF, FGF-2, IL-8, heparin, and TGF-β, to foster the generation of a vasculature-supporting microenvironment and deliver essential nutrients for cancer growth (Sammarco *et al* 2019, Varricchi *et al* 2019). Through the release of VEGF-C and VEGF-D, MCs facilitate the growth and migratory capacity of lymphatic endothelial cells, whose role is essential for metastatic spread of tumors (Skobe *et al* 2001, Detoraki *et al* 2009, Karaman *et al* 2018, Cho *et al* 2022). The interaction of MCs with cancer cells influences their distinct phenotypes, functionality and even viability. In association with anti-tumor immune cells or immunosuppressive cells, MCs modulate the consequences of anti-tumor immunity. Despite comprehensive research being carried out to examine the function and mechanism of MCs in tumors, it is still difficult to reach a consensus. It may be that this issue requires a specific approach in the framework of the TME. This chapter focuses on the microenvironment in which MCs are located to elucidate the distinct characteristics and roles of MCs and reviews the interaction between MCs and various other constituents of the TME. The possible part of MCs in cancer immunotherapy is beginning to be realized and could enhance the efficacy of established cancer treatments.

3.5.2 Effect of MCs on the ECM microenvironment

The characteristics of the MC histotopography phenotype in Marfan syndrome (MFS) comprising a narrower co-localization with elastic fibers, smooth muscle cells, and fibroblasts (Atiakshin *et al* 2024). MFS is an inherited disorder characterized by disturbances in the structural and regulatory characteristics of connective tissue, involving elastic fibers. It is caused by a mutation in the fibrillin-1 protein glycoprotein FBN1 gene (15q21.1), which leads to the production of abnormal fibrillin-1 glycoprotein. Fibrillin-1 assembles into microfibrils in the ECM of connective tissue and forms an essential building block for the elastic fiber processing and physiological function. Fibrillin-1 monomers assemble into a framework that requires tropoelastin to mature into a functional elastic fiber (Wheeler *et al* 2021). Under these conditions, microfibrils function as a scaffold for elastin and assist the assembly mechanism by creating a functional architectural framework of elastic fibers (Wagenseil and Mecham 2007, Kielty 2017). Specifically, a model of elastic fiber assembly has been suggested that takes into account the interplay between elastin, LOXs, fibulins and the microfibril, along with the crucial contribution of the cells in shaping the ultimate functional fiber (Wagenseil and Mecham 2007) that points out to the dynamics nature of the MC-ECM scaffold interaction in terms of mechanical cues. Elastic fibers give the ECM plasticity and resilience to stress in the skin, blood vessel membranes, ligaments, tendons, cartilage, the 'soft skeletons' of parenchymal organs, and in other structures. The skin changes seen in most MFS patients are predominantly caused by a reduced number of fibrillin-1 microfibrils. In the dermal fibroblasts of infants with neonatal MFS, a shortening and fragmentation of the microfibrils has been observed (Godfrey 1994). Studies on fibroblasts have revealed reduced fibrillin-1 synthesis, release, and deposition into the ECM in MFS patients (Milewicz *et al* 1992). In laboratory rodents, simulated models of the condition have revealed a reduced density of elastic fibers and disordered microfibrils inside the murine cornea. In this context, fibrillin-1 plays a regulatory role in the deposition, secretion, and activation of TGF-β, which exerts a variety of biological activities (figure 3.4) (Ramirez *et al* 2004, Ramirez and Sakai 2010, Peng *et al* 2022). Although MCs have a high capacity to reshape the ECM, their pathogenetic importance in MFS remains elusive. MCs are critical in fibrosis and ECM rearrangement (Hügle 2014, Bradding and Pejler 2018, Atiakshin *et al* 2020), pointing to the implication of connective tissue homeostasis as a part of the pathogenesis in MFS. MCs produced multiple intradermal cell accumulations that synchronized cellular functions in the stromal matrix of the tissue microenvironment by means of spatial architectural features, like cell chain assembly and fibrous niche generation. Moreover, MCs may also be involved in the formation of profibrogenic tissue compartments, where permissive circumstances for the evolution of fibrotic alterations are established. The availability of certain MC proteases, like tryptase, chymase and carboxypeptidase A, and other highly biologically active secreted compounds, such as TGF-β and heparin, enables the establishment of appropriate

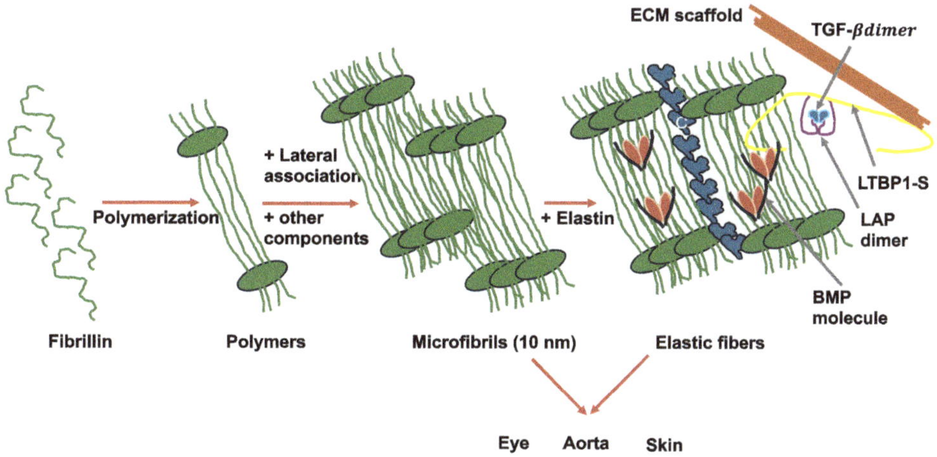

Figure 3.4. Schematic drawing of the key steps of the microfibril assembly and elastic fibers. Examples are eye, aorta and skin tissues.

conditions for the dissolution of fibrous matrix structures and fibrillogenesis, so that new fibrous matrix elements can be handled appropriately (Atiakshin *et al* 2020). In MCs, the increased expression of certain proteases, TGF-β and heparin is associated with the directed release of biologically active substances at the site of the dermal elastic fibers, which in MFS exhibited distinct structural features, such as abnormal variability in thickness along their whole length, areas of alternately thickened and thinned regions, and an irregular surface topography. Thus, a potential role of MCs in strain analysis (tensometry) of the tissue microenvironment has been proposed for MFS.

In particular, the processes of fibrosis and neoformation involving MCs can be carried out at the same time, whereby MCs act as sculptors in a distinct tissue microenvironment (Atiakshin *et al* 2020, Elieh Ali Komi *et al* 2020). The production of a wide range of cytokines and growth factors permits MCs to actively modulate the collagen- and elastin-producing components of the connective tissue and to adjust the constitution of the ECM to specific environmental circumstances (Mukai *et al* 2018, Galli *et al* 2020). This function of MCs is apparent in wound repair in every phase of healing, from inflammation to proliferation and reorganization. Activation of MCs with the subsequent secretion of preformed or *de novo* produced mediators, among them histamine, certain proteases, VEGF, IL-6, IL-8, bFGF, NGF, and PDGF, leads to an enhancement of capillary leakiness, vasodilation, and directional movement of granulocytes, lymphocytes, monocytes, and macrophages. In addition, it stimulates the cell division and enhanced biosynthetic action of fibroblasts, leads to an elevated number of myofibroblasts, improves collagen synthesis and encourages neoangiogenesis (Elieh Ali Komi *et al* 2020). Consequently, the quantitative and qualitative relocations of the dermal MC population in MFS tend to modify the stromal architecture of the connective tissue. MFS at the scale of the stromal landscape of a distinct tissue microenvironment may be auspicious regarding pharmacological targeting. Abnormal microfibrils in MFS to some degree can impair their functioning as so-called tensometers, thereby impairing the capacity of both the stromal and immunocompetent cells of the dermis to properly oversee the integrative metabolic buffering environment, which involves reactions to mechanical and gravitational impacts. Since MCs have been localized in close association with elastic fibers in a variety of pathophysiological settings (Sheu *et al* 1991, Fimiani *et al* 1995), and they secrete enzymes that can effectuate rearrangement of elastic fibers, such as elastase and cathepsin G (Boudier *et al* 1991, Blanco *et al* 2010, Xing *et al* 2011, Schmelzer *et al* 2012, Gudmann *et al* 2018, Thorpe *et al* 2018), it may be speculated that their intrinsic biological characteristics contribute to MFS. In the absence of research on MCs in this hereditary disorder, the aim of the investigation has been to characterize the MC population in MFS and to assess the potential contribution of MCs to the tensometry of the cutaneous local tissue microenvironment, as one of the essential functionalities of elastic fibers (Atiakshin *et al* 2024). It can be speculated that MC function in tensiometry plays a critical role in cancer development and progression and subsequently underpins the bidirectional (MC–ECM scaffold interaction) and the tri-directional (MC–ECM-cancer cell interplay). The MC profiles found in the dermis are likely to be characteristic of connective tissue in general, as there are no structural differences in the macromolecular composition of microfibrils. The elastic reticulum of the skin comprises bundles of microfibrils, such as oxytalan fibers, that enter the dermis from the dermo-epithelial junction, from where they combine with thin elastin-containing fibers, such as elaunin fibers, to create a reticulum of thick elastic fibers in the dermis's more profound layers. Microfibrils in the assembly of elastic fibers are regarded as structural 'tensometers'. Their conformational rearrangements upon mechanical stretching can alter cell migration, functionality and differentiation through integrin receptors in an indirect manner (Kielty 2017). In this regard, the impact of MCs on the elastic compartment of the dermis is of special concern, since the targeted secretory activity in proximity to elastic fibers leads primarily to the secretion, along with heparin and specific proteases, to engage with fibrillin-1 and microfibrils (figure 3.4). Therefore, the ability of the MC to counteract the decrease in the natural interstitial tensinometry of the elastic fibers through the generation of spatial clusters in the connective tissue basement of the skin should be the subject of further investigations. In fact, the microscopic structures that have been identified indicate a more active creation of functional entities in MFS than is usually the case. This process involves large regions of skin and coordinates a large population of cells from both the immune and stromal landscapes.

Moreover, the question of the MC phenomenon in elastic fibers requires careful consideration. The breakdown impact of MC on elastic fibers is recognized and proven in some works in terms of elastinopathy, especially mid-dermal elastolysis marked by selective depletion of elastic fibers in the mid-dermis (Fimiani *et al* 1995). In the formation of skin striae atrophicae, tight co-localization of degranulating MCs with disrupted elastic fibers has been found on the ultrastructural scale (Sheu *et al* 1991). In particular, the elastic fibers became disordered, with irregular borders and/or indications of elastolysis, and they no longer had the specific peripheral organization of microfilaments (Sheu *et al* 1991, Gudmann *et al* 2018). Patterns of elastic fiber stress relaxation alterations have been identified at the light-optical scale (Atiakshin *et al* 2024). Nevertheless, the theoretical possibility that MCs can break down elastic fibers is underpinned through the generation of the corresponding enzymes, like elastase and cathepsin G (Boudier *et al* 1991, Blanco *et al* 2010, Xing *et al* 2011, Schmelzer *et al* 2012, Gudmann *et al* 2018, Thorpe *et al* 2018). The active release of tryptase, chymase, CPA3 and heparin straight onto elastic fibers that has been seen could account for their direct participation in the metabolism of the fibrous element of the dermis. In this context, a logical question poses itself: why is it that the relative proportion of elastic to collagen fibers has not altered considerably? Moreover, the capacity of the MCs to carry out the resynthesis of elastic fibers following their breakdown into specific structural and molecular entities that can then be reused to generate new elastic fibers can be highlighted. This finding may account for the absence of a significant reduction in the integral surface area of elastic fibers found in the skin biopsies. Beforehand, the potential of MCs to take part in collagen fibrillogenesis as a result of direct and indirect actions has been analyzed (Atiakshin *et al* 2020, 2023). This assumption can be used to clarify the mechanisms of elastic fiber reshaping, as the action of heparan sulfate and perhaps heparin, on the activity of the fibrillin-1 molecule's interplay during linear polymerization of microfibrils is clear (Mayer *et al* 2008, Kielty 2017, Herrera-Heredia *et al* 2022). The resemblance between heparan sulfate and heparin increases the potency of MCs when they are engaged in elastic fiber remodeling. Although MCs and basophils represent two primary sources of heparin in humans, MCs are recognized as the major source of heparin in skin because of their long-standing residence and abundance in cutaneous tissues. (Shriver *et al* 2012, Meneghetti *et al* 2015, Maccarana *et al* 2022). In this regard, an elevated number of MCs can be regarded as a compensatory response related to the requirement to provide supplemental ECM strength.

Conversely, the potential contribution of MCs in the progression of fibrotic alterations should be emphasized. Specifically, the development of fibrotic niches in skeletal muscles has been shown to be due to mutations in the FBN1 gene in MFS; this leads to augmented collagen fiber and fibroblast abundance (Tomaz Da Silva *et al* 2023). In the aorta of MFS patients, there was enhanced expression of contractile protein markers, which contained α-SMA, aggregation of type I collagen, fragmentation, disappearance or augmentation of interlaminar elastic fibers, and deposition of acid mucopolysaccharides (Yuan *et al* 2011, Crosas-Molist *et al* 2015). Because MFS characterizes the pathobiology of fibrillin-1, it is likely that both the architecture of the intercellular crosstalk and the mechanisms of fibrillin-1 engagement with ECM constituents are perturbed to some degree. Dysfunctional fibrillin-1 is unable to produce latent TGF-β-binding protein (LTBP) and enriches TGF-β within the ECM, leading to enhanced secretion of the TGF-β (Chaudhry *et al* 2007, Zeigler *et al* 2021). Upon liberation, TGF-β engages with the TGF-β receptor family, leading to the control of physiological processes including angiogenesis, cell proliferation, differentiation, apoptosis, and the restoration and reorganization of the architecture and constituents of the ECM (Massagué 2000, Bertolino *et al* 2005, Chandiran and Cauley 2023, Chen *et al* 2023). TGF-β has been found to have the capacity to cause exaggerated collagen generation and elastic fiber breakdown, and to promote the proliferation and motility of vascular smooth muscle cells (Nataatmadja *et al* 2006, Matt *et al* 2009). These results indicate that TGF-β production is enhanced in MCs. Moreover, MCs were found to actively release the growth factor at tissue target sites, thereby seemingly counterbalancing its deficiency in tissue repositories (niches). Obviously, this leads to enhanced production of TGF-β in MCs. In addition, to enhancing their functionality, MCs can move directly to cellular tissue destinations with no storage, whereupon growth factors are released (Atiakshin *et al* 2024).

The increased frequency of MCs with positive expression of CD117 leads to the hypothesis that the MC population in MFS has a high capacity for continued differentiation, growth, viability, and high levels of effector functions. The elevated frequency of MCs is a hallmark of their active involvement in the metabolic remodeling of the dermal stroma in MFS. Active secretory activity, specifically of proteases, highlights the part played by the ECM in the evolution of the fibrous framework in reply to the diminished strength of the elastic element. In relation to the aforementioned, the compensating actions of MCs turn into pathological effects;

this may account for the appearance of fibrosis in other types of organs (Yuan *et al* 2011, Suarez *et al* 2014, Crosas-Molist *et al* 2015, Karur *et al* 2018, Tomaz Da Silva *et al* 2023).

3.5.3 MC and ECM in the progression of cancer (bidirectional interaction)

ECM governs a wide range of biological processes in both normal and cancer cells, involving the cell's locomotion and adhesion (Rigoni *et al* 2015). In turn, the resident cancer cells, fibroblasts, endothelial cells, mobilized inflammatory cells and pericytes release ECM components (Rigoni *et al* 2015). MC-derived tryptase can induce neovascularization through the activation of MMPs. These enzymes break down the ECM and secrete angiogenic factors (Ammendola *et al* 2017). Tryptase can activate the plasminogen activator and cause the liberation of VEGF and FGF-2 from their tethered state, where they are attached to the ECM (Hu *et al* 2018). The accumulation of MCs in tumors, such as breast cancer, has been detected and MC-derived tryptase can initiate to induce fibroblast differentiation and foster stromal restructuring (Mangia *et al* 2011). MMP-9 belongs to the most important MMPs able to digest and reshape ECM, resulting in a modification of the cellular microenvironment. E-cadherin molecules, linking epithelial cells at adhesion junctions, are split and their expression is reduced under the impact of MC chymase. Chymase is additionally found to enhance MMP-9 expression in cancer cells. Chymase facilitates the disconnection, proliferation and displacement of cell clusters by directly and indirectly acting on the ECM to sustain metastasis (Jiang *et al* 2017).

3.6 Effect of mechanical cues, such as stiffness, on MCs in cancer

The pro-tumorigenic or anti-tumorigenic function of tumor-infiltrating MCs in tumors may vary according to the type of cancer, the level of cancer advancement and the positioning of MCs in the tumor mass. Immunohistochemistry has been used to show that MCs in the tumor stroma are positively linked with the metastasis of ovarian cancer, while this is not the case for MCs located in the tumor parenchyma. To further investigate the effect of the varying stiffnesses of the culture matrix on the biological functionality of MCs in the tumor parenchyma and stroma compartments, a transcriptome analysis of the mouse MC line P815 cultivated in a two-dimensional (2D) or three-dimensional (3D) cell culture set-up was performed. Additional investigations have shown that the softening of the 3D ECM could enhance the mitochondrial activity of MCs to facilitate proliferation via raising the expression rates of mitochondrial-related genes, including Pet100, atp5md, and Cox7a2. Moreover, least absolute shrinkage and selection operator (LASSO) regression analysis identified that Pet100 and Cox7a2 are associated with survival rate in ovarian cancer patients. Subsequently, these two genes have been used to create a risk assessment tool that demonstrated that the high-risk patient group is among the prognostic indicators for ovarian cancer patients. In addition, XCell algorithm analysis indicated that the high-risk group harbored a broader range of immune cell infiltrations. Tumor-infiltrating MCs in the tumor stroma can enhance ovarian cancer metastasis, and the proteins Pet100/Cox7a2 associated with mitochondrial activity could function as prognostic biomarkers for ovarian cancer (Ma *et al* 2024). Consequently, the stiffness of the TME can affect the expression of genes in cancer cells that are candidates for biomarkers of cancer types and the malignant progression of cancer.

3.7 Two ways of MC activation

In the classical sense, MCs can be activated via various membrane receptors that engage with their specific ligand. The non-classical activation of MCs is the mechanical activation via mechanosensory molecules expressed on the cell surface. Apart from the two main activation types, another type of MC activation has been proposed, such as the activation of immune cells, such as MCs is based on the dynamics of the actin cytoskeleton (Colin-York *et al* 2019). Although the molecular assembly of actin is highly regulated, the precise processes that control actin organization in response to activation continue to be elusive. With the help of advanced microscopy, it is demonstrated that rat basophil leukemia (RBL) cells, used as a model for MCs, deploy a coordinated sequence of reorganization steps within cortical actin networks in response to activation. Self-organization demands collective behavior of interacting molecules, which are well away from thermodynamic equilibrium and are propelled by the constant energy influx into a stationary structure, which is typical of reaction–diffusion processes (Misteli 2001, Hannezo *et al* 2015). In fact, cellular order arises as a result of a mixture of complex deterministic interactions, named self-organization, that are triggered by distinct signaling cues and dynamic intermolecular interactions that depend on energy dissipation, named self-organization (Karsenti 2008, Battle *et al* 2016). The actin cortex meets all the requirements of self-organization (Vignaud *et al* 2012, Tan *et al* 2018). The system permanently expends energy to sustain its steady state, and perturbations of local or global biophysical quantities, like mechanical stress, can lead to a sudden symmetry breaking (Van Der Gucht *et al* 2005, Abu Shah and Keren 2014). Symmetry breaking refers to a phenomenon where small fluctuations applied to a system that passes a critical point determine the steady state of the system (Vignaud *et al* 2012, Abu Shah and Keren, 2014). This type of symmetry breaking leads to network instabilities that may quickly lead to new orders like the emergence of distinct filamentous actin (F-actin) networks (Fritzsche *et al* 2017). The application of self-organizing principles permits cells to quickly transform their F-actin meshworks, for example, from isotropic random meshworks into ordered F-actin meshworks that are organized by actin patterning like actin vortices (nodes) and asters (Kruse *et al* 2004). The symmetry breaking of the F-actin network can be seen in reply to the IgE-antigen stimulation of the FcεRI on the plasma membrane of the RBL cells. Subsequently, a reassembly procedure occurred that could be powered by the orchestrated conversion of several nanoscale F-actin structures that resemble self-organizing actin patterning. Actin structures colocalized with areas of Arp2/3 complex assembly, whereas the re-establishment of the actin network was concomitant with areas of myosin II activity. Remarkably, the disassembly of cortical actin co-occurred with granule secretory compartments, implying that cytoskeletal actin organization participates in orchestrating the activation of RBL cells. Although the findings seem to be promising, care needs to be taken that RBL cells have been employed as model system for MCs.

3.7.1 Traditional activation of MCs

In adults, myeloid cells arise from multipotent hematopoietic stem cells located in the bone marrow (St. John *et al* 2023). Human MC precursors (Lin–CD34hiKITmid/hiFcεRI) are a rare subset in circulating blood, comprising only a small fraction of leukocytes in healthy adults (Dahlin *et al* 2016). Circulating MC precursors are recruited by chemokines and cytokines at the interface of the internal and external environment, where they differentiate into resident cells and acquire diverse phenotypes and become functionally heterogeneous (Derakhshan *et al* 2022). Abundant tissue-resident MCs expressing an immature phenotype also multiply in response to infection or chronic inflammation (Rathore and St John 2020).

The activation of MCs is based on their different membrane receptors (figure 3.5). Among the most extensively explored MC receptors is the high-affinity receptor for the Fc region of IgE, referred to as FcεRI or Fc epsilon RI. Various receptors can be found on the surface of MCs that form the basis for the division into different subtypes (Cildir *et al* 2021). FcεRI represents a unique receptor which is detectable in all MCs. The key function of MCs in innate immunity is in allergic diseases and is focused on the activation of FcεRI.

When exposed to xenogeneic antigen, B cells present it toward naive T cells (Wu and Zarrin 2014) and liberate IgE to transfer the immune signal via the blood flow

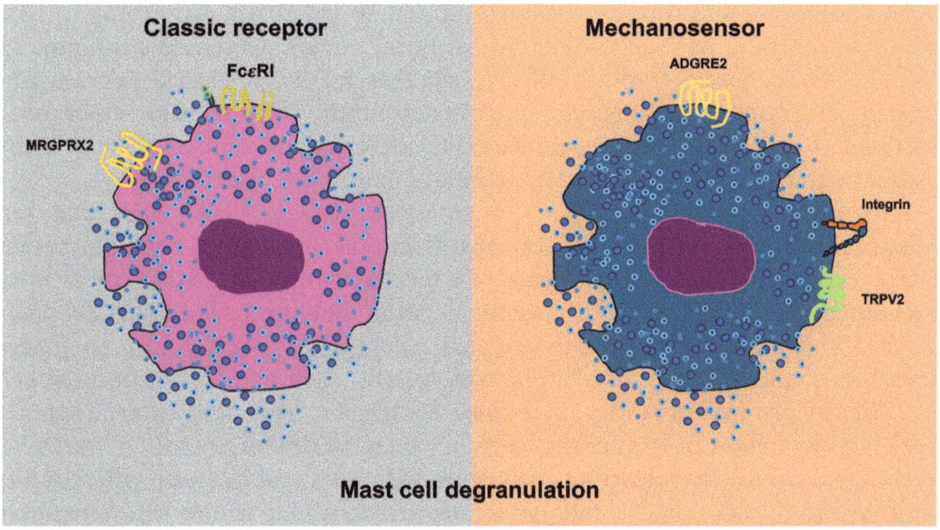

Figure 3.5. Traditional receptors and mechanosensors on human MCs. MCs express multiple receptors and react to various cues. FcεRI and MRGPRX1 represent canonical receptors that govern innate immunity, itch and neuroinflammation. TRPV2, ADGRE2 and integrins are identified as sensors for physical forces on the molecular scale.

toward other cells. Typically, non-attached IgE can remain in peripheral blood for two to four days before it is tethered to FcεRI or broken down. Once the Fc fragment of IgE engages with FcεRI on the surface of MCs in a 1:1 binding stoichiometry (McDonnell *et al* 2023), the activation of an intracellular signal transduction pathway is initiated in MCs, leading to an instant widespread exocytosis of the pre-synthesized allergic effectors via degranulation and accelerating the generation of allergic vesicles (Arthur and Cruse 2022). Moreover, upon the attachment of IgE to MCs, leads to the elevation of MC expansion within the gastrointestinal tract and skin (Kwon *et al* 2022). The immune system is sensitized when FcεRI is coupled with IgE, thereby becoming capable of recognizing bacteria, fungi, protozoa, worms, viruses and every other foreign substance. MRGPRX2 is another highly expressed receptor in human dermatological MCs (Subramanian *et al* 2016). Expression of FcεRI on MCs is widely distributed, whereas MRGPRX2 is primarily expressed in a specific subtype of human MCs referred to as chymase^{+}, which are abundant in connective tissues. In contrary, the mucosal subtype of MCs is deficient in the chymase enzyme and exhibits low MRGPRX2 expression (The Immunological Genome Project Consortium *et al* 2016). This indicates that MRGPRX2 can exert a more prominent role in the chymase^{+} subtype of MCs. In a similar fashion to FcεRI, activation of MRGPRX2 leads to the intracellular recruitment of Ca^{2+} ions (Gaudenzio *et al* 2016). The creation of calcium responses will trigger a downstream reaction in the endoplasmic reticulum and Golgi apparatus, and the induction of chemokines and cytokines synthesis (Ma and Beaven 2011). Nevertheless, the pathophysiological responses triggered through these two receptors vary in their particulars. For instance, activation of Mrgprb2 and FcεRI triggered itch in a mouse model that relies on cutaneous nociceptor afferentation for signal transduction (Tsang and Wong 2020). According to recent studies, in contrast, MC activation by Mrgprb2 causes decreased histamine and serotonin but increased tryptase release, whereas IgE/FcεRI-based degranulation primarily liberates histaminergic pruritogens into the internal environment (Meixiong *et al* 2019). These results help elucidate why several causes can lead to persistent cutaneous pruritus in various inflammatory disorders.

Apart from the mechanosensory behavior of cancer cells, MCs can sense mechanical cues like stiffness. Several investigations suggested a coupling between the activation of MCs and the presence of a mechanical signal. MCs express several mechanoreceptors, including transient receptor potential vanilloid channel 2 (TRPV2) and adhesion G-protein coupled receptor E2 (ADGRE2), to perceive external mechanical forces. Mechanosensitive cells, such as neurons and epithelial cells, release pro-inflammatory cytokines to trigger MC activation when they experience scratches or traumatic stimuli. Integrin-driven signaling paths that perceive mechanical characteristics will cause MCs to travel to areas with variations in stiffness. In this section, the linkage of mechanical stimulus and MC activation is presented and discussed. Moreover, evidence of MC activation through mechanical signals is also seen in fibrosis-driven diseases. The elasticity and linearity of the tissue's mechanical characteristics intensify as fibrosis progresses. The mechanical features of fibrotic microenvironments are therefore both the consequence and the driver of fibrosis advancement (Long *et al* 2022). There is clear experimental evidence that the abundance and density of MCs is increased in pathological fibrotic tissues compared to normal control tissues (Sun *et al* 2021). It was found that mechanical stretching of the lung directly induces the degranulation of MCs and eventually triggers the TGF-β1 signaling cascade, which is the key signaling mechanism in pulmonary fibrosis.

3.7.2 MC activation due to mechanical cues

Mechanical cues have been observed to elevate the density of MCs and their degranulation rate in humans and certain disease models. Mechanical stretching leads to an agglomeration and degranulation of MCs in fibrotic rat lungs, resulting in a heavier fibrosis (Shimbori *et al* 2019). In contrast, the MC stabilizer sodium cromoglicate decreases mechanical stress-induced MC degranulation and abrogates the activation of the TGF-β1 signaling cascade, which is frequently found to be activated in fibrotic disorders. All of this suggests that MCs in fibrotic tissue are highly sensitive to mechanical cues and that the response of MCs to mechanical cues fosters fibrosis. The crosstalk between mechanical stimuli and inflammation turns into a downward spiral due to MCs. Hence, a mechanical cue can be a decisive factor for the activation of MCs in pathological processes.

3.7.2.1 Direct activation through a mechanosensor, such as a mechanoreceptor

Mechanoreceptors can be detected on the membrane of MCs, which transduce mechanical forces into biochemical messages in a chain reaction and regulate several features of cell metabolism. TRPV2 appears to be highly elevated in mechanosensitive MCs (Plum *et al* 2020) and has been shown to attenuate MC activation during physical challenge (Stokes *et al* 2004, Zhang *et al* 2022). TRPV2 belongs to a family of so far six transient receptor potential vanilloid channels (TRPVs). These ion channels react very sensitively to physical alterations like mechanical stress, shear forces, and temperature alterations (Rosenbaum and Islas 2023). In an animal pain research study, a TRPV2-knockout model in male mice has been utilized to implicate TRPV2 in the activation of MCs at acupuncture points and to implicate the liberation of MC mediators in the acupuncture-triggered analgesic response (Huang *et al* 2018). TRPV2 activation enables Ca^{2+} inflow, markedly elevates mobility, and decreases myeloid cell membrane tension, ultimately leading to MC degranulation (Zhang *et al* 2012). Loratadine, the antihistamine, is capable of decreasing TRPV2-mediated calcium inflow, constituting a potential mechanism for the healing of physical urticaria (Van Den Eynde *et al* 2022). In neurons, TRPV2 intimately engages with actin. Mechanical stimulus-driven actin reorganization leads to TRPV2 activation, which in turn increases growth cone motility and actin aggregation, thus facilitating axonal extension. TRPV2 is also necessary for directional motility in various cell types, among them macrophages (Link *et al* 2010), prostate cancers (Monet *et al* 2010), bladder cancer (Liu and Wang 2013), and esophageal squamous cell carcinoma (Kudou *et al* 2019), nevertheless, TRPV2 has so far not been explored in MCs.

Adhesion G-protein coupled receptor E2 (ADGRE2), belonging to the adhesion G-protein coupled receptor subclass, has been implicated in familial vibratory urticaria, a dermatology caused by MCs and a rare form of chronic inducible urticaria where weals result from vibration of the skin (Boyden *et al* 2016). Moreover, the substitution of cysteine by tyrosine in ADGRE2 (pC492Y) renders MCs sensitive to IgE-independent vibration-triggered degranulation (Yang *et al* 2023). As a reaction to local frictional stimulation, vibration urticaria sets off systemic reactions that are initiated by the activation of MCs in the skin and are accompanied by a fast build-up of histamine (McSweeney *et al* 2023). Human MCs that were transfected with pC492Y-ADGRE2 or that originated from patients with familial vibration urticaria underwent destruction when exposed to minimal mechanical force. These data indicate that the encoded adhesion receptor GPCR acts as a mechanosensor in MCs. Subsequent analyses demonstrated that ADGRE2 sheds its extracellular N-terminal fragment in reply to the vibrational cue that is recognized as tethered-ligand activation (Naranjo *et al* 2020). Signal transduction pathways like PLC-β, Akt, MAPK or NF-κB are activated together with $Ca2^{+}$- and K^{+}-mobilization (I *et al* 2021).

While TRPV2, ADGRE2, and other mechanoreceptors have been shown to be involved in MC activation, it is unclear why mechanosensing by MCs would be advantageous for humans. For instance, when MCs react mildly to friction and gradually secrete proteases, a gentle degranulation can serve to loosen stiff tissue and ease tissue healing, rather than cause aggressive and potentially harmful reactions. Evolutionary biology or population genetics are proposed to reveal more substantial evidence to enlighten this issue.

3.7.2.2 Indirect activation through neighboring different cells

The presence of MCs in the vicinity of mechanoreceptors of different cell types may contribute to the development of an inflammatory response via cytokine release. A typical scenario is the activation of intestinal MCs by dermal keratinocytes upon skin scratching. For a long time, it has been established that food allergy with intestinal MC expansion can be due to a leaky skin barrier (Brough *et al* 2020). In a process that is not well understood, the immune system of the dermis picks up antigens and secretes IgE into the bloodstream following scratching that disrupts the normal skin barrier (Bartnikas *et al* 2013). Regrettably, a passive sensitization model in mice using mechanical skin wounding reveals that a hardly mechanical cue can trigger dermal keratinocytes to liberate a lot of IL-33, which diffuses through the circulation towards the digestive tract, where it triggers proliferation and activation of MCs to generate pro-inflammatory cytokines (Leyva-Castillo *et al* 2019). When secreted into the gut, these cytokines would stimulate the tuft cells to release IL-25, which causes the MCs to become further sensitized. The relentless activation of gut MC not only unleashes a cascade of cytokines that directly produce digestive symptoms, but also speeds up the absorption of allergens, setting the stage for debilitating allergies in the future.

Neurons also indirectly contribute to the activation of MCs. MCs are frequently located near sensory nerve fibers and in the vicinity of blood vessels (Mukouyama *et al* 2002, Pundir *et al* 2019), where neuron dendrites and MCs build neuroimmune synapses (Forsythe 2019). Through this specific structure, sensory neurons can sensitively and accurately regulate MCs through adapting neuropeptides and neurotransmitters secreted by them. In general, sensory neurons recognize mechanical cues via TRPV1 and ankyrin 1 (TRPA1) (Moon *et al* 2014). Following depolarization, the substances P. and vasoactive intestinal peptide are liberated at a synapse and trigger MRGPRX2 activation on MCs (Kulka *et al* 2008). These results indicate that MRGPRX2 drives mechanical irritation-driven MC activation through both direct and indirect mechanisms (Yu *et al* 2017, Pundir *et al* 2019). Also, MCs could activate neurons by liberating pro-inflammatory cytokines and cause an activation circuit towards the development of a pathological condition. More worryingly, mechanical cues in the peripheral nervous system can trigger the vagus nerve-innervated MCs in the brain, eventually setting off a neuroinflammatory pathway in the central nervous system (Yang *et al* 2022).

3.8 Effect of mechanical cues, like stiffness, on MC infiltration of solid tumors and TMEs

Early infiltration of the MCs was observed in a variety of human and animal cancers, especially in malignant melanomas, breast and colon cancer. Inside the tumor, interactions arise between MCs and infiltrated immune cells, cancer cells and the extracellular matrix, either via direct cell–cell interactions or the secretion of a variety of mediators that can remodel the TME. MCs actively participate in angiogenesis and drive neovascularization by secreting both classical proangiogenic factors, like VEGF, FGF-2, PDGF, and IL-6, and non-classical proangiogenic factors, primarily proteases including tryptase and chymase. Macrophages enhance invasiveness of cancers through secretion of a wide range of MMPs. The presence of MCs in tumors became even more significant when it appeared that the control of their activation through tyrosine kinase inhibitors like imatinib and masitinib and tryptase inhibitors like gabexat and nafamostat mesylate, or the regulation of their interactions with other cell types, might have certain therapeutic value.

Inside the tumor, MCs secrete angiogenic compounds, among them IL-8, VEGF, FGF-2, NGF, heparin, tryptase, chymase, and TGF-β. Moreover, MMP-2 and MMP-9 liberated from MCs can foster the tumor vascularization and encourage cancer invasiveness (Ribatti and Crivellato 2012). Multiple different cytokines liberated by MCs encompassing IL-1, IL-4, IL-8, IL-6, MCP-3, MCP-4, TNF-α, IFN-γ, LTB4, TGF-β, and chymase promote the development of inflammation, impair the growth of tumors and evoke apoptosis of cancer cells (Ribatti and Crivellato 2012).

A major question is whether the ECM stiffness can modify the migratory capacity of MCs. Apart from the generation of in-site MCs, MC migration out of the peripheral blood and adherence to tissue contribute pronouncedly to the number of activated MCs at pathological regions that underwent mechanical stimulation. The process of cell migration is generally divided into mesenchymal and amoeboid migration modes within the ECM. The majority of immune cells exhibit amoeboid migration modes, with little adhesion to the underlying substrate, a roundish shape, and migration through tissues by fast protrusion and ingression of membrane extensions, which are controlled through G-protein-coupled receptor signal transduction pathways (Lämmermann *et al* 2008, Moreau *et al* 2018). In contrast, MCs depend on mesenchymal migration, an event that relies heavily on the attachment to a substratum (Kaltenbach *et al* 2023). This finding seems to be quite unexpected since MCs are immune cells that have transmigrate out of the vasculature. The expression of integrins at the periphery confers pronounced adhesiveness on MCs. A mechanical cue modifies the microcosmic form of collagen fibers, strands of muscles, nerve fibers and vascular channels within the cell matrix, all of which can connect to integrins on the MC plasma membrane. The unique migratory mode for immune cells indicates that MCs react to altered tissue characteristics in response to mechanical signals and pathophysiological states as potential mechanosensitive immune cells.

MCs inside the peripheral tissue can sense the microenvironmental stiffness through integrins bound to ECM fibers (Saraswathibhatla *et al* 2023). Integrins consist of heterodimeric receptors, which are made up of an alpha and a beta subunit that are not covalently linked. 18 α-subunits (α1–11, αIIb, αD, αE, αL, αM, αV, αX) and 8 β-subunits (β1–8) have been identified, constituting four classes of functional surface receptors, such as leukocyte integrins, collagen-binding integrins, Arg-Gly-Asp (RGD)-binding integrins, and laminin-binding integrins (Kadry and Calderwood 2020). Most integrin subunits have been identified on the plasma membrane of MCs (Krüger-Krasagakes *et al* 1996, Mierke *et al* 2000, Rosbottom *et al* 2002, Berlanga *et al* 2005, Tegoshi *et al* 2005, Abonia *et al* 2006, Sugimoto *et al* 2012, Pastwińska *et al* 2020). The α and β subunit pairing defines the distinct affinity for bioactive amino acid sequences and results in distinct engagement of different ligands like collagen, fibronectin, vitronectin and laminin (Plow *et al* 2000, Hynes 2002). As mechanosensitive receptors, integrin conformation alters in response to external forces and thereby integrins are activated. In particular, the mechanical cue can switch integrins from a bent configuration to an outstretched configuration, which augments the interaction of the integrin with the ligand present on the ECM (Kechagia *et al* 2019). The differential affinity for ECM ligands drives the migration of MCs towards the stiffer end of a substrate featuring a stiffness gradient, which is referred to as durotaxis (Lo *et al* 2000). A research on substrates with different stiffnesses revealed that rat MC lines spread towards regions of high stiffness (Yang *et al* 2018). In another study, a MATLAB program has been used to determine migration speed, persistence index and rotation angles of MC upon culture on polyacrylamide gels exhibiting a gradient in stiffness (Yu *et al* 2021). Ultimately, it has been observed that there is an intermediate level of optimal stiffness where the migration efficiency is the maximum. These findings indicate that the stiffness in the MC migration depends on the stiffness of the ECM.

Primary human small airway epithelial cells (HSAECs) exhibited relative resistance to tryptase. In opposition, chymase broadly affected multiple epithelial cell traits, with a focus on ECM reorganization-related events (Zhao *et al* 2022b). Among these was the inhibited expression of ECM-related genes encoding MMPs like MMP-2 and MMP-9, as verified at the protein level. HSAECs treated with chymase alone or in combination with tryptase released less pro-MMP2 and pro-MMP9 compared to untreated controls or tryptase-treated cells indicating that chymase suppresses the expression of these enzymes. For MMP9, there was a decline in proenzyme and active enzyme concentrations in response to chymase treatment. Conversely, while the amount of pro-MMP2 was decreased in reaction to chymase, active MMP2 was seen only after chymase exposure. The latter observation is consistent with the established ability of chymase to activate pro-MMP2 (Lundequist *et al* 2006, Groschwitz *et al* 2013, Zhao *et al* 2022a). Moreover, chymase induces a slight increase of TIMP-2 release within epithelial cells and the elevated levels of this molecule could therefore aid to decrease metalloprotease levels within tissues. An impact of chymase is the restructuring of the ECM that is aided through the suppression of genes encoding for various collagens, such as COL1A1, COL4A1, COL7A1, and COL22A1, and pro-fibrotic growth factors like TGF-β 1 and FGF1. In addition, chymase also suppressed the fibronectin gene expression and induced the breakdown of fibronectin secreted from epithelial cells. Chymase has also been demonstrated to inhibit the migratory potential of airway epithelial cells and to break down the cell–cell contact protein E-cadherin on the surface of epithelial cells (Zhao *et al* 2022b). In carcinoma, the interaction between MCs and cancer cells may also occur via chymase and its ECM restructuring effects, such as suppression of MMPs and fibronectin and simultaneously breakdown of existent fibronectin.

3.9 Communication between MCs and cancer cells in solid tumors

In TME, the interaction between cancer cells and the non-cancer cells in their vicinity is bidirectional. The mechanisms used by cancer cells to attract and activate MCs and trigger the release of mediators are outlined earlier. In the following section the impact of MCs on the phenotype and functionality of cancer cells. The capacity to sustain continued proliferation is a hallmark of cancer cells (Hanahan and Weinberg 2011). What is important about MCs is that they cause cancer cells to multiply. In one mechanism, activated MCs secrete bioactive mediators like IL-6, IL-17A and exosomes that directly augment the proliferation of cancer cells. In primary cutaneous lymphoma, MCs secrete or trigger the secretion of IL-6 *in vitro*, which enhances the clonogenic proliferation of primary cells (Rabenhorst *et al* 2012). IL-6 has a profound impact on cancer cells, directly inducing the expression of STAT3 target genes, which encode proteins that promote cancer proliferation or survival (Fischer *et al* 2003). Hence, MCs are also accepted as facilitators of multiple myeloma pathogenesis and pathophysiology. Similarly, in an *in vitro* assay, IL-17A released by MCs stimulated gastric cancer cell proliferation and blocked apoptosis, and this effect was abolished by inhibition of IL-17A using neutralizing antibodies. *In vivo* mouse experiments also demonstrated that MC-deficient mice, in which either MCs plus IL-17A neutralizing antibodies or MCs from IL-17A-deficient mice were injected, had significantly less cancer cell proliferation, smaller tumor volume, and slower disease progression compared to control mice, which were injected with either MCs plus control immunoglobulin or wild-type mouse MCs. These findings revealed that IL-17A generated by MCs also fosters the proliferation of cancer cells, whereas the distinct mechanism is still unclear (Yu *et al* 2018). Exosomes released from MCs can have a proliferation-promoting effect. For instance, MCs secrete exosomes carrying the Kit protein, which when captured from lung adenocarcinoma cells activates the PI3K/AKT pathway and enhances the proliferation of cancer cells (Zhu *et al* 2007). In another mechanism, MCs also induce proliferation of cancer cells, by being involved in direct cell-to-cell interactions, in parallel to their effect of secreting active substances. In Hodgkin's lymphoma, MCs represent the most important CD30L-expressing cells, interacting through the CD30-CD30 ligand with Hodgkin and Reed–Sternberg cells to enhance the synthesis of DNA in the latter (Gruber *et al* 1995). Therefore, trials investigating the clinical utility of CD30L-expressing MCs for the future treatment of Hodgkin's lymphoma seem promising.

As the primary tumor advances, additional localized invasion of cancer cells and distant metastases at specific sites can be seen. The transformed cancer cells change their shape and their adhesion to the ECM and to the neighboring cells (Hanahan and Weinberg 2011). MCs have been reported to trigger enhanced invasiveness and metastasis of a multitude of cancer cells. This characteristic could be linked to the developmental control program that causes cancer cells to undergo epithelial-to-mesenchymal transition (EMT). The EMT can be full EMT or hybrid, partial or incomplete EMT, whereby the different stages are reversible (figure 3.6) (Jolly 2015, Brabletz *et al* 2021).

The plasticity of cancer cells, which enables continuous and reversible adjustment to constantly fluctuating conditions, is achieved through the EMT mechanism, which is not genetically determined but instead relies on accumulating mutations. Nevertheless, it is controlled epigenetically by cues from the microenvironment, causing the entire program to be reversible (through activation of the mesenchymal-epithelial transition; MET) and to be subject to strong fluctuations (Brabletz *et al* 2021). The EMT program is executed primarily by a core group of EMT-activating transcription factors (EMT-TFs), among them SNAIL, SNAI1, SLUG, SNAI2, the basic helix-loop-helix factors TWIST1, which is synonymously referred to as TWIST, and TWIST2, and the zinc-finger E-box-binding homeobox factors ZEB1 and ZEB2. All these regulatory factors have the capacity to suppress epithelial genes such as the E-cadherin-encoding gene CDH1 through direct interactions with E-box motifs in their corresponding promoter regions (Nieto *et al* 2016), which has been detected for SNAIL (Batlle *et al* 2000, Cano *et al* 2000), TWIST (Yang *et al* 2004), ZEB1 (Eger *et al* 2005), and ZEB2 (Comijn *et al* 2001). In tandem, EMT-TFs directly or indirectly induce the activation of genes related to a mesenchymal phenotype, comprising vimentin (VIM), fibronectin (FN1) and N-cadherin (CDH2) (Nieto *et al* 2016, Dongre *et al* 2017, Dongre and Weinberg 2019). Many functions, instead of being co-opted, are carried out by different EMT-TFs because they vary in terms of aspects such as expression patterns or widely differing protein size and structure (Stemmler *et al* 2019). The pleiotropic functions of EMT-TFs in cancer biology are demonstrated by their widespread significance, which goes far beyond

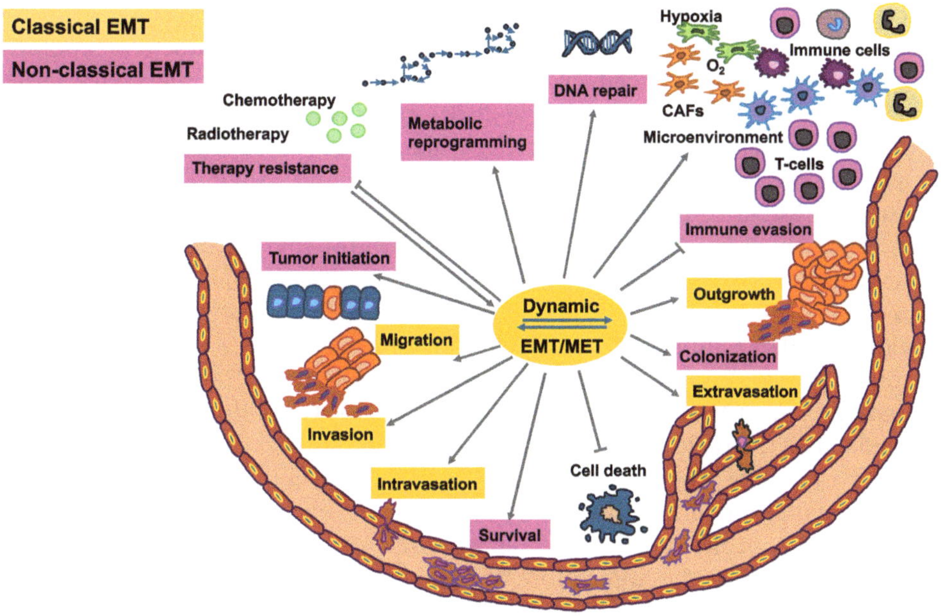

Figure 3.6. Dynamic EMT represents a versatile element in the advancement of cancer. Schematic representation of the different oncogenic features of EMT during tumor advancement. The classic EMT functionalities include enabling cancer cells to migrate, invade, and enter and exit blood and lymph vessels. In the distant sites, MET drives expansion of macrometastases. The non-classical EMT features assist in tumor initiation and subsequent metastatic dissemination. Along the entire process of tumor evolution, they assist cells to adapt to altered circumstances through reprogramming metabolism, improving survival through modified DNA repair and cell death avoidance, escaping immune surveillance and enhancing resistance to chemotherapy and radiotherapy. The important aspect is that EMT not only assists cancer cells in adapting to altered environmental circumstances, but is also triggered by extracellular cues, e.g. from CAFs or immune cells in the TME, or by therapeutic interventions.

the 'classical' EMT characteristics of increased motility and invasive ability (Brabletz *et al* 2018). EMT-TFs have been proven to preserve stem cell characteristics and enhance tumorigenicity, thereby connecting them to the idea of cancer stem cells (CSCs). Moreover, EMT-TFs are implicated in DNA damage repair, evasion of senescence and apoptosis, treatment resistance, and immune escape, leading to a survival-promoting phenotype and conferring an advantage under several types of stress circumstances. Overall, the classic EMT features, along with the highly variable, non-redundant, context-dependent, non-classical features of EMT-TFs, which are also dynamically controlled through TME, enable cancer cells to continually adjust to altered circumstances (Puisieux *et al* 2014, Bar-Hai *et al* 2024). Therefore, therapeutic interference with EMT/plasticity offers the possibility of combating multiple facets of cancer development in just one fell swoop.

In the early stages of cancer development, cancer cells express an epithelial phenotype and progressively acquire more mesenchymal characteristics as the lesion advances. EMT represents an evolutionarily conserved cellular mechanism that transiently displaces epithelial cells in the direction of a mesenchymal cell phenotype. EMT occurs during cancer development and progression. It increases the mobility, invasiveness and capacity of cancer cells to withstand apoptosis, thereby endowing cancer cells with carcinogenic and metastatic characteristics (Ronca *et al* 2017, Komi and Redegeld 2020). In addition, the EMT process confers stem cell characteristics on cancer cells, which results in substantial resistance to therapy. Soluble factors secreted by MCs, like IL-8 and MMP-9, and MC-derived EVs trigger EMT through initiation of a cascade of phosphorylation occurrences in cancer cells, which has been demonstrated *in vitro* for thyroid cancer (Oldford and Marshall 2015, Mukai *et al* 2016), breast cancer (Evans and Hutchinson 2010), gastric cancer (Huang *et al* 2016), non-small cell lung cancer (Chiurchiù and Maccarrone 2016 Chiurchiù *et al* 2018), and bladder cancer (Gupta *et al* 2021). Several drugs have been designed that specifically inhibit the EMT program and are undergoing clinical trials (Takabe and Spiegel 2014). In the future, research would be beneficial that investigates whether targeted manipulation of MCs can prevent tumor advancement and metastasis through modification of EMT. Nevertheless, direct manipulation of EMT effector molecules is pharmacologically difficult in most cases. As recent research has emphasized the various metabolic pathways implicated in EMT, the utilization of metabolism-targeting FDA-approved or clinical-trial agents is proposed as an effort to repurpose drugs to combat EMT in cancer. Metabolism-inhibiting medications could be combined with routine chemotherapy or immunotherapy to target EMT-driven resistant and aggressive cancers. In mouse models of breast cancer with concomitant spontaneous autoimmune arthritis, MCs express the c-kit receptor, which binds to cancer cells via the SCF, thereby enhancing lung and bone metastases (Olivera and Rivera 2011). By inhibiting the SCF-c-Kit interactions with MCs and cancer cells, metastasis of cancer in these two locations is significantly decreased. This result indicates that MC-driven chemotaxis also drives cancer cell metastasis. In certain experimental approaches, MCs have been linked to TGF-β-triggered elevation of cancer cell stem-like characteristics (Jolly *et al* 2004). For instance, a number of *in vitro* experimental analyses have revealed that MCs can decrease gemcitabine/nab-paclitaxel-driven apoptosis of cancer cells through the induction of TGF-β signal transduction in pancreatic cancer cells, which leads to resistance to cancer drug treatments (Oskeritzian *et al* 2010). These investigations have highlighted that MCs encourage the proliferation of cancer cells, trigger the chemotactic behavior of cancer cells and facilitate malignant phenotypes exhibiting the EMT and stem-like characteristics. In opposition to this, several hypotheses have been postulated by some investigators, according to which MCs can suppress the proliferation of cancer cells, cause cellular conversion towards benign phenotypes and could ultimately cause the death of cancer cells. For instance, melanoma cells were found to recruit MCs under *in vitro* co-culture settings, establishing tight contacts when visualized under the microscope. Thereafter, tryptase released by the MCs is transferred to the nucleus of melanoma cells through exosome-driven endocytosis, thereby enhancing histone breakdown, promoting nuclear reshaping and repressing cell proliferation (Kihara *et al* 2014). In principle, it is assumed that tumor-released exosomes have powerful tumor-promoting characteristics and have an important part in sustaining tumor proliferative signaling, providing resistance to apoptosis of cancer cells, facilitating cancer invasion and metastasis, and triggering angiogenesis (Olivera and Rivera 2005). Interestingly, the results presented herein demonstrate a new mechanism for the control of cancer cell proliferation and indicate that MCs in the TME can scavenge exosomes liberated from cancer cells and in turn use them to exert anti-tumor activities. Besides blocking proliferation caused by MC-derived EVs, heparin secreted from MCs in the peri-tumoral fibrous tissue blocks cancer cell proliferation in a CAF-dependent fashion, thus impeding both cancer growth and metastasis (Piñeiro *et al* 2011). These findings indicate that MC-derived compounds also function via the adjacent stromal elements of the tumor. Simultaneously, this phenomenon could provide a new approach for the future development of therapeutics, and could offer an explanation for the efficacy of heparin and its derivatives in preventing the advancement of primary and metastatic cancers in different animal models and cancer patients (Zhou *et al* 2018). MCs in breast cancer drive the luminal phenotype of cancer cells directly through repression of cMET expression and facilitation of estrogen receptor activity. The luminal phenotype usually has a better prognosis; nevertheless, in HER2-positive breast cancer patients, this phenotype is typically linked to treatment setbacks and cancer recurrence (Kargl *et al* 2016). Besides regulating the proliferative potential of cancer cells and their phenotypic characteristics, MCs can also engulf and destroy cancer cells in a direct manner. A microscopic analysis of samples of tumors from breast cancer sufferers enabled to observe the process by which MCs phagocytose and break down cancer cells (Hofmann *et al* 2015). It can be assumed that MCs remain in the breast or connective tissue after the phagocytosis of cancer cells and are finally eliminated. In this process, MCs serve a phagocytic purpose. In addition, MCs can detect tumor-specific antigens through the generation of IgE antibodies and can trigger apoptosis of cancer cells. An *in vitro* study showed that MCs isolated from human adipose tissue could be primed with an anti-HER2/neu-IgE antibody when cocultured with HER2/neu-positive human breast cancer cells, and that this priming led to tumor cell killing. Functional MCs were isolated from adipose tissue of humans who had received cosmetic surgery. This research suggests a new avenue for immunotherapy in breast cancer and possibly for other tumors. In conclusion, adipose tissue can act as a reservoir of MCs and the detection of tumor-specific antigens with IgE antibodies can help develop specific targeted treatments (Ryberg *et al* 2007, Cantarella *et al* 2011).

While the impact of crosstalk between MCs and cancer cells appears to be bidirectional, the results obtained so far are consistent with the hypothesis that the biological properties of both cell types are modified. Thus, targeting MCs in cancer treatment is a new and attractive approach that needs to be evaluated in the context of the specific TME. MCs can actively contribute to cancer cell clearance and cancer rejection through the secretion of IL-1, IL-4, IL-6 and TNF-α (Cimpean *et al* 2017). In turn, tumor-promoting factors secreted by MCs, like FGF-2, NGF, PDGF, VEGF, IL-8 and IL-10, enhance cancer cell proliferation (Ribatti and Ranieri 2015). In addition, histamine promotes cancer cell proliferation through its interaction with H1 receptors (H1R) expressed on the surface of cancer cells (Ribatti and Ranieri

2015). In the TME, MCs are the major producers of S1P, alongside cancer cells. S1P stimulates proliferation, motility and survival of cancer cells (Nakajima *et al* 2017). In solid tumors like thyroid tumors, histamine binding of H1R and H2R leads to proliferation of cancer cells. In addition, CXCL1/GRO-α and CXCL10/IP-10 have been implicated to facilitate cancer cell invasion, proliferation and survival through engagement of CXCR2 and CXCR3, respectively (Visciano *et al* 2015b). Cell–cell interplay between MCs and cancer cells can lead to activation of MCs and the subsequent secretion of mediators. These types of crosstalk lead to the autocrine production of adenosine through MCs by a CD73-driven process. Adenosine then activates adenosine A3R, which activates ERK1/2 MAP kinases and causes IL-8 to be synthesized and secreted by MCs into the TME (Gorzalczany *et al* 2017). The role of MCs in the development of renal cell carcinoma (RCC) has been explored, as has the possibility of mutual crosstalk between MCs and cancer cells (Chen *et al* 2017). First, they introduced conditioned medium from RCC-OSRC-2 cells or from OSRC-2 plus HMC-1 into the lower compartment of a Transwell system, while human umbilical vein endothelial cells (HUVECs) were introduced into the upper compartment. Using this assay, HMC-1-conditioned medium was shown to enhance OSRC-2-facilitated HUVEC engagement. The utilization of bevacizumab or cromolyn to block MC secretion inhibited HUVEC recruitment and the generation of capillary networks *in vitro*. To assess the angiogenic potential of MCs, OSRC-2 cells and HMC-1 cells were subcutaneously co-injected into the dorsal area of nude mice. Simultaneous injection of HMC-1 and OSRC-2 enhances microvessel generation when compared to OSRC-2 injection on its own (Chen *et al* 2017). It has been hypothesized that the PI3K $\rightarrow$ AKT $\rightarrow$ GSK3β signal transduction pathway, a subsequent substrate of the PI3K/Akt signal transduction pathway, causes the expression of adrenomedullin (AM), which in turn recruits MCs into the TME where MCs attract endothelial cells through VEGF and FGF-2 liberation, cause tissue reorganization by MMPs and tryptase release and finally stimulate angiogenesis in RCCs (Chen *et al* 2017).

3.10 Pro-tumoral effects of MCs

Characteristic properties of MCs are the existence of numerous cytoplasmic granules containing a variety of preformed bioactive molecules and the constant expression of FcεRI. Upon activation of FcεRI by IgE and multivalent antigens, clustered receptors initiate biochemical signal transduction pathways that culminate in the secretion of stored and newly generated pro-inflammatory mediators. In addition, MCs can also secrete exosomes either in a constitutive manner or following stimulation. Exosomes have been characterized as nanoscale vesicles of endocytic origin that possess key immunoregulatory characteristics and provide an additional means of intercellular crosstalk. Remarkably, exosomes derived from FcεRI-activation comprise costimulatory and adhesion molecules, lipid factors and MC-specific proteases, and receptor subunits along with IgE and antigens. These results lend credence to the assumption that FcεRI signal transduction exerts an essential role in modulating the constitution and functionalities of exosomes generated by

MCs according to their activation stage (Lecce *et al* 2020). MC exosomes bearing the FcεRI receptor can inhibit allergic responses when they bind to IgE (Xie *et al* 2018). These exosomes derived from bone marrow-derived MCs (BMMCs) can tether to free IgE via FcεRI. Thereby, BMMC exosomes exhibit an anti-IgE effect that reduces IgE abundance and impairs the activation of MCs. Moreover, the liberation of tryptase from MCs promotes the process of metastasis of cancer cells through exosomes (Xiao *et al* 2019). Exosomes produced by lung cancer cells contain SCF for tethering to MCs via their c-KIT receptor. SCF is located on chromosome 12 in humans and in mice on chromosome 10 and is encoded via 9 exons in both species. SCF is generated as two main isoforms, such as SCF^{220} and SCF^{248} (Anderson *et al* 1991, Tsai *et al* 2022), through the process of alternative splicing (figure 3.7) (Mierke 2020). Both isoforms, SCF^{220} and SCF^{248}, encode the membrane-bound SCF that is composed of intracellular, transmembrane and extracellular parts. The isoform SCF^{248} comprises an additional protease cleavage site, which is encoded on exon 6 and that is cleaved through chymase (figure 3.7), MMP-9, and proteases of the ADAMs family to produce the 165 amino acid soluble SCF (Lennartsson and Rönnstrand 2012). Soluble SCF can also be synthesized from ^{SCF220} in a less efficient way by splitting exon 7. The active version of SCF is a non-covalently linked homodimer that engages with the first three extracellular Ig domains of the KIT receptor (Lennartsson and Rönnstrand 2012). The fourth and

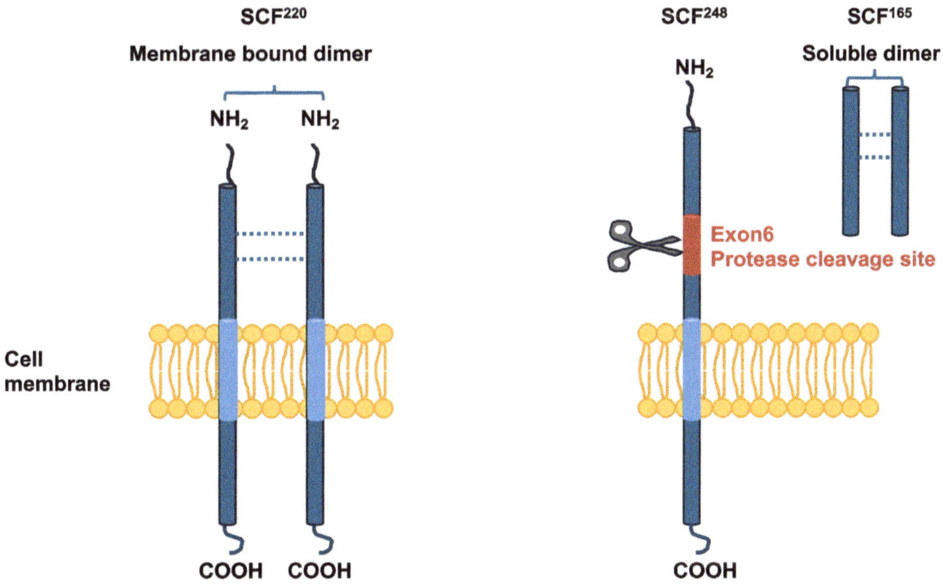

Figure 3.7. Two principal SCF isoforms, SCF^{220} and SCF^{248}, are formed through alternative splicing. Both SCF^{220} and SCF^{248} are composed of an extracellular region, a transmembrane region and an intracellular tail end. Exon 6 in SCF^{248} encodes a protease-sensitive location that can be split to produce soluble SCF^{165}. Small quantities of soluble SCF can also be formed from SCF220 through an alternative protease splitting site that is encoded by exon 7 (not depicted). Biologically active SCF consists of a non-covalent dimer that is present either in membrane-associated shape or in a soluble form.

fifth extracellular Ig domains of KIT contribute to the stability of the homo-dimeric form of the receptors in the presence of ligand engagement (Lennartsson and Rönnstrand 2012). As with other tyrosine kinase receptors, dimerization or oligomerization is necessary for the triggering the activation of the intrinsic tyrosine kinase and the transphosphorylation of KIT receptors (Lennartsson and Rönnstrand 2012). Soluble SCF is usually present in the bloodstream in a monomeric version that fails to activate KIT (Hsu *et al* 1997). The expression of SCF^{220} and SCF^{248} has been shown to be tissue-specific (Broudy 1997). Membrane and soluble SCF homodimers can both engage and activate KIT (Mierke *et al* 2000), but have distinct biological roles in hematopoiesis, where SCF^{248} appears to fulfill a more critical function in MC development and survival (Kapur *et al* 1998, Tajima *et al* 1998).

Cancer cells secrete exosomes and can be internalized by MCs, but the possible functional implications of MCs in cancer metastasis have not yet been elucidated. Interaction of MCs with lung cancer cell membranes triggers the shedding of extracellular vesicles harboring a unique miRNA profile (Shemesh *et al* 2023). Cell–cell contact has been reconstituted when exposing MCs to membranes originating from lung cancer cells, and their activation has been corroborated, as demonstrated by enhanced phosphorylation of the ERK and AKT kinases. MicroRNA profiling of the EV cargo preferentially secreted from lung cancer-activated MCs indicated that they carried significantly elevated levels of miR-100–5p and miR-125b, which are two pro-tumorigenic miRNAs.

Exosomes derived from lung adenocarcinoma cell line A549 have been purified and the internalization of PKH26-labeled exosomes by bone marrow MCs was analyzed using flow cytometry and fluorescence microscopy. Cytokines and tryptase in the MC supernatant have been assayed by ELISA kit, and the existence of tryptase has been proved by Western blot. Cell proliferation and migration have been evaluated by Cell counting Kit-8 (CCK-8) and Transwell experiments. Proteins in the tryptase-JAK-STAT signal transduction route were verified by Western blot. In this work, it was demonstrated that exosomes from A549 cells can be internalized by MCs (Shemesh *et al* 2023). In addition, A549 exosomes incorporate SCF for MCs and then stimulate the activation of MCs via SCF-KIT signal transduction, resulting in MC degranulation and secretory tryptase liberation. Tryptase enhances proliferation and migratory capacity of HUVECs via JAK-STAT signal transduction route.

In summary, these results demonstrate a potential mechanism for metastasis, whereby exosomes can deliver SCF to MCs and thereby activate them, which can impact tryptase secretion and HUVEC angiogenesis. Notably, exosomes originating from lung cancer cells appear to be key players in the tumor burden that occurs during metastasis (Headley *et al* 2016). There is ample evidence that MCs serve as key drivers for metastasis. The cKIT-derived from MCs functions as a protein that functionally interacts with cancer cells through exosomes, and then subsequently induces the cKIT-SCF signaling transduction route, which increases proliferation in lung cancer cells (Xiao *et al* 2014). Nevertheless, the subsequent destiny of exosomes released by lung cancer cells and their first encounter with MCs is poorly understood, and how these exosomes-treated MCs are modified is completely uncharted territory. Moreover, the mechanisms that permit exosomes from early-stage lung cancer cells to undergo transduction from the microenvironment to MCs are unidentified. Exosomes originating from the lung adenocarcinoma cell line A549 were found to harbor the SCF receptor, which can be delivered to MCs via exosomes. Lung cancer cells secreted exosomes with characteristic exosome markers such as TSG101, CD81 and calnexin. Collectively, these findings indicate that exosomes from lung cancer cells promote the proliferation of HUVECs through MC activation and SCF-KIT-induced tryptase liberation, possibly by intensifying JAK-STAT signal transduction in HUVECs. In line with the expected pro-inflammatory role of exosomes, exosomes released from A549 cells activated MCs, inducing the secretion of pro-inflammatory mediators like β-hexosaminidase, tryptase, IL-6, MMP-9 and TNF-α. In specific, protease-activated receptor 2 (PAR2) expressed by HUVECs could be activated by serine proteases like MC-mediator tryptase. The attachment of tryptase to PAR2 could enhance the phosphorylation of JAK and STAT, which is considered to be a part of the signaling pathway in activated HUVECs (Xiao *et al* 2019).

3.10.1 Functional role of MCs in angiogenesis

The tumor's vascular system cannot adequately deliver blood to the expanding tumor cell mass. As a result, the delivery of nutrients to the cancer cells, the exchange of gases and the removal of metabolic compounds generated from the tumor are impaired. This situation therefore causes the development of a hypoxic and acidotic TME, which stimulates angiogenesis and the creation of heterogeneous new vessels to balance the deficiency in blood circulation (Hui and Chen 2015, Jarosz-Biej *et al* 2018, Kabiraj *et al* 2018). In addition, the endothelial cells that comprise the tumor vessels are able to inhibit the enrollment, adhesion and the activity of circulating T cells by functioning as physical protective walls for immune cells (Schaaf *et al* 2018). There are a range of angiogenic mediators secreted from MCs in the TME, among them are IL-8, NGF, TNF-α, TGF-β and urokinase-type plasminogen activator (PA) (De Palma *et al* 2017). MC tryptase enhances endothelial cell proliferation (Guo *et al* 2016), eases the development of vascular tubes *in vitro* and breaks down the connective tissue matrix to provide sufficient room for neovascular development (De Palma *et al* 2017). In addition, histamine, which acts on H1R and H2R, promotes the development of new vascular structures (De Palma *et al* 2017). MCs not only participate in angiogenesis through VEGF-A and VEGF-B (Visciano *et al* 2015b), but are also implicated in lymph angiogenesis through the liberation of VEGF-C and VEGF-D (Detoraki *et al* 2009). In a co-culture model of HMC-1 cells and dermal microvascular endothelial cells (HDMECs), the influence of MC mediators on the development of tubes has been examined. It was observed that calcium ionophore-activated MCs promoted the generation of tubes. Tryptase emerged as the principal facilitator implicated in neovascularization and proliferation of endothelial cells, and tube generation was inhibited with tryptase inhibitors, recombinant leech-derived tryptase inhibitor and bis(5-amidino-2-benzimidazo-lyl) methane (Blair *et al* 1997). In a similar effort, the impact of tryptase secreted by recombinant human lung MC on the proliferation and tube-forming capacity of HUVECs has been examined. Incubation of the cells with tryptase was found to significantly improve their viability and cell multiplication. Treatment of the cells with nafamostat, a tryptase inhibitor, inverted this whole effect. PD98059, an inhibitor of ERK phosphorylation, inhibited the enhancing activity of tryptase. In addition, MC-tryptase not only stimulated the proliferation of HUVECs, but also induced tubular network formation (Guo *et al* 2016). Moreover, MC tryptase has been shown to promote tumorigenesis and angiogenesis *in vivo* after injection of PANC-1 (pancreatic cancer cell line) into nude mice. The cancer cells injected along with tryptase were more numerous and larger than those seen in untreated mice, and nafamostat could block these tumorigenic actions of tryptase (Guo *et al* 2016).

3.10.2 Improved reactivity of MCs

MCs are considered important effector cells in IgE-mediated allergic diseases and are also key actors in the orchestration of innate and adaptive immune reactions, as they can react to a variety of diverse stimuli by secreting pro-inflammatory substances (Galli and Tsai 2012, Voehringer 2013, Olivera *et al* 2018). The presence of MCs is also commonly seen in tumors, indicating that they are involved in the transformation from sustained inflammation to cancer (Varricchi *et al* 2017). The regulatory function of MCs relies on the expression of multiple receptors that, after engagement of ligands, can trigger the secretion of stored effectors and the *de novo* generation of lipid metabolites and several chemokines and cytokines. The prominent FcεRI along with c-Kit (CD117) marks mature MCs in tissues where they undergo final differentiation (Galli *et al* 1995, Kawakami and Galli 2002). SCF appears to be the most thoroughly characterized growth and differentiation factor of MCs, as indicated by the close association of activating Kit mutations and systemic mastocytosis (Metcalfe 2008). In the microenvironment of the tissue, SCF regulates the development of MCs and guarantees the viability of mature MCs (Galli *et al* 1995, Metcalfe 2008, Huber *et al* 2019). Along with SCF, other soluble factors can assist in the activation and expansion of MCs, one of which is a member of the IL-1 superfamily, IL-33 (Cayrol and Girard 2018). When danger signals are involved, IL-33 is the most potent alarm signal of several stimulants for enhancing MC activation (Enoksson *et al* 2011, Saluja *et al* 2016) and is liberated by numerous tissue-resident cells, including epithelial and endothelial cells (Liew *et al* 2016, Martin and Martin 2016). Remarkably, MCs function as critical enhancers of IL-33-driven inflammation not only because of their ample surface expression of ST2, the IL-33-specific subunit of the IL-33 receptor, but also because of their intrinsic ability to liberate IL-33 (Enoksson *et al* 2011, Tung *et al* 2014, Sjöberg *et al* 2015, Saluja *et al* 2016). IL-33 enhances IgE-driven MC reactions and works together with reactivity to SCF via affecting signal transduction events and secretory MC effector capabilities (Drube *et al* 2010, Jung *et al* 2013, Joulia *et al* 2017, Babina *et al* 2019). The ultimate response, nonetheless, may differ according to the environment where the MCs undergo differentiation. At least two subtypes of completely differentiated MCs can

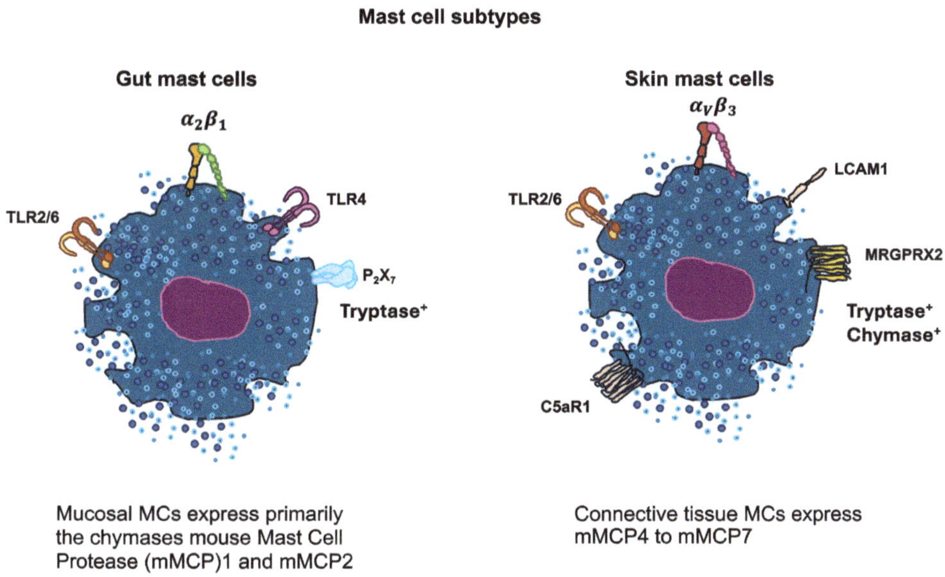

Figure 3.8. There are at least two major human MC subtypes, such as MMCs and CTMCs.

be identified in mice: mucosal MCs (MMCs), which are primarily located in the epithelia of the lungs and gastrointestinal tract, and connective tissue MCs (CTMCs), which are primarily located in the intestinal submucosa, peritoneum, and dermis (Gurish and Austen 2012). Aside from their anatomical position, these two subgroups are differentiated through the expression of certain proteases, whereby MMCs mainly express the chymases mouse mast cell protease (mMCP)1 and mMCP2 and CTMCs express mMCP4-to-7 and carboxypeptidase A (figure 3.8) (Gurish and Austen 2012, Wernersson and Pejler 2014).

Such a classical classification is oversimplified, though, as there is convincing experimental evidence that the microenvironment exerts a prominent influence on the phenotype and functioning of MCs, indicating a high level of plasticity of MCs (Galli *et al* 2011, Xing *et al* 2011). Additionally, the transcription pattern of MCs highlights the heterogeneity of gene expression in various connective tissues (The Immunological Genome Project Consortium *et al* 2016). It is noteworthy that not just the tissue location, but even the disease stage affects the heterogeneity of MCs. For example, an altered MC subtype has been reported in patients with inadequately treated severe Th2-associated asthma, which could be involved in the pathophysiology of this disease (Fajt and Wenzel 2013). In addition, MMCs increase in the acute stage of T. spiralis infection and enhance intestinal barrier permeability via the effects of mMCP-1 to promote parasite excretion (Knight *et al* 2000). In the chronic stage of infection, nevertheless, a primary involvement of CTMC tryptase mMCP-6 has also been proposed (Shin *et al* 2008). MCs are also found in various types of cancers, indicating that they help cancer formation and advancement (Oldford and Marshall 2015). In line with this concept, several lines of evidence have implicated

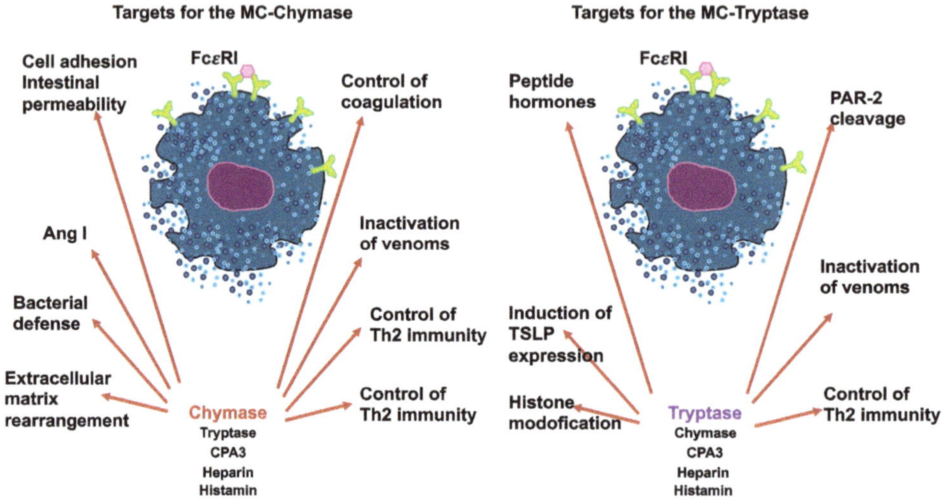

Figure 3.9. Targets of the human chymase (left image) and the human tryptase (right image) of MCs. Several potential aims of the two proteases are illustrated.

MCs and their proteases (figure 3.9) (Hellman *et al* 2022) in governing the transition from chronic intestinal inflammation to colorectal cancer (CRC).

According to the type, grade or staging of cancers, the distinct functions of MCs vary widely and can be either tumor-promoting or tumor-inhibiting (Heijmans *et al* 2012, Oldford and Marshall 2015, Varricchi *et al* 2017, 2019). A comprehensive characterization of how intestinal TME influences the phenotypes and effector functions of MCs is presently lacking. Based on a mouse model for chemically triggered inflammatory CRC, it was shown that the microenvironment of CRC is typified by high quantities of SCF and IL-33 (Molfetta *et al* 2023). As cancer progresses under the effect of SCF, CTMCs enrich in colonic lesions or neoplasms and, although there is marked downregulation of c-Kit, they are still able to secrete large quantities of pro-inflammatory cytokines (Molfetta *et al* 2023). Likewise, when combined to promote primary MC cultures, IL-33 and SCF can co-operate to further elicit pro-inflammatory cytokine generation. There is convincing data that MCs are capable of context-specific and local shifts from one functional subtype to another (Varricchi *et al* 2019). This switch may be the outcome of interactions between MCs and epithelial cells in an acute inflammatory state or with transformed cells in epithelial tumorigenesis. This implies that a pro-fibrotic function of MCs, which is triggered especially during wound healing, can be deregulated in the transition: tumors can foster new pro-inflammatory features in MCs to turn them into a pathogenic phenotype (Rigoni *et al* 2018, Segura-Villalobos *et al* 2022). Nevertheless, it remains to be determined whether the MC phenotype is modified by the TME, and the precise contribution of the unique MC subset in tumor development and progression is far from completely elucidated. Using a colon-type mouse model of CRC, high concentrations of SCF and IL-33 in the TME are associated with the clustering of tumor-associated MCs (TAMCs) with a fibrotic phenotype, which contribute to a pro-inflammatory environment by secreting TNF-α and IL-6. Initial results indicate that mice with a deficiency of MCs are less prone to forming chemically initiated colon tumors (Wedemeyer and Galli 2005) and point to a tumor-promoting effect of MCs in the context of colitis in the development of colon cancer (Rigoni *et al* 2018, Lee *et al* 2022). There are, nonetheless, also trials in which MCs seem to have a protective effect in the development of colorectal cancer (Sinnamon *et al* 2008, Bodduluri *et al* 2018, Song *et al* 2023). In particular, MCs have been demonstrated to either enhance or repress colon tumor development according to microenvironmental cues, exerting a tumorigenic action in colitis-induced CRC and a protective effect in sporadic CRC mouse models (Sakita *et al* 2022). This result is consistent with the current notion that the effect of MCs varies during CRC evolution, according to the genetic background of the mice and the specific cancer models, and can be either protective or detrimental (Molfetta and Paolini 2023). In this experimental context, TAMCs were found to secrete elevated levels of the pro-inflammatory cytokines IL-6 and TNF-α compared to MCs in cancer-free tissue, indicating a pro-tumorigenic activity (Molfetta *et al* 2023). This hypothesis is strengthened by previous evidence demonstrating that intestinal MCs are an essential cellular reservoir of TNF-α, which they secrete to drive progressive epithelial conversion and polyp formation (Bischoff *et al* 1999, Gounaris *et al* 2007, Kim *et al* 2010). It has been demonstrated that in colonic lesions, the principal MC subset is elevated with mMCP4 (Molfetta *et al* 2023), indicating a selective aggregation of connective tissue-like MCs. These findings are consistent with earlier studies documenting a lymphocyte-independent spread of a subset of CTMCs in the transformation of polyps to adenocarcinomas (Gounaris *et al* 2007, Saadalla *et al* 2018, 2019). In addition, by using a combination of *in situ* RNA hybridization and immunofluorescence, for the first time it was possible to prove that mMCP4$^+$ MCs are the principal cell subtype that is accountable for cytokine generation in cancer lesions (Molfetta *et al* 2023). In terms of TME makeup, high levels of SCF and IL-33 were identified (Molfetta *et al* 2023), which is consistent with previous results obtained in human malignant cancers and a genetic mouse model of CRC (Bellone *et al* 2006, Liu *et al* 2014, Cui *et al* 2015, Maywald *et al* 2015, Shah and Van Den Brink 2015). Chronic treatment of TAMCs with both soluble mediators probably accounts for the marked downregulation of c-Kit and the marked upregulation of IL-33R that could be reproducibly detected (Molfetta *et al* 2023). SCF and IL-33 work independently or synergistically to promote the survival, differentiation, and cytokine expression of both human and mouse MCs (Okayama and Kawakami 2006, Drube *et al* 2010, Jung *et al* 2013, Joulia *et al* 2017, Babina *et al* 2019). Their importance is illustrated by the finding that mice missing c-kit or ST2 display a reduced number of MCs in multiple tissues (Wang *et al* 2014, Tsai *et al* 2022), whereas delivery of exogenous SCF or IL-33 augments the MC pool (Huang *et al* 2008, Saluja *et al* 2016). Correspondingly, in intestinal polyposis, IL-33 depletion impacts MC aggregation, and MC-derived secreted proteases and cytokines (Wang *et al* 2014, Cui *et al* 2015). Therefore, it is probable that in this experimental model, the combined effect of SCF and IL-33 is driving the propagation of a fibrotic MC subset inside the tumor, either fostering *de novo* enlistment and maturation of immature MC precursors or promoting proliferation of resident mature MC subsets. While *in vivo* proof of a function of SCF in promoting the build-up of connective tissue-like MC subsets in the tumor has been established (Molfetta *et al* 2023), additional *in vivo* experiments are needed to further underpin the precise function of SCF and IL-33 in MC proliferation and/or attraction and maturation of MC precursors throughout CRC advancement.

Independently, through the establishment of primary cultures from bone marrow- or peritoneum-derived MC precursors, it has been further determined to what extent SCF and IL-33 control the MC phenotype and functionality. Bone marrow-derived precursors cultured in the continuous presence of SCF exhibit a predominant connective tissue-like phenotype marked by high transcriptional abundance of mMCP4, mMCP5, and mMCP6, whereas MCs cultured in the continuous presence of IL-3 alone exhibited substantially higher expression of mMCP1 transcripts (Molfetta *et al* 2023). Upon differentiation, however, SCF appears incapable of further influencing the proteasome level, but this requires confirmation in future experiments. Remarkably, treatment with IL-33 results in a greater selective enhancement of mMCP4, mMCP5, and mMCP6, which is more pronounced in BMMCs cultured with IL-3 and SCF (Molfetta *et al* 2023). Consistent with the present findings, it has been described that SCF mMCP-4 and -6 addition to BMMCs generated with IL-3 alone results in their upregulation (Gurish *et al* 1992), and that IL-33 enhances tryptase mMCP-6 expression in wild-type bone marrow-derived MCs, although not in ST2-deficient MCs (Kaieda *et al* 2010). While BMMCs were undergoing differentiation, SCF-MCs were also found to down-regulate c-kit expression (Molfetta *et al* 2023), reflecting the marked c-kit down-modulation seen in tumor-infiltrating cells *in vivo* (Molfetta *et al* 2023). In addition, they had a reduced tendency to degranulate and secret cytokines when FcεRI engaged with its ligand, in line with previous evidence (Ando *et al* 1993, Ito *et al* 2012). Upon combined SCF and IL-33 stimulation, overriding production of IL-6 and TNF-α was seen in all primary MC cultures, independent of the hypofunctional condition seen in reaction to IgE and antigen in IL-3/SCF-derived BMMCs (Molfetta *et al* 2023). These results are in line with previous work showing a synergistic action of SCF and IL-33 in activating MCs (Jung *et al* 2013) and provide a rationale for the notion that persistent stimulation during colonic carcinogenesis by these soluble factors sculpts not just the MC phenotype but also enhances the secretory output of IL-6 and TNF-α. The molecular signaling events governing cytokine gene expression in MCs are less well characterized than those for granule release. IL-33 has been shown to drive IL-6 generation through ERK1/2 and p38 phosphorylation in a MyD88-driven fashion (Schmitz *et al* 2005, Ho *et al* 2007). In addition, SCF has been reported to increase IL-33-triggered generation of IL-6 (Drube *et al* 2010, 2015). Correspondingly, IL-33 alone was found to be capable of eliciting the phosphorylation of ERK1/2 and p38 along with the activation of NF-κ B (Molfetta *et al* 2023). The activation of these biochemical signal transduction cascades would account for the capacity of IL-33 on its own to stimulate a strong generation of IL-6, which was especially evident in the stimulation of peritoneal-derived MCs (Molfetta *et al* 2023). SCF by itself, in complement to a potent phosphorylation of ERK1/2 seen in both BMMC and PDMC cultures, seems to be capable of activating the PI3 kinase signal transduction pathway and inducing specific STAT3 phosphorylation (Molfetta *et al* 2023). A reasonable explanation for these results is that the enhanced cytokine generation seen by the joint action of SCF and IL-33 relies on the activation of shared or unique but overlapping signal transduction routes. Collectively, these findings emphasize the capacity of SCF in concert with IL-33 to sculpt the phenotype and functioning of MCs, which may promote chronic inflammation and disease in the gut microenvironment via release of IL-6 and TNF-α. Two major subsets of MCs have been identified, as presented before, in the human gastrointestinal tract based on their protease levels: those that contain predominantly tryptase (MCT) and are found mainly in the lamina propria, and those marked by the expression of both tryptase and chymase (MCTC) in the submucosa. Remarkably, *in vitro* research indicates that human MCs are also fine-tunable effector cells that alter their granule levels in response to various inputs such as SCF and multiple cytokines (Galli *et al* 2011). In colorectal cancer patients, no distinction in the ratio of these two subgroups has been found so far. Most investigations on the density of tumor-infiltrating MCs, nevertheless, were conducted using *in situ* detection of tryptase-positive cells, without differentiation between MCTC and MCT subsets (Tan 2005, Ammendola *et al* 2014, Hu *et al* 2018). Therefore, there are no coherent data on the contribution of a specific MC subset in the carcinogenesis mechanism, highlighting the necessity to investigate the implications of MCs and their proteases from multiple viewpoints. In this setting, the question remains whether a shared progenitor gives rise to discrete phenotypic and functional MC subsets and/or whether differentiated MCs may exhibit diverse characteristics due to their sensitivity to SCF, IL-33 and/or some other distinct microenvironmental cues. An in-depth investigation of MC phenotypic and functional plasticity may provide potential therapeutic targets throughout the progression of CRC.

3.11 Anti-tumoral effects of MCs

It has been hypothesized that the stabilization of tumor-resident MCs can re-establish the infiltration of T cells and can sensitize sarcomas to the blocking of PD-L1. To investigate the cellular interplay of tumor-resident MCs in regulating the activity of CAFs to compensate for tumor TME aberrations to improve the efficacy of immune checkpoint blockers in sarcomas. Thus, a co-culture model was employed and subsequently evaluated in mouse models of fibrosarcoma and osteosarcoma with or without treatment with the MC stabilizer and antihistamine ketotifen. To assess the ability of ketotifen to increase tumor sensitivity to treatment, a combination trial with doxorubicin chemotherapy and anti-PD-L1 treatment, such as B7-H1, clone 10F.9G2, has been conducted. The capacity of ketotifen to modify the TME was investigated in human sarcomas in a refocused phase II clinical study. Ketotifen suppression of MC activation effectively reduced CAF proliferation and decreased ECM stiffness, associated with improved vascular perfusion in fibrosarcoma and osteosarcoma, as demonstrated with ultrasound shear wave elastography

(Panagi *et al* 2024). Improved tissue oxygenation enhanced the effectiveness of chemoimmunotherapy, aided by improved T-cell infiltration and the subsequent generation of tumor-specific memory. Significantly, the effectiveness of ketotifen in alleviating the stiffness of tumors in sarcoma patients has been additionally confirmed, underscoring its translational impact. In general, the results of the study indicate that the targeted treatment of MCs with classically administered medications, like antihistamines, is a powerful strategy to break the resistance to immunotherapy observed in sarcomas (Panagi *et al* 2024).

3.12 Conclusions and future perspective

MCs are driving factors of cancer initiation and malignant progression. In contrast they can also contribute to tumor-suppressing functions. The activation and function of MCs relies also on the TME and its especially its mechanobiological characteristics, such as stiffness, viscoelasticity and structural cues like fiber alignment. Therefore, MCs express mechanosensory receptors on their cell surface, such as TRPV2 and ADGRE2 (table 3.1). The communication of cancer cells and MCs occurs via exosomes that display the SCF on the vesicle membrane to stimulate MCs to proliferate and release certain substances that foster the progression of cancer. MCs secrete growth factors for endothelial cells and thereby participate in the induction of angiogenesis. In addition, there exists a direct interaction between MCs and other cell types, such as cancer cells and immune cells. The MC-release proteases can alter the microenvironment of cancers and cleave MCs' own surface expressed receptors, such as the different alternatively spliced isoforms of the membrane-bound SCF and soluble SCF. SCF has been shown to work together with IL-33 to activate MCs and encourage their release of pro-inflammatory cytokines, such as IL-6 and TNF-α. Thereby, it needs to be explored how these cytokines after the mechanical phenotype of cancer cells and immune cells. In addition, MCs have been revealed as a potent player in mediating the switch of cancer cells from a less malignant phenotype to malignant phenotype, whereby this switch is reversible and can be induced multiple times in cancer cells. Moreover, MCs interact with various types of immune cells and promote their transition toward either tumor-promoting or tumor-suppressing phenotypes pf immune cells. Thus, it can be concluded that MCs are the key players in tumor development and the malignant progression. The role of MCs in tumors seems to be similar in multiple tumor types and may rely on the MC subtype, which needs to be addressed in future experiments.

A future perspective is the mechanobiological characterization of MCs in various 3D environments. In specific, it needs to be explored whether the different subtypes of MCs exhibit diverse mechanosensation strategies. As MCs liberate multiple exosomes for the communication with other cell types and to alter the mechanical cues of the TME, it should be explored whether they serve as mechanosensors similar as it has been discussed for endothelial cells (Mierke 2024). In terms of metastatic progression of cancers, the role of MCs in supporting the individual steps of the metastatic cascade would be helpful to get a complete picture of how cancer cells can migrate through the endothelial cell barrier to spread in the vasculature. MCs could also determine the way in which cancer cells migrate through the endothelium, either paracellularly or transcellularly, and the site of transmigration, particularly at the extravasation step. This could represent a new frontier in MC-cancer cell research, which needs to be overcome to elucidate the process of malignant progression of tumors.

References and further reading

Abonia J P, Hallgren J, Jones T, Shi T, Xu Y, Koni P *et al* 2006 Alpha-4 integrins and VCAM-1, but not MAdCAM-1, are essential for recruitment of mast cell progenitors to the inflamed lung *Blood* **108** 1588–94

Abraham S N and St. John A L 2010 Mast cell-orchestrated immunity to pathogens *Nat. Rev. Immunol.* **10** 440–52

Abu Shah E and Keren K 2014 Symmetry breaking in reconstituted actin cortices *eLife* **3** e01433

Almeida J A, Mathur J, Lee Y L, Sarker B and Pathak A 2023 Mechanically primed cells transfer memory to fibrous matrices for invasion across environments of distinct stiffness and dimensionality *MBoC* **34** ar54

Álvarez-Errico D, Oliver-Vila I, Ainsua-Enrich E, Gilfillan A M, Picado C, Sayós J *et al* 2011 CD84 Negatively regulates IgE high-affinity receptor signaling in human mast cells *J. Immunol.* **187** 5577–86

Ammendola M, Gadaleta C D, Frampton A E, Piardi T, Memeo R, Zuccalà V *et al* 2017 The density of mast cells c-Kit$^+$ and tryptase$^+$ correlates with each other and with angiogenesis in pancreatic cancer patients *Oncotarget* **8** 70463–71

Ammendola M, Sacco R, Sammarco G, Donato G, Montemurro S, Ruggieri E *et al* 2014 Correlation between serum tryptase, mast cells positive to tryptase and microvascular density in colo-rectal cancer patients: possible biological-clinical significance *PLoS One* **9** e99512

Anderson D M, Williams D E, Tushinski R, Gimpel S, Eisenman J, Cannizzaro L A *et al* 1991 Alternate splicing of mRNAs encoding human mast cell growth factor and localization of the gene to chromosome 12q22-q24 *Cell Growth Differ.* **2** 373–8

Ando A, Martin T R and Galli S J 1993 Effects of chronic treatment with the c-kit ligand, stem cell factor, on immunoglobulin E-dependent anaphylaxis in mice. Genetically mast cell-deficient Sl/Sld mice acquire anaphylactic responsiveness, but the congenic normal mice do not exhibit augmented responses *J. Clin. Invest.* **92** 1639–49

Arthur G K and Cruse G 2022 Regulation of trafficking and signaling of the high affinity IgE receptor by FcεRIβ and the potential impact of FcεRIβ splicing in allergic inflammation *Int. J. Mol. Sci.* **23** 788

Artuc M, Hermes B, Stckelings U M, Grützkau A and Henz B M 1999 Mast cells and their mediators in cutaneous wound healing—active participants or innocent bystanders? *Exp. Dermatol.* **8** 1–16

Atiakshin D, Buchwalow I and Tiemann M 2020 Mast cells and collagen fibrillogenesis *Histochem. Cell Biol.* **154** 21–40

Atiakshin D, Nikolaeva E, Semyachkina A, Kostin A, Volodkin A, Morozov S *et al* 2024 The contribution of mast cells to the regulation of elastic fiber tensometry in the skin dermis of children with marfan syndrome *Int. J. Mol. Sci.* **25** 9191

Atiakshin D, Soboleva M, Nikityuk D, Alexeeva N, Klochkova S, Kostin A *et al* 2023 Mast cells in regeneration of the skin in burn wound with special emphasis on molecular hydrogen effect *Pharmaceuticals* **16** 348

Babina M, Wang Z, Franke K, Guhl S, Artuc M and Zuberbier T 2019 Yin-Yang of IL-33 in human skin mast cells: reduced degranulation, but augmented histamine synthesis through p38 activation *J. Invest. Dermatol.* **139** 1516–1525.e3

Bagher M, Larsson-Callerfelt A-K, Rosmark O, Hallgren O, Bjermer L and Westergren-Thorsson G 2018 Mast cells and mast cell tryptase enhance migration of human lung fibroblasts through protease-activated receptor 2 *Cell Commun. Signal.* **16** 59

Banafea G H, Bakhashab S, Alshaibi H F, Natesan Pushparaj P and Rasool M 2022 The role of human mast cells in allergy and asthma *Bioengineered* **13** 7049–64

Bar-Hai N, Ben-Yishay R, Arbili-Yarhi S, Herman N, Avidan-Noy V, Menes T *et al* 2024 Modeling epithelial-mesenchymal transition in patient-derived breast cancer organoids *Front. Oncol.* **14** 1470379

Bartnikas L M, Gurish M F, Burton O T, Leisten S, Janssen E, Oettgen H C *et al* 2013 Epicutaneous sensitization results in IgE-dependent intestinal mast cell expansion and food-induced anaphylaxis *J. Allergy Clin. Immunol.* **131** 451–460.e6

Batlle E, Sancho E, Francí C, Domínguez D, Monfar M, Baulida J *et al* 2000 The transcription factor Snail is a repressor of E-cadherin gene expression in epithelial tumour cells *Nat. Cell Biol.* **2** 84–9

Battle C, Broedersz C P, Fakhri N, Geyer V F, Howard J, Schmidt C F *et al* 2016 Broken detailed balance at mesoscopic scales in active biological systems *Science* **352** 604–7

Bellone G, Smirne C, Carbone A, Buffolino A, Scirelli T, Prati A *et al* 2006 KIT/stem cell factor expression in premalignant and malignant lesions of the colon mucosa in relationship to disease progression and outcomes *Int. J. Oncol.* **29** 851–9

Berlanga O, Emambokus N and Frampton J 2005 GPIIb (CD41) integrin is expressed on mast cells and influences their adhesion properties *Exp. Hematol.* **33** 403–12

Bertolino P, Deckers M, Lebrin F and Ten Dijke P 2005 Transforming growth factor-β signal transduction in angiogenesis and vascular disorders *Chest* **128** 585S–90S

Bischoff S C, Lorentz A, Schwengberg S, Weier G, Raab R and Manns M P 1999 Mast cells are an important cellular source of tumour necrosis factor α in human intestinal tissue *Gut* **44** 643–52

Blair R J, Meng H, Marchese M J, Ren S, Schwartz L B, Tonnesen M G *et al* 1997 Human mast cells stimulate vascular tube formation. Tryptase is a novel, potent angiogenic factor *J. Clin. Invest.* **99** 2691–700

Blanco I, Béritze N, Argüelles M, Cárcaba V, Fernández F, Janciauskiene S *et al* 2010 Abnormal overexpression of mastocytes in skin biopsies of fibromyalgia patients *Clin. Rheumatol.* **29** 1403–12

Bodduluri S R, Mathis S, Maturu P, Krishnan E, Satpathy S R, Chilton P M *et al* 2018 Mast cell–dependent CD8+ T-cell recruitment mediates immune surveillance of intestinal tumors in ApcMin/+ mice *Cancer Immunol. Res.* **6** 332–47

Boudier C, Godeau G, Hornebeck W, Robert L and Bieth J G 1991 The elastolytic activity of cathepsin G: an *Ex Vivo* study with dermal elastin *Am. J. Respir. Cell Mol. Biol.* **4** 497–503

Boyden S E, Desai A, Cruse G, Young M L, Bolan H C, Scott L M *et al* 2016 Vibratory urticaria associated with a missense variant in *ADGRE2* *N. Engl. J. Med.* **374** 656–63

Brabletz S, Schuhwerk H, Brabletz T and Stemmler M P 2021 Dynamic EMT: a multi-tool for tumor progression *EMBO J.* **40** e108647

Brabletz T, Kalluri R, Nieto M A and Weinberg R A 2018 EMT in cancer *Nat. Rev. Cancer* **18** 128–34

Bradding P and Pejler G 2018 The controversial role of mast cells in fibrosis *Immunol. Rev.* **282** 198–231

Broudy V C 1997 Stem cell factor and hematopoiesis *Blood* **90** 1345–64

Brough H A, Nadeau K C, Sindher S B, Alkotob S S, Chan S, Bahnson H T *et al* 2020 Epicutaneous sensitization in the development of food allergy: What is the evidence and how can this be prevented? *Allergy* **75** 2185–205

Cano A, Pérez-Moreno M A, Rodrigo I, Locascio A, Blanco M J, Del Barrio M G *et al* 2000 The transcription factor Snail controls epithelial–mesenchymal transitions by repressing E-cadherin expression *Nat. Cell Biol.* **2** 76–83

Cantarella G, Scollo M, Lempereur L, Saccani-Jotti G, Basile F and Bernardini R 2011 Endocannabinoids inhibit release of nerve growth factor by inflammation-activated mast cells *Biochem. Pharmacol.* **82** 380–8

Cayrol C and Girard J 2018 Interleukin-33 (IL-33): a nuclear cytokine from the IL-1 family *Immunol. Rev.* **281** 154–68

Chandiran K and Cauley L S 2023 The diverse effects of transforming growth factor-β and SMAD signaling pathways during the CTL response *Front. Immunol.* **14** 1199671

Chaudhry S S, Cain S A, Morgan A, Dallas S L, Shuttleworth C A and Kielty C M 2007 Fibrillin-1 regulates the bioavailability of TGFβ1 *J. Cell Biol.* **176** 355–67

Chen P-Y, Qin L and Simons M 2023 TGFβ signaling pathways in human health and disease *Front. Mol. Biosci.* **10** 1113061

Chen X, Feng B-S, Zheng P-Y, Liao X-Q, Chong J, Tang S-G *et al* 2008 Fc gamma receptor signaling in mast cells links microbial stimulation to mucosal immune inflammation in the intestine *Am. J. Pathol.* **173** 1647–56

Chen Y, Li C, Xie H, Fan Y, Yang Z, Ma J *et al* 2017 Infiltrating mast cells promote renal cell carcinoma angiogenesis by modulating PI3K → AKT → GSK3β → AM signaling *Oncogene* **36** 2879–88

Cheng S, Li Z, Gao R, Xing B, Gao Y, Yang Y *et al* 2021 A pan-cancer single-cell transcriptional atlas of tumor infiltrating myeloid cells *Cell* **184** 792–809.e23

Chiurchiù V, Leuti A and Maccarrone M 2018 Bioactive lipids and chronic inflammation: managing the fire within *Front. Immunol.* **9** 38

Chiurchiù V and Maccarrone M 2016 Bioactive lipids as modulators of immunity, inflammation and emotions *Curr. Opin. Pharmacol.* **29** 54–62

Cho W, Mittal S K, Elbasiony E and Chauhan S K 2022 Ocular surface mast cells promote inflammatory lymphangiogenesis *Microvasc. Res.* **141** 104320

Cianferoni A 2021 Non-IgE-mediated anaphylaxis *J. Allergy Clin. Immunol.* **147** 1123–31

Cildir G, Yip K H, Pant H, Tergaonkar V, Lopez A F and Tumes D J 2021 Understanding mast cell heterogeneity at single cell resolution *Trends Immunol.* **42** 523–35

Cimpean A M, Tamma R, Ruggieri S, Nico B, Toma A and Ribatti D 2017 Mast cells in breast cancer angiogenesis *Crit. Rev. Oncol. Hematol.* **115** 23–6

Colin-York H, Li D, Korobchevskaya K, Chang V T, Betzig E, Eggeling C *et al* 2019 Cytoskeletal actin patterns shape mast cell activation *Commun. Biol.* **2** 93

Comijn J, Berx G, Vermassen P, Verschueren K, van Grunsven L, Bruyneel E *et al* 2001 The two-handed E box binding zinc finger protein SIP1 downregulates Ecadherin and induces invasion *Mol. Cell.* **7** 1267–78

Crosas-Molist E, Meirelles T, López-Luque J, Serra-Peinado C, Selva J, Caja L *et al* 2015 Vascular smooth muscle cell phenotypic changes in patients with marfan syndrome *ATVB* **35** 960–72

Cui G, Qi H, Gundersen M D, Yang H, Christiansen I, Sørbye S W *et al* 2015 Dynamics of the IL-33/ST2 network in the progression of human colorectal adenoma to sporadic colorectal cancer *Cancer Immunol. Immunother.* **64** 181–90

Dahlin J S, Malinovschi A, Öhrvik H, Sandelin M, Janson C, Alving K *et al* 2016 Lin-CD34hi CD117int/hi FcεRI+ cells in human blood constitute a rare population of mast cell progenitors *Blood* **127** 383–91

De Palma M, Biziato D and Petrova T V 2017 Microenvironmental regulation of tumour angiogenesis *Nat. Rev. Cancer* **17** 457–74

Derakhshan T, Boyce J A and Dwyer D F 2022 Defining mast cell differentiation and heterogeneity through single-cell transcriptomics analysis *J. Allergy Clin. Immunol.* **150** 739–47

Derakhshani A, Vahidian F, Alihasanzadeh M, Mokhtarzadeh A, Lotfi Nezhad P and Baradaran B 2019 Mast cells: a double-edged sword in cancer *Immunol. Lett.* **209** 28–35

Detoraki A, Staiano R I, Granata F, Giannattasio G, Prevete N, De Paulis A *et al* 2009 Vascular endothelial growth factors synthesized by human lung mast cells exert angiogenic effects *J. Allergy Clin. Immunol.* **123** 1142–1149.e5

Dongre A, Rashidian M, Reinhardt F, Bagnato A, Keckesova Z, Ploegh H L *et al* 2017 Epithelial-to-mesenchymal transition contributes to immunosuppression in breast carcinomas *Cancer Res.* **77** 3982–9

Dongre A and Weinberg R A 2019 New insights into the mechanisms of epithelial–mesenchymal transition and implications for cancer *Nat. Rev. Mol. Cell Biol.* **20** 69–84

Drube S, Heink S, Walter S, Löhn T, Grusser M, Gerbaulet A *et al* 2010 The receptor tyrosine kinase c-Kit controls IL-33 receptor signaling in mast cells *Blood* **115** 3899–906

Drube S, Weber F, Göpfert C, Loschinski R, Rothe M, Boelke F *et al* 2015 TAK1 and IKK2, novel mediators of SCF-induced signaling and potential targets for c-Kit-driven diseases *Oncotarget* **6** 28833–50

Eger A, Aigner K, Sonderegger S, Dampier B, Oehler S, Schreiber M *et al* 2005 DeltaEF1 is a transcriptional repressor of E-cadherin and regulates epithelial plasticity in breast cancer cells *Oncogene* **24** 2375–85

Elieh Ali Komi D, Wöhrl S and Bielory L 2020 Mast cell biology at molecular level: a comprehensive review *Clin. Rev. Allerg Immunol.* **58** 342–65

Elpek G O, Gelen T, Aksoy N H, Erdogan A, Dertsiz L, Demircan A *et al* 2001 The prognostic relevance of angiogenesis and mast cells in squamous cell carcinoma of the oesophagus *J. Clin. Pathol.* **54** 940–4

Enoksson M, Lyberg K, Möller-Westerberg C, Fallon P G, Nilsson G and Lunderius-Andersson C 2011 Mast cells as sensors of cell injury through IL-33 recognition *J. Immunol.* **186** 2523–8

Evans J F and Hutchinson J H 2010 Seeing the future of bioactive lipid drug targets *Nat. Chem. Biol.* **6** 476–9

Fajt M L and Wenzel S E 2013 Mast cells, their subtypes, and relation to asthma phenotypes *AnnalsATS* **10** S158–64

Falduto G H, Pfeiffer A, Luker A, Metcalfe D D and Olivera A 2021 Emerging mechanisms contributing to mast cell-mediated pathophysiology with therapeutic implications *Pharmacol. Ther.* **220** 107718

Fernandez-Bravo S, Palacio-Garcia L, Requena-Robledo N, Yuste-Montalvo A, Nuñez-Borque E and Esteban V 2022 Anaphylaxis: mediators, biomarkers, and microenvironments *J. Investig. Allergol. Clin. Immunol.* **32** 419–37

Fimiani M, Mazzatenta C, Alessandrini C, Paola M, Paola C and Andreassi L 1995 Mid-dermal elastolysis: an ultrastructural and biochemical study *Arch. Dermatol. Res.* **287** 152–7

Fischer M, Juremalm M, Olsson N, Backlin C, Sundström C, Nilsson K *et al* 2003 Expression of CCL5/RANTES by Hodgkin and Reed-Sternberg cells and its possible role in the recruitment of mast cells into lymphomatous tissue *Int. J. Cancer* **107** 197–201

Forsythe P 2019 Mast cells in neuroimmune interactions *Trends. Neurosci.* **42** 43–55

Friend D S, Ghildyal N, Austen K F, Gurish M F, Matsumoto R and Stevens R L 1996 Mast cells that reside at different locations in the jejunum of mice infected with Trichinella spiralis exhibit sequential changes in their granule ultrastructure and chymase phenotype *J. Cell Biol.* **135** 279–90

Fritzsche M, Li D, Colin-York H, Chang V T, Moeendarbary E, Felce J H *et al* 2017 Self-organizing actin patterns shape membrane architecture but not cell mechanics *Nat. Commun.* **8** 14347

Fu H, Zhu Y, Wang Y, Liu Z, Zhang J, Wang Z *et al* 2017 Tumor infiltrating mast cells (TIMs) confers a marked survival advantage in nonmetastatic clear-cell renal cell carcinoma *Ann. Surg. Oncol.* **24** 1435–42

Füreder W, Agis H, Willheim M, Bankl H C, Maier U, Kishi K *et al* 1995 Differential expression of complement receptors on human basophils and mast cells. Evidence for mast cell heterogeneity and CD88/C5aR expression on skin mast cells *J. Immunol.* **155** 3152–60

Galli S J, Borregaard N and Wynn T A 2011 Phenotypic and functional plasticity of cells of innate immunity: macrophages, mast cells and neutrophils *Nat. Immunol.* **12** 1035–44

Galli S J, Gaudenzio N and Tsai M 2020 Mast cells in inflammation and disease: recent progress and ongoing concerns *Annu. Rev. Immunol.* **38** 49–77

Galli S J and Tsai M 2012 IgE and mast cells in allergic disease *Nat. Med.* **18** 693–704

Galli S J, Tsai M, Wershil B K, Tam S-Y and Costa J J 1995 Regulation of mouse and human mast cell development, survival and function by stem cell factor, the ligand for the c-*kit* receptor *Int. Arch. Allergy Immunol.* **107** 51–3

Gaudenzio N, Sibilano R, Marichal T, Starkl P, Reber L L, Cenac N *et al* 2016 Different activation signals induce distinct mast cell degranulation strategies *J. Clin. Invest.* **126** 3981–98

Gentek R, Ghigo C, Hoeffel G, Bulle M J, Msallam R, Gautier G *et al* 2018 Hemogenic endothelial fate mapping reveals dual developmental origin of mast cells *Immunity* **48** 1160–1171.e5

Ghouse S M, Polikarpova A, Muhandes L, Dudeck J, Tantcheva-Poór I, Hartmann K *et al* 2018 Although abundant in tumor tissue, mast cells have no effect on immunological micro-milieu or growth of HPV-induced or transplanted tumors *Cell Rep.* **22** 27–35

Godfrey M 1994 From fluorescence to the gene: the skin in the marfan syndrome *J. Invest. Dermatol.* **103** 58S–62S

Godot V, Arock M, Garcia G, Capel F, Flys C, Dy M *et al* 2007 H4 histamine receptor mediates optimal migration of mast cell precursors to CXCL12 *J. Allergy Clin. Immunol.* **120** 827–34

Gorzalczany Y, Akiva E, Klein O, Merimsky O and Sagi-Eisenberg R 2017 Mast cells are directly activated by contact with cancer cells by a mechanism involving autocrine formation of adenosine and autocrine/paracrine signaling of the adenosine A3 receptor *Cancer Lett.* **397** 23–32

Gorzalczany Y, Merimsky O and Sagi-Eisenberg R 2019 Mast cells are directly activated by cancer cell–derived extracellular vesicles by a CD73- and adenosine-dependent mechanism *Transl. Oncol.* **12** 1549–56

Gounaris E, Erdman S E, Restaino C, Gurish M F, Friend D S, Gounari F *et al* 2007 Mast cells are an essential hematopoietic component for polyp development *Proc. Natl Acad. Sci. USA* **104** 19977–82

Groschwitz K R, Wu D, Osterfeld H, Ahrens R and Hogan S P 2013 Chymase-mediated intestinal epithelial permeability is regulated by a protease-activating receptor/matrix metalloproteinase-2-dependent mechanism *Am. J. Physiol. Gastrointest. Liver Physiol.* **304** G479–89

Gruber B, Marchese M and Kew R 1995 Angiogenic factors stimulate mast-cell migration *Blood* **86** 2488–93

Gudmann N S, Manon-Jensen T, Sand J M B, Diefenbach C, Sun S, Danielsen A *et al* 2018 Lung tissue destruction by proteinase 3 and cathepsin G mediated elastin degradation is elevated in chronic obstructive pulmonary disease *Biochem. Biophys. Res. Commun.* **503** 1284–90

Gülen T and Akin C 2022 Anaphylaxis and mast cell disorders *Immunol. Allergy Clin. North Am.* **42** 45–63

Guo X, Zhai L, Xue R, Shi J, Zeng Q and Gao C 2016 Mast cell tryptase contributes to pancreatic cancer growth through promoting angiogenesis via activation of angiopoietin-1 *Int. J. Mol. Sci.* **17** 834

Gupta P, Taiyab A, Hussain A, Alajmi M F, Islam A and Hassan M I 2021 Targeting the sphingosine kinase/sphingosine-1-phosphate signaling axis in drug discovery for cancer therapy *Cancers* **13** 1898

Gurish M F and Austen K F 2012 Developmental origin and functional specialization of mast cell subsets *Immunity* **37** 25–33

Gurish M F, Ghildyal N, McNeil H P, Austen K F, Gillis S and Stevens R L 1992 Differential expression of secretory granule proteases in mouse mast cells exposed to interleukin 3 and c-kit ligand *J. Exp. Med.* **175** 1003–12

Gurish M F, Tao H, Abonia J P, Arya A, Friend D S, Parker C M *et al* 2001 Intestinal mast cell progenitors require CD49dβ7 (α4β7 integrin) for tissue-specific homing *J. Exp. Med.* **194** 1243–52

Hallgren J and Gurish M F 2007 Pathways of murine mast cell development and trafficking: tracking the roots and routes of the mast cell *Immunol. Rev.* **217** 8–18

Hanahan D and Weinberg R A 2011 Hallmarks of cancer: the next generation *Cell* **144** 646–74

Hannezo E, Dong B, Recho P, Joanny J-F and Hayashi S 2015 Cortical instability drives periodic supracellular actin pattern formation in epithelial tubes *Proc. Natl Acad. Sci. USA* **112** 8620–5

Headley M B, Bins A, Nip A, Roberts E W, Looney M R, Gerard A *et al* 2016 Visualization of immediate immune responses to pioneer metastatic cells in the lung *Nature* **531** 513–7

Heijmans J, Büller N V, Muncan V and Van Den Brink G R 2012 Role of mast cells in colorectal cancer development, the jury is still out *Biochim. Biophys. Acta, Mol. Basis Dis.* **1822** 9–13

Hellman L, Akula S, Fu Z and Wernersson S 2022 Mast cell and basophil granule proteases—*in vivo* targets and function *Front. Immunol.* **13** 918305

Hempel H A, Cuka N S, Kulac I, Barber J R, Cornish T C, Platz E A *et al* 2017 Low intratumoral mast cells are associated with a higher risk of prostate cancer recurrence: mast cells and prostate cancer recurrence *Prostate* **77** 412–24
Herrera-Heredia S A, Hsu H-P, Kao C-Y, Tsai Y-H, Yamaguchi Y, Roers A *et al* 2022 Heparin is required for the formation of granules in connective tissue mast cells *Front. Immunol.* **13** 1000405
Higuchi H, Hara M, Yamamoto K, Miyamoto T, Kinoshita M, Yamada T *et al* 2008 Mast cells play a critical role in the pathogenesis of viral myocarditis *Circulation* **118** 363–72
Ho L H, Ohno T, Oboki K, Kajiwara N, Suto H, Iikura M *et al* 2007 IL-33 induces IL-13 production by mouse mast cells independently of IgE-FcεRI signals *J. Leukoc. Biol.* **82** 1481–90
Hofmann N A, Yang J, Trauger S A, Nakayama H, Huang L, Strunk D *et al* 2015 The GPR 55 agonist, L-α-lysophosphatidylinositol, mediates ovarian carcinoma cell-induced angiogenesis *Br. J. Pharmacol.* **172** 4107–18
Hsu Y-R, Wu G-M, Mendiaz E A, Syed R, Wypych J, Toso R *et al* 1997 The majority of stem cell factor exists as monomer under physiological conditions *J. Biol. Chem.* **272** 6406–15
Hu G, Wang S and Cheng P 2018 Tumor-infiltrating tryptase$^+$ mast cells predict unfavorable clinical outcome in solid tumors *Int. J. Cancer* **142** 813–21
Huang B, Lei Z, Zhang G-M, Li D, Song C, Li B *et al* 2008 SCF-mediated mast cell infiltration and activation exacerbate the inflammation and immunosuppression in tumor microenvironment *Blood* **112** 1269–79
Huang D, Duan H, Huang H, Tong X, Han Y, Ru G *et al* 2016 Cisplatin resistance in gastric cancer cells is associated with HER2 upregulation-induced epithelial-mesenchymal transition *Sci. Rep.* **6** 20502
Huang M, Wang X, Xing B, Yang H, Sa Z, Zhang D *et al* 2018 Critical roles of TRPV2 channels, histamine H1 and adenosine A1 receptors in the initiation of acupoint signals for acupuncture analgesia *Sci. Rep.* **8** 6523
Huber M, Cato A C B, Ainooson G K, Freichel M, Tsvilovskyy V, Jessberger R *et al* 2019 Regulation of the pleiotropic effects of tissue-resident mast cells *J. Allergy Clin. Immunol.* **144** S31–45
Hügle T 2014 Beyond allergy: the role of mast cells in fibrosis *Swiss Med. Wkly* **144** 3536
Hui L and Chen Y 2015 Tumor microenvironment: sanctuary of the devil *Cancer Lett.* **368** 7–13
Hynes R O 2002 Integrins *Cell* **110** 673–87
I K-Y, Tseng W-Y, Wang W-C, Gordon S, Ng K-F and Lin H-H 2021 Stimulation of vibratory urticaria-associated adhesion-GPCR, EMR2/ADGRE2, triggers the NLRP3 inflammasome activation signal in human monocytes *Front. Immunol.* **11** 602016
Ito T, Smrž D, Jung M-Y, Bandara G, Desai A, Smržová Š *et al* 2012 Stem cell factor programs the mast cell activation phenotype *J. Immunol.* **188** 5428–37
Jarosz-Biej M, Kamińska N, Matuszczak S, Cichoń T, Pamuła-Piłat J, Czapla J *et al* 2018 M1-like macrophages change tumor blood vessels and microenvironment in murine melanoma *PLoS One* **13** e0191012
Jiang Y, Wu Y, Hardie W J and Zhou X 2017 Mast cell chymase affects the proliferation and metastasis of lung carcinoma cells *in vitro Oncol. Lett.* **14** 3193–8
Jiménez M, Cervantes-García D, Córdova-Dávalos L E, Pérez-Rodríguez M J, Gonzalez-Espinosa C and Salinas E 2021 Responses of mast cells to pathogens: beneficial and detrimental roles *Front. Immunol.* **12** 685865

Johansson A, Rudolfsson S, Hammarsten P, Halin S, Pietras K, Jones J *et al* 2010 Mast cells are novel independent prognostic markers in prostate cancer and represent a target for therapy *Am. J. Pathol.* **177** 1031–41

Jolly M K 2015 Implications of the hybrid epithelial/mesenchymal phenotype in metastasis *Front. Oncol.* **5** 155

Jolly P S, Bektas M, Olivera A, Gonzalez-Espinosa C, Proia R L, Rivera J *et al* 2004 Transactivation of sphingosine-1–phosphate receptors by FcεRI triggering is required for normal mast cell degranulation and chemotaxis *J. Exp. Med.* **199** 959–70

Joulia R, L'Faqihi F-E, Valitutti S and Espinosa E 2017 IL-33 fine tunes mast cell degranulation and chemokine production at the single-cell level *J. Allergy Clin. Immunol.* **140** 497–509.e10

Jung M-Y, Smrž D, Desai A, Bandara G, Ito T, Iwaki S *et al* 2013 IL-33 induces a hyporesponsive phenotype in human and mouse ast cells *J. Immunol.* **190** 531–8

Kabiraj A, Jaiswal R, Singh A, Gupta J, Singh A and Samadi F M 2018 Immunohistochemical evaluation of tumor angiogenesis and the role of mast cells in oral squamous cell carcinoma *J. Cancer Res. Therap.* **14** 495–502

Kadry Y A and Calderwood D A 2020 Chapter 22: structural and signaling functions of integrins *Biochim. Biophys. Acta (BBA)—Biomembr.* **1862** 183206

Kaieda S, Shin K, Nigrovic P A, Seki K, Lee R T, Stevens R L *et al* 2010 Synovial fibroblasts promote the expression and granule accumulation of tryptase via interleukin-33 and its receptor ST-2 (IL1RL1) *J. Biol. Chem.* **285** 21478–86

Kaltenbach L, Martzloff P, Bambach S K, Aizarani N, Mihlan M, Gavrilov A *et al* 2023 Slow integrin-dependent migration organizes networks of tissue-resident mast cells *Nat. Immunol.* **24** 915–24

Kanakura Y, Thompson H, Nakano T, Yamamura T, Asai H, Kitamura Y *et al* 1988 Multiple bidirectional alterations of phenotype and changes in proliferative potential during the *in vitro* and *in vivo* passage of clonal mast cell populations derived from mouse peritoneal mast cells *Blood* **72** 877–85

Kapur R, Majumdar M, Xiao X, McAndrews-Hill M, Schindler K and Williams D A 1998 Signaling through the interaction of membrane-restricted stem cell factor and c-kit receptor tyrosine kinase: genetic evidence for a differential role in erythropoiesis *Blood* **91** 879–89

Karaman S, Leppänen V-M and Alitalo K 2018 Vascular endothelial growth factor signaling in development and disease *Development* **145** dev151019

Kargl J, Andersen L, Hasenöhrl C, Feuersinger D, Stančić A, Fauland A *et al* 2016 GPR55 promotes migration and adhesion of colon cancer cells indicating a role in metastasis *Br. J. Pharmacol.* **173** 142–54

Karsenti E 2008 Self-organization in cell biology: a brief history *Nat. Rev. Mol. Cell Biol.* **9** 255–62

Karur G R, Pagano J J, Bradley T, Lam C Z, Seed M, Yoo S-J *et al* 2018 Diffuse myocardial fibrosis in children and adolescents with Marfan syndrome and Loeys-Dietz syndrome *J. Am. Coll. Cardiol.* **72** 2279–81

Kato Y, Morikawa T, Kato E, Yoshida K, Imoto Y, Sakashita M *et al* 2021 Involvement of activation of mast cells via IgE signaling and epithelial cell–derived cytokines in the pathogenesis of pollen food allergy syndrome in a Murine model *J. Immunol.* **206** 2791–802

Kawakami T and Galli S J 2002 Regulation of mast-cell and basophil function and survival by IgE *Nat. Rev. Immunol.* **2** 773–86

Kechagia J Z, Ivaska J and Roca-Cusachs P 2019 Integrins as biomechanical sensors of the microenvironment *Nat. Rev. Mol. Cell Biol.* **20** 457–73

Kennedy C C, Brown E E, Abutaleb N O and Truskey G A 2021 Development and application of endothelial cells derived from pluripotent stem cells in microphysiological systems models *Front. Cardiovasc. Med.* **8** 625016

Kielty C M 2017 Fell-Muir lecture: fibrillin microfibrils: structural tensometers of elastic tissues? *Int. J. Exp. Pathol.* **98** 172–90

Kiener H P, Baghestanian M, Dominkus M, Walchshofer S, Ghannadan M, Willheim M *et al* 1998 Expression of the C5a receptor (CD88) on synovial mast cells in patients with rheumatoid arthritis *Arthritis Rheum.* **41** 233–45

Kihara Y, Maceyka M, Spiegel S and Chun J 2014 Lysophospholipid receptor nomenclature review: IUPHAR review 8 *Br. J Pharmacol.* **171** 3575–94

Kim Y J, Hong K S, Chung J W, Kim J H and Hahm K B 2010 Prevention of colitis-associated carcinogenesis with infliximab *Cancer Prev. Res.* **3** 1314–33

Kirshenbaum A S, Akin C, Wu Y, Rottem M, Goff J P, Beaven M A *et al* 2003 Characterization of novel stem cell factor responsive human mast cell lines LAD 1 and 2 established from a patient with mast cell sarcoma/leukemia; activation following aggregation of FcεRI or FcγRI *Leuk. Res.* **27** 677–82

Kirshenbaum A S, Petrik A, Walsh R, Kirby T L, Vepa S, Wangsa D *et al* 2014 A ten-year retrospective analysis of the distribution, use and phenotypic characteristics of the LAD2 human mast cell line *Int. Arch. Allergy Immunol.* **164** 265–70

Klein O, Krier-Burris R A, Lazki-Hagenbach P, Gorzalczany Y, Mei Y, Ji P *et al* 2019 Mammalian diaphanous-related formin 1 (mDia1) coordinates mast cell migration and secretion through its actin-nucleating activity *J. Allergy Clin. Immunol.* **144** 1074–90

Knight P A, Wright S H, Lawrence C E, Paterson Y Y W and Miller H R P 2000 Delayed expulsion of the nematode *Trichinella spiralis* in mice lacking the mucosal mast cell–specific granule chymase, mouse mast cell protease-1 *J. Exp. Med.* **192** 1849–56

Ko E-A, Sanders K M and Zhou T 2017 A transcriptomic insight into the impacts of mast cells in lung, breast, and colon cancers *OncoImmunology* **6** e1360457

Komi D E A, Khomtchouk K and Santa Maria P L 2020 A review of the contribution of mast cells in wound healing: involved molecular and cellular mechanisms *Clin. Rev. Allerg Immunol.* **58** 298–312

Komi D E A and Redegeld F A 2020 Role of mast cells in shaping the tumor microenvironment *Clin. Rev. Allerg Immunol.* **58** 313–25

Kovanen P T 2019 Mast cells as potential accelerators of human atherosclerosis—from early to late lesions *Int. J. Mol. Sci.* **20** 4479

Krüger-Krasagakes S, Grützkau A, Baghramian R and Henz B M 1996 Interactions of immature human mast cells with extracellular matrix: expression of specific adhesion receptors and their role in cell binding to matrix proteins *J. Invest. Dermatol.* **106** 538–43

Kruse K, Joanny J F, Jülicher F, Prost J and Sekimoto K 2004 Asters, vortices, and rotating spirals in active gels of polar filaments *Phys. Rev. Lett.* **92** 078101

Kudou M, Shiozaki A, Yamazato Y, Katsurahara K, Kosuga T, Shoda K *et al* 2019 The expression and role of TRPV2 in esophageal squamous cell carcinoma *Sci. Rep.* **9** 16055

Kulka M, Sheen C H, Tancowny B P, Grammer L C and Schleimer R P 2008 Neuropeptides activate human mast cell degranulation and chemokine production *Immunology* **123** 398–410

Kwon D, Park E S, Kim M, Choi Y H, Lee M, Joo S *et al* 2022 Homeostatic serum IgE is secreted by plasma cells in the thymus and enhances mast cell survival *Nat. Commun.* **13** 1418

Lämmermann T, Bader B L, Monkley S J, Worbs T, Wedlich-Söldner R, Hirsch K *et al* 2008 Rapid leukocyte migration by integrin-independent flowing and squeezing *Nature* **453** 51–5

Lecce M, Molfetta R, Milito N D, Santoni A and Paolini R 2020 FcεRI signaling in the modulation of allergic response: role of mast cell-derived exosomes *Int. J. Mol. Sci.* **21** 5464

Lee J-H, Jeon Y-D, Xin M, Lim J, Lee Y-M and Kim D-K 2022 Mast cell modulates tumorigenesis caused by repeated bowel inflammation condition in azoxymethane/dextran sodium sulfate-induced colon cancer mouse model *Biochem. Biophys. Rep.* **30** 101253

Lennartsson J and Rönnstrand L 2012 Stem cell factor receptor/c-Kit: from basic science to clinical implications *Physiol. Rev.* **92** 1619–49

Levi-Schaffer F, Austen K F, Gravallese P M and Stevens R L 1986 Coculture of interleukin 3-dependent mouse mast cells with fibroblasts results in a phenotypic change of the mast cells *Proc. Natl Acad. Sci. USA* **83** 6485–8

Leyva-Castillo J-M, Galand C, Kam C, Burton O, Gurish M, Musser M A *et al* 2019 Mechanical skin injury promotes food anaphylaxis by driving intestinal mast cell expansion *Immunity* **50** 1262–75.e4

Li D and Wu M 2021 Pattern recognition receptors in health and diseases *Signal Transduct. Target. Ther.* **6** 291

Li H, Guo Z and Xiangdong Q 2024 Role of mechanical stimulus in mast cell activation *Digit. Med.* **10** e23-00014

Li P, Cui Y, Song G, Wang Z and Zhang Q 2014 Phenotypic characteristics of nasal mast cells in a mouse model of allergic rhinitis *ORL* **76** 303–13

Li Z, Liu S, Xu J, Zhang X, Han D, Liu J *et al* 2018 Adult tconnective tissue-resident mast cells originate from late erythro-myeloid progenitors *Immunity* **49** 640–53.e5

Lieberman P and Garvey L H 2016 Mast cells and anaphylaxis *Curr. Allergy Asthma Rep.* **16** 20

Liew F Y, Girard J-P and Turnquist H R 2016 Interleukin-33 in health and disease *Nat. Rev. Immunol.* **16** 676–89

Link T M, Park U, Vonakis B M, Raben D M, Soloski M J and Caterina M J 2010 TRPV2 has a pivotal role in macrophage particle binding and phagocytosis *Nat. Immunol.* **11** 232–9

Liu Q and Wang X 2013 Effect of TRPV2 cation channels on the proliferation, migration and invasion of 5637 bladder cancer cells *Exp. Ther. Med.* **6** 1277–82

Liu X, Zhu L, Lu X, Bian H, Wu X, Yang W *et al* 2014 IL-33/ST2 pathway contributes to metastasis of human colorectal cancer *Biochem. Biophys. Res. Commun.* **453** 486–92

Lo C-M, Wang H-B, Dembo M and Wang Y 2000 Cell movement is guided by the rigidity of the substrate *Biophys. J.* **79** 144–52

Long Y, Niu Y, Liang K and Du Y 2022 Mechanical communication in fibrosis progression *Trends Cell Biol.* **32** 70–90

Lundequist A, Åbrink M and Pejler G 2006 Mast cell-dependent activation of pro matrix metalloprotease 2: a role for serglycin proteoglycan-dependent mast cell proteases *Biol. Chem.* **387** 2006

Ma H-T and Beaven M A 2011 Regulators of Ca^{2+} signaling in mast cells: potential targets for treatment of mast cell-related diseases? *Mast Cell Biology* ed A M Gilfillan and D D Metcalfe (Boston, MA: Springer) pp 62–90

Ma X, Ligan C, Huang S, Chen Y, Li M, Cao Y *et al* 2024 Mitochondrial activity related genes of mast cells identify poor prognosis and metastasis of ovarian cancer *Immunobiology* **229** 152831

Maccarana M, Jia J, Li H, Zhang X, Vlodavsky I and Li J-P 2022 Implications of heparanase on heparin synthesis and metabolism in mast cells *Int. J. Mol. Sci.* **23** 4821

Maciel T T, Moura I C and Hermine O 2015 The role of mast cells in cancers *F1000Prime Rep.* **7** 9

Mangia A, Malfettone A, Rossi R, Paradiso A, Ranieri G, Simone G *et al* 2011 Tissue remodelling in breast cancer: human mast cell tryptase as an initiator of myofibroblast differentiation: mast cell as initiator of myofibroblast *Histopathology* **58** 1096–106

Martin N T and Martin M U 2016 Interleukin 33 is a guardian of barriers and a local alarmin *Nat. Immunol.* **17** 122–31

Massagué J 2000 How cells read TGF-β signals *Nat. Rev. Mol. Cell Biol.* **1** 169–78

Matt P, Schoenhoff F, Habashi J, Holm T, Van Erp C, Loch D *et al* 2009 Circulating transforming growth factor-β in Marfan syndrome *Circulation* **120** 526–32

Mayer G, Hamelin J, Asselin M-C, Pasquato A, Marcinkiewicz E, Tang M *et al* 2008 The regulated cell surface zymogen activation of the proprotein convertase PC5A directs the processing of its secretory substrates *J. Biol. Chem.* **283** 2373–84

Maywald R L, Doerner S K, Pastorelli L, De Salvo C, Benton S M, Dawson E P *et al* 2015 IL-33 activates tumor stroma to promote intestinal polyposis *Proc. Natl Acad. Sci. USA* **112** E2487–96

McDonnell J M, Dhaliwal B, Sutton B J and Gould H J 2023 IgE, IgE receptors and anti-IgE biologics: protein structures and mechanisms of action *Annu. Rev. Immunol.* **41** 255–75

McSweeney S M, Christou E A A, Maurer M, Grattan C E, Tziotzios C and McGrath J A 2023 Physical urticaria: clinical features, pathogenesis, diagnostic work-up, and management *J. Am. Acad. Dermatol.* **89** 324–37

Meixiong J, Anderson M, Limjunyawong N, Sabbagh M F, Hu E, Mack M R *et al* 2019 Activation of mast-cell-expressed mas-related G-protein-coupled receptors drives non-histaminergic itch *Immunity* **50** 1163–71.e5

Ménasché G, Longé C, Bratti M and Blank U 2021 Cytoskeletal transport, reorganization, and fusion regulation in mast cell-stimulus secretion coupling *Front. Cell Dev. Biol.* **9** 652077

Méndez-Enríquez E and Hallgren J 2019 Mast cells and their progenitors in allergic asthma *Front. Immunol.* **10** 821

Meneghetti M C Z, Hughes A J, Rudd T R, Nader H B, Powell A K, Yates E A *et al* 2015 Heparan sulfate and heparin interactions with proteins *J. R. Soc. Interface* **12** 20150589

Metcalfe D D 2008 Mast cells and mastocytosis *Blood* **112** 946–56

Mierke C T 2019 The matrix environmental and cell mechanical properties regulate cell migration and contribute to the invasive phenotype of cancer cells *Rep. Prog. Phys.* **82** 064602

Mierke C T 2020 Splicing and alternative splicing and the impact of mechanics *Cellular Mechanics and Biophysics* (Cham: Springer International Publishing) pp 509–93

Mierke C T 2021 Bidirectional mechanical response between cells and their microenvironment *Front. Phys.* **9** 749830

Mierke C T 2024 Mechanosensory entities and functionality of endothelial cells *Front. Cell Dev. Biol. Sec. Cell Adhes. Migrat.* **12** 1446452

Mierke C T, Ballmaier M, Werner U, Manns M P, Welte K and Bischoff S C 2000 Human endothelial cells regulate survival and proliferation of human mast cells *J. Exp. Med.* **192** 801–12

Milewicz D M, Pyeritz R E, Crawford E S and Byers P H 1992 Marfan syndrome: defective synthesis, secretion, and extracellular matrix formation of fibrillin by cultured dermal fibroblasts *J. Clin. Invest.* **89** 79–86

Misteli T 2001 The concept of self-organization in cellular architecture *J. Cell Biol.* **155** 181–6

Molderings G J, Zienkiewicz T, Homann J, Menzen M and Afrin L B 2017 Risk of solid cancer in patients with mast cell activation syndrome: results from Germany and USA *F1000Res* **6** 1889

Molfetta R, Lecce M, Milito N D, Putro E, Pietropaolo G, Marangio C *et al* 2023 SCF and IL-33 regulate mouse mast cell phenotypic and functional plasticity supporting a pro-inflammatory microenvironment *Cell Death Dis.* **14** 616

Molfetta R and Paolini R 2023 The controversial role of intestinal mast cells in colon cancer *Cells* **12** 459

Molin D, Edström A, Glimelius I, Glimelius B, Nilsson G, Sundström C *et al* 2002 Mast cell infiltration correlates with poor prognosis in Hodgkin's lymphoma *Br. J. Haematol.* **119** 122–4

Monet M, Lehen'kyi V, Gackiere F, Firlej V, Vandenberghe M, Roudbaraki M *et al* 2010 Role of cationic channel TRPV2 in promoting prostate cancer migration and progression to androgen resistance *Cancer Res.* **70** 1225–35

Moon T C, Befus A D and Kulka M 2014 Mast cell mediators: their differential release and the secretory pathways involved *Front. Immunol.* **5** 569

Moon T C, St Laurent C D, Morris K E, Marcet C, Yoshimura T, Sekar Y *et al* 2010 Advances in mast cell biology: new understanding of heterogeneity and function *Mucosal Immunol.* **3** 111–28

Moreau H D, Piel M, Voituriez R and Lennon-Duménil A-M 2018 Integrating physical and molecular insights on immune cell migration *Trends Immunol.* **39** 632–43

Mousavizadeh R, Waugh C M, McCormack R G, Cairns B E and Scott A 2024 MRGPRX2-mediated mast cell activation by substance P from overloaded human tenocytes induces inflammatory and degenerative responses in tendons *Sci. Rep.* **14** 13540

Mukai K, Tsai M, Saito H and Galli S J 2018 Mast cells as sources of cytokines, chemokines, and growth factors *Immunol. Rev.* **282** 121–50

Mukai K, Tsai M, Starkl P, Marichal T and Galli S J 2016 IgE and mast cells in host defense against parasites and venoms *Semin. Immunopathol.* **38** 581–603

Mukouyama Y, Shin D, Britsch S, Taniguchi M and Anderson D J 2002 Sensory nerves determine the pattern of arterial differentiation and blood vessel branching in the skin *Cell* **109** 693–705

Nakajima M, Nagahashi M, Rashid O M, Takabe K and Wakai T 2017 The role of sphingosine-1-phosphate in the tumor microenvironment and its clinical implications *Tumour Biol.* **39** 101042831769913

Naranjo A N, Bandara G, Bai Y, Smelkinson M G, Tobío A, Komarow H D *et al* 2020 Critical signaling events in the mechanoactivation of human mast cells through p.C492Y-ADGRE2 *J. Invest. Dermatol.* **140** 2210–20.e5

Nataatmadja M, West J and West M 2006 Overexpression of transforming growth factor-β is associated with increased hyaluronan content and impairment of repair in marfan syndrome aortic aneurysm *Circulation* **114** I-371–7

Ngkelo A, Richart A, Kirk J A, Bonnin P, Vilar J, Lemitre M *et al* 2016 Mast cells regulate myofilament calcium sensitization and heart function after myocardial infarction *J. Exp. Med.* **213** 1353–74

Nieto M A, Huang R Y-J, Jackson R A and Thiery J P 2016 EMT: 2016 *Cell* **166** 21–45

Nishikori Y, Shiota N and Okunishi H 2014 The role of mast cells in cutaneous wound healing in streptozotocin-induced diabetic mice *Arch. Dermatol. Res.* **306** 823–35

Niyonsaba F, Ushio H, Hara M, Yokoi H, Tominaga M, Takamori K *et al* 2010 Antimicrobial peptides human β-defensins and cathelicidin LL-37 induce the secretion of a pruritogenic cytokine IL-31 by human mast cells *J. Immunol.* **184** 3526–34

Okayama Y and Kawakami T 2006 Development, migration, and survival of mast cells *IR* **34** 97–116

Oldford S A and Marshall J S 2015 Mast cells as targets for immunotherapy of solid tumors *Mol. Immunol.* **63** 113–24

Olivera A, Beaven M A and Metcalfe D D 2018 Mast cells signal their importance in health and disease *J. Allergy Clin. Immunol.* **142** 381–93

Olivera A and Rivera J 2005 Sphingolipids and the balancing of immune cell function: lessons from the mast cell *J. Immunol.* **174** 1153–8

Olivera A and Rivera J 2011 An emerging role for the lipid mediator sphingosine-1-phosphate in mast cell effector function and allergic disease *Mast Cell Biology* ed A M Gilfillan and D D Metcalfe (Boston, MA: Springer) pp 123–42

Oskeritzian C A, Price M M, Hait N C, Kapitonov D, Falanga Y T, Morales J K *et al* 2010 Essential roles of sphingosine-1–phosphate receptor 2 in human mast cell activation, anaphylaxis, and pulmonary edema *J. Exp. Med.* **207** 465–74

Oskeritzian C A, Zhao W, Min H-K, Xia H-Z, Pozez A, Kiev J *et al* 2005 Surface CD88 functionally distinguishes the MCTC from the MCT type of human lung mast cell *J. Allergy Clin. Immunol.* **115** 1162–8

Otsu K, Nakano T, Kanakura Y, Asai H, Katz H R, Austen K F *et al* 1987 Phenotypic changes of bone marrow-derived mast cells after intraperitoneal transfer into W/Wv mice that are genetically deficient in mast cells *J. Exp. Med.* **165** 615–27

Panagi M, Mpekris F, Voutouri C, Hadjigeorgiou A G, Symeonidou C, Porfyriou E *et al* 2024 Stabilizing tumor-resident mast cells restores T-cell infiltration and sensitizes sarcomas to PD-L1 inhibition *Clin. Cancer Res.* **30** 2582–97

Paolino G, Belmonte M, Trasarti S, Santopietro M, Bizzoni L, Riminucci M *et al* 2017 Mast cell disorders, melanoma and pancreatic carcinoma: from a clinical observation to a brief review of the literature *Acta Dermatovenerol. Croat.* **25** 112–9

Paruchuri S, Jiang Y, Feng C, Francis S A, Plutzky J and Boyce J A 2008 Leukotriene E4 activates peroxisome proliferator-activated receptor γ and induces prostaglandin D2 generation by human mast cells *J. Biol. Chem.* **283** 16477–87

Pastwińska J, Walczak-Drzewiecka A, Łukasiak M, Ratajewski M and Dastych J 2020 Hypoxia regulates human mast cell adhesion to fibronectin via the PI3K/AKT signaling pathway *Cell Adh. Migr.* **14** 106–17

Peng D, Fu M, Wang M, Wei Y and Wei X 2022 Targeting TGF-β signal transduction for fibrosis and cancer therapy *Mol. Cancer* **21** 104

Pham L, Kennedy L, Baiocchi L, Meadows V, Ekser B, Kundu D *et al* 2022 Mast cells in liver disease progression: an update on current studies and implications *Hepatology* **75** 213–8

Piliponsky A M and Romani L 2018 The contribution of mast cells to bacterial and fungal infection immunity *Immunol. Rev.* **282** 188–97

Piñeiro R, Maffucci T and Falasca M 2011 The putative cannabinoid receptor GPR55 defines a novel autocrine loop in cancer cell proliferation *Oncogene* **30** 142–52

Plow E F, Haas T A, Zhang L, Loftus J and Smith J W 2000 Ligand binding to integrins *J. Biol. Chem.* **275** 21785–8

Plum T, Wang X, Rettel M, Krijgsveld J, Feyerabend T B and Rodewald H-R 2020 Human mast cell proteome reveals unique lineage, putative functions, and structural basis for cell ablation *Immunity* **52** 404–16.e5

Põlajeva J, Bergström T, Edqvist P-H, Lundequist A, Sjösten A, Nilsson G *et al* 2014 Glioma-derived macrophage migration inhibitory factor (MIF) promotes mast cell recruitment in a STAT5-dependent manner *Mol. Oncol.* **8** 50–8

Põlajeva J, Sjösten A M, Lager N, Kastemar M, Waern I, Alafuzoff I *et al* 2011 Mast cell accumulation in glioblastoma with a potential role for stem cell factor and chemokine CXCL12 *PLoS One* **6** e25222

Popescu G, Park Y, Choi W, Dasari R R, Feld M S and Badizadegan K 2008 Imaging red blood cell dynamics by quantitative phase microscopy *Blood Cells Mol. Dis.* **41** 10–6

Puisieux A, Brabletz T and Caramel J 2014 Oncogenic roles of EMT-inducing transcription factors *Nat. Cell Biol.* **16** 488–94

Pundir P, Liu R, Vasavda C, Serhan N, Limjunyawong N, Yee R *et al* 2019 A connective tissue mast-cell-specific receptor detects bacterial quorum-sensing molecules and mediates anti-bacterial immunity *Cell Host Microbe* **26** 114–122.e8

Rabenhorst A, Schlaak M, Heukamp L C, Förster A, Theurich S, Von Bergwelt-Baildon M *et al* 2012 Mast cells play a protumorigenic role in primary cutaneous lymphoma *Blood* **120** 2042–54

Rajput A B, Turbin D A, Cheang M C, Voduc D K, Leung S, Gelmon K A *et al* 2008 Stromal mast cells in invasive breast cancer are a marker of favourable prognosis: a study of 4,444 cases *Breast Cancer Res. Treat.* **107** 249–57

Ramirez F and Sakai L Y 2010 Biogenesis and function of fibrillin assemblies *Cell Tissue Res.* **339** 71–82

Ramirez F, Sakai L Y, Dietz H C and Rifkin D B 2004 Fibrillin microfibrils: multipurpose extracellular networks in organismal physiology *Physiol. Genomics* **19** 151–4

Rao K N and Brown M A 2008 Mast cells: multifaceted immune cells with diverse roles in health and disease *Ann. N.Y. Acad. Sci.* **1143** 83–104

Rathore A P and St John A L 2020 Protective and pathogenic roles for mast cells during viral infections *Curr. Opin. Immunol.* **66** 74–81

Redegeld F A, Yu Y, Kumari S, Charles N and Blank U 2018 Non-IgE mediated mast cell activation *Immunol. Rev.* **282** 87–113

Ribatti D and Crivellato E 2012 Mast cells, angiogenesis, and tumour growth *Biochim. Biophys. Acta (BBA)—Mol. Basis Dis.* **1822** 2–8

Ribatti D and Ranieri G 2015 Tryptase, a novel angiogenic factor stored in mast cell granules *Exp. Cell. Res.* **332** 157–62

Ribatti D, Tamma R and Annese T 2020 Mast cells and angiogenesis in multiple sclerosis *Inflamm. Res.* **69** 1103–10

Ribatti D, Vacca A, Ria R, Marzullo A, Nico B, Filotico R *et al* 2003 Neovascularisation, expression of fibroblast growth factor-2, and mast cells with tryptase activity increase simultaneously with pathological progression in human malignant melanoma *Eur. J. Cancer* **39** 666–74

Rigoni A, Colombo M P and Pucillo C 2015 The role of mast cells in molding the tumor microenvironment *Cancer Microenviron.* **8** 167–76

Rigoni A, Colombo M P and Pucillo C 2018 Mast cells, basophils and eosinophils: from allergy to cancer *Semin. Immunol.* **35** 29–34

Rivellese F, Nerviani A, Rossi F W, Marone G, Matucci-Cerinic M, De Paulis A *et al* 2017 Mast cells in rheumatoid arthritis: friends or foes? *Autoimmun. Rev.* **16** 557–63

Ronca R, Tamma R, Coltrini D, Ruggieri S, Presta M and Ribatti D 2017 Fibroblast growth factor modulates mast cell recruitment in a murine model of prostate cancer *Oncotarget* **8** 82583–92

Rosbottom A, Scudamore C L, Von Der Mark H, Thornton E M, Wright S H and Miller H R P 2002 TGF-β1 regulates adhesion of mucosal mast cell homologues to laminin-1 through expression of integrin α7 *J. Immunol.* **169** 5689–95

Rosenbaum T and Islas L D 2023 Molecular physiology of TRPV channels: controversies and future challenges *Annu. Rev. Physiol.* **85** 293–316

Roy A, Coum A, Marinescu V D, Põlajeva J, Smits A, Nelander S *et al* 2015 Glioma-derived plasminogen activator inhibitor-1 (PAI-1) regulates the recruitment of LRP1 positive mast cells *Oncotarget* **6** 23647–61

Ryberg E, Larsson N, Sjögren S, Hjorth S, Hermansson N, Leonova J *et al* 2007 The orphan receptor GPR55 is a novel cannabinoid receptor *Br. J. Pharmacol.* **152** 1092–101

Saadalla A, Lima M M, Tsai F, Osman A, Singh M P, Linden D R *et al* 2019 Cell intrinsic deregulated ß-catenin signaling promotes expansion of bone marrow derived connective tissue type mast cells, systemic inflammation, and colon cancer *Front. Immunol.* **10** 2777

Saadalla A M, Osman A, Gurish M F, Dennis K L, Blatner N R, Pezeshki A *et al* 2018 Mast cells promote small bowel cancer in a tumor stage-specific and cytokine-dependent manner *Proc. Natl Acad. Sci. USA* **115** 1588–92

Sakita J Y, Elias-Oliveira J, Carlos D, De Souza Santos E, Almeida L Y, Malta T M *et al* 2022 Mast cell-T cell axis alters development of colitis-dependent and colitis-independent colorectal tumours: potential for therapeutically targeting via mast cell inhibition *J. Immunother. Cancer* **10** e004653

Salamon P, Mekori Y A and Shefler I 2020 Lung cancer-derived extracellular vesicles: a possible mediator of mast cell activation in the tumor microenvironment *Cancer Immunol. Immunother.* **69** 373–81

Salomonsson M, Dahlin J S, Ungerstedt J and Hallgren J 2020 Localization-specific expression of CCR1 and CCR5 by mast cell progenitors *Front. Immunol.* **11** 321

Saluja R, Zoltowska A, Ketelaar M E and Nilsson G 2016 IL-33 and thymic stromal lymphopoietin in mast cell functions *Eur. J. Pharmacol.* **778** 68–76

Sammarco G, Varricchi G, Ferraro V, Ammendola M, De Fazio M, Altomare D F *et al* 2019 Mast cells, angiogenesis and lymphangiogenesis in human gastric cancer *Int. J. Mol. Sci.* **20** 2106

Saraswathibhatla A, Indana D and Chaudhuri O 2023 Cell–extracellular matrix mechanotransduction in 3D *Nat. Rev. Mol. Cell Biol.* **24** 495–516

Schaaf M B, Garg A D and Agostinis P 2018 Defining the role of the tumor vasculature in antitumor immunity and immunotherapy *Cell Death Dis.* **9** 115

Schmelzer C E H, Jung M C, Wohlrab J, Neubert R H H and Heinz A 2012 Does human leukocyte elastase degrade intact skin elastin? *FEBS J.* **279** 4191–200

Schmitz J, Owyang A, Oldham E, Song Y, Murphy E, McClanahan T K *et al* 2005 IL-33, an interleukin-1-like cytokine that signals via the IL-1 receptor-related protein ST2 and induces T helper type 2-associated cytokines *Immunity* **23** 479–90

Segura-Villalobos D, Ramírez-Moreno I G, Martínez-Aguilar M, Ibarra-Sánchez A, Muñoz-Bello J O, Anaya-Rubio I *et al* 2022 Mast cell–tumor interactions: molecular mechanisms of recruitment, intratumoral communication and potential therapeutic targets for tumor growth *Cells* **11** 349

Shah Y M and Van Den Brink G R 2015 c-Kit as a novel potential therapeutic target in colorectal cancer *Gastroenterology* **149** 534–7

Shefler I, Salamon P and Mekori Y A 2021 Extracellular vesicles as emerging players in intercellular communication: relevance in mast cell-mediated pathophysiology *Int. J. Mol. Sci.* **22** 9176

Shemesh R, Laufer-Geva S, Gorzalczany Y, Anoze A, Sagi-Eisenberg R, Peled N *et al* 2023 The interaction of mast cells with membranes from lung cancer cells induces the release of extracellular vesicles with a unique miRNA signature *Sci. Rep.* **13** 21544

Sheu H, Yu H and Chang C 1991 Mast cell degranulation and elastolysis in the early stage of striae distensae *J. Cutan. Pathol.* **18** 410–6

Shi S, Ye L, Yu X, Jin K and Wu W 2023 Focus on mast cells in the tumor microenvironment: current knowledge and future directions *Biochim. Biophys. Acta (BBA)—Rev. Cancer* **1878** 188845

Shimbori C, Upagupta C, Bellaye P-S, Ayaub E A, Sato S, Yanagihara T *et al* 2019 Mechanical stress-induced mast cell degranulation activates TGF-β1 signalling pathway in pulmonary fibrosis *Thorax* **74** 455–65

Shin K, Watts G F M, Oettgen H C, Friend D S, Pemberton A D, Gurish M F *et al* 2008 Mouse mast cell tryptase mMCP-6 is a critical link between adaptive and innate immunity in the chronic phase of *Trichinella spiralis* infection *J. Immunol.* **180** 4885–91

Shriver Z, Capila I, Venkataraman G and Sasisekharan R 2012 Heparin and heparan sulfate: analyzing structure and microheterogeneity *Heparin—A Century of Progress* ed R Lever, B Mulloy and C P Page (Berlin: Springer) pp 159–76

Siegel A M, Stone K D, Cruse G, Lawrence M G, Olivera A, Jung M *et al* 2013 Diminished allergic disease in patients with STAT3 mutations reveals a role for STAT3 signaling in mast cell degranulation *J. Allergy Clin. Immunol.* **132** 1388–96.e3

Sinnamon M J, Carter K J, Sims L P, LaFleur B, Fingleton B and Matrisian L M 2008 A protective role of mast cells in intestinal tumorigenesis *Carcinogenesis* **29** 880–6

Sjöberg L C, Gregory J A, Dahlén S-E, Nilsson G P and Adner M 2015 Interleukin-33 exacerbates allergic bronchoconstriction in the mice via activation of mast cells *Allergy* **70** 514–21

Skobe M, Hawighorst T, Jackson D G, Prevo R, Janes L, Velasco P *et al* 2001 Induction of tumor lymphangiogenesis by VEGF-C promotes breast cancer metastasis *Nat. Med.* **7** 192–8

Song F, Zhang Y, Chen Q, Bi D, Yang M, Lu L *et al* 2023 Mast cells inhibit colorectal cancer development by inducing ER stress through secreting Cystatin C *Oncogene* **42** 209–23

Sonoda S, Sonoda T, Nakano T, Kanayama Y, Kanakura Y, Asai H *et al* 1986 Development of mucosal mast cells after injection of a single connective tissue-type mast cell in the stomach mucosa of genetically mast cell-deficient W/Wv mice *J. Immunol.* **137** 1319–22

Soucek L, Lawlor E R, Soto D, Shchors K, Swigart L B and Evan G I 2007 Mast cells are required for angiogenesis and macroscopic expansion of Myc-induced pancreatic islet tumors *Nat. Med.* **13** 1211–8

St. John A L and Abraham S N 2013 Innate immunity and its regulation by mast cells *J. Immunol.* **190** 4458–63

St. John A L, Rathore A P S and Ginhoux F 2023 New perspectives on the origins and heterogeneity of mast cells *Nat. Rev. Immunol.* **23** 55–68

Stankovic B, Bjørhovde H A K, Skarshaug R, Aamodt H, Frafjord A, Müller E *et al* 2019 Immune cell composition in human non-small cell lung cancer *Front. Immunol.* **9** 3101

Stemmler M P, Eccles R L, Brabletz S and Brabletz T 2019 Non-redundant functions of EMT transcription factors *Nat. Cell Biol.* **21** 102–12

Stokes A J, Shimoda L M N, Koblan-Huberson M, Adra C N and Turner H 2004 A TRPV2–PKA signaling module for transduction of physical stimuli in mast cells *J. Exp. Med.* **200** 137–47

Strouch M J, Cheon E C, Salabat M R, Krantz S B, Gounaris E, Melstrom L G *et al* 2010 Crosstalk between mast cells and pancreatic cancer cells contributes to pancreatic tumor progression *Clin. Cancer Res.* **16** 2257–65

Suarez E M, Knackstedt R J and Jenrette J M 2014 Significant fibrosis after radiation therapy in a patient with Marfan syndrome *Radiat. Oncol. J.* **32** 208

Subramanian H, Gupta K and Ali H 2016 Roles of Mas-related G protein–coupled receptor X2 on mast cell–mediated host defense, pseudoallergic drug reactions, and chronic inflammatory diseases *J. Allergy Clin. Immunol.* **138** 700–10

Subramanian H, Gupta K, Guo Q, Price R and Ali H 2011 Mas-related Gene X2 (MrgX2) is a novel G protein-coupled receptor for the antimicrobial peptide LL-37 in human mast cells *J. Biol. Chem.* **286** 44739–49

Sugimoto K, Kudo M, Sundaram A, Ren X, Huang K, Bernstein X *et al* 2012 The αvβ6 integrin modulates airway hyperresponsiveness in mice by regulating intraepithelial mast cells *J. Clin. Invest.* **122** 748–58

Sun K, Li Y and Jin J 2021 A double-edged sword of immuno-microenvironment in cardiac homeostasis and injury repair *Sig. Transduct. Target. Ther.* **6** 79

Tajima Y, Moore M A S, Soares V, Ono M, Kissel H and Besmer P 1998 Consequences of exclusive expression *in vivo* of kit-ligand lacking the major proteolytic cleavage site *Proc. Natl Acad. Sci. USA* **95** 11903–8

Takabe K and Spiegel S 2014 Export of sphingosine-1-phosphate and cancer progression *J. Lipid Res.* **55** 1839–46

Tamma R, Guidolin D, Annese T, Tortorella C, Ruggieri S, Rega S *et al* 2017 Spatial distribution of mast cells and macrophages around tumor glands in human breast ductal carcinoma *Exp. Cell. Res.* **359** 179–84

Tan S-Y 2005 Prognostic significance of cell infiltrations of immunosurveillance in colorectal cancer *WJG* **11** 1210

Tan T H, Malik-Garbi M, Abu-Shah E, Li J, Sharma A, MacKintosh F C *et al* 2018 Self-organized stress patterns drive state transitions in actin cortices *Sci. Adv.* **4** eaar2847

Tegoshi T, Nishida M and Arizono N 2005 Expression and role of E-cadherin and CD103β7 (αEβ7 integrin) on cultured mucosal-type mast cells *APMIS* **113** 91–8

Terada T and Matsunaga Y 2000 Increased mast cells in hepatocellular carcinoma and intrahepatic cholangiocarcinoma *J. Hepatol.* **33** 961–6

The Immunological Genome Project ConsortiumDwyer D F, Barrett N A and Austen K F 2016 Expression profiling of constitutive mast cells reveals a unique identity within the immune system *Nat. Immunol.* **17** 878–87

Theoharides T C, Kempuraj D, Tagen M, Conti P and Kalogeromitros D 2007 Differential release of mast cell mediators and the pathogenesis of inflammation *Immunol. Rev.* **217** 65–78

Thorpe M, Fu Z, Albat E, Akula S, De Garavilla L, Kervinen J *et al* 2018 Extended cleavage specificities of mast cell proteases 1 and 2 from golden hamster: classical chymase and an elastolytic protease comparable to rat and mouse MCP-5 *PLoS One* **13** e0207826

Tomaz Da Silva M, Santos A R, Koike T E, Nascimento T L, Rozanski A, Bosnakovski D *et al* 2023 The fibrotic niche impairs satellite cell function and muscle regeneration in mouse models of Marfan syndrome *Acta Physiol.* **237** e13889

Tsai M, Valent P and Galli S J 2022 KIT as a master regulator of the mast cell lineage *J. Allergy Clin. Immunol.* **149** 1845–54

Tsang M S-M and Wong C K 2020 Functional interaction between sensory neurons and mast cells in the early stage of house dust mite-induced type 2 inflammation and itch: a novel therapeutic target of allergic disease *Cell Mol. Immunol.* **17** 899–900

Tuna B, Yorukoglu K, Unlu M, Mungan M U and Kirkali Z 2006 Association of mast cells with microvessel density in renal cell carcinomas *Eur. Urol.* **50** 530–4

Tung H-Y, Plunkett B, Huang S-K and Zhou Y 2014 Murine mast cells secrete and respond to interleukin-33 *J. Interferon Cytokine Res.* **34** 141–7

Turner H and Kinet J-P 1999 Signalling through the high-affinity IgE receptor FcεRI *Nature* **402** 24–30

Valent P, Akin C, Hartmann K, Nilsson G, Reiter A, Hermine O *et al* 2020 Mast cells as a unique hematopoietic lineage and cell system: from Paul Ehrlich's visions to precision medicine concepts *Theranostics* **10** 10743–68

Van Den Eynde C, Held K, Ciprietti M, De Clercq K, Kerselaers S, Marchand A *et al* 2022 Loratadine, an antihistaminic drug, suppresses the proliferation of endometrial stromal cells by inhibition of TRPV2 *Eur. J. Pharmacol.* **928** 175086

Van Der Gucht J, Paluch E, Plastino J and Sykes C 2005 Stress release drives symmetry breaking for actin-based movement *Proc. Natl Acad. Sci. USA* **102** 7847–52

Varricchi G, De Paulis A, Marone G and Galli S J 2019 Future needs in mast cell biology *Int. J. Mol. Sci.* **20** 4397

Varricchi G, Galdiero M R, Loffredo S, Marone G, Iannone R, Marone G *et al* 2017 Are mast cells MASTers in cancer? *Front. Immunol.* **8** 424

Vignaud T, Blanchoin L and Théry M 2012 Directed cytoskeleton self-organization *Trends Cell Biol.* **22** 671–82

Visciano C, Liotti F, Prevete N, Cali' G, Franco R, Collina F *et al* 2015a Mast cells induce epithelial-to-mesenchymal transition and stem cell features in human thyroid cancer cells through an IL-8–Akt–Slug pathway *Oncogene* **34** 5175–86

Visciano C, Prevete N, Liotti F and Marone G 2015b Tumor-associated mast cells in thyroid cancer *Int. J. Endocrinol.* **2015** 1–8

Voehringer D 2013 Protective and pathological roles of mast cells and basophils *Nat. Rev. Immunol.* **13** 362–75

Vukman K V, Lalor R, Aldridge A and O'Neill S M 2016 Mast cells: new therapeutic target in helminth immune modulation *Parasite Immunol.* **38** 45–52

Wagenseil J E and Mecham R P 2007 New insights into elastic fiber assembly *Birth Defects Res. Pt* C **81** 229–40

Wang J-X, Kaieda S, Ameri S, Fishgal N, Dwyer D, Dellinger A *et al* 2014 IL-33/ST2 axis promotes mast cell survival via BCLXL *Proc. Natl Acad. Sci. USA* **111** 10281–6

Wedemeyer J and Galli S J 2005 Decreased susceptibility of mast cell-deficient Kit/Kit mice to the development of 1, 2-dimethylhydrazine-induced intestinal tumors *Lab. Invest.* **85** 388–96

Wernersson S and Pejler G 2014 Mast cell secretory granules: armed for battle *Nat. Rev. Immunol.* **14** 478–94

Wheeler J B, Ikonomidis J S and Jones J A 2021 Connective tissue disorders and cardiovascular complications: the indomitable role of transforming growth factor-β signaling *Progress in Heritable Soft Connective Tissue Diseases* ed J Halper (Cham: Springer International Publishing) pp 161–84

Wu L C and Zarrin A A 2014 The production and regulation of IgE by the immune system *Nat. Rev. Immunol.* **14** 247–59

Xiao H, He M, Xie G, Liu Y, Zhao Y, Ye X *et al* 2019 The release of tryptase from mast cells promote tumor cell metastasis via exosomes *BMC Cancer* **19** 1015

Xiao H, Lässer C, Shelke G V, Wang J, Rådinger M, Lunavat T R *et al* 2014 Mast cell exosomes promote lung adenocarcinoma cell proliferation—role of KIT-stem cell factor signaling *Cell Commun. Signal.* **12** 64

Xie G, Yang H, Peng X, Lin L, Wang J, Lin K *et al* 2018 Mast cell exosomes can suppress allergic reactions by binding to IgE *J. Allergy Clin. Immunol.* **141** 788–91

Xing W, Austen K F, Gurish M F and Jones T G 2011 Protease phenotype of constitutive connective tissue and of induced mucosal mast cells in mice is regulated by the tissue *Proc. Natl Acad. Sci. USA* **108** 14210–5

Yang B-G, Kim A-R, Lee D, An S B, Shim Y A and Jang M H 2023 Degranulation of mast cells as a target for drug development *Cells* **12** 1506

Yang H-W, Liu X-Y, Shen Z-F, Yao W, Gong X-B, Huang H-X *et al* 2018 An investigation of the distribution and location of mast cells affected by the stiffness of substrates as a mechanical niche *Int. J. Biol. Sci.* **14** 1142–52

Yang J, Dong H-Q, Liu Y-H, Ji M-H, Zhang X, Dai H-Y *et al* 2022 Laparotomy-induced peripheral inflammation activates NR2B receptors on the brain mast cells and results in neuroinflammation in a vagus nerve-dependent manner *Front. Cell. Neurosci.* **16** 771156

Yang J, Mani S A, Donaher J L, Ramaswamy S, Itzykson R A, Come C *et al* 2004 Twist, a master regulator of morphogenesis, plays an essential role in tumor metastasis *Cell* **117** 927–39

Yang Z, Zhang B, Li D, Lv M, Huang C, Shen G-X *et al* 2010 Mast cells mobilize myeloid-derived suppressor cells and treg cells in tumor microenvironment via IL-17 pathway in murine hepatocarcinoma model *PLoS One* **5** e8922

Yano H, Kinuta M, Tateishi H, Nakano Y, Matsui S, Monden T *et al* 1999 Mast cell infiltration around gastric cancer cells correlates with tumor angiogenesis and metastasis *Gastric Cancer* **2** 26–32

Yu Y, Blokhuis B, Derks Y, Kumari S, Garssen J and Redegeld F 2018 Human mast cells promote colon cancer growth via bidirectional crosstalk: studies in 2D and 3D coculture models *OncoImmunology* **7** e1504729

Yu Y, Blokhuis B R, Garssen J and Redegeld F A 2016 Non-IgE mediated mast cell activation *Eur. J. Pharmacol.* **778** 33–43

Yu Y, Ren L-J, Liu X-Y, Gong X-B and Yao W 2021 Effects of substrate stiffness on mast cell migration *Eur. J. Cell Biol.* **100** 151178

Yu Y, Zhang Y, Zhang Y, Lai Y, Chen W, Xiao Z *et al* 2017 LL-37-induced human mast cell activation through G protein-coupled receptor MrgX2 *Int. Immunopharmacol.* **49** 6–12

Yuan S-M, Ma H-H, Zhang R-S and Jing H 2011 Transforming growth factor-beta signaling pathway in Marfans syndrome: a preliminary histopathological study *Vasa* **40** 369–74

Zeigler S M, Sloan B and Jones J A 2021 Pathophysiology and pathogenesis of marfan syndrome *Progress in Heritable Soft Connective Tissue Diseases* ed J Halper (Cham: Springer International Publishing) pp 185–206

Zhang D, Spielmann A, Wang L, Ding G, Huang F, Gu Q *et al* 2012 Mast-cell degranulation induced by physical stimuli involves the activation of transient-receptor-potential channel TRPV2 *Physiol. Res.* **61** 113–24

Zhang L, Simonsen C, Zimova L, Wang K, Moparthi L, Gaudet R *et al* 2022 Cannabinoid non-cannabidiol site modulation of TRPV2 structure and function *Nat. Commun.* **13** 7483

Zhao X O, Lampinen M, Rollman O, Sommerhoff C P, Paivandy A and Pejler G 2022a Mast cell chymase affects the functional properties of primary human airway fibroblasts: implications for asthma *J. Allergy Clin. Immunol.* **149** 718–27

Zhao X O, Sommerhoff C P, Paivandy A and Pejler G 2022b Mast cell chymase regulates extracellular matrix remodeling-related events in primary human small airway epithelial cells *J. Allergy Clin. Immunol.* **150** 1534–44

Zhou X, Guo X, Song Y, Zhu C and Zou W 2018 The LPI/GPR55 axis enhances human breast cancer cell migration via HBXIP and p-MLC signaling *Acta Pharmacol. Sin.* **39** 459–71

Zhu X-Q, Lv J-Q, Lin Y, Xiang M, Gao B-H and Shi Y-F 2007 Expression of chemokines CCL5 and CCL11 by smooth muscle tumor cells of the uterus and its possible role in the recruitment of mast cells *Gynecol. Oncol.* **105** 650–6

IOP Publishing

Physics of Cancer, Volume 6 (Second Edition)
Cellular mechanisms to foster or fight cancer
Claudia Tanja Mierke

Chapter 4

Apoptosis and the two faces of autophagy in cancer on different time scales

4.1 Summary

Every cell is endowed with a conserved clean-up mechanism, referred to as autophagy, to facilitate the recycling of utilized materials and the removal of damaged organelles by lysosomal breakdown. Autophagy serves as an early cellular reaction to stress signals in physiological and pathological conditions. Autophagy can have two opposing effects in cancer development: Autophagy can either block the progression of the disease or, conversely, protect cells, thereby promoting tumor growth. This phenomenon is known as the 'autophagy paradox' and is defined by the fact that the autophagy process supplies the substrates needed for biosynthesis to sustain the energy demands of the cell, and that the excessive programmed activity of this mechanism can cause cell death by apoptosis. Combating cancer is a challenging endeavor due to the high resistance of cancer cells against chemotherapy and radiotherapy. An increasing number of research results suggest that autophagy could play a crucial part in the evolution of resistance by acting as a protector of cancer cells. For this reason, autophagy can act as a sort of double-edged sword in cancer treatment. This chapter aims to explore the impact of autophagy and cancer stem cells on cancer development and to discuss new therapeutic approaches that are aimed at targeting these pathways. Since it is assumed that enhancing autophagy flux protects host cells from damage later by clearing dysfunctional organelles and aberrantly folded proteins, the effect of calorie restriction is discussed. There is growing support for the notion that periodic fasting or caloric limitation can result in the activation of adaptive autophagy and enhance the lifespan of eukaryotic cells. Prolonged caloric restriction, nevertheless, with exaggerated autophagy response is detrimental and can induce type II autophagic cell death.

4.2 Introduction to the process of autophagy in cancer

Autophagy is a cellular self-maintenance process with the primary function of preserving cell metabolism and viability during periods of starvation and stress,

doi:10.1088/978-0-7503-4007-6ch4

thereby avoiding cell death. While apoptosis and autophagy have distinct molecular targets and exhibit different morphological and biochemical features, several interactions between them have been identified (Mizushima and Komatsu 2011). The signaling mechanisms underlying the crosstalk between autophagy and apoptosis have received considerable attention in past years. Notably, special emphasis is placed on how autophagy can be tackled and whether it is beneficial for cancer therapy. The significance of autophagy in cancer treatment is especially worth highlighting. Autophagy is widely regarded as a mechanism for preventing cancer, which takes place in two distinct manners: in a sovereign manner and in a dependent manner (White 2012). It can sustain ordinary cellular homeostasis by clearing oncogenic proteins or defective organelles away from cells. This helps to protect cells from progressing to neoplastic malignancies. Autophagy also interacts extensively with the immune system to offer a non-cellular form of cancer surveillance. Autophagy is a process by which cells scavenge and recycle nutrients, and, regrettably, increased autophagy facilitates the survival and growth of malignant neoplastic lesions. The function of autophagy in cancer is multifaceted: it fosters the survival of cancer cells by providing essential components for growth, but it is also implicated in regulating cancer cell migration and invasion (Rao *et al* 2014). This chapter concentrates on how autophagy can cause cancer, how it impacts cancer stem cells (CSCs), and what autophagy-related therapeutic options are available at present (Chmurska *et al* 2021).

4.3 Apoptosis

The following is a brief historical overview of apoptosis. In 1964, the scientists Lockshin and Williams explored the metamorphosis of the silkworm *Antheraea pernyi* and identified the programmed cell death, which is synonymously referred to as apoptosis. In the silkworm, they analyzed the cytoarchitecture of the degeneration in the intersegmental muscles, whereby the distinct death of specific cells in the course of silkworm metamorphosis was reported. This mode of cell death was termed 'programmed cell death' as the cells died due to the general intrinsic instruction of the insect (Lockshin and Williams 1965). In 1972, the three scientists Kerr, Curie and Wyllie first introduced the term 'apoptosis' (Kerr *et al* 1972, Paweletz 2001, Kerr 2002). Nevertheless, the scientist Flemming detected the mechanism and termed it chromatolysis in the year 1885 (Bursch 2004). Programmed cell death is generally referred to as apoptosis, a term that comes from ancient Greek and translates as falling down (figure 4.1) (Kerr 2002). This mechanism is controlled genetically, beginning with the embryonic phase and extending through the entire ontogenesis. Apoptosis is essential for the reorganization of developing tissues and for the orderly functioning of an immune response (Fulda 2011, Sinha *et al* 2013).

Advances in electron microscopy have made it feasible to identify the distinct morphological alterations that take place when cells undergo apoptosis. The hallmark characteristics of apoptosis are those of pyknosis and cell shrinkage (or cell contraction). During the cell-shrinkage reaction, the cell size is diminished while the cytoplasm and cell organelles condense. The karyopyknosis, which is synonymously known as the shrinkage of cell nuclei and represents a hallmark of apoptotic

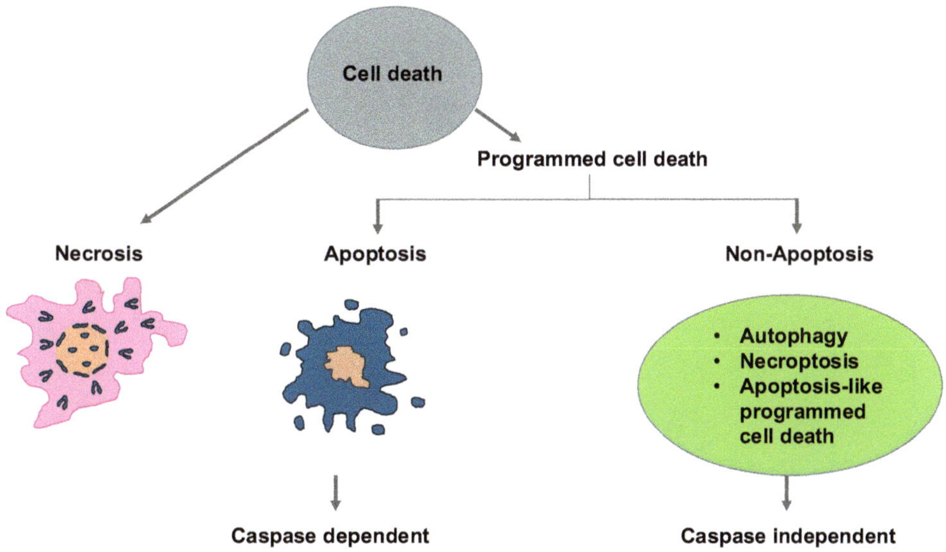

Figure 4.1. Joint mechanism of cancer cell death. Cancer cell death comprises two generally well-defined mechanisms: programmed cell death and necrosis. Programmed cell death primarily deals with apoptosis and cell death based on non-apoptosis, like autophagy, necroptosis and apoptosis-like programmed cell death.

cells, is the result of the condensation of chromatin (Häcker 2000). Hematoxylin-eosin staining revealed that apoptotic cells are like an oval mass containing thick, dark cytoplasm and densely stained, violet, fragmented nuclear chromatin. At the phase of condensation, the chromatin becomes aggregated beneath the nuclear membrane (Joza *et al* 2001). As cytoplasmic condensation and cell shrinkage progress, cell membranes become bleb-like. Simultaneously, two other events proceed, which are karyorrhexis and bud formation, and cell fragments are released. It is also worth mentioning that bud formation is the event that causes cell fragments to be liberated (Elmore 2007). The apoptotic bodies generated in this way comprise cytoplasm with densely organized organelles and nuclear fragments. Apoptotic bodies are subsequently phagocytosed, whereby macrophages, parenchymal cells, and histiocytes are recruited to the sites of cell death and take up (eat) the apoptotic bodies (figure 4.2). The recognition of apoptotic cells have been discussed to be mechanosensory dependent (Sarter *et al* 2007).

The key characteristic of apoptosis is the lack of an inflammatory response in the vicinity of the cells that are dying, which is attributed to the following circumstances:

- cells do not break down when they die of apoptosis;
- due to the tremendous speed of phagocytosis, secondary necrosis is avoided;
- phagocytes like macrophages, histiocytes, interstitial cells are not releasing any anti-inflammatory cytokines (Kurosaka *et al* 2003).

4.3.1 Mechanism of apoptosis

The process of programmed cell death is extremely intricate and comprises a series of molecular events that require ATP. The following phases can be determined throughout apoptosis:

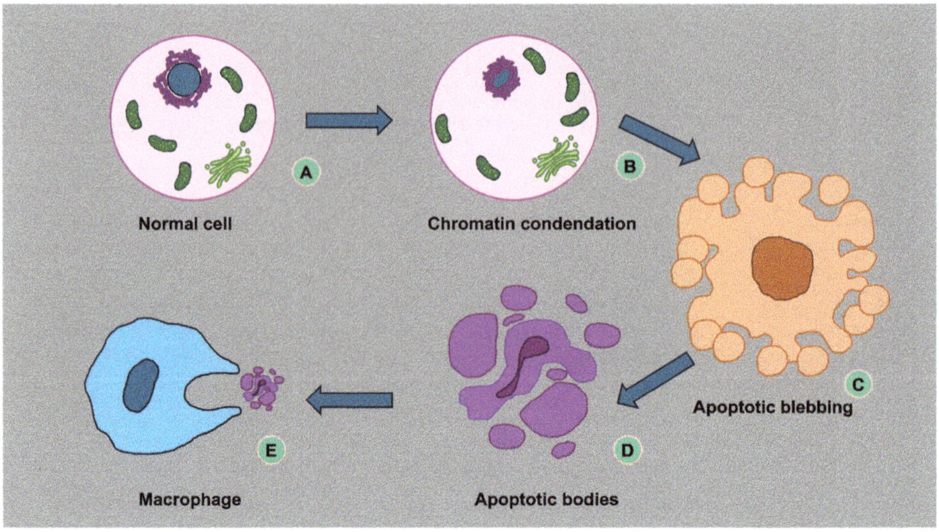

Figure 4.2. The various steps of apoptosis. (A) The first morphological phenomenon of apoptosis is a modification at the scale of the cellular nucleus. (B) Chromatin condenses and deposits beneath the nuclear membrane, causing the nucleus to undergo a shrinking and becoming fragmented. (C) The next phase of 'dying' involves the condensation of the cytoplasm and the appearance of specific bubbles on the cell surface. (D) Apoptotic bodies consist of cell debris and are composed of chromatin, cytoplasm and organelles. (E) The ultimate stage of apoptosis entails the clearance of apoptotic bodies through phagocytosis.

- induction phase;
- effector phase;
- breakdown phase (Elmore 2007).

There are two main routes of apoptosis, such as an extrinsic and intrinsic route. Activation of hydrolytic enzymes, particularly of protease and nuclease, is the immediate reason for the decay of the cell structures occurring during apoptosis. In addition to the stimulation of the activation of hydrolytic enzymes, the cell's own repair systems are also inactivated. Endonuclease activation and fragmentation of DNA are the earliest signs of apoptotic cell death. The production of specific cysteine proteases, such as caspases in a cell is characteristic of a process known as programmed cell death. The highly concentrated phosphatidylserine in the outermost layer of the plasma membrane is another key characteristic of apoptosis (Hengartner 2000). The caspases, a family of cysteine proteases, are triggered irrespective of the apoptosis route. Based on the phase of apoptosis in which they become active, two types of caspases emerge: initiator caspases like caspase-2, caspase-8, caspase-9, caspase-10, caspase-12, and executioner caspases like caspase-3, caspase-6, caspase-7 (figure 4.3) (Rai *et al* 2005). These enzymes are synthesized and stored in an inert state within the cell. Caspases become activated as apoptosis is instigated. They convey a cell signaling message that leads to the activation of transcription factors such as NF-κB and AP-1. As a result, abnormally folded

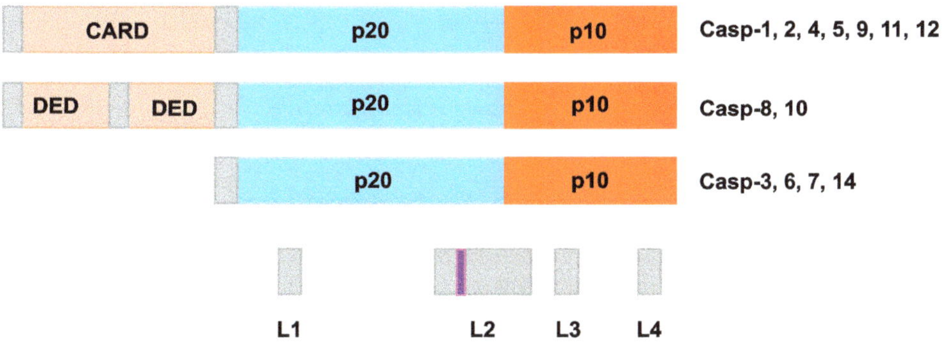

Figure 4.3. Domain architecture of the three main groups of caspases and position of the catalytic center loops (L1–L4). Initiator caspases harbor long prodomains, CARD or DED, while executioner caspases possess short prodomains. Loops have been colored grey. The active Cys residue is indicated by the violet line. The processing that divides the p20 and p10 subunits takes place in L2. The resultant large subunit part of the L2 loop of one monomer and the small subunit part of the L2 loop of the adjacent monomer (L2′) participate in loop bundling.

proteins are generated, which can result in cell death (Li and Yuan 2008, Henshall and Engel 2013).

4.3.1.1 Extrinsic route

The extrinsic apoptosis route is triggered by a restricted range of growth factors or nutrients, and by a rise in the local concentration of hormones and cytokines. In addition, chemical agents, involving physical stressors and cytostatic drugs, are activators of the extrinsic path. The pro-apoptotic message is conveyed through receptors of the TNF family. After the corresponding ligand attaches to the receptor, for example FasL/TNF-α-ligand to FasR/TNFR1, the signal is transferred via the death domain of FADD onto caspase-8. Thereafter, activation of the death-inducing signaling complex (DISC) leads to the activation of caspase-8 and transmits a message to the effector caspases, like caspase-3 and caspase-7 (figure 4.4) (Pop and Salvesen 2009, Sayers 2011).

4.3.1.2 Intrinsic route

The intrinsic route of apoptosis implicates mitochondria. Because of elevated calcium ion (Ca^{2+}) concentration, ROS, hypoxia, hormonal and growth factor deprivation, the process of apoptosis is triggered. The above-mentioned circumstances impact the activation of caspases, permeability of mitochondrial membranes and interference with proteins that lead to cell disruption. This results in the liberation of cytochrome c out of the mitochondrial area and to the triggering of the activation of caspases. The key structural component associated with the leakage of cytochrome c from mitochondria is termed the mitochondrial permeability transition pore (PTP). Cytochrome c attaches to the Apaf-1 protein factor and procaspase-9, thereby creating a tripartite system referred to as apoptosome. The assembly of the apoptosome relies on energy from the hydrolysis of ATP. In addition to cytochrome c, more than 40 proteins are required for the induction of

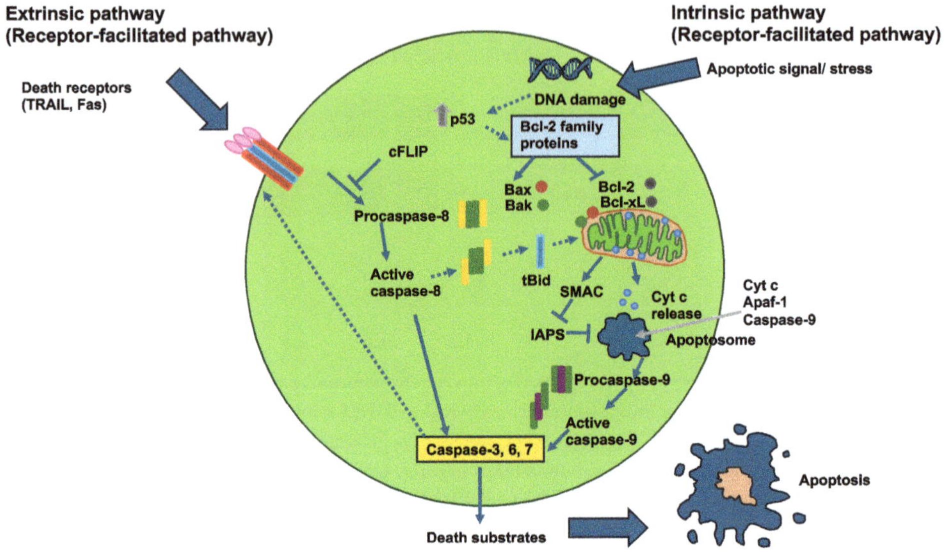

Figure 4.4. Signal transduction routes of apoptosis. Apoptosis primarily comprises two principal routes and a third route represents the executioner way of apoptosis. The extrinsic route is initiated through external cues or ligand molecules and notably concerns the death receptors (DRs). The intrinsic route occurs through the insertion of Bax/Bak into the mitochondrial membrane, leading to the liberation of cytochrome c, which associates with Apaf-1 and procaspase-9 to form an apoptosome, subsequently leading to the activation of the caspase-3 apoptosis signaling cascade. B-cell lymphoma protein 2 (Bcl-2), Bcl-2 homologue splice variants (Bcl-xL), cytochrome C (Cyt C), inhibitor of apoptosis proteins (IAPs), second mitochondrial activator of caspases (SMAC), TNF related apoptosis-inducing ligand (TRAIL), cellular FLICE inhibitory proteins (cFLIP), truncated Bid (tBid)

activation of the intrinsic apoptotic signal transduction pathway, such as apoptosis-inducing factor (AIF). This is transported into the cell nucleus during the initiation phase. AIF and endonuclease G are liberated from the mitochondria and conveyed into the cell nucleus. These proteins influence the apoptotic alterations in the cell nucleus (Li *et al* 2010, Tait and Green 2010, Wu and Bratton 2013). The two apoptosis signaling routes are interconnected and can mutually impact one another through molecules (Igney and Krammer 2002). Intriguingly, extrinsic and intrinsic signal transduction pathways converge in the same phase, namely the effector phase (Pardo *et al* 2007, Jan and Chaudhry 2019) (figure 4.4).

4.4 Autophagy

The term autophagy is derived from the Greek words ‘auto’, meaning ‘self’, and ‘phagy’, meaning ‘to eat’. The first usage of the term ‘autophagy’ was in 1859 by the scientist Anselmier. In his work, Anselmier reported on the impact of fasting on mice. Almost a century had passed before the scientist De Duve finally described the autophage process on lysosomes in 1963 (Klionsky *et al* 2016, Harnett *et al* 2017). Interest in autophagy skyrocketed in 2016 when the Scientist Ohsumi was awarded the Nobel Prize for discovering the ATG genes, which are the genes that govern and orchestrate autophagy in yeast.

4.4.1 Mechanism of autophagy

Autophagy can be broadly defined as a conservative degradation process that is common to all eukaryotic organisms. This process is initiated in response to different kinds of stimuli, like nutrient deprivation, oxidative stress or hypoxia, and is regarded as a cytoprotective mechanism (Zhang *et al* 2017, Bai *et al* 2019). The main task of this mechanism is to sustain the cell's metabolism during starvation and to protect it from the accumulation of damaged proteins or toxins when it is exposed to stress factors (Glick *et al* 2010). Autophagy is classified into three categories based on how cellular components are transported to lysosomes and vacuoles: (1) Microautophagy removes cytoplasmic components, small organelles, and nuclear debris. During microautophagy, the substrate becomes engulfed within a lysosomal membrane and internalized via endocytosis (Reggiori *et al* 2021). (2) Chaperone-mediated autophagy (CMA) involves chaperones, whose role is to catch cytosolic proteins, for example, with an inappropriate conformation. A hallmark of CMA requires each substrate protein to harbor a KFERQ motif somewhere in its amino acid sequence. This sequence is essential for targeting the protein to the lysosome and is identified through heat shock proteins (Arias and Cuervo 2011). (3) Macroautophagy, commonly referred to as autophagy, in which a C shaped double membrane enveloping the cytoplasm demarcates the initiation of autophagy. Both membrane endings elongate and engulf diverse cellular elements, such as damaged mitochondria, ribosomes, and the endoplasmic reticulum. The resulting vesicle is termed an 'autophagosome' (Shao *et al* 2016). Upon fusion of the lysosome with the cargo, catabolism of the cargo is induced through autophagolysosomes. The breakdown products, such as amino acids, carbohydrates, and fatty acids, are fundamental constituents of macromolecular substrates. These substrates are shuttled into the cytoplasm to fuel cellular metabolic demands or are utilized to facilitate damage repair (Tukaj 2013, Zhang *et al* 2018). While the types of autophagy are distinct, substrates are targeted to all kinds of lysosomes. Among the three types of autophagy, macroautophagy has received the most attention (Wu *et al* 2018). The process of autophagy is divided into the subsequent phases (figure 4.5):

- autophagy initiation phase;
- phagophore nucleation phase;
- elongation phase;
- lysosome fusion phase; and
- breakdown phase (Al-Bari 2020, Chmurska *et al* 2021).

The autophagic process is induced through the generation of double-membrane vesicles termed autophagosomes. This biological phenomenon relies on the action of almost 16 Atgs and two different ubiquitin-like conjugation pathways (figure 4.5) (Yorimitsu and Klionsky 2005). Two distinct complexes are necessary to stimulate the generation of autophagosomes. The first complex is made up of PI3K type III, Vps34, Atg14, Atg6/Beclin1 and Vps15/p150.73, and the second is linked to the functional activity of the serine/threonine kinase Atg1 (Hassanpour *et al* 2019). In yeast, Atg8 or Atg13 and Atg17 are indispensable for the serine/threonine kinase

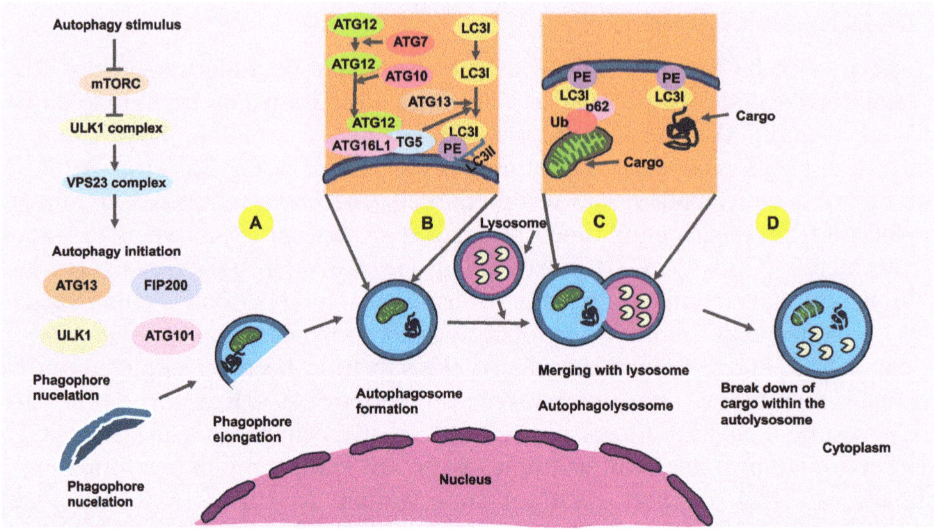

Figure 4.5. Various signals activate the molecular mechanism of autophagy. The cellular process of autophagy comprises a series of sequential molecular and subcellular morphological alterations. In the initial phases, the activation of autophagy upstream molecules results in the generation and elongation of phagophores. The target of these signals is the ULK 1 complex, which includes ULK1, FIP200, ATG13 and ATG101 protein. During autophagy, the fragment of cytoplasm is surrounded by C-shaped double-membrane (A), while autophagosome is being created as the result of connection of two membrane ends (B). These entities scavenge the injured organelles, aggregated or misfolded proteins. Through the activation of various Atgs, phagophores ripen into autophagosomes, which then merge with lysosomes to generate autophagolysosomes. Afterwards autophagosome fuses with lysosome, which is referred to as lysophagy. The resulting vesicle is known as an autophagolysosome (C). In the autophagolysosomes, enzymatic degradation is coordinated by several hydrolases, resulting in extensive enzymatic degradation (D). The hydrolyzed cargo is either recycled back into the cytosol or extruded out of the host cells. Autophagy-related protein (Atg), mammalian target of rapamycin (mTOR).

activity of Atg1, while in mammals, LC3, an ortholog of Atg8, can enhance the Atg1 activity along with GATE-16 and GABARAP (Noda and Fujioka 2015). In the following step, the autophagosomes are enlarged by the activity of ubiquitin-like conjugation systems. For this process, cells can utilize two distinct routes, which comprise LC3/GABARAP/GATE-16 and Atg12 (Varga *et al* 2022). Because of the proteolytic activity of Atg4 on Atg8 (carboxyl terminus), a glycine moiety is uncovered to create the autophagosome (Maruyama and Noda 2018). Phosphatidylethanolamine covalently attaches to Atg8 (lipidized LC3-II of mammals) (Martens and Fracchiolla 2020). As autophagosome generation advances, the Atg16–Atg5–Atg12 complex is dissociated from the autophagosomal membrane and entire autophagosomes can merge with lysosomes (Cui *et al* 2017). The process continues with the generation of autophagolysosomes, where the autophagic load is broken down by lysosomal activity. Alternatively, autophagosomes merge with late endosomes to produce amphisomes (Amini *et al* 2021). Molecular studies have revealed that both amphisomes and autophagosomes ultimately accumulate in lysosomes to degrade and reuse their cargos (Lőrincz and Juhász 2020). A number of components, including Rab and Arf proteins possessing GTPase activity, tethering proteins like SNAREs, binding adaptors, and motor proteins, facilitate the direct engagement of autophagosomes and lysosomes (Lőrincz and Juhász 2020). Autophagocytosed macromolecules undergo recycling within the cytosol to sustain bioenergetic equilibrium or are excreted from the cells (Yu and Klionsky 2022). It is important to remember that the accumulation of autophagosomes, which is otherwise referred to as autophagic flux, can be a reflection of both enhanced autophagic response and aberrant reduction in autophagolysosome biogenesis (Das *et al* 2012). Thus, the accumulation of intracellular autophagosomes may not fully reflect the autophagic reaction.

4.4.1.1 Initiation of autophagy

Activation of the ULK1 complex is the starting point of this signaling pathway. The complex comprises ULK1, ATG13, ATG101 and the FIP200/RB1CC1 protein (Hara *et al* 2008). The master regulator of both complex activity and the overall autophagy program at the same time is mTOR kinase. The mammalian target of rapamycin (mTOR) serves as a sensory mechanism that detects the abundance of nutrients in the surrounding environment. When there is an abundance of nutrients, mTOR blocks autophagy through the phosphorylation of ULK1/2 kinases, thereby impairing the activity of the overall ULK/Atg13/Atg101/FIP200 initiation complex. When cells are starved, mTOR disengages from ULK1/2, leading to their dephosphorylation, which in turn results in the stimulation of cell growth through the suppression of protein catabolism. When mTOR becomes inactive, autophagy activation occurs through the assembly of the ULK complex. The mTOR kinase is not the sole regulator of activity of the ULK complex. When a cell experiences energy deficiency, AMPK kinase can activate it either in a direct or indirect manner by blocking mTOR kinase activity.

4.4.1.2 Nucleation

A complex containing the class III PI3K kinase has an essential function in the nucleation phase, that is, the early phase of phagophore generation. The complex comprises VPS34, VPS15, Beclin-1 and ATG14. The activity of this complex is crucial for the supply of phospholipids, phosphatidylinositol-triphosphate (PI3P), to the site of phagophore formation instead of the generation of isolation membranes. Interaction with the protein UV-radiation-resistance-associated gene (UVRAG) signaling is needed to assemble this complex. The complex generates phosphatidylinositol (3,4,5)-triphosphate, which is required for phagophosphorus elongation (Cuervo 2004, Kondo *et al* 2005, Maria Fimia *et al* 2007).

4.4.1.3 Elongation

The ATG12–ATG5–ATG16L and ATG8 systems are required for the proper progression of the phagophore elongation phase (Kaufmann *et al* 2014). The first complex needs the participation of four proteins: ATG5, 7, 10, 12. Proteins ATG12 and ATG7 create an unstable bond, after which ATG12 is carried over to ATG10, which corresponds to the E2 enzyme. After that, the ATG12 protein connects to

ATG5. In the next phase, the ATG12–ATG5 complex is attached to the ATG16 protein. LC3-II-PE is the second ubiquitous, similar complex required for the creation of autophagosomes. LC3, which is a light chain of microtubule-tethered proteins, is the infant ortholog of the yeast protein ATG8. The proteins ATG3, 4, 7 and phosphatidylethanolamine are involved in the formation of the second ATG8/LC3 conjugation system. The LC3 protein, which is located in the cytosol, is produced as a pro-LC3 precursor, which is converted into the LC3-I form by a proteolytic cleavage through ATG4. The action of ATG7 and ATG3, which function as ubiquitin-like enzymes, and the reversible binding of LC3-I to the amine group phosphatidylethanolamine (PE), creates a mature version LC3-II (Tanida 2011, Nakatogawa 2013).

4.4.1.4 Merging with the lysosome for breakdown

When the extending ends of the phagophore form a bubble, termed an 'autophagosome', the next phase of autophagy begins, namely the coupling of autophagosome and lysosome. The outcome is a bubble enclosed by a singular membrane, where the ultimate destruction of its load is accomplished under the contribution of lysosomal enzymes. The presence of LAMP-2 protein, RAB-GTPases and SNAREs is required for the maturation pathway to proceed orderly (Kaufmann *et al* 2014).

4.5 Two facets of autophagy

Autophagy serves a dual function in cancer development: it can prevent the disease from progressing or protect cancer cells, thereby leading to tumor growth. This phenomenon is referred to as autophagy paradox (Martinez-Outschoorn *et al* 2010). For one thing, autophagy breaks down some cellular components, thereby supplying substrates for biosynthesis. For another thing, overactivity of this pathway can result in excessive breakdown and finally cell death through apoptosis. Apart from this phenomenon on its own, autophagy can facilitate or hinder resistance to drug treatment (Guo *et al* 2016, Huang *et al* 2016). An increasing number of scientists report that autophagy could have a major impact on chemoresistance in several types of cancer. Increased autophagy activity can either enhance or impair resistance to medications, based on the type of cancer and the method of treatment (Sui *et al* 2013, Li *et al* 2017c).

Some cancers utilize autophagy to adjust to hypoxia or nutrient deprivation. Cancer cells have been found to exhibit autophagy initiation as a defense and life-sustaining process during cancer treatments. In addition, hypoxia can initiate autophagy, indicating that this mechanism is implicated in the development of resistance (Bellot *et al* 2009). Autophagy enhances cell survival and the evolution of treatment resistance. In multiple myeloma resistant to DOX, enhanced autophagy activity and reduced apoptosis activity were demonstrated. Cytoprotective autophagy has been proven to be triggered in MDA-MB-231 breast cancer cells exposed to DOX. This work suggests that autophagy may confer protection on cancer cells, leading to decreased sensitivity to drug treatment. Thus, autophagy in cancer treatment is a 'double-edged blade'. Due to its contradictory nature, efforts to block or activate autophagy may be an effective strategy for combating cancer (Tan *et al* 2015).

The link between disrupted autophagy and cancer development is most clearly evident when the latter process is blocked in mice, whereby cancer development was induced by deletion of the BECN1 gene. In the MCF-7 cell line, which has weak expression of Bcl-1, the proliferation was repressed after the transfection of BECN1. Based on the research, the reduced expression of BECN1 is a poor prognosis for the diseased patients (Dalby *et al* 2010). The main factor associating autophagy and apoptosis in the cell is Beclin-1. This protein is found in a multi-protein complex with anti-apoptotic proteins of the BCL-2 family. Beclin-1 is bound by these proteins via its BH3 domain, thereby inhibiting its involvement in the triggering of autophagy (Marquez and Xu 2012, Salminen *et al* 2013). Through autophagy, the cell can re-establish homeostasis and escape cell death. An important point in the interplay between autophagy and apoptosis is that, irrespective of the location in the cell, the association of beclin-1 has no effect on the anti-apoptotic activity of Bcl-2 proteins. Conversely, the location in the cell influences the capacity of Bcl-2 proteins to attach to beclin-1. The nutrient-deficiency autophagy factor 1 can impair this capability. NAF-1 likely functions by stabilizing the Bcl-2-Beclin-1 complex in the endoplasmic reticulum. Under conditions that induce autophagy, proteins like BAD, BNIP3, NIX, NOXA and PUMA tether the anti-apoptotic Bcl2 protein, triggering the liberation of Beclin-1, which in turn permits the assembly of the PI3K complex (Sinha and Levine 2008, Yun and Lee 2018). The formation of Becl-1 complexes with Bcl-2 proteins is controlled at multiple regulatory levels: through JNK kinase in reaction to the presence of nutrients, through DAPK kinase, the HMGB protein, which is activated in the cell in response to oxidative stress, or through the pro-apoptotic proteins Bad and Bax (Wei *et al* 2008, Bovellan *et al* 2010, Mukhopadhyay *et al* 2014).

The aforementioned protein HMGB1 prevents BECLIN-1 and ATG5 from being degraded through calpains. This inhibits the generation of pro-apoptotic cleavage products of these proteins. HMGB1 helps to limit tissue injury as it blocks the activation of apoptosis but preserves autophagy. HMGB1 can thus be regarded as a molecular on–off switch between autophagy and apoptosis (Livesey *et al* 2012). Following calpain cleavage, the ATG5 protein exhibits pro-apoptotic characteristics. In the mitochondria, the N-terminal fragment of ATG5 attaches to the Bcl-XL protein and is therefore involved in the mitochondrial activation of apoptosis. If ATG5 is not translocated into the mitochondrion, autophagy is triggered in the cells (Mizushima and Komatsu 2011). Notably, the ATG5 protein is required for the induction of p53-dependent apoptosis. In addition, ATG5 also binds to the FADD protein, but this interaction does not affect the formation of autophagosomes. It may be related to the suppression of apoptosis regardless of autophagy. Being one of the major proteins of the autophagic signaling cascade, ATG5 may be associated with both the positive and negative control of apoptosis (Ye *et al* 2018).

Caspases involved in the apoptosis process are implicated in the breakdown of autophagy-related proteins, which include Beclin-1, AMBRA1, ATG3, ATG4, ATG5, ATG7 and p62. In addition, some pro-autophagic proteins can be converted into pro-apoptotic proteins after cleavage through caspases, triggering apoptosis (Djavaheri-Mergny *et al* 2010, Ojha *et al* 2015).

The FLIP protein, which inhibits the induction of apoptosis at the stage of the death receptors, also regulates the autophagy pathway. Through its DED domains, the FLIP protein specifically identifies the Atg3 protein and combines with it. This blocks the conjugation of the Atg3 protein with the LC3 protein and inhibits autophagy at the phase of autophagosome assembly (Mizushima *et al* 2008). It has been shown that alterations in other genes influence autophagy. In gastric and colon cancer cells, mutations in the UVRAG gene have been identified. The UVRAG gene codifies a protein that binds to Bcl-1. It has been shown that overexpression of this gene in colon cancer cells leads to decreased proliferation and significantly retarded tumor growth (Liang *et al* 2008). It has been established that genetic changes in ATG genes, seen in human cancer cells, can cause cancer to form in mice. Frameshift mutations in ATG12, ATG2B, ATG9B, and ATG5 are found in 25% of gastric and colon cancers. In addition, ATG5 mutations have been found to result in the impairment of autophagy via the disruption of the ATG5-ATG16L1 interactions (Kang *et al* 2009).

Protein genes with changed expression patterns are potential cancer prognostic markers. High expression rates of several ATG genes have been associated with high patient survival rates, but regrettably, certain ATG proteins have also been found to serve as adverse prognostic indicators (Bortnik and Gorski 2017). According to the type of cancer, there are various steps of ATG gene expression events, high-risk markers for colorectal cancer relapse are upregulated ATG16L2, CAPN2 and TP63, downregulated ATG5 and other genes related to autophagy, comprising SIRT1, RPS6KB1, PEX3, UVRAG and NAF1. Other factors involved in the pathogenesis of gastric cancer include ULK1, BECN1, ATG3, and ATG10 (Cao *et al* 2016, Mo *et al* 2019). There is emerging evidence that the expression of genes associated with autophagy is reduced in cancer cells. This is a big concern since autophagy has a critical function as a cancer suppressor in the event of oncogenesis (Chen and Karantza 2011, Levine and Kroemer 2019). In summary, there is a complex interrelationship between autophagy and apoptosis and the cellular fate is dictated by the cues traveling through this signaling web. Thus, it is important to keep in mind that any dysfunction of either of these processes will result in malfunction of the other.

4.6 Impact on autophagy on cancer treatments

In the past, the effect of autophagy in cancer research was overlooked for a long time. Therefore, the effect of autophagy inducers and autophagy inhibitors in relation to their function in cancer therapy is described and discussed below.

4.6.1 Inducers of autophagy

Drug resistance can be enhanced through apoptosis. For this reason, new approaches to optimize cancer treatment have been pursued. The induction of apoptosis through autophagy via anti-cancer drugs or autophagy inducers could be an intriguing therapeutic approach for eradicating cancer cells (Sui *et al* 2013). It has been shown that cytostatic drugs and radiotherapy may induce autophagy. Autophagy can be triggered by, for instance, imatinib, the tyrosine kinase inhibitor BCR-ABL, or cetuximab, a monoclonal antibody that targets the epidermal growth factor receptor (EGFR), and by proteasome inhibiting compounds. A number of other agents, such as seocalcitol, trigger autophagy in cells (Kanzawa *et al* 2004, Smith *et al* 2010). Moreover, pharmacologically activated autophagy by using rapamycin and metformin analogues could prevent carcinogenesis (Sesen *et al* 2015). Metformin, which is used to control diabetes, belongs to a class of synthetic guanidine derivatives. This chemical compound is considered to be one of the most important medications for the treatment of patients suffering from type 2 diabetes (Pernicova and Korbonits 2014). Metformin can cause a decrease in cell proliferation by activating AMPK, which inhibits mTOR and subsequently activates apoptosis or halts the cell cycle. AMPK activation also stimulates the initiation of autophagy by directly phosphorylating ULK1 and simultaneously blocking the mTORC1 complex through the phosphorylation of TSC2 and Raptor (Kim *et al* 2011, Takahashi *et al* 2014). Moreover, metformin could stimulate autophagy via accumulating LC3-II and decreasing p62 protein expression levels, thus enhancing TRAIL-driven apoptosis in TRAIL-resistant lung cancer cells (Nazim *et al* 2018). Rapamycin, a natural mTOR inhibitory compound, and its analogs have been proven to be effective as potential anti-cancer medications. There are an ever-increasing number of studies indicating that the inhibitory effect of mTOR signal transduction pathway through rapamycin analogs is associated with the activation of autophagy. The results suggest that rapamycin and its analogues may have a low toxicity at specific concentrations and can prevent the development of cancer cells (Blagosklonny 2012). In an interesting note, rapamycin can enhance the anti-cancer effects of doxorubicin, and the combination of the two drugs can suppress cancer cell growth and cause cell death through the mTOR/p70S6K signal transduction route (Li *et al* 2019a). Avoiding disease progression through autophagy activation is a fascinating area of investigation. Whether the enhancement of autophagy is a possible chemopreventive and/or chemotherapeutic mechanism of action of these compounds remains to be elucidated (Dalby *et al* 2010).

4.6.2 Inhibitors of autophagy

Although much progress has been made in the combat against cancer, many forms of cancer do not react well to available treatments. The most recent research indicates that the blocking of autophagy offers new possibilities for improving the effectiveness of anti-cancer medications. Autophagy can halt apoptosis triggered through DNA-damaging substances or hormone treatments. Studies have shown that autophagy can be suppressed using siRNA that blocks the genes that are part of this process, including ATG5, ATG6/BECN1, ATG10 and ATG12, or by using specific inhibitors, such as 3-methyladenine, hydroxychloroquine and bafilomycin A1 (Chen and Karantza 2011). Autophagy inhibitors can work early or late in the autophagy cascade by targeting regulators such as ULK1, VPS34, and ATG4B (table 4.1). The inhibitors that function in the early phase of autophagy comprise

Table 4.1. Categorization of autophagy inhibitors based on their site of function.

Target location	Inhibitory drug	References
PI3K	3-Methyloadenine	Heckmann *et al* (2013)
	LY294002	Wang *et al* (2017)
	SF1126	Qin *et al* (2019)
	Chloroquine	Mauthe *et al* (2018)
	Hydroxychloroquine	Mauthe *et al* (2018)
	Lys05	Cechakova *et al* (2019)
Lysosome	DQ661	Towers and Thorburn (2017)
	VATG-027	Goodall *et al* (2014)
	VATG-032	Goodall *et al* (2014)
	Melfquine	Rodrigues *et al* (2014)
	MRT67307	Petherick *et al* (2015)
ULK	MRT68921	Petherick *et al* (2015)
	SBI-0206965	Pasquier (2016), Chaikuad *et al* (2019)
Vacuolar-ATPase inhibitor	Bafilomycin A1	Mauthe *et al* (2018)
	SAR405	Pasquier (2015)
Vps34	VPS34-IN1	Bago *et al* (2014)
	Compound 13	Pasquier *et al* (2015)
	SB02024	Dyczynski *et al* (2018)

3-methyladenine and wortmannin. These substances target class III PI3K (Thelen *et al* 1994). In comparison to these two substances, Wortmannin is more effective than 3-methyladenine due to its irreversible attachment to the class III PI3K kinase. Other PI3K inhibitors comprise LY294002 and SF1126 which is an analog of LY294002 (Wang *et al* 2017, Qin *et al* 2019).

The serine/threonine kinases ULK1 and ULK2 are potential cancer therapy drug targets. Inhibition of ULK1 has been demonstrated to cause apoptosis and influence tumor size. ULK1 inhibitors, by contrast, impact ULK2. The most extensively investigated compound is SBI-0206965, which has powerful inhibitory characteristics for ULK1 but also possesses inhibitory characteristics outside of the target area (Pasquier 2016, Chaikuad *et al* 2019). It blocks autophagy and triggers apoptosis in NSCLC (Tang *et al* 2017), RCC (Lu *et al* 2018) and neuroblastoma cells (Dower *et al* 2018). Structurally, it is noteworthy that ULK1 inhibitors that have exhibited inhibitory activity can also block Aurora kinase and ULK2. In addition to SBI-0206965, there are additional lead candidates for autophagy inhibitory compounds including MRT68921 and MRT67307 (Petherick *et al* 2015) (table 4.1). The VPS34 protein is an attractive goal for blocking autophagy, as previously discussed. VPS34-IN1 belongs to such inhibitors (Bago *et al* 2014) alongside with SAR405 (Pasquier 2015), compound 13 (Pasquier *et al* 2015) and SB02024 (Dyczynski *et al* 2018). SB02024, which is a potential autophagy blocker, has shown potent inhibitory characteristics, a very favorable pharmacokinetic behavior and synergistic activity with other treatments, which qualifies it for further evaluation as a possible therapeutic candidate for clinical application (Dyczynski *et al* 2018).

In the late phase of autophagy, chloroquine, hydroxychloroquine and bafilomycin A1 blockers are utilized. Bafilomycin A1 works as a selective vacuolar-ATPase blocker, impeding lysosomal proton flux and thus autophagic flux (Chen *et al* 2010). Chloroquine (CQ) and hydroxychloroquine (HCQ) are lysosomotropic agents that inhibit pH acidification in lysomes. These drugs are utilized for the treatment of malaria and rheumatoid arthritis, respectively, as they can cross the blood–brain barrier. CQ and HCQ have been shown to potentiate the cytotoxic activity of chemotherapy drugs (Choi 2012, Verbaanderd *et al* 2017). Moreover, it was demonstrated that the combination of CQ and HCQ treatment significantly enhances cell death. Lys05, an analog of chloroquine, has recently been found to be more effective compared to hydroxychloroquine (Cechakova *et al* 2019). Besides these inhibitors, there are other lysosomal inhibitory proteins like DQ661 (Towers and Thorburn 2017), VATG-027, VATG-032 (Goodall *et al* 2014) and melfquine (Rodrigues *et al* 2014).

4.7 Function of autophagy in cancer treatment

Modulation of autophagy and combined therapy offer new perspectives in the combat against cancer. Combining autophagy inhibitors/activators with chemotherapy or radiotherapy has demonstrated improved cancer therapeutic outcomes. So far, CQ and its derivative HCQ are the sole medications that are currently administered to patients to specifically combat autophagy (Dyczynski *et al* 2018). The scientist Bedoya first proposed CQ as a candidate anti-cancer agent in 1970 after noting its toxicity to lymphoma and melanoma cells (Bedoya 1970). In contrast, Murakami and coworkers were the pioneers in unequivocally identifying the inhibitory effect of CQ on autophagy (Murakami *et al* 1998). Early clinical trials of CQ/HCQ in conjunction with some other therapies have yielded encouraging findings and indicated that the therapeutic objectives of inhibiting autophagy can be attained with only minimal toxicity. For instance, the combination of trastuzumab and chloroquine nearly entirely suppressed HER2-positive breast cancer that had previously exhibited resistance to trastuzumab (Cufí *et al* 2013). In addition, autophagy impairment occurred in 45%–66% of patients receiving concurrent HCQ and radiotherapy and chemotherapy (Rosenfeld *et al* 2014). Moreover, antitumor activity has been seen, for instance, in patients with glioma who underwent chemotherapy in conjunction with radiotherapy and temozolomide (TZD) demonstrated a threefold improvement in median survival in comparison to patients in the control group (Sotelo *et al* 2006). Among melanoma patients who received CQ and TZD, 41% experienced a partial reply or stabilization of the disease; the one-year rate of survival was 84% for patients with cerebral metastases who received CQ and radiotherapy, in comparison with 55% for those who received radiotherapy treatment alone (Rojas-Puentes *et al* 2013, Rangwala *et al* 2014).

The question of autophagy reliance is an important one, as autophagy-dependent cancer cells may experience a synergy of autophagy blockers with some other medications. Conversely, they can have an antagonistic effect in autophagy-independent cancers (Maycotte *et al* 2014). Brain tumors with BRAFV600E mutations are particularly reliant on autophagy (Levy *et al* 2014). Combination treatment with vemurafenib and CQ decreases the viability of cancer cells. These insights have led to CQ being administered to patients with vemurafenib-resistant brain cancers (Mulcahy Levy *et al* 2017). While these studies were initially encouraging, it was determined that blocking autophagy was enough to kill BrafV600E-positive CNS cancer cells. Nonetheless, this was not the situation for their wild-type BRAF-expressing counterparts (Levy *et al* 2014). Regrettably, the suppression of autophagy is contraindicated in certain instances. A recently published study demonstrated that blocking autophagy could reverse the abnormal expression of 6-phosphofructo-2-kinase/fructose-2,6-biphosphatase 3 (PFKFB3) in mouse breast cancer stem cells (BCSC), thereby promoting their revival and proliferation, and leading to subsequent metastasis (La Belle Flynn *et al* 2019). The findings above underscore the need for careful patient and medication choice to maximize the health benefits of autophagy impairment and minimize potential adverse effects. Probably because of several factors, among which the almost total absence of biomarkers to select patients eligible for therapy, the lack of highly specific autophagy inhibitors, and the incompletely characterized possible mechanisms of resistance. Regrettably, the clinical agents presently used, such as CQ and HCQ, are not selective for autophagy (Towers *et al* 2019).

The evolution of nanotechnology potentially allows not merely the direct delivery of anti-cancer agents to the cancer, but also the entrapment of autophagy blockers, which may lead to more potent treatment. For example, CQ administration can provide far superior enrichment of drugs in cancer cells (Lv *et al* 2018, Shao *et al* 2018). By utilizing progress in nanotechnology, it was feasible to encase miR-375, an autophagy inhibitory agent, and sorafenib in lipid-coated calcium carbonate nanoparticles (Zhao *et al* 2018). This strategy significantly suppressed autophagy and enhanced the anti-tumor activity of sorafenib in hepatocellular carcinoma. Dox/Wortmannin-regulated micelles have been designed to achieve a noteworthy anti-cancer efficacy in melanoma and breast cancer through the inhibition of autophagy (Rao *et al* 2019). Since autophagy blockers can be combined with commonly employed chemotherapy agents or other existing cancer treatments, they should be regarded as a highly attractive modality for fighting cancer. Further work is required for future clinical implementation to ascertain the significance of cancer therapy-induced autophagy, to develop a safer and more effective drug delivery system, and to identify prospective new autophagy inhibitor drugs.

4.8 Function of autophagy in stem cells

Stem cells constitute a special cell population within a living organism. They are immortally living, unspecialized cells that exhibit unique plasticity in the appropriate microenvironment and can develop into any type of cell (Daley 2015). The most distinctive property of stem cells is their capacity for self-renewal (multiple symmetrical or asymmetrical cell divisions with no differentiation and no aging and multidirectional differentiation into specialized progeny cells for specific organs. Particularly crucial is the stem cell's capacity for asymmetric cell division, in which one of the daughter cells stays a stem cell, whereas the other starts the program of differentiation (Gaziova and Bhat 2007, Santoro *et al* 2016). Nevertheless, this classification is somewhat oversimplified, as there are in fact several kinds of stem cell, each of which can be distinguished by its proliferative capacity and its potential for differentiation. This demonstrates that it is challenging to find a unique and precise characterization of stem cells when they are heterogeneous (Zheng *et al* 2018).

4.9 Cancer stem cells—a unique kind of stem cells

Over the past few years, it became evident that not all cells in tumors are functionally identical. Significant genetic heterogeneity exists in each individual tumor (Chen *et al* 2016, Bezuidenhout and Shoshan 2019). CSCs are a subset of undifferentiated cells that can be isolated based on the expression of the corresponding surface markers (Zheng *et al* 2018). These cells are typified through their enhanced oncogenesis potential and their capacity to trigger the formation of cancers. They may even be the cause of cancer metastases (Sukowati 2019). CSCs share many characteristics with normal stem cells, including the capacity for self-renewal, proliferation, differentiation, and multiplication (figure 4.6) (Tamazashvili 2019).

The microenvironment of stem cells, referred to as the niche, is an essential factor in keeping stem cell numbers at a constant level (Chen *et al* 2013). Among different functions, the niche determines if a cell self-renews or diversifies into a distinct cell

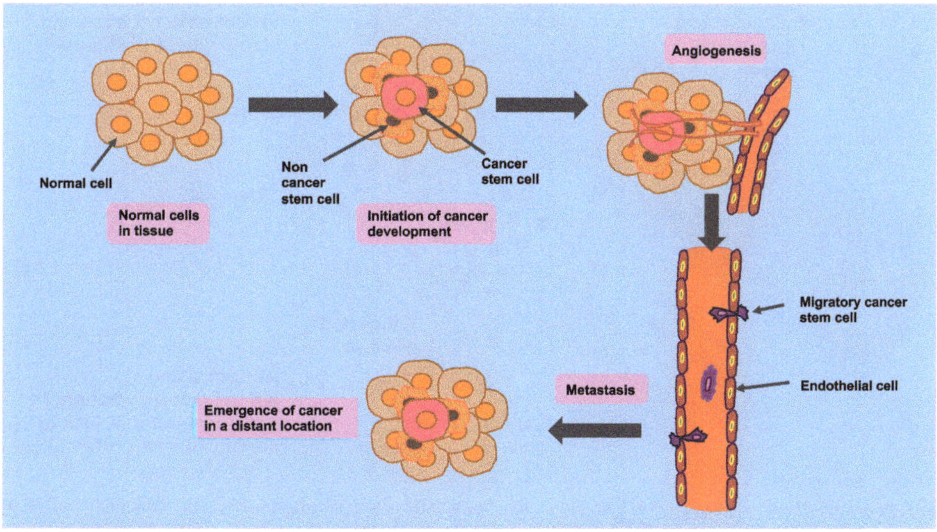

Figure 4.6. The generation of neoplastic stem cells and the mechanism of cancer development and its malignant progression, such as cancer metastases. The initiation and evolution of a tumor is typical performed via sequential phases. Following the emergence of the initial neoplastic cells, that is, both stem and non-stem cells, the development of the tumor is a gradual process. In the next phase, the angiogenesis stage is initiated, and the process of blood vessel development permits the invasive cells to spread to other areas of the organism.

type. The niche consists of a complex architecture formed by the immune system's mesenchymal cells, a framework of blood vessels and a certain component of the extracellular matrix (ECM) (Shen *et al* 2008). To maintain their characteristics and their capacity for regeneration, stem cells need to receive cues from the niche cells in which they reside. The niche also protects against the unrestricted proliferation of stem cells in the organism (So and Cheung 2018). CSCs are likely to be regulated by the identical processes inside the niche (Plaks *et al* 2015). The microenvironment of CSCs is typically found around blood vessels, which promotes their ability to metastasize. It has been hypothesized that the lodging of cells that have metastasized to distant targeted organs may mark the onset of the establishment of a new microenvironment at the site of the metastasis, which will trigger the establishment and growth of secondary cancers in other tissues (figure 4.7) (LaBarge 2010). Stem cells are characterized as having restricted or no capacity for self-renewal, indicating that the CSC pool may be specifically required for cancer generation and propagation.

4.10 Autophagy functions in maintaining of stem cell characteristics

All stem cells must strictly control the intracellular concentration of each protein and the rate of ATP generation to maintain their characteristic and unique traits, such as multipotentiality, differentiation and self-renewal. This kind of strict control

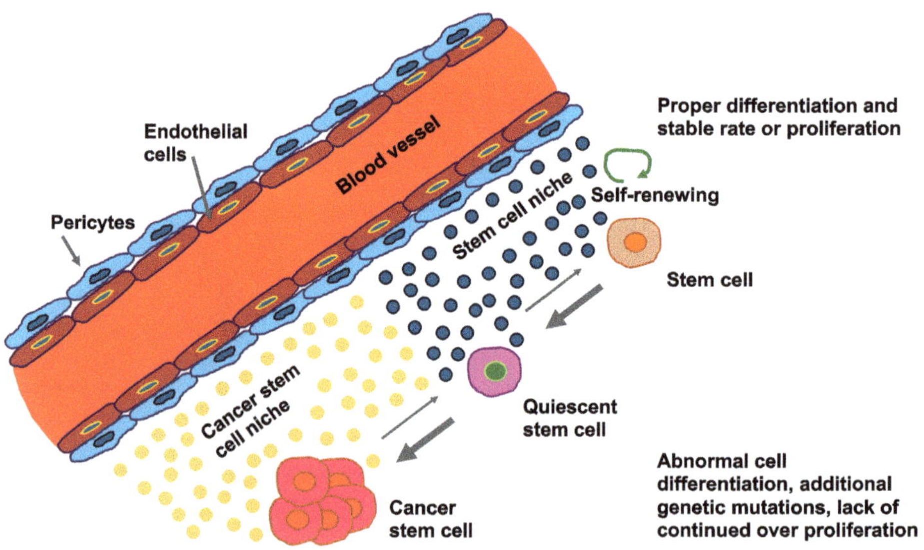

Figure 4.7. Functionality of the normal stem cell niche and the CSC niche. A stem cell niche offers cells physiological conditions in which both cell differentiation and cell proliferation are restricted. A temporary proliferation stimulus is activated solely when the tissue, in which the stem cells are situated, regenerates. If the stem cell acquires suitable mutations in its genetic information, it begins to multiply in an uncontrolled manner. Another hypothesis is that, because of a mutation of this type, the stem cell niche emits continuous cues that trigger stem cell proliferation and expansion. This leads to the generation of a large pool of progenitor cells that carry a genetic mutation.

of cellular metabolism is regarded as a fundamental mechanism for keeping stem cells in a dormant state (García-Prat *et al* 2017). Sustaining the equilibrium between stem cell differentiation processes and their suppression is critical to maintaining stem cells in an orderly condition. An enhanced differentiation program can exhaust the stem cell population and cause stem cell aging. Moreover, enhanced stem cell differentiation can result in the evolution of cancer. Therefore, the implementation of suitable quality check mechanisms is required to maintain homeostasis and the capacity to adequately react to injuries, environmental stresses or signaling cues for differentiation, and to enable adequate tissue regeneration.

4.11 The importance of cancer stem cell autophagy

Autophagy plays a key role in sustaining the activity and aggressive behavior of CSCs. Maintaining homeostasis in the context of autophagy is vital for preserving the pluripotency of stem cells. The phenomenon of pluripotency is a key property of CSCs (figure 4.8); without it, it would not be possible to sustain cancer stem cells in an immature state and to continuously keep the cells in a state of division (Han *et al* 2018b). Proteins such as Beclin-1 or Atg4 appear to be critical to the autophagy process for keeping stem cells working correctly. This is supported by research conducted on CSCs isolated from various types of malignant cancers, such as liver, pancreatic, ovarian or breast cancers (Chaterjee and Van Golen 2011, Gong *et al* 2013, p 1, Song *et al* 2013, Peng *et al* 2017).

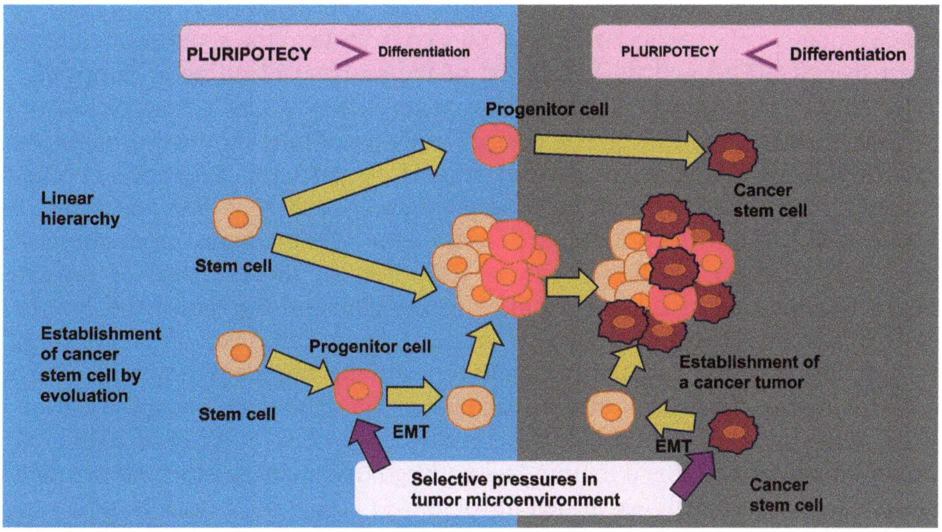

Figure 4.8. Two fundamental ideas on the generation of CSCs. In a proposed hierarchical model of CSC generation, the genetic and epigenetic transformation causes the generation of progenitor cells and CSCs. The cells modified in this manner can be further transformed which subsequently causes the development of a tumor. A different theory is based on the idea of the evolution of CSCs. The entire process starts with the initiation of epithelial-to-mesenchymal transition (EMT).

Some studies have indicated that blocking the autophagy pathway in CSCs can reduce the release of IL-6. This mechanism is very probably realized through the STAT3/JAK2 signaling cascade. It can be inferred from this that the IL-6-JAK2-STAT3 signaling mechanism has a significant part to fulfill in the conversion of non-cancer stem cells into CSCs (Iliopoulos *et al* 2011, Maycotte *et al* 2015). A lot of work is also being conducted on FOXO proteins and the part these transcription factors exert in sustaining the functionality of CSCs (Liang and Ghaffari 2018, Van Doeselaar and Burgering 2018, Yadav *et al* 2018). Evidence indicates that blocking the expression of FOXO3 proteins enhances the self-renewal capacity of CSCs. This phenomenon has been documented in breast, prostate and colon cancers (Dubrovska *et al* 2009, Prabhu *et al* 2015, Smit *et al* 2016). In addition, many investigations point to the central significance of FOXO proteins for the control of the expression of numerous proteins that are linked directly to the autophagy mechanism, such as Beclin 1, LC3, ULK1 or several ATG proteins (Reaganshaw and Ahmad 2007, Van Der Vos and Coffer 2008). In one of the most up-to-date studies, it was demonstrated that the Forkhead box A2 (FOXA2) protein is found overexpressed in CSCs and that its activity is in direct correlation with the level of autophagy. Several studies were carried out to block FOXA2 activity in a cell. In this manner, they were able to significantly decrease the self-renewal capacity of CSCs (Peng *et al* 2017).

4.12 Phenomenon of mitophagy in cancer stem cells

Mitochondria are one of the most essential cell organelles. The mitochondria are the power plants of the cells, they are essential for the generation of cellular ATP through oxidative phosphorylation, and they contribute to the generation of reactive oxygen species (ROS) and the activation of the pathways leading to apoptotic cell fate (Held and Houtkooper 2015, Pickles *et al* 2018). In CSCs, mitochondria are highly elevated in importance. The normal function of mitochondria governs many characteristics of CSCs, such as their migratory capacity, tolerance to chemo-therapeutic drugs, and self-renewal (Peiris-Pagès *et al* 2016, Chae and Kim 2018, Lleonart *et al* 2018). Mitochondria are essential for cellular energy production. Most cancer cells undergo aerobic glycolysis regardless of the presence of oxygen, a phenomenon referred to as the Warburg effect (Pacini and Borziani 2014, Icard *et al* 2018). CSCs are exceptional in this respect. Stem cells are observed to have extraordinary adaptive capabilities that rely on the microenvironment in which the CSCs reside. They can metabolize either via oxidative phosphorylation or aerobic glycolysis (Peiris-Pagès *et al* 2016, Snyder *et al* 2018, Yi *et al* 2018).

Mitochondria are highly dynamic and undergo fusion and fission. The mitochondria splitting is primarily governed through GTPase DRP1 and a series of its associated factors, amongst which is FIS1. The fusion pathway is controlled through GTPase MFN1, GTPase MFN2 and OPA1. Both types of processes are performed in the cell to adjust to the cell's actual requirements and to guarantee the orderly breakdown of defective organelles. Mitochondria that are defective or superfluous are generally broken down by mitophagy (Kasahara and Scorrano 2014). Many lines of evidence point to the fact that sustaining pluripotency of both stem cells and CSCs relies on the orderly operation of DRP1-regulated mitochondrial segregation (Vazquez-Martin *et al* 2012, Son *et al* 2013, Chen and Chan 2017, Seo *et al* 2018). The study also found that high activity of the DRP1 protein can be a poor prognostic indicator in the setting of cancer therapy (Xie *et al* 2015, Kitamura *et al* 2017).

Mitophagy in cancer stem cells is typically carried out via the PINK1-PARKIN protein signaling cascade (figure 4.9) (Pickrell and Youle 2015). The PINK1 protein specifically binds to the membrane of injured or dysfunctional mitochondria and then causes phosphorylation of the E3-PARKIN ligase. Following these modifications, the mitochondria destined for elimination are labeled with ubiquitin, which in turn enables their identification by OPTN, NDP52 and AMBRA1. These proteins

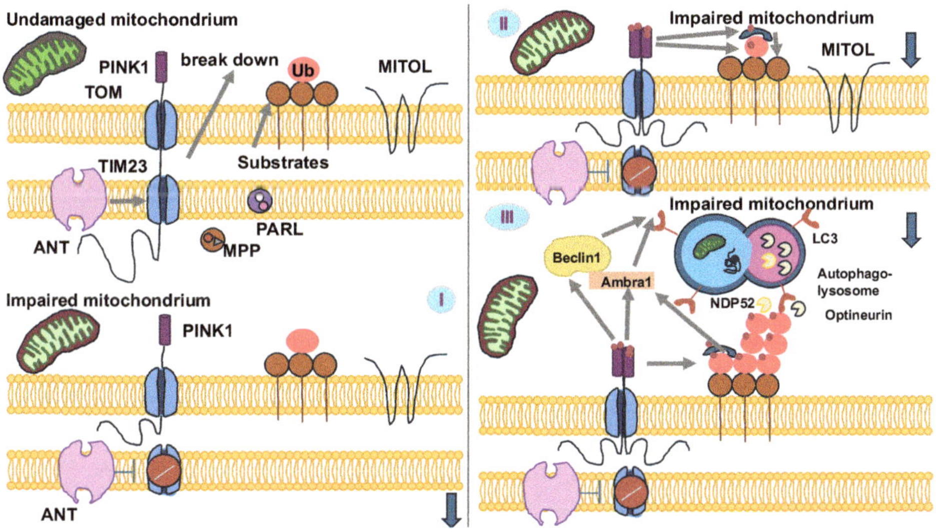

Figure 4.9. Function of the PINK1/Parkin route in the initiation of autophagy in stem cells and differentiated cells. In a functional mitochondrion, PINK1 is translocated to the inner mitochondrial membrane by the TOM/TIM23 translocase machinery. There it is then decomposed by the activity of the two proteases PARL and MPP, before being trafficked to the cytosol and broken down. When mitochondria function improperly as evidenced through mitochondrial membrane potential disruption, ANT impedes the trafficking of PINK1 across the TIM23 translocase. Because of this process, PINK1 is enriched at the outer mitochondrial membrane. The accumulation of PINK1 on the outer mitochondrial membrane leads to the generation of a complex of PINK1 and TOM translocase, which in turn leads to the autophosphorylation of PINK1 and thus its activation. Activated PINK1 phosphorylates the ubiquitinated substrates present on the outer mitochondrial membrane, thereby establishing a link between Parkin and its phosphorylation. Phosphorylated Parkin, for its part, promotes the subsequent ubiquitination of proteins on the outer mitochondrial membrane. The ubiquitinated proteins, in turn, are capable of phosphorylating successive PINK1 molecules, thereby forming a feedback circuit, so that a dysfunctional mitochondrion becomes coated with phospho-ubiquitin. In this case, the mitophagic adaptors optineurin and NDP52 tether to phospho-ubiquitin chains. These proteins interact with the LC3 protein, which triggers the onset of mitophagy. Concurrently, PINK1 binds to the Beclin-1, Parkin and Ambra proteins, which promotes the generation of autophagosomes surrounding the dysfunctional mitochondria.

also assist in the generation of autophagosomal membranes surrounding organelles (Pickles *et al* 2018). These processes appear to be critical for ensuring that CSCs can function correctly (Vazquez-Martin *et al* 2016, Ho *et al* 2017). Research also indicates that mitophagy is upregulated in CSCs through enhanced transcription of the NANOG protein, which seems to be essential for preserving the characteristic traits of stem cells, like the ability to self-renew (Lin and Iacovitti 2015). The results also indicate that CSCs can escape apoptosis activation through the targeted activation of the mitophagy signaling pathways. Under hypoxic circumstances as they are present in most solid tumors, CSCs reprogram their metabolism from oxidative phosphorylation to aerobic glycolysis. This leads to the activation of HIF-1α and HIF-2α which are normally not activated in normoxic cells (Sowter *et al* 2001, Macklin *et al* 2020).

In such conditions, CSCs can utilize the proteins FUNDC1, BNIP3 and BNIP3L/NIX for generating energy and consequently substantially decrease the quantity of mitochondria in the cell via the mitophagy pathway. A low number of mitochondria, in contrast, decreases the likelihood of pro-apoptotic release factors being secreted from the mitochondria and the entire cell being degraded through apoptosis (Peng *et al* 2010, Hamacher-Brady and Brady 2016, Leites and Morais 2018, Naik *et al* 2019).

4.13 NAD+ and nicotinamide phosphoribosyl transferase (NAMPT) can trigger autophagy of cancer stem cells

NAD is a molecule that functions as a cofactor for cellular metabolism and energy generation. It also appears to serve an important function in DNA damage repair and in keeping mitochondria in cancer cells and CSCs in optimal working state (figure 4.10).

NAMPT, by contrast, is a key enzyme in the NAD synthesis pathway. When its activity is blocked, the intracellular NAD^+ reservoir is slowly depleted, which results in the inhibition of ATP synthesis (Garten *et al* 2015). NAMPT shows over-expression in CSCs and the higher the expression level of this transferase, the poorer the prognosis regarding cancer evolution (Gujar *et al* 2016). NAMPT is also emerging as a very critical player in regulating the behavior of CCs, and PARP and SIRTs are also implicated in the entire process (Lucena-Cacace *et al* 2018). It also looks like NAMPT could be one of the determinants of cancer differentiation, as it is involved in the epigenetic reprogramming that characterizes most cancers. The activity of this enzyme is controlled by the intake of NAD in the cell (Jung *et al* 2017, Lucena-Cacace *et al* 2018).

4.14 Impact of autophagy in cancer stem cell resistance to chemotherapy

Notwithstanding the continuous progress in pharmacology and the emergence of new and ever more specific chemotherapies, it is still not feasible to introduce a therapy that is universally effective. Most importantly, pharmacological treatment

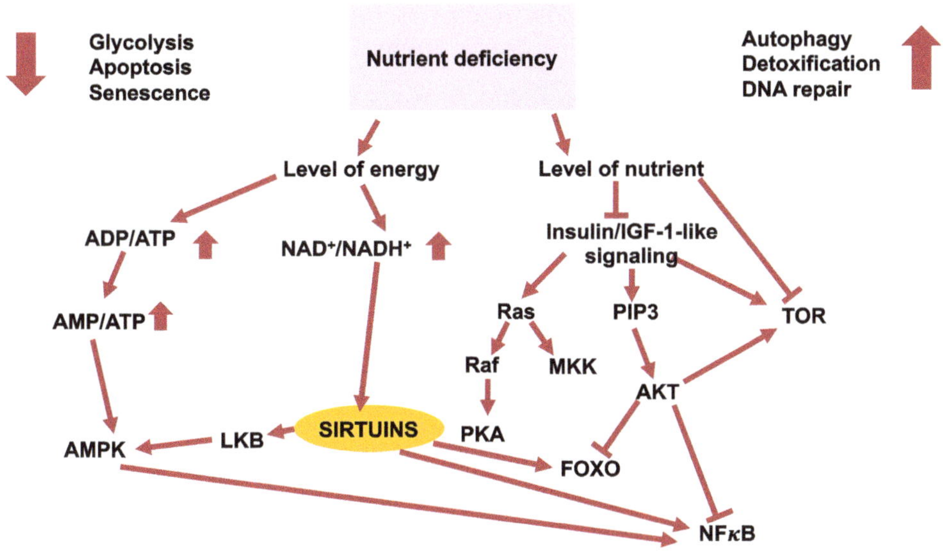

Figure 4.10. The impact of nutrient supply on the initiation of autophagy. The decline in nutrient utilization results in a decline in mitochondrial activity, thereby elevating the AMP/ATP ratio and causing intracellular NAD^+ concentrations to rise. The consequence of the elevation of the AMP/ATP ratio results in AMPK activation. In addition, AMPK can be activated by deacetylation, thereby creating a feedback cycle. Elevated NAD^+ levels enhance the activity of sirtuins (guardians of the genome), which results in the deacetylation of LKB1 and AMPK and the modulation of their respective functions. A downward-pointing arrow represents a reduction in the intensity of the mechanism, while an upward-pointing arrow represents an enhancement in the intensity of the depicted mechanism.

can only destroy cells that proliferate quickly and without restriction, while the quiescent CSCs develop a resistance to the therapy during this time. Such cells are then the cause of the reappearance of the cancer that had seemingly been eradicated. CSCs become drug-resistant through a combination of multiple factors. First, the niche where these cells nest is conducive to this phenomenon. Other important factors involve a high capacity for repairing DNA injury, blocking the initiation of the apoptosis process and high levels of MDR gene expression (Gurusamy *et al* 2018, Askari *et al* 2023). The presence of autophagy in CSCs appears to be critical to sustaining the chemotherapeutic resistance phenomenon. Many trials demonstrate that the pairing of autophagy blockers with traditional chemotherapeutic drugs provides a much stronger therapeutic index (Sui *et al* 2013, Yue *et al* 2013, Golden *et al* 2014, Ojha *et al* 2016, Sun *et al* 2016, Huang *et al* 2017, Naik *et al* 2018). Evidence also suggests that the autophagy mechanism may be involved in the development of resistance to chemotherapy. For instance, blocking the Wnt signaling cascade has been linked to the activity of resveratrol in breast CSCs, which in turn activates the autophagy mechanism within these cells (Fu *et al* 2014). In addition, it has also been proposed that inhibition of mTOR can result in the differentiation of CSCs from the nervous system (Zeng and Zhou 2008, Zhao *et al* 2010, Li *et al* 2017b). Collectively, this experimental evidence indicates that the mechanism of autophagy may finally emerge as a key factor in overcoming cancer multidrug resistance and preventing cancer reappearance.

4.15 Autophagy functions on the motility of cancer stem cells

CSCs exhibit a very high capacity for migration and formation of metastases. This appears to be closely linked to the epithelial-to-mesenchymal transition (EMT), which involves alterations in cell polarization and the mechanism through which cells engage with one another (Al-Hajj *et al* 2003, Morel *et al* 2008, Koren and Bentires-Alj 2015, Donnenberg *et al* 2016). There is growing experimental support for the notion that EMT and autophagy are intimately connected events. Cells that undergo EMT are altered in a distinctive manner. In such cells, the autophagy program is highly induced (Amaravadi 2012, Marcucci *et al* 2017, He *et al* 2019). In addition, there exists also a partial EMT that leads to hybrid EMT states, indicating that plasticity is an inherent feature of CSCs (figure 4.11) (Wang and Unternaehrer 2019, Fernando *et al* 2024). Gaining an awareness of the pivotal elements that govern the transition among these processes, like EMT or mesenchymal-to-epithelial transition (MET), could pave the way for new therapeutic approaches that curb the progression of cancers. Some investigations have also pointed to a totally distinct function of autophagic activity in the case of migration. In certain tumors, a low migratory potential was seen in cells with intense autophagic activity. In these cases, modifying autophagy made it possible to re-establish the display of the mesenchymal phenotype by the cancer stem cells, and therefore these cells became prone to migration and metastasis (Marcucci *et al* 2017, Jia *et al* 2018). This is probably linked to the phenomenon of paracrine secretions from the cells, which stimulates EMT activation. Certain studies also suggest a totally unrelated effect of autophagic activity in relation to migration. In some cancers, despite the high level of activity of the autophagic process, a low migratory capacity of the cells was noted. In such cases, alteration of autophagy made it possible to reconstitute the presentation of the mesenchymal phenotype by the CSCs, and hence these cells were capable of migration and metastatic spread (Marcucci *et al* 2017, Jia *et al* 2018). This is most probably associated with the phenomenon of paracrine secretion from the cells that promotes EMT induction.

4.16 Relationship between autophagy, decrease in calories and cancer

Autophagy is an early cellular reaction to stress signals in physiological and pathological settings. It is hypothesized that enhancing autophagy flux protects host cells from consecutive injuries through clearing dysfunctional organelles and aberrantly folded proteins. Accordingly, manipulation of autophagy is proposed as a therapeutic strategy for various pathological disorders. There is growing experimental evidence that intermittent fasting or calorie limitation can result in the activation of adaptive autophagy and extend the lifespan of eukaryotic cells (Shabkhizan *et al* 2023). Prolonged caloric restraint with exaggerated autophagy reaction is detrimental and can induce type II autophagic cell death. Although there

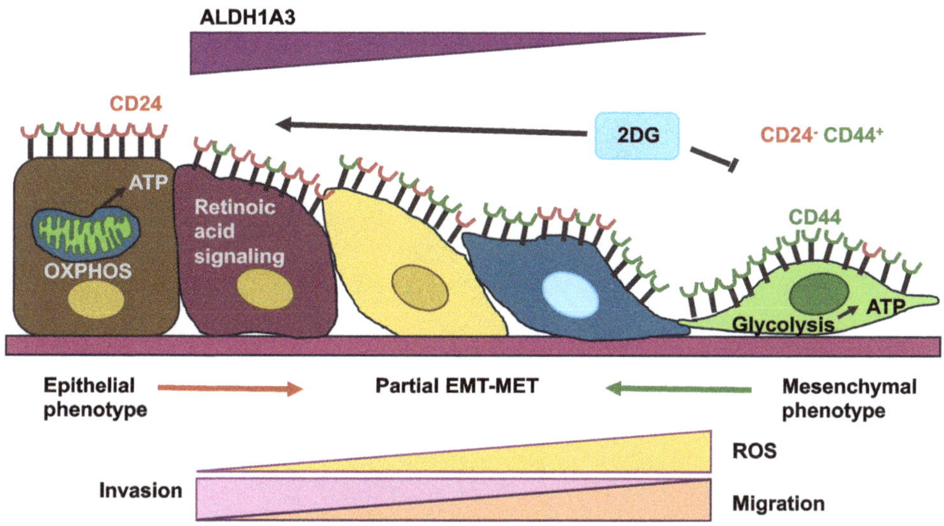

Figure 4.11. Plasticity has been recognized as an intrinsic property of CSCs and governs the equilibrium of critical processes necessary during various steps of breast cancer development, encompassing EMT versus MET and glycolysis versus oxidative phosphorylation. Aldehyde dehydrogenase 1A3 (ALDH1A3) was identified to orchestrate these cancer-promoting events and the balance between the two different breast CSC subpopulations that are characterized according to high ALDH activity and $CD24^{-}CD44^{+}$ cell surface expression (Fernando *et al* 2024). While ALDH1A3 enhances $ALDH^{+}$-breast cancer cells, it conversely represses the subpopulation of $CD24^{-}CD44^{+}$ cells through retinoic acid signaling-facilitated gene expression alterations. This ALDH1A3-induced alteration of CSC subpopulations was associated with reduced migration but enhanced invasion and intermediate (hybrid) EMT phenotype. ALDH1A3 enhances the oxidative phosphorylation and reduces glycolysis and ROS. The consequences of ALDH1A3 decrease were compensated with the glycolysis inhibitor 2-deoxy-D-glucose (2DG). In cell culture and tumor xenograft models, 2DG suppresses the expansion of the $CD24^{-}CD44^{+}$ population and ALDH1A3-knockdown triggered ROS generation. Combined blockade of ALDH1A3 and glycolysis most potently impairs mammary tumor growth and tumor-initiating cell growth, implicating the dual targeting of ALDH1A3 and glycolysis as possessing therapeutic potency to curtail CSCs and tumor advancement. In summary, these results identify ALDH1A3 as a master regulator of pathways necessary for breast cancer advancement, and elimination of ALDH1A3 renders breast cancer cells much more sensitive to glycolytic inhibitory treatment.

is a strong correlation between caloric restriction and the autophagic reaction in various cell types, the exact molecular mechanisms underlying this process are still elusive. The following section discusses the possible implications of long-term and short-term calorie limitation on autophagic reaction and the homeostasis of cells.

The term autophagy relates to cellular mechanisms involved in self-eating and various homeostatic functions. The activation of autophagy assists host cells in the clearance of aberrantly folded, aggregated proteins and dysfunctional organelles (González-Rodríguez *et al* 2022). In addition to its function in cellular homeostasis, autophagy is actively engaged in the development, differentiation and regenerative capacity of a variety of embryonic and adult cells (Hassanpour *et al* 2018, 2020). In eukaryotes, the autophagy mechanism and the proteasome represent the two

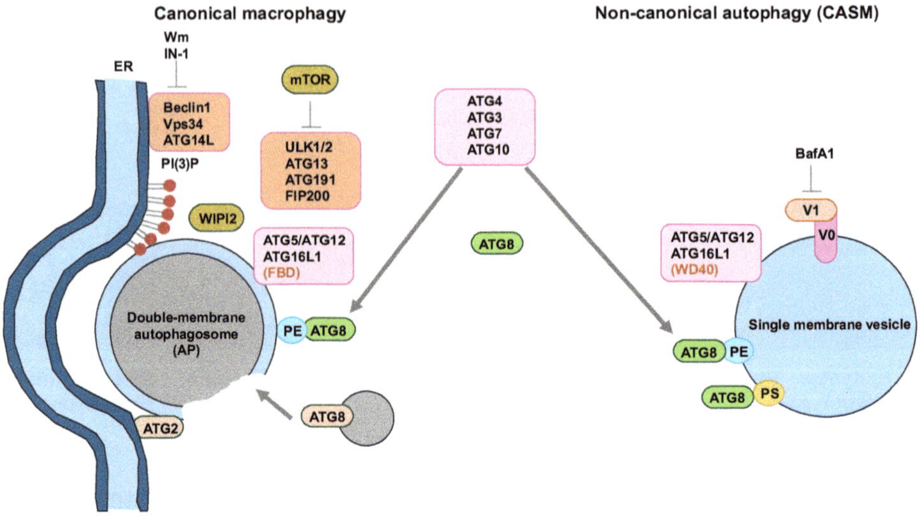

Figure 4.12. Molecular characteristics of autophagy-related signal transduction routes. The schematic sketch points out to the differences in molecular signal transduction mechanisms and the membrane target of canonical macroautophagy and noncanonical autophagy. autophagosome (AP), bafilomycin A1 (BafA1), wortmannin (Wm), inhibitor 1 (IN-1), conjugation of ATG8s to single membranes (CASM).

principal catabolic routes for the disposal of degraded molecular debris (figure 4.12) (Li *et al* 2022).

In distinction to proteasomal clearance, autophagy is the sole mechanism to dispose of whole damaged organelles (Hassanpour *et al* 2018). So far, three different types of autophagy mechanisms have been identified in eukaryotic cells, such as macroautophagy, microautophagy and chaperone-based autophagy (figure 4.13) (Hassanpour *et al* 2018). More recent work has shown that macroautophagy plays the most important function in autophagy and will be referred to as autophagy in the following. From an ultrastructural perspective, the autophagic reaction is triggered through the generation of double-membrane vesicles referred to as autophagosomes (Shahabad *et al* 2022).

In the subsequent steps, the autophagosomes are delivered to and merge with lysosomes, where they undergo enzymatic breakdown driven by the activity of multiple acidic hydrolases (Fang *et al* 2022). In contrast to macroautophagy, microautophagy is considered a non-selective mechanism that is triggered through the direct uptake and sequestration of cytosolic components across the lysosomal membrane. In the third type of autophagy, the chaperone-based autophagy, LAMP-2A activity guides a complex of chaperone proteins and targeted proteins toward the lysosomes (Schnebert *et al* 2022).

Besides the protective function of autophagy in damaged cells, abnormal or overly active autophagic reactions can lead to multiple pathological consequences, such as apoptosis and pyroptosis, through the induction of programmed cell death (Rezabakhsh *et al* 2017, An *et al* 2022, Duan *et al* 2022, Wang *et al* 2022). Similarly,

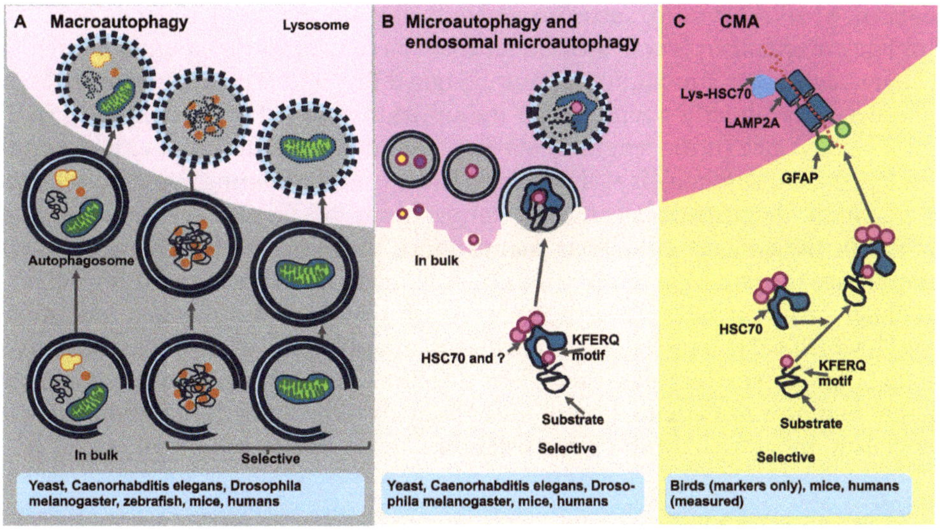

Figure 4.13. There are three different types of autophagy, namely macrophagy, mitophagy and chaperone-mediated autophagy (CMA). (A) Macroautophagy involves the elimination of cytosolic cargo through a sealing membrane that assembles through the combination of certain proteins and lipids in a intricate and multistep mechanism The membrane forms a closure around an autophagosome that is being transported along microtubules. Merging of autophagosomes and lysosomes results in the breakdown of the enclosed material. Macroautophagy can occur in bulk or in a selective manner, based on the cargo it is engulfing. (B) Microautophagy encloses cytosolic cargo in small vesicles that are produced through invagination of the lysosomal membrane either in bulk or in a selective manner through detection and targeting via heat shock-related 71 kDa protein (HSC70; synonymously referred to as HSPA8) and co-chaperones to be identified. (C) CMA comprises the selective breakdown of KFERQ-like motifs in proteins, which are subsequently broken down via the chaperone HSC70 and co-chaperones like the carboxyl terminal of HSC70-interacting protein (CHIP), heat shock protein 40 (HSP40; synonymously referred to as DNABJ1) and HSP70–HSP90 organizing protein (HOP) and their uptake into lysosomes through the receptor lysosome-associated membrane protein type 2 A (LAMP2A). The grey fields indicate that each of the three mechanisms is evolutionary conservated. chaperone-assisted selective autophagy (CASA), glial fibrillary acidic protein (GFAP), lysosomal HSC70 (lys-HSC70), and ubiquitin (Ub).

a strong decrease in autophagy activity can lead to different pathological states like genomic instability, anaplastic alterations, infections, accelerated aging and various metabolic disorders (Cassidy and Narita 2022, Chang *et al* 2022, Chao *et al* 2022, Zhou *et al* 2022). Because of the reciprocal influence of various autophagy components and other signaling pathways, the autophagy mechanism has been linked to pleiotropic phenomena (Jiang *et al* 2019, Xie *et al* 2020). The presence of sophisticated autophagy complexes and overlapping molecular effectors results in the simultaneous execution of diverse reaction pathways (Su *et al* 2020). In line with these observations, it is reasonable to assume that certain metabolic circumstances can modulate the autophagic reaction in either physiological or pathological settings (Frisardi *et al* 2021). Molecular studies have found that low-energy states or the lack of essential nutrients like amino acids and glucose can cause a drop in ATP and an enhanced AMP/ATP ratio, subsequently inducing autophagy (Kim *et al* 2022). The autophagy mechanism is activated under these circumstances to counteract the ATP deficiency, while the stimulation of autophagic efflux and the consequent biological phenomena are directly connected to the utilization of ATP (Li *et al* 2017a). The precise significance of the length of starvation and the activation of either adaptive or exaggerated autophagy is still obscure. There is accumulating evidence related to the potential therapeutic/detrimental implications of fasting on the autophagic reaction in the context of cellular maintenance of homeostasis. This section aims to emphasize the potential implications of sustained and temporary caloric deficit on autophagic reaction and cellular homeostatic responses. The molecular mechanisms through which caloric restraint can elicit both adaptive and exaggerated autophagic reactions are extensively discussed. The question at the heart of the discussion is whether and how the attenuation of autophagy through calorie limitation can assist host cells in avoiding harmful circumstances.

4.16.1 General physical health advantages of intermittent fasting

The concept of intermittent fasting encompasses various short- and long-term dietary strategies that consist of cycles of regular eating and fasting periods (Harvie and Howell 2017, Mattson *et al* 2017, Sun *et al* 2017a). Calorie limitation programs typically range from 24 h to 3 weeks and supply 30% of overall energy requirements (Teong *et al* 2023). In addition to various programs endorsed by nutritionists, intermittent fasting involves total or semi-abstinence from food and drink in religious practices (Lessan and Ali 2019). For a number of decades, intermittent fasting has been utilized in different guises in various scientific studies (Różański *et al* 2021). In the alternate day fasting regimen, fasting is imposed every other day and eating of energy-dense foods is discouraged, while in the feeding-dosed groups, access to food is permitted *ad libitum* throughout the feeding period (d). It is noteworthy that the modified fasting is a form of fasting every other day where the overall energy supply is between 25% and 30% and takes place over a restricted period of time with an open eating window of two to four hours. A modified fasting cure is a different type of intermittent fasting in which 20%–25% of the daily energy intake is covered by periodic fasting days. This technique is rooted in the 5:2 weight-loss diet, in which you fast for two days a week and eat normally on the other days. With limited time eating, there are certain time slots for eating and fasting. There is also another type of fasting with time-limited eating schedules, such as eating only from sunrise to sunset. During the eating window of four to ten hours, the quantity of energy is not restricted (Różański *et al* 2021, Teong *et al* 2023). Several results are in line with the idea that calorie limitation is associated with longer life and improves prevention of chronic disease through the regulation of anti-inflammatory reactions (Longo and Mattson 2014). Intermittent fasting could be an alternative approach to common therapies as it has anti-cancer characteristics and even improves cancer patients' tolerability to chemotherapy. Supporting this assumption, healthy cells are more robust to chemotherapy-induced damage; hence, the severity of off-target adverse effects is reduced in cancer patients who have fasted

(Lee *et al* 2012). More specifically, subjecting normal cells to calorie limitation causes energy (ATP) to be expended on purposes of preservation and restoration instead of proliferation.

In cancer cells, the activation of multiple oncogenes, like the IGF-1R, Ras and AKT/mTOR signaling routes, enhances proliferation levels and thereby increases the sensitivity of the cells to chemo-/radiotherapy treatment schemes (Sadeghian *et al* 2021). A number of publications have indicated massive alterations in metabolic profiles under regularly experienced hunger episodes (Sun *et al* 2017a). During the first 10 h of fasting, glycogen is depleted from liver tissue, followed by the degradation of fatty acids in the muscles and adipose tissue. As a result, considerable amounts of free fatty acids and amino acids are generated (Sun *et al* 2017a). Notably, intermittent fasting has been found to alleviate inflammatory reactions and has yielded multiple positive effects in individuals suffering from asthma or rheumatoid arthritis (Longo and Mattson 2014). An example of this mechanism is the induction of IL-4, IL-5 and IL-13 by type 2 $CD4^+$ lymphocytes to enhance the polarization of macrophages towards M2 phenotypes, thereby conferring anti-inflammatory characteristics (Kim *et al* 2017). The beneficial impacts of intermittent fasting on gut microbiota and immune system activity may occur in individuals suffering from multiple sclerosis (Carbone *et al* 2016). The molecular characterization of peripheral mononuclear cells obtained from 21 volunteers after 24 h of fasting and 3 h of food intake demonstrated the induction of $CD4^+$ Th2 cells in both healthy and asthmatic subjects (Traba *et al* 2015, Han *et al* 2018a).

The precise significance of fasting and immunomodulatory actions is related to the activation of FOXO4 and its classical target, FK506-binding protein 5. These compounds are regarded as fasting reaction items with regard to the immune system, which can modulate the generation and secretion of pro-inflammatory cytokines from Th1 and Th17 lymphocytes (Han *et al* 2021). In a related experiment, intermittent fasting substantially decreased the number of $CD16^+$ and $CD56^+$ natural killer cells, with no impact on circulating $CD3^+$, $CD4^+$ and $CD8^+$ lymphocytes (Gasmi *et al* 2018). Fasting can enhance the activity of the antioxidant system to counter oxidative stress. For instance, SOD, which is an enzyme participating in the neutralization of ROS, is induced in cells undergoing intermittent fasting (Li *et al* 2019b). The data demonstrated that absolute protein levels of soluble intracellular adhesion molecule-1 (ICAM-1), LDL, and T3 hormone are decreased in middle-aged subjects who undergo short- and long-term alternate days fasting (Longo *et al* 2021). Biochemical research indicated that 49 metabolites exhibited a 20% decrease, and that values for amino acids were about 44.9%. The apparent decrease of specific amino acids like methionine, a factor that contributes to aging, may delay the aging progression (Stekovic *et al* 2019). Intermittent fasting's anti-aging advantage is also associated with the control of dynamic growth and stem cell performance (Yin *et al* 2013, Mihaylova *et al* 2014). These cells can replace lost cells and re-establish the functionality of injured cells. Implementing fasting regimens in animal models can result in the eradication of age-related alterations like Alzheimer's and an increased capacity for recovery in ischemic stroke (Longo and Mattson 2014). Histological studies have revealed enhanced proliferation of neural stem cells in the hippocampus of rodents about three weeks following intermittent fasting. These alterations are in line with the local distribution of BDNF (Mana *et al* 2017). It is noteworthy that the stimulation of the mTOR pathway decreases telomere shortening and preserves pluripotency capacity in intermittent fasting regimes (Fontana and Partridge 2015, Iglesias *et al* 2019).

4.16.2 Intermittent fasting and autophagy

It has turned out that there exists a relationship between intermittent fasting and autophagy that finally may impact the outcome of cancer treatments and the malignant progression of cancers. This phenomenon seems to be universal.

4.16.2.1 Molecular mechanisms of intermittent fasting-triggered autophagy

New data indicates a strong relationship between intermittent fasting and autophagy reaction. The way intermittent fasting can modulate autophagy and the mechanisms through which it can achieve this are the topic of ongoing discussions (Chaudhary *et al* 2022). A close relationship between starvation and autophagy induction has been seen through the deacetylation of ATG4B and further interaction with pro-LC3 (Sun *et al* 2022). Starvation in mice has been verified to cause a simultaneous decrease in P300 and an increase in SIRT2 activity, which leads to the deacetylation of ATG4B (Sun *et al* 2022). Acetyltransferase activity of P300 probably plays a role in nuclear targeting of ATG8–PE ubiquitin system and suppression of autophagy in Bombyx mori (Wu *et al* 2021, Shabkhizan *et al* 2023). On the basis of the findings, a two-dimensional fasting period in Sirt2/mice resulted in a decrease in LC3-II, an accumulation of acetylated ATGB4 in the liver tissue and a repression of the autophagic reaction, indicating a tight relationship of autophagy and the activity of SIRT2 (Sun *et al* 2022, Shabkhizan *et al* 2023). When intracellular ATP and glucose levels fall below normal values, the increase in AMP favors allosteric modifications in AMPK, thereby enhancing its enzymatic activity (Zong *et al* 2019). Thus, AMPK activity limits mTORC1 and protein biosynthesis to attenuate ATP expenditure through the coordination of anabolic and catabolic programs (Zhang *et al* 2022). A subsequent decrease in fructose-1,6-bisphosphate concentrations results in the activation of lysosomal AMPK through the vacuolar H-ATPase–aldolase complex. In parallel with these alterations, the AXIN vacuolar H-ATPase complex modulates the activity of LKB1, leading to the phosphorylation of AMPK (Zhang *et al* 2022). To reduce fatty acid synthesis to a minimum, AMPK recruits acetyl-CoA carboxylases to induce fatty acid oxidation. In addition, phosphorylation of SREBP-1c through AMPK inhibits fatty acid production at the expression level (Zhang *et al* 2022). AMPK-dependent direct phosphorylation of ULK1, which is the orthologue of yeast Atg1, and BCLN1 triggers multiple Atgs involved in the autophagy signaling cascade (Kim *et al* 2013).

4.16.2.2 Autophagy control through insulin receptor signaling

The insulin signaling cascade is a key regulator of anabolic and catabolic processes in health and disease (Poloz and Stambolic 2015). The hormone insulin is able to attach to membrane-associated receptors INSR-1, -2 and IGF1R, which exhibit tyrosine kinase activity, thereby triggering messages connected with energy homeostasis (Saltiel and Kahn 2001). In contrast to hypoglycemic states, activation of INSR-1, -2 and IGF1R enhances glucose utilization through the PI3K/Akt signaling route (Saltiel and Kahn 2001, Poloz and Stambolic 2015). Together with these alterations, the mechanisms of glycolysis, glycogenesis and lipogenesis are triggered through the activation of phosphofructokinase-2 and the blocking of GSK3β (Saltiel and Kahn 2001, Poloz and Stambolic 2015). Gluconeogenesis is decreased by the availability of insulin after phosphorylation of FOXO1 (Kim and Spiegelman 1996, Porstmann *et al* 2008). Evidence shows that insulin impairs the autophagy machinery through mTORC1 activation, ULK1 phosphorylation, and repression of FOXO1 expression (Frendo-Cumbo *et al* 2021). Activation has been shown to modulate phosphorylation of mTORC1 and FOXO1/3 factors, suggesting a reciprocal crosstalk between autophagy and the insulin signaling cascades (Frendo-Cumbo *et al* 2021). The results revealed a significant upsurge in the expression of p53, p21, p16, and β-galactosidase in chondrocytes that were exposed to hyperglycemic circumstances (Wang *et al* 2021). Elevating glucose levels enhances concentrations of intracellular p62 and ROS, which occurs at the same time as the repression of LC3II/I and the phosphorylated AMPK/AMPK ratio (Wang *et al* 2021). Naturally, it is important to note that the autophagy mechanism is activated during the early phases after subjecting cells to high glucose concentrations. Nevertheless, it is conceivable that the persistence of hyperglycemia could trigger an aberrant (excessive) autophagic reaction (Rezabakhsh *et al* 2017, Nakai *et al* 2020). Collectively, these findings indicated shared activators of glucose metabolism, the insulin signaling cascade, and the autophagy reaction, as well as their mutual interplay across various circumstances.

4.16.2.3 Function of intermittent fasting on autophagic reaction in pathological settings

The obvious and early activity of the autophagy mechanism has been observed in a number of pathological conditions (Lim *et al* 2021). In accordance with this assertion, indications of the involvement of autophagy as a protective reaction in the early phases have been found in various neurodegenerative disorders like Huntington's disease, Alzheimer's disease and Parkinson's disease (Chandran and Rochet 2022). As neurodegenerative diseases advance, the build-up of aberrantly folded proteins and peptides leads to an elevated risk of proteotoxicity (Bishop *et al* 2010, Cuervo *et al* 2010). The physiological role of autophagy is associated with the removal of harmful proteins and the prevention of neuronal proteotoxicity (Fleming *et al* 2011). According to these observations, it is assumed that the activation of autophagy by various signals, including regular food shortages and different exogenous substances, is a therapeutic option for patients (Alirezaei *et al* 2010, Krishnan *et al* 2020). It has been revealed that short-term fasting of 24 to 48 h triggers an autophagic reaction in hepatic tissues and neuronal cells in murine models (Alirezaei *et al* 2010). On the basis of the findings, the amount of autophagosomes in liver cells, which are identified as GFP-LC3 spots, rose in the first 24 h following food limitation, reaching maximum concentrations levels after

48 h (Alirezaei *et al* 2010). A similar trend has been seen in the cortical neurons of mice with limited food intake compared to those of the control group. Autophagosome numbers and sizes were elevated in mice with limited food consumption. Remarkably, these circumstances can foster the retrograde transport of autophagosomes away from the neurites back to the cell soma, pointing to the generation of autophagolysosomes and their lysosomal breakdown (Alirezaei *et al* 2010). Equally, the presence of perinuclear reticular patterns in Purkinje cells pointed to active autophagosome-lysosome merging after nutritional deprivation (Alirezaei *et al* 2010). Therefore, it has been hypothesized that in the case of non-fed mice, decreased p70S6 kinase activity and phosphorylated S6RP are related to the suppression of mTOR and the autophagic reaction (Alirezaei *et al* 2010). These results demonstrate that dietary caloric limitation can ameliorate neuronal damage linked to proteotoxicity and alterations in intracellular inclusion clearance through the lysosomal catabolic pathway. In favor of this assumption, SIRT1-driven deacetylation of FoxO influences the activity of ROCK1, thereby enhancing α-secretase and decreasing Aβ plaque deposition. (Yang and Zhang, 2020). It has been proven that α-secretase, a metalloproteinase, can cleave the amyloid precursor protein inside the Aβ domain and insoluble Aβ can be aggregated within the ECM (Lichtenthaler 2012). These findings suggest that calorie limitation may have neuroprotective potential in neurological disorders through the effects of autophagy modulation. Toxic protein deposits can be cleared through therapeutic drugs or specific diets that aim at the autophagy pathway. Since calorie limitation is a cost-effective and convenient alternative to medication and available drug treatments, it appears to be a viable approach for both prophylactic and therapeutic purposes in patients with neurological disturbances. Notwithstanding the positive impact of food limitations in mitigating neurodegenerative diseases, there are some contradictory and inconclusive findings (Le Bras 2020, Milošević *et al* 2021). Feeding 5XFAD mice (AD model) every other day not only failed to decrease plaque deposition in the brain parenchyma, but also enhanced cortical neuroinflammatory reaction and decreased protein levels of factors linked to synaptic plasticity. (Le Bras 2020). This obvious inconsistency may be due to multiple causes, including food ingredients, experimental procedures, duration of dietary constraints, genetic characteristics, and age of the groups examined (Rodriguez-Navarro *et al* 2012). A large number of studies have pointed to impaired autophagy and decreased lysosomal degradation during calorie limitation (Sun *et al* 2020). Admittedly, a calorie limitation strategy may not be a suitable approach for reducing intracellular inclusion bodies in elderly people due to the impairment of the autophagy mechanism. Moreover, cells can adjust by controlled energy decline, acute depletion in the short term or extensive ATP decline in the long term, thereby preventing vacuolar hydrolysis. In these situations, macroautophagy is an essential part of the advancement of pathological alterations (Lang *et al* 2014). Self-extensive and hostile ATP breakdown pushes cells towards necrotic alterations rather than apoptosis and/or autophagic cell death (Pajuelo *et al* 2018). Excessive autophagy flow activated through fast ATP breakdown overwhelms lysosomal clearance capacity, leading to autophagosome accumulation and subsequently cell death (Zhu *et al* 2020,

Duan *et al* 2022). The fact that certain treatment programs are not suitable for use in the case of long-term fasting under *in vivo* circumstances should not be overlooked. For instance, in rats that fasted overnight, local calcium levels in the brain increased at an abnormal rate when administered systemically, with the reintroduction of glucose leading to a restoration of brain calcium concentrations to normal levels (Hanahisa and Yamaguchi 1996). A cause for this effect could be that chronic hunger can result in a leaky blood–brain barrier due to the fast ATP degradation and the disruption of cellular homeostasis. Therefore, caution is advised when recommending starvation diets for conventionally treated patients.

As with the central nervous system, the autophagy flow in the heart is critical to cardiac performance and homeostatic maintenance of cardiomyocytes (Kirshenbaum 2012). The induction of AMPK in ischemic cardiomyocytes represents a protective strategy to rescue damaged cells (Sciarretta *et al* 2011, Bagherniya *et al* 2018). Intermittent fasting in endothelial cells confers protective functional activity and decreases the level of damage in obese individuals through the modulation of specific biomarkers and the autophagic reaction (Maday and Holzbaur 2016). The cardioprotective effect of intermittent fasting has been shown in a mouse model of ischemia-reperfusion damage (Godar *et al* 2015). Six weeks of fasting every other day enhanced autophagy flow, whereby the autophagosomes are increased, and decreased infarct size relative to non-fasted mice. It was hypothesized that the impairment of autophagosome-lysosome merging caused by chloroquine treatment would cause the build-up of LC3-II and SQSTM1/p62 (Godar *et al* 2015). Lysosomal Lamp2 repression enhanced punctate GFP-LC3 and SQSTM1/p62, thus intermittent fasting exacerbates ventricular restructuring and cardiomyocyte toxicity in Lamp2$^{-/-}$ mice (Godar *et al* 2015). These features suggest that activating the lysosomal clearance mechanism through autophagy after intermittent fasting may have regenerative effects on cardiac tissue with the onset of ischemic alterations. It is noteworthy that in fasting cardiomyocytes, active Atg5 expression and autophagic flow help to confer protection against ischemic challenges (El Agaty *et al* 2022). In elderly rats exposed to acute myocardial infarction induced by isoproterenol, four weeks of intermittent fasting resulted in the induction of Atg5 and a decrease in cardiac creatine kinase, MDA, TNF-α, and FBS (El Agaty *et al* 2022). In liver tissue, autophagy is of great significance due to its special metabolic characteristics and its specific function (Amir and Czaja 2011). It is thus logical to assume that mildly insulting circumstances and metabolic disturbances can result in the activation of autophagy in hepatocytes (Hazari *et al* 2020). An inadequate autophagic reaction can cause various disorders including non-alcoholic steatohepatitis, viral illnesses and certain carcinomas to advance (Amir and Czaja 2011). It has been proposed that an 8-week fasting treatment of 72 h per week can decrease hepatic fat content in mice maintained on a high-fat diet through the activation of LC3, LAMP1 and BCLN1 in comparison to mice fed on food pellets (Chaudhary *et al* 2022). There were no significant variations in the autophagy levels in skeletal muscles (Chaudhary *et al* 2022). A reason for this could be that the liver is the principal location for ketone body generation and build-up in starvation and there is a strong correlation between the autophagy level and ketone body concentrations

(Takagi *et al* 2016). The transcriptional downregulation of Atg7 is linked to the intracellular clustering of NCoR1, which is a PPAR-α coregulator, the impairment of PPAR and the attenuation of fatty acid β-oxidation (Saito *et al* 2019). Thus, the sensitivity of tissues toward intermittent fasting can be distinct, whereby the autophagic reactions may differ (Chaudhary *et al* 2022). It is noteworthy that intermittent fasting engages a variety of cues to trigger adaptive autophagy in reaction to ischemic circumstances (Rickenbacher *et al* 2014). In fasted mice exposed to ischemic-reperfusion damage to the liver tissue, the time-dependent protective action of intermittent fasting cannot be overlooked (Rickenbacher *et al* 2014). The transcription of pro-inflammatory genes like TNF-α, IL1-β, IL-6 and Ccl2 and elevated systemic HMGB1 levels were attenuated in mice with lesioned hepatocytes that fasted for 1 day, while 2- and 3-day fasting regimens had no beneficial impact (Rickenbacher *et al* 2014). These actions are linked to the modulation of resident Kupffer cells in liver tissue and infiltration of neutrophils in reaction to ischemic damage. Together with these alterations, the levels of ROS and TNF-α were not elevated in LPS-exposed Raw 264–7 mouse macrophages subjected to declining fetal calf serum levels (Rickenbacher *et al* 2014). The data showed an elevation of LC3, BCLN1, SIRT1, and HMGB1 intracellular accumulation in the fasting mice in comparison with the control groups (Rickenbacher *et al* 2014). It is assumed that the autophagy mechanism breaks down the pro-inflammatory HMGB1 and prevents its continued release into the circulation and the subsequent activation of phagocytes, such as macrophages (Rickenbacher *et al* 2014, Yang *et al* 2020). Thus, it can be proposed that regular fasting is able to trigger the activation of the autophagic reaction in ischemic hepatocytes, whereby direct breakdown of pro-inflammatory substances and impairment of their liberation into the ECM is permitted. In addition, there is another potential impact of intermittent fasting on hormonal control of the breakdown performance of lysosomes (Han *et al* 2021). Increased systemic FGF21 has been implicated to increase β-oxidation rate of fatty acids and autophagic reaction (increased GFP-LC3 punctate structures) in mice undergoing 12- and 24-h fasting procedures (Chen *et al* 2017). When studying $Fgf21^{-/-}$ mice, it has been found that intracellular LC3-II and p62 levels were markedly increased, concurrent with impairment of Lamp1 and lysosomal-autophagic signaling cascade (Chen *et al* 2017). In wild-type mice, fasting can block mTORC1, resulting in a decline in phosphorylated TFEB and subsequent accumulation in the cytosol. Dephosphorylated TFEB enters the nucleus and induces autophagy and stimulates lysosomal activity. It is remarkable to point out that the removal of Fgf21 can enhance the activity of Mid1 and increase the phosphorylation of TFEB. Due to the complex coupling of fasting and lipid metabolism, it can be assumed that intermittent fasting can direct the autophagic flux to govern lysosomal functionality. It has recently been postulated that calorie limitation together with the induction of autophagy may modify the mode of cell death and attenuate the resulting pathological consequences (Sharma *et al* 2022). It was pointed out that stimulating autophagy through rapamycin can alleviate the severity of acute pancreatitis caused by caerulein in mice by raising caspase-3 levels, which is a protein linked to apoptosis, and lowering HMGB1 protein levels, which is a protein coupled to necrosis. Amylase activity and the systemic amounts of pro-inflammatory factors including IL-6, TNF-α, MCP-1 and GM-CSF were also decreased.

Myocytes are fundamentally distinct from other cells because they have a specialized program to generate energy through the breakdown of proteins into amino acids when they are under stress (Masiero *et al* 2009). Autophagolysosomes are hence fundamental sites of catabolism for the generation of essential amino acids (Masiero *et al* 2009). The results are in line with the notion that fasting can promote the liberation of phenylalanine from skeletal muscle along an mTOR-dependent route. In addition, fasting can inhibit mTORC1 activity, thereby decreasing phosphorylated ULK1 and 4EBP1 proteins. These changes are accompanied by a simultaneous rise in intracellular amounts of LC3B, p62 and FOXO3a (Vendelbo *et al* 2014). Other results have revealed opposite impacts of the presence of insulin that can reverse these circumstances and consequently lead to near-to-normal situations (Shabkhizan *et al* 2023).

4.16.2.4 Intermittent fasting and role of autophagy in cancers

Numerous experiments have pointed to the critical importance of calorie limitation and intermittent fasting in the onset and growth of cancer (Castejón *et al* 2020, Gray *et al* 2020). Building on this, various types of caloric restraints, including dietary restrictions, intermittent fasting, and caloric limitation, have been used to curb the development and advancement of various kinds of cancer. In addition, the 'calorie restriction mimetics' with restricted calories provide a low-calorie diet that does not require cancer patients to follow stringent diets (Vidoni *et al* 2021). Several lines of evidence indicated that low-energy nutrition per se can impact the metabolism and microenvironment of cancer cells, thereby causing a decline in growth and metastasis levels (Vidoni *et al* 2021). It is worth noting that low-carbohydrate, high-fat, and high-protein nutrition has several advantages for cancer patients with regard to survival rate and the tolerability of chemotherapy adverse effects (Klement *et al* 2020). Amino acids like methionine, leucine, glutamine and arginine, as well as glucose, are crucial for the dynamic growth of cancer cells and motile behavior in relation to normal cell types. These characteristics transform cancer cells into frontier cells when subjected to calorie limitation and hunger (Vidoni *et al* 2021). Nevertheless, the precise impact of intermittent fasting on anaplastic diseases is still largely unclear and is the subject of current extensive investigations (Wilhelmi De Toledo *et al* 2013, Nencioni *et al* 2018). Naturally, calorie limitation by itself is not a definitive cancer therapy and is regarded as a complementary approach to enhance the therapeutic effectiveness of chemotherapeutic agents (Tang *et al* 2021), although numerous animal model studies have demonstrated the cancer-preventing benefits of intermittent fasting (Caccialanza *et al* 2018). Because of the indispensable part intermittent fasting programs perform in the dynamic proliferation of cancer cells, recent clinical trials have concentrated on the implementation of specific fasting programs in cancer patients using specially designed and tolerated diets (Emmons and Colditz, 2017, Jaffee *et al* 2017, Kerr *et al* 2017). Clinical findings have indicated that insulin and certain other compounds including IGF-1 and glucose are decreased in individuals subjected to 5 days of fasting imitating nutrition (Wei *et al* 2017). In a mouse model, such a metabolic phenotype can decrease cancer incidence, obesity and inflammatory disorders (Brandhorst *et al* 2015). As mentioned earlier, intermittent fasting and calorie limitation after cellular stress caused by harmful circumstances can trigger autophagy signals, causing mTORC1 to be disabled (Antunes *et al* 2018, Chung and Chung 2019, Chung *et al* 2020). Anti-tumor mechanisms like E-cadherin, adiponectin/leptin relationship, metalloproteinases, chemokines and cytokines, caspase-3 and histone H2AX, which represents an enzyme linked to DNA damage repair, are increased upon fasting (Messaoudi *et al* 2006, Lu *et al* 2008, Orgel and Mittelman 2013, Dong *et al* 2014, Dogan *et al* 2017). In addition, *in vitro* and *in vivo* diet plans can tailor the sensitivity of the host cell toward insulin (Dao *et al* 2019).

Data from preclinical studies have demonstrated that short-term caloric deprivation of cancer patients prior to or during radiation treatment can potentiate anticancer effects (Icard *et al* 2020). Notably, non-calorie-restricted ketogenic dietary regimens appear to lead to contradictory results in patients receiving caloric restraint and radiotherapy when compared to patients receiving isocaloric dietary regimens (Icard *et al* 2020). *In vitro* analyses showed that culturing various cancer cells including HepG2 [hepatocellular carcinoma] and HuH6-clone5 [hepatoblastoma cells] in a serum-free environment for 6–24 h enhanced radiosensitivity by activating mTOR and causing an overabundance of ROS (Murata *et al* 2015). Exposure of human A549 cells to hypoglycemic levels (2.8 mmol l^{-1}) for 24 h decreased clonogenicity and augmented radiation-induced DNA fragmentation, while normal HSF7 fibroblasts displayed no susceptibility to glucose shortage (Ampferl *et al* 2018). The reason for this difference could be that the baseline metabolic state is carefully monitored by repressive checkpoints like SIRT3 and PTEN. in reaction to energy stress. This gives normal cells flexibility, while these factors are often silent in cancer cells (Haigis *et al* 2012). It is noteworthy that the kind of nutrition and the duration of calorie limitation can impact the dynamic proliferation of cancer cells (Icard *et al* 2020). Notwithstanding the stimulating impact of ketogenic diets on the induction of the autophagy reaction and curative processes (Wang *et al* 2018), the application of calorie limitation and therapeutic ketogenic regimens to cancer sufferers may lead to even more contradictory results (Icard *et al* 2020). In the majority of cancer cells, the protein concentrations of enzymes that break down ketone bodies like SCOT and 3β-OHBD are lower compared to normal cells (Branco *et al* 2016), and ketogenic diets have been utilized to avoid weight reduction in cancer patients receiving medical treatment (Weber *et al* 2020). It has been hypothesized that certain cancer cells from various tissues including blood, breast and brain, can enhance the activity of the enzymes SCOT and 3β-OHBD to break down fatty acids and thereby promote tumor growth. Therefore, incidental caloric deprivation in cancer patients receiving therapy may introduce extra distortion (Fine *et al* 2012). A 6 h fasting experiment in rats receiving a ketogenic meal for 4 weeks caused a decrease in LC3II and ULK and a rise in serum creatinine and kidney fibrosis (Jia *et al* 2021). High levels of triglycerides and cholesterol can interfere with the merging of autophagosomes and lysosomes, leading to autophagosome aggregation and an unfinished autophagic reaction. In addition to these consequences, an excess of lipids in the presence of inadequate glucose intake can result in impaired glycolipid-protein metabolism and perhaps cause abnormalities in vascular tissue (Jia *et al* 2021). The activation of autophagy in cancer cells after intermittent fasting can have both harmful and positive effects. The autophagy mechanism prevents cell injury through the stimulation of DNA damage repair mechanisms and antioxidant mechanisms, while under certain circumstances it can speed up the cell death of cancer cells. Importantly, autophagy activation in specific cancer cells is also linked to proliferation and metastatic spread (Rodríguez-Vargas *et al* 2012, Zhang *et al* 2015). These findings suggest that autophagy has a dual impact on dynamic tumor evolution. It should be clarified as to whether and how autophagy regulates the protecting and impairing mechanisms in cancer cells. The suppression of BCLN1 is associated with the development of several types of cancer in tissues like the ovaries, breasts and testis. The removal of BCLN1 in mice resulted in the development of liver and lung cancer due to a partly inadequate autophagic reaction (Antunes *et al* 2018). It is remarkable that BCLN1 and BRCA1, which is a tumor suppressor gene, are in close proximity to one another and that a BRCA1 deletion in breast cancer patients can impact BCLN1 (Laddha *et al* 2014). Notably, specific cancer cells, like several Ras-transformed anaplastic alterations in the bladder, pancreas, and colon, display stimulated autophagic reactions which are linked to normal cells. Consequently, controlling autophagy is a key element in combating these cancers (Guo *et al* 2011). The activation of the GTPase KRAS is closely linked to autophagy. In certain cancer cells, the mutation of Ras genes, like H-RasV12 or K-RasV12, can cause enhanced autophagic flux notwithstanding of the activation of mTORC1 (Guo *et al* 2011). It has been found that a limited calorie supply can predispose cancer cells to apoptosis through the triggering of $CD8^+$ lymphocytes (Ibrahim *et al* 2021). In a fasting state, blocking mTORC1 via AMP/ATP↑ upregulation, can result in the stimulation of autophagy, the advancement of ketogenesis and the blocking of glycolysis and glutaminolysis within cancer cells (Okawa *et al* 2021). Calorie limitation can suppress the PI3K/Akt/mTOR pathway and thereby glycolysis in cancer cells by decreasing serum IGF-1 levels (Okawa *et al* 2021). It is striking that the decrease in glycolytic activity in cancer cells is a major factor in reducing the proliferation rate (Pouyafar *et al* 2019). Together with these alterations, the degree of differentiation in the direction of Th1, Th17 lymphocytes and M1 macrophages is elevated (Okawa *et al* 2021). Intermittent fasting can reduce the frequency of regulatory $CD3^+/CD4^+/CD25^+$ T cells within the tumor stroma (Di Biase *et al* 2016). In reaction to the calorie limitation, the sensitivity of NK cells towards cancer cells is increased.

Remarkably, the activation of specific oncogenes like PI3K, Akt1 and Bcl2 inhibits the autophagic reaction. The repression of certain proteins (DAPK, PTEN, TSC1, TSC2 and LKB1/STK11) is linked to the lysosomal breakdown capacity in cancer cells (Maiuri *et al* 2009). When cells are subjected to calorie limitation and a shortage of adequate ATP, they attempt to accommodate harmful situations, such as chemotherapy, through the induction of autophagy, a process that is not induced in cancer cells, leaving them vulnerable to cell damage (Lum *et al* 2005). The effect of calorie limitation of 70% of normal food consumption on the dynamic proliferative activity of human colorectal HCT116 cancer cells in a mouse xenograft tumor model was examined (Fu *et al* 2019). The inhibition of the activity of the cysteine protease ATG4B through S130 together with calorie limitation in tumor-bearing mice was hypothesized to cause the aggregation of LC3-II, which is an increased delipidated LC3, and p62 in the cancer cells, which led to apoptotic cell death through caspase-3 activity (Fu *et al* 2019, Shabkhizan *et al* 2023). *In vitro* analyses showed that culturing MCF-7 and HCT116 cells under calorie-limited circumstances resulted in the induction and activation of the protein kinase CK2, such as the isoforms CK2α and CK2β, SIRT1 and phosphorylated AMPK (Park *et al* 2022). These characteristics are concordant with the activation of the molecular autophagy mechanism, such as Atg5, Atg7, LC3BII, BCLIN1 and Ulk1. Calorie limitation can enhance the siRNA inhibitory efficiency on protein kinase CK2 and the suppression of autophagic response in these cancer cell lines (Park *et al* 2022). The data corroborated that Atg5þ/þ kidney epithelial cell growth was decreased in calorie-limited diet-fed nude mice as compared to the standard diet group (Lashinger *et al* 2016). In this context, it has been suggested that the decrease in systemic glucose and amino acids and the concomitant rise in ketone bodies can override the normal autophagic pathway (Lashinger *et al* 2016). Strikingly, the dynamic expansion of Atg5$^{-/-}$ renal epithelial cells was severely attenuated in conditions mimicking calorie limitation. These results demonstrate that the combination of calorie limitation and autophagy regulation can influence cancer cell proliferation even under *in vivo* circumstances. While calorie limitation can attenuate the proliferation of Atg5-expressing cancer cells, these beneficial outcomes are exacerbated by the concomitant impairment of the molecular autophagy pathway. It has been found that fasting every other day for a period of 14 days can exhibit tumoricidal activity against the CT26 cell line of murine colorectal carcinoma (Sun *et al* 2017b). The data demonstrated that tumor volume and M2 polarization of tumor-associated macrophages were decreased with caloric restraint. Together with the triggering of autophagy, such as increased Atg5 and enhanced LC3II/I ratio, in fasting, the expression of CD73 and the generation of adenosine were additionally reduced (Sun *et al* 2017b). Induction of excessive ROS and nitrogen species generation using cold atmospheric pressure plasma was recently shown to result in autophagic and apoptotic cell death of B16 melanoma cells following subcutaneous inoculation into mice (Golpour *et al* 2022). A 24 h starvation of mice, such as three times per five days, resulted in marked tumoricidal activity in comparison to mice subjected to cold atmospheric pressure plasma. This finding indicates that mechanical stimulation can cause a similar effect on cancer cells as conventional stimuli, such as starvation excessive autophagy, which is indicated by increased levels of LC3 and ATG5, triggered apoptotic alterations that occurred concurrently with the induction of pro-apoptotic proteins like Bax and caspase-3 (Golpour *et al* 2022). The overactivity of Atg5 favors its splitting by calpain and subsequent interaction with Bcl-X, leading to apoptosis (Codogno and Meijer 2006).

4.17 Conclusions and future directions

Many trials have demonstrated that the efficacy of existing treatments can be enhanced by influencing autophagy. Nevertheless, there are some remaining challenges. The first thing is to fully comprehend the interaction between autophagy and cancer resistance (Lorente *et al* 2018, Li *et al* 2020). There are established cases in which the stimulation of autophagy impeded the expansion of a tumor. The issue is whether blocking autophagy is the correct strategy. Is the type of tumor relevant for the success of blocking autophagy? A deeper insight into the dual function of autophagy could pave the way for the design of new anti-cancer therapies (Vega-Rubín-de-Celis *et al* 2018).

Another problem is the design of new compounds that target autophagy in the face of drug-resistant development. Polytherapies, which utilize established chemotherapeutic agents in combination with new compounds, are an approach that may help address this issue (Bhat *et al* 2018, Pérez-Hernández *et al* 2019). Therapy with autophagic inhibiting drugs is also an outstanding problem, as the side reactions of autophagy blocking are not well characterized. Tissue injury was seen after autophagy inhibitor administration, but the suitable dosage of the drug is difficult to determine as it varies from patient to patient (Karsli-Uzunbas *et al* 2014).

Another question is whether the early stage of autophagy or the later stage of autolysosome breakdown might be more useful to block. Another issue is the paucity of suitable tools to track autophagy in cancer patients. The LC3 and SQSTM1/p62 are the sole markers that enable the surveillance of autophagy, but they are part of the process (Morgan and Thorburn, 2016). Currently, it is not possible to visualize the activity of the autophagy breakdown phase. In addition, the post-translational changes that have a major influence on the control of autophagy cannot be neglected. Hence, it is highly critical to identify new autophagy hallmarks that are not part of the autophagy pathway and incorporate possible post-translational modifications (Bortnik and Gorski 2017). CSCs are a further area of concern. CSCs may be strongly associated with the development of drug resistance in cancer cells. In theory, it should be feasible to sensitize cancer cells by simultaneously targeting autophagy and cancer stem cells, and this might even result in overcoming drug resistance (Nazio *et al* 2019, Smith and Macleod 2019). Although many questions and uncertainties remain, it can be assumed that autophagy is crucial for several mechanisms in cancer. Autophagy is involved in both the maintenance of cell proliferation following chemotherapy and their removal. Thus, it is crucial to fine-tune these functions of autophagy to block cell proliferation. There are high expectations for the utilization of autophagy to block the conversion of stem cells into CSCs, which could be the keystone to a successful anti-cancer treatment. For this reason, it is so crucial to seek out new approaches that can modulate this process to ensure that autophagy only acts as an anti-proliferative mechanism during effective chemotherapy.

Finally, the protective/detrimental impacts of autophagy modulated by calorie limitation were debated. The data demonstrate that adaptive and excessive autophagy can be triggered according to the intensity and length of calorie limitation regimes. It appears that a supervised calorie decline, or intermittent fasting can

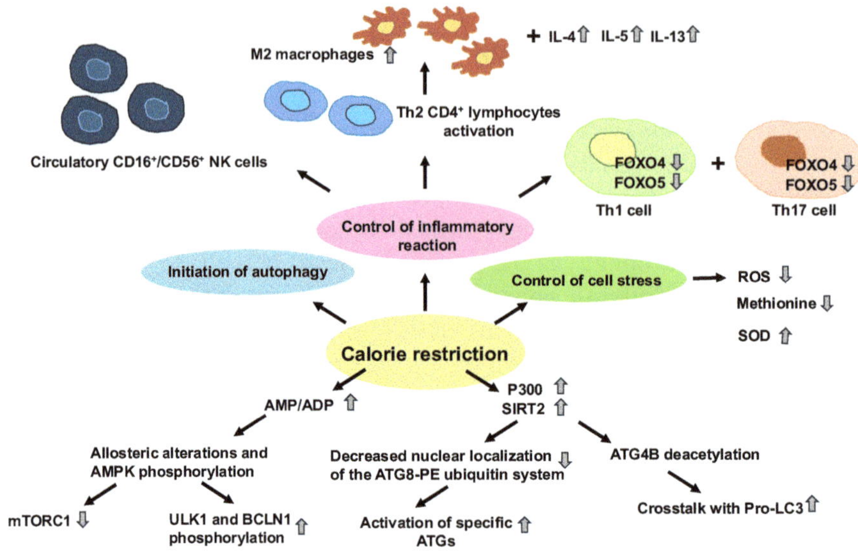

Figure 4.14. Potential effects of calorie restriction on autophagic response, cellular stress, and the activity and predominant subtype state of immune cells.

shield various cellular processes through an adjustment of autophagy as an answer to multiple pathological situations (figure 4.14). Through the activation of adaptive autophagy and subsequent important downstream events, intermittent fasting or calorie limitation can result in life-prolonging and anti-aging benefits through the establishment of an appropriate balance of homeostasis. A moderate calorie limitation for short- and long-term or intermittent fasting can be implemented without significant side effects (Napoleão *et al* 2021). Despite the benefits, insufficient energy consumption and dehydration are still potential risk factors in calorie-restricted volunteers and need to be thoroughly controlled.

Accordingly, an aberrant (excessive) autophagy reaction not only fails to protect the host cells, but also drives the cells into autophagic cell death and apoptosis. Many investigations are required to examine the fundamental mechanisms related to the effects of intermittent fasting on autophagy regulation. Although calorie restriction exerts a regulatory action on the autophagy cascade, the mechanisms underpinning the activation of adaptive/excessive autophagy have not been explored. This represents a frontier in cancer research. Advances in molecular biology will make it possible to observe alterations in the autophagy reaction in real time in individuals participating in various intermittent fasting routines. Due to the paucity of adequate data on the safety and side effects of intermittent fasting programs in patients, caloric limitation needs to be implemented with great prudence in clinical environments. Subsequently, pairing autophagy inducers and specific calorie limitations in certain pathological settings may assist clinicians in developing *de novo* therapeutic intervention strategies.

References and further reading

Al-Bari M A A 2020 A current view of molecular dissection in autophagy machinery *J. Physiol. Biochem.* **76** 357–72

Al-Hajj M, Wicha M S, Benito-Hernandez A, Morrison S J and Clarke M F 2003 Prospective identification of tumorigenic breast cancer cells *Proc. Natl Acad. Sci. USA* **100** 3983–8

Alirezaei M, Kemball C C, Flynn C T, Wood M R, Whitton J L and Kiosses W B 2010 Short-term fasting induces profound neuronal autophagy *Autophagy* **6** 702–10

Amaravadi R K 2012 Autophagy and tumor cell invasion *Cell Cycle* **11** 3718–8

Amini H, Rezabakhsh A, Heidarzadeh M, Hassanpour M, Hashemzadeh S, Ghaderi S *et al* 2021 An examination of the putative role of melatonin in exosome biogenesis *Front. Cell Dev. Biol.* **9** 686551

Amir M and Czaja M J 2011 Autophagy in nonalcoholic steatohepatitis *Expert Rev. Gastroenterol. Hepatol.* **5** 159–66

Ampferl R, Rodemann H P, Mayer C, Höfling T T A and Dittmann K 2018 Glucose starvation impairs DNA repair in tumour cells selectively by blocking histone acetylation *Radiother. Oncol.* **126** 465–70

An W, Zhang Y, Lai H, Zhang Y, Zhang H, Zhao G *et al* 2022 *Alpinia katsumadai Hayata* induces growth inhibition and autophagy-related apoptosis by regulating the AMPK and Akt/mTOR/p70S6K signaling pathways in cancer cells *Oncol. Rep.* **48** 142

Antunes F, Erustes A G, Costa A J, Nascimento A C, Bincoletto C, Ureshino R P *et al* 2018 Autophagy and intermittent fasting: the connection for cancer therapy? *Clinics* **73** e814s

Arias E and Cuervo A M 2011 Chaperone-mediated autophagy in protein quality control *Curr. Opin. Cell Biol.* **23** 184–9

Askari H, Sadeghinejad M and Fancher I S 2023 Mechanotransduction and the endothelial glycocalyx: interactions with membrane and cytoskeletal proteins to transduce force *Current Topics in Membranes* (Amsterdam: Elsevier) pp 43–60

Bagherniya M, Butler A E, Barreto G E and Sahebkar A 2018 The effect of fasting or calorie restriction on autophagy induction: a review of the literature *Ageing Res. Rev.* **47** 183–97

Bago R, Malik N, Munson M J, Prescott A R, Davies P, Sommer E *et al* 2014 Characterization of VPS34-IN1, a selective inhibitor of Vps34, reveals that the phosphatidylinositol 3-phosphate-binding SGK3 protein kinase is a downstream target of class III phosphoinositide 3-kinase *Biochem. J.* **463** 413–27

Bai Y, Chen Y, Chen X, Jiang J, Wang X, Wang L *et al* 2019 Trichostatin A activates FOXO1 and induces autophagy in osteosarcoma *Arch. Med. Sci.* **15** 204–13

Bedoya V 1970 Effect of chloroquine on malignant lymphoreticular and pigmented cells *in vitro* *Cancer Res.* **30** 1262–75

Bellot G, Garcia-Medina R, Gounon P, Chiche J, Roux D, Pouysségur J *et al* 2009 Hypoxia-induced autophagy is mediated through hypoxia-inducible factor induction of BNIP3 and BNIP3L via their BH3 domains *Mol. Cell. Biol.* **29** 2570–81

Bezuidenhout N and Shoshan M 2019 A shifty target: tumor-initiating cells and their metabolism *Int. J. Mol. Sci.* **20** 5370

Bhat P, Kriel J, Shubha Priya B, Basappa , Shivananju N S and Loos B 2018 Modulating autophagy in cancer therapy: advancements and challenges for cancer cell death sensitization *Biochem. Pharmacol.* **147** 170–82

Bishop N A, Lu T and Yankner B A 2010 Neural mechanisms of ageing and cognitive decline *Nature* **464** 529–35

Blagosklonny M V 2012 Rapalogs in cancer prevention: anti-aging or anticancer? *Cancer Biol. Ther.* **13** 1349–54

Bortnik S and Gorski S M 2017 Clinical applications of autophagy proteins in cancer: from potential targets to biomarkers *Int. J. Mol. Sci.* **18** 1496

Bovellan M, Fritzsche M, Stevens C and Charras G 2010 Death-associated protein kinase (DAPK) and signal transduction: blebbing in programmed cell death *FEBS J.* **277** 58–65

Branco A F, Ferreira A, Simões R F, Magalhães-Novais S, Zehowski C, Cope E *et al* 2016 Ketogenic diets: from cancer to mitochondrial diseases and beyond *Eur. J. Clin. Invest.* **46** 285–98

Brandhorst S, Choi I Y, Wei M, Cheng C W, Sedrakyan S, Navarrete G *et al* 2015 A periodic diet that mimics fasting promotes multi-system regeneration, enhanced cognitive performance, and healthspan *Cell Metab.* **22** 86–99

Bursch W 2004 Multiple cell death programs: Charon's lifts to Hades *FEMS Yeast Res.* **5** 101–10

Caccialanza R, Cereda E, De Lorenzo F, Farina G and Pedrazzoli POn behalf of the AIOM-SINPE-FAVO Working Group 2018 To fast, or not to fast before chemotherapy, that is the question *BMC Cancer* **18** 337

Cao Q-H, Liu F, Yang Z-L, Fu X-H, Yang Z-H, Liu Q *et al* 2016 Prognostic value of autophagy related proteins ULK1, Beclin 1, ATG3, ATG5, ATG7, ATG9, ATG10, ATG12, LC3B and p62/SQSTM1 in gastric cancer *Am. J. Transl. Res.* **8** 3831–47

Carbone F, La Rocca C, De Candia P, Procaccini C, Colamatteo A, Micillo T *et al* 2016 Metabolic control of immune tolerance in health and autoimmunity *Semin. Immunol.* **28** 491–504

Cassidy L D and Narita M 2022 Autophagy at the intersection of aging, senescence, and cancer *Mol. Oncol.* **16** 3259–75

Castejón M, Plaza A, Martinez-Romero J, Fernandez-Marcos P J, De Cabo R and Diaz-Ruiz A 2020 Energy restriction and colorectal cancer: a call for additional research *Nutrients* **12** 114

Cechakova L, Ondrej M, Pavlik V, Jost P, Cizkova D, Bezrouk A *et al* 2019 A potent autophagy inhibitor (Lys05) enhances the impact of ionizing radiation on human lung cancer cells H1299 *Int. J. Mol. Sci.* **20** 5881

Chae Y C and Kim J H 2018 Cancer stem cell metabolism: target for cancer therapy *BMB Rep.* **51** 319–26

Chaikuad A, Koschade S E, Stolz A, Zivkovic K, Pohl C, Shaid S *et al* 2019 Conservation of structure, function and inhibitor binding in UNC-51-like kinase 1 and 2 (ULK1/2) *Biochem. J.* **476** 875–87

Chandran A and Rochet J-C 2022 Shining a light on autophagy in neurodegenerative diseases *J. Biol. Chem.* **298** 101437

Chang K-C, Liu P-F, Chang C-H, Lin Y-C, Chen Y-J and Shu C-W 2022 The interplay of autophagy and oxidative stress in the pathogenesis and therapy of retinal degenerative diseases *Cell Biosci.* **12** 1

Chao X, Wang S, Fulte S, Ma X, Ahamed F, Cui W *et al* 2022 Hepatocytic p62 suppresses ductular reaction and tumorigenesis in mouse livers with mTORC1 activation and defective autophagy *J. Hepatol.* **76** 639–51

Chaterjee M and Van Golen K L 2011 Breast cancer stem cells survive periods of farnesyl-transferase inhibitor-induced dormancy by undergoing autophagy *Bone Marrow Res.* **2011** 1–7

Chaudhary R, Liu B, Bensalem J, Sargeant T J, Page A J, Wittert G A *et al* 2022 Intermittent fasting activates markers of autophagy in mouse liver, but not muscle from mouse or humans *Nutrition* **101** 111662

Chen C-L, Uthaya Kumar D B, Punj V, Xu J, Sher L, Tahara S M *et al* 2016 NANOG Metabolically reprograms tumor-initiating stem-like cells through tumorigenic changes in oxidative phosphorylation and fatty acid metabolism *Cell Metab.* **23** 206–19

Chen H and Chan D C 2017 Mitochondrial dynamics in regulating the unique phenotypes of cancer and stem cells *Cell Metab.* **26** 39–48

Chen L, Wang K, Long A, Jia L, Zhang Y, Deng H *et al* 2017 Fasting-induced hormonal regulation of lysosomal function *Cell Res.* **27** 748–63

Chen N and Karantza V 2011 Autophagy as a therapeutic target in cancer *Cancer Biol. Ther.* **11** 157–68

Chen S, Lewallen M and Xie T 2013 Adhesion in the stem cell niche: biological roles and regulation *Development* **140** 255–65

Chen S, Rehman S K, Zhang W, Wen A, Yao L and Zhang J 2010 Autophagy is a therapeutic target in anticancer drug resistance *Biochim. Biophys. Acta (BBA)—Rev. Cancer* **1806** 220–9

Chmurska A, Matczak K and Marczak A 2021 Two faces of autophagy in the struggle against cancer *Int. J. Mol. Sci.* **22** 2981

Choi K S 2012 Autophagy and cancer *Exp. Mol. Med.* **44** 109

Chung H Y, Kim D H, Bang E and Yu B P 2020 Impacts of calorie restriction and intermittent fasting on health and diseases: current trends *Nutrients* **12** 2948

Chung K W and Chung H Y 2019 The effects of calorie restriction on autophagy: role on aging intervention *Nutrients* **11** 2923

Codogno P and Meijer A J 2006 Atg5: more than an autophagy factor *Nat. Cell Biol.* **8** 1045–7

Cuervo A M 2004 Autophagy: many paths to the same end *Mol. Cell. Biochem.* **263** 55–72

Cuervo A M, Wong E S P and Martinez-Vicente M 2010 Protein degradation, aggregation, and misfolding *Mov. Disord.* **25** S49–54

Cufí S, Vazquez-Martin A, Oliveras-Ferraros C, Corominas-Faja B, Cuyàs E, López-Bonet E *et al* 2013 The anti-malarial chloroquine overcomes Primary resistance and restores sensitivity to Trastuzumab in HER2-positive breast cancer *Sci. Rep.* **3** 2469

Cui C, Merritt R, Fu L and Pan Z 2017 Targeting calcium signaling in cancer therapy *Acta Pharm. Sin.* B **7** 3–17

Dalby K, Tekedereli I, Lopez-Berestein G and Ozpolat B 2010 Targeting the pro-death and pro-survival functions of autophagy as novel therapeutic strategies in cancer *Autophagy* **6** 322–9

Daley G Q 2015 Stem cells and the evolving notion of cellular identity *Phil. Trans. R. Soc.* B **370** 20140376

Dao M C, Sokolovska N, Brazeilles R, Affeldt S, Pelloux V, Prifti E *et al* 2019 A data integration multi-omics approach to study calorie restriction-induced changes in insulin sensitivity *Front. Physiol.* **9** 1958

Das G, Shravage B V and Baehrecke E H 2012 Regulation and function of autophagy during cell survival and cell death *Cold Spring Harb. Perspect. Biol.* **4** a008813–a3

Di Biase S, Lee C, Brandhorst S, Manes B, Buono R, Cheng C-W *et al* 2016 Fasting-mimicking diet reduces HO-1 to promote T cell-mediated tumor cytotoxicity *Cancer Cell* **30** 136–46

Djavaheri-Mergny M, Maiuri M C and Kroemer G 2010 Cross talk between apoptosis and autophagy by caspase-mediated cleavage of Beclin 1 *Oncogene* **29** 1717–9

Dogan S, Ray A and Cleary M P 2017 The influence of different calorie restriction protocols on serum pro-inflammatory cytokines, adipokines and IGF-I levels in female C57BL6 mice: short term and long term diet effects *Meta Gene.* **12** 22–32

Dong S, Khoo A, Wei J, Bowser R K, Weathington N M, Xiao S *et al* 2014 Serum starvation regulates E-cadherin upregulation via activation of c-Src in non-small-cell lung cancer A549 cells *Am. J. Physiol. Cell Physiol.* **307** C893–9

Donnenberg V S, Huber A, Basse P, Rubin J P and Donnenberg A D 2016 Neither epithelial nor mesenchymal circulating tumor cells isolated from breast cancer patients are tumorigenic in NOD-scid Il2rgnull mice *npj Breast Cancer* **2** 16004

Dower C M, Bhat N, Gebru M T, Chen L, Wills C A, Miller B A *et al* 2018 Targeted inhibition of ULK1 promotes apoptosis and suppresses tumor growth and metastasis in neuroblastoma *Mol. Cancer Ther.* **17** 2365–76

Duan Y, Wang J, Wang J, Yang Q, Zhang Q, Lu S-Y *et al* 2022 Dual-enzyme catalytic nanosystem-mediated ATP depletion strategy for tumor elimination via excessive autophagy pathway *Chem. Eng. J.* **446** 136795

Dubrovska A, Kim S, Salamone R J, Walker J R, Maira S-M, García-Echeverría C *et al* 2009 The role of PTEN/Akt/PI3K signaling in the maintenance and viability of prostate cancer stem-like cell populations *Proc. Natl Acad. Sci. USA* **106** 268–73

Dyczynski M, Yu Y, Otrocka M, Parpal S, Braga T, Henley A B *et al* 2018 Targeting autophagy by small molecule inhibitors of vacuolar protein sorting 34 (Vps34) improves the sensitivity of breast cancer cells to Sunitinib *Cancer Lett.* **435** 32–43

El Agaty S M, Nassef N A, Abou-Bakr D A and Hanafy A A 2022 Chronic activation of cardiac Atg-5 and pancreatic Atg-7 by intermittent fasting alleviates acute myocardial infarction in old rats *Egypt Heart J.* **74** 31

Elmore S 2007 Apoptosis: a review of programmed cell death *Toxicol. Pathol.* **35** 495–516

Emmons K M and Colditz G A 2017 Realizing the potential of cancer prevention—the role of implementation science *N. Engl. J. Med.* **376** 986–90

Lichtenthaler S F 2012 Alpha-secretase cleavage of the amyloid precursor protein: proteolysis regulated by signaling pathways and protein trafficking *Curr. Alzheimer Res.* **9** 165–77

Fang Q, Liu X, Ding J, Zhang Z, Chen G, Du T *et al* 2022 Soluble epoxide hydrolase inhibition protected against diabetic cardiomyopathy through inducing autophagy and reducing apoptosis relying on Nrf2 upregulation and transcription activation *Oxid. Med. Cell. Longev.* **2022** 1–20

Fernando W, Cruickshank B M, Arun R P, MacLean M R, Cahill H F, Morales-Quintanilla F *et al* 2024 ALDH1A3 is the switch that determines the balance of ALDH+ and CD24–CD44+ cancer stem cells, EMT-MET, and glucose metabolism in breast cancer *Oncogene* **43** 3151–69

Fine E J, Segal-Isaacson C J, Feinman R D, Herszkopf S, Romano M C, Tomuta N *et al* 2012 Targeting insulin inhibition as a metabolic therapy in advanced cancer: a pilot safety and feasibility dietary trial in 10 patients *Nutrition* **28** 1028–35

Fleming A, Noda T, Yoshimori T and Rubinsztein D C 2011 Chemical modulators of autophagy as biological probes and potential therapeutics *Nat. Chem. Biol.* **7** 9–17

Fontana L and Partridge L 2015 Promoting health and longevity through diet: from model organisms to humans *Cell* **161** 106–18

Frendo-Cumbo S, Tokarz V L, Bilan P J, Brumell J H and Klip A 2021 Communication between autophagy and insulin action: at the crux of insulin action-insulin resistance? *Front. Cell Dev. Biol.* **9** 708431

Frisardi V, Matrone C and Street M E 2021 Metabolic syndrome and autophagy: focus on HMGB1 protein *Front. Cell Dev. Biol.* **9** 654913

Fu Y, Chang H, Peng X, Bai Q, Yi L, Zhou Y *et al* 2014 Resveratrol inhibits breast cancer stem-like cells and induces autophagy via suppressing Wnt/β-catenin signaling pathway *PLoS One* **9** e102535

Fu Y, Hong L, Xu J, Zhong G, Gu Q, Gu Q *et al* 2019 Discovery of a small molecule targeting autophagy via ATG4B inhibition and cell death of colorectal cancer cells *in vitro* and *in vivo* *Autophagy* **15** 295–311

Fulda S 2011 Targeting apoptosis signaling pathways for anticancer therapy *Front. Oncol.* **1** 23

García-Prat L, Sousa-Victor P and Muñoz-Cánoves P 2017 Proteostatic and metabolic control of stemness *Cell Stem Cell* **20** 593–608

Garten A, Schuster S, Penke M, Gorski T, De Giorgis T and Kiess W 2015 Physiological and pathophysiological roles of NAMPT and NAD metabolism *Nat. Rev. Endocrinol.* **11** 535–46

Gasmi M, Sellami M, Denham J, Padulo J, Kuvacic G, Selmi W *et al* 2018 Time-restricted feeding influences immune responses without compromising muscle performance in older men *Nutrition* **51–52** 29–37

Gaziova I and Bhat K M 2007 Generating asymmetry: ith and without self-renewal *Asymmetric Cell Division* ed A Macieira-Coelho (Berlin: Springer) pp 143–78

Glick D, Barth S and Macleod K F 2010 Autophagy: cellular and molecular mechanisms *J. Pathol.* **221** 3–12

Godar R J, Ma X, Liu H, Murphy J T, Weinheimer C J, Kovacs A *et al* 2015 Repetitive stimulation of autophagy-lysosome machinery by intermittent fasting preconditions the myocardium to ischemia-reperfusion injury *Autophagy* **11** 1537–60

Golden E B, Cho H-Y, Jahanian A, Hofman F M, Louie S G, Schönthal A H *et al* 2014 Chloroquine enhances temozolomide cytotoxicity in malignant gliomas by blocking autophagy *Neurosurg. Focus* **37** E12

Golpour M, Alimohammadi M, Sohbatzadeh F, Fattahi S, Bekeschus S and Rafiei A 2022 Cold atmospheric pressure plasma treatment combined with starvation increases autophagy and apoptosis in melanoma *in vitro* and *in vivo* *Exp. Dermatol.* **31** 1016–28

Gong C, Bauvy C, Tonelli G, Yue W, Deloménie C, Nicolas V *et al* 2013 Beclin 1 and autophagy are required for the tumorigenicity of breast cancer stem-like/progenitor cells *Oncogene* **32** 2261–72

González-Rodríguez P, Klionsky D J and Joseph B 2022 Autophagy regulation by RNA alternative splicing and implications in human diseases *Nat. Commun.* **13** 2735

Goodall M L, Wang T, Martin K R, Kortus M G, Kauffman A L, Trent J M *et al* 2014 Development of potent autophagy inhibitors that sensitize oncogenic BRAF V600E mutant melanoma tumor cells to vemurafenib *Autophagy* **10** 1120–36

Gray A, Dang B N, Moore T B, Clemens R and Pressman P 2020 A review of nutrition and dietary interventions in oncology *SAGE Open Med.* **8** 2050312120926877

Gujar A D, Le S, Mao D D, Dadey D Y A, Turski A, Sasaki Y *et al* 2016 An NAD^+-dependent transcriptional program governs self-renewal and radiation resistance in glioblastoma *Proc. Natl Acad. Sci. USA* **113** E8247–56

Guo J Y, Chen H-Y, Mathew R, Fan J, Strohecker A M, Karsli-Uzunbas G *et al* 2011 Activated ras requires autophagy to maintain oxidative metabolism and tumorigenesis *Genes Dev* **25** 460–70

Guo J Y, Teng X, Laddha S V, Ma S, Van Nostrand S C, Yang Y *et al* 2016 Autophagy provides metabolic substrates to maintain energy charge and nucleotide pools in Ras-driven lung cancer cells *Genes Dev* **30** 1704–17

Gurusamy N, Alsayari A, Rajasingh S and Rajasingh J 2018 Adult stem cells for regenerative therapy *Prog. Mol. Biol. Transl. Sci.* **160** 1–22

Häcker G 2000 The morphology of apoptosis *Cell Tissue Res.* **301** 5–17

Haigis M C, Deng C-X, Finley L W S, Kim H-S and Gius D 2012 SIRT3 is a mitochondrial tumor suppressor: a scientific tale that connects aberrant cellular ROS, the warburg effect, and carcinogenesis *Cancer Res.* **72** 2468–72

Hamacher-Brady A and Brady N R 2016 Mitophagy programs: mechanisms and physiological implications of mitochondrial targeting by autophagy *Cell. Mol. Life Sci.* **73** 775–95

Han K, Nguyen A, Traba J, Yao X, Kaler M, Huffstutler R D *et al* 2018a A pilot study to investigate the immune-modulatory effects of fasting in steroid-naive mild asthmatics *J. Immunol.* **201** 1382–8

Han K, Singh K, Rodman M J, Hassanzadeh S, Wu K, Nguyen A *et al* 2021 Fasting-induced FOXO4 blunts human $CD4^+$ T helper cell responsiveness *Nat. Metab.* **3** 318–26

Han Y, Fan S, Qin T, Yang J, Sun Y, Lu Y *et al* 2018b Role of autophagy in breast cancer and breast cancer stem cells (Review) *Int. J. Oncol.* **52** 1057–70

Hanahisa Y and Yamaguchi M 1996 Characterization of calcium accumulation in the brain of rats administered orally calcium: the significance of energy-dependent mechanism *Mol. Cell. Biochem.* **158** 1–7

Hara T, Takamura A, Kishi C, Iemura S, Natsume T, Guan J-L *et al* 2008 FIP200, a ULK-interacting protein, is required for autophagosome formation in mammalian cells *J. Cell Biol.* **181** 497–510

Harnett M M, Pineda M A, Latré De Laté P, Eason R J, Besteiro S, Harnett W *et al* 2017 From Christian de Duve to Yoshinori Ohsumi: more to autophagy than just dining at home *Biomed. J.* **40** 9–22

Harvie M and Howell A 2017 Potential benefits and harms of intermittent energy restriction and intermittent fasting amongst obese, overweight and normal weight subjects—a narrative review of human and animal evidence *Behav. Sci.* **7** 4

Hassanpour M, Rahbarghazi R, Nouri M, Aghamohammadzadeh N, Safaei N and Ahmadi M 2019 Role of autophagy in atherosclerosis: foe or friend? *J. Inflamm.* **16** 8

Hassanpour M, Rezabakhsh A, Pezeshkian M, Rahbarghazi R and Nouri M 2018 Distinct role of autophagy on angiogenesis: highlights on the effect of autophagy in endothelial lineage and progenitor cells *Stem Cell Res. Ther.* **9** 305

Hassanpour M, Rezaie J, Darabi M, Hiradfar A, Rahbarghazi R and Nouri M 2020 Autophagy modulation altered differentiation capacity of CD146+ cells toward endothelial cells, pericytes, and cardiomyocytes *Stem Cell Res. Ther.* **11** 139

Hazari Y, Bravo-San Pedro J M, Hetz C, Galluzzi L and Kroemer G 2020 Autophagy in hepatic adaptation to stress *J. Hepatol.* **72** 183–96

He R, Wang M, Zhao C, Shen M, Yu Y, He L *et al* 2019 TFEB-driven autophagy potentiates TGF-β induced migration in pancreatic cancer cells *J. Exp. Clin. Cancer Res.* **38** 340

Heckmann B L, Yang X, Zhang X and Liu J 2013 The autophagic inhibitor 3-methyladenine potently stimulates PKA-dependent lipolysis in adipocytes *Br. J Pharmacol.* **168** 163–71

Held N M and Houtkooper R H 2015 Mitochondrial quality control pathways as determinants of metabolic health *BioEssays* **37** 867–76

Hengartner M O 2000 The biochemistry of apoptosis *Nature* **407** 770–6

Henshall D C and Engel T 2013 Contribution of apoptosis-associated signaling pathways to epileptogenesis: lessons from Bcl-2 family knockouts *Front. Cell. Neurosci.* **7** 110

Ho T T, Warr M R, Adelman E R, Lansinger O M, Flach J, Verovskaya E V *et al* 2017 Autophagy maintains the metabolism and function of young and old stem cells *Nature* **543** 205–10

Huang H, Song J, Liu Z, Pan L and Xu G 2017 Autophagy activation promotes bevacizumab resistance in glioblastoma by suppressing Akt/mTOR signaling pathway *Oncol Lett.* **15** 1487–94

Huang Z, Zhou L, Chen Z, Nice E C and Huang C 2016 Stress management by autophagy: implications for chemoresistance *Int. J. Cancer* **139** 23–32

Ibrahim E M, Al-Foheidi M H and Al-Mansour M M 2021 Energy and caloric restriction, and fasting and cancer: a narrative review *Support. Care Cancer* **29** 2299–304

Icard P, Ollivier L, Forgez P, Otz J, Alifano M, Fournel L *et al* 2020 Perspective: do fasting, caloric restriction, and diets increase sensitivity to radiotherapy? A literature review *Adv. Nutr.* **11** 1089–101

Icard P, Shulman S, Farhat D, Steyaert J-M, Alifano M and Lincet H 2018 How the Warburg effect supports aggressiveness and drug resistance of cancer cells? *Drug Resist. Updat.* **38** 1–11

Iglesias M, Felix D A, Gutiérrez-Gutiérrez Ó, De Miguel-Bonet M D M, Sahu S, Fernández-Varas B *et al* 2019 Downregulation of mTOR signaling increases stem cell population telomere length during starvation of immortal planarians *Stem Cell Rep.* **13** 405–18

Igney F H and Krammer P H 2002 Death and anti-death: tumour resistance to apoptosis *Nat. Rev. Cancer* **2** 277–88

Iliopoulos D, Hirsch H A, Wang G and Struhl K 2011 Inducible formation of breast cancer stem cells and their dynamic equilibrium with non-stem cancer cells via IL6 secretion *Proc. Natl Acad. Sci. USA* **108** 1397–402

Jaffee E M, Dang C V, Agus D B, Alexander B M, Anderson K C, Ashworth A *et al* 2017 Future cancer research priorities in the USA: a lancet oncology commission *Lancet Oncol.* **18** e653–706

Jan R and Chaudhry G-S 2019 Understanding apoptosis and apoptotic pathways targeted cancer therapeutics *Adv. Pharm. Bull.* **9** 205–18

Jia L, Huang S, Yin X, Zan Y, Guo Y and Han L 2018 Quercetin suppresses the mobility of breast cancer by suppressing glycolysis through Akt-mTOR pathway mediated autophagy induction *Life Sci.* **208** 123–30

Jia P, Huang B, You Y, Su H and Gao L 2021 Ketogenic diet aggravates kidney dysfunction by exacerbating metabolic disorders and inhibiting autophagy in spontaneously hypertensive rats *Biochem. Biophys. Res. Commun.* **573** 13–8

Jiang G-M, Tan Y, Wang H, Peng L, Chen H-T, Meng X-J *et al* 2019 The relationship between autophagy and the immune system and its applications for tumor immunotherapy *Mol. Cancer* **18** 17

Joza N, Susin S A, Daugas E, Stanford W L, Cho S K, Li C Y J *et al* 2001 Essential role of the mitochondrial apoptosis-inducing factor in programmed cell death *Nature* **410** 549–54

Jung J, Kim L J Y, Wang X, Wu Q, Sanvoranart T, Hubert C G *et al* 2017 Nicotinamide metabolism regulates glioblastoma stem cell maintenance *JCI Insight* **2** e90019

Kang M R, Kim M S, Oh J E, Kim Y R, Song S Y, Kim S S *et al* 2009 Frameshift mutations of autophagy-related genes *ATG2B, ATG5, ATG9B* and *ATG12* in gastric and colorectal cancers with microsatellite instability *J. Pathol.* **217** 702–6

Kanzawa T, Germano I M, Komata T, Ito H, Kondo Y and Kondo S 2004 Role of autophagy in temozolomide-induced cytotoxicity for malignant glioma cells *Cell Death Differ.* **11** 448–57

Karsli-Uzunbas G, Guo J Y, Price S, Teng X, Laddha S V, Khor S *et al* 2014 Autophagy is required for glucose homeostasis and lung tumor maintenance *Cancer Discov.* **4** 914–27

Kasahara A and Scorrano L 2014 Mitochondria: from cell death executioners to regulators of cell differentiation *Trends Cell Biol.* **24** 761–70

Kaufmann A, Beier V, Franquelim H G and Wollert T 2014 Molecular mechanism of autophagic membrane-scaffold assembly and disassembly *Cell* **156** 469–81

Kerr J, Anderson C and Lippman S M 2017 Physical activity, sedentary behaviour, diet, and cancer: an update and emerging new evidence *Lancet Oncol.* **18** e457–71

Kerr J F R 2002 History of the events leading to the formulation of the apoptosis concept *Toxicology* **181–182** 471–4

Kerr J F R, Wyllie A H and Currie A R 1972 Apoptosis: a basic biological phenomenon with wideranging implications in tissue kinetics *Br. J. Cancer* **26** 239–57

Kim D, Kim J, Yu Y S, Kim Y R, Baek S H and Won K-J 2022 Systemic approaches using single cell transcriptome reveal that C/EBPγ regulates autophagy under amino acid starved condition *Nucleic Acids Res.* **50** 7298–309

Kim J B and Spiegelman B M 1996 ADD1/SREBP1 promotes adipocyte differentiation and gene expression linked to fatty acid metabolism *Genes Dev* **10** 1096–107

Kim J, Kim Y C, Fang C, Russell R C, Kim J H, Fan W *et al* 2013 Differential regulation of distinct Vps34 complexes by AMPK in nutrient stress and autophagy *Cell* **152** 290–303

Kim J, Kundu M, Viollet B and Guan K-L 2011 AMPK and mTOR regulate autophagy through direct phosphorylation of Ulk1 *Nat. Cell Biol.* **13** 132–41

Kim K-H, Kim Y H, Son J E, Lee J H, Kim S, Choe M S *et al* 2017 Intermittent fasting promotes adipose thermogenesis and metabolic homeostasis via VEGF-mediated alternative activation of macrophage *Cell Res.* **27** 1309–26

Kirshenbaum L A 2012 Regulation of autophagy in the heart in health and disease *J. Cardiovasc. Pharmacol.* **60** 109

Kitamura S, Yanagi T, Imafuku K, Hata H, Abe R and Shimizu H 2017 Drp1 regulates mitochondrial morphology and cell proliferation in cutaneous squamous cell carcinoma *J. Dermatol. Sci.* **88** 298–307

Klement R J, Brehm N and Sweeney R A 2020 Ketogenic diets in medical oncology: a systematic review with focus on clinical outcomes *Med. Oncol.* **37** 14

Klionsky D J, Abdelmohsen K, Abe A, Abedin M J, Abeliovich H, Acevedo Arozena A *et al* 2016 Guidelines for the use and interpretation of assays for monitoring autophagy (3rd edition) *Autophagy* **12** 1–222

Kondo Y, Kanzawa T, Sawaya R and Kondo S 2005 The role of autophagy in cancer development and response to therapy *Nat. Rev. Cancer* **5** 726–34

Koren S and Bentires-Alj M 2015 Breast tumor heterogeneity: source of fitness, hurdle for therapy *Mol. Cell* **60** 537–46

Krishnan S, Shrestha Y, Jayatunga D P W, Rea S, Martins R and Bharadwaj P 2020 Activate or inhibit? Implications of autophagy modulation as a therapeutic strategy for Alzheimer's disease *Int. J. Mol. Sci.* **21** 6739

Kurosaka K, Takahashi M, Watanabe N and Kobayashi Y 2003 Silent cleanup of very early apoptotic cells by macrophages *J. Immunol.* **171** 4672–9

La Belle Flynn A, Calhoun B C, Sharma A, Chang J C, Almasan A and Schiemann W P 2019 Autophagy inhibition elicits emergence from metastatic dormancy by inducing and stabilizing Pfkfb3 expression *Nat. Commun.* **10** 3668

LaBarge M A 2010 The difficulty of targeting cancer stem cell niches *Clin. Cancer Res.* **16** 3121–9

Laddha S V, Ganesan S, Chan C S and White E 2014 Mutational landscape of the essential autophagy gene *BECN1* in human cancers *Mol. Cancer Res.* **12** 485–90

Lang M J, Martinez-Marquez J Y, Prosser D C, Ganser L R, Buelto D, Wendland B *et al* 2014 Glucose starvation inhibits autophagy via vacuolar hydrolysis and induces plasma membrane internalization by down-regulating recycling *J. Biol. Chem.* **289** 16736–47

Lashinger L M, O'Flanagan C H, Dunlap S M, Rasmussen A J, Sweeney S, Guo J Y *et al* 2016 Starving cancer from the outside and inside: separate and combined effects of calorie restriction and autophagy inhibition on Ras-driven tumors *Cancer Metab.* **4** 18

Le Bras A 2020 Dietary restriction exacerbates Alzheimer's features in mice *Lab Anim.* **49** 76–6

Lee C, Raffaghello L, Brandhorst S, Safdie F M, Bianchi G, Martin-Montalvo A *et al* 2012 Fasting cycles retard growth of tumors and sensitize a range of cancer cell types to chemotherapy *Sci. Transl. Med.* **4** 124ra27

Leites E P and Morais V A 2018 Mitochondrial quality control pathways: PINK1 acts as a gatekeeper *Biochem. Biophys. Res. Commun.* **500** 45–50

Lessan N and Ali T 2019 Energy metabolism and intermittent fasting: the ramadan perspective *Nutrients* **11** 1192

Levine B and Kroemer G 2019 Biological functions of autophagy genes: a disease perspective *Cell* **176** 11–42

Levy J M M, Thompson J C, Griesinger A M, Amani V, Donson A M, Birks D K *et al* 2014 Autophagy inhibition improves chemosensitivity in BRAFV600E brain tumors *Cancer Discov.* **4** 773–80

Li F-J, Xu Z-S, Soo A D S, Lun Z-R and He C Y 2017a ATP-driven and AMPK-independent autophagy in an early branching eukaryotic parasite *Autophagy* **13** 715–29

Li J, Liu W, Hao H, Wang Q and Xue L 2019a Rapamycin enhanced the antitumor effects of doxorubicin in myelogenous leukemia K562 cells by downregulating the mTOR/p70S6K pathway *Oncol Lett.* **18** 2694–703

Li J and Yuan J 2008 Caspases in apoptosis and beyond *Oncogene* **27** 6194–206

Li J, Zhou J, Li Y, Qin D and Li P 2010 Mitochondrial fission controls DNA fragmentation by regulating endonuclease G *Free Radical Biol. Med.* **49** 622–31

Li L, Liu W-L, Su L, Lu Z-C and He X-S 2020 The role of autophagy in cancer radiotherapy *CMP* **13** 31–40

Li T, Qi M, Gatesoupe F-J, Tian D, Jin W, Li J *et al* 2019b Adaptation to fasting in crucian carp (*Carassius auratus*): gut microbiota and its correlative relationship with immune function *Microb. Ecol.* **78** 6–19

Li X, Wu X-Q, Deng R, Li D-D, Tang J, Chen W-D *et al* 2017b CaMKII-mediated Beclin 1 phosphorylation regulates autophagy that promotes degradation of Id and neuroblastoma cell differentiation *Nat. Commun.* **8** 1159

Li Y, Li S and Wu H 2022 Ubiquitination-Proteasome System (UPS) and autophagy two main protein degradation machineries in response to cell stress *Cells* **11** 851

Li Y-Y, Feun L, Thongkum A, Tu C-H, Chen S-M, Wangpaichitr M *et al* 2017c Autophagic mechanism in anti-cancer immunity: its pros and cons for cancer therapy *Int. J. Mol. Sci.* **18** 1297

Liang C, Lee J, Inn K-S, Gack M U, Li Q, Roberts E A *et al* 2008 Beclin1-binding UVRAG targets the class C Vps complex to coordinate autophagosome maturation and endocytic trafficking *Nat. Cell Biol.* **10** 776–87

Liang R and Ghaffari S 2018 Stem cells seen through the FOXO lens: an evolving paradigm *Curr. Topics Dev. Biol.* **127** 23–47
Lim S M, Mohamad Hanif E A and Chin S-F 2021 Is targeting autophagy mechanism in cancer a good approach? The possible double-edge sword effect *Cell Biosci.* **11** 56
Lin R and Iacovitti L 2015 Classic and novel stem cell niches in brain homeostasis and repair *Brain Res.* **1628** 327–42
Livesey K M, Kang R, Vernon P, Buchser W, Loughran P, Watkins S C *et al* 2012 p53/HMGB1 complexes regulate autophagy and apoptosis *Cancer Res.* **72** 1996–2005
Lleonart M E, Abad E, Graifer D and Lyakhovich A 2018 Reactive oxygen species-mediated autophagy defines the fate of cancer stem cells *Antioxid. Redox Signal.* **28** 1066–79
Lockshin R A and Williams C M 1965 Programmed cell death—I. Cytology of degeneration in the intersegmental muscles of the Pernyi silkmoth *J. Insect Physiol.* **11** 123–33
Longo V D, Di Tano M, Mattson M P and Guidi N 2021 Intermittent and periodic fasting, longevity and disease *Nat. Aging* **1** 47–59
Longo V D and Mattson M P 2014 Fasting: molecular mechanisms and clinical applications *Cell Metab.* **19** 181–92
Lorente J, Velandia C, Leal J A, Garcia-Mayea Y, Lyakhovich A, Kondoh H *et al* 2018 The interplay between autophagy and tumorigenesis: exploiting autophagy as a means of anticancer therapy *Biol. Rev.* **93** 152–65
Lőrincz P and Juhász G 2020 Autophagosome-lysosome fusion *J. Mol. Biol.* **432** 2462–82
Lu C, Shi Y, Wang Z, Song Z, Zhu M, Cai Q *et al* 2008 Serum starvation induces H2AX phosphorylation to regulate apoptosis via p38 MAPK pathway *FEBS Lett.* **582** 2703–8
Lu J, Zhu L, Zheng L, Cui Q, Zhu H, Zhao H *et al* 2018 Overexpression of ULK1 represents a potential diagnostic marker for clear cell renal carcinoma and the antitumor effects of SBI-0206965 *EBioMedicine* **34** 85–93
Lucena-Cacace A, Otero-Albiol D, Jiménez-García M P, Muñoz-Galvan S and Carnero A 2018 *NAMPT* is a potent oncogene in colon cancer progression that modulates cancer stem cell properties and resistance to therapy through sirt1 and PARP *Clin. Cancer Res.* **24** 1202–15
Lum J J, Bauer D E, Kong M, Harris M H, Li C, Lindsten T *et al* 2005 Growth factor regulation of autophagy and cell survival in the absence of apoptosis *Cell* **120** 237–48
Lv T, Li Z, Xu L, Zhang Y, Chen H and Gao Y 2018 Chloroquine in combination with aptamer-modified nanocomplexes for tumor vessel normalization and efficient erlotinib/Survivin shRNA co-delivery to overcome drug resistance in EGFR-mutated non-small cell lung cancer *Acta Biomater.* **76** 257–74
Macklin P S, Yamamoto A, Browning L, Hofer M, Adam J and Pugh C W 2020 Recent advances in the biology of tumour hypoxia with relevance to diagnostic practice and tissue-based research *J. Pathol.* **250** 593–611
Maday S and Holzbaur E L F 2016 Compartment-specific regulation of autophagy in primary neurons *J. Neurosci.* **36** 5933–45
Maiuri M C, Tasdemir E, Criollo A, Morselli E, Vicencio J M, Carnuccio R *et al* 2009 Control of autophagy by oncogenes and tumor suppressor genes *Cell Death Differ.* **16** 87–93
Mana M D, Kuo E Y-S and Yilmaz Ö H 2017 Dietary regulation of adult stem cells *Curr. Stem Cell Rep.* **3** 1–8
Marcucci F, Ghezzi P and Rumio C 2017 The role of autophagy in the cross-talk between epithelial-mesenchymal transitioned tumor cells and cancer stem-like cells *Mol. Cancer* **16** 3

Maria Fimia G, Stoykova A, Romagnoli A, Giunta L, Di Bartolomeo S, Nardacci R *et al* 2007 Ambra1 regulates autophagy and development of the nervous system *Nature* **447** 1121–5

Marquez R T and Xu L 2012 Bcl-2:Beclin 1 complex: multiple, mechanisms regulating autophagy/apoptosis toggle switch *Am. J. Cancer Res.* **2** 214–21

Martens S and Fracchiolla D 2020 Activation and targeting of ATG8 protein lipidation *Cell Discov.* **6** 23

Martinez-Outschoorn U E, Trimmer C, Lin Z, Whitaker-Menezes D, Chiavarina B, Zhou J *et al* 2010 Autophagy in cancer associated fibroblasts promotes tumor cell survival: role of hypoxia, HIF1 induction and NFκB activation in the tumor stromal microenvironment *Cell Cycle* **9** 3515–33

Maruyama T and Noda N N 2018 Autophagy-regulating protease Atg4: structure, function, regulation and inhibition *J. Antibiot.* **71** 72–8

Masiero E, Agatea L, Mammucari C, Blaauw B, Loro E, Komatsu M *et al* 2009 Autophagy is required to maintain muscle mass *Cell Metab.* **10** 507–15

Mattson M P, Longo V D and Harvie M 2017 Impact of intermittent fasting on health and disease processes *Ageing Res. Rev.* **39** 46–58

Mauthe M, Orhon I, Rocchi C, Zhou X, Luhr M, Hijlkema K-J *et al* 2018 Chloroquine inhibits autophagic flux by decreasing autophagosome-lysosome fusion *Autophagy* **14** 1435–55

Maycotte P, Gearheart C M, Barnard R, Aryal S, Mulcahy Levy J M, Fosmire S P *et al* 2014 STAT3-mediated autophagy dependence identifies subtypes of breast cancer where autophagy inhibition can be efficacious *Cancer Res.* **74** 2579–90

Maycotte P, Jones K L, Goodall M L, Thorburn J and Thorburn A 2015 Autophagy supports breast cancer stem cell maintenance by regulating IL6 secretion *Mol. Cancer Res.* **13** 651–8

Messaoudi I, Warner J, Fischer M, Park B, Hill B, Mattison J *et al* 2006 Delay of T cell senescence by caloric restriction in aged long-lived nonhuman primates *Proc. Natl Acad. Sci. USA* **103** 19448–53

Mihaylova M M, Sabatini D M and Yilmaz Ö H 2014 Dietary and metabolic control of stem cell function in physiology and cancer *Cell Stem Cell* **14** 292–305

Milošević M, Arsić A, Cvetković Z and Vučić V 2021 Memorable food: fighting age-related neurodegeneration by precision nutrition *Front. Nutr.* **8** 688086

Mizushima N and Komatsu M 2011 Autophagy: renovation of cells and tissues *Cell* **147** 728–41

Mizushima N, Levine B, Cuervo A M and Klionsky D J 2008 Autophagy fights disease through cellular self-digestion *Nature* **451** 1069–75

Mo S, Dai W, Xiang W, Li Y, Feng Y, Zhang L *et al* 2019 Prognostic and predictive value of an autophagy-related signature for early relapse in stages I–III colon cancer *Carcinogenesis* **40** 861–70

Morel A-P, Lièvre M, Thomas C, Hinkal G, Ansieau S and Puisieux A 2008 Generation of breast cancer stem cells through epithelial-mesenchymal transition *PLoS One* **3** e2888

Morgan M J and Thorburn A 2016 Measuring autophagy in the context of cancer *Tumor Microenvironment* ed C Koumenis, L M Coussens, A Giaccia and E Hammond (Cham: Springer International Publishing) pp 121–43

Mukhopadhyay S, Panda P K, Sinha N, Das D N and Bhutia S K 2014 Autophagy and apoptosis: where do they meet? *Apoptosis* **19** 555–66

Mulcahy Levy J M, Zahedi S, Griesinger A M, Morin A, Davies K D, Aisner D L *et al* 2017 Autophagy inhibition overcomes multiple mechanisms of resistance to BRAF inhibition in brain tumors *eLife* **6** e19671

Murakami N, Oyama F, Gu Y, McLennan I S, Nonaka I and Ihara Y 1998 Accumulation of tau in autophagic vacuoles in chloroquine myopathy *J. Neuropathol. Exp. Neurol.* **57** 664–73

Murata Y, Uehara Y and Hosoi Y 2015 Activation of mTORC1 under nutrient starvation conditions increases cellular radiosensitivity in human liver cancer cell lines, HepG2 and HuH6 *Biochem. Biophys. Res. Commun.* **468** 684–90

Naik P P, Birbrair A and Bhutia S K 2019 Mitophagy-driven metabolic switch reprograms stem cell fate *Cell. Mol. Life Sci.* **76** 27–43

Naik P P, Mukhopadhyay S, Panda P K, Sinha N, Das C K, Mishra R *et al* 2018 Autophagy regulates cisplatin-induced stemness and chemoresistance via the upregulation of CD44, ABCB1 and ADAM17 in oral squamous cell carcinoma *Cell Prolif.* **51** e12411

Nakai N, Kitai S, Iida N, Inoue S and Higashida K 2020 Autophagy under glucose starvation enhances protein translation initiation in response to re-addition of glucose in C2C12 myotubes *FEBS Open Bio.* **10** 2149–56

Nakatogawa H 2013 Two ubiquitin-like conjugation systems that mediate membrane formation during autophagy *Essays Biochem.* **55** 39–50

Napoleão A, Fernandes L, Miranda C and Marum A P 2021 Effects of calorie restriction on health span and insulin resistance: classic calorie restriction diet vs. ketosis-inducing diet *Nutrients* **13** 1302

Nazim U M, Jeong J-K and Park S-Y 2018 Ophiopogonin B sensitizes TRAIL-induced apoptosis through activation of autophagy flux and downregulates cellular FLICE-like inhibitory protein *Oncotarget* **9** 4161–72

Nazio F, Bordi M, Cianfanelli V, Locatelli F and Cecconi F 2019 Autophagy and cancer stem cells: molecular mechanisms and therapeutic applications *Cell Death Differ.* **26** 690–702

Nencioni A, Caffa I, Cortellino S and Longo V D 2018 Fasting and cancer: molecular mechanisms and clinical application *Nat. Rev. Cancer* **18** 707–19

Noda N N and Fujioka Y 2015 Atg1 family kinases in autophagy initiation *Cell. Mol. Life Sci.* **72** 3083–96

Ojha R, Ishaq M and Singh S K 2015 Caspase-mediated crosstalk between autophagy and apoptosis: mutual adjustment or matter of dominance *J. Cancer Res. Ther.* **11** 514–24

Ojha R, Singh S K and Bhattacharyya S 2016 JAK-mediated autophagy regulates stemness and cell survival in cisplatin resistant bladder cancer cells *Biochim. Biophys. Acta (BBA)—Gen. Subj.* **1860** 2484–97

Okawa T, Nagai M and Hase K 2021 Dietary intervention impacts immune cell functions and dynamics by inducing metabolic rewiring *Front. Immunol.* **11** 623989

Orgel E and Mittelman S D 2013 The links between insulin resistance, diabetes, and cancer *Curr. Diab. Rep.* **13** 213–22

Pacini N and Borziani F 2014 Cancer stem cell theory and the warburg effect, two sides of the same coin *Int. J. Mol. Sci.* **15** 8893–930

Pajuelo D, Gonzalez-Juarbe N, Tak U, Sun J, Orihuela C J and Niederweis M 2018 NAD+ depletion triggers macrophage necroptosis, a cell death pathway exploited by *Mycobacterium tuberculosis* *Cell Rep.* **24** 429–40

Pardo J, Wallich R, Ebnet K, Iden S, Zentgraf H, Martin P *et al* 2007 Granzyme B is expressed in mouse mast cells *in vivo* and *in vitro* and causes delayed cell death independent of perforin *Cell Death Differ.* **14** 1768–79

Park J-W, Jeong J and Bae Y-S 2022 Protein kinase CK2 is upregulated by calorie restriction and induces autophagy *Mol. Cells* **45** 112–21

Pasquier B 2015 SAR405, a PIK3C3/Vps34 inhibitor that prevents autophagy and synergizes with MTOR inhibition in tumor cells *Autophagy* **11** 725–6

Pasquier B 2016 Autophagy inhibitors *Cell. Mol. Life Sci.* **73** 985–1001

Pasquier B, El-Ahmad Y, Filoche-Rommé B, Dureuil C, Fassy F, Abecassis P-Y *et al* 2015 Discovery of (2*S*)-8-[(3*R*)-3-Methylmorpholin-4-yl]-1-(3-methyl-2-oxobutyl)-2-(trifluoromethyl)-3,4-dihydro-2*H*-pyrimido[1,2-*a*]pyrimidin-6-one: a novel potent and selective inhibitor of Vps34 for the treatment of solid tumors *J. Med. Chem.* **58** 376–400

Paweletz N 2001 Walther Flemming: pioneer of mitosis research *Nat. Rev. Mol. Cell Biol.* **2** 72–5

Peiris-Pagès M, Martinez-Outschoorn U E, Pestell R G, Sotgia F and Lisanti M P 2016 Cancer stem cell metabolism *Breast Cancer Res.* **18** 55

Peng Q, Qin J, Zhang Y, Cheng X, Wang X, Lu W *et al* 2017 Autophagy maintains the stemness of ovarian cancer stem cells by FOXA2 *J. Exp. Clin. Cancer Res.* **36** 171

Peng W-X, Pan F-Y, Liu X-J, Ning S, Xu N, Meng F-L *et al* 2010 Hypoxia stabilizes microtubule networks and decreases tumor cell chemosensitivity to anticancer drugs through Egr-1 *Anat Rec.* **293** 414–20

Pérez-Hernández M, Arias A, Martínez-García D, Pérez-Tomás R, Quesada R and Soto-Cerrato V 2019 Targeting autophagy for cancer treatment and tumor chemosensitization *Cancers* **11** 1599

Pernicova I and Korbonits M 2014 Metformin—mode of action and clinical implications for diabetes and cancer *Nat. Rev. Endocrinol.* **10** 143–56

Petherick K J, Conway O J L, Mpamhanga C, Osborne S A, Kamal A, Saxty B *et al* 2015 Pharmacological inhibition of ULK1 kinase blocks mammalian target of rapamycin (mTOR)-dependent autophagy *J. Biol. Chem.* **290** 11376–83

Pickles S, Vigié P and Youle R J 2018 Mitophagy and quality control mechanisms in mitochondrial maintenance *Curr. Biol.* **28** R170–85

Pickrell A M and Youle R J 2015 The roles of PINK1, parkin, and mitochondrial fidelity in Parkinson's disease *Neuron* **85** 257–73

Plaks V, Kong N and Werb Z 2015 The cancer stem cell niche: how essential is the niche in regulating stemness of tumor cells *Cell Stem Cell* **16** 225–38

Poloz Y and Stambolic V 2015 Obesity and cancer, a case for insulin signaling *Cell Death Dis.* **6** e2037–7

Pop C and Salvesen G S 2009 Human caspases: activation, specificity, and regulation *J. Biol. Chem.* **284** 21777–81

Porstmann T, Santos C R, Griffiths B, Cully M, Wu M, Leevers S *et al* 2008 SREBP activity is regulated by mTORC1 and contributes to Akt-dependent cell growth *Cell Metab.* **8** 224–36

Pouyafar A, Heydarabad M Z, Abdolalizadeh J, Rahbarghazi R and Talebi M 2019 Modulation of lipolysis and glycolysis pathways in cancer stem cells changed multipotentiality and differentiation capacity toward endothelial lineage *Cell Biosci.* **9** 30

Prabhu V V, Allen J E, Dicker D T and El-Deiry W S 2015 Small-molecule ONC201/TIC10 targets chemotherapy-resistant colorectal cancer stem–like cells in an Akt/Foxo3a/TRAIL–dependent manner *Cancer Res.* **75** 1423–32

Qin A-C, Li Y, Zhou L-N, Xing C-G and Lu X-S 2019 Dual PI3K-BRD4 inhibitor SF1126 inhibits colorectal cancer cell growth *in vitro* and *in vivo* *Cell. Physiol. Biochem.* **52** 758–68

Rai N K, Tripathi K, Sharma D and Shukla V K 2005 Apoptosis: a basic physiologic process in wound healing *Int. J. Low. Extrem. Wounds* **4** 138–44

Rangwala R, Leone R, Chang Y C, Fecher L A, Schuchter L M, Kramer A *et al* 2014 Phase I trial of hydroxychloroquine with dose-intense temozolomide in patients with advanced solid tumors and melanoma *Autophagy* **10** 1369–79

Rao J, Mei L, Liu J, Tang X, Yin S, Xia C *et al* 2019 Size-adjustable micelles co-loaded with a chemotherapeutic agent and an autophagy inhibitor for enhancing cancer treatment via increased tumor retention *Acta Biomater.* **89** 300–12

Rao S, Tortola L, Perlot T, Wirnsberger G, Novatchkova M, Nitsch R *et al* 2014 A dual role for autophagy in a murine model of lung cancer *Nat. Commun.* **5** 3056

Reaganshaw S and Ahmad N 2007 The role of Forkhead-box Class O (FoxO) transcription factors in cancer: a target for the management of cancer *Toxicol. Appl. Pharmacol.* **224** 360–8

Reggiori F, Gabius H-J, Aureli M, Römer W, Sonnino S and Eskelinen E-L 2021 Glycans in autophagy, endocytosis and lysosomal functions *Glycoconj. J.* **38** 625–47

Rezabakhsh A, Cheraghi O, Nourazarian A, Hassanpour M, Kazemi M, Ghaderi S *et al* 2017 Type 2 diabetes inhibited human mesenchymal stem cells angiogenic response by over-activity of the autophagic pathway *J.Cell. Biochem.* **118** 1518–30

Rickenbacher A, Jang J H, Limani P, Ungethüm U, Lehmann K, Oberkofler C E *et al* 2014 Fasting protects liver from ischemic injury through Sirt1-mediated downregulation of circulating HMGB1 in mice *J. Hepatol.* **61** 301–8

Rodrigues F A R, Bomfim I D S, Cavalcanti B C, Pessoa C, Goncalves R S B, Wardell J L *et al* 2014 Mefloquine–oxazolidine derivatives: a new class of anticancer agents *Chem. Biol. Drug Des.* **83** 126–31

Rodriguez-Navarro J A, Kaushik S, Koga H, Dall'Armi C, Shui G, Wenk M R *et al* 2012 Inhibitory effect of dietary lipids on chaperone-mediated autophagy *Proc. Natl Acad. Sci. USA* **109** E705–14

Rodríguez-Vargas J M, Ruiz-Magaña M J, Ruiz-Ruiz C, Majuelos-Melguizo J, Peralta-Leal A, Rodríguez M I *et al* 2012 ROS-induced DNA damage and PARP-1 are required for optimal induction of starvation-induced autophagy *Cell Res.* **22** 1181–98

Rojas-Puentes L L, Gonzalez-Pinedo M, Crismatt A, Ortega-Gomez A, Gamboa-Vignolle C, Nuñez-Gomez R *et al* 2013 Phase II randomized, double-blind, placebo-controlled study of whole-brain irradiation with concomitant chloroquine for brain metastases *Radiat. Oncol.* **8** 209

Rosenfeld M R, Ye X, Supko J G, Desideri S, Grossman S A, Brem S *et al* 2014 A phase I/II trial of hydroxychloroquine in conjunction with radiation therapy and concurrent and adjuvant temozolomide in patients with newly diagnosed glioblastoma multiforme *Autophagy* **10** 1359–68

Różański G, Pheby D, Newton J L, Murovska M, Zalewski P and Słomko J 2021 Effect of different types of intermittent fasting on biochemical and anthropometric parameters among patients with metabolic-associated fatty liver disease (MAFLD)—A systematic review *Nutrients* **14** 91

Sadeghian M, Rahmani S, Khalesi S and Hejazi E 2021 A review of fasting effects on the response of cancer to chemotherapy *Clin. Nutr.* **40** 1669–81

Saito T, Kuma A, Sugiura Y, Ichimura Y, Obata M, Kitamura H *et al* 2019 Autophagy regulates lipid metabolism through selective turnover of NCoR1 *Nat. Commun.* **10** 1567

Salminen A, Kaarniranta K and Kauppinen A 2013 Beclin 1 interactome controls the crosstalk between apoptosis, autophagy and inflammasome activation: impact on the aging process *Ageing Res. Rev.* **12** 520–34

Saltiel A R and Kahn C R 2001 Insulin signalling and the regulation of glucose and lipid metabolism *Nature* **414** 799–806

Santoro A, Vlachou T, Carminati M, Pelicci P G and Mapelli M 2016 Molecular mechanisms of asymmetric divisions in mammary stem cells *EMBO Rep.* **17** 1700–20

Sarter K, Mierke C, Beer A, Frey B, Führnrohr B G, Schulze C *et al* 2007 Sweet clearance: involvement of cell surface glycans in the recognition of apoptotic cells: minireview *Autoimmunity* **40** 345–8

Sayers T J 2011 Targeting the extrinsic apoptosis signaling pathway for cancer therapy *Cancer Immunol. Immunother.* **60** 1173–80

Schnebert S, Goguet M, Vélez E J, Depincé A, Beaumatin F, Herpin A *et al* 2022 Diving into the evolutionary history of HSC70-linked selective autophagy pathways: endosomal micro-autophagy and chaperone-mediated autophagy *Cells* **11** 1945

Sciarretta S, Hariharan N, Monden Y, Zablocki D and Sadoshima J 2011 Is autophagy in response to ischemia and reperfusion protective or detrimental for the heart? *Pediatr. Cardiol.* **32** 275–81

Seo B J, Yoon S H and Do J T 2018 Mitochondrial dynamics in stem cells and differentiation *Int. J. Mol. Sci.* **19** 3893

Sesen J, Dahan P, Scotland S J, Saland E, Dang V-T, Lemarié A *et al* 2015 Metformin inhibits growth of human glioblastoma cells and enhances therapeutic response *PLoS One* **10** e0123721

Shabkhizan R, Haiaty S, Moslehian M S, Bazmani A, Sadeghsoltani F, Saghaei Bagheri H *et al* 2023 The beneficial and adverse effects of autophagic response to caloric restriction and fasting *Adv. Nutr.* **14** 1211–25

Shahabad Z A, Avci C B, Bani F, Zarebkohan A, Sadeghizadeh M, Salehi R *et al* 2022 Photothermal effect of albumin-modified gold nanorods diminished neuroblastoma cancer stem cells dynamic growth by modulating autophagy *Sci. Rep.* **12** 11774

Shao M, Zhu W, Lv X, Yang Q, Liu X, Xie Y *et al* 2018 Encapsulation of chloroquine and doxorubicin by MPEG-PLA to enhance anticancer effects by lysosomes inhibition in ovarian cancer *Int. J. Nanomedicine* **13** 8231–45

Shao X, Lai D, Zhang L and Xu H 2016 Induction of autophagy and apoptosis via PI3K/AKT/TOR pathways by Azadirachtin A in *Spodoptera litura* cells *Sci. Rep.* **6** 35482

Sharma M K, Priyam K, Kumar P, Garg P K, Roy T S and Jacob T G 2022 Effect of calorie-restriction and rapamycin on autophagy and the severity of caerulein-induced experimental acute pancreatitis in mice *Front. Gastroenterol.* **1** 977169

Shen Q, Wang Y, Kokovay E, Lin G, Chuang S-M, Goderie S K *et al* 2008 Adult SVZ stem cells lie in a vascular niche: a quantitative analysis of niche cell–cell interactions *Cell Stem Cell* **3** 289–300

Sinha K, Das J, Pal P B and Sil P C 2013 Oxidative stress: the mitochondria-dependent and mitochondria-independent pathways of apoptosis *Arch. Toxicol.* **87** 1157–80

Sinha S and Levine B 2008 The autophagy effector Beclin 1: a novel BH3-only protein *Oncogene* **27** S137–48

Smit L, Berns K, Spence K, Ryder W D, Zeps N, Madiredjo M *et al* 2016 An integrated genomic approach identifies that the PI3K/AKT/FOXO pathway is involved in breast cancer tumor initiation *Oncotarget* **7** 2596–610

Smith A G and Macleod K F 2019 Autophagy, cancer stem cells and drug resistance *J. Pathol.* **247** 708–18

Smith D M, Patel S, Raffoul F, Haller E, Mills G B and Nanjundan M 2010 Arsenic trioxide induces a beclin-1-independent autophagic pathway via modulation of SnoN/SkiL expression in ovarian carcinoma cells *Cell Death Differ.* **17** 1867–81

Snyder V, Reed-Newman T C, Arnold L, Thomas S M and Anant S 2018 Cancer stem cell metabolism and potential therapeutic targets *Front. Oncol.* **8** 203

So W-K and Cheung T H 2018 Molecular regulation of cellular quiescence: a perspective from adult stem cells and its niches *Cellular Quiescence* ed H D Lacorazza (New York: Springer) pp 1–25

Son M-Y, Choi H, Han Y-M and Sook Cho Y 2013 Unveiling the critical role of REX1 in the regulation of human stem cell pluripotency *Stem Cells* **31** 2374–87

Song Y, Zhang S, Guo X, Sun K, Han Z, Li R *et al* 2013 Autophagy contributes to the survival of CD133+ liver cancer stem cells in the hypoxic and nutrient-deprived tumor microenvironment *Cancer Lett.* **339** 70–81

Sotelo J, Briceño E and López-González M A 2006 Adding chloroquine to conventional treatment for glioblastoma multiforme: a randomized, double-blind, placebo-controlled trial *Ann. Intern Med* **144** 337

Sowter H M, Ratcliffe P J, Watson P, Greenberg A H and Harris A L 2001 HIF-1-dependent regulation of hypoxic induction of the cell death factors BNIP3 and NIX in human tumors *Cancer Res.* **61** 6669–73

Stekovic S, Hofer S J, Tripolt N, Aon M A, Royer P, Pein L *et al* 2019 Alternate day fasting improves physiological and molecular markers of aging in healthy, non-obese humans *Cell Metab.* **30** 462–476.e6

Su T, Li X, Yang M, Shao Q, Zhao Y, Ma C *et al* 2020 Autophagy: an intracellular degradation pathway regulating plant survival and stress response *Front. Plant Sci.* **11** 164

Sui X, Chen R, Wang Z, Huang Z, Kong N, Zhang M *et al* 2013 Autophagy and chemotherapy resistance: a promising therapeutic target for cancer treatment *Cell Death Dis.* **4** e838–8

Sukowati C H C 2019 Heterogeneity of hepatic cancer stem cells *Stem Cells Heterogeneity in Cancer* ed A Birbrair (Cham: Springer International Publishing) pp 59–81

Sun L, Li Y-J, Yang X, Gao L and Yi C 2017a Effect of fasting therapy in chemotherapy-protection and tumor-suppression: a systematic review *Transl. Cancer Res.* **6** 354–65

Sun L, Xiong H, Chen L, Dai X, Yan X, Wu Y *et al* 2022 Deacetylation of ATG4B promotes autophagy initiation under starvation *Sci. Adv.* **8** eabo0412

Sun P, Wang H, He Z, Chen X, Wu Q, Chen W *et al* 2017b Fasting inhibits colorectal cancer growth by reducing M2 polarization of tumor-associated macrophages *Oncotarget* **8** 74649–60

Sun R, Shen S, Zhang Y-J, Xu C-F, Cao Z-T, Wen L-P *et al* 2016 Nanoparticle-facilitated autophagy inhibition promotes the efficacy of chemotherapeutics against breast cancer stem cells *Biomaterials* **103** 44–55

Sun Y, Li M, Zhao D, Li X, Yang C and Wang X 2020 Lysosome activity is modulated by multiple longevity pathways and is important for lifespan extension in C. elegans *eLife* **9** e55745

Tait S W G and Green D R 2010 Mitochondria and cell death: outer membrane permeabilization and beyond *Nat. Rev. Mol. Cell Biol.* **11** 621–32

Takagi A, Kume S, Kondo M, Nakazawa J, Chin-Kanasaki M, Araki H *et al* 2016 Mammalian autophagy is essential for hepatic and renal ketogenesis during starvation *Sci. Rep.* **6** 18944

Takahashi A, Kimura F, Yamanaka A, Takebayashi A, Kita N, Takahashi K *et al* 2014 Metformin impairs growth of endometrial cancer cells via cell cycle arrest and concomitant autophagy and apoptosis *Cancer Cell Int.* **14** 53

Tamazashvili T 2019 Systemic stem cells equilibrium theory—the law of life *Georgian Med. News.* **296** 144–9

Tan Q, Wang H, Hu Y, Hu M, Li X, A *et al* 2015 Src/ STAT 3-dependent heme oxygenase-1 induction mediates chemoresistance of breast cancer cells to doxorubicin by promoting autophagy *Cancer Sci.* **106** 1023–32

Tang C-C, Huang T-C, Tien F-M, Lin J-M, Yeh Y-C and Lee C-Y 2021 Safety, feasibility, and effects of short-term calorie reduction during induction chemotherapy in patients with diffuse large B-cell lymphoma: a pilot study *Nutrients* **13** 3268

Tang F, Hu P, Yang Z, Xue C, Gong J, Sun S *et al* 2017 SBI0206965, a novel inhibitor of Ulk1, suppresses non-small cell lung cancer cell growth by modulating both autophagy and apoptosis pathways *Oncol. Rep.* **37** 3449–58

Tanida I 2011 Autophagosome formation and molecular mechanism of autophagy *Antioxid. Redox Signal.* **14** 2201–14

Teong X T, Liu K, Vincent A D, Bensalem J, Liu B, Hattersley K J *et al* 2023 Intermittent fasting plus early time-restricted eating versus calorie restriction and standard care in adults at risk of type 2 diabetes: a randomized controlled trial *Nat. Med.* **29** 963–72

Thelen M, Wymann M P and Langen H 1994 Wortmannin binds specifically to 1-phosphatidylinositol 3-kinase while inhibiting guanine nucleotide-binding protein-coupled receptor signaling in neutrophil leukocytes *Proc. Natl Acad. Sci. USA* **91** 4960–4

Towers C G and Thorburn A 2017 Targeting the lysosome for cancer therapy *Cancer Discov.* **7** 1218–20

Towers C G, Wodetzki D and Thorburn A 2019 Autophagy and cancer: modulation of cell death pathways and cancer cell adaptations *J. Cell Biol.* **219** e201909033

Traba J, Kwarteng-Siaw M, Okoli T C, Li J, Huffstutler R D, Bray A *et al* 2015 Fasting and refeeding differentially regulate NLRP3 inflammasome activation in human subjects *J. Clin. Invest.* **125** 4592–600

Tukaj C 2013 The significance of macroautophagy in health and disease *Folia Morphol.* **72** 87–93

Van Der Vos K E and Coffer P J 2008 FOXO-binding partners: it takes two to tango *Oncogene* **27** 2289–99

Van Doeselaar S and Burgering B M T 2018 FOXOs maintaining the equilibrium for better or for worse *Curr. Top. Dev. Biol.* **127** 49–103

Varga V B, Keresztes F, Sigmond T, Vellai T and Kovács T 2022 The evolutionary and functional divergence of the Atg8 autophagy protein superfamily *Biol. Futura.* **73** 375–84

Vazquez-Martin A, Cufí S, Corominas-Faja B, Oliveras-Ferraros C, Vellon L and Menendez J A 2012 Mitochondrial fusion by pharmacological manipulation impedes somatic cell reprogramming to pluripotency: new insight into the role of mitophagy in cell stemness *Aging* **4** 393–401

Vazquez-Martin A, Den Haute C V, Cufí S, Faja B C, Cuyàs E, Lopez-Bonet E *et al* 2016 Mitophagy-driven mitochondrial rejuvenation regulates stem cell fate *Aging* **8** 1330–52

Vega-Rubín-de-Celis S, Zou Z, Fernández Á F, Ci B, Kim M, Xiao G *et al* 2018 Increased autophagy blocks HER2-mediated breast tumorigenesis *Proc. Natl Acad. Sci. USA* **115** 4176–81

Vendelbo M H, Møller A B, Christensen B, Nellemann B, Clasen B F F, Nair K S *et al* 2014 Fasting increases human skeletal muscle net phenylalanine release and this is associated with decreased mTOR signaling *PLoS One* **9** e102031

Verbaanderd C, Maes H, Schaaf M B, Sukhatme V P, Pantziarka P, Sukhatme V *et al* 2017 Repurposing drugs in oncology (ReDO)—chloroquine and hydroxychloroquine as anti-cancer agents *ecancer* **11** 781

Vidoni C, Ferraresi A, Esposito A, Maheshwari C, Dhanasekaran D N, Mollace V *et al* 2021 Calorie restriction for cancer prevention and therapy: mechanisms, expectations, and efficacy *J. Cancer Prev.* **26** 224–36

Wang B, Shi Y, Chen J, Shao Z, Ni L, Lin Y *et al* 2021 High glucose suppresses autophagy through the AMPK pathway while it induces autophagy via oxidative stress in chondrocytes *Cell Death Dis.* **12** 506

Wang B-H, Hou Q, Lu Y-Q, Jia M-M, Qiu T, Wang X-H *et al* 2018 Ketogenic diet attenuates neuronal injury via autophagy and mitochondrial pathways in pentylenetetrazol-kindled seizures *Brain Res.* **1678** 106–15

Wang G, Zhang C, Jiang F, Zhao M, Xie S and Liu X 2022 NOD2-RIP2 signaling alleviates microglial ROS damage and pyroptosis via ULK1-mediated autophagy during *Streptococcus pneumonia* infection *Neurosci. Lett.* **783** 136743

Wang H and Unternaehrer J J 2019 Epithelial-mesenchymal Transition and cancer stem cells: at the crossroads of differentiation and dedifferentiation *Dev. Dyn.* **248** 10–20

Wang Y, Kuramitsu Y, Baron B, Kitagawa T, Tokuda K, Akada J *et al* 2017 PI3K inhibitor LY294002, as opposed to wortmannin, enhances AKT phosphorylation in gemcitabine-resistant pancreatic cancer cells *Int. J. Oncol.* **50** 606–12

Weber D D, Aminzadeh-Gohari S, Tulipan J, Catalano L, Feichtinger R G and Kofler B 2020 Ketogenic diet in the treatment of cancer—where do we stand? *Mol. Metab.* **33** 102–21

Wei M, Brandhorst S, Shelehchi M, Mirzaei H, Cheng C W, Budniak J *et al* 2017 Fasting-mimicking diet and markers/risk factors for aging, diabetes, cancer, and cardiovascular disease *Sci. Transl. Med.* **9** eaai8700

Wei Y, Sinha S C and Levine B 2008 Dual role of JNK1-mediated phosphorylation of Bcl-2 in autophagy and apoptosis regulation *Autophagy* **4** 949–51

White E 2012 Deconvoluting the context-dependent role for autophagy in cancer *Nat. Rev. Cancer* **12** 401–10

Wilhelmi De Toledo F, Buchinger A, Burggrabe H, Hölz G, Kuhn C, Lischka E *et al* 2013 Fasting therapy—an expert panel update of the 2002 consensus guidelines *Complement Med. Res.* **20** 434–43

Wu C-C and Bratton S B 2013 Regulation of the intrinsic apoptosis pathway by reactive oxygen species *Antioxid. Redox Signal.* **19** 546–58

Wu W, Li K, Guo S, Xu J, Ma Q, Li S *et al* 2021 P300/HDAC1 regulates the acetylation/deacetylation and autophagic activities of LC3/Atg8-PE ubiquitin-like system *Cell Death Discov* **7** 128

Wu Y, Yao J, Xie J, Liu Z, Zhou Y, Pan H *et al* 2018 The role of autophagy in colitis-associated colorectal cancer *Signal Transduct Target. Ther.* **3** 31

Xie Q, Liu Y and Li X 2020 The interaction mechanism between autophagy and apoptosis in colon cancer *Transl. Oncol.* **13** 100871

Xie Q, Wu Q, Horbinski C M, Flavahan W A, Yang K, Zhou W *et al* 2015 Mitochondrial control by DRP1 in brain tumor initiating cells *Nat. Neurosci.* **18** 501–10

Yadav R K, Chauhan A S, Zhuang L and Gan B 2018 FoxO transcription factors in cancer metabolism *Semin. Cancer Biol.* **50** 65–76

Yang H, Wang H and Andersson U 2020 Targeting inflammation driven by HMGB1 *Front. Immunol.* **11** 484

Yang Y and Zhang L 2020 The effects of caloric restriction and its mimetics in Alzheimer's disease through autophagy pathways *Food Funct* **11** 1211–24

Ye X, Zhou X-J and Zhang H 2018 Exploring the role of autophagy-related gene 5 (ATG5) yields important insights into autophagy in autoimmune/autoinflammatory diseases *Front. Immunol.* **9** 2334

Yi M, Li J, Chen S, Cai J, Ban Y, Peng Q *et al* 2018 Emerging role of lipid metabolism alterations in Cancer stem cells *J. Exp. Clin. Cancer Res.* **37** 118

Yin H, Price F and Rudnicki M A 2013 Satellite cells and the muscle stem cell niche *Physiol. Rev.* **93** 23–67

Yorimitsu T and Klionsky D J 2005 Autophagy: molecular machinery for self-eating *Cell Death Differ.* **12** 1542–52

Yu G and Klionsky D J 2022 Life and death decisions—the many faces of autophagy in cell survival and cell death *Biomolecules* **12** 866

Yue W, Hamaï A, Tonelli G, Bauvy C, Nicolas V, Tharinger H *et al* 2013 Inhibition of the autophagic flux by salinomycin in breast cancer stem-like/progenitor cells interferes with their maintenance *Autophagy* **9** 714–29

Yun C W and Lee S H 2018 The roles of autophagy in cancer *Int. J. Mol. Sci.* **19** 3466

Zeng M and Zhou J-N 2008 Roles of autophagy and mTOR signaling in neuronal differentiation of mouse neuroblastoma cells *Cell. Signal.* **20** 659–65

Zhang C-S, Li M, Wang Y, Li X, Zong Y, Long S *et al* 2022 The aldolase inhibitor aldometanib mimics glucose starvation to activate lysosomal AMPK *Nat. Metab.* **4** 1369–401

Zhang D, Tang B, Xie X, Xiao Y-F, Yang S-M and Zhang J-W 2015 The interplay between DNA repair and autophagy in cancer therapy *Cancer Biol. Ther.* **16** 1005–13

Zhang G, Wang Z, Du Z and Zhang H 2018 mTOR regulates phase separation of PGL granules to modulate their autophagic degradation *Cell* **174** 1492–1506.e22

Zhang M, Liu D and Ge L 2017 In vitro dissection of autophagy *CP Cell Biol.* **77** 11.23.1–17

Zhao P, Li M, Wang Y, Chen Y, He C, Zhang X *et al* 2018 Enhancing anti-tumor efficiency in hepatocellular carcinoma through the autophagy inhibition by miR-375/sorafenib in lipid-coated calcium carbonate nanoparticles *Acta Biomater.* **72** 248–55

Zhao Y, Huang Q, Yang J, Lou M, Wang A, Dong J *et al* 2010 Autophagy impairment inhibits differentiation of glioma stem/progenitor cells *Brain Res.* **1313** 250–8

Zheng H, Pomyen Y, Hernandez M O, Li C, Livak F, Tang W *et al* 2018 Single-cell analysis reveals cancer stem cell heterogeneity in hepatocellular carcinoma *Hepatology* **68** 127–40

Zhou L, Haiyilati A, Li J, Li X, Gao L, Cao H *et al* 2022 Gga-miR-30c-5p Suppresses avian reovirus (ARV) replication by inhibition of ARV-induced autophagy via targeting ATG5 *J. Virol.* **96** e00759-22

Zhu Y-X, Jia H-R, Gao G, Pan G-Y, Jiang Y-W, Li P *et al* 2020 Mitochondria-acting nanomicelles for destruction of cancer cells via excessive mitophagy/autophagy-driven lethal energy depletion and phototherapy *Biomaterials* **232** 119668

Zong Y, Zhang C-S, Li M, Wang W, Wang Z, Hawley S A *et al* 2019 Hierarchical activation of compartmentalized pools of AMPK depends on severity of nutrient or energy stress *Cell Res.* **29** 460–73

IOP Publishing

Physics of Cancer, Volume 6 (Second Edition)
Cellular mechanisms to foster or fight cancer
Claudia Tanja Mierke

Chapter 5

Mechanobiological aspects of autophagy in cancer and various subtypes of selective autophagy

5.1 Summary

The orderly conduct of cellular processes, the sustenance of cellular homeostasis and the survival of cells rely on the effective coordination of cellular responses to mechanical stimuli such as stress. The development of cancer is a frequent detrimental outcome of dysregulation of the synchronized reactions of the cell. In this scheme, the process of autophagy plays a central role by maintaining the equilibrium between synthesis, breakdown and recycling of cytosolic elements like proteins, lipids and organelles. There are different types of selective autophagy subtypes, such as pexophagy, lysophagy, mitophagy. ER-phagy, aggrephagy, xenophagy, ribophagy, ferritinophagy, glycophagy and lipophagy, that fulfill different tasks and may be able to sense mechanical cues. Various kinds of stress from the surroundings trigger autophagy, among them hypoxia, damage to DNA, inflammation and metabolic stresses like starvation. In complement to these chemical demands, cells also must deal with mechanical stresses emanating from their own surrounding microenvironment (Mierke 2019, 2020a, Hernández-Cáceres *et al* 2021). Cells fulfill this purpose through the activation of intrinsic mechanical reactions conveyed through active cytoskeletal mechanisms and mechanosensitive protein constructs that interface cells toward their mechanical surroundings. Although it is recognized that autophagy and cell mechanics are critical in oncogenesis and malignancy propagation, their interacting mechanisms are widely overseen. In this chapter, the contribution of physical forces to the control of autophagy and their potential consequences in both physiological and pathological states will be highlighted. By considering mechanics, there are new questions to advance the investigation of the mechanical constraints of autophagy and to comprehend the degree to what extent mechanical cues impact this fundamental process in physiological and pathological contexts.

doi:10.1088/978-0-7503-4007-6ch5

5.2 Introduction to the coupling of autophagy and mechanical stress

Comprehension of the homeostatic processes during carcinogenesis and how cancer cells react to chemical and physical environmental impacts is one of the most fascinating areas of cancer research. In this scenario, the catabolic activity of autophagy appears to be the key mechanism for preserving the equilibrium between synthesis, breakdown and reutilization of cytosolic constituents (Eskelinen and Saftig 2009). These common housekeeping functions constitute a cellular mechanism that maintains steady-state homeostasis, increases resistance to stress, and ensures survival. Autophagy is activated by several harmful environmental factors, including hypoxia, DNA injury, inflammatory conditions, and metabolic stresses like fasting. Apart from facing chemical threats, cells are also subject to mechanical stress originating from the environment (Mierke 2019, 2021, 2023, 2024a). Mechanical sensing is conferred through force-induced conformational alterations of mechanosensitive proteins that are directly or indirectly linked to the cytoskeleton, and through the activation of mechanosensitive ion channels or cellular entities (Holle and Engler 2011 Iskratsch *et al* 2014 Mierke 2024b). Mechanosensory perception leads to a change in intracellular tension by rearranging the cytoskeleton and actomyosin contraction, thereby integrating mechanical cues into biochemical cascades, which is referred to as mechanotransduction, and, in the longer perspective, causing an alteration in gene expression (Iskratsch *et al* 2014). Therefore, the physical characteristics of the microenvironment, such as the constitution, stiffness, viscoelasticity and architecture of the extracellular matrix (ECM), have a pervasive influence on the cellular genotype, phenotype, processes, organization of tissues and organismal general biological functionality (Iskratsch *et al* 2014). This link between mechanical and biological reactions is also relevant to cancer in terms of cancer cell transformation and advancement, during which the local physical microenvironment experiences profound alterations. These changes in the tumor microniche are caused by increased cell contractility, elevated pressure due to abnormal cell proliferation and expansion of the tumor mass, and changes in the makeup, architectural design and rheological characteristics of the encompassing ECM (Chaudhuri *et al* 2018, Northcott *et al* 2018). These mechanical alterations were reported to parallel the activation of autophagy, that could be an integrated reaction to mechanical stress utilized by cancer cells to evade programmed cell death and promote their adaptive reaction to the new mechanical landscape (Das *et al* 2020). Moreover, there is convincing data indicating that autophagy affects multiple cancer hallmarks, such as cell motility and invasion, cancer stem cell survival and differentiation, epithelial-to-mesenchymal transition (EMT), resilience to apoptosis and anoikis, evasion from immune supervision, and cancer cell quiescence (Mowers *et al* 2017, Das *et al* 2020). Nevertheless, the cause-and-effect relationship of cellular mechanics and autophagy and their mutual influence on carcinogenesis are patchy and largely unproven. In this chapter the autophagic process is outlined from a mechanical perspective and the crosstalk between mechanotransduction and cellular catabolism is presented to gain an understanding of their possible contribution to the regulation of cancer transformation and survival.

5.3 The cytoskeleton controls cell mechanical properties

Essential life functions of eukaryotic cells, like resistance to deformation, regulation of cell form, cell migration and transportation of intracellular freight, rely on the activity of the cytoskeleton, an intertwined assembly of filamentous polymers, motor proteins and modulating proteins (Banerjee *et al* 2020). This cytoskeletal network consists of three interconnected structural elements, that is, microtubules, intermediate filaments, and microfilaments (actin), which form the motor of the cells through the transduction of chemical energy into mechanical energy by ATP-dependent polymerization and the function of motor proteins. This mechanical energy is utilized to generate forces that dislocate cellular components, such as the generation of cellular protrusions and the transportation of freight, and/or can cause the storage of elastic energy in these components, like cortical tension and cellular contractility. The entire procedure of autophagy, which is a succession of membrane reorganization incidents, is mechanically achieved and orchestrated through ATP-dependent cytoskeletal dynamics inducing mechanical distortion and transportation (Aguilera *et al* 2012, Kast and Dominguez 2017). The cytoskeleton serves as an essential scaffold for orchestrating and directing the proper positioning, anchoring, attachment, priming, merging, and trafficking of organelles like autophagosomes and lysosomes. The actin cytoskeleton consists of actin filaments and fibers, which assemble and disassemble to form mesh-like subcellular networks with Arp2/3-facilitated branching, and filament bundles with formin-facilitated crosslinking. These structural architectures and bundles provide structural stability to cell membranes and govern membrane dynamics (Chhabra and Higgs 2007). What is crucial is that the activity of the molecular motors of the myosin family places the actin filaments in a state of tension. Analogous to a stretch spring, the liberation of this tension generates kinetic energy that is utilized for vesicle trafficking and membrane reorganization involved in autophagosome generation (Kast *et al* 2015, Zientara-Rytter and Subramani 2016). Moreover, certain myosins, such as myosin VI, are directly implicated in the trafficking of several cargoes, among them autophagosomes (Tumbarello *et al* 2012). In addition, microtubule dynamics through assembly and disassembly and the activity of their associated motor proteins, like kinesins and dyneins (Geeraert *et al* 2010, Nakamura and Yoshimori 2017) coordinate the trafficking of preautophagosomal particles and autophagosomes throughout the cytoplasm during maturation of autophagosomes (Jordens *et al* 2001, Kimura *et al* 2008) and bidirectional autolysosome trafficking (Yang *et al* 2011). The collaboration and competitiveness between actin and microtubules are accountable for a major portion of cellular mechanics. Collectively, these ATP-dependent cytoskeletal events constitute the mechanism to surmount the energy hurdles provided by membrane elasticity and resistance to deformation, that govern each single step of the autophagic pathway (Bahrami *et al* 2017). Ultimately, it is assumed that intermediate filaments, such as keratins and vimentin, which show no obvious dynamics and possess no motor proteins, confer mechanical stability on the cell and on its organelles (Goldman *et al* 2012). Intermediate filaments serve as a key determinant of the positioning of autophagosomes and lysosomes through their capacity to constrain their free, random motions (Biskou *et al* 2019). For example, it has been found that interconnecting networks of vimentin fibers form cages around cellular organelles such as the nucleus, the endoplasmic reticulum, and the mitochondria (Lowery *et al* 2015). Accordingly, disruption of the Vimentin meshwork through pharmacological intervention leads to a disrupted flow of the autophagic vesicle network, such as the autophagic flux, to the perinuclear localization of autophagic vesicles and to a mislocalization of autophagic vesicles in distinct phases of the pathway (Biskou *et al* 2019).

5.4 Connection of mechanics and autophagy

From a mechanical perspective, the autophagic event can be broken down into a series of seven major phases, as illustrated in figure 5.1: initiation, nucleation, elongation, closure, maturation of the autophagosome and transport towards the perinuclear zone of the cell, merging with the lysosome, and ultimately breakdown and recirculation of the cargo (Al-Bari 2020).

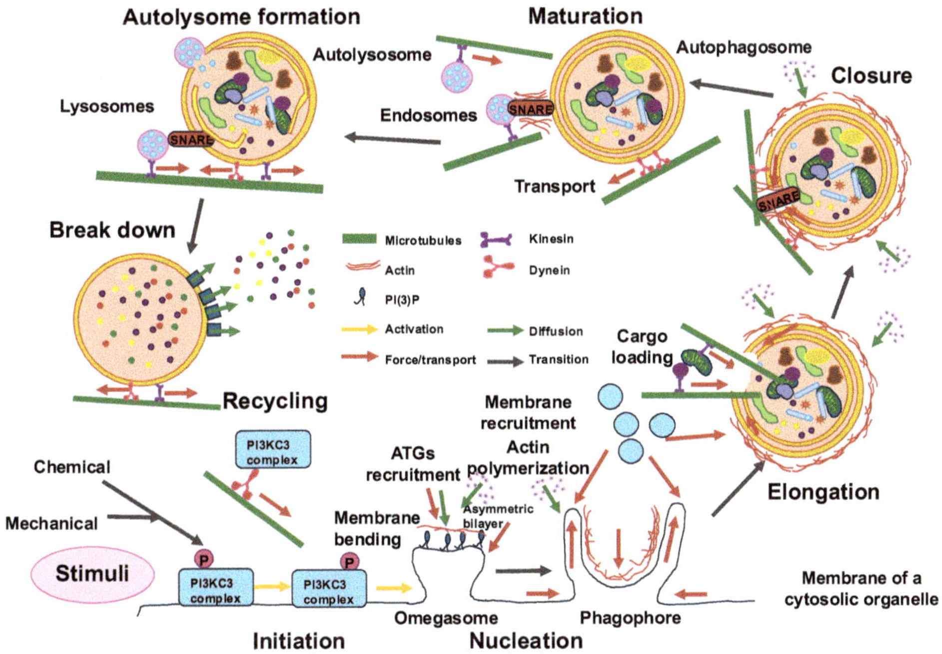

Figure 5.1. The mechanisms of the autophagic pathway. From a mechanical perspective, the autophagic event can be dissected into seven major phases: initiation, nucleation, elongation, closure, maturation of the autophagosome, autolysosome generation, and ultimately breakdown and cargo recirculation. Active processes of the cytoskeleton and membranous organization throughout the consecutive steps of autophagy are emphasized.

5.4.1 Initiation step

In the event of chemical or mechanical cell stimulation, the autophagic process is initiated by the enrollment of nuclear autophagy elements (figure 5.1, initiation step). This phase is equivalent to the activation of the Unc-51-like kinase 1 (ULK1) protein complex (Hurley and Young 2017). As illustrated in the scheme in figure 5.2, the regulation of the ULK1 complex is accomplished by increasing the activity of AMPK, which is induced by ATP depletion (Kim *et al* 2011) and/or by blocking the mechanistic target of rapamycin complex 1 (mTORC1), which serves as a repression factor for autophagy and retains ULK1 in an inactive conformational structure under basal physiological conditions (Kim *et al* 2011). In the canonical induction of autophagy, metabolic stresses or chemical stimuli, like nutrient deprivation, trigger uncoupling of mTORC1 from ULK1, which is then activated and complexes with ATG13 and FIP200. The ULK1 complex can be built by chemical or mechanical stimuli that is illustrated in figure 5.2.

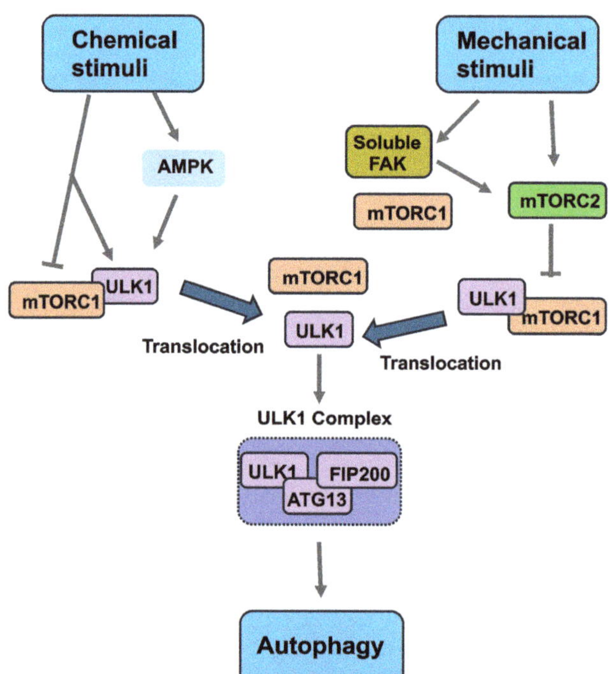

Figure 5.2. Schematic illustration of the signaling events for the activation of the ULK1 complex upon induction of autophagy. Autophagy is triggered through the ULK1 signal transduction pathway, leading to the dissociation of mTORC1 from ULK1, which then undergoes activation and attaches to ATG13 and FIP200 to generate the ULK1 complex. This uncoupling and the subsequent cascade of signals can be triggered through chemical cues, such as an increase in AMPK activity, and/or through mechanical cues. This second route is attained through the blocking of mTORC1 through the mechanosensitive mTORC2, which responds to various mechanical signals. For example, mTORC2 can be triggered directly through the soluble version of focal adhesion kinase (FAK), which is liberated from focal adhesions under low-traction conditions such as cell disengagement from the substrate.

Signaling at an early step is required for the subsequent events of autophagosome generation (figure 5.1, initiation step). It is still uncertain whether mechanical stress and stimuli have a direct part to play in the activation of ULK1. It has been proposed that mechanosensitive mTORC2 (Case *et al* 2011, Sen *et al* 2014) operates in a negative regulatory feedback circuit with mTORC1 (Tsuji-Tamura and Ogawa 2016, Jhanwar-Uniyal *et al* 2017), thus indirectly trigger the activation of ULK1 to induce autophagy through the inactivation of mTORC1 repressor action (Conciatori *et al* 2018). Mechanosensitive FAK can mechanically stimulate mTORC2 activation (Thompson *et al* 2013) (figure 5.2). Within adherent cells, FAK localizes to focal adhesions, a protein complex that regulates cell-substrate adhesion. A reduction in the mechanical forces acting on the focal adhesions, which can arise when cells become dislodged from the substrate or due to alterations in the rheological characteristics of the ECM, leads to a de-coupling of FAK from the focal adhesion complex (Kuo 2014). Soluble FAK can phosphorylate and thereby activate mTORC2 and subsequently trigger the induction of the autophagy process (Thompson *et al* 2013). Remarkably, mTORC2 is also able to activate AKT, thereby counterbalancing the inhibitory activity of mTORC1 via an indirect chain of signaling events (Case *et al* 2011). FAK, mTORC2 and AKT could therefore represent a potential negative regulation or an off switch to delay the autophagic degradation pathway (figure 5.2).

5.4.2 Nucleation step

The formation of the initial membrane complex, which extends and matures into an autophagosome, starts with the attachment of the activated ULK1 complex at or near the phagophore generation sites (Karanasios *et al* 2013) (figure 5.1, nucleation step). In yeast, the phagophore is localized in a restricted and specialized compartment between the exit point of the endoplasmic reticulum (ER) (Graef *et al* 2013, Hollenstein and Kraft 2020) and the vacuole, which is the lysosomal breakdown compartment in yeast (Yim and Mizushima 2020). Remarkably, the autophagosome stays in this compartment during the entire autophagic event in yeast. In contrast, in mammalian cells, phagophore assembly foci can be localized at several cytoplasmic sites, like the ER, ER-mitochondria contacts or the ER-plasma membrane, and specific subdomains of the plasma membrane harboring the primary cilium (Hamasaki *et al* 2013, Pampliega *et al* 2013, Nascimbeni *et al* 2017). Therefore, the autophagosome must be translocated through the cytosol for orderly ripening. These discrepancies between mammalian and yeast cells could reflect variations in the structural intricacy and spatial organization of the stress signaling pathways. This raises the tantalizing possibility that the higher level of spatial organization exhibited by the autophagic network in mammalian cells may reflect the higher level of mechanical organization exhibited by mammalian cells compared to the walled and sessile yeast cells. The generation of the autophagosome starts with a curved membrane formation, which is referred to as an omegasome due to its shape, which is like the Greek letter omega (Axe *et al* 2008). The omegasom forms a double-membrane structure that takes up lipids from most of the cell's interior compartments (Axe *et al* 2008, Yu *et al* 2018). The omegasome expands into a beaker-shaped double membrane, which is referred to as the phagophore or isolation membrane, that is generally connected at its foundation to the ER membrane (Carlsson and Simonsen 2015). Finally, the connecting point between the ER and the omegasome is fused, forming an organelle with two distinct membranes (Axe *et al* 2008). To accomplish this, multiple mechanical and energetic demands must be satisfied. This involves the recruiting of specialized ATGs, actin cytoskeleton to assist and orient the curved membrane, and the acquisition of the required material, especially phospholipids, to enable the *de novo* assembly of the phagophore (Mi *et al* 2015) (figure 5.1, nucleation step). The ULK1 complex is essential for the activation of all these processes. As a first event, ULK1 localizes to the phagocytic compartment, where it engages and activates PI3KC3, a kinase complex that comprises VPS34, Beclin-1, VPS15 and ATG14 (Yang and Klionsky 2010). PI3KC3 is activated in response to ULK1-based phosphorylation of Ambra1 (Di Bartolomeo *et al* 2010), a protein that interfaces with Beclin-1. The PI3KC3 complex, which is bound to the cytoskeleton by an interplay between the Ambra1 and dynein light chains (Di Bartolomeo *et al* 2010), causes the detachment of PI3KC3 from the dynein light chain and the microtubule meshwork, thereby permitting the activation of the complex and its trafficking to the omegasome. At this point, PI3KC3 phosphorylates phosphatidylinositol to produce phosphatidylinositol-3-phosphate (PI(3)P), which stimulates membrane bending and the subsequent recruiting of the additional ATG proteins needed in the subsequent phases of autophagosome generation (Axe *et al* 2008) (figure 5.1, nucleation step, ATGs enrolment). In addition to PI(3)P, membrane curvature is also maintained through Atg17, which is the yeast counterpart of FIP200 and a specific scaffolding protein that can also offer a curvature sensing mechanism (Ragusa *et al* 2012, Wang *et al* 2013). The Atg17 dimer has several hydrophobic residues that facilitate its interaction with the membrane. Atg17 dimers assemble to link fused vesicles, acquiring a distinctive biconcave moon shape (Ragusa *et al* 2012) hat is optimal for generating and propagating a membrane bending. In addition, PI(3)P recruits specific membrane-associated nucleation promoting factors (NPFs), such as WHAMM (Mathiowetz *et al* 2017), JMY (Coutts and La Thangue 2015), and WASH (Xia *et al* 2013). In reply to the specific juxtaposition of these factors, Arp2/3 and CapZ both polymerize a branching network of actin at the ER (figure 5.1, nucleation step, actin polymerization). This ATP-dependent and spatio-regulated polymerization of actin exerts compressive forces opposing the membrane, thereby maintaining the dome-like curvature of the isolation membrane (Mi *et al* 2015). Moreover, this ramified actin meshwork offers a structural framework to maintain the distinctive curvature of the membranes that give rise first to the omegasome and then to the adjoining phagophore. Due to their high, curved edges, the latter would be more likely to open energetically into a spherical vacuole than to maintain their characteristic cup shape. The favored shape of a vesicle is determined through the minimization of the membrane bending energy for a specified confined volume (Sackmann 1994). To surmount this energy hurdle, cells utilize various auxiliaries, including the asymmetric distribution of lipids and proteins between the two sides of the bilayer, such as PI(3)P and cholesterol, and the action of scaffold proteins like Atg1) and cytoskeletal scaffolds, such as actin filaments (Sackmann 1994, Chen and Rand 1997, Stachowiak *et al* 2012, Martens *et al* 2016) (figure 5.1, nucleation step). In an area that is over-proportionally concentrated on protein-mediated signaling events, the significance of the physical characteristics of the phospholipid bilayers has been widely neglected. While PI(3)P and actin polymerization prepare the physical landscape, ULK1 also triggers a second critical cascade that results in the replenishment of phospholipids to reconstitute the preautophagosomal double membrane, which is accomplished by the sorting of vesicles that acquire inputs from diverse membrane origins, comprising mitochondria-associated ER membrane, ER, Golgi, plasma membrane, and recycling endosomes (Staiano and Zappa 2019, Melia *et al* 2020). This appears to be achieved by two mechanisms: ATG9-mediated vesicle trafficking and merging with the omegasome (Karanasios *et al* 2013) and ATG2-facilitated transportation of lipids from a donor pool to the omegasome (Chowdhury *et al* 2018). Different signaling routes such as EGF/Src cause the uptake and phosphorylation of cytosolic ATG9 at the target membrane and the subsequent generation of ATG9-containing vesicles (Jia *et al* 2017, Zhou *et al* 2017). The issue of selectivity of the membrane source in relation to the type of autophagy and the type of cargo to be quarantined is a matter of ongoing discussion (Axe *et al* 2008). In essence, intracellular membrane transport is controlled by the Rab family of small monomeric GTPases (Stenmark 2009). In their GTP-bound form, Rab proteins sequester effector proteins to orchestrate vesicle traffic, whereas hydrolysis of coupled GTP to GDP results in the dissociation of effectors and their removal from the membranes. After activation of the autophagic pathway, the activated Rab11/Ypt11 GTPase governs the delivery of ATG9 vesicles to the omegasome due to the physical interaction of ATG9 with ULK1 (Longatti *et al* 2012, Ragusa *et al* 2012, Wang *et al* 2013). Actomyosin contractility appears to be essential for ATG9 vesicle trafficking. Activation of myosin IIA through the MLCK-like protein Sqa, which lies downstream of UKL1, was demonstrated to trigger the transportation of ATG9 vesicles toward the phagophore (Tang *et al* 2011). Although the suggested mechanism of cargo trafficking through myosin IIA may appear outlandish, as myosin IIA itself cannot be regarded as a cargo carrier (Bialik *et al* 2011), it is conceivable that actin wires under tension direct the flux of vesicles toward the phagophore. Recent work has proposed a novel mechanism for the translocation of phospholipids of the donor membrane toward the nascent autophagosome (Sawa-Makarska *et al* 2020). In fact, based on experimental findings, Atg9 defines membrane-targeting motifs with a donor compartment. In this case, phospholipids are exchanged between compartments through lipid transfer proteins such as Atg2, which leads to a net flux of lipids out of the vesicles toward the autophagosome in the absence of vesicle merging (Sawa-Makarska *et al* 2020).

5.4.3 Elongation and closure steps

After preparing the physical surroundings, the membrane of the emerging phagophore expands into an open, bowl-shaped structure through the merging of additional membranes (figure 5.1, elongation step). This newly formed structure encapsulates a small fraction of the cytosol, which can accept the material to be recycled (cargo loading), and ultimately closes by SNARE-mediated merging (figure 5.1, elongation step and closure step). From a mechanical point of view, the growth of the phagophore double membrane has the identical mechanical requirements as the step before. Therefore, this step continues the identical dynamics, whereby lipids are attracted by ATG9-driven merging and/or translocation (figure 5.1, elongation step, membrane recruitment) and the polymerization of the actin cytoskeleton to form a structural framework that assists in the growth of the double membrane and retains its morphology (Monastyrska *et al* 2008) (figure 5.1, elongation step, actin polymerization). These processes are tightly governed by various ATG proteins including ATG3, ATG7, ATG5, ATG12, and ATG16L1, and the lipidation of LC3/GABARAP family proteins (Al-Bari 2020). The lipidation pathway is the process by which the cytosolic LC3-I protein is fused with phosphatidylethanolamine (PE), resulting in the membrane-tethered LC3-II (Tanida *et al* 2004, Parzych and Klionsky 2014 Brier *et al* 2019). The LC3 lipidation needs the curved edge of the phagophore, because ATG3, the E2-like enzyme required for LC3 lipidation, functions exclusively on highly bent membranes (Nath *et al* 2014). In addition, it was found that the phagophore's localized curvature rises after LC3 integration, which points to the curvature-inducing characteristics of LC3 (Dall'Armi *et al* 2013). Therefore, the localization and accumulation of PE on the inner membrane of the phagophore is essential for the advancement of autophagy. In fact, it has also been suggested that phospholipid translocation (PE precursor) from the ER toward the acceptor membrane on neighboring organelles such as mitochondria may be the mechanism that triggers phagophore generation at locations distinct from the ER (Hailey *et al* 2010, McEwan and Dikic 2010). Besides other cargo receptors, LC3 is essential for the choice and incorporation of specific material into the autophagosome, as outlined in a review article (Birgisdottir *et al* 2013) (figure 5.1, elongation step, cargo loading). LC3 is also recognized to orchestrate cytoskeletal dynamics. In one scenario, LC3 engages NPFs such as WHAMM and JMY to facilitate Arp2/3-faciliated extension of the membrane-proximal actin meshwork to enable phagophore elongation and morphogenesis (Kast *et al* 2015, Hu and Mullins 2019). Conversely, it has been postulated that microtubule-interacting LC3 may orchestrate the transportation and sorting of malfunctioning organelles (Hanna *et al* 2012), the extension of the phagophore and, at a later stage, the closing of the autophagosome (Fass *et al* 2006, Geeraert *et al* 2010, Weidberg *et al* 2010). After the phagophore has been filled with its cargo, it seals to create a double-membrane organelle, the autophagosome, to enclose its internal breakdown compartment (Ravikumar *et al* 2010, Zhao and Zhang 2019) (figure 5.1, closure step). Before being sealed, all ATG proteins attached to the PI(3)P platform are dislodged from the outer surface of the autophagosome. This process involves elimination of PI(3)P by phosphoinositide phosphatases and potentially additional factors (Cebollero *et al* 2012, Nakatogawa *et al* 2012, Lee and Lee 2016). It should be emphasized that the elimination of PI(3)P is an integral mechanism for the disassembly of the nucleating-elongating ATG network that is necessary for the generation of the mature autophagosome (Cebollero *et al* 2012). Ultimately, the phagophore is terminated through a fission mechanism of the inner and outer membrane of the phagophore to form an autophagosome enclosed by a double membrane (Knorr *et al* 2015). This mechanism, although not yet fully elucidated, is primarily dependent on the endosomal sorting complex required for transport (ESCRT) (Melia *et al* 2020, Li *et al* 2021a) and displays a similar membrane topology to canonical ESCRT-dependent cellular membrane separation events, like cytokinesis, plasma membrane healing, and multivesicular body formation (Yu and Melia 2017, Takahashi *et al* 2019b). The ESCRT pathway consists of several conserved protein complexes (ESCRT-I, -II and -III) and additional accessory proteins, including the ATPase protein VPS4, which breaks down and reuses the ESCRT-III complex (Christ *et al* 2017, Schöneberg *et al* 2017). During the course of the event, ESCRT-III subunits organize into helical filaments, which are the driving factor for initiating membrane deformation, whereas VPS4 recruitment powers membrane tethering (Takahashi *et al* 2019b) and ultimate cleavage (Adell *et al* 2014, Chiaruttini *et al* 2015). Besides the ESCRT complex, the motor protein Myo6 and the actin cytoskeleton are involved in sealing the phagophore (Yu and Melia 2017, Takahashi *et al* 2019b). Collectively, these elements pull the free ends of the autophagosome together, facilitating SNARE-based merging (Corona and Jackson 2018).

5.5 The autolysosome

When an autophagosome or amphisome merges with a lysosome, the emerging compartment is termed an autolysosome (Klionsky *et al* 2014). Certain authors of scientific articles on autophagy use the terms ‘autolysosome’ and ‘autophagolysosome’ synonymously, which is not precisely. These terms should be used precisely to denote two distinct compartments. It is necessary to keep this distinction of compartments because the autophagolysosome has a special origin in the process of xenophagy that differentiates it from an autolysosome. The direct merging of an autophagosome with a lysosome creates an autolysosome, a term most researchers in the field of autophagy are in agreement on (Ericsson 1969). The convergence of macroautophagy with endocytosis and phagocytosis is a possible cause of confusion about nomenclature. When an endosome, which is a compartment enclosed with a single membrane, merges with an autophagosome, an amphisome is generated (figure 5.3) (Gordon and Seglen 1988) this is also a compartment with a single membrane in which the inner autophagosome membrane is broken down. The subsequent merging of an amphisome with a lysosome produces an autolysosome, and the application of the term autolysosome in this context is universally accepted. Phagocytosis involves the formation of a single-membrane phagosome, which can merge straight away with a lysosome to form a phagolysosome (Armstrong and Hart 1971). There is also broad agreement on the usage of this term to describe the merging entity. Importantly, this is the result of a traditional phagosome-lysosome merging event without autophagy.

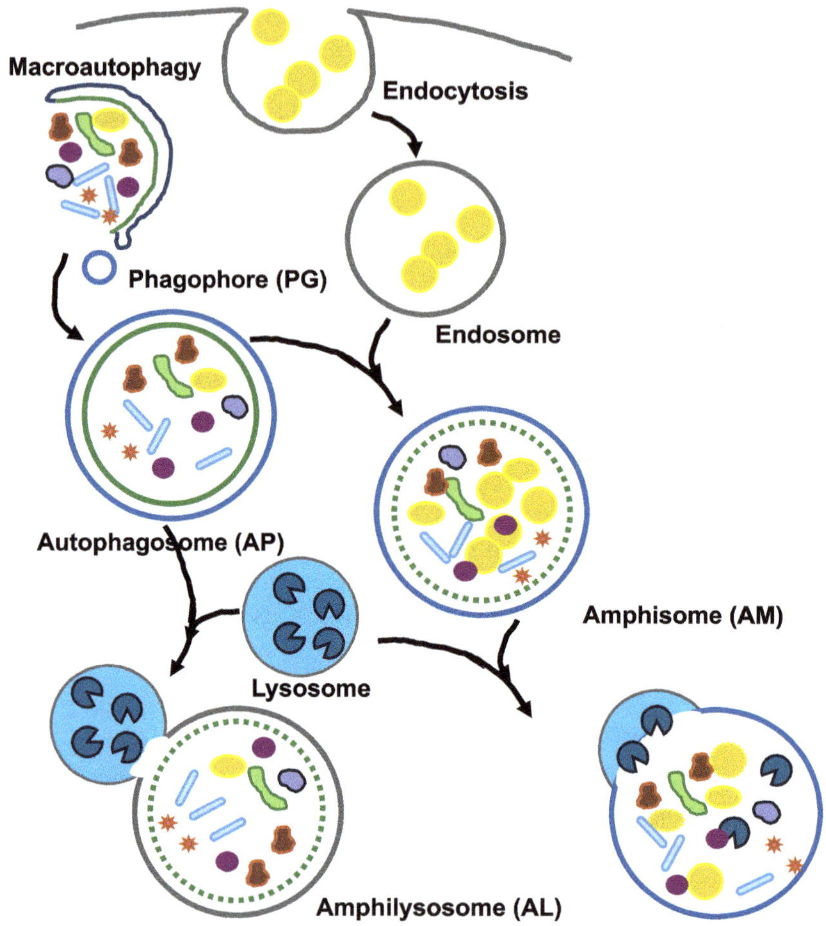

Figure 5.3. Schematic drawing of the emergence of an autolysosome. The merging of a double-membrane autophagosome with a lysosome results in the formation of an autolysosome. Likewise, the convergence of macroautophagy and endocytosis results in a single-membrane amphisome, which also leads to an autolysosome when it merges with a lysosome.

An example particularly relevant to autophagy is the merging of a phagosome decorated with LC3 with a lysosome, which takes place during LC3-driven phagocytosis (Martinez *et al* 2011). In this case, the fusion product is termed an autophagolysosome (figure 5.4). It is worth emphasizing this clear-cut concept, because some microbes, such as *Mycobacterium tuberculosis*, survive within their hosts through the prevention of phagosome-lysosome merging, thereby avoiding the establishment of a conventional phagolysosome (Armstrong and Hart 1971, Deretic *et al* 2013). The phagosome may also be incorporated into an autophagosome (Amer and Swanson 2005). In this scenario, when the autophagosome merges with a lysosome, the emerging compartment is referred to as an autophagolysosome once again (figure 5.4). The autophagolysosome thus refers to a very distinct event that is the outcome of certain

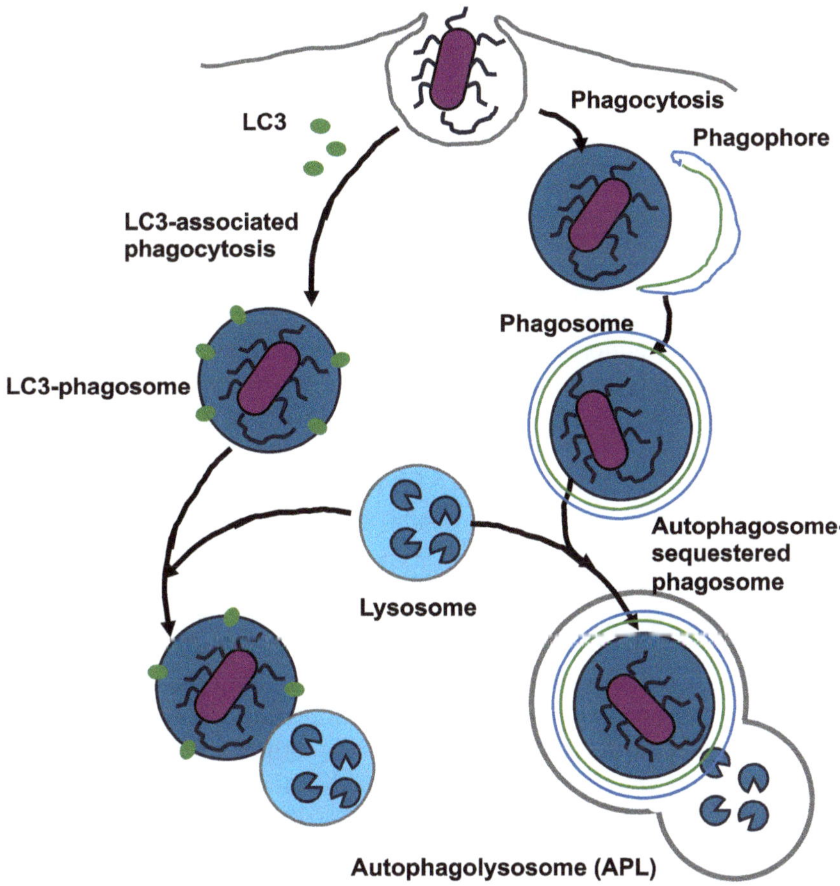

Figure 5.4. The emergence of autophagolysosomes. There are different kinds of xenophagy, the selective engulfment and breakdown of invasive microbes through macroautophagy. When a phagosome marked with LC3 or an autophagosome that has engulfed a phagosome merges with a lysosome, an autophagolysosome is created. For clarity, LC3 is not depicted on the phagophore or autophagosome.

types of xenophagy, especially when the event comprises a phagosome and elements of the autophagic system. Consequently, the distinction between the terms autolysosome, phagolysosome and autophagolysosome must be retained.

5.5.1 Maturation and generation of the autolysosome

As soon as the double membrane has merged, the autophagosomes start to mature. This process involves the merging of the autophagosome with early/late endosomes (figure 5.1, maturation step) and its transportation toward a perinuclear area, which is facilitated through microtubules and dynein (figure 5.1, maturation step, transport) (Jordens *et al* 2001, Kimura *et al* 2008). Subsequently, the mature autophagosome merges with the lysosome to produce the autolysosome, in which the breakdown of the freight takes place (figure 5.1, autolysosome formation step). Merging of autophagosomes with endosomes/lysosomes can proceed through a number of mechanisms, such as kiss-and-run, full merging or tubule-based merging (Trivedi *et al* 2020). These processes appear to involve two distinct regulated events: docking and merging. When the autophagosome and the lysosome come together, the outer membrane of the autophagosome merges with the lysosome to create an autolysosome. Merging of endo-lysosomal vesicles with autophagosomes generally involves Rab GTPases for vesicle transport and tethering (notably Rab7), membrane-binding complexes, and SNAREs to transduce vesicle merging in a distinct fashion (Itakura *et al* 2012b, Diao *et al* 2015, Guerra and Bucci 2016). The molecular mechanism that regulates the merging of autophagosomes with lysosomes is not yet completely elucidated. Recent findings indicate that elevated PI(4)P concentrations on late endosomes/lysosomes promote the renrolment of the multisubunit homotypic merging and vacuole protein sorting complex (HOPS) (Miao *et al* 2020). The HOPS complex interferes with the binding of the lysosomal protein LC3 to the autophagosome membrane (Manil-Ségalen *et al* 2014, McEwan *et al* 2015) and by direct interference with the STX17 localized in the autophagosome, the assembly of the SNARE complexes (Jiang *et al* 2014) between STX17 and its interaction partners, the ubiquitous SNAP29 and the lysosomal VAMP3 (Itakura *et al* 2012b, Miao *et al* 2020). In addition to the HOPS complex, TECPR1, a protein found on lysosomal membranes, has also been suggested as a docking factor that promotes autophagosome-lysosome merging (Chen *et al* 2012) by attracting LC3-matured autophagosomes toward lysosomes and facilitating the clearance of protein aggregations (Wetzel *et al* 2020). The entire docking and fusion process is guided by the ubiquitous actin cytoskeleton, which acts to maintain curvature and supply the mechanical energy required to position the membranes of different organelles in proximity for merging (Kast and Dominguez 2017, Nakamura and Yoshimori 2017). The latter event is conferred by WHAMM-dependent polymerization of the ramified actin meshwork via cortactin and Arp2/3, which results in the occurrence of stress-bearing actin comets (Kast *et al* 2015), and by the unorthodox myosin motor protein Myo1C (Zhang *et al* 2019). In contrast to the canonical fast-twitch myosins (e.g. myosin IIA), Myo1C is a low-speed monomeric actin-based motor protein that is suitable for the slow translocation of large cargos. Although its mechanistic effect is not yet fully understood, it is assumed that it connects the PI(3,5)P2-enriched membrane freight, which has been generated through a PIKfyve-dependent PI(3)P dissipative process (Kim *et al* 2014) with the actin cytoskeleton (Kruppa *et al* 2016) and subsequently serves to stabilize membrane folds (Bose *et al* 2004).

5.5.2 Breakdown and recycling of cargo

After fusion of the lysosomes with the autophagosome and breakdown of the inner membrane, the process of breaking down the autophagosomal load is initiated, as illustrated in figure 5.1—degradation and recycling (Yu *et al* 2018). During this phase, autolysosomes are considerably diminished in size (Dai *et al* 2019), as a result of the breakdown of their load and the transportation of small solutes, such as amino acids, monosaccharides and nucleosides, which is facilitated by the solute carrier translocator (Freeman and Grinstein 2018). The transport of solutes throughout the autolysosome membrane is trailed by the succeeding osmotic forces that create a water flow towards the outside. Autolysosome retraction is necessary for the downstream event of lysosomal membrane reclamation. The strong membrane curvature that results from the contraction of the autolysosome causes the enrichment of the protein complexes needed for the processes of vesiculation and tubulation, which enable lysosomal vesicles to rebuild themselves (McMahon and Boucrot 2015). Autolysosome tubulation is also aided through the protein WHAMM, which, once attracted to the surface of the autolysosome, encourages the generation of a ramified actin scaffolding that eases the operation (Dai *et al* 2019).

5.6 Interplay between cellular mechanics and autophagy

A multitude of biophysical cues triggers cellular reactions and governs cellular fate (figure 5.5(A)) (Sainio and Järveläinen 2014). Many of these cues arise from the short-term (figure 5.5(A), blue boxes) interplay between cells and their physico-chemical microsurroundings.

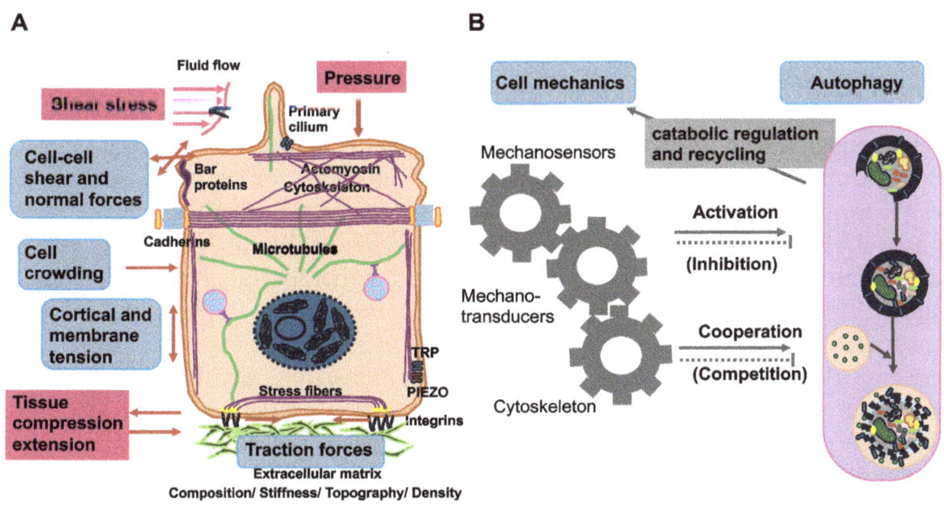

Figure 5.5. Schematic illustration of cell mechanics and their interaction with autophagy. Cells exposed to a wide range of mechanical forces (red arrows) from the surroundings that produce cell-autonomous forces that are transmitted through the cytoskeleton. (A) Forces perceived by the cells fall into two main classes: those on the short length scale (blue boxes) and those on the long length scale (red boxes). Short- and long-range forces are sensed by cells through different mechanosensory receptors, which can be categorized into cell–cell adhesion complexes (integrin- and cadherin-mediated adhesions), mechanosensitive ion channels (TRP, Piezo), tension and bending sensors localized at the plasma membrane and the actin cortical skeleton, such as BAR proteins and filamin, and the primary cilium. (B) Mechanical stimuli are converted into biochemical signals (mechanotransducers) like Ca^{2+}, transcription factors (YAP/TAZ) and signaling molecules like phosphatases and kinases, which influence the cytoskeleton, transcription and other essential cellular processes. Autophagy is induced through mechanical processes directly through cellular signal transduction and indirectly through cooperation with the cytoskeleton. Although it is probable that negative feedback through inhibition and competition for cytoskeletal elements, is present, this has not yet been sufficiently investigated in the scientific literature. Autophagy is a mechanism that governs different mechanical events by assuring the recirculation of cellular elements and supplying energy in the catabolic pathway.

Cells have been found to firstly perceive the physical nature of the ECM, such as its makeup, stiffness, topography and thickness, and respond through applying traction forces toward the underlying substrate (Engler *et al* 2006, Vogel and Sheetz 2006, Doss *et al* 2020), secondly geometric information, such as size, boundary, and curvedness, which modulate cortical and membrane tension (Vedula *et al* 2012, Ravasio *et al* 2015a), thirdly the existence of neighboring cells, such as cell compactness (synonymously referred to as crowding), and their physical behavior, such as cell traction and pushing, which generate cell–cell shear forces and normal forces, and the chemical makeup of the interstitial and luminal liquids, such as osmotic pressure, which causes cell swelling or retraction and a resulting change in membrane tension (Baumgarten and Feher 2001). Cells also sense and react to strong mechanical forces (figure 5.5(A), red boxes) like shear stress and fluid pressure resulting from the flux of fluids or solid material within the lumen of tubular organs such gut, blood vessels, and urinary tract, and the lateral stretching and compression of tissues that are necessary for the physiological operation of the lungs, muscles, and gastrointestinal system (Dietl *et al* 2004, Jennings *et al* 2005, Ravasio *et al* 2011, Shivashankar *et al* 2015). Short- and long-range forces (figure 5.5(A)) trigger cellular responses that are both responsive and adaptive, and primarily involve active signaling processes that are facilitated by the cytoskeleton (Iskratsch *et al* 2014). Activation can be initiated through the direct exposure of the cytoskeleton to external mechanical stimuli through recognition mechanisms that engage different cell-surface mechanosensors, like mechanically activated ion channels such as TRP and Piezo, proteins that detect the tension and warping of the plasma membrane, such as BAR proteins, and of the cytoskeleton, such as filamin, as well as adhesion protein complexes such as focal adhesions and adherens junctions (Fletcher and Mullins 2010). These mechanosensors convert mechanical stimuli via mechano-transducers, which is otherwise known as the mechanotransduction process, into biochemical cues that govern cytoskeletal organization, membranous transport, expression profiles of genes and, in the end, the entire cellular functionality (Iskratsch *et al* 2014, Chen *et al* 2017, Wolfenson *et al* 2019) (figure 5.5(B)). Mechanosensation usually involves a force-driven conformational rearrangement of the sensor protein that can result in the aperture of a channel, such as usually calcium channels, which then triggers a cellular reaction through an electrochemical impulse or through the uncoupling of proteins, the mechanotransducers, that are part of the sensor complex. In its freely diffusing state, the mechanotransducer is involved in enzymatic processes such as phosphorylation, either as an enzyme or as a metabolite. In both cases, the increase in calcium and/or the triggering of protein phosphorylation pathways results in short- and long-term adjustment to mechanical signals. As a key component of the cell's innate adaptation mechanisms, the autophagic reaction helps to overcome mechanical hurdles and allows the cell to accommodate the constantly fluctuating physical landscape (figure 5.5(B)). In general, mechanical stimuli can influence the autophagic pathway in two principal modes: first, through specific interactions between mechanotransduction and autophagy control proteins, such as mTORC and AMPK These regulatory proteins are involved in the induction and/or repression of the autophagic program and/or recruit cytoskeletal elements and phospholipid membranes through the non-specific crosstalk/competition mechanisms existing between mechanical events and autophagy (Dower *et al* 2018). There is growing experimental support for the notion that mechanical signals actually integrate into the signals needed to activate autophagy (King 2012, Das *et al* 2018, 2019, 2020). In contrast, the rivalry for cellular elements between the two processes and the implications of this rivalry, although very plausible, have yet to be discussed in the relevant literature. A last point of convergent evolution is the importance of the regulatory function of autophagic degradation and reuse of biological compounds in the control of the turnover of cellular constituents required for the correct execution of mechanical tasks. In the following sections, crosstalk and interplay between cell mechanics and autophagy will be elucidated. Next, some of the most prominent and established connections between the mechanotransduction machinery and the autophagic process will be thoroughly discussed.

5.6.1 Extracellular matrix environment and focal adhesions

The macromolecular assembly, three-dimensional architecture and rheological features of the ECM are subject to continuous reorganization as a result of enzymatic and mechanical activities of the cells (Vogel and Sheetz 2006, Vogel 2018). These changes and the restructuring processes provide a multifaceted microniche, which in turn influences the phenotype and performance of the cells and can lead to the development of diseases like fibrosis and cancer if incorrectly regulated (Sainio and Järveläinen 2014). The capacity of cells to perceive the mechanical characteristics of the ECM under physiologic and pathologic circumstances can be ascribed to integrin-mediated adhesions, commonly referred to as focal adhesions (Jiang *et al* 2006, Ross *et al* 2013). Focal adhesions consist of several mechanosensors, such as talin and vinculin, signaling molecules, such as FAK, Src and PI3K, adaptor proteins, such as paxillin, and actin linker proteins like filamin and alpha-actinin, which physically attach the integrins to the cytoskeleton and are positioned in specific layers with different dynamic remodeling rates (Wozniak *et al* 2004, Kechagia *et al* 2019, Legerstee and Houtsmuller 2021). Engagement of ECM ligands to integrin heterodimers drives stress-triggered conformational alterations in the cytoplasmic tail of integrin, which results in the enrollment of talin and paxillin (Kim *et al* 2003, Luo *et al* 2007). As tension rises and focal adhesion ripens, protein tyrosine kinase 2 and Src are enlisted, thereby contributing enzymatic kinase activity to drive subsequent signal transduction, which encompasses Rho-GTPase signal transduction pathway, anoikis signaling, mitogenic signal routes, and ECM metabolism (Caswell *et al* 2009). In this way, integrin-facilitated adhesions engage with the ECM and perceive its stiffness, which in its turn regulates cell response, such as migratory behavior and invasion (Seetharaman and Etienne-Manneville 2018). Multiple investigations focus on how the ECM and integrin-based adhesions may induce autophagy through FAK and integrin linked kinase (ILK) via detachment-induced cell death and therefore be associated with anoikis and in turn promote cancer advancement (Lock and Debnath 2008, Vlahakis and Debnath 2017 Anlaş and Nelson 2020). Crucially, these emerging links between integrin-mediated adhesion pathways and autophagy are important for immune surveillance (Gubbiotti and Iozzo 2015) and hence impact on the occurrence of specific diseases, such as cancer. It has been demonstrated that matrix components can control autophagy in both positive (activators) and negative (inhibitors) directions. Decorin, collagen VI, kringle 5, perlecan and endostatin act as activators (Nguyen *et al* 2007 2009 Gubbiotti and Iozzo 2015 Castagnaro *et al* 2018), while laminin a2 functions as an inhibitor of the autophagic pathway (Carmignac *et al* 2011). The ECM, displaying various physical and structural properties, can trigger biochemical signaling pathways that engage membrane receptors such as integrins, VEGFR2, GRP78 (Nguyen *et al* 2007, 2009, Neill *et al* 2014), regulatory proteins (AKT, mTORC1 and 2) and autophagy-specific enhancers, comprising VPS34, Beclin-1 and lipidated LC3 (Neill *et al* 2012, Gubbiotti and Iozzo 2015). Conversely, autophagy controls integrin-based adhesion and thereby cell migration through regulating the metabolism of focal adhesion through a process engaging LC3, paxillin and Src (Sharifi *et al* 2016).

5.6.2 Intercellular adhesions

Besides the ECM, the cells of a tissue engage in physical interactions to other cells like epithelial cells and muscle cells through transmembrane receptors that establish extracellular connections with receptors on nearby cells to maintain the integrity of the tissue and govern collective cellular dynamics (Angulo-Urarte *et al* 2020). Cell–cell junctions are facilitated by a variety of adhesion complexes including adherens junctions, tight junctions and desmosomes, all of which exhibit different functionality and molecular properties. Adherens junctions act as force sensor complexes. Tight junctions seem to function only through a physical bond between the two complexes in parallel with adherens junctions. The involvement of the desmosome during mechanotransduction events at the junctions continues to be obscure. In adherens junctions, the linkage between the cadherin transmembrane receptor and the actin cytoskeleton is facilitated by a protein complex referred to as the cadhesome network (Zaidel-Bar 2013). This complex, like the aforementioned integrin-based adhesion, exhibits a clearly defined spatial arrangement in which force transmission is facilitated through protein conformation, which itself regulates the linkage of the cadherins with the actin cytoskeleton (Bertocchi *et al* 2017, Han and De Rooij 2017). Tensions at adherens junctions cause a conformational switch of α-catenin with the consequence that previously concealed attachment sites for vinculin are uncovered, which causes an enhanced functional incorporation of the complex into the actin dynamics. Tension-activated conformational modifications of vinculin can activate the signaling layer of the actomyosin contraction apparatus in various manners and enable selective actin polymerization through the vinculin-coupled Mena-VASP complex (Bertocchi *et al* 2017, Han and De Rooij 2017). Vinculin thus acts as a 'molecular coupler' that combines mechanical and biochemical cues to connect and disconnect the cell–cell junction with both internal and external forces. This striking spatial arrangement and the underlying molecular mechanism endow cells with the strength and plasticity that highly dynamic epithelial tissues require for biological functions such as collective cell migration, tissue repair, and tissue stretching. Autophagy exerts a critical function in the homeostasis of the junctions through the active control of the recycling of the junctional complexes in reply to various intra- and extracellular stimuli that has been outlined in a review article (Nighot and Ma 2016). Experimental findings demonstrate an autophagy-driven relocalization of cadherin (Damiano *et al* 2020) and claudin (Hu *et al* 2015) away from the cell membrane toward the cytosol, from which they are ultimately cleared through the autophagosome or lysosome. The impact of cell–cell adhesion on the autophagic pathway has received little attention. Nonetheless, it has been demonstrated that the action of force on E-cadherin triggers autophagy through the activation of liver kinase B1 (LKB1), which leads to the attraction of the autophagy initiator factor AMPK toward the E-cadherin complex (Bays *et al* 2017).

5.6.3 Yes-associated protein/transcriptional co-activator with PDZ-binding motif signal transduction

The Yes-associated protein (YAP, alternatively referred to as YAP1) and its paralogue TAZ (containing a PDZ-binding motif) are transcriptional coactivators that shuttle between the cytoplasm and nucleus and govern organ size and maintain tissue homeostasis. Autophagy and mechanosensation are intertwined through the YAP/TAZ signaling system, in addition to what was mentioned in the preceding sections. YAP and the TAZ control the expression of genes in a force-dependent fashion. The groundbreaking work of Piccolo and coworkers demonstrated that mechanical forces control the cytosolic distribution and nuclear translocation of YAP/TAZ (Dupont *et al* 2011). By examining YAP subcellular localization and transcriptional activation, YAP activity was found to be modulated through ECM stiffness, cell density, and cellular geometry. When cells have low density or reside on stiff ECM, YAP and TAZ are active and translocate to the nucleus to engage the DNA-binding transcription factor TEAD and drive the expression of several growth-related genes, thereby ultimately leading to cell proliferation (Zhao *et al* 2008, Zhou *et al* 2016). In contrast, YAP/TAZ is not active in the cytoplasm when cells are at high cell density or when seeded on a soft matrix (Dupont *et al* 2011, Wada *et al* 2011, Aragona *et al* 2013) which results in blockage of proliferation due to contact impairment. This force-dependent regulation of proliferation is a key mechanism for sustaining tissue equilibrium and promoting tissue repair. Disruption of this system can result in out-of-control cell growth, which is a hallmark of cancer. Intriguingly, among the transcriptional targets of YAP/TAZ is Armus (Totaro *et al* 2019), which belongs to the Rab-GAP family of proteins that regulates autophagosome-lysosome merging (Carroll *et al* 2013). As a result, the effectiveness of the autophagic flow has been found to rely on the physical characteristics of the cell's micromilieu through the mechanical signaling reaction of YAP/TAZ (Totaro *et al* 2019). Moreover, mTORC1 controls YAP by promoting its autophagic destruction

(Liang *et al* 2014), thereby providing another connection between cellular nutrient levels and the activity of YAP (Pocaterra *et al* 2020).

5.6.4 Ion channels sense mechanical cues

Mechanosensitive calcium influx has been linked to the initialization and elongation steps of autophagy (Kondratskyi *et al* 2018). ER-resident channels display the capacity to govern autophagy at various phages, such as initiation and the autophagosome-lysosome merging steps, because of its distinct function as a hub for the nucleation of autophagosomes. Nevertheless, these channels, located in the ER, were not associated with mechanosensory perception. Alternatively, plasma membrane channels have been implicated in the induction process through AMPK and mTOR signal pathways. Intriguingly, two major families of calcium channels, the osmo-mechano and voltage-gated transient receptor potential (TRP) channels and the pore-forming piezo channels (Moroni *et al* 2018) are recognized as mechanosensitive. These two channel families are controlled by alterations in membrane tension, as can occur through the stretching and compression of the plasma membrane caused by cell migration or the shearing of cells. These channels also react to osmotic stimuli that raise membrane tension levels in the course of cell swelling (Liedtke and Heller 2007).

5.6.5 Plasma membrane and cytoskeletal tensions

As the physical barrier separating the cell from its surroundings, the plasma membrane plays a crucial part in mechanosensation and mechanotransduction (Ayad *et al* 2019, Le Roux *et al* 2019). The lipid bilayer, whose extensibility is very low, because it cracks with an extension of only 3%–5%, is mechanically reinforced through the actin cortex, which, due to its active and dynamic behavior, is capable of absorbing a large part of the generated tension, controlling the folding and unfolding of the plasma membrane into and out of membrane pools, and assisting in vesicular transportation and vesicular merging. Mechanical cues at the plasma membrane can be categorized as tensile, such as cell stretching and hypoosmotic expansion; compressive, such as cell compression and hyperosmotic retraction; shear stress, like fluid flow over adherent cells; and topographically generated stimuli, such as restriction through the physical microenvironment. It has been proposed that the shear stress of fluids triggers the initiation of autophagy through activating Rho-GTPases, like Rac1, RhoA, and Cdc42, which leads to upregulation of Beclin-1, ATG5, ATG7, and LC3 (Yan *et al* 2019). In addition, cells react to mechanical stress through the fast generation of autophagosomes via an mTOR-independent route (King *et al* 2011). The autophagic reaction shows a high degree of specificity for mechanical stress with a transient and progressive reaction to the impulse, such as half-maximal reactions at about 0.2 kPa (King *et al* 2011). The exact recognition and signaling pathways are not fully understood, but it is possible that BAR proteins, which have been characterized as primary membrane tension sensors, are implicated (Peter *et al* 2004). Another mechanism of tension measurement concerns the actin-skeleton protein filamin. Filamin A regulates the tension level of the actin cytoskeleton through its ability to crosslink actin filaments at wide angles (Glogauer *et al* 1998, Razinia *et al* 2012). When cells are subjected to stress by mere flow, filamin enriches in the entire cell and enhances the mechanical stability of the cytoskeleton as a whole (Jackson *et al* 2008). Moreover, filamin A connects integrin with actin, thereby providing force-dependent strengthening at the focal adhesions (Gehler *et al* 2009). In reaction to tension, filamin A experiences conformational alterations that enhance its ubiquitination and ultimate targeting through chaperone-mediated autophagy (Ulbricht *et al* 2013b).

5.6.6 Possible mechanosensing mechanism at the primary cilium

Key cellular functions including cell migration, differentiation, cell cycle restart and apoptosis critically involve the function of the primary cilium (Satir *et al* 2010). The primary cilium, present in most cell types, is a non-motile microtubule-based appendage that perceives extracellular chemical and mechanical cues (Praetorius, 2015, Orhon *et al* 2016). For example, the cilium in kidney cells functions as a flow sensing device. Shear forces that induce curvature of the cilium trigger calcium influx into the cell through polycystin-2 (PC2) and transient receptor potential vanilloid 4 (TRPV4) (Praetorius and Spring 2001). This purely sheer-stress-dependent signal transmission induces autophagy and regulates cell size (Orhon *et al* 2016, Zemirli *et al* 2019) via the LKB1-AMPK-mTOR signaling cascade (Boehlke *et al* 2010). In contradistinction to autophagy caused through hunger, autophagy is initiated through mechanical cues from the cilium in a way that is separate from ULK1, Beclin-1 and PI3K/VPS34 (Boukhalfa *et al* 2020a, 2020b). Recently, it has been proposed that PI3KC2α lipid kinase (PI3K class II), which is necessary for ciliogenesis and cilia perfomance, can enhance the assembly of a local pool of PI (3)P in reply to shear stress (Boukhalfa *et al* 2020b). PI(3)P, in contrast, is essential for the membrane trafficking and activation of Rab11a (Franco *et al* 2014, Campa *et al* 2018), and acts as a scaffold for the recruitment of the autophagosome assembly machinery and its subsequent activation (Puri *et al* 2018, Vicinanza *et al* 2019). Conversely, the size and assembly of primary cilia (ciliogenesis) are controlled through autophagy. This includes the breakdown of ciliogenesis regulatory proteins (Pierce and Nachury 2013 Tang *et al* 2013), as it has been shown that several members of the autophagic pathway, such as ATG16L1 accumulate at the basal body of the cilium upon fasting (Pampliega *et al* 2013).

5.7 Mechanics of autophagy in the malignant progression of cancer

Malignant conversion is associated with a continuing deterioration in tissue homeostasis and disturbances in the architecture of the tissue. It is now well accepted that a crucial feature of this transition comprises alterations in the cell's mechanical phenotype and the concomitant remodeling of the local microenvironment, thereby establishing a distinct mechanical niche populated mainly with cancer cells embedded in a dense ECM scaffold (Kumar and Weaver 2009, Ravasio *et al* 2015b, Northcott *et al* 2018). Moreover, a number of additional cells can be identified in the tumor microenvironment, such as blood and lymphatic endothelial cells, lymphocytes, inflammatory cells, and CAFs (Kumar and Weaver 2009, Northcott *et al* 2018, Sharma *et al* 2020). According to the cancer stage and development context, autophagy was identified as a 'double-edged sword' because it can function as either a tumor-suppressing or tumor-promoting process, based on the cellular environment in which it operates (White and DiPaola 2009, Bhutia *et al* 2013). In line with its supportive function in cell survival and anti-aging strategies, autophagy acts as a major quality control mechanism that recognizes alterations in organelle structure and protein folding, thereby inhibiting tumorigenesis. Conversely, the identical mechanisms favor the viability of cancer cells, preventing them from succumbing to apoptosis. This takes place by supporting the reactions to environmental stress and producing the energy required for unrestrained growth and metastasis through the restoration and breakdown of cell organelles (figure 5.6) (Wang 1998, Morselli *et al* 2009, White and DiPaola 2009, Panda *et al* 2015), which has been reported to alter cellular mechanical cues, such as Golgi, endoplasmatic reticulum, mitochondria, vacuoles, lysosomes and peroxysomes (Fischer *et al* 2020, Mierke 2020b, 2020c, 2020d, 2020e).

In the case of solid tumors, multiple mechanical features of the tumor stroma promote the tumor-promoting effect of autophagy (figure 5.6, right panel). When encapsulated by the ECM scaffold, cancer spheroids are subjected to forces imparted by the expanding tumor mass because of unrestrained proliferation and the resilience of the circumjacent stromal tissue toward deformation (Nia *et al* 2017, Jain *et al* 2020). This results in elevated interstitial pressure (Jain *et al* 2020) and produces shear forces inside the micromilieu of the tumor (Shieh and Swartz 2011,

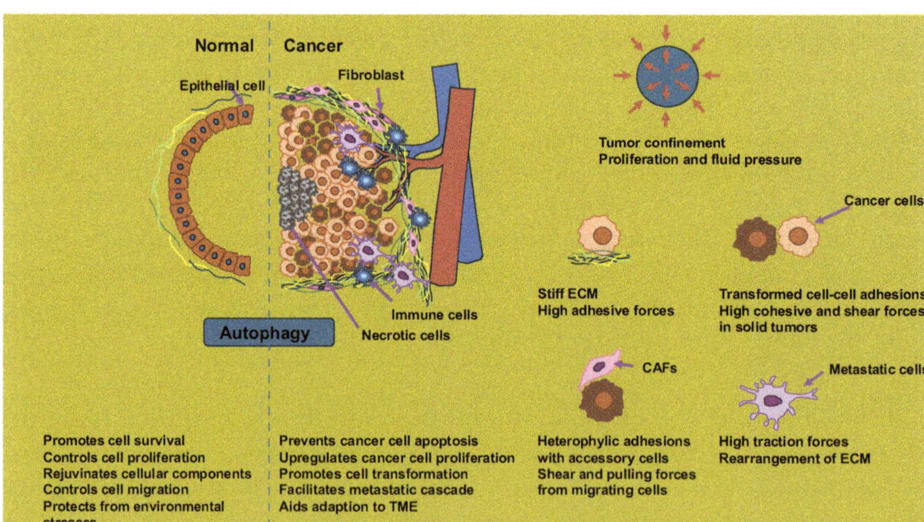

Figure 5.6. Autophagy and mechanical features in malignant transition of cancer. The maintenance of cellular homeostasis and the defense against mechanical stress are illustrated and described in the left panel (the dashed line divides the normal and cancer settings). Cancer cells use autophagy to adjust to the tumor microenvironment and drive malignant disease progression. The panel on the right depicts the mechanical elements of the tumor microenvironment.

Hyler *et al* 2018, Das *et al* 2020). Finally, this mechanical stress has a direct impact on cell growth as it causes compression of the cancer cells and/or an indirect impact as it causes compression of the surrounding blood and lymph vessels (Stylianopoulos 2017). Due to the persistent compression of the vessels inside the tumor, insufficient blood flow in the tissue leads to hypoxia and ultimately possibly to necrosis inside the tumor (Stylianopoulos *et al* 2013). Hypoxia enhances EMT, cytoskeletal rearrangement, and loss of epithelial cell–cell junctions via disassembly. This results in dynamic cell extension, directional migration (Lamouille *et al* 2014), and, in consequence, an enhanced metastatic capability of the cancer cells. Moreover, the molecular composition of the intermediate filaments undergoes alterations during EMT, whereby there is a shift from keratin to vimentin (Kumar and Weaver 2009, Wirtz *et al* 2011). In addition, the actin cytoskeleton undergoes a hypercontractile rearrangement in this process via TGF-β-driven activation of signaling routes including Rho-GTPases, p38MAPK, and ERK1/2 (Xie *et al* 2004). Activation of this signaling cascade leads to actin rearrangement and the generation of cellular protrusions, which comprise lamellipodia and filopodia (Derynck and Zhang 2003, Ridley 2011). In addition, TGF-β and hypoxia also drive the generation of CAFs (Gilkes *et al* 2014, Stylianou *et al* 2018), which communicate with all cellular constituents of the tumor microenvironment.

The ECM stiffness enables the fibroblasts associated with cancer to modulate the cytoskeleton of the cancer cells (Gaggioli *et al* 2007, García-Palmero *et al* 2016, Najafi *et al* 2019). These alterations of the cellular cytoskeleton throughout transformation or EMT radically modify their mechanical phenotype, especially the level of tension applied to adjacent cells and the ECM, resulting in enhanced migratory, invasive, and dispersive capacities (Yu *et al* 2011a, Li and Wang 2020). To metastasize effectively, cancer cells travel from the local site of disease to adjacent tissues and then into the vasculature. They then enter the bloodstream through the basal lamina and detach from the ECM to become circulating cancer cells, which is referred to as intravasation (Plaks *et al* 2013, Massagué and Obenauf 2016). Ultimately, circulating tumor cells (CTCs) that persist in the circulation can exit the vascular bed and seed secondary sites, thereby establishing metastases (Strilic and Offermanns 2017). All cells that move in the circulation are exposed to varying shear forces. The hemodynamic shear force that arises from the motion of blood flowing along the cell surface is affected by both fluid viscosity and fluid flow rate (Wirtz *et al* 2011). Shear stress can also be induced through frictional interplay with endothelial cells (Fan *et al* 2016, Northcott *et al* 2018). Likewise, cancer cells in the circulation must survive challenging circumstances, such as apoptosis caused by shedding of the ECM, for instance through anoikis, attacks by the immune system, and fluctuations in shear forces (Strilic and Offermanns 2017). The physiological shear stress (0.5–3 Pa) generated due to fluid flow can inhibit cancer cell proliferation, while also enhancing their migration and adhesion (Avvisato *et al* 2007, Mitchell and King, 2013, Ma *et al* 2017, Xiong *et al* 2017). There is clear evidence that mechanical stress, such as pressure and shear forces in the vicinity of tumors, promote malignant progress by causing autophagy (Das *et al* 2018, Wang *et al* 2018a, Das *et al* 2019, Yan *et al* 2019). Accordingly, cervical cancer cells subjected to pulsatile laminar shear stress of 2 Pa for 3 and 6 min exhibit autophagy through a lipid raft-driven p38MAPK-dependent pathway and retarded apoptotic cell death (Das *et al* 2018). Nevertheless, shear stress is not automatically favorable for cancer cells. In contrast, it has been demonstrated that increased shear stress of 6 Pa, as it appears during vigorous exercise, can cause cancer cell death (Regmi *et al* 2017). In addition, it has been proven that fluid shear stress around 0.05–1.2 Pa causes apoptosis and autophagy in multiple cancer cell lines, among them hepatocarcinoma, osteosarcoma, oral squamous cell carcinoma and carcinomatous alveolar basal epithelia. Remarkably, this type of death caused by fluid shear stress was not seen in non-cancerous cells (Lien *et al* 2013). It seems, therefore, that autophagy can function as either a survival or anti-survival mechanism, according to the magnitude and length of time shear stress is applied. In addition, it has been demonstrated that shear stress- or pressure-induced autophagy is involved in cytoskeletal rearrangement and protein recirculation, which are vital for cancer advancement (Sharifi *et al* 2016, Yan *et al* 2019). In fact, enhanced tissue stiffness is involved in regulating multiple cancer hallmarks, including growth, invasion, and metastatic spread (Chaudhuri *et al* 2014, Tung *et al* 2015, Jain *et al* 2020). Similarly, it has been found that the stiffness of cancer tissue taken from the breast and liver is significantly increased compared to the respective physiological counterpart (Masuzaki *et al* 2007, Lopez *et al* 2011, Pang *et al* 2016, 2017). ECM stiffening in cancers is caused by the rearrangement of the stroma following the hyperactivity of ECM proteins and enzymes that covalently crosslink collagen fibrils and other ECM constituents (Egeblad *et al* 2010, Lu *et al* 2012). Crosslinking of collagen improves activation of integrins, maturation of focal adhesions and intracellular contraction, and consequently leads to a resulting rise in stiffness of the actin cytoskeleton, which can facilitate the migration and invasion of cancer cells (Levental *et al* 2009, Pickup *et al* 2013, Gilkes *et al* 2014, Rubashkin *et al* 2014). The enhanced stiffness of the ECM also contributes to the malignant phenotypic transition: cytoskeletal tension results in enhanced cell attachment to the ECM and perturbation of cell–cell junctions (Paszek *et al* 2005). Increased collagen accumulation in the ECM results in Hippo pathway activation (Zhao *et al* 2007, 2008, Aragona *et al* 2013) with a resulting disruption of the contact inhibitory circuit. Autophagy is also known to centrally mediate these events. Autophagy is known to be impaired in contactin-inhibited cells in both 2D and 3D culture systems using soft ECM substrates. In these cells, YAP/TAZ, which were discussed before in the context of their modulation through mechanical forces, cannot co-transcriptionally control the expression of myosin II genes, which leads to the disassembly of F-actin stress fibers and, in turn, to an attenuation of autophagosome generation. This depletion of F-actin stress fibers is also accompanied by a decrease in the number of ATG16L1 puncta per cell and a reduced co-localization of ATG9A-LC3, indicating an impaired transport of major autophagy proteins and therefore an impaired autophagic reaction (Pavel *et al* 2018). In addition, pressure-induced autophagy can enhance the release of matrix metalloproteinase-2 and the conversion of focal adhesion paxillin, thereby increasing the invasiveness of the commonly known HeLa cervical cancer cell line (Das *et al* 2019). In accordance with these findings, it was hypothesized that paxillin directly interacts with LC3 to trigger the disassembly of focal adhesions in the human breast cancer cell lines MDA-MB-231 and the murine melanoma cell line B16.F10, and to enhance metastasis *in vivo* in mice bearing 4T1 mammary tumors (Sharifi *et al* 2016). A further mechanism of force sensing in cancer comes from filamin A. This actin and actin-integrin crosslinker is diminished in human bladder cancer, thereby minimizing autophagy in these cancer cells, as evidenced by the reduction in LC3-II concentrations and LC3-I abundance (Wang *et al* 2018b). It has also been shown that, when overexpressed, filamin A reduces autophagy and inhibits the invasive potential of cancer cells. The mechanism of activity may comprise blocking the expression of matrix metalloproteinases, modulating integrin activity and increasing apoptosis (Sun *et al* 2014, Krebs *et al* 2015, Wang *et al* 2018b). Intriguingly, YAP/TAZ signal transduction has been found to induce transcription of filamin A to sustain actin tethering and crosslinking during mechanical tension (Ulbricht *et al* 2013a). This may be a possible mode of action for cancer cells to gain control of autophagy via a crosstalk involving YAP/TAZ and cytoskeletal components. Minor mechanical stress has been found to activate caveolin-1, thereby initiating the FAK/Src and ROCK/p-MLC signal transduction routes, which are implicated in cytoskeletal rearrangement, cell motility, focal adhesion dynamics and breast cancer cell attachment (Xiong *et al* 2017). PI3K/AKT activation and caveolin-1- dependent β-catenin-TCF/LEF activity also appear to be associated with enhanced VEGF expression and thus increased angiogenic capacity of the tumor (Sanhueza *et al* 2015). Shear stress-induced activation of caveolin-1 can trigger PI3K/AKT/mTOR signal transduction and metalloprotease activity, which have been demonstrated to increase cell migration and metastasis of breast cancer cells (Yang *et al* 2022). Reciprocally, phosphorylated caveolin-1 was observed to activate autophagy through engagement with the Beclin-1/VPS34 complex during oxidative stress and provide protective effects toward ischemic injury (Nah *et al* 2017). These data indicate that the role of caveolin-1 may rely on cell-type context (Quest *et al* 2008), leading to variable autophagic results. Notably, like autophagy, caveolin-1 has also been linked both mechanisms, such as cancer impairment and cancer advancement (Bender *et al* 2000, Simón *et al* 2020). While potentially beneficial in the case of emerging tumors, elevated amounts of caveolin-1 mRNA or protein have been observed in several types of cancer, and these strongly coincide with unfavorable survival rates in patients with advanced-stage cancer (Xiong *et al* 2017). These data also suggest that caveolin-1 is involved in the metastatic cascade, as indicated by enhanced migration, invasion, and attachment-independent proliferation (Diaz-Valdivia *et al* 2015). Most recent findings have implicated the primary cilium and autophagy in crosstalk to control cancer initiation and progression (Fabbri *et al* 2019, Ko *et al* 2019, Liu *et al* 2019a). In addition to the fact that it is regarded as a survival mechanism in tumorigenesis, an over-accumulation of autophagosomes can lead to autophagic cell death or apoptosis (Shimizu *et al* 2004, Espert 2006, Mukhopadhyay *et al* 2014, Zhou *et al* 2017), which, with regard to cancer, inhibits tumor growth and the spread of metastases. It has been shown that the accumulation of autophagosomes is enhanced under acute shear stress, such as 10 Pa for 60 min, which is associated with enhanced merging of autophagic vesicles with multivesicular bodies and a decrease in autophagosome-lysosome merging in HeLa and MDA-MB-231 cancer cell lines (Wang *et al* 2019). Moreover, inhibition of autophagosome breakdown caused by mechanical stress is linked to enhanced liberation of autophagic compounds in extracellular nanovesicles, perhaps through a Ca^{2+}-dependent signal transduction route that engages autophagy, multivesicular bodies, and exosomes (Wang *et al* 2019). Hence, exosome release may offer an alternative way to ensure cellular homeostasis when the autophagy route is impaired or inadequate to eliminate large quantities of defective proteins and avoid cell death (Wang *et al* 2019). Under mechanical stress, these findings point to a potential interplay between degradative and secretory autophagy to sustain cellular equilibrium and cancer cell viability (Gonzalez *et al* 2020). In addition, after pathological stress, detrimental nucleic acids, molecular chaperones, cytosolic factors and aberrantly folded proteins are secreted into the extracellular compartment through exosomes and may promote tumor growth and metastatic spread (Zhang *et al* 2015b, Zomer *et al* 2015).

The previous sections have indicated that the relationship between cell mechanics and autophagy is bidirectional. In the setting of cancer, autophagy modulates multiple metastasis-related signaling cascades, which are linked to cell mechanics according to the cell type and tumor surrounding environment. The autophagic protein LC3-II conveys the targeted breakdown of focal adhesion proteins like Src and paxillin (Sandilands *et al* 2012, Sharifi *et al* 2016) to enhance the dissolution and catabolism of focal adhesions and facilitate cell migratory movements. Integrins are recycled and shed in various ways, which differ based on their conformation, their activation through ECM proteins, and the attachment of effector proteins like talins and FERMTs/Kindlins. The transport, recycling, and breakdown of integrins influence their exposure at the plasma membrane, the dynamics of focal adhesions, and Rho-GTPase-driven cytoskeletal rearrangements to drive cell movement (Arjonen *et al* 2012, Dower *et al* 2018). Deficiency in nutrients (Rainero *et al* 2015) enhances integrin endocytosis and ECM breakdown (Colombero *et al* 2021), whereas hypoxia (Yoon *et al* 2005) stimulates the intracellular recycling of certain integrins. Because nutrient deprivation and hypoxia are common hallmarks of the tumor microenvironment, continued study of how autophagy regulates integrin transport has the potential to reveal the general impact of autophagy on cancer metastasis. In addition, it may elucidate how stress in the microenvironment affects cell mechanics to promote cancer cell escape from the primary tumor for metastatic spread.

5.8 Mechanobiology of autophagy in cancer therapy and prevention of chemoresistance

Over the last ten years, a variety of new treatments have been implemented that have considerably raised the survivability rates of cancer patients. Nevertheless, highly aggressive cancers frequently become resistant to treatment, either initially or over time, and ultimately can no longer be cured. Therefore, new therapeutic approaches are warranted to circumvent drug tolerance and enhance the responsiveness to treatment. The mechanobiology of autophagy could represent a new and promising approach for future treatments. Although there seem to be no approved drugs or drugs being tested in clinical trials that consider the mechanobiology of autophagy in their mechanism of action, there are examples of proteins/signaling pathways that are regulated through mechanical forces and thereby influence autophagy. As mentioned earlier, the Hippo-YAP/TAZ signaling pathway is activated in cancer cells and controlled through mechanical forces. Inhibition of this pathway has yielded encouraging findings in combating drug resistance, as reviewed in (Nguyen and Yi 2019, Thompson 2020). Intriguingly, inhibition of this signaling path also decreases autophagy (Pavel *et al* 2018), which is antagonized through various medications that are presently being evaluated in clinical studies. This indicates that dual blocking of the YAP-TAZ signal transduction pathway and autophagy may augment treatment responsiveness. Mechanisms also govern the epidermal growth factor receptor (EGFR), a protein that is frequently augmented or mutated within glioblastomas and in which autophagy fosters cell viability (Jutten *et al* 2013). Blocking autophagy as an adjunct to radiotherapy has yielded positive outcomes, and these could be further optimized based on an understanding of the mechanics of cancer. In line with these results, it has recently been shown that the drug ruxolitinib, which is currently employed to treat myeloproliferative neoplasms and blocks cell contractility in these cells to prevent signaling after focal adhesions, inhibits Janus-associated kinase (JAK) and triggers the initiation of autophagy (Machado-Neto *et al* 2020). Therefore, a cancer treatment must involve a combination of ruxolitinib with pharmacological inhibitors of autophagy. Another medication undergoing clinical investigation is losartan, an angiotensin II receptor blockage agent that decreases intratumoral interstitial fluid pressure in solid tumors (Chauhan *et al* 2013). Notably, this medication has also been found to block autophagy and induce autophagic cell death specifically in cancer cells (Woo and Jung 2017), which further highlights the relevance of manipulating autophagy and mechanobiology to combat cancer. For example, bladder cancer cells have exceptional adaptive abilities: molded by their environment, these cells are exposed to a cocktail of metabolites and xenobiotics, along with physiological mechanical stimuli. These reactions can lead to resistance to chemotherapy and are mostly attributable to autophagy. In the face of molecules able to rewire the plasticity of tumors, naturally derived compounds hold potential as beneficial treatment choices. Fungal-derived metabolites like bafilomycin and wortmannin are largely accepted as autophagy-inhibiting agents. Their potential to modulate the adaptability of bladder cancer cells when exposed to chemical and physical stimuli was investigated (Jobst *et al* 2024). In addition, mycotoxins present in food were also analyzed, such as deoxynivalenol with a concentration between 0.1 and 10 μM and fusarin acid with a concentration between 0.1 and 1 mM. Deoxynivalenol and fusarium acid, which exhibit Janus-faced characteristics, are reported as harmful toxins on the side of health risks. At the same time, they are also being experimentally investigated for potential selective pharmacological uses, among other things for their effectiveness against cancer. In non-cytotoxic concentrations, bafilomycin (1–10 nM) and wortmannin (1 μM) altered cell shape and decreased cancer cell migratory behavior in 2D environments.

The cellular sensitivity to treatment with bafilomycin (1 nM) was decreased under conditions of shear force exposure and mechanically controlled PIEZO channel blockade. Likewise, for fusarinic acid (0.5 mM), PIEZO1 expression and inhibitory effects were broadly consistent with the compound's modulatory effect on cancer cell motility. Moreover, this work revealed that the activity pattern of compounds with comparable cytotoxic capabilities, such as co-incubation of deoxynivalenol with bafilomycin or fusarin acid with wortmannin, can differ considerably in their impact on cell mechanotransduction control. Taking into account the correlation between cancer advancement and reaction to mechanical cues, these data suggest a new perspective for exploring chemoresistance and the signaling pathways connected with it (Jobst *et al* 2024). Autophagy is a cornerstone of the adaptive plasticity of the bladder cell: this sophisticated mechanism enables it to deal with the environmental constraints of a continuously fluctuating environment, which is characterized by permanent mechanical stress and a challenging chemical environment, composed of bioactive xenobiotic and dietary metabolic products (Bolt and Klimecki 2012, Lin and Hwang 2017, Aman *et al* 2021). Autophagy modulators can enter the body inadvertently as urinary impurities (Klionsky *et al* 2016, Yin *et al* 2020, Lee *et al* 2022), or intentionally as medications, like novel cures for cancer (Lin and Hwang 2017, Marinković *et al* 2018) or neurodegenerative illnesses (Martinez-Vicente 2015, Corti *et al* 2020, Dong *et al* 2023). Their effect is currently underestimated and their part in the adaptation of bladder cancer cells to physical and chemical stress factors is not fully recognized. Most crucially, the autophagy pathway is deregulated in bladder cancer (Zhou *et al* 2021) which could lead to chemoresistance (Li *et al* 2019a). It has been previously established that autophagy regulation affects the morphometric adjustment to shear stress (Del Favero *et al* 2021) and can impair the migratory capacity of bladder cancer cells (Jobst *et al* 2023). A weakness of this investigation is the usage of 2D gap closure migration assays and solely two cancer cell lines, such as the human urinary bladder carcinoma T24 cell line and the human colon carcinoma HCT116 cell line. Nevertheless, these results are consistent with the scientific literature, which reports on the correlation between autophagy and mechanotransduction in renal cells (Orhon *et al* 2016, Boukhalfa *et al* 2020a, Miceli *et al* 2020). This also applies to other autophagy-inhibiting compounds, for instance, to natural products (Abbas and Mirocha 1988, Yu *et al* 2011b), such as bafilomycin and wortmannin. Although the susceptibility of bladder cells to autophagy regulators has been investigated to promote treatment efficacy (Lin and Hwang 2017, Li *et al* 2019a), little is understood about the possible implications for mechanotransduction of bladder cells. Subsequently, T24 cells appeared to be more resistant to the chemical blockage of autophagy when exposed to shear forces, reflecting a response that has been found for shear-stress-induced chemoresistance (Ip *et al* 2016, Novak *et al* 2019). Deoxynivalenol and fusariin acid were investigated in the search for molecules that could additionally modify these reactions and are currently being re-examined for potentially positive impacts, including an evaluation of anti-cancer effects (Fernandez-Pol *et al* 1993, Stack *et al* 2004, Ye *et al* 2013, Habrowska-Górczyńska *et al* 2023). Because cancer cells can mitigate the subsequent harm via autophagy (Pagotto *et al* 2017, Wang *et al* 2020a,

Wu *et al* 2021a, Debnath *et al* 2023), but also since autophagy and mechanotransduction can reciprocally affect each other (King 2012, Hernández-Cáceres *et al* 2021), one of the aims was to examine whether the mycotoxins could be coupled with autophagy blockers to enhance their biological activity. In addition to this goal, the possible inhibitory effects of toxins as food contaminants may be evaluated, assuming a possible impairment of pharmacological therapy. As described in the following, the cytotoxicity of fusariin acid and deoxynivalenol was exacerbated to some extent in conjunction with the autophagy-inhibiting agents, confirming that the adaptability of cancer cells relies on a properly functioning autophagy. Interestingly, mechanobiology did not completely agree with cell viability readouts; cell biomass readouts yielded relatively robust values across treatment groups, such as deoxynivalenol versus deoxynivalenol and bafilomycin/wortmannin combinations or fusarium acid /bafilomycin/wortmannin combinations, which rendered it improbable that discrepancies in proliferation rates were the cause of the discrepancies seen in the migration assays. Likewise, cells incubated with deoxynivalenol exhibited decreased cell motility at similar cytotoxic potency, while fusarium acid markedly enhanced migration velocities, which is possibly an undesirable consequence when assessing prospective anti-cancer compounds. Along these notes, the co-treatment of fusarium acid with wortmannin substantially reduced cell viability while keeping migration abilities like those of solvent controls. Moreover, the joint incubation of deoxynivalenol and bafilomycin had hardly any impact on cell viability, but the migration potential was significantly diminished. It is evident that complex signaling pathways could lead to these phenotypes; among other things, a role for cytoskeletal components is probable and is an open research question. For instance, earlier studies have demonstrated that deoxynivalenol exposure, at non-cytotoxic levels, strongly affects cytoskeletal constituents, rendering cells susceptible to the deformability of the ECM (Del Favero *et al* 2018). Two different hypotheses were pursued in the investigation of the mechanisms of action, which focused on the cell membranes. First, that the candidate compounds may alter cell motility and mechanotransduction and modify plasma membrane morphology. Although some important effects were detected, membrane shape alone barely correlated with the reaction to shear stress/migration capacity. In line with this, because cells need more than just lipid membrane functionality (Tanaka *et al* 2017, Sackmann and Tanaka 2021) to migrate in an efficient manner, the ability of calcium regulation was examined as an auxiliary factor that could potentially influence cell motility (Wei *et al* 2012). In T24 cells, intracellular Ca^{2+} ion values recorded after 24-h treatment with the autophagy inhibitor compounds were consistent with previous published reports; therefore, bafilomycin alone enhanced calcium signaling, potentially through blocking the ER calcium ATPase Ca-P60A/SERCA (Mauvezin and Neufeld 2015). Notably, fusarin acid and deoxynivalenol attenuated this reaction, despite producing rather minimal responses as sole agents. It is established that Wortmannin acts by modifying Ca^{2+}-transporters in that it inhibits the kinase function of the mammalian target of rapamycin (mTOR) (Brunn *et al* 1996). The function of inositol 1,4,5-trisphosphate sensitive Ca^{2+} channels (IP (3)R), which represent the major Ca^{2+} liberation channels of the ER, relies on the phosphorylation via mTOR and thus they are impaired through wortmannin (Frégeau *et al* 2011). To strengthen the connection between mechanotransduction and calcium signaling, the intracellular levels of the divalent ion were determined after treatment with YODA1, which is a PIEZO1 inducer (Davies *et al* 2019). In cells exposed to bafilomycin, the magnitude of the calcium fluxes triggered by PIEZO was lower than in the control cells. The pre-incubation with wortmannin resulted in a weaker inhibition. The mycotoxins deoxynivalenol and fusarin acid also markedly altered YODA1-induced Ca^{2+} ion reaction; although this could be due in part to their mitochondrial toxicity (Bensassi *et al* 2012, Sheik Abdul *et al* 2016), these mycotoxins are also recognized to directly interfere with Ca^{2+} kinetics (Bouizgarne *et al* 2006, Ren *et al* 2016). The reaction of fusarinic acid and deoxynivalenol was largely diminished when co-incubated with the autophagy regulators as compared to only being exposed to the toxin, which highlights the capacity of autophagy impairment to adjust chemosensitivity. As the kinetics of calcium measures and migration assays are apparently dissimilar, additional investigations were conducted to examine the participation of PIEZO channels in the control of mechanotransduction arising from autophagy regulation. To dissect the role of mechanosensitive ion channels, PIEZOs have been blocked through tarantula toxin GsMTx-4 (Bowman *et al* 2007). Remarkably, PIEZOs impairment on its own decreased migration but when coupled with autophagy impairment partially recovered the migration defect, reflecting the impact of shear stress on cell motile behavior. Therefore, it is probable that shear forces induce autophagy in bladder cells, as has been observed in other cell types (Liu *et al* 2015, Meng *et al* 2022). Interference with the mechanotransduction system, which is frequently dysregulated in bladder cancer, for example, due to PIEZO1/2 overexpression (Etem *et al* 2018, Yu and Liao 2021), may be a useful new approach to address this chemoresistance pathway and improve existing chemotherapies (Ip *et al* 2016, De Felice and Alaimo 2020, Kim *et al* 2022). In this context, several natural products are recognized to block PIEZO1 activation, including the aforementioned GsMTx-4 (Bowman *et al* 2007), or the plant alkaloid jatrorrhizine, which is employed in classical Chinese medicine (Hong *et al* 2023). Since a connection between metastatic potency, shear stress and PIEZO1 in PC3 prostate cancer cell lines has already been proven (Kim *et al* 2022), it seemed highly unlikely that the sensitivity of bladder cells could be an exceptional phenomenon. It can hence be hypothesized that the linkage between metastatic capacity, shear stress and PIEZO1 seems to be of universal nature and even not restricted to cancer cells. Consequently, the motility experiments were performed again on the colon cancer cell line HCT116. In contrast to T24 cells, the motile behavior of HCT116 cells was not impaired by equimolar amounts of GsMTx-4. When the cells were exposed to bafilomycin, the use of an inhibitor blocked the motility of the former but not the motility of the latter cancer cell type. These findings indicate that there exist inhibitor- and hence substance-specific effects that mask the universality phenomenon of the metastatic capacity, shear stress and PIEZO1 connection. Apparently, several substance-specific responses were evident, but notable was the sensitiveness of the colon cancer cells toward wortmannin treatment. The expression of PIEZO1 channels was analyzed in the attempt to identify potential sources of variation among cell lines and between individual treatments. These experiments showed that the baseline expression of PIEZO1 was found to be markedly distinct between T24 and HCT116. In addition, HT116 expressed higher levels of PIEZO1 compared to other colon cancer cell lines, such as SW480 and pancreatic cancer cells (Sun *et al* 2020b). These data suggest that the differential impact of 100 nM GsMTx-4 on the control of migratory behavior of the two cell types. Due to the increased expression levels of PIEZO1 in HCT116 cells, it could be due to the amount of the inhibitory substance, which was simply not high enough to efficiently block PIEZO1. Moreover, in line with the migration experiments, PIEZO1 immunolocalization was more potently reprogrammed upon bafilomycin incubation in T24 cells, whereas wortmannin restored higher expression values of the mechanosensitive channel in HCT116 cells.

5.9 Short classification of the autophagy process in its major subgroups

Autophagy refers to a mechanism that is well conserved in eukaryotes, ranging from yeast to humans, and is implicated in upholding cellular homeostasis through breaking down organelles, proteins, lipids, and nucleic acids, and in the recycling of their building elements when cells are confronted with nutrient shortages. In addition, autophagy is used to clear pathogenic microorganisms that have entered the cytoplasm and to get rid of defective organelles and aberrant proteins. Autophagy dysregulation has been linked to a wide range of diseases, among them neurodegenerative disorders, infectious diseases, inflammatory disorders, metabolic disorders, cancer and the aging process (Dikic and Elazar 2018). There exist two primary types of autophagy, which are distinguished by their specificity. The first is 'bulk' or 'non-selective' autophagy, which randomly removes and breaks down portions of the cytoplasm, which comprise organelles and macromolecular complexes. The second is 'selective' autophagy, which targets certain, frequently dangerous cargo for breakdown (figure 5.7). Selective autophagy comprises various pathways that have been classified based on their targeted objects: mitophagy (mitochondria), lysophagy (lysosomes), aggrephagy (protein and RNA aggregates), xenophagy (intracellular pathogens), ER-phagy (endoplasmic reticulum (ER)), pexophagy (peroxisomes), ribophagy (ribosomes), ferritinophagy (ferritin), glycophagy (glycogen), lipophagy (lipid droplets) and fluidophagy (droplets). The many target cargoes of selective autophagy are associated with various physiological functions, and preventing these cargoes from being broken down causes various diseases (Jiang and Mizushima 2014). These two types of autophagy require the generation of spherical, double-walled organelles termed autophagosomes, which enclose cellular constituents intended for breakdown. The generation of autophagosomes requires autophagy-related (ATG) proteins, which are well conserved from yeast to mammalian cells. Over 40 ATG genes have been characterized, among them the core ATG genes, such as ATG1 to ATG18, which are most often required for both non-selective and selective autophagy modes. Non-selective autophagy permits

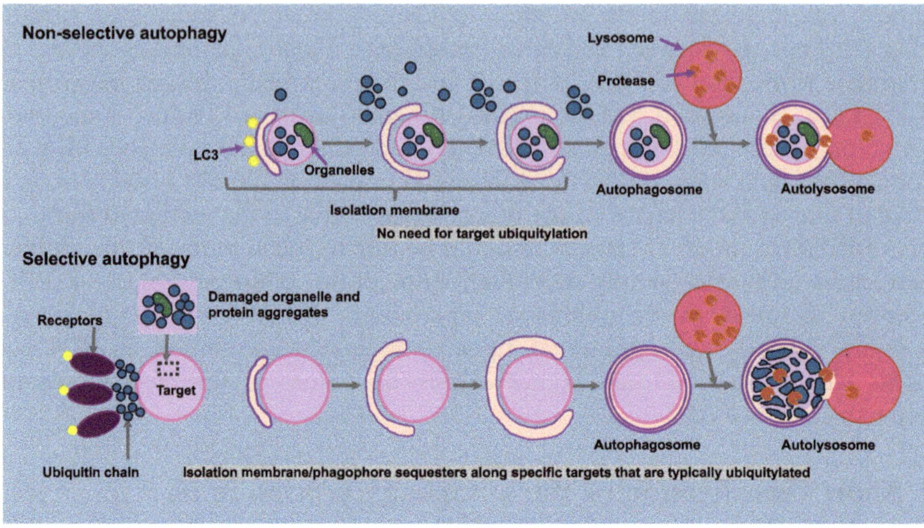

Figure 5.7. Schematic drawing of non-selective and selective autophagy processes. Non-selective and selective autophagy are both forms of autophagic break down of cytoplasmic components sequestered by a double-membrane structure, which is referred to as an autophagosome. In non-selective autophagy, an isolation membrane/phagophore forms at the autophagosome assembly sites when a few signals are received, for example, a lack of nutrients. In this event, the generation of the isolation membrane is triggered in the endoplasmic reticulum (ER), which serves to stimulate activation of the ULK complex, which consisted of the autophagy-related protein 13 (ATG13), FIP200, ATG101 and ULK1 or ULK2 and acts as a framework for the subsequent engagement of downstream factors and the phosphorylation-dependent control of their functions. The enrolment of ATG9-containing vesicles that can deliver lipids is then followed through the expansion of the membrane. This results in the bulk of the cytosol and organelles to be enclosed in autophagosomes, which eventually merge with the lysosome so that the load can be degraded using hydrolytic enzymes. In selective autophagy, autophagosomes are generated *de novo* on distinct targets. Selective autophagy target detection varies depending on the type of selective autophagy; discontinuation of the membrane that collects galectins at the site of target detection is important in ubiquitination of the selective autophagy freight. Ubiquitylation represents a critical but not exclusive factor for the identification of targets for elimination through selective autophagy. It is facilitated by a variety of distinct E3 ubiquitin ligases and facilitates the engagement of receptor proteins that accumulate on the surface of the targeted protein or organelle. These receptor clusters can attract the ULK1 complex and offer a site to connect the isolation membranes with the targets by attaching them to lipidized LC3 proteins or their homologs, the GABARAP proteins. This is succeeded with the extension of the isolation membrane and the selective capture of the targeted cargo through autophagosomes. It is noteworthy that these autophagosomes tend to be enlarged when compared to normal autophagosomes, which are formed via the non-selective autophagy route to take up large loads like complete organelles.

cells to withstand nutrient deprivation and survive long enough for the next nutrient source to become accessible (Ohsumi 1999). When cells sense nutrient deprivation, autophagy is triggered, resulting in the generation of beaker-shaped membrane structures referred to as isolation membranes or phagophores, the origin of which is not yet clear. There are, nevertheless, many reports supporting the participation of different organelles, including the ER, mitochondria, and plasma membranes (Hamasaki *et al* 2013). These membrane protrusions are marked with lipidated, covalently attached to phosphatidylethanolamine LC3/GABARAP proteins, which are homologs of yeast Atg8 that is frequently employed as an autophagosome membrane marker.

Auxiliary membranes are delivered to the expanding isolation membrane to enable its stretching and orderly wrapping. The isolation membrane eventualy shuts and creates autophagosomes, which may have a diameter of up to 1 μm (Lamb *et al* 2013). Selective autophagy relies on cargo-specific recruitment, which is often accomplished through ubiquitylation of the respective cargo (Gubas and Dikic 2022), which differs from non-selective autophagy. The cargo may then be targeted for sequestration through specific receptor proteins comprising p62 (synonymously referred to as SQSTM1), TAX1BP1, NDP52 (synonymously referred to as CALCOCO2), NBR1, and Optineurin (OPTN) (Johansen and Lamark 2020). These receptor proteins, which are characterized by the possession of LC3-interacting regions (LIR) and ubiquitin-tethering domains, enable the association of autophagic cargo and lipidation of LC3 for autophagic cargo entrapment (Kirkin and Rogov 2019). Selective autophagosomes can differ in their size from 1 to 10 μm based on their target (Nakagawa *et al* 2004). In addition to the ubiquitination of target proteins for selective autophagy, some receptor proteins undergo ubiquitination themselves to modulate their own function. The lysine-7 residue of p62 is subject to ubiquitylation through the E3 ligase TRIM21, thereby inhibiting its oligomerization (Pan *et al* 2016), which is necessary for cargo sorting events in the autophagosome. Ubiquitination of lysine 420 of p62 within the UBA domain that attaches ubiquitin to the freight via KEAP1-Cullin 3 enhances the ability of p62 to attach freight (Lee *et al* 2017b). Ubiquitination thus functions in both target identification and regulatory control of receptor proteins. In this part, the focus is on the various types of selective autophagy in mammalian cells, with an eye to mitophagy, aggrephagy, lysophagy, and xenophagy, which have been most extensively investigated up to now. The latest advances in cargo labeling, cargo detection, cargo selective removal, and cargo breakdown were highlighted, with a particular focus on the importance of these pathways in health and disease (Vargas *et al* 2023).

5.10 Mitophagy

Mitochondrial networking is vital for the maintenance of proper fitness in most eukaryotic cells. Malfunctioning proteins in the respiratory chain complex can lead to a shortage of energy and to the build-up of reactive oxygen species (ROS), which are damaging and toxic to the cell. To avoid the build-up of defective mitochondria, they are selectively eliminated through autophagy in a process referred to as mitophagy (Pickles *et al* 2018). PTENinduced putative kinase 1 (PINK1) and Parkin can act as a monitoring mechanism for mitochondria that have been injured. A major mechanism ensuring the specificity of injury-triggered mitophagy involves the ubiquitination of outer mitochondrial membrane (OMM) proteins, which stimulates the engagement of autophagy receptors solely for the organelles that require elimination (figure 5.8) (Tanaka 2020). PINK1 and Parkin act as master controllers of this ubiquitin-tagging pathway. PINK1 serves as a monitor for

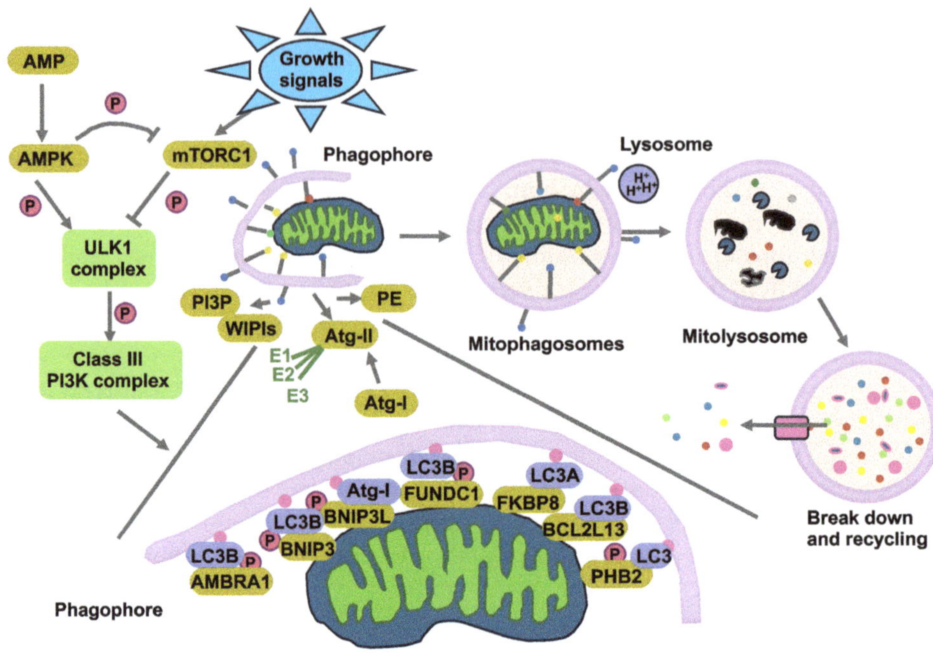

Figure 5.8. Mitophagy independent of Parkin. The outer mitochondrial membrane (OMM) contains a plethora of proteins comprising BNIP3, BNIP3L, FUNDC1, BCL2L13, FKBP8, and PHB2, as well as non-ubiquitinated autophagy receptors with LIR motifs. Direct association of lipidated LC3 and GABARAP family members with the phagosomal membrane triggers mitophagy.

mitochondrial health by exclusively aggregating in dysfunctional mitochondria (Narendra *et al* 2008, Jin *et al* 2010). PINK1 is transported into healthy mitochondria via the TOM and TiM complexes and then proteolytically processed in the lumen by the protease PARL and, to a lesser degree, through Oma1, both of which are found on the inner mitochondrial membrane (Jin *et al* 2010, Sekine and Youle 2018). PARL splitting of PINK1 leads to the splitting of a short N-terminal region, thereby uncovering N-degron in the transmembrane domain and exposing it to proteasome-driven breakdown (Deas *et al* 2011, Yamano and Youle 2013). When mitochondrial membrane potential is dissipated, TIM complex-based entry is compromised and PINK1 cannot access the inner membrane, thereby preventing it from accessing PARL (Sekine and Youle 2018). This results in the clustering of PINK1 in the OMM, especially at injured mitochondria, where it may then phosphorylate serine 65 (Ser65) on ubiquitin chains linked to a plethora of OMM proteins (Kane *et al* 2014, Kazlauskaite *et al* 2014, Koyano *et al* 2014, Shiba-Fukushima *et al* 2014).

It is unresolved which E3 ligases are engaged in basal mitochondrial protein ubiquitylation. USP30, by contrast, localizes to mitochondria, where it regulates the basal levels of OMM ubiquitylation (Bingol *et al* 2014). Parkin, which is an E3 ubiquitin ligase, directly interacts with Ser65-phospho-ubiquitin, and thus activated PINK1 can selectively promote the translocation of Parkin from the cytosol toward dysfunctional mitochondria (Narendra *et al* 2008, Kondapalli *et al* 2012). In addition, PINK1 also phosphorylates Parkin in its ubiquitin-like domain, likewise at Ser65 (Kondapalli *et al* 2012), which frees Parkin from its autoinhibited conformation (Gladkova *et al* 2018, Sauvé *et al* 2018). Parkin, formerly active within mitochondria, ubiquitinates countless OMM proteins (Sarraf *et al* 2013, Ordureau *et al* 2014 2018). These emerging ubiquitin chains can then be additionally phosphorylated through PINK1, which results in an even more pronounced attachment of Parkin to phospho-ubiquitin and an activated state of Parkin at the mitochondria (Narendra *et al* 2008, Kondapalli *et al* 2012, Kane *et al* 2014, Koyano *et al* 2014, Ordureau *et al* 2014). PINK1-Parkin dependent feedforward ubiquitylation of OMM proteins results in the subsequent enrichment of numerous proteins like p97 (Tanaka *et al* 2010), which is synonymously referred to as VCP, and Rab GTPases (Yamano *et al* 2014, 2018, Heo *et al* 2018), that are crucial for the proper execution of mitophagy. p97 promotes the removal of OMM proteins like mitofusins to increase mitophagy rates upon Parkin activation (Tanaka *et al* 2010), while RAB7 trafficking to mitochondria governs LC3-labelled membrane biogenesis and extension for autophagosome generation (Yamano *et al* 2014, 2018). Important is that the phospho-ubiquitin chains produced by PINK1-Parkin can attract autophagy receptors (Heo *et al* 2015, Lazarou *et al* 2015, Moore and Holzbaur 2016). Intriguingly, PINK1 conjugates with Parkin mono- and short phospho-ubiquitin chains at injured mitochondria to induce mitophagy (Swatek *et al* 2019), which may bring Parkin close to the OMM surface to ubiquitinate additional OMM proteins.

NDP52 and OPTN function as the primary ubiquitin-driven mitophagy receptors. A systematic approach utilizing combinatorial CRISPR-Cas9 knockout (KO) lines of five autophagy receptors identified OPTN and NDP52 as the two ubiquitin-independent receptors that are most instrumental for Parkin-independent mitophagy (Heo *et al* 2015, Lazarou *et al* 2015, Moore and Holzbaur 2016, Evans and Holzbaur 2020). OPTN and NDP52 are targeted to mitochondria through their individual ubiquitin-binding domains. More subtly, accumulation of matrix--localized protein aggregates at mitochondria also leads to focal enrichment of receptor proteins at these aggregates and their clearance, which relies on Parkin (Burman *et al* 2017). As is outlined in the following xenophagy section, NDP52 and OPTN are also implicated in the elimination of invading bacteria (Thurston *et al* 2009, Wild *et al* 2011, Ravenhill *et al* 2019). Considering the bacterial origin of mitochondria, the coincidence of xenophagic and mitophagic ubiquitin-binding receptors could be associated with the bacterial origin of mitochondria. In fact, TBK1 kinase, which is a key regulator of innate immune responses, also plays an essential role in the proper timing of mitophagy (Moore and Holzbaur 2016, Richter *et al* 2016, Heo *et al* 2018, Vargas *et al* 2019). Both NDP52 and OPTN directly engage TBK1 and serve as substrates for TBK1 (Thurston *et al* 2009 Wild *et al* 2011, Richter *et al* 2016). Phosphorylation of NDP52 and OPTN through TBK1 supports the attachment of these receptors to the mitochondria by influencing their ability to connect with ubiquitin chains and thereby positively controlling the speed of mitophagy (Heo *et al* 2015, Moore and Holzbaur 2016, Richter *et al* 2016).

Moreover, phosphorylation of OPTN inside its LIR domain through TBK1 enhances the affinity of OPTN toward lipidated LC3 (Wild *et al* 2011) to promote cargo retrieval on the isolation membrane.

There exist additional mitophagy receptors that work independently of ubiquitin (Terešak *et al* 2022). Mitochondria are the site of localization for numerous receptors, such as NIX (19 kDa protein X, which interacts with protein-3 (NIP3)) and BNIP3 (BCL2/adenovirus E1B 19 kDa protein-interacting with protein-3) (Zhang and Ney 2009). NIX was originally identified as a key mitophagic receptor in reticulocyte maturation (Sandoval *et al* 2008, Novak *et al* 2010). It has been established that BNIP3, which is homologous to NIX, governs both mitophagy and ER-phagy (Hanna *et al* 2012). Despite having LIR domains, NIX and BNIP3 are mitochondrial receptors lacking the ubiquitin-binding domains that typify OPTN and NDP52. Among other non-ubiquitin receptors for mitophagy, NIPSNAP1/2 proteins, located in the mitochondrial matrix, can act in concomitance with NIX and BNIP. In fact, NIPSNAP1/2 aggregates on the OMM following mitochondrial depolarization and can enlist LC3. Interestingly, NIPSNAP1/2 also binds the NDP52 protein directly through its zinc finger domain, which is the same domain involved in recognizing Parkin-generated ubiquitin chains (Princely Abudu *et al* 2019). In this way, mitochondrial receptors can interact with ubiquitin-binding receptors and thereby facilitate their recruitment, which then triggers the autophagic pathway by attracting other early autophagy proteins. Ultimately, the ubiquitin-independent mitophagy receptors NIX and BNIP3 are controlled by distinct post-translational modifications, including phosphorylation and ubiquitylation, which modify their action in mitophagy (Li *et al* 2021b).

OPTN and NDP52 convey *de novo* autophagosome formation upon mitophagy. Earlier models of the role of mitophagy receptors emphasized their function in targeting preformed LC3-lipidized membranes to injured mitochondria originating at other sites within the cell. Nevertheless, after Parkin is activated, a mitophagosome can selectively swallow mitochondria without the aid of LC3/GABARAP family proteins (Nguyen *et al* 2016). In this case, the mitophagosome expansion rate is compromised and the merging of the mitophagosome with the lysosome is impeded. In fact, both aTG9a and the UlK1 complex, which are necessary for the initiation of autophagosomes, are localized to mitochondria in cells deficient in ATG3, a protein that acts as a key factor in LC3 lipidation, during PINK1-Parkin-facilitated mitophagy (Itakura *et al* 2012a). These outcomes indicate strongly that LC3/GABARAP proteins are not necessary for the induction of Parkin-mediated mitophagy but are nevertheless required for the enlargement of the emerging autophagosome and its consecutive merging with the lysosome. These findings also point to the fact that mitophagy receptors promote the biogenesis of isolation membranes directly on the surface of mitochondria to be catabolized. Consistent with this model, it was earlier shown that in the lack of NDP52 and OPTN, enrollment of ULK1 to mitochondria is compromised, indicating that receptor proteins are able to enlist the upstream autophagy mechanism toward mitochondria (Lazarou *et al* 2015). More recent work has demonstrated that NDP52 interacts with FIP200, a central scaffold member of the ULK1 complex, and that this engagement is essential for the generation of the isolation membrane through direct activation of ULK1 on load substrates like injured mitochondria and intruding bacteria (Ravenhill *et al* 2019, Vargas *et al* 2019, Fu *et al* 2021). In addition, the activity of TBK1 eases the interplay between NDP52 and FIP200 (Vargas *et al* 2019). Accordingly, the impact of the engagement between NDP52 and FIP200 has been highlighted, and it has been revealed that NDP52 allosterically activates the membrane affinity of FIP200 and ULK1 (Shi *et al* 2020b). In a similar vein, OPTN has recently been demonstrated to combine with ATG9A vesicles (O'Loughlin *et al* 2020, Yamano *et al* 2020) and with FIP200 (O'Loughlin *et al* 2020). It has been demonstrated that the engagement of OPTN via its leucine zipper domain with ATG9A is crucial for the initiation of mitophagy (Yamano *et al* 2020). A screening for new mitophagy activators revealed that the antiparasitic agent ivermectin can stimulate mitophagy (Zachari *et al* 2019). The ubiquitin ligases cIAP1, cIAP2 and TRAF2 are implicated in ivermectin-triggered mitophagy. Moreover, ivermectin stimulates TBK1, which supports the enrollment of OPTN into the mitochondria (Zachari *et al* 2019). Like NDP52, OPTN and TAX1BP1, which represents a homolog of NDP52, were identified to be linked to different ATG components in mitophagy based on proximal proteomics (Heo *et al* 2019). OPTN has also been demonstrated to associate with the ATG16L1-ATG5-ATG12 complex (Bansal *et al* 2018), which powers progression of LC3 lipidation. Finally, OPTN was found to engage with and enroll ATG9A vesicles toward mitochondria in mitophagy (Yamano *et al* 2020). Therefore, OPTN combines with upstream ATG compounds and is probably capable of inducing mitophagy via localizing upstream factors, comparable to NDP52, on injured mitochondria. Consistent with these findings, subsequent work on NDP52, OPTN and TAX1BP1 utilizing a fully recapitulated *in vitro* autophagy system has revealed that all these receptors are competent to enlist upstream autophagy proteins, but with differing degrees of affinity (Chang *et al* 2021). It is notable that the speed at which the receptors can fulfill this role is enhanced when ubiquitin chains are appended (Shi *et al* 2020a, Chang *et al* 2021). ULK1 becomes activated through AMPK upon nutrient depletion in the process of bulk autophagy (Kim *et al* 2011). It is noteworthy that for selective autophagy, the ULK1 localization on freight is all that is needed to facilitate its activation (Vargas *et al* 2019, Shi *et al* 2020a). In fact, ubiquitin strands on the surface of freight items facilitate the concentration of autophagy receptors (Turco *et al* 2019, 2021, Shi *et al* 2020a). The enrollment of receptors on ubiquitinated loads acts as a kind of hub to enhance the local concentrations of early autophagy constituents, involving ULK1, thereby encouraging its activation by autophosphorylation (Vargas *et al* 2019, Shi *et al* 2020a, Turco *et al* 2021). Consistent with these findings, ULK1 can be directly targeted to mitochondria for autoinactivation even in the complete lack of AMPK (Vargas *et al* 2019), implying that ULK1 activation through localized aggregation can occur on freight with no upstream control. Hence, PINK1-Parkin-driven phospho-ubiquitylation of mitochondrial membrane proteins represents a decisive first step in mitophagy for the localization of OPTN and NDP52, leading to the sequestration of upstream autophagy proteins directly to mitochondria, where autophagosome biogenesis is initiated through autoactivation of ULK1. Finally, an unorthodox activation route for mitophagy through OPTN and TBK1 regardless of ULK1/2 kinases and FIP200 has recently been uncovered. Moreover, TBK1 is directly linked to class III phosphatidylinositol-3-kinase complex I. Hence, TBK1 is an additional selective autophagy-activating kinase that can function in the absence of the ULK1-FIP200 signaling complex (Nguyen *et al* 2023). While not strictly required for the initiation step, the engagement of the LIR domains of the autophagy receptor with LC3/GABARAP proteins is instrumental for mitophagy. For example, it has been found that as soon as emerging autophagosomes assemble at the mitochondria, lipated LC3 can utilize the LIR domain to attract NDP52 and OPTN in a way that is independent of ubiquitin (Padman *et al* 2019). It is assumed that this ubiquitin-independent but LC3-dependent sorting of NDP52 and OPTN causes the recruitment of additional upstream autophagy complexes to the ripening autophagosome, thereby boosting its rate of expansion (Padman *et al* 2019). In addition, ubiquitin-independent mitophagy receptors utilize LIR-tethering to LC3 to facilitate mitophagy. Together, these results provide the model that three receptor proteins, NDP52, its homolog TAX1BP1 and OPTN, work together to trigger mitophagy through promoting autophagosome biogenesis by directly interacting with key components of autophagy at ubiquitinated injured mitochondria. It is currently unknown how exactly ubiquitin-dependent and ubiquitin-independent mitophagy routes can mutually balance each other in physiological conditions. Moreover, it is presently not understood whether certain cell types utilize more ubiquitin-dependent or ubiquitin-independent mitophagy.

5.10.1 Relevance of mitophagy under physiological conditions

The effects of deregulated mitophagy on disease development are underscored by the fact that mutations in genes encoding PINK1 and Parkin lead to familial Parkinson's disease (Kitada *et al* 1998, Valente *et al* 2004). Research on flies showed an epistatic interaction between PINK1 and Parkin, with PINK1 acting as an upstream factor (Clark *et al* 2006, Heo *et al* 2019). The question is how does disruption of mitophagy lead to neurodegeneration? Mitochondria not only perform a central role in energy generation but are also known to serve as signaling centers for diverse cellular processes including apoptosis and innate immunity. For example, RNA viruses trigger the mitochondrial antiviral signaling protein (MAVS), a protein situated at the OMM (Cai *et al* 2017). Mitochondria mediate apoptosis via the secretion of different cytotoxic proteins, which is dependent on OMM permeabilization (Wang and Youle 2009), and Parkin-mediated ubiquitination of the pore-forming proteins BAK and BAX can precisely adjust apoptosis by impeding the oligomerization of BAX/BAK (Johnson *et al* 2012, Bernardini *et al* 2019). In addition, VDaC1, which has been well characterized as a Parkin substrate, is implicated in the differentiation between mitophagy and apoptosis (Ham *et al* 2020). Notably, polyubiquitylation and monoubiquitylation of VDAC1 through Parkin, which is carried out at specific lysine residues, was demonstrated to enhance mitophagy and prevent apoptosis, respectively. In this way, mitophagy controls physiological signal transduction routes that rely on mitochondria as signaling sites through the regulation of mitochondrial cargo inside cells.

The innate immune routes in eukaryotes can act on invading pathogens like bacteria, viruses and fungi (Akira *et al* 2006). The effectiveness of innate immunity is based on the capacity to accurately distinguish pathogenic signature molecules and peptides. Due to their α-proteobacterial origin, mitochondria pose a challenge for the innate immune system. Mitochondria-derived damage-associated molecular patterns (DAMPs) potently activate the innate immune reactions (Youle 2019). When liberated into the cytosol, a DAMP, the mitochondrial DNA (mtDNA), triggers the activation of the cGAS-STING signal transduction pathway. STING is a dimeric, ER-resident protein that induces pro-inflammatory reaction through the activation of interferon-stimulated gene expression (West *et al* 2015). STING becomes activated through cGAMP, which is produced by the coupling of cGAS to cytosolic double-stranded DNA and functions as a key hub in the defense response to double-stranded DNA viruses (Wu and Chen 2014). In this way, mitochondrial injury can result in the liberation of mtDNA (West *et al* 2015, Sliter *et al* 2018) and other DAMPs into the cytosol, thereby causing inflammation (Moehlman and Youle 2020).

Impaired mitophagy *in vivo* leads to STING activation, causing elevated IL6 concentrations (Sliter *et al* 2018). It is noteworthy that the removal of STING in Parkin null mice (Pickrell and Youle 2015) not just alleviates the inflammation seen in these mice, but a variety of symptoms related to Parkinson's disease, such as the depletion of dopaminergic neurons in the substantia nigra and motor dysfunction (Sliter *et al* 2018). In Parkinson's disease patients carrying mutations in genes that code for PINK1 and Parkin, the amount of circulating mtDNA is elevated in comparison to healthy controls (Borsche *et al* 2020). In addition, IL6 is also elevated in the blood serum of these patients (Borsche *et al* 2020). Therefore, a defect in mitophagy, which is linked to mutations in Parkinson's disease, enhances the inflammatory reaction, probably by failing to prevent the spread of mtDNA from injured mitochondria. These findings suggest that persistent activation of the innate immune system resulting from mitophagy disorders is a potential pathological characteristic of Parkinson's disease, which is the cause of neurodegeneration. The dopaminergic neurons destroyed in Parkinson's disease have an elevated need for mitochondria in comparison to other neurons due to their rhythm-generating activity and widespread axonal branching, which may explain why this neuronal cell type is unusually reliant on the quality management of mitophagy (Surmeier 2018). Besides PINK1 and Parkin, OPTN and TBK1 are also involved in neurodegenerative diseases like amyotrophic lateral sclerosis (Evans and Holzbaur 2019). Perhaps neurons are inherently susceptible to mitochondrial disequilibrium because neuronal activity relies on the preservation of chemical gradients across the plasma membrane, a bioenergetically expensive process that relies on the preservation of healthy mitochondria (Van Laar and Berman 2013). Moreover, the complex morphological organization of neurons provides another level of spatial intricacy for mitochondrial sustenance—mitochondrial fine-tuning demands concerted expression of both nuclear and mitochondrial genomes, which, like the synaptic mitochondria, may be located tens of centimeters apart (Aschrafi *et al* 2016, Mandal and Drerup 2019, Youle 2019). These requirements can work in combination to make certain neuronal subpopulations sensitive to deficiencies in mitophagy. In particular, dopaminergic neurons which are a neuronal subtype depleted in Parkinson's disease, require more mitochondria than other neurons due to their rhythm-generating activity and extensive axonal branching, which may render this neuronal cell type particularly reliant on the quality assurance of mitophagy (Surmeier 2018). In addition, dysfunctional mitophagy leads to the liberation of mtDNA and DAMPs, inducing inflammatory reactions that are detrimental to neuronal viability and further increase the stress of preserving aging neurons (Borsche *et al* 2020, Moehlman and Youle 2020).

5.10.2 Mitophagy, mechanical cues and cancer

Hypoxia is a hallmark of the tumor microenvironment and can speed up the dissemination of cancer. In the early phase of carcinogenesis, a microenvironment created by fast-multiplying cancer cells and characterized by localized hypoxia and nutrient deprivation leads to mitochondrial malfunction and, consequently, to mitophagy. Cancer cells require increased nutrients, and mitophagy supplies amino acids for cell growth through lysosomal recycling. Mitophagy delivers not merely nutrients for ATP generation and biogenesis, but also meets the metabolic demands of cancer cells through the breakdown of carbohydrates, proteins, lipids, and nucleotides (Kroemer *et al* 2010). To sustain unrestrained growth levels, cancer cells use unorthodox mechanisms to gain energy from the outside environment. Mitochondrial oxidative phosphorylation (OXPHOS) is impaired in cancer cells due to mitochondrial disorders. This repurposing of energy metabolism is referred to as the Warburg effect (Icard *et al* 2018). Mitosis enhances the glycolytic metabolic route and diminishes the utilization of the OXPHOS mechanism, allowing OXPHOS to supply the fast-increasing energy requirements. Mitochondrial OXPHOS and glycolysis function in synergy to preserve the equilibrium of energy metabolism within cancer cells (Wu *et al* 2021b). In addition, mitophagy can prevent the generation of ROS and the unproductive consumption of precious nutrients like oxygen that fuels the rapid proliferation of cancer cells (figure 5.9) (Poole and Macleod 2021).

Mitochondria are a key controller of metabolism. Mitochondria are composed of four distinct compartments, namely the outer mitochondrial membrane (OMM), the intermembrane space, the inner mitochondrial membrane (IMM), and the mitochondrial matrix (Green 1983). The IMM comprises a several-fold larger surface compared to the OMM, which leads to an invagination of cristae membranes that contain the oxidative phosphorylation (OXPHOS) system, nvolving respiratory complexes I to IV and F1F0-ATP synthase for the generation of ATP (Pfanner *et al* 2019). In addition, mitochondria harbor a 16-kb circular genome, which is referred to as mitochondrial DNA (mtDNA) and encodes tRNAs, rRNAs, and intramembrane proteins that are needed for respiration (Frey and Mannella 2000). Recently, nanotube-based mitochondrial trafficking from immune cells to tumor

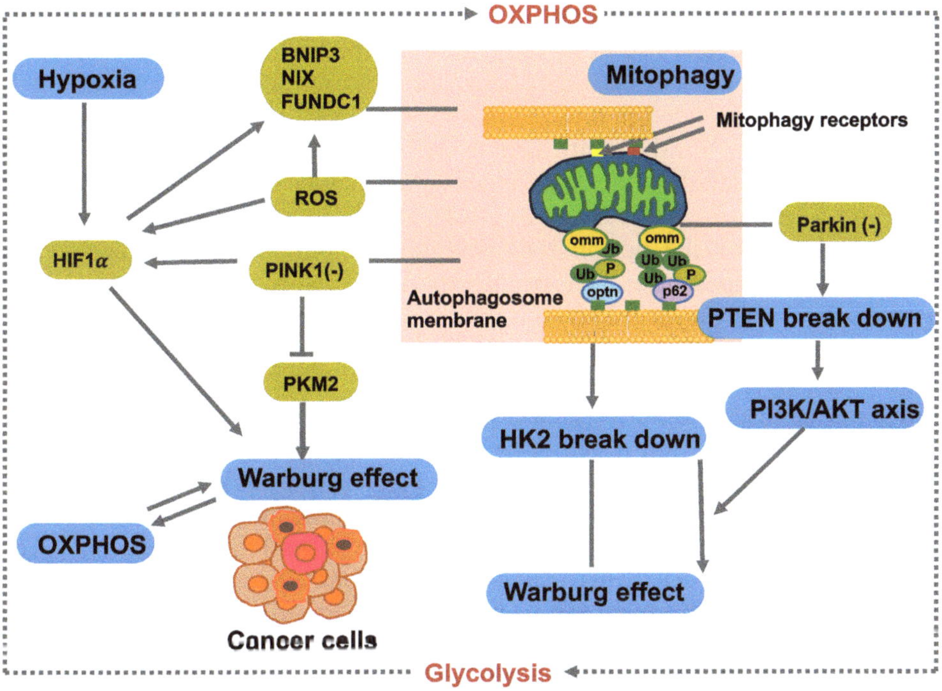

Figure 5.9. Mitophagy supplies cancer cells with an adequate supply of nutrition, energy and oxygen. The Warburg effect refers to the process whereby glycolysis, which represents a process in which glucose is processed into lactic acid, is enhanced through mitophagy. Mitochondrial OXPHOS and glycolysis work together to keep the energy metabolism in equilibrium in cancer cells. The absence of Parkin triggers the breakdown of PTEN, which in turn stimulates the characteristic carcinogenic signal transduction route referred to as the PI3K/AKT signaling cascade. This pathway accelerates aerobic glycolysis in cancer cells. Likewise, the lack of PINK1 can induce the Warburg effect by decreasing the activity of pyruvate kinase M2 (PKM2) and promoting the stabilization of HIF1a, thereby maintaining the fast proliferation rate of cancer cells. PKM2 plays a key regulatory role in glycolysis, and a decrease in PKM2 activity can enhance the fast proliferation of cancer cells through enhancement of the pentose phosphate route. In addition, hexokinase 2 (HK2) is broken down in a selective manner through p62/SQSTM1-dependent mitophagy to regulate glycolysis values. HIF1a transcription undergoes upregulation, thereby inducing glycolytic metabolism, and its target genes undergo enhanced expression, thereby guiding mitophagy. These effects are also induced through hypoxia stimulus and the resulting rise in mitochondrial ROS.

cells was identified (Saha *et al* 2022). Actually, it has been hypothesized that there is a close interrelationship between the morphodynamics, mechanics, and metabolism of mitochondria (Senft and Ronai 2016, Su *et al* 2023). Mitochondria themselves are extremely dynamic organelles. In general, mitochondria are capable of sensing and responding to a wide range of mechanical, physical and metabolic cues (Senft and Ronai 2016, Giacomello *et al* 2020, Chen *et al* 2023, Su *et al* 2023). The intracellular transfer of these physical signals leads to alterations in the dynamic and metabolic features of the mitochondria (Senft and Ronai 2016, Giacomello *et al* 2020, Chen *et al* 2023, Su *et al* 2023). The process of mitochondrial unification and division affects mitochondrial performance and balances between mitochondrial energy generation and carrying out various cell death pathways (Senft and Ronai 2016, Giacomello *et al* 2020, Chen *et al* 2023, Su *et al* 2023). The processes that control mitochondrial dynamics are tightly controlled in normal physiological settings but are frequently dysregulated in diverse tumors. The processes that control mitochondrial dynamics are tightly controlled in normal physiological settings but are frequently dysregulated in diverse tumors (Senft and Ronai 2016, Giacomello *et al* 2020, Chen *et al* 2023, Su *et al* 2023). It is important to note that there is growing support for the notion that deregulated mitochondrial dynamics exert profound effects on cancer cell proliferation, metastasis, drug resistance, and modification of the tumor microenvironment (Senft and Ronai 2016, Giacomello *et al* 2020, Su *et al* 2020, Chen *et al* 2023). Cellular metabolism controlled in a dynamic fashion through mitochondria has recently emerged as a process affected through mechanical cues during cancer development. While the general architecture of metabolic routes within a cell is well understood, the governing of metabolic rearrangement in this framework is still obscure. In particular, the interrelations between cellular mechanics and mitochondria-driven metabolism are only partially elucidated. Mitochondria display striking plasticity and dynamics and play a decisive part in cell metabolism, responses to stress and the upkeep of homeostasis. These organelles act as a focal intracellular site for essential biochemical activities including ATP synthesis, fatty acid synthesis, intracellular ROS generation, OXPHOS, thermogenesis, and control of calcium homeostasis (Wellen and Thompson 2012, Mookerjee *et al* 2015, Chen *et al* 2023). Metabolically active mitochondria generate signaling intermediates that have a decisive influence on the regulation of cell performance and phenotype (Weinberg *et al* 2015). In particular, mt-ROS have been shown to be essential for the proper control of the intracellular signaling cascade of inflammatory reactions (Weinberg *et al* 2015). Most importantly, impaired mitochondria are closely associated with a variety of diseases and pathologies, ranging from neurodegenerative disorders and metabolic diseases to cancer. These disorders are generally typified through impaired mitochondrial functioning (Mishra *et al* 2014, Lightowlers *et al* 2015, Chen *et al* 2023).

It is known that mitochondria continuously alter their shape and level of activity by going through processes of division, merging, mitophagy and trafficking cycles (Mishra *et al* 2014, Lightowlers *et al* 2015, Chen *et al* 2023). These pathways govern and monitor the morphology, quality, quantity and dispersion of mitochondria and their functioning (Mishra *et al* 2014, Lightowlers *et al* 2015, Chen *et al* 2023). A novel quality control mechanism of mitochondria, which is referred to as mitocytosis, during the migration of cells was identified (Jiao *et al* 2021). Migratory cells utilize specialized organelles, referred to as migrasomes, to scavenge for defective mitochondria. This scavenging mechanism affects cellular viability (Jiao *et al* 2021). Mitochondrial constituents that are defective can be eliminated through the process of mitochondrial dynamics. Mitochondria that are seriously impaired and no longer functional are generally removed through mitophagy, thereby avoiding a potential risk of cell damage (Youle and Van Der Bliek 2012, Chen *et al* 2023). An optimal mitochondrial functioning in healthy cells demands an equilibrium of mitochondrial dynamics (Tilokani *et al* 2018). Collectively, these findings suggest that partitioning serves a key function in preserving mitochondrial quality through the elimination of defective or damaged mitochondria, especially in the face of intense cellular stress that can cause apoptosis. In turn, merging promotes intermixing and the sharing of intramitochondrial material between mitochondria, thereby preserving mitochondrial functionality (Youle and Van Der Bliek 2012, Chen *et al* 2023).

Emerging evidence indicates that the processes of mitochondrial partitioning and unification are dysregulated in cancer (Wallace 2012, Chen *et al* 2023). As the key regulators of mitochondrial dynamics relate to cancer advancement, it is evident that mitochondrial dynamics controllers perform a pivotal function in tumorigenesis and cancer advancement in various cancers, because the abnormal expression of master regulators of mitochondrial dynamics in cancer results in striking consequences in mitochondrial functionality, frequently linked to enhanced proliferation, migration, invasion, and cancer cell survival. Mitochondria perform a variety of critical tasks during tumorigenesis and carcinogenesis, involving bioenergetics, calcium homeostasis, cancer anabolism, redox control, transcription of genes, and control of fundamental cellular activities (Wallace 2012 Chen *et al* 2023). In addition, the proper performance of immune processes depends on proper mitochondrial metabolic efficiency within immune cells, indicating that interfering with mitochondrial dynamics could be a powerful potential treatment approach to fight cancer (Breda *et al* 2019).

Cellular dysfunction is marked by unrestrained growth of cells, defective cell cycle control and anomalies in programmed cell death, which are well-recognized hallmarks of cancer (Hanahan 2022). Mitochondrial dynamics appears to drive these processes. In various cancers, like lung cancer (Rehman *et al* 2012), metastatic breast cancer (Tang *et al* 2018, Humphries *et al* 2020), glioblastoma (Kriel *et al* 2018), neuroblastoma (Qiao *et al* 2017), colorectal cancer (Abdelmaksoud *et al* 2023), hepatocellular carcinoma (HCC) (Zhang *et al* 2018, Wang *et al* 2022), pancreatic cancer (Liang *et al* 2020b, Carmona-Carmona *et al* 2022), and melanoma (Wieder *et al* 2015), cancer cells frequently exhibit a high frequency of fragmented mitochondria. This phenotype is often linked to enhanced expression or elevated activation of dynamin-1-like protein (Drp1) and/or the decreased expression of mitofusin 2 (MFN2) (Senft and Ronai 2016). The association between increased fission or decreased fusion has been linked to cancer advancement. Drp1 inhibition or MFN2 overexpression reverses cancer progression, leading to cell cycle halt and increased spontaneous apoptosis (Senft and Ronai 2016). Overexpression of survivin has been shown to cause fragmentation of mitochondria, which is coupled to a decrease in the activity of complex I (Hagenbuchner *et al* 2013). This change enhances glycolysis, impedes the build-up of ROS and boosts resistance to chemotherapeutic agents (Hagenbuchner *et al* 2013). Conversely, blocking glycolysis using the glucose analog 2-deoxy-D-glucose rendered survivin-overexpressing neuroblastoma cells more susceptible toward chemotherapeutic agents (Hagenbuchner *et al* 2013).

Altered mitochondrial dynamics can affect signal transduction events during tumorigenesis and cancer advancement. Phosphorylation of Drp1 at Ser616, which is activated by mitogen-activated protein kinase 1 (MAPK1), leads to mitochondrial fragmentation, which is connected to the growth of tumors (Kashatus *et al* 2015). Moreover, decreased Drp1 expression inhibits the growth of tumors derived from MAPK-facilitated malignant tumors (Kashatus *et al* 2015). In HCC, ECM-associated protein, collagen and calcium-binding EGF domains 1 (CCBE1) exerts an significant function in augmenting mitochondrial fusion and suppressing HCC advancement through restraining HCC cell proliferation and metastasis (Tian *et al* 2023). Notably, reduced CCBE1 expression has been linked to unfavorable prognosis and worse patient outcomes in HCC (Tian *et al* 2023). CCBE1 mechanistically impedes mitochondrial division through inhibiting Drp1 phosphorylation at Ser616, consequently blocking Drp1 targeting to mitochondria (Tian *et al* 2023). MFN2 has been found to inhibit the proliferation and cell cycle of cervical cancer cells of the Hela cell line by blocking the expression of key proteins like NF-κB p65, Myc and mechanistic targets of rapamycin kinase (mTOR) (Liu *et al* 2019b). Moreover, mitochondrial fragmentation is closely associated with the advancement of cancer metastasis. For instance, increased levels of active Drp1 and reduced expression of mitofusin 1 (MFN1) result in enhanced mitochondrial fragmentation in metastasizing breast cancer cells (Zhao *et al* 2013). Actually, mitochondrial fission has been established to be vital for the migration and invasion of breast cancer cells (Zhao *et al* 2013). In contrast, lengthening or bundling mitochondria has the capacity to substantially decrease the metastatic abilities of breast cancer cells. This result was achieved by either Drp1 depletion or MFN1 overexpression. In contrast, the knockout of the MFN1 gene causes mitochondrial fragmentation within breast cancer cells, thereby increasing their metastatic capacity (Zhao *et al* 2013).

In addition, dysregulated mitochondria have been found to play a critical role in the acquisition of resistance to drugs in cancer cells (Vasan *et al* 2019). For example, resistance to chemotherapy and breast cancer recurrence are mainly caused by breast cancer stem cells (BCSCs) (Vasan *et al* 2019, Chen *et al* 2023). Recent evidence indicates that BCSCs exhibit significantly increased expression of the proteins Fission Mitochondrial 1 (Fis1) and MFN1 (Fu *et al* 2023). AZD5363 (capivasertib) therapy was found to impact mitochondrial dynamics in BCSCs via the suppression of MFN1 expression, thus enhancing BCSC chemotherapeutic sensitivity to doxorubicin (Fu *et al* 2023). In addition, breast cancer cells were found to metastasize to organs with a more permissive microenvironment. This type of metastasis is linked to elevated mitochondrial fission, which is promoted through Drp1 and the mitochondrial elongation factor 1/2 (MIEF1/2) (Romani *et al* 2022). The impairment of Drp1 was reversed after chemotherapy involving cisplatin (Romani *et al* 2022). These insights form the foundation for possible treatment approaches to avoid chemotherapy resistance in metastases through regulated and adjustment of mitochondrial dynamics (Romani *et al* 2022).

In common, proteins that induce mitotic division are frequently elevated in a variety of tumors when compared to normal tissues. Increased expression of these proteins is often associated with adverse clinical prognoses and influences tumor growth, motility, invasion, and the acquisition of resistance to chemotherapeutic agents (Chen *et al* 2023). Characteristic traits of cancer cells involve hyperproliferation and enhanced resilience to apoptosis (Hanahan 2022). These properties should be promoted to some level of energetic support. It is thus logical that cancer cells show abnormalities in mitochondrial functionality, especially a shift from oxidative metabolism to aerobic glycolysis (Archer 2013). The contribution of mitochondrial dynamics to cancer is related to the requirement for mitochondrial fission to occur during the mitotic cell division process. The synchronized manner in which this occurs, termed mitotic splitting, ensures that mitochondria are not distributed unevenly among the daughter cells that arise (Archer 2013). Mitochondrial fragmentation therefore appears to have an important function in cancer progression through several mechanisms. In particular, it can increase mitotic division and interfere with intramitochondrial calcium waves, thus impeding apoptosis induced through calcium signal transduction (Archer 2013).

5.10.3 Mechanical signal controls mitochondrial features in cancer

ECM stiffness is a key physical parameter that delivers a universal signaling cue to orchestrate cell proliferation, differentiation, and death (Petridou *et al* 2017, Vining and Mooney 2017, Tschumperlin *et al* 2018, Min and Schwartz 2019). Deviating ECM stiffness is linked to cancer advancement and aids in controlling the destiny of cancer cells. Notably, mechanical forces emanating from a stiffer cancer ECM substantially increase both cancer growth and invasiveness (Mohammadi and Sahai 2018). Somewhat unsurprisingly, physical perturbations of the tumor ECM have been demonstrated to mechanically control metabolism in a variety of distinct cancer types (Evers *et al* 2021, Romani *et al* 2021, Zanotelli *et al* 2021). In addition, mitochondria, a key site of intracellular metabolism, are also markedly changed in cancers (Wallace 2012, Archer 2013, Chen *et al* 2023). Taken together, these observations support the notion that mechanical signals emanating from the tumor microenvironment could control mitochondrial dynamics and ultimately functionality in cancer cells. Physical cues emanating from substrate stiffness can influence the functionality and dynamics of mitochondria inside cancer cells (Youle and Van Der Bliek 2012). Increased substrate stiffness fosters the extension of mitochondria leading to elongated mitochondria (Youle and Van Der Bliek 2012, Romani *et al* 2022). In particular, pancreatic cancer cells exhibited elongated and merged mitochondria when grown on 2–4 GPa highly stiff material like glass and 38 kPa stiff substrates in comparison to 0.7 kPa soft substrates (Youle and Van Der Bliek 2012). Metastatic breast cancer cells exposed to a soft ECM displayed Drp1-driven mitochondrial fragmentation (Romani *et al* 2022). Breast cancer cells grown on soft (0.5 kPa) fibronectin-coated acrylamide hydrogels exhibited fragmented mitochondria, in contrast to cells cultured on stiff (15 kPa) substrates (Romani *et al* 2022). A weakness of these experiments is that the acrylamide hydrogels are purely elastic and homogenous materials, whereas the commonly employed collagen fiber matrices and natural environments exhibit viscoelastic characteristics and are inhomogeneous features (Hayn *et al* 2020 2023). Another shortcoming is that the cells are grown on even 2D acrylamide gels and not in a 3D environment as is the case in tissues. Mechanistically, this increased mitochondrial division was regulated by peri-mitochondrial F-actin, whose assembly was in turn governed through Spire1C and Arp2/3 (Romani *et al* 2022). The growing ECM rigidity in conjunction with pancreatic cancer cells encouraged the emergence of more elongated and merged mitochondria. (Youle and Van Der Bliek 2012). Remarkably, these alterations in mitochondrial dynamics were coupled with alterations in arginine metabolism and creatine biosynthetic pathways (Youle and Van Der Bliek 2012). It was determined that creatinine enrichment occurred on a soft matrix, while a rigid matrix favored high creatine and phosphocreatine levels (Youle and Van Der Bliek 2012). All in all, the pancreatic cancer cells adapted their metabolic processes in line with matrix stiffness alterations, diverting L-arginine metabolism to the creatine biosynthetic route (Youle and Van Der Bliek 2012). Moreover, ECM stiffening (an elevation of the elastic modulus from 0.35 to 40 kPa) led to an elongation of mitochondria by enhancing merging and restraining fission within lung adenocarcinoma A549 cells and fibrosarcoma HT1080 cells (Chen *et al* 2021). Kindlin-2 was determined to be required for the upregulation of merging mitochondria. The repression of Drp1 expression was controlled by PINCH-1, which led to a blockade of mitochondrial fission (Chen *et al* 2021). It is important to mention that some other research points to the fact that stiff (and not soft) substrates can enhance the fragmentation of mitochondria (Tharp *et al* 2021, Xie *et al* 2021, Guo *et al* 2023). For example, breast epithelial cells were grown on polyacrylamide hydrogel surfaces with different elasticities (rigidities) ranging from normal breast (400 Pa), where they presented elongated mitochondria, to tumors (6–60 kPa) ECM, where they presented mitochondrial fragmentation (Tharp and Weaver 2018, Tharp *et al* 2021). Moreover, mesenchymal stem cells cultured on soft of 1 kPa substrates exhibited a filamentous architecture of mitochondria. In contrast, fragmented mitochondrial organization was detected on stiff of 20 kPa substrates (Xie *et al* 2021). Subsequently, it was observed that human lung fibroblasts perceive progressively rising matrix stiffness, such as from soft 1 kPa to stiff 20 kPa, and acquire mitochondrial dynamics conducive to mitochondrial fission and elevated ATP generation (Guo *et al* 2023). While these investigations attempted to establish a connection between the dynamics of mitochondria and cancer, they used non-cancerous cell models, which may account for the contradictory findings (Chen *et al* 2021, Romani *et al* 2022). In addition, the stiffness measurements in various scientific investigations are not consistent, which results in varying findings.

Adhesion-based mechanical signaling converts physical signals from the altered ECM, which cause alterations in mitochondrial dynamics, ultimately contributing to metabolic rerouting of cancer cells (Youle and Van Der Bliek 2012, Romani *et al* 2021, 2022). New findings imply that in cancer progression, stiffening of the ECM triggers integrin- and cadherin-facilitated adhesions, which results in cytoskeletal reorganization and focal adhesions. Cytoskeletal changes in turn affect the expression and activity of metabolic enzymes and mitochondrial fusion/fission events, leading to oncogenic reprogramming of the metabolism of cancer cells (Youle and Van Der Bliek 2012, Romani *et al* 2021, 2022, Yanes and Rainero 2022). For instance, cells undergo stiffness fluctuations throughout the entire metastasis cascade when they switch from the stiff environment of primary tumors toward the frequently more compliant environment of secondary metastatic niches (Youle and Van Der Bliek 2012, Romani *et al* 2021, 2022, Yanes and Rainero 2022). In metastatic breast cancer cells of the MDA-MB-231 type, changes in both collagen matrix densities and fiber alignment are crucial in influencing the intracellular ATP:ADP ratio (Zanotelli *et al* 2018). When cells are embedded in denser collagen matrices that impede cell migration, an increase in the ATP:ADP ratio is seen compared to lower-density gels. Moreover, increased cell migration in aligned collagen matrices is associated with a concomitant decrease in the ATP:ADP ratio (Zanotelli *et al* 2018). Ultimately, inhibition of contractile myofibrils and their migration led to a decline in intracellular ATP:ADP levels. These findings indicate that a local ECM can actively modulate three-dimensional metastatic invasion of cancer cells through an intracellular energy redistribution (Zanotelli *et al* 2018). Increased ECM stiffening causes pancreatic cancer cells to undergo metabolic rearrangement to enhance cell invasion and motility (Youle and Van Der Bliek 2012). Soft substrates have been found to favor glycolysis, while the stiffer substrates promote ATP generation through the TCA cycle and OXPHOS (Youle and Van Der Bliek 2012). In fact, cancer cell invasion was decreased through pharmacological blocking of mitochondrial ATP synthase and OXPHOS by oligomycin A (Youle and Van Der Bliek 2012). In addition, metastatic breast cancer cells grown on soft ECM enhanced cytoplasmic, mitochondrial, and membrane localized lipid ROS generation (Romani *et al* 2022). Increased ROS led to the induction of NRF2 (Nuclear Factor Erythroid 2-Related Factor) transcription factor. Moreover, soft ECM enhanced the activation of Drp1, which was subsequently involved in the modulation of ROS values and the attenuation of oxidative stress (Romani *et al* 2022). Moreover, ECM-driven mechanotransduction appears to contribute to chemoresistance. Cells cultured on a soft ECM displayed markedly enhanced resistance to cisplatin and arsenic trioxide (As_2O_3) (Romani *et al* 2022). The situation in 3D tumor microenvironments, nonetheless, is likely more intricate compared to 2D ECM surfaces. A recent publication demonstrated that soft 3D collagen matrices can increase glycolysis within hepatocellular carcinoma cells (Frtús *et al* 2023). Mitochondrial depolarization, coinciding with decreased expression of the mitochondrial cytochrome c oxidase I, occurred (Frtús *et al* 2023). This adjustment of mitochondrial activity and functioning resulted in a decelerated proliferation rate and a resting state in hepatocarcinoma cells (Frtús *et al* 2023). This research hypothesized that there may be multiple physical signals in the 3D tumor microenvironment that impact mitochondrial functionality (Frtús *et al* 2023). It was hypothesized that cells could be exposed to concurrent mechanical stimuli, such as adhesion and pressure, when grown in 3D collagen scaffolds, which affect cell function through mitochondrial dynamics adjustment (Frtús *et al* 2023). Collagen fibers offered attachment sites, while the porous nature of the collagen scaffold facilitated cell–cell communication and increased pressure on adjacent cells (Frtús *et al* 2023).

5.11 Lysophagy

Lysosomes are filled with approximately 50 hydrolytic enzymes that can break down proteins, carbohydrates, lipids and polynucleotides. The interior of the lysosome is acidic at a pH of about 5 and serves as an important site for intracellular digestive

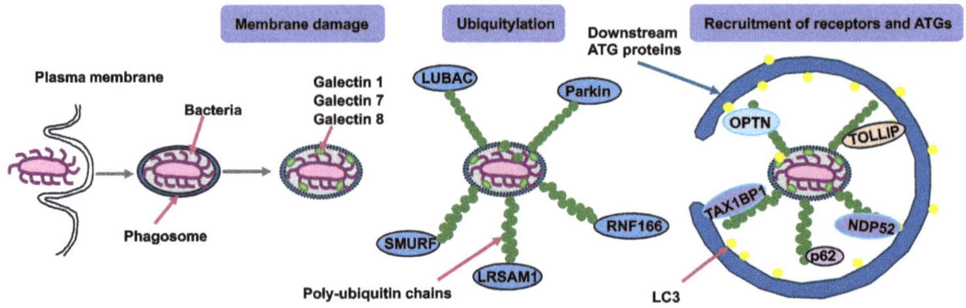

Figure 5.10. Schematic drawing of the process of lysophagy, which is the elimination of injured lysosomes through the process of autophagy. TRIM16, UBE2QL1, SCFFBXO27, LRSAM1 and other lysosomal autophagy proteins are recruited to the ubiquitinated lysosomal membrane proteins in case of an injury of the lysosomal membrane or also under physiological conditions. Autophagy adapters are then attracted to trigger autophagy. Galectin-3 usually exists in the cytoplasm and nucleus, but when lysosomes are damaged, it can be recruited to them. Phagosome assembly is initiated through the clustering of autophagy initiating proteins, which is enabled through the TRIM16-galectin-3 complex. In contrast, galectin-8 attracts LC3-positive phagocytes to induce lysophagy through direct linkage to the autophagy receptor NDP52, in an ubiquitin-independent manner.

processes, which may culminate in autophagy (Coffey and De Duve 1968). Damage to the lysosome causes hydrolytic enzymes to be released into the cytoplasm, thereby inducing cell death (Boya 2012). Lysosomal membranes are vulnerable to injury caused by extracellular materials that penetrate cells, among them cholesterol, uric acid crystals, human amyloid β aggregates, and ultrafine particles like silica and asbestos (Boya 2012, Maejima *et al* 2013). With lysosomal membrane disruption, cells seek to isolate and/or try to fix the lysosomal membrane disruption (Stahl-Meyer *et al* 2021). Damaged lysosomes can therefore be specifically broken down by selective autophagy, which is referred to as lysophagy (Hung *et al* 2013, Maejima *et al* 2013) (figure 5.10). Lysosomal membrane disruption has been implicated in causing or exacerbating lifestyle-related disorders like type 2 diabetes, atherosclerosis, arthritis and neurodegenerative disorders (Aman *et al* 2021). Hence, the mechanism for repairing and clearing defective lysosomes deserves special emphasis.

5.11.1 Cellular reaction to lysosomal membrane injury

Lysosomes can be destroyed synthetically with a medication referred to as ILOM, which becomes membrane-lytic when split in lysosomes through cathepsin D (Maejima *et al* 2013). These membrane defects are sensed through galectins, primarily galectin-3 (Gal3), which travel from the cytoplasm into the lysosomal compartment, where they attach to the Nglycans of proteins. Correspondingly, lysosomal disruption by LLOMe is evidenced through the co-localization of lysosomes with Gal3, which is concomitant with ubiquitin and lapidated LC3. Upon LLOMe washing out, Gal3, ubiquitin and lipidated LC3 relocalization reverts to the cytoplasmic distribution seen in the pre-treatment condition (Maejima *et al* 2013). In cells with an autophagy deficiency, no corresponding reduction in the number of Gal3-positive lysosomes could be discerned following LLOMe removal, suggesting that autophagy participates in the cellular reaction to lysosomal membrane injury through lysophagy. In autophagy-defective cells, by contrast, the percentage of Gal3-positive lysosomes is reduced after LLOMe washing out, indicating that lysosomal reparation is accomplished by a mechanism distinct from autophagy.

5.11.2 Mechanistic insight into lysophagy

Lysophagy regulated through ubiquitylation and ubiquitin chain sensing via selective autophagy receptors. Ubiquitination takes place in the luminal space of the lysosome. Two receptors that are presumed to participate in lysophagy are p62 (Papadopoulos *et al* 2017) and TAX1BP1 (Eapen *et al* 2021). Analogous to Gal3 recruitment, ubiquitylation on injured lysosomes appeared only after 30 min of LLOMe exposure (Maejima *et al* 2013). TRIM16 was identified as one of over 600 E3 ubiquitin ligases in humans. It is specifically targeted to defective lysosomes through its interaction with Gal3 (Chauhan *et al* 2016). Because of its interaction with ULK1, BECLIN 1 and ATG16, TRIM16 acts as a conduit linking impaired lysosomes with ATG proteins and is required in the initial steps of lysophagy to direct ATG proteins to the injured organelle and to facilitate autophagosome membrane generation. It is noteworthy that Gal3 is merely a hallmark of dysfunctional lysosomes and not a required player in lysophagy; the necessity of TRIM16 is not obvious. FBXO27, which is a substrate recognition branch of the SKP1-CUL1-Fbox (SCF) complex, is a further E3 ubiquitin ligase that has been implicated in lysophagy (Yoshida *et al* 2017). Both FBXO27 and Gal3 show co-localization after LLOMe application, and the defective lysosome repairing capacity is decreased about 20% in FBXO27-KO cells in comparison to control cells. In FBXO27-overexpressing cells, the most important lysosomal membrane proteins LAMP1 and, especially, LAMP2, are heavily ubiquitinated when there is a lysosomal defect. Nevertheless, FBXO27 is not expressed universally (its expression is mainly limited to muscle and adipose tissue), indicating that other E3 ubiquitin ligases, yet to be identified, are involved in lysophagy. Lys48 and Lys63 ubiquitin chains are found on disrupted lysosomes when they are subjected to lysophagy (Papadopoulos *et al* 2017). Lys63 ubiquitin chains are evident in the earliest phases of injury, while the Lys48-ubiquitin chain reaches a maximum level 2–4 h following LLOMe challenge. Upon ubiquitylation, endo-lysosomal damage response (ELDR) complexes comprising the deubiquitinating enzymes YOD1 and p97 attach to injured Lys48-ubiquitinated lysosomes, leading to Lys48-specificubiquitin removal and LC3 attraction to trigger the initiation of lysophagy. Mutations in p97 have been implicated in diverse neurodegenerative disorders, and accumulations of defective lysosomes with Lys48 ubiquitylation are detected in tissues of patients suffering from multisystem proteinopathy. The function of this specific deubiquitylation, however, is still elusive. In a topical study of the ubiquitylated proteome in disrupted lysosomes, a new protein, calponin 2 (CNN2), has been found to be a target for ubiquitylation/deubiquitylation. CNN2 is an actin filament stabilizer that is sequestered to injured lysosomes with the actin-nucleating factor Arp2/3 to facilitate the assembly of branched actin filaments and drive lysophagy. CNN2 undergoes ubiquination and then eliminated through p97 to permit extension of the isolation membrane/phagophore surrounding compromised lysosomes. The small heat shock protein HSPB1, that binds to CNN2 and supports its clearance from dysfunctional lysosomes, was also identified in the screen. Similar to p97, HSPB1 is linked to neurodegeneration (Kravić *et al* 2022). UBE2QL1 was identified as a lysophagy-required E2 enzyme during the characterization of about 40 human E2 ubiquitin-conjugating enzymes (Koerver *et al* 2019). UBE2QL1 is required for Lys48-ubiquitin chain recruitment and localizes to injured lysosomes during the 2–3 h following LLOMe challenge. UBE2Q1 deficiency severely impairs the sequestration of p97, p62 and LC3 to the injured lysosomes. Nevertheless, as the targeting of UBE2Q1 to injured lysosomes takes place only at subsequent steps following LLOMe challenge, UBE2Q1 may be implicated in the disposal of severely injured lysosomes. Collectively, additional work is needed to elucidate the mechanisms underpinning ubiquitylation/deubiquitylation dynamism at the lysosomal membrane and to establish its part in the reaction to lysosomal injury.

5.11.3 Role of lysophagy under physiological conditions

A disruption of lysophagy can have significant pathological implications. For instance, it has been proven that the autophagic elimination of lysosomes is crucial for preventing the emergence of acute hyperuricemic nephropathy in mice, which is caused by harm to the lysosomes through uric acid crystals (Maejima *et al* 2013). Moreover, lysophagy may be crucial as a defense against lifestyle-related disorders (Aman *et al* 2021). In addition, lysosomal injuries may also be generated through cytotoxic protein aggregation, which is a hallmark of neurodegenerative diseases like α-synuclein, amyloid-β, tau, and abnormal huntingtin (Zhang *et al* 2009). In fact, when there is a breach in the integrity of the lysosomal membrane, these proteins build up in the cell and create cytotoxic aggregates that can propagate to neighboring cells and cause additional pathology. Therefore, when not properly treated, lysosomal defects can give rise to neurodegenerative diseases. Likewise, injury to lysosomes can result in the release of cathepsin D. This release, in turn, has been demonstrated to trigger the liberation of cytochrome c from the mitochondria, which leads to apoptosis (Boya 2012). Remarkably, cathepsin D has been found to be liberated into the cytoplasm of aging rat neurons, indicating yet another connection between disturbed lysophagy and neuronal degeneration. As lysosomes are key calcium repositories, lysosomal membrane disruption can also lead to calcium efflux. Calcium outflow leads to activation of the calcium-dependent, non-lysosomal cysteine protease calpain, which interferes with lysosomal functioning, impairs autophagy and causes additional lysosomal injury, thereby triggering cell death (Papadopoulos and Meyer 2017). This has been implicated in the breakdown of calcium homeostasis that is linked to Alzheimer's disease (McBrayer and Nixon 2013).

5.12 Aggrephagy

Cells possess multiple avenues to clear protein aggregates inside themselves, which comprise the ubiquitin-proteasome pathway (UPS), chaperone-mediated autophagy, and selective autophagy referred to as aggrephagy (figure 5.11) (Lamark and Johansen 2012, Menzies *et al* 2015). Misfolded proteins are subject to aggregation due to the presence of unstructured segments. They usually occur as soluble monomers, small aggregates or large, non-soluble inclusion bodies, which are presumably cytotoxic in that they overload the protein degradation system, disturb membrane structure or combine with different proteins. The aggregation state of proteins is a hallmark of many human diseases, and the solubility of aggregated proteins and the size of the aggregates can govern which route is activated for their

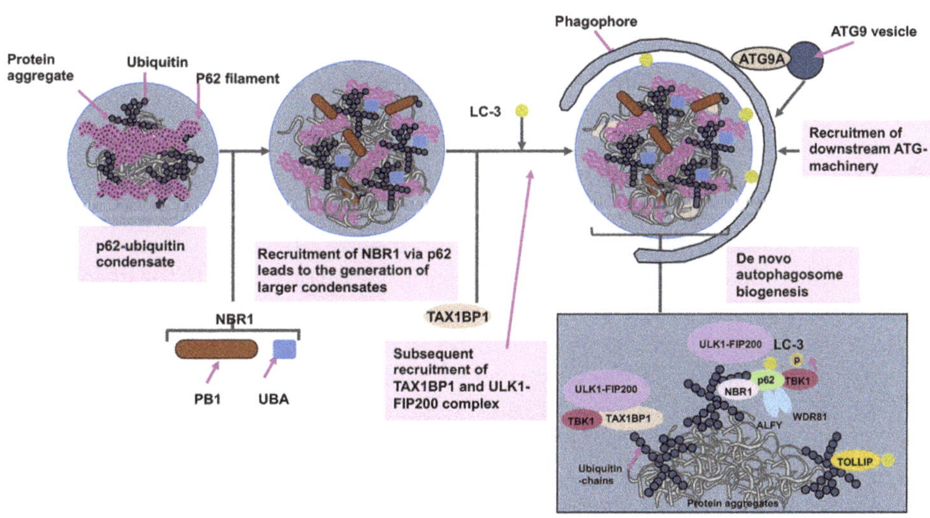

Figure 5.11. General principle of receptor enrolment in aggrephagy. Upon aggrephagy, p62 molecules scavenge ubiquitinated aberrantly folded proteins and oligomerize, leading to the generation of p62 biomolecular condensates, which are fueled by liquid–liquid phase partitioning owing to multivalent interplay between p62 and ubiquitin chains. The PB1 domain of p62 plays a vital part in its phase separation. Two proteins, namely WDR81 and ALFY, ease the p62-based removal of aberrant proteins, enhance the interaction of p62 with ubiquitin and recruit LC3 and GABARAP toward ubiquitinated proteins, respectively. NBR1 can then be directly engaged to p62 filaments through its PB1 domain, which leads to a superior recruitment efficiency of p62 to NBR1's UBA domain. In addition, the additional recruitment of the receptor TAX1BP1 to these condensates is aided through NBR1. The p62 and TAX1BP1 proteins are recruited to these aggregations through their interaction with FIP200. FIP200 also acts as a site for the selective recruitment of distinct autophagy-related (ATG) protein constituents, like ATG9A-containing vesicles and the PI3K complex. The resulting complex in turn stimulates the *de novo* biogenesis of autophagosomes on top of these aggregated protein substrates. p62 connects the aggregates with the isolation membrane through linking to the lipidated LC3. In addition, ubiquitin tethering also causes the clustering of a further receptor protein, which is named TOLLIP. An additional LC3 adapter is needed to promote aggrephagy and probably works in tandem with the various receptors. The kinase TBK1 phosphorylates receptor proteins and is bound to freight substrates to enhance receptor functionality when freight, lipidation of LC3 and/or interaction with different upstream ATG compounds are engaged.

clearance, with aggrephagy being a key mechanism for the elimination of possibly cytotoxic deposits (Scotter *et al* 2014, Dikic 2017).

5.12.1 p62 and additional ubiquitin-dependent receptors of aggrephagy

p62 functions as a key aggrephagy receptor, and p62's function was uncovered together with the initial identification of the LIR motif (Bjørkøy *et al* 2005, p. 62; Pankiv *et al* 2007, p. 62; Wurzer *et al* 2015, Zaffagnini *et al* 2018, Turco *et al* 2021). p62 recruits ubiquitin that is attached to misfolded proteins, thereby enabling the assembly of the autophagic compartment and the internalization of the misfolded proteins. Besides ubiquitin, the members of the BEACH domain-containing protein family, WDR81 and ALFY, enhance p62-facilitated clearance of aberrantly folded proteins through enhancing p62 conjugation with ubiquitin and promoting LC3 and GABARAP sequestration to ubiquitinated proteins (Clausen *et al* 2010, Liu *et al* 2017). A pivotal function of p62 in aggregate autophagy involves the engagement of the ULK1 complex with ubiquitinated proteins by p62's ability to interact in a direct manner with FIP200 (Turco *et al* 2019). Consequently, it leads to the generation of autophagosomes localized straight on protein aggregates to convey specific freight elimination. The interaction between FIP200 and p62 is conferred through the C-terminal claw domain of FIP200, that connects to the disorganized region of p62, which coincides with the LIR motif (Turco *et al* 2019).

Intriguingly, unlike NDP52, p62 necessitates an unperturbed LIR for its interaction with FIP200, which indicates that p62 may need to attach to ripening autophagosomes with lipidated LC3 for its efficient engagement of the ULK1 complex. In addition, the interaction of p62 with FIP200 is surpassed through LC3, pointing to a temporal directionality in the interaction of p62 with the ULK1 complex and nascent autophagosomes (Turco *et al* 2019). Finally, the FIP200 interaction domain of p62 is phosphorylated in several sites, which strengthens the physical connection between p62 and FIP200 (Turco *et al* 2019), even though the identity of the kinases that phosphorylate p62 at these sites remains uncertain. The kinase TBK1 also participates in promoting aggrephagia by phosphorylating p62 at Ser403 to increase its interaction with ubiquitin and transduce receptor homo-oligomerization toward filamentous polymers (Matsumoto *et al* 2015). Moreover, TBK1 can engage several ubiquitin-binding receptors, among them OPTN45, NDP52 (Thurston *et al* 2009) and TAX1BP1 (Heo *et al* 2019), which could enable localized activation at aggregate nucleates. Whether TBK1 is implicated in modulating the interface between p62 and FIP200, nevertheless, remains to be elucidated.

Oligomerization of p62 is essential for the correct targeting of ubiquitylated cargos to the isolation membrane (Wurzer *et al* 2015, Zaffagnini *et al* 2018, Savova *et al* 2020) which is consistent with the earlier observation that p62 oligomerization is important for its receptor activity (Pankiv *et al* 2007). In particular, the ubiquitin-facilitated oligomerization of p62 powers the liquid–liquid phase separation of the p62-misfolded protein complex, which leads to the generation of liquid-like membraneless condensates. This process is driven by multivalent ubiquitin-p62 chain interferences (Sun *et al* 2018, 2020a). In addition, NBR1, formerly characterized as an aggrephagy receptor (Kirkin *et al* 2009), supports p62 oligomerization and phase separation through its PB1 and UBA domains (Savova *et al* 2020). The heterooligomeric complex of p62 and NBR1 may exhibit a higher affinity toward ubiquinated substrates compared to p62 oligomers on their own (Savova *et al* 2020). These results are consistent with the earlier results that the UBA domain of NBR1 exhibits a tighter binding to ubiquitin compared to the UBA domain of p62 (Long *et al* 2010, Walinda *et al* 2014). Therefore, heterooligomerization of p62 and NBR1 could enable more potent condensate assembly via enhanced binding of ubiquitin.

Finally, how autophagosomal membranes enclose p62 condensates has been recently elucidated, demonstrating the role of surface tension in this event. It emerged from this study that p62 condensates with high surface tension became fully embedded in the autophagosome, while droplets of low surface tension were partially broken down (Agudo-Canalejo *et al* 2021). Moreover, larger p62 condensates of high surface tension were not completely enveloped due to restricted autophagosomal membrane availability, whereas smaller ones were easily encapsidated, which indicated that the size of protein condensates matters in regulating autophagosomal membrane dynamics (Agudo-Canalejo *et al* 2021). In fact, the autophagic clearance mechanism for protein condensates may be separate from those that give rise to non-soluble aggregates. The researchers therefore proposed the concept of fluidophagy as a selective clearance mechanism for liquid condensates (Agudo-Canalejo *et al* 2021, Schultz *et al* 2021).

In addition to p62, it has been demonstrated that TAX1BP1 is also essential for removing mutant huntingtin protein with polyglutamine expansion (PolyQ Htt) aggregates in several cell models, like induced pluripotent stem cells that have not been derived cortical neurons (Sarraf *et al* 2020). TAX1BP1 has also been found to be critical for the removal of NBR1-positive protein aggregates in cells in which autophagy flux is impaired (Ohnstad *et al* 2020). In addition, TAX1BP1, similar to NDP52, can bind to FIP200 through its SKICH domain (Ohnstad *et al* 2020). The connection between TAX1BP1 and FIP200 enables the removal of NBR1-positive protein condensates regardless of LC3 lipidation in cells where autophagy flux is impeded (Ohnstad *et al* 2020). Remarkably, LC3-independent elimination of NBR1-positive protein condensates/aggregates through TAX1BP1 seems not to necessitate the ubiquitin-binding ability of TAX1BP1, since removal of the protein's UBZ domain leaves its function unaffected (Ohnstad *et al* 2020). Hence, TAX1BP1, similar to p62, is able to tether the ULK1 complex onto protein aggregates to facilitate their elimination through its connection with FIP200. TOLLIP represents another receptor for the removal of protein aggregates and was initially characterized in the clearance of insoluble PolyQ Htt (Lu *et al* 2014). Additional findings propose that TOLLIP also plays a role in mitophagy (Ryan *et al* 2020) and endosomal microautophagy (Zellner *et al* 2021).

5.12.2 Chaperone-assisted selective autophagy and alternative aggrephagy paths

Chaperone-assisted selective autophagy (CASA) accounts for the clearance of aggresomes, which are formations comprised of aberrantly folded proteins that

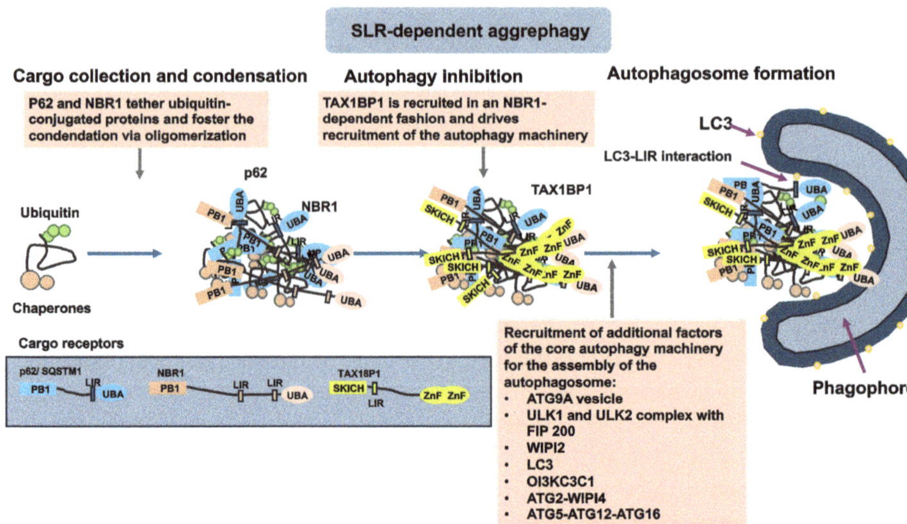

Figure 5.12. SLR-dependent autophagy route. There are three major steps in the SLR-dependent autophagic cascade. The first step is the collection of cargo and the condensation of the aggregate through oligomerization, The second step is the triggering of the initiation of the autophagy process, whereby the major action is the attraction of the TAX1BP1 protein in an NBR1-dependent fashion. The third step is the creation of the autophagosome that involves the enlistment of various proteins of the main autophagy mechanism of autophagosome formation, such as ATG9A vesicles, ULK1 and UKL2 complexes, including FIP200, WIPI2, LC3, PI3C3C1, ATG2-WIP14, and ATG5-ATG12-ATG16.

accumulate when proteasome activity is blocked (Kawaguchi *et al* 2003, Tedesco *et al* 2023) (figure 5.12). CASA could be viewed as a specialized form of aggrephagy, whereby the HSP70 family and HSPB8 chaperones, together with the ubiquitin E3 ligase CHIP, which is synonymously referred to as STUB1, identify aberrantly folded proteins and label these with ubiquitin (Arndt *et al* 2010). The decisive factor in this process is the function of BAG3, which links all the factors concerned and guides the complex of proteins toward the microtubule organizing center (Klimek *et al* 2017). Aggregates are first deposited in a cage-like structure formed by vimentin, which are referred to as vimentin cages, before being transported into the lysosome for piece-by-piece elimination (Kawaguchi *et al* 2003). An alternative signaling route has recently been discovered that incorporates the TRiC subunit CCT2 and transports stiff end-stage aggregates toward the lysosome (Ma *et al* 2022) (figure 5.12). In this degradation route, CCT2 binds a protein aggregate to LC3 in a direct manner, thereby promoting its clearance. The binding is uncoupled from ubiquitylation, p62, NBR1 and TAX1BP1. Unexpectedly, it is found to be separate from the remainder of the TRiC complex. TRiC normally performs the folding of a variety of substrates, among them proteins of the cytoskeleton (Gestaut *et al* 2022).

Cells could also reduce some of the burden of aggregation through liberation of protein aggregates. This exocytotic secretion can be pharmacologically enhanced, for example through the inhibition of PIKFYVE, which is a kinase implicated in vesicle merging events (Hung *et al* 2023). Disabling PIKFYVE exacerbates pathologies in model systems for amyotrophic lateral sclerosis (ALS), which represents another devastating disease linked to aggregated proteins (Hung *et al* 2023). Cells consequently have adopted a variety of strategies to get rid of protein aggregates. An important challenge is to decipher how these signaling routes are orchestrated and how they operate in diverse cell types, involving postmitotic cells like neurons and muscle cells.

5.12.3 Function of aggrephagy in neurodegenerative processes

Many neurodegenerative diseases are marked by the build-up of protein aggregates due to aging. Among the aggrephagy proteins are hyperphosphorylated tau fibrils (Inami *et al* 2011), amyloidβ (Pickford *et al* 2008), polyQ Htt (Ravikumar *et al* 2004), αsynuclein (Winslow *et al* 2010), as well as the RNAbinding proteins transactive TDP43 (Buchan *et al* 2013, Scotter *et al* 2014) and fUs (Buchan *et al* 2013). The postmitotic character of neurons in combination with their longevity may render them vulnerable to aggregation of pathological proteins. A key pathogenic mechanism in neurodegeneration is therefore the aggregation of proteins, a process that can be intensified because of the inefficient autophagic removal of such proteins that takes place during aging. Protein misfolding diseases may be associated with sequence changes resulting from genetic defects, such as in cystic fibrosis. Several approaches to the removal of aggregated proteins utilize receptor-like compounds. A protein-targeting chimeric molecule (PROTAC) molecule consists of an E3 recruitment ligand, a POI-targeting domain, and a flexible linker that connects the two ligands (Zhao *et al* 2022). The supplementation of PROTAC enhances the generation of the ternary complex POI-PROTAC-E3, promotes the ubiquitination of the POI and its consequent clearance through the UPS (Sun *et al* 2019, Wang *et al* 2020b, Yang *et al* 2021). The first PROTAC molecule has been reported in 2001 (Sakamoto *et al* 2001). Protac-1 was engineered to attach to the target protein methionine aminopeptidase-2 (MetAP-2) and attract the SCF ubiquitin ligase complex for proteasomal clearance (Sakamoto *et al* 2001). The Protac-1 comprises two domains: one domain is composed of a phosphopeptide derived from IkBα (IPP) that interacts with SCF, and the other domain, which is made up of ovalicin, engages with MetAP-2 (Sakamoto *et al* 2001). This same group then proceeded to show that a chimeric molecule consisting of the IκB phosphopeptide and small molecules is able to break down the estrogen receptor (ER) and the androgen receptor (AR), which stimulate the growth of breast and prostate cancer, respectively (Sakamoto *et al* 2003).

In 2008, the scientific team of Crew showed the first prototype of PROTACs based on small molecules (Schneekloth *et al* 2008). This PROTAC, composed of a non-steroidal androgen receptor ligand (SARM), an MDM2 ligand that works by targeting the ubiquitin ligase murine double minute 2 (MDM2), and a PEG-based linker, has been utilized to break down AR (Schneekloth *et al* 2008). Compared to peptide-based PROTACs, small-molecule PROTACs tend to be more easily absorbed by cells and are more rapidly advanced to the stage where they can be turned into medications (Toure and Crews 2016). Besides MDM2, several other E3 ligases were exploited in the PROTAC technology, involving Von-Hippel-Lindau (VHL) (Bond *et al* 2020), cereblon (CRBN) (Wang *et al* 2020b), and cell inhibitor of apoptosis protein (cIAP) (Yang and Sun 2015).

PROTACs offer numerous benefits over conventional small-molecule inhibitor drugs (Zeng *et al* 2021). In the first place, PROTACs significantly broaden the scope of proteins that can be targeted by pharmaceutical drugs. More than 4000 disease-associated proteins have been identified. Of these, only about 400 proteins have been successfully used in current therapeutic approaches. Many of them could not be treated with conventional inhibitors because of their structural intricacy and side effects. In the second palce, while conventional inhibitors disrupt only partially the functioning of the protein, PROTACs break down the protein, thereby eliminating all of its functions. Thirdly, traditional kinase inhibitors frequently result in resistance to medication due to mutations or overexpression of the drug-targeted proteins. PROTACs, by contrast, could minimize drug resistance through long-term selection pressure because of the breakdown of target proteins. Finally, PROTACs are sub-stoichiometrically and catalytically active, which enables them to act at low concentrations, thus minimizing potential toxic adverse effects.

5.13 Xenophagy

Xenophagy refers to a form of selective autophagy where autophagosomes scission and eliminate pathogens that invade the cytoplasm. While the first barrier against pathogens is a coordinated reaction of the immune system, non-phagocytic cells, such as epithelial cells, can also antagonize pathogens by xenophagy (figure 5.13) (Shibutani *et al* 2015). In the same way, it can also fight a wide range of viruses that cause infection through a process termed virophagy (Dong and Levine 2013).

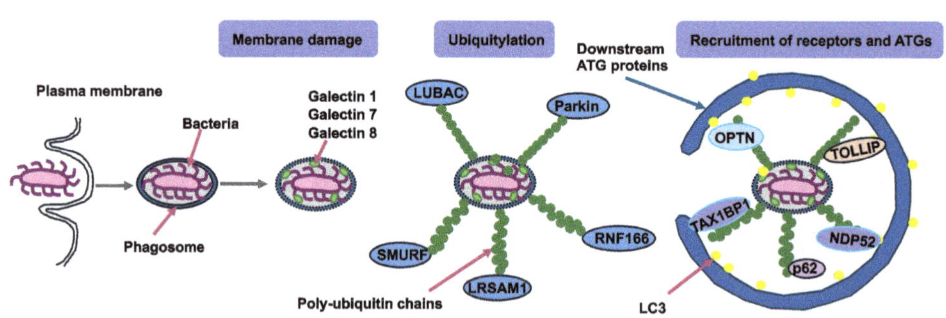

Figure 5.13. Receptor enrolment in xenophagy. Bacteria that invade host cells are enclosed in phagosomes from the host membranes, sometimes creating niche environments for bacterial growth, like *Salmonella*-containing vacuoles. *Salmonella* evades these vacuoles through vacuole membrane permeabilization and enters the cytoplasm. Within the cytosol, bacteria are flagged through galectins, which specifically recognize bacterial surface polysaccharides, and ubiquitination, through the activity of a panel of ubiquitin ligases, leading to the engagement of receptor proteins and mechanisms that favor autophagosome formation. Receptor proteins such as TAX1BP1, NDP52, p62, optineurin (OPTN) and TOLLIP attach to bacteria and isolation membranes through tethering of both LC3 on the isolation membrane and ubiquitin on the bacteria.

5.13.1 Aims for xenophagy

While the mechanism whereby host cells identify xenophagy substrates is common to other types of selective autophagy, xenophagy is distinct from other forms of selective autophagy in that it seeks out invaders that challenge host cells, instead of elements within the cell itself. Whereas xenophagy restricts bacterial proliferation within host cells, several pathogens have the ability to block autophagosome formation or inactivate lysosomal enzymes to escape destruction, such as *Listeria* RavZ protein, which impairs the recycling of LC3, *Shigella* IcsB protein, which interferes with the detection of the bacterial VirG protein using ATG5 and *Salmonella* SopF, which disturbs the interaction of VATPaseATG16L1 during infection (Ogawa *et al* 2005, Choy *et al* 2012, Xu *et al* 2019, 2022). In certain circumstances, these pathogens co-opt and exploit the xenophagy system to enhance their proliferation (Cemma and Brumell 2012, Dong and Levine 2013). Nevertheless, xenophagy represents an essential survival mechanism, as it acts against many lethal pathogens, like group A *Streptococcus* (GAS) (Nakagawa *et al* 2004) and *Salmonella* spp. (Birmingham *et al* 2006), which are frequently highly resistant to antibiotics. In this subsection, the mechanistic knowledge of xenophagy for bacteria is discussed in somewhat more depth than that of virophagy, since the mechanism of autophagy-related proteins for virophagy is very different, whereas the mechanism for detecting invading bacteria has some similarity to the mechanism of other selective autophagies. For example, the LC3 conjugation system enhances the interruption of the replication complex of murine norovirus, while other autophagy-related genes, like ATG14, are not essential for the process, indicating that the LC3 conjugation system has another antiviral effect beyond autophagy-mediated elimination of viral elements (Biering *et al* 2017). Moreover, while ubiquitin-facilitated clearance certainly represents an important mechanism for the xenophagy of bacteria. Multiple cases of autophagy-driven clearance of viral proteins may not necessarily depend on ubiquitination. For instance, the autophagic decomposition of the HIV-1 viral protein Tat inside T-lymphocytes is caused through its non-ubiquitin-dependent association with p62 (Sagnier *et al* 2015). The E3 ubiquitin protein ligase SMURF1 and p62 enhance the breakdown of the capsid protein of the Sindbis virus, although the E3 ubiquitin ligase domain of SMURF1 is not necessary for the process of virophagy. In addition, it must be considered that the antiviral effect of autophagy *in vitro* is not the same in all cell types, even though numerous investigations provide substantial support for the protective effect of autophagy *in vivo*. The controversy may originate from the involvement of autophagy in antigen presentation to enhance the host defense toward viral infection *in vivo* by activating systemic immune reactions (Dong and Levine 2013).

5.13.2 Identification of bacteria as part of xenophagy

When bacteria enter cells, they are enveloped in endosomal membranes that are broken down through the endosomal-lysosomal system. The bacteria can, nevertheless, hinder this breakdown. For instance, *Salmonella* hijacks the host endocytic route to create *Salmonella*-filled vacuoles, where they multiply (Birmingham *et al* 2006). Nonetheless, a minor but significant portion of the invading *Salmonella* is liberated into the cytoplasm through disruption of the endosomal membrane enclosing the bacteria, and these are tagged with polyubiquitin (Perrin *et al* 2004). This subset of *Salmonella* is labeled for LC3 and is trafficked to an autophagosome (Xu *et al* 2022). It has been demonstrated that the inclusion of polystyrene beads carrying a membrane-disrupting reagent inside endosomes is usually enough to trigger the assembly of autophagosome-like membranes around the beads (Fujita *et al* 2013). Therefore, the laceration of host membranes may act as a universal alarm cue to induce xenophagy. The presentation of glycans, which are molecules located in the surface of cell membranes or within the lumen of endosomal membranes, at sites of endosomal membrane disruption appears to be a critical event in the signaling that accompanies this process, since galectins attach to the glycans and rapidly accumulate at the points of disruption. The concentration of galectins not only serves as a marker for the disrupted membrane, which is the case with the lysosomal lesions discussed above, but also has an important function in pathogen detection. Galectins tether to polysaccharides on the bacterial surface and thereby attract NDP52 to enable the generation of autophagosomes. This initial engagement of NDP52 with the invading bacteria through galectin-8 primes NDP52 for ubiquitin tethering on the bacteria. This enhances the percentage of LC3-positively taggged *Salmonella* and elevates the autophagy-facilitated growth restriction of invading *Salmonella* (Thurston *et al* 2012). Some other galectins, like galectin 1 and galectin 7, are attracted in a TOLLIP-dependent fashion following the invasion of GAS and inhibit the spread of GAS (Lin *et al* 2020). Galectins thus do not merely function as warning signals, instead they also regulate the process of autophagosome membrane biogenesis in a multileveled fashion throughout xenophagy.

5.13.3 Bacterial polyubiquitylation and enrolment of receptor proteins

As with other selective autophagy routes, the ubiquitination of bacterial constituents represents an essential feature for the detection of pathogens for xenophagy. The *M. tuberculosis* protein Rv1468c becomes attached directly to ubiquitin to be shielded from the autophagosomal membrane (Chai *et al* 2019). The N-acetylglucosamine residues of GAS surface carbohydrates can be detected with FBXO2 and enhance the ubiquitylation of the invading GAS (Yamada *et al* 2021). Ubiquitination of *Salmonella* lipopolysaccharide carried out by the ubiquitin ligase RNF213 is necessary for limiting bacterial growth in host cells. This supports the notion that xenophagy is driven by non-pathogen-derived, non-proteinaceous ubiquitylation targets (Otten *et al* 2021).

Xenophagy relies on the binding of bacteria to autophagosomal compartments through receptor proteins that are able to concurrently engage LC3 and ubiquitin (von Muhlinen *et al* 2012). p62 is attracted through invading *Salmonella* by its ubiquitin-binding domain 1 (Zheng *et al* 2009). Ubiquitin-binding domain-dead p62 is unable to enhance the recruitment of LC3 to invading *Salmonella*, indicating that the association of p62 with ubiquitinated bacteria is necessary for the linking of autophagosomal membranes together with bacteria. NDP52 plays a unique and integral part in xenophagy as it also has galectin-binding domains on top of the aforementioned ubiquitin-binding motif (Thurston *et al* 2009). In addition, it plays a part in the extrusion of intracellular bacteria; NDP52 is able to interact not merely with LC3 but additionally with myosin VI, which enhances the intracellular transportation of bacterium-carrying autophagosomes to the minus end of the actin filaments to ease their merging with the lysosome (Verlhac *et al* 2015). In addition, NDP52 is needed to target the ULK1 complex to bacteria within the cytosol, thereby reinforcing the concept that autophagosomal structures for selective autophagy are generated *de novo* on the targeted substrates (Ravenhill *et al* 2019, Shi *et al* 2020a). NDP52 and p62 can be separately targeted by *Salmonella* for their own benefit, but they function through the identical pathway, as simultaneous knockdown of both receptors compared to the knockdown of each individual receptor causes no additional enhancement of *Salmonella* growth (Cemma *et al* 2011). Moreover, OPTN has been demonstrated to facilitate xenophagy as a receptor protein and to inhibit the multiplication of *Salmonella* (Bansal *et al* 2018). The depletion of TAX1BP1 also leads to an accumulation of ubiquitin-positive *Salmonella* and promotes their hyperproliferation (Tumbarello *et al* 2015). In conjunction with the preceding regulatory factors LAMTOR1 and LAMTOR2, TAX1BP1 promotes the ripening of autophagosomes that harbor invading GAS and inhibits the survival of GAS (Lin *et al* 2019). TOLLIP can also serve an important function in the xenophagy of GAS because it enhances the ability of invading GAS to attract galectin 7 and additional receptor proteins (Lin *et al* 2020).

Ubiquitination of the substrates for xenophagy, so that they can be identified by the receptors, involves several E3 ligases that attach various types of polyubiquitin chains, among them polyubiquitin chains connected by Lys6, Lys27, Lys33, Lys48 and Lys63, along with linear (Met1) polyubiquitin linkages. Each E3 ligase can exert different mechanisms to limit the proliferation of invading bacteria. Parkin, an E3 ligase necessary for mitophagy, is required for Lys63-tethered ubiquitylation and the growth impairment of *M. tuberculosis* (Manzanillo *et al* 2013). Conversely, in the E3 ligase, SMURF1 promotes Lys48-coupled ubiquitination (Franco *et al* 2017). Parkin is required for the enrollment of p62 toward the invading *M. tuberculosis*, while SMURF1 can be disregarded for this process. In contrast, SMURF1 is required to localize the proteasome to the bacteria, while Parkin is not required. The LRR-containing RING E3 ligase LRSAM1, that exhibits *in vitro* E3 ligase activity for Lys6- and Lys27-connected polyubiquitin chains, is necessary for the ubiquitination of multiple bacterial species, such as *Salmonella* (Huett *et al* 2012). RNF166 becomes engaged with bacteria and eases the following engagement of p62, that is also a target of Lys33-connected ubiquitination. Moreover, RNF166 is a slightly more difficult to grasp, unusual type of ubiqutination string that can foster intracellular transportation (Heath *et al* 2016). LUBAC produces linear polyubiquitin chains and is induced upon infection with *Salmonella* (Fiskin *et al* 2016, Noad *et al* 2017, Fuseya *et al* 2020). Remarkably, LUBAC targets bacteria that are pre-coated with ubiquitin, indicating that it augments ubiquitin coverage (Noad *et al* 2017).

These linear polyubiquitin strings on invading bacteria trigger the assembly of the OPTN/NeMO-dependent autophagosome-lysosome complex for xenophagy, but also promotes the activation of the pro-inflammatory NFκB signal transduction pathway. In this way, LUBAC-dependent detection of bacteria orchestrates the steps of the antibacterial reaction within higher eukaryotes (Noad *et al* 2017).

Importantly, while the majority of ubiquitin signal transduction normally takes place on membrane-like structures enveloping invading bacteria throughout xenophagy, there is emerging experimental evidence that bacterial carbohydrates and host proteins are ubiquitinated in the process of bacterial xenophagy. However, the identity of these host factors remains elusive (Manzanillo *et al* 2013). Based on the mechanistic resemblance of xenophagy to other selective autophagy routes, several host proteins that are recognized for ubiquitylation to be sequestered within an autophagosome for some other selective autophagy, such as LAMP1 and LAMP2 during lysophagy, represent powerful potential candidates for factors implicated in xenophagy. Besides, ubiquitination is possibly not indispensable for xenophagy. For instance, *Salmonella* can be sorted onto autophagosomes either by ubiquitylation of the bacterial surface, such as discussed before, or by generation of the lipid second messenger diacylglycerol on *Salmonella*-carrying vacuoles (Shahnazari *et al* 2010). Therefore, identifying ubiquitination target sites and elucidating the distinction between ubiquitin-dependent and ubiquitin-independent xenophagy hold the key to exploring the working mechanism of xenophagy. In the following subsections, a concise outline of some of the other arising paths of selective autophagy is offered, with a special emphasis on the receptor proteins participating in each of the processes.

5.14 ER-phagy

The ER critically contributes to protein expression, and one mode of ER quality assurance involves the selective autophagy of ER membrane subregions that are malfunctioning due to, for instance, aberrantly folded proteins that give rise to ER stress. This mechanism is referred to as ER-phagy or reticulophagy, whereby, the former is more commonly used (figure 5.14). Mammalian cells express several ER-phagy receptor proteins. FAM134B localizes to the ER and possesses a C-terminal LIR motif to promote autophagic membrane docking at the ER (Khaminets *et al* 2015). RTN3, a family member of reticulon proteins, represents another ER-phagy receptor that harbors several N-terminal LIR motifs and acts in an independent manner to FAM134B (Grumati *et al* 2017). Moreover, SEC62 (Fumagalli *et al* 2016), TEX264 (An *et al* 2019, Chino *et al* 2019), atlastin 3 (Chen *et al* 2019), CCPG1 (Smith *et al* 2018) and CALCOCO1 (Nthiga *et al* 2020) have each been characterized as ER-phagy receptors. ER receptors with different specificities may react to various ER stressors. In addition, the ER constitutes a multicomponent organelle that consists of heterogeneous subcellular structures comprising tubules and cisternae that extend throughout the entire cell. It is still unknown whether the different ER-phagy receptors are involved in the clearance of distinct subcellular ER patterns.

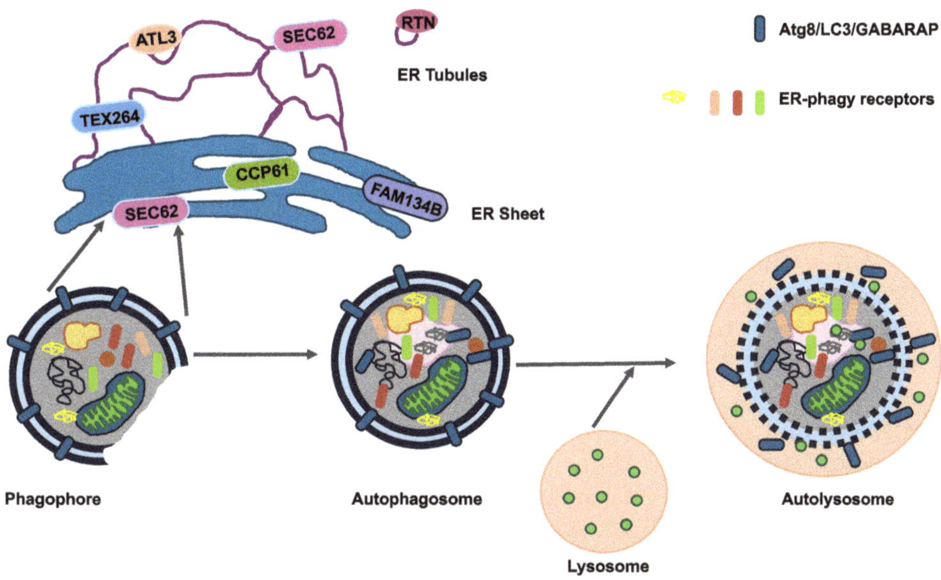

Figure 5.14. The catabolic process of the enodplasmatic reticulums (ER)'-phagy's, which is referred to as ER-phagy. As described in LC3/GABARAP/Atg8, ER phagocytosis receptors are concentrated in ER micro-domains that must be broken down. The ER and the autophagy process are linked by this mechanism. The isolation membrane combines with the ER assemblies and enlarges in phagocytes. The phagosome then engulfs the ER piece and closes to create an autophagosome. The autophagosome and lysosome then merge to generate a new organelle, the vacuole in yeast and plants, or an autolysosome in mammalian cells. Finally, the constituents absorbed by the autophagosome are metabolized using lysosome/vacuole hydrolases.

Of the different ER-phagy receptors, CCPG1 appears to be unique in its ability to engage both LC3 proteins and FIP200 through different motifs; the engagement of both ATG proteins is required for CCPG1-facilitated ER-phagy (Smith *et al* 2018). The binding of CCPG1 to FIP200 matches the emerging model of receptor proteins that interact with the upstream ATG mechanism. Importantly, the ER-phagy receptors mentioned so far are already resident at the ER and therefore need no ubiquitin to become functional receptors. In addition, p62 promotes the clearance of ER out of hepatocytes, as has been seen in p62-deficient mice, which display reduced levels of ER-phagy in comparison to wild-type mice (Yang *et al* 2016). Moreover, p62 has been demonstrated to interact with the transmembrane E3 ligase TRIM13, which is labeled with Lys63-linked ubiquitin, to mediate ER-phagy in cells suffering from proteotoxic stress, thus participating in ER protein quality assurance and ameliorating ER stress (Ji *et al* 2019). A genome-wide CRISPR-Cas9 screening identified ubiquitin-like post-translational modification, which is referred to as UFMylation, in which ubiquitin fold modifier 1 (UFM1) protein is linked, as a critical modulator of ER-phagy. UFL1 ligase migrates to the ER under stress to add UFMylate to ER-resident proteins (Liang *et al* 2020a). The role of UFL1 is similar to the function of PINK1-Parkin in labeling defective mitochondria for mitophagy. The UFL1 ligase then enlists an evolutionarily conserved cytosolic protein, such as

C53, that functions as an ER-phagy receptor and engages with ATG8 proteins through a noncanonical LIR motif, thus facilitating delivery of the isolation membrane toward the ER to mediate ER-phagy (Stephani *et al* 2020).

5.15 Pexophagy

Peroxisomes serve as a site for both ROS generation and detoxification. Hence, cells need to sustain peroxisome homeostasis, otherwise they risk suffering oxidative injury. Peroxisome numbers are also affected through metabolic requirements. In yeast, for instance, peroxisomes multiply rapidly when cultured with nutrients that peroxisomes need for metabolism. Once they are converted to other carbon nutrients, the high number of peroxisomes are no longer needed and are eliminated through pexophagy. Pexophagy is the selective breakdown of excess or defective peroxisomes and acts as the primary mechanism for decreasing the number of peroxisomes within a cell. Both p62 and NBR1 have been demonstrated to be involved in pexophagy in mammalian cells (Kim *et al* 2008, Deosaran *et al* 2012, Yamashita *et al* 2014). PEX2, which is a peroxisomal E3 ligase, is supposed to ubiquitinate peroxisomal membrane proteins when there is a lack of food, thereby causing NBR1 to trigger pexophagy (Sargent *et al* 2016). Increased cellular ROS levels trigger the activation of Ataxia-telangiectasia-mutated kinase (ATM), which suppresses mTOR and stimulates the activation of ULK1 to encourage autophagy. ATM also specifically triggers pexophagy via its transport to peroxisomes, where ATM can phosphorylate the peroxisomal import receptor PEX5, resulting in its ubiquitylation, thereby promoting p62 target selection for autophagosome biogenesis (Zhang *et al* 2015a). In a recent study, MARCH5 has been shown to be a ubiquitin ligase that is required for pexophagy and is localized to peroxisomes via its association with PEX19 (Zheng *et al* 2022). Finally, PEX14 can directly tether with LC3, indicating that this protein serves as a pexophagy receptor via a mechanism that is not dependent on ubiquitination (Hara-Kuge and Fujiki 2008).

5.15.1 What is the regulatory role of pexophagy in selective autophagy?

Selective autophagy represents an integral process for sustaining cellular homeostasis through the continuous reutilization of compromised or redundant constituents. More than a dozen selective autophagy routes control the breakdown of various cellular substrates, but it is still unclear whether these routes can affect one another. This problem was addressed as a model in the context of pexophagy, which is the autophagic breakdown of peroxisomes (figure 5.15).

5.15.2 The ACBD5 protein can promote the efficient recruitment of receptors and adapters that are specific for pexophagy

Upregulated pexophagy was observed to interfere with selective autophagy of both mitochondria and protein aggregates inside cells through its depletion of the autophagy induction factor ULK1. This result was validated in cell models of the pexophagic variant of peroxisome dysfunction, which is referred to as Zellweger Spectrum Disorder, a disorder caused mainly due to peroxisomal malfunction.

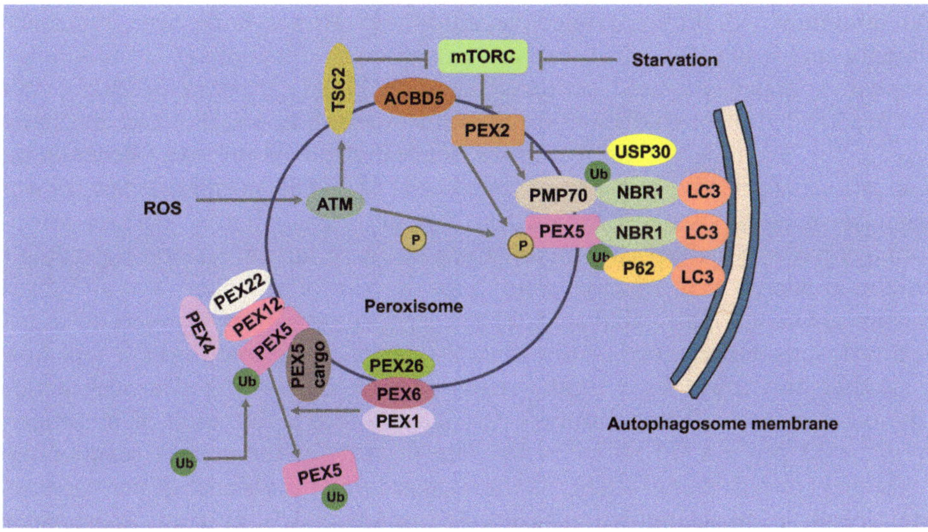

Figure 5.15. Schematic drawing of key proteins of pexophagy. Normally, mTORC1-sriven proteasome signaling keeps PEX2 expression at a low level. During starvation, an accumulation of PEX2 results in PEX5 and PMP70 being ubiquitinated, thereby inducing pexophagy in an NBR1-dependent manner. USP30 counterbalances PEX2 through deubiquitination of its substrate to impede pexophagy. The initial reaction pathway of peroxisomal ROS involves ATM serine/threonine kinase. TSC2 is activated through ATM kinase, which then causes the inactivation of mTORC1. Moreover, ATM can phosphorylate PEX5 at Ser141, thereby causing PEX5 to be ubiquitinated at Lys209. The proteophagic targeting of the ubiquitin-dependent peroxisome involves monobituted PEX5 at Cys11. PEX1 and PEX6, that are connected to the peroxisome through PEX26, cause PEX5 to ubiquitinate and subsequently detach it from the membrane after transit. To date, ACBD5 is the only known protein that is exclusive to phagocytes. Thereafter, ubiquitinated PEX5 interacts with p62/NBR1 to induce ROS-driven autophagy. Monobitized PEX5 at Cys11 serves as the cleavage target of the ubiquitin-dependent peroxisome proteophage. PEX1 and PEX6, that are attached through PEX26 to the peroxisome, cause PEX5 to ubiquinate and subsequently degrade it after its passage through the membrane.

Moreover, this result has been generalized to cell models of Parkinson's disease and Huntington's disease through the identification that enhanced clearance of protein aggregates reciprocally attenuates pexophagy, thereby extending the generalizability of restricted selective autophagy. These results indicate that the clearance capacity of selective autophagy can be restricted through an augmentation of a substrate. Macroautophagy, later referred to as autophagy, is a well-conserved pathway that is essential for the clearance of cytoplasmic components, such as malfunctioning or excess organelles. (Vargas *et al* 2023). Autophagy involves a double-walled vesicle, which is referred to as the autophagosome, that initially encapsulates cytoplasmic material and later merges with a lysosome to break it down (Vargas *et al* 2023). Selective autophagy arises from a specific cytoplasmic substrate being targeted for clearance through an assemblage of specialized autophagy receptors across its outer surface (Vargas *et al* 2023). In mammalian cells, autophagy receptors can either home to substrates at the point of expression or engage an 'eat-me' tag on substrates, which is typically ubiquitin (Kirkin and Rogov 2019).

At minimum, 13 different selective autophagy routes have been reported in mammals, involving the clearance of peroxisomes (pexophagy), mitochondria (mitophagy), and aggregated proteins (aggrephagia) (Vargas *et al* 2023). While they are activated through distinct mechanisms, they all utilize the same mechanism to create the machinery that forms the sequestration membrane. Because most investigations of selective autophagy have been conducted as single events in which specific harm is imposed on organelles, it is not clear whether various selective autophagy pathways affect each other. Even less is so far understood regarding the conditions under which selective autophagy is controlled when several substrate pathways are concurrently activated. Zellweger spectrum disorder refers to a group of relatively uncommon inherited disorders that are characterized through depletion of peroxisomes and result in a multisystemic disease comprising neurodegeneration (Argyriou *et al* 2016). The Zellweger spectrum disorder is the result of mutations in one of the 13 peroxin (PEX) genes that are needed for the correct functioning of peroxisomes (Argyriou *et al* 2016). At the cellular level, the absence of peroxisomes in humans results in an accumulation of defective mitochondria, aberrant lysosomes and lipid droplets (Goldfischer *et al* 1973, Mooi *et al* 1983, Yakunin *et al* 2010, Salpietro *et al* 2015, Berendse *et al* 2019). In three mouse models of Zellweger spectrum disorder, there was an increase in autophagy substrates, such as dysfunctional mitochondria and alpha-synuclein (αS) oligomers in the hepatic and cerebral tissues (Baumgart *et al* 2001, Rahim *et al* 2016). It is not clear, however, why cells with Zellweger Spectrum Disorder are unable to eliminate these cytotoxins, although there are no obvious deficiencies in autophagy. PEX13 was previously assumed to be the protein that governs mitophagy and virophagy, since knocking down this gene impairs both selective autophagy routes (Lee *et al* 2017a). Nevertheless, it has recently been established that the depletion of PEX13 results in the upregulation of pexophagy in cells, indicating that pexophagy, and not the depletion of PEX13, could affect other selective autophagy routes (Demers *et al* 2023). A related increase in pexophagy takes place in cells lacking PEX1, which is the gene most frequently mutated in Zellweger Spectrum Disorders (Law *et al* 2017). Since different selective autophagy pathways require overlapping mechanisms, it is possible that enhanced pexophagy restricts the activity of other selective autophagy routes in Zellweger Spectrum Disorder. A study was conducted to determine whether different selective autophagy signaling routes can mutually influence each other. Consequently, cells with enhanced pexophagy exhibit compromised aggrephagy and mitophagy. The compromised selective autophagy resulted not from the depletion of peroxisomes or their functionality, but from the exhaustion of basal autophagic capacity due to the excessive pexophagic pathway activation. Moreover, these investigations revealed ULK1 as a restricted factor that impedes selective autophagy within cells with enhanced pexophagy. In addition, the data obtained in cell models of Parkinson's and Huntington's disease are indicative of the generalizability of these results. In summary, these investigations have demonstrated that pexophagy-dependent restriction of mitophagy and aggrephagy fosters the pathological aggregation of proteins and the malfunction of mitochondria in models of Zellweger Spectrum Disorder, leading to the realization that one type of selective autophagy can impact on others.

5.16 Ribophagy

Proteostasis is critical for cellular integrity, and the capacity of a cell to regulate the amount of ribosomes that are accessible to translation by ribophagy may have broad implications for multiple molecular signaling routes (Nakatogawa 2018). Autophagy has been suggested as a mechanism for controlling the number of ribosomes. Pharmacological blockade of mTOR, starvation, and arsenite were demonstrated to lead to ribophagy (An and Harper 2018). Nuclear FMR1-interacting protein 1 (NUFIP1) has been found to serve as a ribophagy receptor for mammals. NUFIP1 may physically interact with LC3B and ribosomes to promote ribophagy, and NUFIP1 decrease impedes ribophagy (Wyant *et al* 2018). Nevertheless, recent studies have shown that the elimination of NUFIP1 has no effect on ribophagy, which points to the existence of other receptors. In addition, ribosomal export to lysosomes accounted for only a small fraction of ribosomal frequency when cells were starved or after mTOR blockade, implying that ribophagy has negligible effects on bulk ribosome functioning and protein biosynthesis (An *et al* 2020). In summary, much additional investigation is needed to elucidate the molecular constituents and function of mammalian ribophagy.

5.17 Golgiphagy

A new kind of autophagy has been revealed that is termed Golgiphagy (Li *et al* 2012, Lu *et al* 2020, Nthiga *et al* 2021a, 2021b, Jetto *et al* 2022), which regulates the amount of Golgi membranes through the soluble reticulophagy receptor CALCOCO1. Golgiphagy represents a selective type of autophagy that governs the turnover of the Golgi complex (Rahman *et al* 2022). The Golgi apparatus consists of flat membrane bags situated all around the nucleus. Its main function is to modify proteins and fat molecules for transport and utilization in other locations inside and outside the cell. Selective autophagy receptors and adaptors incorporate short linear motifs, referred to as LIR (LC3-interacting regions) motifs, which are required for interaction with Atg8 family proteins. LIR motifs attach to the hydrophobic cavities of the LIR motif docking site (LDS) of the corresponding Atg8 family proteins. The LDS attachment site of Atg8a plays an important part in this selective autophagy pathway. The finding that the Golgi microtubule-associated (GMAP) protein associates with Atg8a and the LIR motif in position 320–325 is critical for the coupling between the phagophore and the Golgi apparatus. Hence, the function of this form of selective autophagy involves modulating Golgi size and shape. Therefore, Golgi stress has been implicated in the evolution of age-related diseases; this process of autophagy may be essential for sustaining cellular equilibrium.

5.18 Selective autophagy linked to cellular processes or molecules

Apart from the selective autophagy of whole key organelles, there can also be selective autophagy governing the iron homeostasis, named ferritinophagy, the regulation of glycogen, which is known as glycophagy, or smaller organelles, such as lipid droplets that is referred to as lipophagy.

5.18.1 Ferritinophagy

Selective autophagy may also regulate iron homeostasis via the specific breakdown of ferritin, which functions as an iron-tethering protein. This phenomenon is fittingly referred to as ferritinophagy. While iron is needed for a variety of biological functions, excessive levels of free iron can produce ROS and can lead to a mode of cell death referred to as ferroptosis. Ferritin can bind free iron and keep intracellular iron balance within acceptable limits (Maio and Rouault 2020). In contrast, whenever iron concentrations are in the low range, ferritinophagy is induced to liberate iron (Kaur and Debnath 2015). Nuclear receptor co-activator 4 (NCOA4) acts as the receptor protein that transduces ferritinophagy (Mancias *et al* 2014). NCOA4 binds to the heavy and light chains of ferritin and simultaneously to LC3 proteins (Mancias *et al* 2014) and it is essential during erythropoiesis (Mancias *et al* 2015). Intriguingly, NCOA4 has been demonstrated to cooperate with TAX1BP1 to promote ferritin targeting to the lysosome despite the absence of FIP200, indicating that the ULK 1 complex is not necessary for NCOA4-TAX1BP1-driven ferritinophagy (Goodwin *et al* 2017), although the question remains as to how membrane isolation is produced. Consistent with this concept, the kinase TBK1 was identified as being involved in iron level control and, together with TAX1BP1 and ATG9A, is required to convey lysosomal sorting of ferritin into cells devoid of FIP200 (Goodwin *et al* 2017).

5.18.2 Glycophagy

Glycogen, consisting of glucose, is a polysaccharide with multiple branches. It is an integral part of muscle and liver cells in the storage of energy. The control of glycogen levels has a critical function in energy metabolism. The catalysis of glycogen returning to glucose, referred to as glycogenolysis, is carried out by glycogen phosphorylase. In addition, cells can degrade glycogen via autophagy, which is referred to as glycophagy (Koutsifeli *et al* 2022). Glycophagy clears ill-branched glycogen strings that are ineffective in glucose storing and recirculating and acts as a homeostatic mechanism. Moreover, muscle glycophagy is associated with energy metabolism and insulin signal transduction, implying a compensatory function in metabolic disorders (Heden *et al* 2022). Starch binding domain-containing protein 1 (STBD1) is abundant in the liver and skeletal muscle and interacts with both glycogen and the LC3 family protein GABARAPL1 through an LIR motif (Jiang *et al* 2010, 2011). Consequently, STBPD1 functions as a receptor protein within the process of glycophagy. It is important to note that a mutation in the LIR motif of STBD1 was recently identified in tissue specimens from colorectal adenocarcinoma patients (Han *et al* 2021). In fact, overexpression of the wild-type, and not the LIR mutant STBD1, blocked the proliferation of several cancer cells, indicating a far-reaching effect of this receptor protein and glycophagy in the tumorigenesis of several types of cancer (Han *et al* 2021). In cardiomyocytes, elevated insulin and glucose concentrations caused an increase in STBD1 (Mellor *et al* 2014). Moreover, depletion of GABARAPL1 *in vivo* has been reported to interfere with cardiac glycophagy and diastolic functioning, thereby phenotypically imitating diabetic cardiac disease (Mellor *et al* 2021). Notably, a gene therapy strategy to enhance GABARAPL1 expression solely in the heart of mice receiving a high-fat dietary intervention ameliorated glycogen build-up in the heart muscle of these mice (Mellor *et al* 2021), indicating that enhancing cardiac glycophagy may alleviate cardiac impairment induced by diabetes. Additional research is needed to determine whether there exist other receptors for glycophagy besides STBD1 and to examine whether potential receptors for glycophagy can also engage with upstream autophagy factors.

5.18.3 Lipophagy

Lipid droplets serve as a lipid reservoir in a variety of cell types. Dysregulation of these organelles has been linked to several metabolic diseases (Olzmann and Carvalho 2019, Herker *et al* 2021). The shape of the lipid droplets differs considerably between and within individual cells, implying that the biogenesis and breakdown of these organelles are controlled in a dynamic manner (Olzmann and Carvalho 2019). In contrast to other organelles, which are mostly enclosed in a lipid bilayer, lipid droplets are characterized by a single-layered phospholipid structure surrounding a nucleus of neutral lipids, mainly triacylglycerols and sterol esters (Tauchi-Sato *et al* 2002). Biogenesis of lipid droplets takes place in the ER, whereas neutral lipids aggregate in the organelle (Szymanski *et al* 2007, Choudhary *et al* 2015). The breakdown of lipid droplets through autophagy is referred to as lipophagy (Singh *et al* 2009, Zechner *et al* 2017, Schott *et al* 2019). It seems that there is a threshold for the size below which lipid droplets can be taken up by autophagosomes; lipophagy only targets smaller lipid droplets (Singh *et al* 2009, Schott *et al* 2019), whereas larger droplets are broken down by lipolysis, which is triggered by lipases. Therefore, these two mechanisms probably work together to reduce the lipid droplet concentration (Zechner *et al* 2017). In a recent functional screening for genes involved in lipophagy in macrophage foam cells, small interfering RNAs were used to target genes involved in lipophagy. This study resulted in the identification of various autophagy-related genes that may govern lipophagy, among them ATG5 and UVRAG, and the selective autophagy receptors SQSTM1, NBR1, and OPTN (Robichaud *et al* 2021). The protein makeup of the lipid droplet surface comprises about 150 proteins, which were determined using proximity-based proteomics analysis in U20S cells. In the course of the same investigation, p62 was found to be a protein concentrated in lipid droplets, together with WIPI2 and ATG2A (Bersuker *et al* 2018). In a previous work, p62 was found to be necessary for ethanol-triggered lipophagy (Wang *et al* 2017). Ubiquitylation's involvement in the acquisition of autophagy receptors to lipid droplets is uncertain and similarly the participation of E3 ligases, which are associated with lipophagy, is also debatable. Autophagy inhibition, nevertheless, enhances the enrollment of p62 and ubiquitin in liquid droplets, which implies that a ubiquitination route is necessary for perpetual lipophagy (Robichaud *et al* 2021). Ultimately, lipid droplets establish junctions with countless organelles (Herker *et al* 2021) and BNIP3, which acts as a mitophagy receptor protein, has been found to contribute to lipophagy in that it stimulates the lysosomal transfer of mitochondria-linked lipid droplets toward hepatocytes (Berardi *et al* 2022). Therefore, it is conceivable that different selective autophagy routes give rise to lipophagy via the ubiquitous interorganellar interfaces constituted through lipid droplet formation.

5.19 Conclusion and future directions

Autophagy has emerged as one of the most important regulators of cellular, tissue, and organismic homeostasis, and the mechanistic regulation of autophagy has been shown to be of great importance in understanding the mechanisms of autophagy. This plays a particularly crucial role in carcinogenesis, cancer growth and malignant progression. The vibrant research in this discipline has revealed the intricacies of the molecular mechanism of autophagy and its biochemical control and the biomedical implications involved in its impairment. The role of mechanobiology in autophagy is not limited to physiological processes; it is even more important in pathological processes such as cancer. The complex mechanobiology that governs mechanical and biochemical events at the cellular and/or tissue level is also an emerging field of research that is attracting more and more attention. In this chapter, the purpose was to highlight the part played by physical forces in the regulation of autophagy and its possible impact on both physiological and pathological settings. More crucially, it was anticipated that this work would generate questions that would help elucidate the mechanical demands of autophagy and the degree to which mechanical cues can influence this process. For example, a diet high in saturated fat can adversely impact the autophagic flux within neurons (Hernández-Cáceres *et al* 2019, 2020). Notably, the steric nature of these phospholipids is recognized to mechanically diminish membrane curvature and could thus interfere with autophagy via hindering the merging of vesicles. The mechanical function of phospholipids, in contrast, is widely absent in the published literature and may be an interesting avenue for future investigations. Likewise, research into cytoskeletal dynamics, the mechanical cross-talk between cellular processes, and the importance of environmental cues are areas that need to be explored to set new frontiers for the research. To accomplish this, a change of paradigm is required, applying modern interdisciplinary strategies that couple cell biology, physics and bioengineering (Grenci *et al* 2019). Towards this purpose, state-of-the-art techniques like super-resolution microscopy and tunability of the mechanochemical environment (Bertocchi *et al* 2017), such as by integrating biomimetic substrates and microfluidics will offer intriguing possibilities and perspectives. Together, these emerging technological and new conceptual avenues will result in a deeper comprehension of autophagy and the mechanisms implicated in the onset of diseases, thereby smoothing the path towards the elucidation of new pharmacological drug targets. After deciphering the molecular mechanisms of starvation-induced non-selective autophagy, the emphasis in recent times has been on investigating the function and routes of selective autophagy. Selective autophagy involves the uptake and lysosomal digestion of protein aggregates like ferritin, organelle sub-compartments or complete organelles, and even pathogenic bacteria. Current efforts to understand how selective autophagy targets are recognized are in the early stages. Two principal avenues are being investigated: the first concerns how cells identify which material should be destroyed, and the second how cargo encounters the autophagy mechanism to trigger its incorporation. A widely accepted concept is that defective cargo is labeled with ubiquitin chains via the ubiquitin pathway. Selective autophagy receptors can then interact with the ubiquitin chains on the freight and with the canonical non-selective autophagy mechanism, like the lipidated LC3 on the isolation membranes. Receptor proteins may also specifically govern isolation membrane activation on cargo substrates through direct attraction and activation of ULK1 on the target site for diverse selective autophagy routes (Smith *et al* 2018, Ravenhill *et al* 2019, Turco *et al* 2019, Vargas *et al* 2019, Shi *et al* 2020a, Sankar *et al* 2024). *In vitro* autophagic receptor reconstitution experiments and structural investigations (Sawa-Makarska *et al* 2020, Shi *et al* 2020b, 2020a, Chang *et al* 2021, Fu *et al* 2021, Turco *et al* 2021) have provided a more in-depth comprehension of selective autophagy receptor functioning and selective autophagy process activation. Moreover, selective autophagy investigations utilizing the yeast model have unveiled mechanisms that are conserved within mammalian cells (Farré and Subramani 2016). For example, it has been shown that Atgl, a ULK1 yeast homologue, is autoactivated through autophosphorylation upon cargo acquisition in selective autophagy (Kamber *et al* 2015, Torggler *et al* 2016, Pan *et al* 2020). There are, nevertheless, several unresolved questions regarding selective autophagy and its receptors. For non-ubiquitin-dependent selective autophagy, it is not understood how such autophagy receptors are able to be mobilized and to become selective for the type of cargo to be discarded. Another feature of selective autophagy which is poorly characterized is whether distinct autophagy receptors, utilized for the disposal of the same cargo, can enable context-dependent regulation of selective autophagy in that they are triggered specifically by distinct biological cues. The question of how various ubiquitin-independent receptors are organized in space and time, how their functioning is limited to specific cargo and what physiological importance the redundant functionality of certain receptors has, still needs to be resolved. The membrane origin for selective autophagosome biogenesis is obscure. Plasma membrane-derived ATG9 vesicles and plasma membrane-derived ATG16L1 vesicles have been proposed to constitute membrane reservoirs. The ER, in contrast, has received the most substantial evidence as a membrane reservoir for the nascent autophagosome (Nakatogawa 2020). The finding that receptors can trigger selective autophagy of freight substrates has prompted a drive to identify multispecific pharmaceuticals that can imitate the action of receptor proteins to break down pathogenic freight like malfunctioning mitochondria (Takahashi *et al* 2019a) and aggregates of proteins (Li *et al* 2019b). Deciphering the fundamental processes that drive selective autophagy has the potential to provide pharmacologically tenable approaches to the treatment of several diseases, especially neurodegeneration and aging, which are commonly linked to protein/organelle quality control insufficiency. In summary, blocking autophagy with BAFI and WORT can impair motility and alter the shape of bladder cancer cells. Interestingly, the effect of autophagy-inhibiting agents can be attenuated through physical stimulation and mechanosensitive signal transduction routes, providing a rationale that physical stimuli may help to activate key physiological signaling pathways that modulate cellular chemosensitivity. Although additional mechanistic investigations are warranted to thoroughly examine the molecular mechanisms that maintain these responses, it is conceivable that this could potentially maintain cancer cell survival mechanisms like those that confer chemoresistance (Clausen *et al* 2010, Zaffagnini *et al* 2018). Autophagy inducers can alter the mechanical-chemical reactions of T24 and HCT116 in a cell-type-specific fashion. Importantly, motility assays yielded different results than cytotoxicity assays, highlighting the potential of physical stimuli to remodel cancer physiology. Nevertheless, the translation of these cascades and the implementation of biomechanical activation in *in vitro* drug testing strategies pose a formidable challenge.

References and further reading

Abbas H K and Mirocha C J 1988 Isolation and purification of a hemorrhagic factor (wortmannin) from *Fusarium oxysporum* (N17B) *Appl. Environ. Microbiol.* **54** 1268–74

Abdelmaksoud N M, Abulsoud A I, Abdelghany T M, Elshaer S S, Rizk S M and Senousy M A 2023 Mitochondrial remodeling in colorectal cancer initiation, progression, metastasis, and therapy: a review *Pathol—. Res. Pract.* **246** 154509

Adell M A Y, Vogel G F, Pakdel M, Müller M, Lindner H, Hess M W *et al* 2014 Coordinated binding of Vps4 to ESCRT-III drives membrane neck constriction during MVB vesicle formation *J. Cell Biol.* **205** 33–49

Agudo-Canalejo J, Schultz S W, Chino H, Migliano S M, Saito C, Koyama-Honda I *et al* 2021 Wetting regulates autophagy of phase-separated compartments and the cytosol *Nature* **591** 142–6

Aguilera M O, Berón W and Colombo M I 2012 The actin cytoskeleton participates in the early events of autophagosome formation upon starvation induced autophagy *Autophagy* **8** 1590–603

Akira S, Uematsu S and Takeuchi O 2006 Pathogen recognition and innate immunity *Cell* **124** 783–801

Al-Bari M A A 2020 A current view of molecular dissection in autophagy machinery *J. Physiol. Biochem.* **76** 357–72

Aman Y, Schmauck-Medina T, Hansen M, Morimoto R I, Simon A K, Bjedov I *et al* 2021 Autophagy in healthy aging and disease *Nat Aging.* **1** 634–50

Amer A O and Swanson M S 2005 Autophagy is an immediate macrophage response to Legionella pneumophila *Cell Microbiol.* **7** 765–78

An H and Harper J W 2018 Systematic analysis of ribophagy in human cells reveals bystander flux during selective autophagy *Nat. Cell Biol.* **20** 135–43

An H, Ordureau A, Körner M, Paulo J A and Harper J W 2020 Systematic quantitative analysis of ribosome inventory during nutrient stress *Nature* **583** 303–9

An H, Ordureau A, Paulo J A, Shoemaker C J, Denic V and Harper J W 2019 TEX264 is an endoplasmic reticulum-resident ATG8-interacting protein critical for ER remodeling during nutrient stress *Mol. Cell* **74** 891–908.e10

Angulo-Urarte A, van der Wal T and Huveneers S 2020 Cell-cell junctions as sensors and transducers of mechanical forces *Biochim. Biophys. Acta (BBA)—Biomembr.* **1862** 183316

Anlaş A A and Nelson C M 2020 Soft microenvironments induce chemoresistance by increasing autophagy downstream of integrin-linked kinase *Cancer Res.* **80** 4103–13

Aragona M, Panciera T, Manfrin A, Giulitti S, Michielin F, Elvassore N *et al* 2013 A mechanical checkpoint controls multicellular growth through YAP/TAZ regulation by actin-processing factors *Cell* **154** 1047–59

Archer S L 2013 Mitochondrial dynamics—mitochondrial fission and fusion in human diseases *N. Engl. J. Med.* **369** 2236–51

Argyriou C, D'Agostino M D and Braverman N 2016 Peroxisome biogenesis disorders *TRD* **1** 111–44

Arjonen A, Alanko J, Veltel S and Ivaska J 2012 Distinct recycling of active and inactive β1 integrins *Traffic* **13** 610–25

Armstrong J A and Hart P D 1971 Response of cultured macrophages to *Mycobacterium Tuberculosis*, with observations on fusion of lysosomes with phagosomes *J. Exp. Med.* **134** 713–40

Arndt V, Dick N, Tawo R, Dreiseidler M, Wenzel D, Hesse M *et al* 2010 Chaperone-assisted selective autophagy is essential for muscle maintenance *Curr. Biol.* **20** 143–8

Aschrafi A, Kar A N, Gale J R, Elkahloun A G, Vargas J N S, Sales N *et al* 2016 A heterogeneous population of nuclear-encoded mitochondrial mRNAs is present in the axons of primary sympathetic neurons *Mitochondrion* **30** 18–23

Avvisato C L, Yang X, Shah S, Hoxter B, Li W, Gaynor R *et al* 2007 Mechanical force modulates global gene expression and β-catenin signaling in colon cancer cells *J. Cell Sci.* **120** 2672–82

Axe E L, Walker S A, Manifava M, Chandra P, Roderick H L, Habermann A *et al* 2008 Autophagosome formation from membrane compartments enriched in phosphatidylinositol 3-phosphate and dynamically connected to the endoplasmic reticulum *J. Cell Biol.* **182** 685–701

Ayad N M E, Kaushik S and Weaver V M 2019 Tissue mechanics, an important regulator of development and disease *Phil. Trans. R. Soc.* B **374** 20180215

Bahrami A H, Lin M G, Ren X, Hurley J H and Hummer G 2017 Scaffolding the cup-shaped double membrane in autophagy *PLoS Comput. Biol.* **13** e1005817

Banerjee S, Gardel M L and Schwarz U S 2020 The actin cytoskeleton as an active adaptive material *Annu. Rev. Condens. Matter Phys.* **11** 421–39

Bansal M, Moharir S C, Sailasree S P, Sirohi K, Sudhakar C, Sarathi D P *et al* 2018 Optineurin promotes autophagosome formation by recruiting the autophagy-related Atg12-5-16L1 complex to phagophores containing the Wipi2 protein *J. Biol. Chem.* **293** 132–47

Baumgart E, Vanhorebeek I, Grabenbauer M, Borgers M, Declercq P E, Fahimi H D *et al* 2001 Mitochondrial alterations caused by defective peroxisomal biogenesis in a mouse model for Zellweger syndrome (PEX5 knockout mouse) *Am. J. Pathol.* **159** 1477–94

Baumgarten C M and Feher J J 2001 Osmosis and regulation of cell volume *Cell Physiology Source Book* (Amsterdam: Elsevier) pp 319–55

Bays J L, Campbell H K, Heidema C, Sebbagh M and DeMali K A 2017 Linking E-cadherin mechanotransduction to cell metabolism through force-mediated activation of AMPK *Nat. Cell Biol.* **19** 724–31

Bender F C, Reymond M A, Bron C and Quest A F 2000 Caveolin-1 levels are down-regulated in human colon tumors, and ectopic expression of caveolin-1 in colon carcinoma cell lines reduces cell tumorigenicity *Cancer Res.* **60** 5870–8

Bensassi F, Gallerne C, Sharaf El Dein O, Lemaire C, Hajlaoui M R and Bacha H 2012 Involvement of mitochondria-mediated apoptosis in deoxynivalenol cytotoxicity *Food Chem. Toxicol.* **50** 1680–9

Berardi D E, Bock-Hughes A, Terry A R, Drake L E, Bozek G and Macleod K F 2022 Lipid droplet turnover at the lysosome inhibits growth of hepatocellular carcinoma in a BNIP3-dependent manner *Sci. Adv.* **8** eabo2510

Berendse K, Koot B G P, Klouwer F C C, Engelen M, Roels F, Lacle M M *et al* 2019 Hepatic symptoms and histology in 13 patients with a Zellweger spectrum disorder *J. Inherit. Metab. Dis.* **42** 955–65

Bernardini J P, Brouwer J M, Tan I K, Sandow J J, Huang S, Stafford C A *et al* 2019 Parkin inhibits BAK and BAX apoptotic function by distinct mechanisms during mitophagy *EMBO J.* **38** e99916

Bersuker K, Peterson C W H, To M, Sahl S J, Savikhin V, Grossman E A *et al* 2018 A proximity labeling strategy provides insights into the composition and dynamics of lipid droplet proteomes *Dev. Cell* **44** 97–112.e7

Bertocchi C, Wang Y, Ravasio A, Hara Y, Wu Y, Sailov T *et al* 2017 Nanoscale architecture of cadherin-based cell adhesions *Nat. Cell Biol.* **19** 28–37

Bhutia S K, Mukhopadhyay S, Sinha N, Das D N, Panda P K, Patra S K *et al* 2013 Autophagy *Advances in Cancer Research* (Amsterdam: Elsevier) pp 61–95

Bialik S, Pietrokovski S and Kimchi A 2011 Myosin drives autophagy in a pathway linking Atg1 to Atg9: myosin drives autophagy in a pathway linking Atg1 to Atg9 *EMBO J.* **30** 629–30

Biering S B, Choi J, Halstrom R A, Brown H M, Beatty W L, Lee S *et al* 2017 Viral replication complexes are targeted by LC3-guided interferon-inducible GTPases *Cell Host Microbe.* **22** 74–85.e7

Bingol B, Tea J S, Phu L, Reichelt M, Bakalarski C E, Song Q *et al* 2014 The mitochondrial deubiquitinase USP30 opposes parkin-mediated mitophagy *Nature* **510** 370–5

Birgisdottir Å B, Lamark T and Johansen T 2013 The LIR motif—crucial for selective autophagy *J. Cell Sci.* **126** 3237–47

Birmingham C L, Smith A C, Bakowski M A, Yoshimori T and Brumell J H 2006 Autophagy controls salmonella infection in response to damage to the salmonella-containing vacuole *J. Biol. Chem.* **281** 11374–83

Biskou O, Casanova V, Hooper K M, Kemp S, Wright G P, Satsangi J *et al* 2019 The type III intermediate filament vimentin regulates organelle distribution and modulates autophagy *PLoS One* **14** e0209665

Bjørkøy G, Lamark T, Brech A, Outzen H, Perander M, Øvervatn A *et al* 2005 p62/SQSTM1 forms protein aggregates degraded by autophagy and has a protective effect on huntingtin-induced cell death *J. Cell Biol.* **171** 603–14

Boehlke C, Kotsis F, Patel V, Braeg S, Voelker H, Bredt S *et al* 2010 Primary cilia regulate mTORC1 activity and cell size through Lkb1 *Nat. Cell Biol.* **12** 1115–22

Bolt A M and Klimecki W T 2012 Autophagy in toxicology: self-consumption in times of stress and plenty *J. Appl. Toxicol.* **32** 465–79

Bond M J, Chu L, Nalawansha D A, Li K and Crews C M 2020 Targeted degradation of oncogenic $KRAS^{G12C}$ by VHL-recruiting PROTACs *ACS Cent. Sci.* **6** 1367–75

Borsche M, König I R, Delcambre S, Petrucci S, Balck A, Brüggemann N *et al* 2020 Mitochondrial damage-associated inflammation highlights biomarkers in PRKN/PINK1 parkinsonism *Brain* **143** 3041–51

Bose A, Robida S, Furcinitti P S, Chawla A, Fogarty K, Corvera S *et al* 2004 Unconventional myosin myo1c promotes membrane fusion in a regulated exocytic pathway *Mol. Cell. Biol.* **24** 5447–58

Bouizgarne B, El-Maarouf-Bouteau H, Frankart C, Reboutier D, Madiona K, Pennarun A M *et al* 2006 Early physiological responses of *Arabidopsis thaliana* cells to fusaric acid: toxic and signalling effects *New Phytol.* **169** 209–18

Boukhalfa A, Nascimbeni A C, Dupont N, Codogno P and Morel E 2020a Primary cilium-dependent autophagy drafts PIK3C2A to generate PtdIns3P in response to shear stress *Autophagy* **16** 1143–4

Boukhalfa A, Nascimbeni A C, Ramel D, Dupont N, Hirsch E, Gayral S *et al* 2020b PI3KC2α-dependent and VPS34-independent generation of PI3P controls primary cilium-mediated autophagy in response to shear stress *Nat. Commun.* **11** 294

Bowman C L, Gottlieb P A, Suchyna T M, Murphy Y K and Sachs F 2007 Mechanosensitive ion channels and the peptide inhibitor GsMTx-4: history, properties, mechanisms and pharmacology *Toxicon* **49** 249–70

Boya P 2012 Lysosomal function and dysfunction: mechanism and disease *Antioxid. Redox Signal.* **17** 766–74

Breda C N D S, Davanzo G G, Basso P J, Saraiva Câmara N O and Moraes-Vieira P M M 2019 Mitochondria as central hub of the immune system *Redox Biol.* **26** 101255

Brier L W, Ge L, Stjepanovic G, Thelen A M, Hurley J H and Schekman R 2019 Regulation of LC3 lipidation by the autophagy-specific class III phosphatidylinositol-3 kinase complex *Mol. Biol. Cell* **30** 1098–107

Brunn G J, Williams J, Sabers C, Wiederrecht G, Lawrence J C and Abraham R T 1996 Direct inhibition of the signaling functions of the mammalian target of rapamycin by the phosphoinositide 3-kinase inhibitors, wortmannin and LY294002 *EMBO J.* **15** 5256–67

Buchan J R, Kolaitis R-M, Taylor J P and Parker R 2013 Eukaryotic stress granules are cleared by autophagy and Cdc48/VCP function *Cell* **153** 1461–74

Burman J L, Pickles S, Wang C, Sekine S, Vargas J N S, Zhang Z *et al* 2017 Mitochondrial fission facilitates the selective mitophagy of protein aggregates *J. Cell Biol.* **216** 3231–47

Cai X, Xu H and Chen Z J 2017 Prion-like polymerization in immunity and inflammation *Cold Spring Harb. Perspect. Biol.* **9** a023580

Campa C C, Margaria J P, Derle A, Del Giudice M, De Santis M C, Gozzelino L *et al* 2018 Rab11 activity and PtdIns(3)P turnover removes recycling cargo from endosomes *Nat. Chem. Biol.* **14** 801–10

Carlsson S R and Simonsen A 2015 Membrane dynamics in autophagosome biogenesis *J. Cell Sci.* **128** 193–205

Carmignac V, Svensson M, Körner Z, Elowsson L, Matsumura C, Gawlik K I *et al* 2011 Autophagy is increased in laminin $\alpha 2$ chain-deficient muscle and its inhibition improves muscle morphology in a mouse model of MDC1A *Hum. Mol. Genet.* **20** 4891–902

Carmona-Carmona C A, Dalla Pozza E, Ambrosini G, Errico A and Dando I 2022 Divergent roles of mitochondria dynamics in pancreatic ductal adenocarcinoma *Cancers (Basel)* **14** 2155

Carroll B, Mohd-Naim N, Maximiano F, Frasa M A, McCormack J, Finelli M *et al* 2013 The TBC/RabGAP armus coordinates Rac1 and Rab7 functions during autophagy *Dev. Cell* **25** 15–28

Case N, Thomas J, Sen B, Styner M, Xie Z, Galior K *et al* 2011 Mechanical regulation of glycogen synthase kinase 3β (GSK3β) in mesenchymal stem cells is dependent on Akt protein serine 473 phosphorylation via mTORC2 protein *J. Biol. Chem.* **286** 39450–6

Castagnaro S, Chrisam M, Cescon M, Braghetta P, Grumati P and Bonaldo P 2018 Extracellular collagen VI has prosurvival and autophagy instructive properties in mouse fibroblasts *Front. Physiol.* **9** 1129

Caswell P T, Vadrevu S and Norman J C 2009 Integrins: masters and slaves of endocytic transport *Nat. Rev. Mol. Cell Biol.* **10** 843–53

Cebollero E, van der Vaart A, Zhao M, Rieter E, Klionsky D J, Helms J B *et al* 2012 Phosphatidylinositol-3-phosphate clearance plays a key role in autophagosome completion *Curr. Biol.* **22** 1545–53

Cemma M and Brumell J H 2012 Interactions of pathogenic bacteria with autophagy systems *Curr. Biol.* **22** R540–5

Cemma M, Kim P K and Brumell J H 2011 The ubiquitin-binding adaptor proteins p62/SQSTM1 and NDP52 are recruited independently to bacteria-associated microdomains to target Salmonella to the autophagy pathway *Autophagy* **7** 341–5

Chai Q, Wang X, Qiang L, Zhang Y, Ge P, Lu Z *et al* 2019 A *Mycobacterium tuberculosis* surface protein recruits ubiquitin to trigger host xenophagy *Nat. Commun.* **10** 1973

Chang C, Shi X, Jensen L E, Yokom A L, Fracchiolla D, Martens S *et al* 2021 Reconstitution of cargo-induced LC3 lipidation in mammalian selective autophagy *Sci. Adv.* **7** eabg4922

Chaudhuri O, Koshy S T, Branco da Cunha C, Shin J-W, Verbeke C S, Allison K H *et al* 2014 Extracellular matrix stiffness and composition jointly regulate the induction of malignant phenotypes in mammary epithelium *Nat. Mater.* **13** 970–8

Chaudhuri P K, Low B C and Lim C T 2018 Mechanobiology of tumor growth *Chem. Rev.* **118** 6499–515

Chauhan S, Kumar S, Jain A, Ponpuak M, Mudd M H, Kimura T *et al* 2016 TRIMs and galectins globally cooperate and TRIM16 and galectin-3 co-direct autophagy in endomembrane damage homeostasis *Dev. Cell* **39** 13–27

Chauhan V P, Martin J D, Liu H, Lacorre D A, Jain S R, Kozin S V *et al* 2013 Angiotensin inhibition enhances drug delivery and potentiates chemotherapy by decompressing tumour blood vessels *Nat. Commun.* **4** 2516

Chen D, Fan W, Lu Y, Ding X, Chen S and Zhong Q 2012 A mammalian autophagosome maturation mechanism mediated by TECPR1 and the Atg12-Atg5 conjugate *Mol. Cell* **45** 629–41

Chen K, Wang Y, Deng X, Guo L and Wu C 2021 Extracellular matrix stiffness regulates mitochondrial dynamics through PINCH-1- and kindlin-2-mediated signalling *Curr. Res. Cell Biol.* **2** 100008

Chen Q, Xiao Y, Chai P, Zheng P, Teng J and Chen J 2019 ATL3 is a tubular ER-phagy receptor for GABARAP-mediated selective autophagy *Curr. Biol.* **29** 846–855.e6

Chen W, Zhao H and Li Y 2023 Mitochondrial dynamics in health and disease: mechanisms and potential targets *Sig. Transduct Target. Ther.* **8** 333

Chen Y, Ju L, Rushdi M, Ge C and Zhu C 2017 Receptor-mediated cell mechanosensing *Mol. Biol. Cell* **28** 3134–55

Chen Z and Rand R P 1997 The influence of cholesterol on phospholipid membrane curvature and bending elasticity *Biophys. J.* **73** 267–76

Chhabra E S and Higgs H N 2007 The many faces of actin: matching assembly factors with cellular structures *Nat. Cell Biol.* **9** 1110–21

Chiaruttini N, Redondo-Morata L, Colom A, Humbert F, Lenz M, Scheuring S *et al* 2015 Relaxation of loaded ESCRT-III spiral springs drives membrane deformation *Cell* **163** 866–79

Chino H, Hatta T, Natsume T and Mizushima N 2019 Intrinsically disordered protein TEX264 mediates ER-phagy *Mol. Cell* **74** 909–921.e6
Choudhary V, Ojha N, Golden A and Prinz W A 2015 A conserved family of proteins facilitates nascent lipid droplet budding from the ER *J. Cell Biol.* **211** 261–71
Chowdhury S, Otomo C, Leitner A, Ohashi K, Aebersold R, Lander G C *et al* 2018 Insights into autophagosome biogenesis from structural and biochemical analyses of the ATG2A-WIPI4 complex *Proc. Natl Acad. Sci. USA* **115** E9792–801
Choy A, Dancourt J, Mugo B, O'Connor T J, Isberg R R, Melia T J *et al* 2012 The *Legionella* effector RavZ inhibits host autophagy through irreversible Atg8 deconjugation *Science* **338** 1072–6
Christ L, Raiborg C, Wenzel E M, Campsteijn C and Stenmark H 2017 Cellular functions and molecular mechanisms of the ESCRT membrane-scission machinery *Trends Biochem. Sci.* **42** 42–56
Clark I E, Dodson M W, Jiang C, Cao J H, Huh J R, Seol J H *et al* 2006 Drosophila pink1 is required for mitochondrial function and interacts genetically with parkin *Nature* **441** 1162–6
Clausen T H, Lamark T, Isakson P, Finley K D, Larsen K B, Brech A *et al* 2010 p62/SQSTM1 and ALFY interact to facilitate the formation of p62 bodies/ALIS and their degradation by autophagy *Autophagy* **6** 330–44
Coffey J W and De Duve C 1968 Digestive activity of lysosomes. I. The digestion of proteins by extracts of rat liver lysosomes *J. Biol. Chem.* **243** 3255–63
Colombero C, Remy D, Antoine S, Macé A-S, Monteiro P, ElKhatib N *et al* 2021 mTOR repression in response to amino acid starvation promotes ECM degradation through MT1-MMP endocytosis arrest *Adv. Sci.* **8** 2101614
Conciatori F, Ciuffreda L, Bazzichetto C, Falcone I, Pilotto S, Bria E *et al* 2018 mTOR cross-talk in cancer and potential for combination therapy *Cancers* **10** 23
Corona A K and Jackson W T 2018 Finding the middle ground for autophagic fusion requirements *Trends Cell Biol.* **28** 869–81
Corti O, Blomgren K, Poletti A and Beart P M 2020 Autophagy in neurodegeneration: new insights underpinning therapy for neurological diseases *J. Neurochem.* **154** 354–71
Coutts A S and La Thangue N B 2015 Actin nucleation by WH2 domains at the autophagosome *Nat. Commun.* **6** 7888
Dai A, Yu L and Wang H-W 2019 WHAMM initiates autolysosome tubulation by promoting actin polymerization on autolysosomes *Nat. Commun.* **10** 3699
Dall'Armi C, Devereaux K A and Di Paolo G 2013 The role of lipids in the control of autophagy *Curr. Biol.* **23** R33–45
Damiano V, Spessotto P, Vanin G, Perin T, Maestro R and Santarosa M 2020 The autophagy machinery contributes to E-cadherin turnover in breast cancer *Front. Cell Dev. Biol.* **8** 545
Das J, Agarwal T, Chakraborty S and Maiti T K 2019 Compressive stress-induced autophagy promotes invasion of HeLa cells by facilitating protein turnover *in vitro Exp. Cell. Res.* **381** 201–7
Das J, Chakraborty S and Maiti T K 2020 Mechanical stress-induced autophagic response: a cancer-enabling characteristic? *Semin. Cancer Biol.* **66** 101–9
Das J, Maji S, Agarwal T, Chakraborty S and Maiti T K 2018 Hemodynamic shear stress induces protective autophagy in HeLa cells through lipid raft-mediated mechanotransduction *Clin. Exp. Metastasis* **35** 135–48

Davies J E, Lopresto D, Apta B H R, Lin Z, Ma W and Harper M T 2019 Using Yoda-1 to mimic laminar flow *in vitro*: a tool to simplify drug testing *Biochem. Pharmacol.* **168** 473–80

De Felice D and Alaimo A 2020 Mechanosensitive piezo channels in cancer: focus on altered calcium signaling in cancer cells and in tumor progression *Cancers* **12** 1780

Deas E, Plun-Favreau H, Gandhi S, Desmond H, Kjaer S, Loh S H Y *et al* 2011 PINK1 cleavage at position A103 by the mitochondrial protease PARL *Hum. Mol. Genet.* **20** 867–79

Debnath J, Gammoh N and Ryan K M 2023 Autophagy and autophagy-related pathways in cancer *Nat. Rev. Mol. Cell Biol.* **24** 560–75

Del Favero G, Woelflingseder L, Janker L, Neuditschko B, Seriani S, Gallina P *et al* 2018 Deoxynivalenol induces structural alterations in epidermoid carcinoma cells A431 and impairs the response to biomechanical stimulation *Sci. Rep.* **8** 11351

Del Favero G, Zeugswetter M, Kiss E and Marko D 2021 Endoplasmic reticulum adaptation and autophagic competence shape response to fluid shear stress in T24 bladder cancer cells *Front. Pharmacol.* **12** 647350

Demers N D, Riccio V, Jo D S, Bhandari S, Law K B, Liao W *et al* 2023 PEX13 prevents pexophagy by regulating ubiquitinated PEX5 and peroxisomal ROS *Autophagy* **19** 1781–802

Deosaran E, Larsen K B, Hua R, Sargent G, Wang Y, Kim S *et al* 2012 NBR1 acts as an autophagy receptor for peroxisomes *J. Cell Sci.* **126** 939–52

Deretic V, Saitoh T and Akira S 2013 Autophagy in infection, inflammation and immunity *Nat. Rev. Immunol.* **13** 722–37

Derynck R and Zhang Y E 2003 Smad-dependent and Smad-independent pathways in TGF-β family signalling *Nature* **425** 577–84

Di Bartolomeo S, Corazzari M, Nazio F, Oliverio S, Lisi G, Antonioli M *et al* 2010 The dynamic interaction of AMBRA1 with the dynein motor complex regulates mammalian autophagy *J. Cell Biol.* **191** 155–68

Diao J, Liu R, Rong Y, Zhao M, Zhang J, Lai Y *et al* 2015 ATG14 promotes membrane tethering and fusion of autophagosomes to endolysosomes *Nature* **520** 563–6

Diaz-Valdivia N, Bravo D, Huerta H, Henriquez S, Gabler F, Vega M *et al* 2015 Enhanced caveolin-1 expression increases migration, anchorage-independent growth and invasion of endometrial adenocarcinoma cells *BMC Cancer* **15** 463

Dietl P, Frick M, Mair N, Bertocchi C and Haller T 2004 Pulmonary consequences of a deep breath revisited *Neonatology* **85** 299–304

Dikic I 2017 Proteasomal and autophagic degradation systems *Annu. Rev. Biochem.* **86** 193–224

Dikic I and Elazar Z 2018 Mechanism and medical implications of mammalian autophagy *Nat. Rev. Mol. Cell Biol.* **19** 349–64

Dong X and Levine B 2013 Autophagy and viruses: adversaries or allies? *J. Innate Immun.* **5** 480–93

Dong Y, Zhuang X-X, Wang Y-T, Tan J, Feng D, Li M *et al* 2023 Chemical mitophagy modulators: drug development strategies and novel regulatory mechanisms *Pharmacol. Res.* **194** 106835

Doss B L, Pan M, Gupta M, Grenci G, Mège R-M, Lim C T *et al* 2020 Cell response to substrate rigidity is regulated by active and passive cytoskeletal stress *Proc. Natl Acad. Sci. USA* **117** 12817–25

Dower C M, Wills C A, Frisch S M and Wang H-G 2018 Mechanisms and context underlying the role of autophagy in cancer metastasis *Autophagy* **14** 1110–28

Dupont S, Morsut L, Aragona M, Enzo E, Giulitti S, Cordenonsi M *et al* 2011 Role of YAP/TAZ in mechanotransduction *Nature* **474** 179–83

Eapen V V, Swarup S, Hoyer M J, Paulo J A and Harper J W 2021 Quantitative proteomics reveals the selectivity of ubiquitin-binding autophagy receptors in the turnover of damaged lysosomes by lysophagy *eLife* **10** e72328

Egeblad M, Rasch M G and Weaver V M 2010 Dynamic interplay between the collagen scaffold and tumor evolution *Curr. Opin. Cell Biol.* **22** 697–706

Engler A J, Sen S, Sweeney H L and Discher D E 2006 Matrix elasticity directs stem cell lineage specification *Cell* **126** 677–89

Ericsson J L E 1969 Studies on induced cellular autophagy *Exp. Cell. Res.* **55** 95–106

Eskelinen E-L and Saftig P 2009 Autophagy: a lysosomal degradation pathway with a central role in health and disease *Biochim. Biophys. Acta (BBA)—Mol. Cell Res.* **1793** 664–73

Espert L 2006 Autophagy is involved in T cell death after binding of HIV-1 envelope proteins to CXCR4 *J. Clin. Invest.* **116** 2161–72

Etem E Ö, Ceylan G G, Özaydın S, Ceylan C, Özercan I and Kuloğlu T 2018 The increased expression of Piezo1 and Piezo2 ion channels in human and mouse bladder carcinoma *Adv. Clin. Exp. Med.* **27** 1025–31

Evans C S and Holzbaur E L 2020 Degradation of engulfed mitochondria is rate-limiting in Optineurin-mediated mitophagy in neurons *eLife* **9** e50260

Evans C S and Holzbaur E L F 2019 Autophagy and mitophagy in ALS *Neurobiol. Dis.* **122** 35–40

Evers T M J, Holt L J, Alberti S and Mashaghi A 2021 Reciprocal regulation of cellular mechanics and metabolism *Nat. Metab.* **3** 456–68

Fabbri L, Bost F and Mazure N M 2019 Primary cilium in cancer hallmarks *Int. J. Mol. Sci.* **20** 1336

Fan R, Emery T, Zhang Y, Xia Y, Sun J and Wan J 2016 Circulatory shear flow alters the viability and proliferation of circulating colon cancer cells *Sci. Rep.* **6** 27073

Farré J-C and Subramani S 2016 Mechanistic insights into selective autophagy pathways: lessons from yeast *Nat. Rev. Mol. Cell Biol.* **17** 537–52

Fass E, Shvets E, Degani I, Hirschberg K and Elazar Z 2006 Microtubules support production of starvation-induced autophagosomes but not their targeting and fusion with lysosomes *J. Biol. Chem.* **281** 36303–16

Fernandez-Pol J A, Klos D J and Hamilton P D 1993 Cytotoxic activity of fusaric acid on human adenocarcinoma cells in tissue culture *Anticancer Res.* **13** 57–64

Fischer T, Hayn A and Mierke C T 2020 Effect of nuclear stiffness on cell mechanics and migration of human breast cancer cells *Front. Cell Dev. Biol.* **8** 393

Fiskin E, Bionda T, Dikic I and Behrends C 2016 Global analysis of host and bacterial ubiquitinome in response to *Salmonella typhimurium* infection *Mol. Cell* **62** 967–81

Fletcher D A and Mullins R D 2010 Cell mechanics and the cytoskeleton *Nature* **463** 485–92

Franco I, Gulluni F, Campa C C, Costa C, Margaria J P, Ciraolo E *et al* 2014 PI3K class II α controls spatially restricted endosomal PtdIns3P and Rab11 activation to promote primary cilium function *Dev. Cell* **28** 647–58

Franco L H, Nair V R, Scharn C R, Xavier R J, Torrealba J R, Shiloh M U *et al* 2017 The ubiquitin ligase smurf1 functions in selective autophagy of *Mycobacterium tuberculosis* and anti-tuberculous host defense *Cell Host Microbe* **21** 59–72

Freeman S A and Grinstein S 2018 Resolution of macropinosomes, phagosomes and autolysosomes: osmotically driven shrinkage enables tubulation and vesiculation *Traffic* **19** 965–74

Frégeau M, Régimbald-Dumas Y and Guillemette G 2011 Positive regulation of inositol 1,4,5-trisphosphate-induced Ca^{2+} release by mammalian target of rapamycin (mTOR) in RINm5F cells *J.Cell. Biochem.* **112** 723–33

Frey T G and Mannella C A 2000 The internal structure of mitochondria *Trends Biochem. Sci.* **25** 319–24

Frtús A, Smolková B, Uzhytchak M, Lunova M, Jirsa M, Petrenko Y *et al* 2023 Mechanical regulation of mitochondrial dynamics and function in a 3D-engineered liver tumor microenvironment *ACS Biomater. Sci. Eng.* **9** 2408–25

Fu T, Zhang M, Zhou Z, Wu P, Peng C, Wang Y *et al* 2021 Structural and biochemical advances on the recruitment of the autophagy-initiating ULK and TBK1 complexes by autophagy receptor NDP52 *Sci. Adv.* **7** eabi6582

Fu Y, Dong W, Xu Y, Li L, Yu X, Pang Y *et al* 2023 Targeting mitochondrial dynamics by AZD5363 in triple-negative breast cancer MDA-MB-231 cell–derived spheres *Naunyn-Schmiedeberg's Arch. Pharmacol.* **396** 2545–53

Fujita N, Morita E, Itoh T, Tanaka A, Nakaoka M, Osada Y *et al* 2013 Recruitment of the autophagic machinery to endosomes during infection is mediated by ubiquitin *J. Cell Biol.* **203** 115–28

Fumagalli F, Noack J, Bergmann T J, Cebollero E, Pisoni G B, Fasana E *et al* 2016 Translocon component Sec62 acts in endoplasmic reticulum turnover during stress recovery *Nat. Cell Biol.* **18** 1173–84

Fuseya Y, Fujita H, Kim M, Ohtake F, Nishide A, Sasaki K *et al* 2020 The HOIL-1L ligase modulates immune signalling and cell death via monoubiquitination of LUBAC *Nat. Cell Biol.* **22** 663–73

Gaggioli C, Hooper S, Hidalgo-Carcedo C, Grosse R, Marshall J F, Harrington K *et al* 2007 Fibroblast-led collective invasion of carcinoma cells with differing roles for RhoGTPases in leading and following cells *Nat. Cell Biol.* **9** 1392–400

García-Palmero I, Torres S, Bartolomé R A, Peláez-García A, Larriba M J, Lopez-Lucendo M *et al* 2016 Twist1-induced activation of human fibroblasts promotes matrix stiffness by upregulating palladin and collagen α1(VI) *Oncogene* **35** 5224–36

Geeraert C, Ratier A, Pfisterer S G, Perdiz D, Cantaloube I, Rouault A *et al* 2010 Starvation-induced hyperacetylation of tubulin is required for the stimulation of autophagy by nutrient deprivation *J. Biol. Chem.* **285** 24184–94

Gehler S, Baldassarre M, Lad Y, Leight J L, Wozniak M A, Riching K M *et al* 2009 Filamin A–β1 integrin complex tunes epithelial cell response to matrix tension *Mol. Biol. Cell* **20** 3224–38

Gestaut D, Zhao Y, Park J, Ma B, Leitner A, Collier M *et al* 2022 Structural visualization of the tubulin folding pathway directed by human chaperonin TRiC/CCT *Cell* **185** 4770–87.e20

Giacomello M, Pyakurel A, Glytsou C and Scorrano L 2020 The cell biology of mitochondrial membrane dynamics *Nat. Rev. Mol. Cell Biol.* **21** 204–24

Gilkes D M, Semenza G L and Wirtz D 2014 Hypoxia and the extracellular matrix: drivers of tumour metastasis *Nat. Rev. Cancer* **14** 430–9

Gladkova C, Maslen S L, Skehel J M and Komander D 2018 Mechanism of parkin activation by PINK1 *Nature* **559** 410–4

Glogauer M, Arora P, Chou D, Janmey P A, Downey G P and McCulloch C A G 1998 The role of actin-binding protein 280 in integrin-dependent mechanoprotection *J. Biol. Chem.* **273** 1689–98

Goldfischer S, Moore C L, Johnson A B, Spiro A J, Valsamis M P, Wisniewski H K *et al* 1973 Peroxisomal and mitochondrial defects in the cerebro-hepato-renal syndrome *Science* **182** 62–4

Goldman R D, Cleland M M, Murthy S N P, Mahammad S and Kuczmarski E R 2012 Inroads into the structure and function of intermediate filament networks *J. Struct. Biol.* **177** 14–23

Gonzalez C D, Resnik R and Vaccaro M I 2020 Secretory autophagy and its relevance in metabolic and degenerative disease *Front. Endocrinol.* **11** 266

Goodwin J M, Dowdle W E, DeJesus R, Wang Z, Bergman P, Kobylarz M *et al* 2017 Autophagy-independent lysosomal targeting regulated by ULK1/2-FIP200 and ATG9 *Cell Rep.* **20** 2341–56

Gordon P B and Seglen P O 1988 Prelysosomal convergence of autophagic and endocytic pathways *Biochem. Biophys. Res. Commun.* **151** 40–7

Graef M, Friedman J R, Graham C, Babu M and Nunnari J 2013 ER exit sites are physical and functional core autophagosome biogenesis components *Mol. Biol. Cell* **24** 2918–31

Green D E 1983 Mitochondria—structure, function, and replication *N. Engl. J. Med.* **309** 182–3

Grenci G, Bertocchi C and Ravasio A 2019 Integrating microfabrication into biological investigations: the benefits of interdisciplinarity *Micromachines* **10** 252

Grumati P, Morozzi G, Hölper S, Mari M, Harwardt M-L I, Yan R *et al* 2017 Full length RTN3 regulates turnover of tubular endoplasmic reticulum via selective autophagy *eLife* **6** e25555

Gubas A and Dikic I 2022 A guide to the regulation of selective autophagy receptors *FEBS J.* **289** 75–89

Gubbiotti M A and Iozzo R V 2015 Proteoglycans regulate autophagy via outside-in signaling: an emerging new concept *Matrix Biol.* **48** 6–13

Guerra F and Bucci C 2016 Multiple roles of the small GTPase Rab7 *Cells* **5** 34

Guo T, Jiang C, Yang S-Z, Zhu Y, He C, Carter A B *et al* 2023 Mitochondrial fission and bioenergetics mediate human lung fibroblast durotaxis *JCI Insight* **8** e157348

Habrowska-Górczyńska D E, Kowalska K, Urbanek K A, Domińska K, Kozieł M J and Piastowska-Ciesielska A W 2023 Effect of the mycotoxin deoxynivalenol in combinational therapy with TRAIL on prostate cancer cells *Toxicol. Appl. Pharmacol.* **461** 116390

Hagenbuchner J, Kuznetsov A V, Obexer P and Ausserlechner M J 2013 BIRC5/Survivin enhances aerobic glycolysis and drug resistance by altered regulation of the mitochondrial fusion/fission machinery *Oncogene* **32** 4748–57

Hailey D W, Rambold A S, Satpute-Krishnan P, Mitra K, Sougrat R, Kim P K *et al* 2010 Mitochondria supply membranes for autophagosome biogenesis during starvation *Cell* **141** 656–67

Ham S J, Lee D, Yoo H, Jun K, Shin H and Chung J 2020 Decision between mitophagy and apoptosis by Parkin via VDAC1 ubiquitination *Proc. Natl Acad. Sci. USA* **117** 4281–91

Hamasaki M, Furuta N, Matsuda A, Nezu A, Yamamoto A, Fujita N *et al* 2013 Autophagosomes form at ER–mitochondria contact sites *Nature* **495** 389–93

Han M K L and De Rooij J 2017 Resolving the cadherin–F-actin connection *Nat. Cell Biol.* **19** 14–6

Han Z, Zhang W, Ning W, Wang C, Deng W, Li Z *et al* 2021 Model-based analysis uncovers mutations altering autophagy selectivity in human cancer *Nat. Commun.* **12** 3258

Hanahan D 2022 Hallmarks of cancer: new dimensions *Cancer Discov.* **12** 31–46

Hanna R A, Quinsay M N, Orogo A M, Giang K, Rikka S and Gustafsson Å B 2012 Microtubule-associated protein 1 light chain 3 (LC3) interacts with Bnip3 protein to selectively remove endoplasmic reticulum and mitochondria via autophagy *J. Biol. Chem.* **287** 19094–104

Hara-Kuge S and Fujiki Y 2008 The peroxin Pex14p is involved in LC3-dependent degradation of mammalian peroxisomes *Exp. Cell. Res.* **314** 3531–41

Hayn A, Fischer T and Mierke C T 2020 Inhomogeneities in 3D collagen matrices impact matrix mechanics and cancer cell migration *Front. Cell Dev. Biol.* **8** 593879

Hayn A, Fischer T and Mierke C T 2023 The role of ADAM8 in mechanophenotype of cancer cells in 3D extracellular matrices *Front. Cell Dev. Biol.* **11** 1148162

Heath R J, Goel G, Baxt L A, Rush J S, Mohanan V, Paulus G L C *et al* 2016 RNF166 determines recruitment of adaptor proteins during antibacterial autophagy *Cell Rep.* **17** 2183–94

Heden T D, Chow L S, Hughey C C and Mashek D G 2022 Regulation and role of glycophagy in skeletal muscle energy metabolism *Autophagy* **18** 1078–89

Heo J-M, Harper N J, Paulo J A, Li M, Xu Q, Coughlin M *et al* 2019 Integrated proteogenetic analysis reveals the landscape of a mitochondrial-autophagosome synapse during PARK2-dependent mitophagy *Sci. Adv.* **5** eaay4624

Heo J-M, Ordureau A, Paulo J A, Rinehart J and Harper J W 2015 The PINK1-PARKIN mitochondrial ubiquitylation pathway drives a program of OPTN/NDP52 recruitment and TBK1 activation to promote mitophagy *Mol. Cell* **60** 7–20

Heo J-M, Ordureau A, Swarup S, Paulo J A, Shen K, Sabatini D M *et al* 2018 RAB7A phosphorylation by TBK1 promotes mitophagy via the PINK-PARKIN pathway *Sci. Adv.* **4** eaav0443

Herker E, Vieyres G, Beller M, Krahmer N and Bohnert M 2021 Lipid droplet contact sites in health and disease *Trends Cell Biol.* **31** 345–58

Hernández-Cáceres M P, Cereceda K, Hernández S, Li Y, Narro C, Rivera P *et al* 2020 Palmitic acid reduces the autophagic flux in hypothalamic neurons by impairing autophagosome-lysosome fusion and endolysosomal dynamics *Mol. Cell. Oncol.* **7** 1789418

Hernández-Cáceres M P, Munoz L, Pradenas J M, Pena F, Lagos P, Aceiton P *et al* 2021 Mechanobiology of autophagy: the unexplored side of cancer *Front. Oncol.* **11** 632956

Hernández-Cáceres M P, Toledo-Valenzuela L, Díaz-Castro F, Ávalos Y, Burgos P, Narro C *et al* 2019 Palmitic acid reduces the autophagic flux and insulin sensitivity through the activation of the free fatty acid receptor 1 (FFAR1) in the Hypothalamic Neuronal Cell Line N43/5 *Front. Endocrinol.* **10** 176

Holle A W and Engler A J 2011 More than a feeling: discovering, understanding, and influencing mechanosensing pathways *Curr. Opin. Biotechnol.* **22** 648–54

Hollenstein D M and Kraft C 2020 Autophagosomes are formed at a distinct cellular structure *Curr. Opin. Cell Biol.* **65** 50–7

Hong T, Pan X, Xu H, Zheng Z, Wen L, Li J *et al* 2023 Jatrorrhizine inhibits Piezo1 activation and reduces vascular inflammation in endothelial cells *Biomed. Pharmacother.* **163** 114755

Hu C-A A, Hou Y, Yi D, Qiu Y, Wu G, Kong X *et al* 2015 Autophagy and tight junction proteins in the intestine and intestinal diseases *Anim. Nutr.* **1** 123–7

Hu X and Mullins R D 2019 LC3 and STRAP regulate actin filament assembly by JMY during autophagosome formation *J. Cell Biol.* **218** 251–66

Huett A, Heath R J, Begun J, Sassi S O, Baxt L A, Vyas J M *et al* 2012 The LRR and RING Domain protein LRSAM1 Is an E3 ligase crucial for ubiquitin-dependent autophagy of intracellular salmonella typhimurium *Cell Host Microbe* **12** 778–90

Humphries B A, Cutter A C, Buschhaus J M, Chen Y-C, Qyli T, Palagama D S W *et al* 2020 Enhanced mitochondrial fission suppresses signaling and metastasis in triple-negative breast cancer *Breast Cancer Res.* **22** 60

Hung S-T, Linares G R, Chang W-H, Eoh Y, Krishnan G, Mendonca S *et al* 2023 PIKFYVE inhibition mitigates disease in models of diverse forms of ALS *Cell* **186** 786–802.e28

Hung Y-H, Chen L M-W, Yang J-Y and Yuan Yang W 2013 Spatiotemporally controlled induction of autophagy-mediated lysosome turnover *Nat. Commun.* **4** 2111

Hurley J H and Young L N 2017 Mechanisms of autophagy initiation *Annu. Rev. Biochem.* **86** 225–44

Hyler A R, Baudoin N C, Brown M S, Stremler M A, Cimini D, Davalos R V *et al* 2018 Fluid shear stress impacts ovarian cancer cell viability, subcellular organization, and promotes genomic instability *PLoS One* **13** e0194170

Icard P, Shulman S, Farhat D, Steyaert J-M, Alifano M and Lincet H 2018 How the Warburg effect supports aggressiveness and drug resistance of cancer cells? *Drug Resist. Updat.* **38** 1–11

Inami Y, Waguri S, Sakamoto A, Kouno T, Nakada K, Hino O *et al* 2011 Persistent activation of Nrf2 through p62 in hepatocellular carcinoma cells *J. Cell Biol.* **193** 275–84

Ip C K M, Li S-S, Tang M Y H, Sy S K H, Ren Y, Shum H C *et al* 2016 Stemness and chemoresistance in epithelial ovarian carcinoma cells under shear stress *Sci. Rep.* **6** 26788

Iskratsch T, Wolfenson H and Sheetz M P 2014 Appreciating force and shape—the rise of mechanotransduction in cell biology *Nat. Rev. Mol. Cell Biol.* **15** 825–33

Itakura E, Kishi-Itakura C, Koyama-Honda I and Mizushima N 2012a Structures containing Atg9A and the ULK1 complex independently target depolarized mitochondria at initial stages of Parkin-mediated mitophagy *J. Cell Sci.* **125** 1488–99

Itakura E, Kishi-Itakura C and Mizushima N 2012b The hairpin-type tail-anchored SNARE syntaxin 17 targets to autophagosomes for fusion with endosomes/lysosomes *Cell* **151** 1256–69

Jackson W M, Jaasma M J, Tang R Y and Keaveny T M 2008 Mechanical loading by fluid shear is sufficient to alter the cytoskeletal composition of osteoblastic cells *Am. J. Physiol. Cell Physiol.* **295** C1007–15

Jain S, Cachoux V M L, Narayana G H N S, de Beco S, D'Alessandro J, Cellerin V *et al* 2020 The role of single-cell mechanical behaviour and polarity in driving collective cell migration *Nat. Phys.* **16** 802–9

Jennings P, Bertocchi C, Frick M, Haller T, Pfaller W and Dietl P 2005 Ca^{2+} Induced surfactant secretion in alveolar type II cultures isolated from the H-2K^b-tsA58 transgenic mouse *Cell. Physiol. Biochem.* **15** 159–66

Jetto C T, Nambiar A and Manjithaya R 2022 Mitophagy and neurodegeneration: between the knowns and the unknowns *Front. Cell Dev. Biol.* **10** 837337

Jhanwar-Uniyal M, Amin A G, Cooper J B, Das K, Schmidt M H and Murali R 2017 Discrete signaling mechanisms of mTORC1 and mTORC2: connected yet apart in cellular and molecular aspects *Adv. Biol. Regul.* **64** 39–48

Ji C H, Kim H Y, Heo A J, Lee S H, Lee M J, Kim S B *et al* 2019 The N-degron pathway mediates ER-phagy *Mol. Cell* **75** 1058–1072.e9

Jia R, Guardia C M, Pu J, Chen Y and Bonifacino J S 2017 BORC coordinates encounter and fusion of lysosomes with autophagosomes *Autophagy* **13** 1648–63

Jiang G, Huang A H, Cai Y, Tanase M and Sheetz M P 2006 Rigidity sensing at the leading edge through $\alpha v\beta 3$ integrins and RPTPα *Biophys. J.* **90** 1804–9

Jiang P and Mizushima N 2014 Autophagy and human diseases *Cell Res.* **24** 69–79
Jiang P, Nishimura T, Sakamaki Y, Itakura E, Hatta T, Natsume T *et al* 2014 The HOPS complex mediates autophagosome–lysosome fusion through interaction with syntaxin 17 *Mol. Biol. Cell* **25** 1327–37
Jiang S, Heller B, Tagliabracci V S, Zhai L, Irimia J M, DePaoli-Roach A A *et al* 2010 Starch binding domain-containing protein 1/genethonin 1 is a novel participant in glycogen metabolism *J. Biol. Chem.* **285** 34960–71
Jiang S, Wells C D and Roach P J 2011 Starch-binding domain-containing protein 1 (Stbd1) and glycogen metabolism: Identification of the Atg8 family interacting motif (AIM) in Stbd1 required for interaction with GABARAPL1 *Biochem. Biophys. Res. Commun.* **413** 420–5
Jiao H, Jiang D, Hu X, Du W, Ji L, Yang Y *et al* 2021 Mitocytosis, a migrasome-mediated mitochondrial quality-control process *Cell* **184** 2896–2910.e13
Jin S M, Lazarou M, Wang C, Kane L A, Narendra D P and Youle R J 2010 Mitochondrial membrane potential regulates PINK1 import and proteolytic destabilization by PARL *J. Cell Biol.* **191** 933–42
Jobst M, Hossain M, Kiss E, Bergen J, Marko D and Del Favero G 2024 Autophagy modulation changes mechano-chemical sensitivity of T24 bladder cancer cells *Biomed. Pharmacother.* **170** 115942
Jobst M, Kiss E, Gerner C, Marko D and Del Favero G 2023 Activation of autophagy triggers mitochondrial loss and changes acetylation profile relevant for mechanotransduction in bladder cancer cells *Arch. Toxicol.* **97** 217–33
Johansen T and Lamark T 2020 Selective autophagy: ATG8 family proteins, LIR motifs and cargo receptors *J. Mol. Biol.* **432** 80–103
Johnson B N, Berger A K, Cortese G P and LaVoie M J 2012 The ubiquitin E3 ligase parkin regulates the proapoptotic function of Bax *Proc. Natl Acad. Sci. USA* **109** 6283–8
Jordens I, Fernandez-Borja M, Marsman M, Dusseljee S, Janssen L, Calafat J *et al* 2001 The Rab7 effector protein RILP controls lysosomal transport by inducing the recruitment of dynein-dynactin motors *Curr. Biol.* **11** 1680–5
Jutten B, Keulers T G, Schaaf M B E, Savelkouls K, Theys J, Span P N *et al* 2013 EGFR overexpressing cells and tumors are dependent on autophagy for growth and survival *Radiother. Oncol.* **108** 479–83
Kamber R A, Shoemaker C J and Denic V 2015 Receptor-bound targets of selective autophagy use a scaffold protein to activate the Atg1 kinase *Mol. Cell* **59** 372–81
Kane L A, Lazarou M, Fogel A I, Li Y, Yamano K, Sarraf S A *et al* 2014 PINK1 phosphorylates ubiquitin to activate Parkin E3 ubiquitin ligase activity *J. Cell Biol.* **205** 143–53
Karanasios E, Stapleton E, Manifava M, Kaizuka T, Mizushima N, Walker S A *et al* 2013 Dynamic association of the ULK1 complex with omegasomes during autophagy induction *J. Cell Sci.* **132415**
Kashatus J A, Nascimento A, Myers L J, Sher A, Byrne F L, Hoehn K L *et al* 2015 Erk2 phosphorylation of Drp1 promotes mitochondrial fission and MAPK-driven tumor growth *Mol. Cell* **57** 537–51
Kast D J and Dominguez R 2017 The cytoskeleton–autophagy connection *Curr. Biol.* **27** R318–26
Kast D J, Zajac A L, Holzbaur E L F, Ostap E M and Dominguez R 2015 WHAMM directs the Arp2/3 complex to the ER for autophagosome biogenesis through an actin comet tail mechanism *Curr. Biol.* **25** 1791–7

Kaur J and Debnath J 2015 Autophagy at the crossroads of catabolism and anabolism *Nat. Rev. Mol. Cell Biol.* **16** 461–72

Kawaguchi Y, Kovacs J J, McLaurin A, Vance J M, Ito A and Yao T-P 2003 The deacetylase HDAC6 regulates aggresome formation and cell viability in response to misfolded protein stress *Cell* **115** 727–38

Kazlauskaite A, Kondapalli C, Gourlay R, Campbell D G, Ritorto M S, Hofmann K *et al* 2014 Parkin is activated by PINK1-dependent phosphorylation of ubiquitin at Ser65 *Biochem. J.* **460** 127–41

Kechagia J Z, Ivaska J and Roca-Cusachs P 2019 Integrins as biomechanical sensors of the microenvironment *Nat. Rev. Mol. Cell Biol.* **20** 457–73

Khaminets A, Heinrich T, Mari M, Grumati P, Huebner A K, Akutsu M *et al* 2015 Regulation of endoplasmic reticulum turnover by selective autophagy *Nature* **522** 354–8

Kim G H E, Dayam R M, Prashar A, Terebiznik M and Botelho R J 2014 PIKFYVE inhibition interferes with phagosome and endosome maturation in macrophages *Traffic* **15** 1143–63

Kim J, Kundu M, Viollet B and Guan K-L 2011 AMPK and mTOR regulate autophagy through direct phosphorylation of Ulk1 *Nat. Cell Biol.* **13** 132–41

Kim M, Carman C V and Springer T A 2003 Bidirectional transmembrane signaling by cytoplasmic domain separation in integrins *Science* **301** 1720–5

Kim O-H, Choi Y W, Park J H, Hong S A, Hong M, Chang I H *et al* 2022 Fluid shear stress facilitates prostate cancer metastasis through Piezo1-Src-YAP axis *Life Sci.* **308** 120936

Kim P K, Hailey D W, Mullen R T and Lippincott-Schwartz J 2008 Ubiquitin signals autophagic degradation of cytosolic proteins and peroxisomes *Proc. Natl Acad. Sci. USA* **105** 20567–74

Kimura S, Noda T and Yoshimori T 2008 Dynein-dependent movement of autophagosomes mediates efficient encounters with lysosomes *Cell Struct. Funct.* **33** 109–22

King J S 2012 Mechanical stress meets autophagy: potential implications for physiology and pathology *Trends Mol. Med.* **18** 583–8

King J S, Veltman D M and Insall R H 2011 The induction of autophagy by mechanical stress *Autophagy* **7** 1490–9

Kirkin V, Lamark T, Sou Y-S, Bjørkøy G, Nunn J L, Bruun J-A *et al* 2009 A role for NBR1 in autophagosomal degradation of ubiquitinated substrates *Mol. Cell* **33** 505–16

Kirkin V and Rogov V V 2019 A diversity of selective autophagy receptors determines the specificity of the autophagy pathway *Mol. Cell* **76** 268–85

Kitada T, Asakawa S, Hattori N, Matsumine H, Yamamura Y, Minoshima S *et al* 1998 Mutations in the parkin gene cause autosomal recessive juvenile parkinsonism *Nature* **392** 605–8

Klimek C, Kathage B, Wördehoff J and Höhfeld J 2017 BAG3-mediated proteostasis at a glance *J. Cell Sci.* **130** 2781–8

Klionsky D J, Abdelmohsen K, Abe A, Abedin M J, Abeliovich H, Acevedo Arozena A *et al* 2016 Guidelines for the use and interpretation of assays for monitoring autophagy (3rd edition) *Autophagy* **12** 1–222

Klionsky D J, Eskelinen E-L and Deretic V 2014 Autophagosomes, phagosomes, autolysosomes, phagolysosomes, autophagolysosomes … Wait, I'm confused *Autophagy* **10** 549–51

Knorr R L, Lipowsky R and Dimova R 2015 Autophagosome closure requires membrane scission *Autophagy* **11** 2134–7

Ko J Y, Lee E J and Park J H 2019 Interplay between primary cilia and autophagy and its controversial roles in cancer *Biomol. Ther.* **27** 337–41

Koerver L, Papadopoulos C, Liu B, Kravic B, Rota G, Brecht L *et al* 2019 The ubiquitin-conjugating enzyme UBE 2 QL 1 coordinates lysophagy in response to endolysosomal damage *EMBO Rep.* **20** e48014

Kondapalli C, Kazlauskaite A, Zhang N, Woodroof H I, Campbell D G, Gourlay R *et al* 2012 PINK1 is activated by mitochondrial membrane potential depolarization and stimulates Parkin E3 ligase activity by phosphorylating Serine 65 *Open Biol.* **2** 120080

Kondratskyi A, Kondratska K, Skryma R, Klionsky D J and Prevarskaya N 2018 Ion channels in the regulation of autophagy *Autophagy* **14** 3–21

Koutsifeli P, Varma U, Daniels L J, Annandale M, Li X, Neale J P H *et al* 2022 Glycogen-autophagy: molecular machinery and cellular mechanisms of glycophagy *J. Biol. Chem.* **298** 102093

Koyano F, Okatsu K, Kosako H, Tamura Y, Go E, Kimura M *et al* 2014 Ubiquitin is phosphorylated by PINK1 to activate parkin *Nature* **510** 162–6

Kravić B, Bionda T, Siebert A, Gahlot P, Levantovsky S, Behrends C *et al* 2022 Ubiquitin profiling of lysophagy identifies actin stabilizer CNN2 as a target of VCP/p97 and uncovers a link to HSPB1 *Mol. Cell* **82** 2633–2649.e7

Krebs K, Ruusmann A, Simonlatser G and Velling T 2015 Expression of FLNa in human melanoma cells regulates the function of integrin $\alpha 1\beta 1$ and phosphorylation and localisation of PKB/AKT/ERK1/2 kinases *Eur. J. Cell Biol.* **94** 564–75

Kriel J, Müller-Nedebock K, Maarman G, Mbizana S, Ojuka E, Klumperman B *et al* 2018 Coordinated autophagy modulation overcomes glioblastoma chemoresistance through disruption of mitochondrial bioenergetics *Sci. Rep.* **8** 10348

Kroemer G, Mariño G and Levine B 2010 Autophagy and the integrated stress response *Mol. Cell* **40** 280–93

Kruppa A J, Kendrick-Jones J and Buss F 2016 Myosins, actin and autophagy *Traffic* **17** 878–90

Kumar S and Weaver V M 2009 Mechanics, malignancy, and metastasis: the force journey of a tumor cell *Cancer Metastasis Rev.* **28** 113–27

Kuo J-C 2014 Focal adhesions function as a mechanosensor *Prog. Mol. Biol. Transl. Sci.* **126** 55–73

Lamark T and Johansen T 2012 Aggrephagy: selective disposal of protein aggregates by macroautophagy *Int. J. Cell Biol.* **2012** 1–21

Lamb C A, Yoshimori T and Tooze S A 2013 The autophagosome: origins unknown, biogenesis complex *Nat. Rev. Mol. Cell Biol.* **14** 759–74

Lamouille S, Xu J and Derynck R 2014 Molecular mechanisms of epithelial–mesenchymal transition *Nat. Rev. Mol. Cell Biol.* **15** 178–96

Law K B, Bronte-Tinkew D, Di Pietro E, Snowden A, Jones R O, Moser A *et al* 2017 The peroxisomal AAA ATPase complex prevents pexophagy and development of peroxisome biogenesis disorders *Autophagy* **13** 868–84

Lazarou M, Sliter D A, Kane L A, Sarraf S A, Wang C, Burman J L *et al* 2015 The ubiquitin kinase PINK1 recruits autophagy receptors to induce mitophagy *Nature* **524** 309–14

Le Roux A-L, Quiroga X, Walani N, Arroyo M and Roca-Cusachs P 2019 The plasma membrane as a mechanochemical transducer *Phil. Trans. R. Soc.* B **374** 20180221

Lee M Y, Sumpter R, Zou Z, Sirasanagandla S, Wei Y, Mishra P *et al* 2017a Peroxisomal protein PEX 13 functions in selective autophagy *EMBO Rep.* **18** 48–60

Lee R, Kim D-W, Lee W-Y and Park H-J 2022 Zearalenone induces apoptosis and autophagy in a spermatogonia cell line *Toxins* **14** 148

Lee Y, Chou T-F, Pittman S K, Keith A L, Razani B and Weihl C C 2017b Keap1/Cullin3 modulates p62/SQSTM1 activity via UBA domain ubiquitination *Cell Rep.* **20** 1994

Lee Y-K and Lee J-A 2016 Role of the mammalian ATG8/LC3 family in autophagy: differential and compensatory roles in the spatiotemporal regulation of autophagy *BMB Rep.* **49** 424–30

Legerstee K and Houtsmuller A 2021 A layered view on focal adhesions *Biology* **10** 1189

Levental K R, Yu H, Kass L, Lakins J N, Egeblad M, Erler J T *et al* 2009 Matrix crosslinking forces tumor progression by enhancing integrin signaling *Cell* **139** 891–906

Li F, Guo H, Yang Y, Feng M, Liu B, Ren X *et al* 2019a Autophagy modulation in bladder cancer development and treatment (Review) *Oncol. Rep.*

Li L, Tong M, Fu Y, Chen F, Zhang S, Chen H *et al* 2021a Lipids and membrane-associated proteins in autophagy *Protein Cell* **12** 520–44

Li W, Li J and Bao J 2012 Microautophagy: lesser-known self-eating *Cell. Mol. Life Sci.* **69** 1125–36

Li X and Wang J 2020 Mechanical tumor microenvironment and transduction: cytoskeleton mediates cancer cell invasion and metastasis *Int. J. Biol. Sci.* **16** 2014–28

Li Y, Zheng W, Lu Y, Zheng Y, Pan L, Wu X *et al* 2021b BNIP3L/NIX-mediated mitophagy: molecular mechanisms and implications for human disease *Cell Death Dis.* **13** 14

Li Z, Wang C, Wang Z, Zhu C, Li J, Sha T *et al* 2019b Allele-selective lowering of mutant HTT protein by HTT–LC3 linker compounds *Nature* **575** 203–9

Liang J R, Lingeman E, Luong T, Ahmed S, Muhar M, Nguyen T *et al* 2020a A genome-wide ER-phagy screen highlights key roles of mitochondrial metabolism and ER-resident UFMylation *Cell* **180** 1160–1177.e20

Liang J, Yang Y, Bai L, Li F and Li E 2020b DRP1 upregulation promotes pancreatic cancer growth and metastasis through increased aerobic glycolysis *J. Gastro. Hepatol.* **35** 885–95

Liang N, Zhang C, Dill P, Panasyuk G, Pion D, Koka V *et al* 2014 Regulation of YAP by mTOR and autophagy reveals a therapeutic target of tuberous sclerosis complex *J. Exp. Med.* **211** 2249–63

2007 *TRP Ion Channel Function in Sensory Transduction and Cellular Signaling Cascades* ed W B Liedtke and S and Heller (Boca Raton, FL: CRC Press/Taylor and Francis)

Lien S-C, Chang S-F, Lee P-L, Wei S-Y, Chang M D-T, Chang J-Y *et al* 2013 Mechanical regulation of cancer cell apoptosis and autophagy: roles of bone morphogenetic protein receptor, Smad1/5, and p38 MAPK *Biochim. Biophys. Acta (BBA)——Mol. Cell Res.* **1833** 3124–33

Lightowlers R N, Taylor R W and Turnbull D M 2015 Mutations causing mitochondrial disease: what is new and what challenges remain? *Science* **349** 1494–9

Lin C, Nozawa T, Minowa-Nozawa A, Toh H, Aikawa C and Nakagawa I 2019 LAMTOR2/LAMTOR1 complex is required for TAX1BP1-mediated xenophagy *Cell. Microbiol.* **21** e12981

Lin C-Y, Nozawa T, Minowa-Nozawa A, Toh H, Hikichi M, Iibushi J *et al* 2020 Autophagy receptor tollip facilitates bacterial autophagy by recruiting galectin-7 in response to group A streptococcus infection *Front. Cell. Infect. Microbiol.* **10** 583137

Lin J-F and Hwang T I S 2017 Autophagy regulation in bladder cancer as the novel therapeutic strategy *Transl. Cancer Res.* **6** S708–19

Liu J, Bi X, Chen T, Zhang Q, Wang S-X, Chiu J-J *et al* 2015 Shear stress regulates endothelial cell autophagy via redox regulation and Sirt1 expression *Cell Death Dis.* **6** e1827–7

Liu L, Sheng J-Q, Wang M-R, Gan Y, Wu X-L, Liao J-Z *et al* 2019a Primary cilia blockage promotes the malignant behaviors of hepatocellular carcinoma via induction of autophagy *BioMed Res. Int.* **2019** 1–14

Liu X, Li Y, Wang X, Xing R, Liu K, Gan Q *et al* 2017 The BEACH-containing protein WDR81 coordinates p62 and LC3C to promote aggrephagy *J. Cell Biol.* **216** 1301–20

Liu X, Sun J, Yuan P, Shou K, Zhou Y, Gao W *et al* 2019b Mfn2 inhibits proliferation and cell-cycle in Hela cells via Ras-NF-κB signal pathway *Cancer Cell Int.* **19** 197

Lock R and Debnath J 2008 Extracellular matrix regulation of autophagy *Curr. Opin. Cell Biol.* **20** 583–8

Long J, Garner T P, Pandya M J, Craven C J, Chen P, Shaw B *et al* 2010 Dimerisation of the UBA domain of p62 inhibits ubiquitin binding and regulates NF-κB signalling *J. Mol. Biol.* **396** 178–94

Longatti A, Lamb C A, Razi M, Yoshimura S, Barr F A and Tooze S A 2012 TBC1D14 regulates autophagosome formation via Rab11- and ULK1-positive recycling endosomes *J. Cell Biol.* **197** 659–75

Lopez J I, Kang I, You W-K, McDonald D M and Weaver V M 2011 *In situ* force mapping of mammary gland transformation *Integr. Biol.* **3** 910–21

Lowery J, Kuczmarski E R, Herrmann H and Goldman R D 2015 Intermediate filaments play a pivotal role in regulating cell architecture and function *J. Biol. Chem.* **290** 17145–53

Lu K, Psakhye I and Jentsch S 2014 Autophagic clearance of PolyQ proteins mediated by ubiquitin-Atg8 adaptors of the conserved CUET protein family *Cell* **158** 549–63

Lu L, Tang M, Qi Z, Huang S, He Y, Li D *et al* 2020 Regulation of the Golgi apparatus via GOLPH3-mediated new selective autophagy *Life Sci.* **253** 117700

Lu P, Weaver V M and Werb Z 2012 The extracellular matrix: a dynamic niche in cancer progression *J. Cell Biol.* **196** 395–406

Luo B-H, Carman C V and Springer T A 2007 Structural basis of integrin regulation and signaling *Annu. Rev. Immunol.* **25** 619–47

Ma S, Fu A, Chiew G G Y and Luo K Q 2017 Hemodynamic shear stress stimulates migration and extravasation of tumor cells by elevating cellular oxidative level *Cancer Lett.* **388** 239–48

Ma X, Lu C, Chen Y, Li S, Ma N, Tao X *et al* 2022 CCT2 is an aggrephagy receptor for clearance of solid protein aggregates *Cell* **185** 1325–45.e22

Machado-Neto J A, Coelho-Silva J L, Santos F P D S, Scheucher P S, Campregher P V, Hamerschlak N *et al* 2020 Autophagy inhibition potentiates ruxolitinib-induced apoptosis in JAK2V617F cells *Invest. New Drugs.* **38** 733–45

Maejima I, Takahashi A, Omori H, Kimura T, Takabatake Y, Saitoh T *et al* 2013 Autophagy sequesters damaged lysosomes to control lysosomal biogenesis and kidney injury *EMBO J.* **32** 2336–47

Maio N and Rouault T A 2020 Outlining the complex pathway of mammalian Fe-S cluster biogenesis *Trends Biochem. Sci.* **45** 411–26

Mancias J D, Pontano Vaites L, Nissim S, Biancur D E, Kim A J, Wang X *et al* 2015 Ferritinophagy via NCOA4 is required for erythropoiesis and is regulated by iron dependent HERC2-mediated proteolysis *eLife* **4** e10308

Mancias J D, Wang X, Gygi S P, Harper J W and Kimmelman A C 2014 Quantitative proteomics identifies NCOA4 as the cargo receptor mediating ferritinophagy *Nature* **509** 105–9

Mandal A and Drerup C M 2019 Axonal transport and mitochondrial function in neurons *Front. Cell. Neurosci.* **13** 373

Manil-Ségalen M, Lefebvre C, Jenzer C, Trichet M, Boulogne C, Satiat-Jeunemaitre B *et al* 2014 The C. elegans LC3 acts downstream of GABARAP to degrade autophagosomes by interacting with the HOPS subunit VPS39 *Dev. Cell* **28** 43–55

Manzanillo P S, Ayres J S, Watson R O, Collins A C, Souza G, Rae C S *et al* 2013 The ubiquitin ligase parkin mediates resistance to intracellular pathogens *Nature* **501** 512–6

Marinković M, Šprung M, Buljubašić M and Novak I 2018 Autophagy modulation in cancer: current knowledge on action and therapy *Oxid. Med. Cell. Longev.* **2018** 8023821

Martens S, Nakamura S and Yoshimori T 2016 Phospholipids in autophagosome formation and fusion *J. Mol. Biol.* **428** 4819–27

Martinez J, Almendinger J, Oberst A, Ness R, Dillon C P, Fitzgerald P *et al* 2011 Microtubule-associated protein 1 light chain 3 alpha (LC3)-associated phagocytosis is required for the efficient clearance of dead cells *Proc. Natl Acad. Sci. USA* **108** 17396–401

Martinez-Vicente M 2015 Autophagy in neurodegenerative diseases: from pathogenic dysfunction to therapeutic modulation *Semin. Cell Dev. Biol.* **40** 115–26

Massagué J and Obenauf A C 2016 Metastatic colonization by circulating tumour cells *Nature* **529** 298–306

Masuzaki R, Tateishi R, Yoshida H, Sato T, Ohki T, Goto T *et al* 2007 Assessing liver tumor stiffness by transient elastography *Hepatol. Int.* **1** 394–7

Mathiowetz A J, Baple E, Russo A J, Coulter A M, Carrano E, Brown J D *et al* 2017 An Amish founder mutation disrupts a PI(3)P-WHAMM-Arp2/3 complex–driven autophagosomal remodeling pathway *Mol. Biol. Cell* **28** 2492–507

Matsumoto G, Shimogori T, Hattori N and Nukina N 2015 TBK1 controls autophagosomal engulfment of polyubiquitinated mitochondria through p62/SQSTM1 phosphorylation *Hum. Mol. Genet.* **24** 4429–42

Mauvezin C and Neufeld T P 2015 Bafilomycin A1 disrupts autophagic flux by inhibiting both V-ATPase-dependent acidification and Ca-P60A/SERCA-dependent autophagosome-lysosome fusion *Autophagy* **11** 1437–8

McBrayer M and Nixon R A 2013 Lysosome and calcium dysregulation in Alzheimer's disease: partners in crime *Biochem. Soc. Trans.* **41** 1495–502

McEwan D G and Dikic I 2010 Not all autophagy membranes are created equal *Cell* **141** 564–6

McEwan D G, Popovic D, Gubas A, Terawaki S, Suzuki H, Stadel D *et al* 2015 PLEKHM1 regulates autophagosome-lysosome fusion through HOPS complex and LC3/GABARAP proteins *Mol. Cell* **57** 39–54

McMahon H T and Boucrot E 2015 Membrane curvature at a glance *J. Cell Sci.* **128** 1065–70

Melia T J, Lystad A H and Simonsen A 2020 Autophagosome biogenesis: from membrane growth to closure *J. Cell Biol.* **219** e202002085

Mellor K M, Varma U, Koutsifeli P, Curl C L, Janssens J V, Daniels L J *et al* 2021 Protective role of the Atg8 homologue Gabarapl1 in regulating cardiomyocyte glycophagy in diabetic heart disease bioXriv: 2021.06.21.449174 https://www.biorxiv.org/content/10.1101/2021.06.21.449174v1

Mellor K M, Varma U, Stapleton D I and Delbridge L M D 2014 Cardiomyocyte glycophagy is regulated by insulin and exposure to high extracellular glucose *Am. J. Physiol. Heart Circ. Physiol.* **306** H1240–5

Meng Q, Pu L, Qi M, Li S, Sun B, Wang Y *et al* 2022 Laminar shear stress inhibits inflammation by activating autophagy in human aortic endothelial cells through HMGB1 nuclear translocation *Commun. Biol.* **5** 425

Menzies F M, Fleming A and Rubinsztein D C 2015 Compromised autophagy and neurodegenerative diseases *Nat. Rev. Neurosci.* **16** 345–57

Mi N, Chen Y, Wang S, Chen M, Zhao M, Yang G *et al* 2015 CapZ regulates autophagosomal membrane shaping by promoting actin assembly inside the isolation membrane *Nat. Cell Biol.* **17** 1112–23

Miao G, Zhang Y, Chen D and Zhang H 2020 The ER-localized transmembrane protein TMEM39A/SUSR2 regulates autophagy by controlling the trafficking of the PtdIns(4)P phosphatase SAC1 *Mol. Cell* **77** 618–632.e5

Miceli C, Roccio F, Penalva-Mousset L, Burtin M, Leroy C, Nemazanyy I *et al* 2020 The primary cilium and lipophagy translate mechanical forces to direct metabolic adaptation of kidney epithelial cells *Nat. Cell Biol.* **22** 1091–102

Mierke C T 2019 The matrix environmental and cell mechanical properties regulate cell migration and contribute to the invasive phenotype of cancer cells *Rep. Prog. Phys.* **82** 064602

Mierke C T 2020a Mechanical cues affect migration and invasion of cells from three different directions *Front. Cell Dev. Biol.* **8** 583226

Mierke C T 2020b Mechanical view on the endoplasmatic reticulum and golgi *Cellular Mechanics and Biophysics* (Cham: Springer International Publishing) pp 191–262

Mierke C T 2020c Mechanical view on the mitochondria *Cellular Mechanics and Biophysics* (Cham: Springer International Publishing) pp 163–89

Mierke C T 2020d Mechanical view on vacuoles *Cellular Mechanics and Biophysics* (Cham: Springer International Publishing) pp 263–75

Mierke C T 2020e The cell nucleus and its compartments *Cellular Mechanics and Biophysics* (Cham: Springer International Publishing) pp 333–414

Mierke C T 2021 The pertinent role of cell and matrix mechanics in cell adhesion and migration *Front. Cell Dev. Biol.* **9** 720494

Mierke C T 2023 The versatile roles of ADAM8 in cancer cell migration, mechanics, and extracellular matrix remodeling *Front. Cell Dev. Biol.* **11** 1130823

Mierke C T 2024a Extracellular matrix cues regulate mechanosensing and mechanotransduction of cancer cells *Cells* **13** 96

Mierke C T 2024b Mechanosensory entities and functionality of endothelial cells *Front. Cell Dev. Biol. Sec. Cell Adh. Migr.* **12** 1446452

Min E and Schwartz M A 2019 Translocating transcription factors in fluid shear stress-mediated vascular remodeling and disease *Exp. Cell. Res.* **376** 92–7

Mishra P, Carelli V, Manfredi G and Chan D C 2014 Proteolytic cleavage of Opa1 stimulates mitochondrial inner membrane fusion and couples fusion to oxidative phosphorylation *Cell Metab.* **19** 630–41

Mitchell M J and King M R 2013 Fluid shear stress sensitizes cancer cells to receptor-mediated apoptosis via trimeric death receptors *New J. Phys.* **15** 015008

Moehlman A T and Youle R J 2020 Mitochondrial quality control and restraining innate immunity *Annu. Rev. Cell Dev. Biol.* **36** 265–89

Mohammadi H and Sahai E 2018 Mechanisms and impact of altered tumour mechanics *Nat. Cell Biol.* **20** 766–74

Monastyrska I, He C, Geng J, Hoppe A D, Li Z and Klionsky D J 2008 Arp2 links autophagic machinery with the actin cytoskeleton *Mol. Biol. Cell* **19** 1962–75

Mooi W J, Dingemans K P, Van Den Bergh Weerman M A, Jobsis A C, Heymans H S A and Barth P G 1983 Ultrastructure of the liver in the cerebrohepatorenal syndrome of Zellweger *Ultrastruct. Pathol.* **5** 135–44

Mookerjee S A, Goncalves R L S, Gerencser A A, Nicholls D G and Brand M D 2015 The contributions of respiration and glycolysis to extracellular acid production *Biochim. Biophys. Acta (BBA)——Bioenerg.* **1847** 171–81

Moore A S and Holzbaur E L F 2016 Dynamic recruitment and activation of ALS-associated TBK1 with its target optineurin are required for efficient mitophagy *Proc. Natl Acad. Sci. USA* **113** E3349–58

Moroni M, Servin-Vences M R, Fleischer R, Sánchez-Carranza O and Lewin G R 2018 Voltage gating of mechanosensitive PIEZO channels *Nat. Commun.* **9** 1096

Morselli E, Galluzzi L, Kepp O, Vicencio J-M, Criollo A, Maiuri M C *et al* 2009 Anti- and pro-tumor functions of autophagy *Biochim. Biophys. Acta (BBA)——Mol. Cell Res.* **1793** 1524–32

Mowers E E, Sharifi M N and Macleod K F 2017 Autophagy in cancer metastasis *Oncogene* **36** 1619–30

Mukhopadhyay S, Panda P K, Sinha N, Das D N and Bhutia S K 2014 Autophagy and apoptosis: where do they meet? *Apoptosis* **19** 555–66

Nah J, Yoo S-M, Jung S, Jeong E I, Park M, Kaang B-K *et al* 2017 Phosphorylated CAV1 activates autophagy through an interaction with BECN1 under oxidative stress *Cell Death Dis.* **8** e2822–2

Najafi M, Farhood B and Mortezaee K 2019 Extracellular matrix (ECM) stiffness and degradation as cancer drivers *J. Cell. Biochem.* **120** 2782–90

Nakagawa I, Amano A, Mizushima N, Yamamoto A, Yamaguchi H, Kamimoto T *et al* 2004 Autophagy defends cells against invading group A *streptococcus Science* **306** 1037–40

Nakamura S and Yoshimori T 2017 New insights into autophagosome–lysosome fusion *J. Cell Sci.* **130** 1209–16

Nakatogawa H 2018 Spoon-feeding ribosomes to autophagy *Mol. Cell* **71** 197–9

Nakatogawa H 2020 Mechanisms governing autophagosome biogenesis *Nat. Rev. Mol. Cell Biol.* **21** 439–58

Nakatogawa H, Ishii J, Asai E and Ohsumi Y 2012 Atg4 recycles inappropriately lipidated Atg8 to promote autophagosome biogenesis *Autophagy* **8** 177–86

Narendra D, Tanaka A, Suen D-F and Youle R J 2008 Parkin is recruited selectively to impaired mitochondria and promotes their autophagy *J. Cell Biol.* **183** 795–803

Nascimbeni A C, Giordano F, Dupont N, Grasso D, Vaccaro M I, Codogno P *et al* 2017 ER – plasma membrane contact sites contribute to autophagosome biogenesis by regulation of local PI 3P synthesis *EMBO J.* **36** 2018–33

Nath S, Dancourt J, Shteyn V, Puente G, Fong W M, Nag S *et al* 2014 Lipidation of the LC3/GABARAP family of autophagy proteins relies on a membrane-curvature-sensing domain in Atg3 *Nat. Cell Biol.* **16** 415–24

Neill T, Schaefer L and Iozzo R V 2012 Decorin *Am. J. Pathol.* **181** 380–7

Neill T, Schaefer L and Iozzo R V 2014 Instructive roles of extracellular matrix on autophagy *Am. J. Pathol.* **184** 2146–53

Nguyen C D K and Yi C 2019 YAP/TAZ signaling and resistance to cancer therapy *Trends Cancer* **5** 283–96

Nguyen T M B, Subramanian I V, Kelekar A and Ramakrishnan S 2007 Kringle 5 of human plasminogen, an angiogenesis inhibitor, induces both autophagy and apoptotic death in endothelial cells *Blood* **109** 4793–802

Nguyen T M B, Subramanian I V, Xiao X, Ghosh G, Nguyen P, Kelekar A *et al* 2009 Endostatin induces autophagy in endothelial cells by modulating Beclin 1 and β-catenin levels *J. Cell. Mol. Med.* **13** 3687–98

Nguyen T N, Padman B S, Usher J, Oorschot V, Ramm G and Lazarou M 2016 Atg8 family LC3/GABARAP proteins are crucial for autophagosome–lysosome fusion but not autophagosome formation during PINK1/Parkin mitophagy and starvation *J. Cell Biol.* **215** 857–74

Nguyen T N, Sawa-Makarska J, Khuu G, Lam W K, Adriaenssens E, Fracchiolla D *et al* 2023 Unconventional initiation of PINK1/parkin mitophagy by optineurin *Mol. Cell* **83** 1693–1709.e9

Nia H T, Liu H, Seano G, Datta M, Jones D, Rahbari N *et al* 2017 Solid stress and elastic energy as measures of tumour mechanopathology *Nat. Biomed. Eng.* **1** 0004

Nighot P and Ma T 2016 Role of autophagy in the regulation of epithelial cell junctions *Tissue Barriers* **4** e1171284

Noad J, Von Der Malsburg A, Pathe C, Michel M A, Komander D and Randow F 2017 LUBAC-synthesized linear ubiquitin chains restrict cytosol-invading bacteria by activating autophagy and NF-κB *Nat Microbiol.* **2** 17063

Northcott J M, Dean I S, Mouw J K and Weaver V M 2018 Feeling stress: the mechanics of cancer progression and aggression *Front. Cell Dev. Biol.* **6** 17

Novak C M, Horst E N, Taylor C C, Liu C Z and Mehta G 2019 Fluid shear stress stimulates breast cancer cells to display invasive and chemoresistant phenotypes while upregulating *PLAU* in a 3D bioreactor *Biotech Bioeng.* **116** 3084–97

Novak I, Kirkin V, McEwan D G, Zhang J, Wild P, Rozenknop A *et al* 2010 Nix is a selective autophagy receptor for mitochondrial clearance *EMBO Rep.* **11** 45–51

Nthiga T M, Kumar Shrestha B, Lamark T and Johansen T 2021a The soluble reticulophagy receptor CALCOCO1 is also a Golgiphagy receptor *Autophagy* **17** 2051–2

Nthiga T M, Kumar Shrestha B, Sjøttem E, Bruun J, Bowitz Larsen K, Bhujabal Z *et al* 2020 CALCOCO 1 acts with VAMP -associated proteins to mediate ER-phagy *EMBO J.* **39** e103649

Nthiga T M, Shrestha B K, Bruun J-A, Larsen K B, Lamark T and Johansen T 2021b Regulation of golgi turnover by CALCOCO1-mediated selective autophagy *J. Cell Biol.* **220** e202006128

Ogawa M, Yoshimori T, Suzuki T, Sagara H, Mizushima N and Sasakawa C 2005 Escape of intracellular *Shigella* from autophagy *Science* **307** 727–31

Ohnstad A E, Delgado J M, North B J, Nasa I, Kettenbach A N, Schultz S W *et al* 2020 Receptor-mediated clustering of FIP200 bypasses the role of LC3 lipidation in autophagy *EMBO J.* **39** e104948

Ohsumi Y 1999 Molecular mechanism of autophagy in yeast, *Saccharomyces cerevisiae Phil. Trans. R. Soc. Lond.* B **354** 1577–81

O'Loughlin T, Kruppa A J, Ribeiro A L R, Edgar J R, Ghannam A, Smith A M *et al* 2020 OPTN recruitment to a Golgi-proximal compartment regulates immune signalling and cytokine secretion *J. Cell Sci.* **133** jcs239822

Olzmann J A and Carvalho P 2019 Dynamics and functions of lipid droplets *Nat. Rev. Mol. Cell Biol.* **20** 137–55

Ordureau A, Paulo J A, Zhang W, Ahfeldt T, Zhang J, Cohn E F *et al* 2018 Dynamics of PARKIN-dependent mitochondrial ubiquitylation in induced neurons and model systems revealed by digital snapshot proteomics *Mol. Cell* **70** 211–27.e8

Ordureau A, Sarraf S A, Duda D M, Heo J-M, Jedrychowski M P, Sviderskiy V O *et al* 2014 Quantitative proteomics reveal a feedforward mechanism for mitochondrial PARKIN translocation and ubiquitin chain synthesis *Mol. Cell* **56** 360–75

Orhon I, Dupont N and Codogno P 2016 Primary cilium and autophagy: the avengers of cell-size regulation *Autophagy* **12** 2258–9

Otten E G, Werner E, Crespillo-Casado A, Boyle K B, Dharamdasani V, Pathe C *et al* 2021 Ubiquitylation of lipopolysaccharide by RNF213 during bacterial infection *Nature* **594** 111–6

Padman B S, Nguyen T N, Uoselis L, Skulsuppaisarn M, Nguyen L K and Lazarou M 2019 LC3/GABARAPs drive ubiquitin-independent recruitment of Optineurin and NDP52 to amplify mitophagy *Nat. Commun.* **10** 408

Pagotto A, Pilotto G, Mazzoldi E L, Nicoletto M O, Frezzini S, Pastò A *et al* 2017 Autophagy inhibition reduces chemoresistance and tumorigenic potential of human ovarian cancer stem cells *Cell Death Dis.* **8** e2943–3

Pampliega O, Orhon I, Patel B, Sridhar S, Díaz-Carretero A, Beau I *et al* 2013 Functional interaction between autophagy and ciliogenesis *Nature* **502** 194–200

Pan J-A, Sun Y, Jiang Y-P, Bott A J, Jaber N, Dou Z *et al* 2016 TRIM21 ubiquitylates SQSTM1/p62 and suppresses protein sequestration to regulate redox homeostasis *Mol. Cell* **61** 720–33

Pan Z-Q, Shao G-C, Liu X-M, Chen Q, Dong M-Q and Du L-L 2020 Atg1 kinase in fission yeast is activated by Atg11-mediated dimerization and cis-autophosphorylation *eLife* **9** e58073

Panda P K, Mukhopadhyay S, Das D N, Sinha N, Naik P P and Bhutia S K 2015 Mechanism of autophagic regulation in carcinogenesis and cancer therapeutics *Semin. Cell Dev. Biol.* **39** 43–55

Pang M, Teng Y, Huang J, Yuan Y, Lin F and Xiong C 2017 Substrate stiffness promotes latent TGF-β1 activation in hepatocellular carcinoma *Biochem. Biophys. Res. Commun.* **483** 553–8

Pang M-F, Siedlik M J, Han S, Stallings-Mann M, Radisky D C and Nelson C M 2016 Tissue stiffness and hypoxia modulate the integrin-linked kinase ILK to control breast cancer stem-like cells *Cancer Res.* **76** 5277–87

Pankiv S, Clausen T H, Lamark T, Brech A, Bruun J-A, Outzen H *et al* 2007 p62/SQSTM1 binds directly to Atg8/LC3 to facilitate degradation of ubiquitinated protein aggregates by autophagy *J. Biol. Chem.* **282** 24131–45

Papadopoulos C, Kirchner P, Bug M, Grum D, Koerver L, Schulze N *et al* 2017 VCP /p97 cooperates with YOD 1, UBXD 1 and PLAA to drive clearance of ruptured lysosomes by autophagy *EMBO J.* **36** 135–50

Papadopoulos C and Meyer H 2017 Detection and clearance of damaged lysosomes by the endo-lysosomal damage response and lysophagy *Curr. Biol.* **27** R1330–41

Parzych K R and Klionsky D J 2014 An overview of autophagy: morphology, mechanism, and regulation *Antioxid. Redox Signal.* **20** 460–73

Paszek M J, Zahir N, Johnson K R, Lakins J N, Rozenberg G I, Gefen A *et al* 2005 Tensional homeostasis and the malignant phenotype *Cancer Cell* **8** 241–54

Pavel M, Renna M, Park S J, Menzies F M, Ricketts T, Füllgrabe J *et al* 2018 Contact inhibition controls cell survival and proliferation via YAP/TAZ-autophagy axis *Nat. Commun.* **9** 2961

Perrin A J, Jiang X, Birmingham C L, So N S Y and Brumell J H 2004 Recognition of bacteria in the cytosol of mammalian cells by the ubiquitin system *Curr. Biol.* **14** 806–11
Peter B J, Kent H M, Mills I G, Vallis Y, Butler P J G, Evans P R *et al* 2004 BAR domains as sensors of membrane curvature: the amphiphysin BAR structure *Science* **303** 495–9
Petridou N I, Spiró Z and Heisenberg C-P 2017 Multiscale force sensing in development *Nat. Cell Biol.* **19** 581–8
Pfanner N, Warscheid B and Wiedemann N 2019 Mitochondrial proteins: from biogenesis to functional networks *Nat. Rev. Mol. Cell Biol.* **20** 267–84
Pickford F, Masliah E, Britschgi M, Lucin K, Narasimhan R, Jaeger P A *et al* 2008 The autophagy-related protein beclin 1 shows reduced expression in early Alzheimer disease and regulates amyloid β accumulation in mice *J. Clin. Invest.* **118** 2190–9
Pickles S, Vigié P and Youle R J 2018 Mitophagy and quality control mechanisms in mitochondrial maintenance *Curr. Biol.* **28** R170–85
Pickrell A M and Youle R J 2015 The roles of PINK1, Parkin, and mitochondrial fidelity in Parkinson's disease *Neuron* **85** 257–73
Pickup M W, Laklai H, Acerbi I, Owens P, Gorska A E, Chytil A *et al* 2013 Stromally derived lysyl oxidase promotes metastasis of transforming growth factor-β–deficient mouse mammary carcinomas *Cancer Res.* **73** 5336–46
Pierce N W and Nachury M V 2013 Cilia grow by taking a bite out of the cell *Dev. Cell* **27** 126–7
Plaks V, Koopman C D and Werb Z 2013 Circulating tumor cells *Science* **341** 1186–8
Pocaterra A, Romani P and Dupont S 2020 YAP/TAZ functions and their regulation at a glance *J. Cell Sci.* **133** jcs230425
Poole L P and Macleod K F 2021 Mitophagy in tumorigenesis and metastasis *Cell. Mol. Life Sci.* **78** 3817–51
Praetorius H A 2015 The primary cilium as sensor of fluid flow: new building blocks to the model. A review in the theme: cell signaling: proteins, pathways and mechanisms *Am. J. Physiol. Cell Physiol.* **308** C198–208
Praetorius H A and Spring K R 2001 Bending the MDCK cell primary cilium increases intracellular calcium *J. Membr. Biol.* **184** 71–9
Princely Abudu Y, Pankiv S, Mathai B J, Håkon Lystad A, Bindesbøll C, Brenne H B *et al* 2019 NIPSNAP1 and NIPSNAP2 act as 'Eat Me' signals for mitophagy *Dev. Cell* **49** 509–525.e12
Puri C, Vicinanza M, Ashkenazi A, Gratian M J, Zhang Q, Bento C F *et al* 2018 The RAB11A-positive compartment is a primary platform for autophagosome assembly mediated by WIPI2 recognition of PI3P-RAB11A *Dev. Cell* **45** 114–31.e8
Qiao P, Zhao F, Liu M, Gao D, Zhang H and Yan Y 2017 Hydrogen sulfide inhibits mitochondrial fission in neuroblastoma N2a cells through the Drp1/ERK1/2 signaling pathway *Mol. Med. Rep.* **16** 971–7
Quest A F G, Gutierrez-Pajares J L and Torres V A 2008 Caveolin-1: an ambiguous partner in cell signalling and cancer *J Cell. Mol. Med.* **12** 1130–50
Ragusa M J, Stanley R E and Hurley J H 2012 Architecture of the Atg17 complex as a scaffold for autophagosome biogenesis *Cell* **151** 1501–12
Rahim R S, Chen M, Nourse C C, Meedeniya A C B and Crane D I 2016 Mitochondrial changes and oxidative stress in a mouse model of Zellweger syndrome neuropathogenesis *Neuroscience* **334** 201–13
Rahman A, Lőrincz P, Gohel R, Nagy A, Csordás G, Zhang Y *et al* 2022 GMAP is an Atg8a-interacting protein that regulates Golgi turnover in Drosophila *Cell Rep.* **39** 110903

Rainero E, Howe J D, Caswell P T, Jamieson N B, Anderson K, Critchley D R *et al* 2015 Ligand-occupied integrin internalization links nutrient signaling to invasive migration *Cell Rep.* **10** 398–413

Ravasio A, Cheddadi I, Chen T, Pereira T, Ong H T, Bertocchi C *et al* 2015a Gap geometry dictates epithelial closure efficiency *Nat. Commun.* **6** 7683

Ravasio A, Hobi N, Bertocchi C, Jesacher A, Dietl P and Haller T 2011 Interfacial sensing by alveolar type II cells: a new concept in lung physiology? *Am. J. Physiol. Cell Physiol.* **300** C1456–65

Ravasio A, Le A P, Saw T B, Tarle V, Ong H T, Bertocchi C *et al* 2015b Regulation of epithelial cell organization by tuning cell–substrate adhesion *Integr. Biol.* **7** 1228–41

Ravenhill B J, Boyle K B, Von Muhlinen N, Ellison C J, Masson G R, Otten E G *et al* 2019 The cargo receptor NDP52 initiates selective autophagy by recruiting the ULK complex to cytosol-invading bacteria *Mol. Cell* **74** 320–329.e6

Ravikumar B, Sarkar S, Davies J E, Futter M, Garcia-Arencibia M, Green-Thompson Z W *et al* 2010 Regulation of mammalian autophagy in physiology and pathophysiology *Physiol. Rev.* **90** 1383–435

Ravikumar B, Vacher C, Berger Z, Davies J E, Luo S, Oroz L G *et al* 2004 Inhibition of mTOR induces autophagy and reduces toxicity of polyglutamine expansions in fly and mouse models of Huntington disease *Nat. Genet.* **36** 585–95

Razinia Z, Mäkelä T, Ylänne J and Calderwood D A 2012 Filamins in mechanosensing and signaling *Annu. Rev. Biophys.* **41** 227–46

Regmi S, Fu A and Luo K Q 2017 High shear stresses under exercise condition destroy circulating tumor cells in a microfluidic system *Sci. Rep.* **7** 39975

Rehman J, Zhang H J, Toth P T, Zhang Y, Marsboom G, Hong Z *et al* 2012 Inhibition of mitochondrial fission prevents cell cycle progression in lung cancer *FASEB J.* **26** 2175–86

Ren Z, Wang Y, Deng H, Deng Y, Deng J, Zuo Z *et al* 2016 Effects of deoxynivalenol on calcium homeostasis of concanavalin A—stimulated splenic lymphocytes of chickens *in vitro Exp. Toxicol. Pathol.* **68** 241–5

Richter B, Sliter D A, Herhaus L, Stolz A, Wang C, Beli P *et al* 2016 Phosphorylation of OPTN by TBK1 enhances its binding to Ub chains and promotes selective autophagy of damaged mitochondria *Proc. Natl Acad. Sci. USA* **113** 4039–44

Ridley A J 2011 Life at the leading edge *Cell* **145** 1012–22

Robichaud S, Fairman G, Vijithakumar V, Mak E, Cook D P, Pelletier A R *et al* 2021 Identification of novel lipid droplet factors that regulate lipophagy and cholesterol efflux in macrophage foam cells *Autophagy* **17** 3671–89

Romani P, Nirchio N, Arboit M, Barbieri V, Tosi A, Michielin F *et al* 2022 Mitochondrial fission links ECM mechanotransduction to metabolic redox homeostasis and metastatic chemotherapy resistance *Nat. Cell Biol.* **24** 168–80

Romani P, Valcarcel-Jimenez L, Frezza C and Dupont S 2021 Crosstalk between mechanotransduction and metabolism *Nat. Rev. Mol. Cell Biol.* **22** 22–38

Ross T D, Coon B G, Yun S, Baeyens N, Tanaka K, Ouyang M *et al* 2013 Integrins in mechanotransduction *Curr. Opin. Cell Biol.* **25** 613–8

Rubashkin M G, Cassereau L, Bainer R, DuFort C C, Yui Y, Ou G *et al* 2014 Force engages vinculin and promotes tumor progression by enhancing PI3K activation of phosphatidylinositol (3,4,5)-triphosphate *Cancer Res.* **74** 4597–611

Ryan T A, Phillips E O, Collier C L, Jb Robinson A, Routledge D, Wood R E *et al* 2020 Tollip coordinates Parkin-dependent trafficking of mitochondrial-derived vesicles *EMBO J.* **39** e102539

Sackmann E 1994 Membrane bending energy concept of vesicle- and cell-shapes and shape-transitions *FEBS Lett.* **346** 3–16

Sackmann E and Tanaka M 2021 Critical role of lipid membranes in polarization and migration of cells: a biophysical view *Biophys. Rev.* **13** 123–38

Sagnier S, Daussy C F, Borel S, Robert-Hebmann V, Faure M, Blanchet F P *et al* 2015 Autophagy restricts HIV-1 infection by selectively degrading tat in $CD4^+$ T lymphocytes *J. Virol.* **89** 615–25

Saha T, Dash C, Jayabalan R, Khiste S, Kulkarni A, Kurmi K *et al* 2022 Intercellular nanotubes mediate mitochondrial trafficking between cancer and immune cells *Nat. Nanotechnol.* **17** 98–106

Sainio A and Järveläinen H 2014 Extracellular matrix macromolecules: potential tools and targets in cancer gene therapy *Mol. Cell Ther.* **2** 14

Sakamoto K M, Kim K B, Kumagai A, Mercurio F, Crews C M and Deshaies R J 2001 Protacs: Chimeric molecules that target proteins to the Skp1–Cullin–F box complex for ubiquitination and degradation *Proc. Natl Acad. Sci. USA* **98** 8554–9

Sakamoto K M, Kim K B, Verma R, Ransick A, Stein B, Crews C M *et al* 2003 Development of protacs to target cancer-promoting proteins for ubiquitination and degradation *Mol. Cell. Proteomics* **2** 1350–8

Salpietro V, Phadke R, Saggar A, Hargreaves I P, Yates R, Fokoloros C *et al* 2015 Zellweger syndrome and secondary mitochondrial myopathy *Eur. J. Pediatr.* **174** 557–63

Sandilands E, Serrels B, McEwan D G, Morton J P, Macagno J P, McLeod K *et al* 2012 Autophagic targeting of Src promotes cancer cell survival following reduced FAK signalling *Nat. Cell Biol.* **14** 51–60

Sandoval H, Thiagarajan P, Dasgupta S K, Schumacher A, Prchal J T, Chen M *et al* 2008 Essential role for Nix in autophagic maturation of erythroid cells *Nature* **454** 232–5

Sanhueza C, Wehinger S, Castillo Bennett J, Valenzuela M, Owen G I and Quest A F G 2015 The twisted survivin connection to angiogenesis *Mol. Cancer* **14** 198

Sankar D S, Kaeser-Pebernard S, Vionnet C, Favre S, De Oliveira Marchioro L, Pillet B *et al* 2024 The ULK1 effector BAG2 regulates autophagy initiation by modulating AMBRA1 localization *Cell Rep.* **43** 114689

Sargent G, Van Zutphen T, Shatseva T, Zhang L, Di Giovanni V, Bandsma R *et al* 2016 PEX2 is the E3 ubiquitin ligase required for pexophagy during starvation *J. Cell Biol.* **214** 677–90

Sarraf S A, Raman M, Guarani-Pereira V, Sowa M E, Huttlin E L, Gygi S P *et al* 2013 Landscape of the PARKIN-dependent ubiquitylome in response to mitochondrial depolarization *Nature* **496** 372–6

Sarraf S A, Shah H V, Kanfer G, Pickrell A M, Holtzclaw L A, Ward M E *et al* 2020 Loss of TAX1BP1-directed autophagy results in protein aggregate accumulation in the brain *Mol. Cell* **80** 779–95.e10

Satir P, Pedersen L B and Christensen S T 2010 The primary cilium at a glance *J. Cell Sci.* **123** 499–503

Sauvé V, Sung G, Soya N, Kozlov G, Blaimschein N, Miotto L S *et al* 2018 Mechanism of parkin activation by phosphorylation *Nat. Struct. Mol. Biol.* **25** 623–30

Savova A, Romanov J and Martens S 2020 NBR1 directly promotes the formation of p62—ubiquitin condensates via its PB1 and UBA domains bioXriv:2020.09.18.303552 https://www.biorxiv.org/content/10.1101/2020.09.18.303552v1

Sawa-Makarska J, Baumann V, Coudevylle N, Von Bülow S, Nogellova V, Abert C *et al* 2020 Reconstitution of autophagosome nucleation defines Atg9 vesicles as seeds for membrane formation *Science* **369** eaaz7714

Schneekloth A R, Pucheault M, Tae H S and Crews C M 2008 Targeted intracellular protein degradation induced by a small molecule: en route to chemical proteomics *Bioorg. Med. Chem. Lett.* **18** 5904–8

Schöneberg J, Lee I-H, Iwasa J H and Hurley J H 2017 Reverse-topology membrane scission by the ESCRT proteins *Nat. Rev. Mol. Cell Biol.* **18** 5–17

Schott M B, Weller S G, Schulze R J, Krueger E W, Drizyte-Miller K, Casey C A *et al* 2019 Lipid droplet size directs lipolysis and lipophagy catabolism in hepatocytes *J. Cell Biol.* **218** 3320–35

Schultz S W, Agudo-Canalejo J, Chino H, Migliano S M, Saito C, Koyama-Honda I *et al* 2021 Should I bend or should I grow: the mechanisms of droplet-mediated autophagosome formation *Autophagy* **17** 1046–8

Scotter E L, Vance C, Nishimura A L, Lee Y-B, Chen H-J, Urwin H *et al* 2014 Differential roles of the ubiquitin proteasome system (UPS) and autophagy in the clearance of soluble and aggregated TDP-43 species *J. Cell Sci.* **127** 1263–78

Seetharaman S and Etienne-Manneville S 2018 Integrin diversity brings specificity in mechanotransduction *Biol. Cell* **110** 49–64

Sekine S and Youle R J 2018 PINK1 import regulation; a fine system to convey mitochondrial stress to the cytosol *BMC Biol.* **16** 2

Sen B, Xie Z, Case N, Thompson W R, Uzer G, Styner M *et al* 2014 mTORC2 regulates mechanically induced cytoskeletal reorganization and lineage selection in marrow-derived mesenchymal stem cells *J. Bone Miner. Res.* **29** 78–89

Senft D and Ronai Z A 2016 Regulators of mitochondrial dynamics in cancer *Curr. Opin. Cell Biol.* **39** 43–52

Shahnazari S, Yen W-L, Birmingham C L, Shiu J, Namolovan A, Zheng Y T *et al* 2010 A diacylglycerol-dependent signaling pathway contributes to regulation of antibacterial autophagy *Cell HostMicrobe* **8** 137–46

Sharifi M N, Mowers E E, Drake L E, Collier C, Chen H, Zamora M *et al* 2016 Autophagy promotes focal adhesion disassembly and cell motility of metastatic tumor cells through the direct interaction of paxillin with LC3 *Cell Rep.* **15** 1660–72

Sharma A, Seow J J W, Dutertre C-A, Pai R, Blériot C, Mishra A *et al* 2020 Onco-fetal reprogramming of endothelial cells drives immunosuppressive macrophages in hepatocellular carcinoma *Cell* **183** 377–94.e21

Sheik Abdul N, Nagiah S and Chuturgoon A A 2016 Fusaric acid induces mitochondrial stress in human hepatocellular carcinoma (HepG2) cells *Toxicon* **119** 336–44

Shi X, Chang C, Yokom A L, Jensen L E and Hurley J H 2020a The autophagy adaptor NDP52 and the FIP200 coiled-coil allosterically activate ULK1 complex membrane recruitment *eLife* **9** e59099

Shi X, Yokom A L, Wang C, Young L N, Youle R J and Hurley J H 2020b ULK complex organization in autophagy by a C-shaped FIP200 N-terminal domain dimer *J. Cell Biol.* **219** e201911047

Shiba-Fukushima K, Arano T, Matsumoto G, Inoshita T, Yoshida S, Ishihama Y *et al* 2014 Phosphorylation of mitochondrial polyubiquitin by PINK1 promotes parkin mitochondrial tethering *PLoS Genet.* **10** e1004861

Shibutani S T, Saitoh T, Nowag H, Münz C and Yoshimori T 2015 Autophagy and autophagy-related proteins in the immune system *Nat. Immunol.* **16** 1014–24

Shieh A C and Swartz M A 2011 Regulation of tumor invasion by interstitial fluid flow *Phys. Biol.* **8** 015012

Shimizu S, Kanaseki T, Mizushima N, Mizuta T, Arakawa-Kobayashi S, Thompson C B *et al* 2004 Role of Bcl-2 family proteins in a non-apoptotic programmed cell death dependent on autophagy genes *Nat. Cell Biol.* **6** 1221–8

Shivashankar G V, Sheetz M and Matsudaira P 2015 Mechanobiology *Integr. Biol.* **7** 1091–2

Simón L, Campos A, Leyton L and Quest A F G 2020 Caveolin-1 function at the plasma membrane and in intracellular compartments in cancer *Cancer Metastasis Rev.* **39** 435–53

Singh R, Kaushik S, Wang Y, Xiang Y, Novak I, Komatsu M *et al* 2009 Autophagy regulates lipid metabolism *Nature* **458** 1131–5

Sliter D A, Martinez J, Hao L, Chen X, Sun N, Fischer T D *et al* 2018 Parkin and PINK1 mitigate STING-induced inflammation *Nature* **561** 258–62

Smith M D, Harley M E, Kemp A J, Wills J, Lee M, Arends M *et al* 2018 CCPG1 is a non-canonical autophagy cargo receptor essential for ER-phagy and pancreatic ER proteostasis *Dev. Cell* **44** 217–232.e11

Stachowiak J C, Schmid E M, Ryan C J, Ann H S, Sasaki D Y, Sherman M B *et al* 2012 Membrane bending by protein–protein crowding *Nat. Cell Biol.* **14** 944–9

Stack B C, Hansen J P, Ruda J M, Jaglowski J, Shvidler J and Hollenbeak C S 2004 Fusaric acid: a novel agent and mechanism to treat HNSCC *Otolaryngol. Head Neck Surg.* **131** 54–60

Stahl-Meyer J, Stahl-Meyer K and Jäättelä M 2021 Control of mitosis, inflammation, and cell motility by limited leakage of lysosomes *Curr. Opin. Cell Biol.* **71** 29–37

Staiano L and Zappa F 2019 Hijacking intracellular membranes to feed autophagosomal growth *FEBS Lett.* **593** 3120–34

Stenmark H 2009 Rab GTPases as coordinators of vesicle traffic *Nat. Rev. Mol. Cell Biol.* **10** 513–25

Stephani M, Picchianti L, Gajic A, Beveridge R, Skarwan E, Sanchez De Medina Hernandez V *et al* 2020 A cross-kingdom conserved ER-phagy receptor maintains endoplasmic reticulum homeostasis during stress *eLife* **9** e58396

Strilic B and Offermanns S 2017 Intravascular survival and extravasation of tumor cells *Cancer Cell* **32** 282–93

Stylianopoulos T 2017 The solid mechanics of cancer and strategies for improved therapy *J. Biomech. Eng.* **139** 021004

Stylianopoulos T, Martin J D, Snuderl M, Mpekris F, Jain S R and Jain R K 2013 Coevolution of solid stress and interstitial fluid pressure in tumors during progression: implications for vascular collapse *Cancer Res.* **73** 3833–41

Stylianou A, Gkretsi V and Stylianopoulos T 2018 Transforming growth factor-β modulates pancreatic cancer associated fibroblasts cell shape, stiffness and invasion *Biochim. Biophys. Acta (BBA)—Gen. Subj.* **1862** 1537–46

Su C, Li J, Zhang L, Wang H, Wang F, Tao Y *et al* 2020 The biological functions and clinical applications of integrins in cancers *Front. Pharmacol.* **11** 579068

Su É, Villard C and Manneville J 2023 Mitochondria: at the crossroads between mechanobiology and cell metabolism *Biol. Cell* **115** e2300010

Sun D, Wu R, Li P and Yu L 2020a Phase separation in regulation of aggrephagy *J. Mol. Biol.* **432** 160–9

Sun D, Wu R, Zheng J, Li P and Yu L 2018 Polyubiquitin chain-induced p62 phase separation drives autophagic cargo segregation *Cell Res.* **28** 405–15

Sun G G, Lu Y F, Zhang J and Hu W N 2014 Filamin A regulates MMP-9 expression and suppresses prostate cancer cell migration and invasion *Tumor Biol.* **35** 3819–26

Sun X, Gao H, Yang Y, He M, Wu Y, Song Y *et al* 2019 PROTACs: great opportunities for academia and industry *Sig. Transduct. Target. Ther.* **4** 64

Sun Y, Li M, Liu G, Zhang X, Zhi L, Zhao J *et al* 2020b The function of Piezo1 in colon cancer metastasis and its potential regulatory mechanism *J. Cancer Res. Clin. Oncol.* **146** 1139–52

Surmeier D J 2018 Determinants of dopaminergic neuron loss in Parkinson's disease *FEBS J.* **285** 3657–68

Swatek K N, Usher J L, Kueck A F, Gladkova C, Mevissen T E T, Pruneda J N *et al* 2019 Insights into ubiquitin chain architecture using Ub-clipping *Nature* **572** 533–7

Szymanski K M, Binns D, Bartz R, Grishin N V, Li W-P, Agarwal A K *et al* 2007 The lipodystrophy protein seipin is found at endoplasmic reticulum lipid droplet junctions and is important for droplet morphology *Proc. Natl Acad. Sci. USA* **104** 20890–5

Takahashi D, Moriyama J, Nakamura T, Miki E, Takahashi E, Sato A *et al* 2019a AUTACs: cargo-specific degraders using selective autophagy *Mol. Cell* **76** 797–810.e10

Takahashi Y, Liang X, Hattori T, Tang Z, He H, Chen H *et al* 2019b VPS37A directs ESCRT recruitment for phagophore closure *J. Cell Biol.* **218** 3336–54

Tanaka A, Cleland M M, Xu S, Narendra D P, Suen D-F, Karbowski M *et al* 2010 Proteasome and p97 mediate mitophagy and degradation of mitofusins induced by parkin *J. Cell Biol.* **191** 1367–80

Tanaka K 2020 The PINK1–parkin axis: an overview *Neurosci. Res.* **159** 9–15

Tanaka M, Kikuchi T, Uno H, Okita K, Kitanishi-Yumura T and Yumura S 2017 Turnover and flow of the cell membrane for cell migration *Sci. Rep.* **7** 12970

Tang H, Peng S, Dong Y, Yang X, Yang P, Yang L *et al* 2018 MARCH5 overexpression contributes to tumor growth and metastasis and associates with poor survival in breast cancer *CMAR* **11** 201–15

Tang H-W, Wang Y-B, Wang S-L, Wu M-H, Lin S-Y and Chen G-C 2011 Atg1-mediated myosin II activation regulates autophagosome formation during starvation-induced autophagy: Atg1 regulates myosin II activation in autophagy *EMBO J.* **30** 636–51

Tang Z, Lin M G, Stowe T R, Chen S, Zhu M, Stearns T *et al* 2013 Autophagy promotes primary ciliogenesis by removing OFD1 from centriolar satellites *Nature* **502** 254–7

Tanida I, Ueno T and Kominami E 2004 Human light chain 3/MAP1LC3B is cleaved at its carboxyl-terminal Met121 to expose Gly120 for lipidation and targeting to autophagosomal membranes *J. Biol. Chem.* **279** 47704–10

Tauchi-Sato K, Ozeki S, Houjou T, Taguchi R and Fujimoto T 2002 The surface of lipid droplets is a phospholipid monolayer with a unique fatty acid composition *J. Biol. Chem.* **277** 44507–12

Tedesco B, Vendredy L, Timmerman V and Poletti A 2023 The chaperone-assisted selective autophagy complex dynamics and dysfunctions *Autophagy* **19** 1619–41

Terešak P, Lapao A, Subic N, Boya P, Elazar Z and Simonsen A 2022 Regulation of PRKN-independent mitophagy *Autophagy* **18** 24–39

Tharp K M, Higuchi-Sanabria R, Timblin G A, Ford B, Garzon-Coral C, Schneider C *et al* 2021 Adhesion-mediated mechanosignaling forces mitohormesis *Cell Metab.* **33** 1322–41.e13

Tharp K M and Weaver V M 2018 Modeling tissue polarity in context *J. Mol. Biol.* **430** 3613–28

Thompson B J 2020 YAP/TAZ: drivers of tumor growth, metastasis, and resistance to therapy *BioEssays* **42** 1900162

Thompson W R, Guilluy C, Xie Z, Sen B, Brobst K E, Yen S S *et al* 2013 Mechanically activated fyn utilizes mTORC2 to regulate RhoA and adipogenesis in mesenchymal stem cells *Stem Cells* **31** 2528–37

Thurston T L M, Ryzhakov G, Bloor S, Von Muhlinen N and Randow F 2009 The TBK1 adaptor and autophagy receptor NDP52 restricts the proliferation of ubiquitin-coated bacteria *Nat. Immunol.* **10** 1215–21

Thurston T L M, Wandel M P, Von Muhlinen N, Foeglein Á and Randow F 2012 Galectin 8 targets damaged vesicles for autophagy to defend cells against bacterial invasion *Nature* **482** 414–8

Tian G-A, Xu W-T, Zhang X-L, Zhou Y-Q, Sun Y, Hu L-P *et al* 2023 CCBE1 promotes mitochondrial fusion by inhibiting the TGFβ-DRP1 axis to prevent the progression of hepatocellular carcinoma *Matrix Biol.* **117** 31–45

Tilokani L, Nagashima S, Paupe V and Prudent J 2018 Mitochondrial dynamics: overview of molecular mechanisms *Essays Biochem.* **62** 341–60

Torggler R, Papinski D, Brach T, Bas L, Schuschnig M, Pfaffenwimmer T *et al* 2016 Two independent pathways within selective autophagy converge to activate Atg1 kinase at the vacuole *Mol. Cell* **64** 221–35

Totaro A, Zhuang Q, Panciera T, Battilana G, Azzolin L, Brumana G *et al* 2019 Cell phenotypic plasticity requires autophagic flux driven by YAP/TAZ mechanotransduction *Proc. Natl Acad. Sci. USA* **116** 17848–57

Toure M and Crews C M 2016 Small-molecule PROTACS: new approaches to protein degradation *Angew. Chem. Int. Ed.* **55** 1966–73

Trivedi P C, Bartlett J J and Pulinilkunnil T 2020 Lysosomal biology and function: modern view of cellular debris bin *Cells* **9** 1131

Tschumperlin D J, Ligresti G, Hilscher M B and Shah V H 2018 Mechanosensing and fibrosis *J. Clin. Invest.* **128** 74–84

Tsuji-Tamura K and Ogawa M 2016 Inhibition of the PI3K–Akt and mTORC1 signaling pathways promotes the elongation of vascular endothelial cells *J. Cell Sci.* **129** 1165–78

Tumbarello D A, Manna P T, Allen M, Bycroft M, Arden S D, Kendrick-Jones J *et al* 2015 The autophagy receptor TAX1BP1 and the molecular motor myosin VI are required for clearance of *Salmonella typhimurium* by autophagy *PLoS Pathog.* **11** e1005174

Tumbarello D A, Waxse B J, Arden S D, Bright N A, Kendrick-Jones J and Buss F 2012 Autophagy receptors link myosin VI to autophagosomes to mediate Tom1-dependent autophagosome maturation and fusion with the lysosome *Nat. Cell Biol.* **14** 1024–35

Tung J C, Barnes J M, Desai S R, Sistrunk C, Conklin M W, Schedin P *et al* 2015 Tumor mechanics and metabolic dysfunction *Free Radical Biol. Med.* **79** 269–80

Turco E, Savova A, Gere F, Ferrari L, Romanov J, Schuschnig M *et al* 2021 Reconstitution defines the roles of p62, NBR1 and TAX1BP1 in ubiquitin condensate formation and autophagy initiation *Nat. Commun.* **12** 5212

Turco E, Witt M, Abert C, Bock-Bierbaum T, Su M-Y, Trapannone R *et al* 2019 FIP200 claw domain binding to p62 promotes autophagosome formation at ubiquitin condensates *Mol. Cell* **74** 330–346.e11

Ulbricht A, Arndt V and Höhfeld J 2013a Chaperone-assisted proteostasis is essential for mechanotransduction in mammalian cells *Commun. Integr. Biol.* **6** e24925

Ulbricht A, Eppler F J, Tapia V E, van der Ven P F M, Hampe N, Hersch N *et al* 2013b Cellular mechanotransduction relies on tension-induced and chaperone-assisted autophagy *Curr. Biol.* **23** 430–5

Valente E M, Abou-Sleiman P M, Caputo V, Muqit M M K, Harvey K, Gispert S *et al* 2004 Hereditary early-onset Parkinson's disease caused by mutations in *PINK1 Science* **304** 1158–60

Van Laar V S and Berman S B 2013 The interplay of neuronal mitochondrial dynamics and bioenergetics: implications for Parkinson's disease *Neurobiol. Dis.* **51** 43–55

Vargas J N S, Hamasaki M, Kawabata T, Youle R J and Yoshimori T 2023 The mechanisms and roles of selective autophagy in mammals *Nat. Rev. Mol. Cell Biol.* **24** 167–85

Vargas J N S, Wang C, Bunker E, Hao L, Maric D, Schiavo G *et al* 2019 Spatiotemporal control of ULK1 activation by NDP52 and TBK1 during selective autophagy *Mol. Cell* **74** 347–62.e6

Vasan N, Baselga J and Hyman D M 2019 A view on drug resistance in cancer *Nature* **575** 299–309

Vedula S R K, Leong M C, Lai T L, Hersen P, Kabla A J, Lim C T *et al* 2012 Emerging modes of collective cell migration induced by geometrical constraints *Proc. Natl Acad. Sci. USA* **109** 12974–9

Verlhac P, Grégoire I P, Azocar O, Petkova D S, Baguet J, Viret C *et al* 2015 Autophagy receptor NDP52 regulates pathogen-containing autophagosome maturation *Cell Host Microbe* **17** 515–25

Vicinanza M, Puri C and Rubinsztein D C 2019 Coincidence detection of RAB11A and PI(3)P by WIPI2 directs autophagosome formation *Oncotarget* **10** 2579–80

Vining K H and Mooney D J 2017 Mechanical forces direct stem cell behaviour in development and regeneration *Nat. Rev. Mol. Cell Biol.* **18** 728–42

Vlahakis A and Debnath J 2017 The interconnections between autophagy and integrin-mediated cell adhesion *J. Mol. Biol.* **429** 515–30

Vogel V 2018 Unraveling the mechanobiology of extracellular matrix *Annu. Rev. Physiol.* **80** 353–87

Vogel V and Sheetz M 2006 Local force and geometry sensing regulate cell functions *Nat. Rev. Mol. Cell Biol.* **7** 265–75

von Muhlinen N, Akutsu M, Ravenhill B J, Foeglein Á, Bloor S, Rutherford T J *et al* 2012 LC3C, bound selectively by a noncanonical LIR motif in NDP52, is required for antibacterial autophagy *Mol. Cell* **48** 329–42

Wada K-I, Itoga K, Okano T, Yonemura S and Sasaki H 2011 Hippo pathway regulation by cell morphology and stress fibers *Development* **138** 3907–14

Walinda E, Morimoto D, Sugase K, Konuma T, Tochio H and Shirakawa M 2014 Solution structure of the ubiquitin-associated (UBA) domain of human autophagy receptor NBR1 and its interaction with ubiquitin and polyubiquitin *J. Biol. Chem.* **289** 13890–902

Wallace D C 2012 Mitochondria and cancer *Nat. Rev. Cancer* **12** 685–98

Wang C and Youle R J 2009 The role of mitochondria in apoptosis *Annu. Rev. Genet.* **43** 95–118

Wang J, Menon S, Yamasaki A, Chou H-T, Walz T, Jiang Y *et al* 2013 Ypt1 recruits the Atg1 kinase to the preautophagosomal structure *Proc. Natl Acad. Sci. USA* **110** 9800–5

Wang K, Wei Y, Liu W, Liu L, Guo Z, Fan C *et al* 2019 Mechanical stress-dependent autophagy component release via extracellular nanovesicles in tumor cells *ACS Nano* **13** 4589–602

Wang L, Zhou J, Yan S, Lei G, Lee C-H and Yin X-M 2017 Ethanol-triggered lipophagy requires SQSTM1 in AML12 hepatic cells *Sci. Rep.* **7** 12307

Wang N 1998 Mechanical interactions among cytoskeletal filaments *Hypertension* **32** 162–5

Wang Q, Yu P, Liu C, He X and Wang G 2022 Mitochondrial fragmentation in liver cancer: emerging player and promising therapeutic opportunities *Cancer Lett.* **549** 215912

Wang X, Jiang Y, Zhu L, Cao L, Xu W, Rahman S U *et al* 2020a Autophagy protects PC12 cells against deoxynivalenol toxicity via the Class III PI3K/beclin 1/Bcl-2 pathway *J. Cell. Physiol.* **235** 7803–15

Wang X, Zhang Y, Feng T, Su G, He J, Gao W *et al* 2018a Fluid shear stress promotes autophagy in hepatocellular carcinoma cells *Int. J. Biol. Sci.* **14** 1277–90

Wang Y, Jiang X, Feng F, Liu W and Sun H 2020b Degradation of proteins by PROTACs and other strategies *Acta Pharm. Sin. B.* **10** 207–38

Wang Z, Li C, Jiang M, Chen J, Yang M and Pu J 2018b Filamin A (FLNA) regulates autophagy of bladder carcinoma cell and affects its proliferation, invasion and metastasis *Int. Urol. Nephrol.* **50** 263–73

Wei C, Wang X, Zheng M and Cheng H 2012 Calcium gradients underlying cell migration *Curr. Opin. Cell Biol.* **24** 254–61

Weidberg H, Shvets E, Shpilka T, Shimron F, Shinder V and Elazar Z 2010 LC3 and GATE-16/GABARAP subfamilies are both essential yet act differently in autophagosome biogenesis *EMBO J.* **29** 1792–802

Weinberg S E, Sena L A and Chandel N S 2015 Mitochondria in the regulation of innate and adaptive immunity *Immunity* **42** 406–17

Wellen K E and Thompson C B 2012 A two-way street: reciprocal regulation of metabolism and signalling *Nat. Rev. Mol. Cell Biol.* **13** 270–6

West A P, Khoury-Hanold W, Staron M, Tal M C, Pineda C M, Lang S M *et al* 2015 Mitochondrial DNA stress primes the antiviral innate immune response *Nature* **520** 553–7

Wetzel L, Blanchard S, Rama S, Beier V, Kaufmann A and Wollert T 2020 TECPR1 promotes aggrephagy by direct recruitment of LC3C autophagosomes to lysosomes *Nat. Commun.* **11** 2993

White E and DiPaola R S 2009 The double-edged sword of autophagy modulation in cancer *Clin. Cancer Res.* **15** 5308–16

Wieder S Y, Serasinghe M N, Sung J C, Choi D C, Birge M B, Yao J L *et al* 2015 Activation of the mitochondrial fragmentation protein DRP1 correlates with BRAF V600E melanoma *J. Investig. Dermatol.* **135** 2544–7

Wild P, Farhan H, McEwan D G, Wagner S, Rogov V V, Brady N R *et al* 2011 Phosphorylation of the autophagy receptor optineurin restricts *Salmonella* Growth *Science* **333** 228–33

Winslow A R, Chen C-W, Corrochano S, Acevedo-Arozena A, Gordon D E, Peden A A *et al* 2010 α-Synuclein impairs macroautophagy: implications for Parkinson's disease *J. Cell Biol.* **190** 1023–37

Wirtz D, Konstantopoulos K and Searson P C 2011 The physics of cancer: the role of physical interactions and mechanical forces in metastasis *Nat. Rev. Cancer* **11** 512–22

Wolfenson H, Yang B and Sheetz M P 2019 Steps in mechanotransduction pathways that control cell morphology *Annu. Rev. Physiol.* **81** 585–605

Woo Y and Jung Y-J 2017 Angiotensin II receptor blockers induce autophagy in prostate cancer cells *Oncol. Lett.* **13** 3579–85

Wozniak M A, Modzelewska K, Kwong L and Keely P J 2004 Focal adhesion regulation of cell behavior *Biochim. Biophys. Acta (BBA)——Mol. Cell Res.* **1692** 103–19

Wu H, Gao Y, Li S, Bao X, Wang J and Zheng N 2021a Lactoferrin alleviated AFM1-induced apoptosis in intestinal NCM 460 cells through the autophagy pathway *Foods* **11** 23

Wu J and Chen Z J 2014 Innate immune sensing and signaling of cytosolic nucleic acids *Annu. Rev. Immunol.* **32** 461–88

Wu Z, Zuo M, Zeng L, Cui K, Liu B, Yan C *et al* 2021b OMA1 reprograms metabolism under hypoxia to promote colorectal cancer development *EMBO Rep.* **22** e50827

Wurzer B, Zaffagnini G, Fracchiolla D, Turco E, Abert C, Romanov J *et al* 2015 Oligomerization of p62 allows for selection of ubiquitinated cargo and isolation membrane during selective autophagy *Elife* **4** e08941

Wyant G A, Abu-Remaileh M, Frenkel E M, Laqtom N N, Dharamdasani V, Lewis C A *et al* 2018 NUFIP1 is a ribosome receptor for starvation-induced ribophagy *Science* **360** 751–8

Xia P, Wang S, Du Y, Zhao Z, Shi L, Sun L *et al* 2013 WASH inhibits autophagy through suppression of Beclin 1 ubiquitination *EMBO J.* **32** 2685–96

Xie J, Bao M, Hu X, Koopman W J H and Huck W T S 2021 Energy expenditure during cell spreading influences the cellular response to matrix stiffness *Biomaterials* **267** 120494

Xie L, Law B K, Chytil A M, Brown K A, Aakre M E and Moses H L 2004 Activation of the Erk pathway is required for TGF-β1-induced EMT *In Vitro Neoplasia* **6** 603–10

Xiong N, Li S, Tang K, Bai H, Peng Y, Yang H *et al* 2017 Involvement of caveolin-1 in low shear stress-induced breast cancer cell motility and adhesion: roles of FAK/Src and ROCK/p-MLC pathways *Biochim. Biophys. Acta, Mol. Cell Res.* **1864** 12–22

Xu Y, Cheng S, Zeng H, Zhou P, Ma Y, Li L *et al* 2022 ARF GTPases activate Salmonella effector SopF to ADP-ribosylate host V-ATPase and inhibit endomembrane damage-induced autophagy *Nat. Struct. Mol. Biol.* **29** 67–77

Xu Y, Zhou P, Cheng S, Lu Q, Nowak K, Hopp A-K *et al* 2019 A bacterial effector reveals the V-ATPase-ATG16L1 axis that initiates xenophagy *Cell* **178** 552–566.e20

Yakunin E, Moser A, Loeb V, Saada A, Faust P, Crane D I *et al* 2010 α-Synuclein abnormalities in mouse models of peroxisome biogenesis disorders *J. Neurosci. Res.* **88** 866–76

Yamada A, Hikichi M, Nozawa T and Nakagawa I 2021 FBXO2/SCF ubiquitin ligase complex directs xenophagy through recognizing bacterial surface glycan *EMBO Rep.* **22** e52584

Yamano K, Fogel A I, Wang C, Van Der Bliek A M and Youle R J 2014 Mitochondrial Rab GAPs govern autophagosome biogenesis during mitophagy *eLife* **3** e01612

Yamano K, Kikuchi R, Kojima W, Hayashida R, Koyano F, Kawawaki J *et al* 2020 Critical role of mitochondrial ubiquitination and the OPTN–ATG9A axis in mitophagy *J. Cell Biol.* **219** e201912144

Yamano K, Wang C, Sarraf S A, Münch C, Kikuchi R, Noda N N *et al* 2018 Endosomal Rab cycles regulate Parkin-mediated mitophagy *eLife* **7** e31326

Yamano K and Youle R J 2013 PINK1 is degraded through the N-end rule pathway *Autophagy* **9** 1758–69

Yamashita S, Abe K, Tatemichi Y and Fujiki Y 2014 The membrane peroxin PEX3 induces peroxisome-ubiquitination-linked pexophagy *Autophagy* **10** 1549–64

Yan Z, Su G, Gao W, He J, Shen Y, Zeng Y *et al* 2019 Fluid shear stress induces cell migration and invasion via activating autophagy in HepG2 cells *Cell Adh. Migr.* **13** 152–63

Yanes B and Rainero E 2022 The interplay between cell-extracellular matrix interaction and mitochondria dynamics in cancer *Cancers* **14** 1433

Yang H, Ni H-M, Guo F, Ding Y, Shi Y-H, Lahiri P *et al* 2016 Sequestosome 1/p62 protein is associated with autophagic removal of excess hepatic endoplasmic reticulum in mice *J. Biol. Chem.* **291** 18663–74

Yang Q, Zhao J, Chen D and Wang Y 2021 E3 ubiquitin ligases: styles, structures and functions *Mol. Biomed.* **2** 23

Yang T, Park C, Rah S-H and Shon M J 2022 Nano-precision tweezers for mechanosensitive roteins and beyond *Mol. Cells* **45** 16–25

Yang X and Sun S 2015 Targeting signaling factors for degradation, an emerging mechanism for TRAF functions *Immunol. Rev.* **266** 56–71

Yang Y, Feng L-Q and Zheng X-X 2011 Microtubule and kinesin/dynein-dependent, bi-directional transport of autolysosomes in neurites of PC12 cells *Int. J. Biochem. Cell Biol.* **43** 1147–56

Yang Z and Klionsky D J 2010 Mammalian autophagy: core molecular machinery and signaling regulation *Curr. Opin. Cell Biol.* **22** 124–31

Ye J, Montero M and Stack Jr B C 2013 Effects of fusaric acid treatment on HEp2 and docetaxel-resistant HEp2 laryngeal squamous cell carcinoma *Chemotherapy* **59** 121–8

Yim W W-Y and Mizushima N 2020 Lysosome biology in autophagy *Cell Discov.* **6** 6

Yin H, Han S, Chen Y, Wang Y, Li D and Zhu Q 2020 T-2 Toxin induces oxidative stress, apoptosis and cytoprotective autophagy in chicken hepatocytes *Toxins* **12** 90

Yoon S-O, Shin S and Mercurio A M 2005 Hypoxia stimulates carcinoma invasion by stabilizing microtubules and promoting the Rab11 trafficking of the $\alpha6\beta4$ Integrin *Cancer Res.* **65** 2761–9

Yoshida Y, Yasuda S, Fujita T, Hamasaki M, Murakami A, Kawawaki J *et al* 2017 Ubiquitination of exposed glycoproteins by SCF^{FBXO27} directs damaged lysosomes for autophagy *Proc. Natl Acad. Sci. USA* **114** 8574–9

Youle R J 2019 Mitochondria—striking a balance between host and endosymbiont *Science* **365** eaaw9855

Youle R J and Van Der Bliek A M 2012 Mitochondrial fission, fusion, and stress *Science* **337** 1062–5

Yu H, Mouw J K and Weaver V M 2011a Forcing form and function: biomechanical regulation of tumor evolution *Trends Cell Biol.* **21** 47–56

Yu J-L and Liao H-Y 2021 Piezo-type mechanosensitive ion channel component 1 (Piezo1) in human cancer *Biomed. Pharmacother.* **140** 111692

Yu L, Chen Y and Tooze S A 2018 Autophagy pathway: cellular and molecular mechanisms *Autophagy* **14** 207–15

Yu S and Melia T J 2017 The coordination of membrane fission and fusion at the end of autophagosome maturation *Curr. Opin. Cell Biol.* **47** 92–8

Yu Z, Zhao L-X, Jiang C-L, Duan Y, Wong L, Carver K C *et al* 2011b Bafilomycins produced by an endophytic actinomycete Streptomyces sp. YIM56209 *J. Antibiot.* **64** 159–62

Zachari M, Gudmundsson S R, Li Z, Manifava M, Cugliandolo F, Shah R *et al* 2019 Selective autophagy of mitochondria on a ubiquitin-endoplasmic-reticulum platform *Dev. Cell* **50** 627–643.e5

Zaffagnini G, Savova A, Danieli A, Romanov J, Tremel S, Ebner M *et al* 2018 p62 filaments capture and present ubiquitinated cargos for autophagy *EMBO J.* **37** e98308
Zaidel-Bar R 2013 Cadherin adhesome at a glance *J. Cell Sci.* **126** 373–8
Zanotelli M R, Goldblatt Z E, Miller J P, Bordeleau F, Li J, VanderBurgh J A *et al* 2018 Regulation of ATP utilization during metastatic cell migration by collagen architecture *Mol. Biol. Cell* **29** 1–9
Zanotelli M R, Zhang J and Reinhart-King C A 2021 Mechanoresponsive metabolism in cancer cell migration and metastasis *Cell Metab.* **33** 1307–21
Zechner R, Madeo F and Kratky D 2017 Cytosolic lipolysis and lipophagy: two sides of the same coin *Nat. Rev. Mol. Cell Biol.* **18** 671–84
Zellner S, Schifferer M and Behrends C 2021 Systematically defining selective autophagy receptor-specific cargo using autophagosome content profiling *Mol. Cell* **81** 1337–1354.e8
Zemirli N, Boukhalfa A, Dupont N, Botti J, Codogno P and Morel E 2019 The primary cilium protein folliculin is part of the autophagy signaling pathway to regulate epithelial cell size in response to fluid flow *CST* **3** 100–9
Zeng , Huang S, Zheng W, Liyan Cheng X, Zhang Z, Wang J *et al* 2021 Proteolysis targeting chimera (PROTAC) in drug discovery paradigm: recent progress and future challenges *Eur. J. Med. Chem.* **210** 112981
Zhang J and Ney P A 2009 Role of BNIP3 and NIX in cell death, autophagy, and mitophagy *Cell Death Differ.* **16** 939–46
Zhang J, Tripathi D N, Jing J, Alexander A, Kim J, Powell R T *et al* 2015a ATM functions at the peroxisome to induce pexophagy in response to ROS *Nat. Cell Biol.* **17** 1259–69
Zhang L, Sheng R and Qin Z 2009 The lysosome and neurodegenerative diseases *Acta Biochim Biophys Sin (Shanghai)* **41** 437–45
Zhang L, Zhang S, Yao J, Lowery F J, Zhang Q, Huang W-C *et al* 2015b Microenvironment-induced PTEN loss by exosomal microRNA primes brain metastasis outgrowth *Nature* **527** 100–4
Zhang Y, Jiang X, Deng Q, Gao Z, Tang X, Fu R *et al* 2019 Downregulation of MYO1C mediated by cepharanthine inhibits autophagosome-lysosome fusion through blockade of the F-actin network *J. Exp. Clin. Cancer Res.* **38** 457
Zhang Y, Li H, Chang H, Du L, Hai J, Geng X *et al* 2018 MTP18 overexpression contributes to tumor growth and metastasis and associates with poor survival in hepatocellular carcinoma *Cell Death Dis.* **9** 956
Zhao B, Wei X, Li W, Udan R S, Yang Q, Kim J *et al* 2007 Inactivation of YAP oncoprotein by the Hippo pathway is involved in cell contact inhibition and tissue growth control *Genes Dev* **21** 2747–61
Zhao B, Ye X, Yu J, Li L, Li W, Li S *et al* 2008 TEAD mediates YAP-dependent gene induction and growth control *Genes Dev* **22** 1962–71
Zhao J, Zhang J, Yu M, Xie Y, Huang Y, Wolff D W *et al* 2013 Mitochondrial dynamics regulates migration and invasion of breast cancer cells *Oncogene* **32** 4814–24
Zhao L, Zhao J, Zhong K, Tong A and Jia D 2022 Targeted protein degradation: mechanisms, strategies and application *Sig. Transduct. Target. Ther.* **7** 113
Zhao Y G and Zhang H 2019 Autophagosome maturation: an epic journey from the ER to lysosomes *J. Cell Biol.* **218** 757–70
Zheng J, Chen X, Liu Q, Zhong G and Zhuang M 2022 Ubiquitin ligase MARCH5 localizes to peroxisomes to regulate pexophagy *J. Cell Biol.* **221** e202103156

Zheng Y T, Shahnazari S, Brech A, Lamark T, Johansen T and Brumell J H 2009 The adaptor protein p62/SQSTM1 targets invading bacteria to the autophagy pathway *J. Immunol.* **183** 5909–16

Zhou C, Li A H, Liu S and Sun H 2021 Identification of an 11-autophagy-related-gene signature as promising prognostic biomarker for bladder cancer patients *Biology* **10** 375

Zhou C, Ma K, Gao R, Mu C, Chen L, Liu Q *et al* 2017 Regulation of mATG9 trafficking by Src- and ULK1-mediated phosphorylation in basal and starvation-induced autophagy *Cell Res.* **27** 184–201

Zhou Y, Huang T, Cheng A, Yu J, Kang W and To K 2016 The TEAD family and its oncogenic role in promoting tumorigenesis *Int. J. Mol. Sci.* **17** 138

Zientara-Rytter K and Subramani S 2016 Role of actin in shaping autophagosomes *Autophagy* **12** 2512–5

Zomer A, Maynard C, Verweij F J, Kamermans A, Schäfer R, Beerling E *et al* 2015 *In Vivo* imaging reveals extracellular vesicle-mediated phenocopying of metastatic behavior *Cell* **161** 1046–57

IOP Publishing

Physics of Cancer, Volume 6 (Second Edition)
Cellular mechanisms to foster or fight cancer
Claudia Tanja Mierke

Chapter 6

Cancer as a mechanobiological disease: targeting the mechanism of mechanical plasticity and dynamical mechanoregulation

6.1 Summary

Since the incidence of cancer is expected to increase considerably over time, non-traditional areas of research are being scrutinized by scientists to identify new therapies. The field is developing rapidly, and many avenues initially appear logical and promising. On closer analysis, however, it becomes clear that the phenomenon is quite complicated and that not just one perspective is sufficient to fully understand it. An up-and-coming field that is receiving more and more attention is that of cellular mechanical mechanisms. In a general discussion of the physical characteristics of cancer, it was discussed whether specific types of cancer could be characterized as stiffer or softer. There are many articles that advocate for both perspectives, pointing out that cancer is not overly governed. This assumption is not shared by many scientists; in fact, the hypothesis is put forward that the progression of cancers is highly regulated and that most cancers stick to general schemes of malignant progression. A new perspective is that cancer is instead highly adaptable so that it can withstand the ever-changing microenvironments that cancer cells experience, like tumor compression and shear forces within the vasculature and body. What enables cancer cells to adapt in this way is the special proteins that comprise the mechanical framework that gives rise to a specific mechanical program within the cancer cell. Serendipitously, several of these proteins, including myosin II, α-actinin, filamin, and actin, have either changed expression in cancer and/or have been directly implicated in carcinogenesis. For this purpose, orientation towards the mechanical properties of the system could be a promising therapeutic strategy for the future. Nevertheless, the focus on the mechanical part is anything but trivial and is still being developed. It seems clear, nevertheless, that the mechanical characteristics play a crucial role in cancers and cannot be neglected, since there is an interaction between biological and mechanical

doi:10.1088/978-0-7503-4007-6ch6

characteristics that is of a complicated nature. As much as the mechanical program is engaged in cancer development and metastasis, it also drives a plethora of other essential cellular processes, like cell division, cell adhesion, metabolism, autophagy and motility. Consequently, the approach of targeting cancer based on the mechanical phenotype must carefully circumvent potential adverse effects and, more importantly, must elucidate how a change in the mechanical phenotype will translate into a benefit for cancer patients. This chapter presents and discusses the potential of focusing on the mechanical system, while also highlighting the hurdles and deficiencies of this type of strategy for cancer therapy.

6.2 Introduction to the mechanobiology of cancer

Cancers are a unique class of diseases where the pathology is primarily driven not by environmental, pathogenic or parasitic insults, but rather by unfavorable mutations that transform a normal cell into a malignant one. The adaption of these genetic mutations of cancer cells is termed the somatic mutations theory (Weinberg 2013). This genetic alteration of cells is not uniform in cancers and can take many forms. Not negligible is the epileptic alteration of cancer cells, which is also strongly influenced by mechanical alterations (Lin *et al* 2015, Mierke 2019, 2020, Nguyen *et al* 2022). When cell work and the various influencing variables such as 'biological' energy, information, matter and cell mechanics are considered as the central characteristic of the cell, specific gene mutations cannot be considered as the sole underlying causal mechanism for the development and progression of cancer (Hanselmann and Welter 2022). This kind of reductionist, monocausal perspective fails to consider the dynamic and highly intricate system of a cell. It can thus be stated that each of the factors, such as energy, information, matter and cell mechanics is able to alter the cell's functioning and therefore the cell's order to such an extent that it can become a cancer cell. Complex biological and medical issues need to be resolved to fully grasp how cancer develops, metastasizes and which factors enhance or repress its cancerous traits. Pathologists can identify various stages of cancer by examining cancerous tissue and identifying the cellular shape and tissue architecture. These analyses only provide information about a specific tumor condition at a specific point in time, but not about its dynamic progression. They are also subject to fluctuations based on the sample collection site within the broad tumor. The morphology and form of cancer cells is governed by multiple factors like intracellular tension and a modified nuclear function. The cytoskeleton is mainly controlled by the cortical cytoskeleton, which aids in determining the cancer development stage. These components are governed by mechanical cues. Notably, a mechanical program gives these cancer cells the capacity to adjust to and maneuver in the ever-changing mechanical environments, such as the tumor microenvironment (TME). The mechanical environment and the mechanical stresses that the TME exerts on cancer cells undergo continuous evolution throughout cancer progression (figure 6.1). These signals, for their part, influence gene expression and the progression of the cancer. The mechanical program endows the cancer cells with the capacity to perpetually sense and react

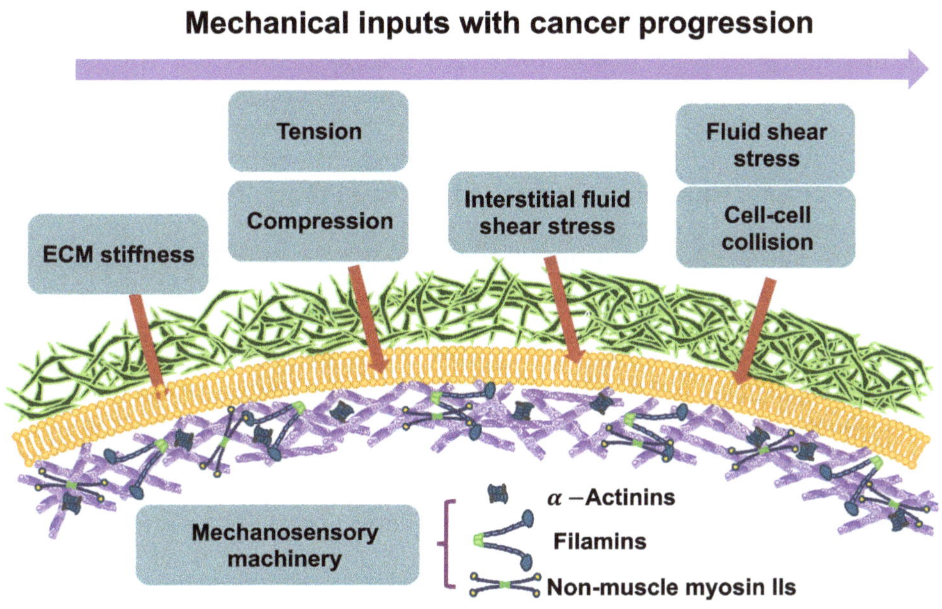

Figure 6.1. The crosstalk between the extrinsic mechanical milieu and the intrinsic mechanosensitive mechanisms. Throughout the various stages of cancer development, cancer cells are exposed to different external mechanical cues, beginning with the stiffness and tension of the extracellular matrix (ECM), but these mechanical cues broaden progressively to incorporate fluid shear stress and cell–cell collisions and cell–cell interactions during cancer development. These forces exert a significant influence on gene expression and the progression of cancer cells. Intrinsically, these cells also possess a mechanosensitive mechanism that equips them with the capability to perceive and react to these mechanical signals.

to mechanical signals, and possibly functions as a communication hub with other cellular signal transduction routes that also fuel cancer evolution.

For the scope of this chapter, it is important to note that the mechanical program involves two classes of actin-associated mechanical proteins. One protein group can carry mechanical loads without assembling in reaction to mechanical stresses, which is referred to as the non-mechanoresponsive group. Conversely, a separate group of proteins carries the load and assembles locally in reaction to mechanical stresses, which is referred to as the mechanoresponsive group (figure 6.1). An example of these distinct functions is Dictyostelium, in which myosin II, α-actinin, cortexillin, and filamin perceive and react to mechanical stresses. Conversely, other actin-associated proteins like dynacortin, coronin, and fimbrin can withstand strain and are not subject to stress-dependent accumulation (Luo *et al* 2013). Deleting these load-bearing actin crosslinks relocates the applied stresses on the mechanoresponsive proteins like myosin II, enabling them to build up more in reaction to lower stresses. In humans, certain paralogs of α-actinins (ACTN4), filamins, such as filamin B (FLNB), and non-muscle myosin II proteins (NMII) are mechanoresponsive, whereas certain sister paralogs are not mechanoresponsive. The mechanoresponsivity of NMIIB is a unique case in that it is cell-type specific. Remarkably, the expression rates of at least one of these mechanoresponsive proteins are increased in numerous types of cancer (Surcel *et al* 2019, Parajón *et al* 2021). In multiple cancers, the expression profiles of mechanosensitive protein families are frequently and profoundly modified in a highly paralogue-specific fashion, such as mechanosensitive isoforms exhibiting a tendency towards elevated expression levels, while non-mechanosensitive isoforms frequently exhibit reduced or constant expression patterns that correlate with cancer progression. Alterations in the expression levels of very low abundance, even mechanoresponsive isoforms mirror a complete reprogramming of cancer cells that promotes the enhanced adaptive flexibility (plasticity) needed for efficient growth and cancer metastasis (Surcel *et al* 2019). Consequently, these alterations recapitulate a repurposing of cancer to foster elevated mechanical adaptability, which is required for growth and metastasis. Mechanoresponsive isoforms are predicted on the basis of their distinct biochemistry, including actin-binding affinities and the dynamic nature of the bipolar myosin filament arrangement (Luo *et al* 2012, 2013, Schiffhauer *et al* 2016). This differential biochemistry can be utilized to design cancer-fighting chemical assays that act isoform-specifically. While most cancer therapies are developed to interfere with target structures, the mechanoresponsive apparatus offers a unique target space, where activation of key players can decrease metastatic burden without affecting healthy tissues. This targeted displacement of the adaptive system from its optimal (so-called sweet spot) activity range, which is the state that maximizes adaptability and consequently proliferation, invasion and metastasis capacity, demands a deep mechanobiome comprehension. This chapter explains how cancer can be viewed as a biophysical disease in which the mechanical characteristics of the environment and the cells themselves impose various stages of cancer development. The emphasis is on the mechanical behavior that provides the level of adaptability required for cancer cells to travel through intricate mechanical environments, and the targeting accuracy of this mechanism, as well as possible hurdles for future development of cancer therapies, are addressed.

6.3 The tumor mechanical microenvironment is subject to dynamic changes

The TME comprises a dynamic ecosystem of cancer cells enveloped by the ECM and contiguous stromal cells. The TME harbors chemical cues that can drive cancer cell proliferation and invasion through modifying their mechanical characteristics and modes of behavior. It has been demonstrated in multiple investigations that the incidence of invasion and metastasis is elevated in the setting of intratumoral hypoxia (Brahimi-Horn *et al* 2007, Vaupel *et al* 2007). Cancer cells react to hypoxia via elevation of hypoxia-inducible factors (HIFs) 1 and 2 levels. The upward regulation of HIFs is positively associated with metastases and a bad prognosis in breast cancer (Bos *et al* 2003, Semenza 2010). Chronic hypoxia also leads to HIF-dependent mortality of lung endothelial cells and enhances pulmonary microvascular permeability, thereby making the lung susceptible to metastases (Reiterer *et al* 2019). An additional investigation demonstrated that the increased movement and invasion of breast cancer cells under hypoxic circumstances are encouraged through the HIF-driven RhoA-ROCK1 route. These conditions enhance phosphorylation of myosin phosphatase target subunit 1 (MYPT1) and myosin light chain (MLC), resulting in alterations in the integral mechanical responses of these cells, involving cell contraction, the establishment of focal adhesions, and matrix contraction (Kitazawa *et al* 2003). In addition, other chemical cues, such as the signal transduction of stress hormones via β-adrenergic receptors, may also modify the mechanics of cells participating in the tumor and TME. Soluble stress hormones can interact with Arp2/3 and cause actin filament restructuring, which consequently leads to enhanced stiffness and diminished deformability of macrophages. These alterations may impact the migratory and phagocytic capacity of these cells (Kim *et al* 2019). Remarkably, activation of β-adrenergic signal transduction paths also decreases deformability in several cancer cell lines, encompassing ovarian, prostate, melanoma, and leukemia cells. This decrease is linked to the reorganization of the actin cytoskeleton and the activity of myosin II (Kim *et al* 2016b). These results explain the consistent finding that β-adrenergic signaling is highly involved in increasing metastasis and the progression of cancer (Sloan *et al* 2010, Creed *et al* 2015, Le *et al* 2016).

Aside from chemical cues, the TME also contains constant physical and mechanical influences, thereby constituting the mechanical microenvironment. Notably, ECM inside the TME is usually regarded as a by-product of tumorigenesis and is actively involved in facilitating cancer growth and proliferation, increasing invasion and spreading, and altering the gene expression signature of cancer cells, which in turn has a far-reaching impact on cancer advancement (Northcott *et al* 2018).

When a tumor forms, specific cues stimulate enhanced matrix laydown, thereby elevating the total stiffness of the ECM enveloping the tumor. The growing tumor progressively sustains the forces exerted by its environment, and in turn, the tumor is subjected to compression and tension, resulting in the activity of several growth factors favoring tumor growth (Yang *et al* 2000, Kalli *et al* 2022). At the same time, the ECM can be restructured due to tumor compression, further increasing the tension on the cancer cells (Levental *et al* 2009). In the event of an infiltration, the invading cancer cells have to battle against a progressively stiffened ECM as they make their way towards the vascular system (Wyckoff *et al* 2000, Condeelis and Pollard 2006). During this process, these cells remain subject to various mechanical forces, which vary from interstitial compression up to shear forces. When cancer cells migrate and intravasate into the circulatory system, they are subjected to shear forces exerted by other neighboring cells in the vascular system and also through hydrodynamic flow (Kumar and Weaver 2009, Yu *et al* 2011, Chen *et al* 2013, Mierke 2024a). Thereby the endothelial cell lining of the vessels fulfills mechanosensory and mechanoregulating functions (Mierke 2024b). There is growing evidence that the tumor's mechanical microenvironment can control and redirect the development, proliferation and invasion of cancers. Moreover, the mechanical environmental cues can impact the mechanical characteristics of cancer cells and tumor-associated immune cells. There are several lines of evidence that support the notion that the mechanical niche can substantially drive cancer progression (Das *et al* 2020, Amos and Choi 2021, Riehl *et al* 2021). In addition, the mechanical microenvironment is a continuously changing terrain during cancer advancement, which constantly prompts the cancer cells to adjust to these mechanical stimuli.

6.4 The stiffness of the substrate impacts carcinogenesis

Substrate stiffness ranks among the most important mechanical cues for cells and is mainly determined by the constitution of the ECM enclosing the primary tumor (Swaminathan *et al* 2011, Jain *et al* 2020, Liu *et al* 2020a; Riehl *et al* 2021). There is growing experimental support for the concept that substrate stiffness regulates multiple signaling events and functions as a driving force in carcinogenesis. For instance, elevated ECM stiffness causes increased invasion and metastasis within breast cancer cells in three-dimensional (3D) models (Levental *et al* 2009, Chaudhuri *et al* 2014, Peela *et al* 2016). Additional investigations of colorectal cancer support the idea that matrix stiffness fosters tumorigenesis and the advancement of cancer (Kim 2009, Krndija *et al* 2010, Baker *et al* 2013). In particular, enhanced activity of lysyl oxidase, which acts as a catalyst for collagen crosslinking, results in elevated substrate stiffness and promotes the advancement of colorectal cancer (Baker *et al* 2013). The researchers also made similar observations in patients with glioblastoma, pancreatic cancer and ovarian cancer (Ulrich *et al* 2009, Swaminathan *et al* 2011, Rice *et al* 2017). Importantly, substrate stiffness can affect ECM mechanical characteristics through modifying cellular gene expression. The expression of integrins, which is controlled by YAP signal transduction, matrix metalloproteases and ECM proteins, rises with ascending substrate stiffness (Nukuda *et al* 2015). The important point to note is that YAP signal transduction alone may not convey its impact on substrate stiffness in each type of cancer, and YAP exhibits a cell-context-dependent function in mechanotransduction. For instance, in a mechanotransduction investigation utilizing adjustable 3D cell cultures, it was determined that breast cancer mechanotransduction proceeds irrespective of YAP signal transduction (Lee *et al* 2019b). Independently, this positive coupling between substrate stiffness and ECM mechanics emphasizes the crosstalk between substrate stiffness, external mechanical forces, and the expression of genes. Consequently, the bi-directional interaction between cancer cells and their microenvironment can alter the ECM mechanics in a possibly dynamic interactive manner. Thus, not only do the cancer cells themselves advance to a malignant state, but also the microenvironment of cancers undergoes significant changes contributing to the malignant progression of cancers.

Many cytoskeletal signaling and regulation pathways are directly altered by substrate stiffness during tumor progression. For instance, ROCK1, a modulator of myosin II activity, and ROCK2, a modulator of cofilin, are involved in controlling the motility of breast cancer cells in a substrate stiffness-dependent way (Peng *et al* 2019). Utilizing polyacrylamide gels that imitate various breast tissue substrate stiffnesses, it has been established that cancer cells cultivated on stiff substrates exhibit greater migration and invasiveness than cancer cells grown on soft substrates. These variations are accompanied with increased expression of ROCK1 and

ROCK2, as well as increased activation of their respective downstream targets. Additional work highlights the key function of myosin II, which is a central cytoskeletal protein, in enhancing oncogenic transformation on stiff substrates. In this transformation scenario, the transforming growth factor-β1 (TGF-β1) signal transduction route reacts to increased substrate stiffness and turns out to be a key factor for oncogenic proliferation in hepatocellular carcinomas and breast cancer cells. In addition, the treatment with blebbistatin, an inhibitor of myosin II, inhibits the increase in the TGF-β1 signal transduction route (Pang *et al* 2017, Lin *et al* 2018).

6.5 Several mechanical forces contribute to cancer advancement

Besides substrate stiffness, carcinogenesis is also significantly impacted by a number of other mechanical factors in the environment (Tang *et al* 2020, Amos and Choi 2021, Riehl *et al* 2021). A topical investigation has shown that the invasion of cancer cells increases with the degree of mechanical restriction. In particular, the migration velocity of colon and breast cancer cells is faster through tight microchannels compared to wide channels, both of which imitate the different dimensions of the TME restriction (Holle *et al* 2019). In addition, hemodynamic forces, immunological stress, collisional forces, and fluid forces have a profound effect on the gene expression signature of cells throughout invasion. Shear forces were observed to work through ligand-dependent activation and phosphorylation of the MAPK/PI3K/Akt signal transduction route, which plays a key part in the proliferation of cancer cells (Laakkonen *et al* 2007, Maimari *et al* 2016). However, it is classically assumed that shear forces cause death of cancer cells circulating in the bloodstream, intermediate shear forces can ease the adhesion and extravasation of circulating tumor cells (CTCs) and thus increase their metastatic capacity (Headley *et al* 2016, Follain *et al* 2018).

In addition to the shear forces that occur in the vascular system, the fluidic forces in the mechanical microenvironment of the tumor can also influence the advancement of cancer, particularly regarding invasion and metastasis. It was revealed that hydrodynamic forces generated within microfluidic platforms promote the epithelial-mesenchymal transition (EMT), a crucial metastatic step, in ovarian cancer cells (Rizvi *et al* 2013). Fluid forces enhance the invasion of glioblastoma cells via CXCR4 signal transduction, which is a widely examined modulator of glioma invasion. Cells exposed to high fluid pressure closely match the aggressiveness of cancer in patient-derived glioma stem cells (Cornelison *et al* 2018). In addition, a related investigation indicates that the tissue fluidity of glioblastoma cells confers the capacity to penetrate effectively into normal brain matter, which promotes their aggressive potential. The fluid characteristic of the cancer mass as determined using magnetic resonance elastography shows a significant positive relation with the proliferative capacity of these cells (Streitberger *et al* 2020). Overall, cancer cells experience a wide variety of mechanical forces as they transform and as the disease progresses. These external forces exert a powerful effect on cancer cell development, advancement and aggressiveness. Yet there is also an interplay between the mechanical microenvironment and intracellular gene expression and signal transduction events. These cancer cells depend not only on their physical environment. There is growing support for the idea that an intrinsic mechanical response program endows these cells with the capacity to continuously perceive and adjust to their environments.

6.6 Are tumors made up of soft cancer cells that promote the malignant progression of cancers?

Stiffness refers to how deformable a particular material becomes under a specified amount of applied force (Baumgart 2000, Guimarães *et al* 2020). The modulus of elasticity or Young's modulus (E) serves as a unit of measurement for this characteristic. The corresponding unit used is Pa (or N m^{-2}), since it is obtained from stress divided by strain. Stress and stiffness thus have the identical units, since stress is a force per unit area (N m^{-2} or Pa) and strain refers to the normalized deformation and is unitless. Even though Young's modulus represents the most important measure of elasticity in biological systems, the shear modulus (G), storage modulus (G') and loss modulus (G'') are also frequently analyzed (Guimarães *et al* 2020). The approximate conversion from Young's modulus to the shear modulus can be calculated as $E \approx 3G$. The storage modulus equals the shear modulus at low frequencies, therefore $E \approx 3G \approx 3G'$. The loss modulus refers to the viscous (non-elastic) characteristics of a substance.

Cancer cells subjected to softer microenvironments are better at populating secondary tissues (Filipe *et al* 2024). Tissue stiffness is primarily determined through the ECM. An increased accumulation of ECM fibers and enhanced crosslinking of these fibers are associated with stiffer tissues (Butcher *et al* 2009). Matrix density depends on fibroblasts, which either deposit or break down collagen fibers or reorganize the surrounding ECM in response to microenvironmental stimuli. The level of crosslinking is controlled through enzymes like lysyl oxidases (LOX), which catalyze the formation of covalent bonds linking ECM proteins (Baker *et al* 2013). Non-enzymatic collagen crosslinking like glycation causes stiffening of the matrix as well (Ansari and Rasheed 2010). Additional mechanisms of tissue stiffening involve ECM fiber orientation, the interstitial fluid pressure, strain stiffening arising from forces applied by cells, and cell jamming (Mohammadi and Sahai 2018). Tissue stiffness values vary from 500 Pa for brain tissue to 13–20 GPa for cortical bone tissue (Rho *et al* 1993). Stromal tissue lies in the middle range between 1 and 10 kPa (Butcher *et al* 2009). Increased cell survival is revealed to be fueled through augmented fatty acid metabolism in triple-negative breast cancer (TNBC) cells that are placed in softer extracellular microenvironments. Evidence is presented that disengagement of cellular mechanosensory function through an integrin β1-blocking antibody prompts stiff, preconditioned TNBC cells to adopt the behavior of their soft equivalents, as demonstrated both *in vitro* and *in vivo*. This is the first piece of work to reveal that softer tumor microenvironments can modify disease progression by imparting increased metabolic flexibility and subtle survival benefits to TNBC cells.

ECM is recognized to be heavily deregulated in cancer, with aberrant expression, laydown, and remodeling of a panoply of ECM constituents that collectively change the conduct of both cancer and stromal cells and drive disease advancement (Cox 2021). In addition to the biochemical consequences of these ECM alterations in cells, the tumor ECM also has important biophysical implications that modify cellular behavior inside the primary tumor. Several solid tumors display substantial alterations in local forces and viscoelastic characteristics of the tissue, either spatially across a single tumor or over time as the disease advances (Butcher *et al* 2009, Er *et al* 2022). In fact, enhanced breast tissue stiffness is known to coincide with the initiation and advancement of breast cancers (Levental *et al* 2009, Boyd *et al* 2014), which is mainly due to the enhanced synthesis and deposition of ECM constituents (Cox 2021). In this context, research has also uncovered important biologically meaningful consequences of alterations in biophysical characteristics, ranging from the impact of stiffness on the cancerous phenotypic state of cells (Paszek *et al* 2005, Wu *et al* 2023), to modulation of EMT (Wei *et al* 2015, Rice *et al* 2017), increased invasiveness of cancer cells (Levental *et al* 2009, Pathak and Kumar 2012, Haage and Schneider 2014), and acquisition of resistance to chemotherapy (Shin and Mooney 2016, Rice *et al* 2017). Frequently, the cellular contractility is also altered in concert with the environmental stiffness, which additionally promotes the malignant progression of cancers (Haage and Schneider 2014). Most existing results, nevertheless, pertain to the impact of stiffness on cells within the primary tumor, and the more long-term implications of primary tumor biomechanics on cancer cells as they exit the primary tumor, proceed to metastatic dissemination, survive, and obviously colonize at secondary locations remain to be explored.

Over the past few years, the importance of cellular energy during cancer evolution has been gaining more and more attention. In fact, cancer cells are recognized to adapt their metabolic program to best meet the needs of every distinct phase of the metastatic dissemination cascade (Vander Heiden *et al* 2009, Faubert *et al* 2020, Bergers and Fendt 2021, Martínez-Reyes and Chandel 2021). Specifically, in TNBC, although there is diversity in the prevailing metabolic phenotype of cancer cells across the primary tumor, the rate of glycolytic reliance is more predominant in the triple-negative subtype group compared to other breast cancer subtypes (Choi *et al* 2013, Kim *et al* 2013, Pelicano *et al* 2014, Lanning *et al* 2017, Gong *et al* 2021). Like other types of cancers, glycolysis is frequently employed in the proliferative phases of tumor growth, where glycolytic intermediates support the generation of essential building components for cell proliferation such as lipids, nucleotides and amino acids, thereby promoting the enhanced biomass generation that is characteristic and essential to the advancement of cancer (Vander Heiden *et al* 2009, Faubert *et al* 2020, Martínez-Reyes and Chandel 2021). Particularly interesting, nevertheless, are the metabolic alterations that probably arise throughout the actions of metastatic spreading, secondary tissue seeding and survival (LeBleu *et al* 2014, Kim and DeBerardinis 2019, Faubert *et al* 2020, Parida *et al* 2022). In this context, the metabolic plasticity characteristics of TNBC (Lanning *et al* 2017, Roshanzamir *et al* 2022) permit the cells to adjust their cellular metabolism. In fact, metastatic TNBC lesions have been demonstrated to alter their metabolic signatures to become more like their target tissues, with a large accumulation of energy-producing pathways, such as oxidative phosphorylation and fatty acid metabolism, which are especially abundant in lung and liver metastatic lesions (Elia *et al* 2018, Roshanzamir *et al* 2022). These metabolic switches have been found to favor cell survival in allowing dispersed cancer cells to access alternative sources of energy, particularly in low-nutrient niches, while also generating oxidative stress that drives the generation of detoxification cofactors like nicotinamide adenine dinucleotide phosphate (NADPH) (Carracedo *et al* 2013, Koundouros and Poulogiannis 2020). This concept of metabolic plasticity is proposed to confer a decisive benefit during the phases of spread and survival of cancer cells in TNBC and even in other types of cancer in a broader sense (LeBleu *et al* 2014, Simões *et al* 2015). The underlying indicators that initiate and control this adjustment are, nonetheless, still being examined thoroughly. Employing experimental models of metastasis, survival, and outgrowth, the data reveal that the biomechanical characteristics of the primary TME crucially determine the metastatic potential of TNBC cells. Importantly, in both *in vitro* and *in vivo* models, a softer TME has been observed to equip cancer cells with improved survival mechanisms, which subsequently augments their metastatic colonization of secondary niches. Moreover, it turned out that this is tightly connected to the metabolic signature of cells that were preconditioned in soft or stiff microenvironments, which captures the divergence between healthy and tumor tissue as well as the heterogeneity inside tumors, with biomechanically driven remodeling of fatty acid oxidation. Consequently, this altered metabolic signature leads to an increased capacity to populate the secondary metastatic niche following their departure from the primary tumor. Ultimately, by blocking β1 integrin-facilitated mechanosensing, the priming actions of soft microenvironments in cancer cells can be mimicked in stiff microenvironments, resulting in enhanced fatty acid metabolism and improved survival in *in vitro* and *in vivo* models of metastatic engraftment. Consequently, there exists an overall benefit for cancer cells, such as those primed to become metastatic cancer cells, to face a mechanically softer environment in terms of malignant progression and survival.

Gaining an insight into the intricate mechanical heterogeneity of the primary TME and the manifold ways it affects cancer cell growth, metastasis, and secondary site dissemination in both the shorter and longer time scales is crucial for the design of new cancer therapeutics (Martinez-Outschoorn *et al* 2017). The work aimed to tackle this problem through a systematic evaluation of the behavior and viability of biomechanically preconditioned breast cancer cells in multiple *in vitro* and *in vivo* models that replicate the different phases of the metastatic dissemination cascade. Abundant evidence over the past decades has implicated that microenvironment stiffness, powered by modified matrix rearrangement including enhanced collagen density, crosslinking and stiffening (Cox 2021, Koorman *et al* 2022) is a determining feature for disease advancement (Wei *et al* 2015, Rice *et al* 2017). Pioneering work carried out nearly two decades earlier revealed for the first time that non-tumorigenic cells are able to adopt tumorigenic characteristics simply as a result of alterations in extracellular stiffness (Paszek *et al* 2005). Subsequent work revealed that high-density collagen inside the mammary fat pad of Col1a1 mutant mice is linked to enhanced stiffness and results in augmented tumor initiation and progression in comparison to wild-type mice (Provenzano *et al* 2008). This malignant transformation in reply to stiffness was subsequently traced to the mechanical activation of the Rho/ROCK signal transduction route (Wozniak *et al* 2003, Boyle *et al* 2020), resulting in a wide range of downstream signaling alterations involving YAP/TAZ translocation and Src-driven phosphorylation signaling alterations across a range of cellular signaling routes (Butcher *et al* 2009, Pickup *et al* 2014, Mohammadi and Sahai 2018). Recently, studies have even identified the effects of tumor stiffening on stromal cell populations, with two-way pro-tumorigenic impacts on cancer cells that subsequently promote disease advancement (Reid *et al* 2017, Hupfer *et al* 2021).

In contrast, there exist merely a handful of studies addressing the potentially tumor-promoting implications of soft microenvironments on disease advancement, demonstrating increased dedifferentiation, tumorigenicity, and chemoresistance of soft microenvironments in a panel of various cancers *in vitro*, comprising breast cancer (Romani *et al* 2022), neuroblastoma (Lam *et al* 2010), hepatocarcinoma (Schrader *et al* 2011), melanoma, lymphoma and ovarian cancer (Liu *et al* 2012, McGrail *et al* 2014). Increased formation of tumors *in vivo* was observed in cultured soft melanoma cells when they were inoculated into the tail veins of mice (Liu *et al* 2012). Most impressive is a report describing the emergence of invasive orthotopic breast tumors, which is followed by a surgical resection and a comprehensive mechanical characterization of the primary tumors, and the consequential implications on the metastatic load (Fenner *et al* 2014). After an observation period, it has been seen that mice with softer, more compliant primary tumors developed considerably more far-reaching metastatic lesions compared to mice bearing stiffer primary tumors. The hypothesis was put forward that the mice with softer tumors possibly had a higher number of tumor-causing cells, which led to increased local relapse and more extensive metastases (Fenner *et al* 2014). Notwithstanding these compelling and important reports, the specific biological alterations that arise in reaction to softened TMEs and in turn fuel pro-tumorigenic pathways in cancer cells are not clearly delineated. The study highlights the differential effect of primary tumor biomechanical cues on the various phases of the metastatic process and highlights the pleiotropic nature of β1 integrin in cancer evolution and metastasis. Significant metabolic changes were detected in breast carcinoma cells in reaction to alterations in the stiffness of the microenvironment, with alterations in mitochondrial metabolism and use of substrates (Filipe *et al* 2024). In studying the downstream biological implications of these metabolic alterations, it was revealed that softer microenvironments may endow breast cancer cells with superior survival mechanisms when subjected to high-stress situations *in vitro* and *in vivo*. In fact, these findings are consistent with previously published data demonstrating that treatment of triple-negative breast cancer cells with compounds that inhibit mitochondrial metabolism before intravenous administration of the cells decreased their metastatic capacity (Davis *et al* 2020). Moreover, it was found that soft primed cells not only store considerably more lipids in the shape of lipid droplets, but are also metabolically more active, with an elevated lipid metabolism through the citric acid cycle. Finally, elevated fatty acid oxidation capacity is associated with improved colonization potential. The importance of fatty acid oxidation in tumor advancement is progressively being acknowledged (Koundouros and Poulogiannis 2020, Li *et al* 2022, Lumaquin-Yin *et al* 2023), with many attributing the remarkable survival mechanisms activated in FAO-rich cells, especially under high-stress circumstances like anoikis, to the fact that cells rely progressively on non-glucose sources for ATP and antioxidant generation (Schafer *et al* 2009, Carracedo *et al* 2012, Menard *et al* 2016, Wang *et al* 2018b).

Ultimately, by using an integrin β1-inhibiting antibody to break cell–matrix interactions and mimic a softer microenvironment, it is apparent that cellular energetics of cells cultured on stiff matrices can be altered to resemble those on softer matrices. These findings are consistent with others that have shown β1 integrin blockade to have pro-metastatic effects in *in vivo* models of breast cancer metastasis (Truong *et al* 2014, Moritz *et al* 2021), albeit these trials lacked assessment of cellular energetics. There are, nonetheless, other reports that point to anti-tumorigenic impacts of β1 integrin blockage in breast cancer (Weaver *et al* 1997, Park *et al* 2006, 2008). These discrepancies are probably attributable to the time of drug administration and the specific disease phase studied, with the latter trials concentrating mainly on primary tumor growth and proliferation rather than metastatic colonization of secondary locations. Of particular note was the finding that the integrin β1-inhibiting antibody, while decreasing the presence of extensive proliferative cell colonies *in vitro*, elevated the frequency of smaller colonies three- to six-fold (Park *et al* 2006), which corroborates the results that integrin β1 blockade results in enhanced tumor generation. Approaches aimed at targeting β1 integrin therapeutically have not advanced in the clinic, and this work highlights the pleiotropic role of β1 integrin in cancer advancement and metastasis.

Gaining an insight into both the drivers and implications of biomechanical heterogeneity in tumors will provide a more holistic view of how cancer ecosystems operate. Together with the work of others (Plodinec *et al* 2012), this work demonstrates that remarkable biomechanical heterogeneity is inherent to any single cancers at any moment in time (Filipe *et al* 2024). The biomechanical heterogeneity is perceived by all cells inside the tumor, causing widespread cellular repurposing that finally determines the course of tumor advancement. The long-term implications of biomechanical reprogramming influencing cancer cell functions after they have exited the primary tumor site were illuminated.

The identification and sorting of highly cancerogenic and metastatic cancer cells within a heterogeneous population of cells is a formidable endeavor. Using microfluidic technologies, marker-based heterogeneous cancer stem cells (CSCs) can be separated into mechanically stiff and soft subpopulations (Lv *et al* 2021). The separated soft cancer cells (less than 400 Pa), but not the stiff cells (above 700 Pa), can establish a tumor in 100 immunocompetent mice per injection. Remarkably, while only the soft, but not the stiff cells, derived from $CD133^+$, $ALDH^+$ or side population CSCs can generate tumors in NOD-SCID or immunocompetent mice when only 100 cells are inoculated. The Wnt signal transduction molecule B-cell CLL/lymphoma 9-like (BCL9L) is elevated in soft cancer cells and modulates their stem cell characteristics and oncogenicity. In clinical studies, BCL9L expression levels are coupled with a poorer outcome. Based on these results, intrinsic softness can be considered a universal hallmark of highly cancerogenic and metastatic cancer cells (Lv *et al* 2021).

Cell stiffness has been linked to stem cell-like characteristics in both development and cancer. In this context, mechanical softness, which is linked to a high expression of the transcription factor BCL9L, is characterized as a universal marker for CSCs. Microfluidic sorting enables the isolation of cancer cells based on their mechanical stiffness:

- Soft cancer cells exhibit elevated tumorigenicity *in vivo*.
- BCL9L acts as a controller of stemness within soft cancer cells.
- BCL9L expression levels and softness of cancer cells are related to disease advancement in melanoma, breast and colorectal cancer patients.

The idea of CSCs or cancerogenic cells stems from the finding that a very small fraction of cells in a cancer can give rise to cancer in severe combined immunodeficiency (SCID) mice (Lapidot *et al* 1994, Al-Hajj *et al* 2003, Singh *et al* 2003, Hope *et al* 2004, O'Brien *et al* 2007, Quintana *et al* 2008, Schatton *et al* 2008). These tumorigenic cells seem to be largely responsible for therapy resistance and cancer recurrence, rendering them a compelling therapeutic target. Nevertheless, despite intensive investigations, the characteristics of this crucial subgroup of cancer cells continue to be largely elusive. Moreover, there are no strict methodologies for separating these cells out of a tumor, as the traditional cell surface biomarkers are inconsistent and vary widely across various types of cancer (Hope *et al* 2004, Dieter *et al* 2011). Therefore, the development of a technique that allows the accurate sorting and identification of cells with tumorigenic potential is highly warranted. The significance of the mechanical characteristics of a living cell for the performance and functionality of cells has been emphasized (figure 6.2) (Engler *et al* 2006, Chowdhury *et al* 2010, Urbanska *et al* 2017).

It is established that cells exert actomyosin-dependent contractile forces in reaction to progressive stiffening of the ECM (Discher *et al* 2005, Mierke *et al* 2008, Irianto *et al* 2016). Endogenous contraction of this kind can in turn increase cell stiffness (Wang *et al* 1993). In addition, to correctly perceive and react to the mechanical stimuli of the environment, the stiffness of a cell needs to be like that of the ECM (Discher *et al* 2005, 2009, Wu *et al* 2018), indicating that soft cells can survive in a soft stroma and stiff cells perform at their best in a stiff niche. In support, soft 3D fibrin matrices were found to promote H3K9 demethylation and enhance Sox2 expression and melanoma stem cell self-renewal, while stiff matrices exhibited contrary actions (Tan *et al* 2014). The H3K9 demethylation and increased Sox2 expression contribute to the softening of cells (Tan *et al* 2014). This finding raises the question of how soft cells can arise in a stiff tumor and tumor environment. At first glance, this seems to contradict the finding that a subset of soft cancer cells within a primary tumor can promote metastasis.

In line with this concept, cells with different levels of stiffness can cohabit in the same tumor tissue because of the heterogeneity of the tumor's mechanical

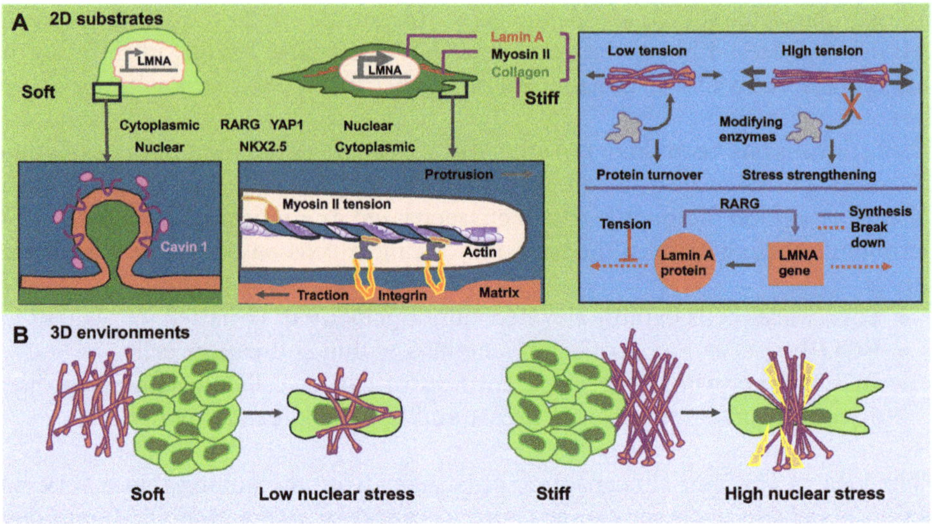

Figure 6.2. (A) Cell shape arises downstream of matrix stiffness and involves the concerted activity of multiple drivers. In F-actin-rich protrusions, F-actin anchors the matrix through integrin-tethered complexes, and more distally within the cell body, the F-actin scaffold also tethers myosin II filaments (Giannone *et al* 2007). Myosin II exerts tension on the surrounding matrix, and if the matrix is stiff and the adhesive ligand is present in abundance, then (1) integrin adhesions become stabilized, (2) myosin II tractions on the matrix continue to rise, and (3) a flattened lamellipodium adheres and propagates on a stiff matrix. When the matrix is soft, events (1) through (3) go awry and cells are unable to spread. The rearrangement of the plasma membrane in relation to the local environment is also supported by several lines of evidence, among them an elevated frequency of the cavin 1 caveolae factor due to elevated tissue stiffness (Swift *et al* 2013, Mierke 2024a). Regulatory signal transduction pathways also react to and mirror the surrounding local matrix microenvironment. For instance, important mesenchymal stem cell (MSC) transcription factors are subject to distinct subcellular localization according to the micro-stiffness of the underlying substrate. In cells cultivated on soft gels, RARG and YAP1 localize in the cytoplasm, while NKX2.5 can be found in the nucleus. When the cells are cultivated on stiff gels, instead, these factors translocate from the cytoplasm to the nucleus and vice versa. (Top right image) For ordinary control of lamin A, high tension in cells on stiff matrices places a stress on the nucleus and causes stabilization of the lamin A protein, which attracts a large quantity of RARG from the cytoplasm toward the nucleus (through a joint binding factor), allowing RARG to ramp up expression of lamin A both on stiff gels and within stiff tissues (Swift *et al* 2013). Mechanobiological gene circuitry can therefore be regulated by tension-suppressed dissociation and breakdown of filamentous systems in a 'use-it-or-lose-it' mechanistic fashion. Epigenetic mechanisms implicated in matrix mechanosensing are still an active area of research. (B) Invasive cancer cell migration into adjacent healthy tissue is a feature of multiple processes, and soft tissues with low collagen matrix content and wider pores are presumably more susceptible to this colonization. In fact, invasion across small pores in stiff matrices can induce high nuclear tension that is severe enough to breach the nuclear lamina and exacerbate DNA harm and cell death (Harada *et al* 2014). Cancerous genomic alterations are surprisingly well categorized in terms of the micro-stiffness scale of normal or host tissue. Changes in chromosome copy number and translocations in pediatric cancers (Chen *et al* 2014) tend to rise with tissue micro-stiffness, in the same way that somatic mutations increase in various cancers (Alexandrov *et al* 2013).

microenvironments (Plodinec *et al* 2012, Elosegui-Artola *et al* 2014). Despite this appreciation, direct demonstration that cellular softness acts as a fundamental characteristic for cancerogenic cells continues to be challenging. Therefore, a technique to segregate soft cells from stiff ones has been devised in a study and the proof that these soft cells are highly cancerous and have the capability to metastasize.

6.6.1 Stiffness and epithelial cancer invasion

The epithelial cancer invasion evolution and related physical clues are visualized in figure 6.3. The primary tumor releases matrix-modifying enzymes and activates stromal cells to communicate with its microenvironment. The tumor–stroma barrier stiffens, thereby protecting the cancer from outside factors like immune cells, chemical stimuli and medications. During the malignant progression of cancers, the cancer cells detach from their cell–cell adhesions and disrupt the basement membrane. Cancer-associated stromal cells reorganize the ECM to establish channels or tracks for the invasion of cancer cells. The matrix is also broken

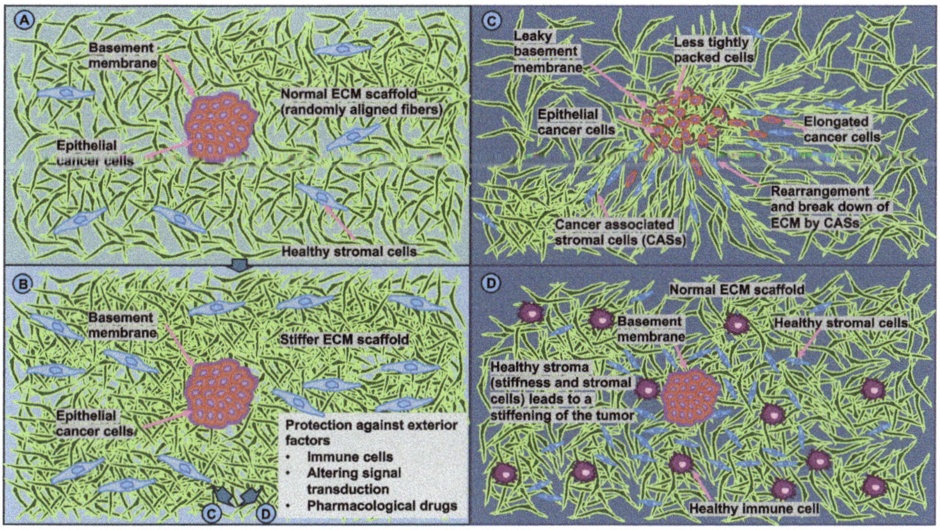

Figure 6.3. Development of incipient epithelial cancer invasion. (A) The cancer cells are embedded within a basement membrane. The healthy stroma surrounds the primary tumor and comprises randomly arranged collagen fibers and a variety of healthy stromal cells, including fibroblasts, healthy epithelial cells, adipocytes and additional organ-specific cells. (B) The cancer cells instruct the stromal cells to stiffen the ECM across the tumor–stroma boundary. This shields the primary tumor from external factors like immune cells, chemical stimuli and therapeutics. A stiffness gradient is created, with the ECM being stiffer in the vicinity than the ECM at a distance. The cancer can then develop either into (C) or (D). This relies on several factors that are not completely elucidated. (C) The cancer cells detach from their cell–cell adhesions and penetrate the basement membrane. Cancer-associated stromal cells aid in this process through restructuring the ECM to provide a trail for the cancer cells to infiltrate along. The matrix is also broken down at the local level to ease cell migration. The cancer cells alter their morphology, expand and increase in flexibility. The entire tumor increases in size and ceases to be round. The cancer cells then enter the vascular system and establish metastases. (D) The nearby ECM is disintegrated to decrease stiffness, whereby a direct decrease of crosslinking and breakdown of fibers through cancer cells occurs or through the impact of healthy stromal cells. The healthy environment suffocates the tumor. Immunocytes can then attack the tumor to keep it suppressed.

down on a local basis to facilitate cell migration and invasion. The cancer cells expand, polarize and become more plastic. The primary tumor becomes more elongated, and its rounded shape becomes irregular. Cancer cells invade the vasculature and begin to metastasize. Reintroduction of a healthy microenvironment, in contrast, through targeted manipulation of matrix stiffness or reprogramming or repopulation of healthy stromal cells has been demonstrated to effectively suppress tumor growth. Immune cells can then attack the remaining tumor cells to suppress it more effectively. A closer look at the biophysical and chemical interactions between cancer cells and the TME reveals the following (figure 6.4): First, the matrix stiffens in response to the activation of TGFβ within newly mutated cancer cells. TGFβ in turn can activate LOX enzymes, thus enhancing the level of crosslinking. TGFβ also influences intracellular signal transduction via the alteration of integrins (Baker *et al* 2013, Amendola *et al* 2019) and by inducing the mesenchymal marker Snail, leading to the repression of E-cadherin (Peinado *et al* 2005), which in turn results in the activation of EMT. EMT enables the cells to adopt a mesenchymal behavior, such as increased motility, and thus promotes invasiveness (Wei *et al* 2015). Stromal cells play a major role in reorganizing the TME. Cancer-associated fibroblasts (CAFs) are involved in the desmoplastic reaction, which causes extensive fibrotic reaction within tumors. CAFs guide the invasion of cancer cells in that they establish tracks of a reorganized matrix that cancer cells can migrate along (Gaggioli *et al* 2007). Vertically oriented collagen

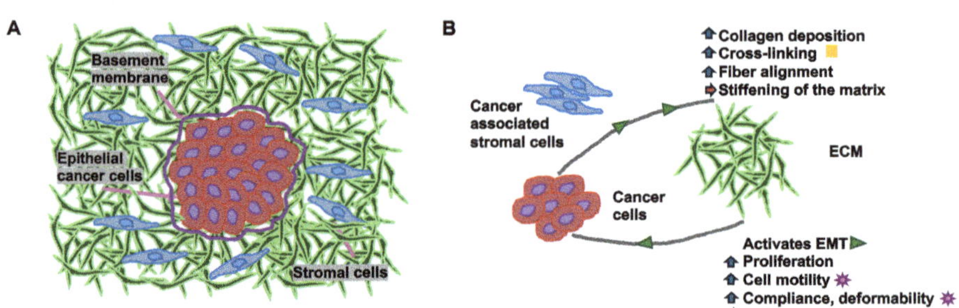

C

Proteins	Functions
Talin, Vinculin, Focal adhesion kinase (FAK)	Act in generation of focal adhesions underneath integrins enabling cells to perceive ambient mechanical signals
Lysyl oxidase (LOX)	Elevates crosslinking between the ECM fibers and thus decreases the pore size and the matrix stiffening
Matrixmetalloproteinase (MMPs)	Degrades ECM scaffold and facilitates the invasion of cancer cells with ease thrugh the fibrillar collagen network
Transforming growth factor-β (TGF-β)	Induces generation of matrix proteins and matrix-modifying enzymes like LOX and MMPs, which crosslink and deposit the ECM scaffold
Snail, Twist, Zeb	Act as mesenchymal markers and are increased during EMT
E-cadherin	Expressed in epithelial cells and is decrease is a hallmark of EMT
Caveolin-1 (Cav-1)	Induces the Rho GTPase
Rho GTPase	Governs cytoskeletal dynamics and intracellular force, therefore elevated expression is associated with soft, deformable cells

Figure 6.4. Stiffness and cancer invasion. (A) Schematic representation of an epithelial tumor, which is enclosed by a basement membrane and encased in a stroma. (B) Schematic representation of the interaction between cancer cells and the adjacent stroma. Both cancer cells and stromal cells cause stiffening of the matrix via enhanced collagen secretion, crosslinking, reshaping and orientation, resulting in stiffening of the matrix. The stiffened ECM stimulates cancer invasion through activation of EMT, thereby favoring proliferation and cellular motility, compliance and deformability. (C) Proteins associated with this metastatic process and their functions are listed. They can be linked to the schematic via the symbols.

fibers are associated with enhanced invasion because they provide a more efficient route for cells to migrate alongside them (Conklin *et al* 2011). It is a paradox that cancer cells seem to become softer and more deformable as a result of the overexpression of Rho GTPase (Fritz *et al* 1999). The deformability enhances the invasive capacity, as it allows the cells to move more easily within the matrix.

6.6.2 Microfluidic chip-based sorting of soft cancer cells

In fact, analysis by atomic force microscopy (AFM) revealed that the stiffness of cancer cells from mice (4T1), human breast cancer (MCF-7), mouse B16 melanoma and primary human melanoma (MP-1) differed widely and fell between 0.2 and 1.3 kPa. Remarkably, over 60% of cancer cells exhibited a stiffness of at least 0.7 kPa, and less than 10% of cancer cells displayed a stiffness of less than 0.4 kPa. As softness (the reciprocal of stiffness) confers a higher level of deformability on a cell, this has been evaluated as a possibility to isolate soft cancer cells from stiff ones (Mohamed *et al* 2009, Zhang *et al* 2012).

The unique microfluidic chip for label-free cell sorting is developed based on the physical characteristics of the cells, such as stiffness/deformability (figure 6.5). The designed microfluidic device is composed of two main parts, the fluidic channels and the micro-weir structures. The gap created between the micro-weir and the main channel can be used as a passive and selective isolation barrier for cells with varying stiffness. Since cancer cells are typically around 20–25 μm in size (Hosokawa *et al* 2010, Hvichia *et al* 2016), the microfluidic chips in this investigation were manufactured with a gap size of 15 μm by adjusting the height of the micro-weir and flow channels to 25 and 40 μm, respectively. Four kinds of chips with various lengths, widths of micro-weir structure, distances between micro-weir structures and overall numbers of micro-weir structures were developed and evaluated to conduct the sorting, based on previous studies (Mohamed *et al* 2009, Zhang *et al* 2012).

Cells with a density of 1×10^4 to 2×10^4 ml^{-1} with varying levels of stiffness were inserted into the chip by a syringe pump at a flow rate of 10 μl min^{-1}. The cells were

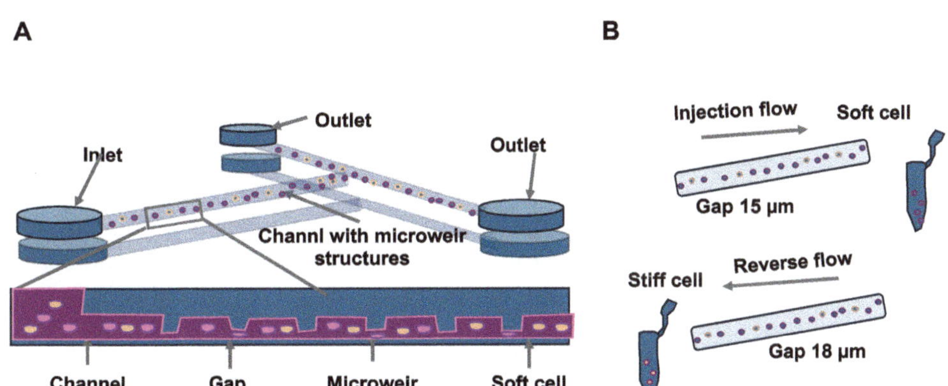

Figure 6.5. Schematic drawing of the microfluidic chip utilized for sorting of stiff and soft cells. The chip contains a channel with gaps and micro-weirs that is required for the cell sorting application.

collected as soft cells while streaming out of the outlet. Additionally, cancer cells were perfused in large quantities through the microfluidic channels that had a larger gap of 18 μm and the channels were rinsed in reverse. The cells then flowed out of the backflow inlet and were captured as stiff cells. For confirming whether these cells flowing out of the outlet of each type of microtube were soft or stiff cells, the stiffness of the cells was analyzed by AFM. In fact, the cells coming out of the outlet of each type of microtube were far softer compared to the cells coming out of the inlet. Microfluidic tube No. 2 (260 μm distance between the ribs and 11.44 mm channel length) showed the most efficient sorting performance and was therefore selected for the following experiments. The melanoma, such as B16F1 and MP-1, and breast cancer, such as 4T1 and MCF-7, cells separated by the microfluidic channel all exhibited the soft property. It was noted that the isolated stiff and soft cells had a comparable cell size. Additionally, the stiff and soft cancer cells *in vitro* grew along comparable growth curves. Since the stiffness of the separated soft cancer cells was lower than 0.4 kPa and the stiffness of the stiff cells was higher than 0.65 kPa, soft cells were those with a stiffness of below 0.4 kPa and stiff cells were those with a stiffness of above 0.7 kPa. In addition, for re-separation with the microfluidic chip, the isolated soft and stiff cells were re-mixed. The distribution of soft cells prior to re-separation was observed to perfectly match the distribution after re-separation, thus indicating that this microfluidic tube does not change the initial stiffness of the cells. The sorted cells can even remember their original stiffness value, which is referred to as cellular memory. F-actin is a key contributor toward cellular stiffness (Wang *et al* 1993). Cytochalasin D (Cyto), which blocks the polymerization of actin, can reduce cell stiffness; in contrast, jasplakinolide (Jas), which is a natural cyclodepsipeptide that potently promotes the polymerization of actin, can promote cell stiffness. Upon the treatment of cancer cells, like 4T1, MCF-7, B16 or MP-1, with Cyto or Jas, the amount of soft cancer cells at the Cyto outlet was augmented, whereas it was diminished by Jas. In this way, a microfluidic chip can be used to separate a small population of cancer cells with the mechanical characteristic of softness from the main cell population.

6.6.3 Highly tumorigenic and metastatic cancer cells are soft

Subsequently, the biological properties of the soft cancer cells were examined. Previous studies had shown that cancerogenic cells rather than differentiated cancer cells are isolated and grow in soft 3D fibrin gels of 90 Pa (Liu *et al* 2012, 2018). These isolated cells are substantially softer than their differentiated peers and can be portrayed by $CD133^{hi}$ melanoma cells or $ALDH^{+}$-breast cancer cells *in vivo* (Liu *et al* 2018). These results led us to propose the hypothesis that mechanical softness could be a unifying characteristic for tumorigenic cells. To verify this hypothesis, soft or stiff cancer cells, such as B16, MP-1, 4T1 and MCF-7 were inoculated into the soft 3D fibrin gels of 90 Pa after being sorted through a microfluidic channel. More than 95% of soft cancer cells were observed to form colonies, whereas stiff cancer cells produced only a few colonies of far smaller size. In the meantime, softness of isolated soft MCF-7 cells could be preserved in soft fibrin gels, while when soft cells were seeded in stiff culture plates, they began to stiffen after 4 h and attained the stiff peak after 12 h. Subsequently, 100 separate soft or stiff cells of 4T1 and MCF-7 were injected into the mammary fat pads of NOD/SCID IL-2Rγ-null (NSG) mice. After 12 weeks, 60% of the soft cells (6 out of 10 for 4T1, 4 out of 10 for MCF-7) were able to generate a primary tumor *in situ*, whereas none of the 100 injected stiff cells generated a primary tumor. Importantly, 100 soft 4T1 cells were even able to establish a tumor in immunocompetent wild-type BALB/c mice at a tumor-forming rate of 3 out of 8, thus implying that the soft cancer cells possess a high tumorigenic potential. Moreover, by conducting dilution experiments to determine tumor frequency for every soft or stiff cell inoculation setting (O'Brien *et al* 2007), soft cell inoculation resulted in the highest tumor frequency. Additionally, when performing serial transplants with 100 soft tumor cells, the tumor size and phenotype of the first implant of the second and third generation of passaged cancers in mice was similar. A hallmark of malignant melanoma is its propensity to metastasize at the lungs. Eight weeks after an intravenous injection of 100 soft or stiff cells sorted from B16-F1 or MP-1 in NSG mice, metastatic lung tumors were apparent in the soft cell population. Remarkably, only 10 soft cells were able to produce metastatic tumors, such as 2 out of 12 for B16 cells or 2 out of 8 for MP-1 cells, whereas no metastatic lung tumors were seen after injection of 100 stiff cells. Subsequently, the 4T1 cell lung metastasis model was utilized to confirm this result. Injection of soft or stiff 4T1 cells were made into the chest fat pads of wild-type BALB/C mice. After eight weeks, the animals were killed and H&E staining of the lungs confirmed metastatic tumors of 4 out of 8 in the soft cell group. No metastatic lung tumors occurred in the stiff cell group zero out of 8. Consistent with these *in vivo* findings, the soft cancer cells showed greater capacity for migration and invasion *in vitro* when compared to the stiff cancer cells. Collectively, these results indicate that soft cancer cells are highly cancerogenic due to their capacity to give rise to tumors at either primary or metastatic locations.

6.6.4 Softness as a hallmark of tumor-forming cells

In the next step, the question arose whether this mechanical softness could serve as a reliable biomarker for oncogenic cells. Despite their unreliability and low specificity, chemical molecules on the cell surface or in the cytosol, like the enzyme ALDH1, have often been utilized to designate CSCs (Ginestier *et al* 2007, Douville *et al* 2009). Using softness to screen ALDH1$^+$- and ALDH1$^-$-breast cancer cells, such as 4T1 and MCF-7, revealed that about 65% of ALDH1$^+$-cells are soft and about 6% of ALDH1$^-$ cells are soft. Strikingly, 100 soft ALDH1$^+$ or ALDH1$^-$ cancer cells could establish tumors in both NSG and WT mice when injected, but 100 stiff ALDH1$^+$ or ALDH1$^-$ cancer cells could not generate tumors even in NSG mice. Additionally, inoculation of 100 unsorted ALDH1$^+$ cancer cells showed only half the tumorgenicity of the sorted soft cells; nevertheless, these sorted soft cells had similar tumorgenicity to the sorted ALDH1$^-$ soft cancer cells. Coherently, only soft ALDH1$^-$ (100) but not stiff ALDH1$^+$-cells could develop microscopic lung metastases. Apart from 4T1 and MCF-7, akin findings were observed from both melanoma cell lines, B16 and MP-1. As a result, about 60% of $CD133^+$ cells and about 4% of $CD133^-$ cells exhibited softness, and 100 soft $CD133^-$ cells had the capacity to establish subcutaneous tumor and lung metastases, while this was not detected in the 100 stiff $CD133^+$ cells. Therefore, surface marker-defined cancer stem cells can comprise both the soft and stiff subpopulations, while solely the soft subpopulation harbors tumorigenic activity. To corroborate this interpretation, a similar strategy was employed to separate a stem-cell-like side population SP from cancer cells based on Hoechst 33342 dye efflux (Golebiewska *et al* 2011). Notably, approximately 50%–60% of SP cells exhibited a soft characteristic. In addition, the isolated SP and non-SP (NSP) cells from 4T1 or MCF-7 were categorized in more detail into stiff and soft subpopulations utilizing microfluidic technologies. Injection of 100 soft NSP cancer cells into the mammary fat pads of NSG mice resulted in about 50% incidence of visible tumor development within 12 weeks, in line with expectations. Comparable results were achieved with the stiff SP or soft NSP B16 or MP-1 cells. Moreover, 80 metastatic micronodules were detected in the lungs of the soft NSP group, although not in the stiff SP group, on day 80 using H&E staining. Therefore, SP cells also comprise a soft subpopulation and use it to allow a tumor to establish at the primary and metastatic locations. Collectively, these findings indicate that intrinsic softness could be a suitable feature for labeling tumorigenic cells.

6.6.5 Soft cancer cells switch on the Wnt-BCL9L signaling axis to develop stem cells

In addition, the biological distinction between soft and stiff cells at the genetic level was explored to elucidate the utility of softness as a surrogate marker for tumorigenic cells. For this purpose, the entire genome of the soft, tumorigenic, ALDH $^+$-cells and the stiff, differentiated cells were first analyzed. There was nothing different about the DNA values in the three groups. Afterwards, gene expression pattern was evaluated using mRNA sequencing. With the help of principal component analysis (PCA), the difference between the soft and stiff populations was much greater than that between the CSC and stiff populations, whereas the internal variances were lower in the soft population compared to the CSC population. Gene ontology analysis revealed that these variously expressed gene families are associated with stem cell proliferation, migration of cells and immune system responses. Moreover, the 2000 genes that were found to be variably expressed in soft, stiff, and CSC cells were plotted to generate a graphical heat map, which displays a closer cluster of genes between the soft and stiff groups as compared to those between the CSC and stiff groups, indicating that softness serves as a more suitable molecular marker to depict the inherent characteristics of tumorigenic cells. Applying a similar strategy, chromatin access analysis using ATAC-seq identified many variably accessible peak regions in both soft and stiff cancer cells. Remarkably, BCL9L, WNT2B, and WNT3A were some of the most highly expressed genes and were also more pronounced when chromatin was open in soft cancer cells. Additionally, BCL9L emerged as a leading candidate from the analysis of stem cell-associated genes obtained from RNA sequencing. Strikingly,

Wnt signaling is crucial for the maintenance of pluripotency and self-renewal of stem cells, while BCL9/BCL9L functions as an extra transcriptional coactivator and is a component of the Wnt enhanceosome (Van Tienen *et al* 2017). Therefore, the focus was on BCL9L, which is a homolog of BCL9 exhibiting functional redundancy.

6.6.6 BCL9L acts as a cellular stemness biomarker for soft cancer cells

In fact, as demonstrated through qPCR, Western blot, and immunostaining, the expression of BCL9L was markedly increased in soft breast cancer (4T1 and MCF-7) and melanoma (B16 and MP-1) cells. Consistent with this BCL9L upregulation, the expression of total β-catenin and its nuclear version was also increased in the soft cancer cells, and increased nuclear accumulation was evident. In particular, the elimination of BCL9L using CRISPR/Cas9 led to a reduction in β-catenin expression at both the entire cell and nuclear levels of the soft cancer cells. In addition, this BCL9L knockout reduced the number of metastatic colonies and their size within the soft 3D fibrin gels. Moreover, stem cell gene expression like Nestin, OCT3/4, and SOX2 for 4T1/MCF-7; Nanog, OCT3/4, SOX2, and CD133 for B16F1/MP-1 was increased in the soft cancer cells, whereas BCL9L knockout reverted this phenomenon. Subsequently, mice were injected with 100 soft SGGFP or BCL9L-SGs cancer cells. After 12 weeks, mice that received SGGFP$^+$ cells developed tumors, whereas no tumors developed in the BCL9L-knockout mice. Moreover, BCL9L knockout inhibited lung metastasis of soft cancer cells. Consistent with this *in vivo* finding, BCL9L knockout also blocked soft cancer cell invasion *in vitro*. Altogether, these findings indicate that BCL9L is an intrinsic molecular constituent in soft cells that serves to sustain their stem cell characteristics.

6.6.7 Soft cancer cells from patients exhibited increased expression of BCL9L

Ultimately, to confirm these results, clinical patient samples were examined. Microfluidic devices were utilized to isolate soft cancer cells from breast or colon cancer tissues. Soft primary breast or colon cancer cells were observed to have a cell stiffness of about 0.3 kPa, whereas their isolated stiff variants were above 0.7 kPa. Additionally, the percentage of soft primary cancer cells obtained from patients with colorectal ($n = 23$) or breast ($n = 11$) cancer strongly associated with the patient's pathological tumor grade. Next, expression of BCL9L was verified in soft and stiff cancer cells that were derived from breast cancer patients. The BCL9L levels were observed to be much elevated in soft cancer cells than in stiff ones. Additional BCL9L testing in 16 melanoma patients demonstrated that BCL9L was strongly expressed in melanoma tissue and that BCL9L levels in the tumor showed a positive relationship with melanoma stage. The high expression of BCL9L across tumor tissues was consistently linked to a worse prognosis in patients with melanoma, breast, pancreatic, lung and liver cancer. These findings indicated that mechanical softness could be a possible prognostic marker for patients suffering from cancer.

The theory of CSCs continues to be debated in terms of their origin, frequency, phenotype and functionality (Clevers 2011, Kaiser 2015, Wang *et al* 2015). Despite recent CSC biomarkers facilitating the establishment of clinical diagnoses and CSC-driven therapies (Keysar and Jimeno 2010, Medema 2013), such biomarkers may be more or less present in normal stem cells, in differentiated cancer cells, or perhaps even in normal tissues (Ginestier *et al* 2007, Gregorieff *et al* 2015). Moreover, the markers for CSCs are very fragile. For instance, both $CD133^+$ and $CD133^-$ cells were found to be CSCs (Singh *et al* 2003, Beier *et al* 2007, Shmelkov *et al* 2008). Moreover, both $CD34^+$ $CD38^-$ and $CD34^+$ $CD38^+$ AML cells have shown cancerogenic potential (Taussig *et al* 2008, 2010). Another conundrum is how to quantify cancer stem cells in a biopsy. Another conundrum is how to quantify CSCs in a biopsy. As there exist many extracellular vesicles with the tumor that carry cell surface receptors, these vesicles can merge with receiving cells that can acquire altered cell biological characteristics and thus mimic CSCs. The gold standard for demonstrating that suspected CSCs are, in fact, CSCs would be to inoculate them at various densities into immunodeficient mice and monitor for tumors to emerge. Another issue is the enormous animal testing for every patient and consequently, the ethical issues of this therapeutic approach. Yet these limited dilution tests usually involve four or five cell seeding densities, multiple test groups of mice, and at least half a dozen mice per group. Therefore, these types of tests are both costly and labor intensive. Moreover, a difficulty regarding stem cells is that ordinary cancer cells can occasionally ‘revert’ toward CSCs. In contrast to the versatility of biochemical molecules, the physical characteristic of a cell is far more stable. The evidence that the softness of cells can serve as a universal hallmark to characterize CSCs has been demonstrated (Lv *et al* 2021). Softness gives cells the capacity to deform. In contrast to microfluidic chips utilized for soft cell isolation presented so far (Mohamed *et al* 2009, Guo *et al* 2012), a chip with single-stream characteristics, like a longer channel in a confined region, efficient cell collection by forward and reverse streaming, and comfortable single-cell monitoring by microscopy, has been designed. The utilization of a silicone form for device production guarantees the process precision and the comfortable manufacture of several duplicates of polydimethylsiloxane microfluidic chips, both of which are essential for effective experimental control. This device was utilized to isolate both soft and stiff cancer cells. Remarkably, colonies were only generated by the isolated soft cancer cells in the soft 3D fibrin gels, while the stiff cells failed to generate colonies. In addition, traditional marker-based CSCs, like $CD133^+$, $ALDH1^+$ or SP^+ CSCs isolated from melanoma or breast cancer cells, comprise both soft and stiff subpopulations, and solely the soft subpopulation has the capacity to establish a tumor. Apart from tumor generation, the ability of CSCs to metastasize to remote organs is an essential property of CSCs. Soft-cell tumor cells generally exhibit the ability to invade and metastasize, while stiff cancer cells have only a low ability to metastasize.

A positive correlation between elevated tissue stiffness and aggressive cancer performance has been reported to result in a cancer development model that relies on either static or dynamic stiffening of cancer tissue (Paszek *et al* 2005, Wei *et al* 2015, Ondeck *et al* 2019). There exists no proof for the assumption that the cells emerging from the stiffened tumor stroma exhibit tumorigenic and metastatic activity. Moreover, it is uncertain whether the tumorigenic cells are stiff or soft. Despite the overall stiffness, the local microenvironments are very heterogeneous in terms of tumor stiffness (Plodinec *et al* 2012). Enhanced tissue stiffness can be accounted for more ECMs, which probably restrict blood vessel disposition and result in tumor hypoxia, which is a frequent feature of the TMEs. It is recognized that hypoxia within primary tumors is linked to enhanced metastasis and a poorer prognosis in cancer patients (Gilkes *et al* 2014, Brooks *et al* 2016, Rankin and Giaccia 2016). More recently, hypoxia was found to enhance mammary tumor recurrence in humans (Tang *et al* 2019). Ischemic-like conditions, such as hypoxia, encourage the investigation of how tumor spheroids migrate, invade, and interfere with stromal cells under various metabolic conditions (Anandi *et al* 2025). It is noteworthy that hypoxic regions can be extremely soft because of localized tissue necrosis and breakdown of the ECM. Increased tissue stiffness can therefore lead to more soft tumor cells being located at the hypoxic sites, which promotes aggressive cancer progression. Cancer cells with the highest migration and invasion capacity are five times less stiff compared to cells with the lowest migration and invasion capacity (Swaminathan *et al* 2011). The results strongly indicate that just the naturally soft cancer cells are highly tumorigenic and metastasizing when tested in animal models (figure 6.6) (Lv *et al* 2021). These results are in line with reports showing that metastasizing cancer cells are considerably softer compared to non-metastasizing cancer cells (Guck *et al* 2005, Cross *et al* 2007, Plodinec *et al* 2012, Xu

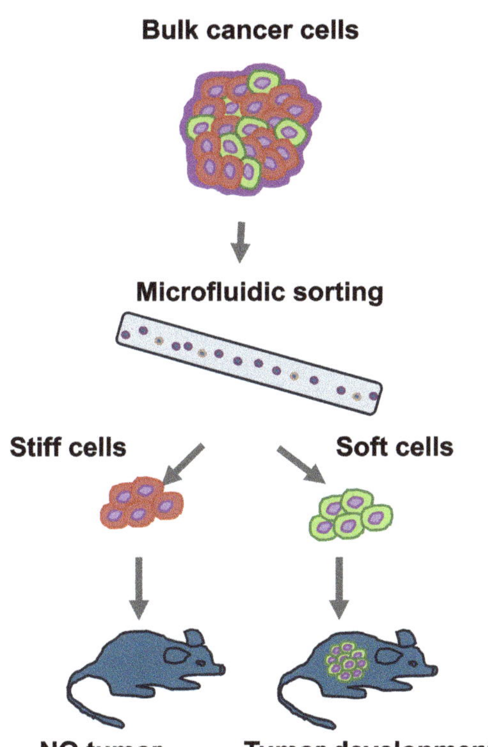

Figure 6.6. Tumorigenic capability of injected stiff and soft cancer cell subgroups into mice after microfluidic-based cell sorting from the bulk primary tumor population.

et al 2012). A weakness of the mechanical analysis of the cells is that the cells have been analyzed as non-adherent cells and no parallel metastatic analysis has been performed with these passages cancer cell lines (Guck *et al* 2005). In another experimental approach pleural effusions of patients have been analyzed using AFM, which is a technique that is adhesion-dependent (Cross *et al* 2007), whereby the adhesion of these cells toward the new substrates is not discussed at all, but may have an impact on cellular stiffness, such as different levels of cell-matrix adhesion receptors and adaption of these cells toward their new (very stiff) environment.

Moreover, these data on soft cancer cells are in line with the finding that less adherent cancer cells are generally more migratory and metastatic (Hope *et al* 2004). It is possible that, on top of hypoxia, the local stiffening of tumor tissue facilitates the EMT, and that the stiffening of cancer cells of a subpopulation promotes the differentiation and migration of this subpopulation out of the tumor stroma. Nevertheless, only the soft, undifferentiated cancer cells, which also originate from the stroma, can create metastases and multiply at the secondary sites, a fact that has been postulated (Tan *et al* 2014). Nonetheless, it will be important to thoroughly investigate in the future whether the soft or stiff cancer cells display a metastatic and tumorigenic potency in human patients.

The CSC theory predicts the utilization of CSC markers to determine patient outcomes, but the available evidence is contradictory. There are reports demonstrating that CSC markers are associated with unfavorable prognosis and worse overall survival (Ginestier *et al* 2007, Iinuma *et al* 2011, Stavropoulou *et al* 2016), but there are also reports demonstrating that CSC markers like CD133 and ALDH1 are not linked to patient prognosis (Lugli *et al* 2010, Wakamatsu *et al* 2012, Kapucuoğlu *et al* 2015, Miller *et al* 2017). This could be attributed to the assumption that marker-based CSCs harbor both soft and stiff cancer cells. Therefore, the utilization of cellular softness could be a prospective physical biomarker that can be employed to forecast the outcome of cancer patients. An exciting finding of this work is that BCL9L was found to be a biological hallmark for soft cancer cells, differentiating them from their stiff equivalents. Wnt/β-catenin signaling has been recognized to be critical for maintaining pluripotency and self-renewal of stem cells. To carry out its role, β-catenin translocates to the nucleus, where β-catenin, in conjunction with the transcription factors TCF/Lef, triggers Wnt-facilitated transcription. BCL9L, by contrast, serves as an additional transcriptional coactivator and is a component of the Wnt enhanceosome. BCL9L expression is elevated in soft cancer cells but is only slightly expressed in stiff cancer cells. More importantly, even in a small cohort analysis, BCL9L expression was found to be associated with the outcome of cancer patients. Based on these analyzes, it is hypothesized that BCL9L may serve as a valuable prognostic marker for cancer patients.

6.7 Dynamical mechanoregulation of cancer metastasis

Cellular transformation encompasses genetic and biochemical alterations that reconfigure cell metabolism, increase cell proliferation, and favor cell resistance to death (Hanahan 2022). Proto-oncogene occurrences that result in epithelial transformation also fundamentally change the mechanical characteristics of cancer cells (Plodinec *et al* 2012, Panciera *et al* 2020). These mechanical alterations within the cell do not take place in a physical vacuum. Cancer growth is physically housed in the primary originating tissue, giving rise to distinct solid, liquid and tension stresses that exert physical forces on adjacent epithelial cells, stromal fibroblasts, vascular endothelial and smooth muscle cells, neurons, and immune cell subsets like macrophages and cytotoxic lymphocytes. In this turbulent mechanical scenery, cancer cells infiltrate locally into the surrounding tissue and gain access to the lymphatic or blood circulatory system, thereby escaping their primary organ of origin. Cancer cells are exposed to shear forces as they spread hematogenously, which affects their interplay with platelets, neutrophils, and endothelial cells and finally influences their ability to survive (Follain *et al* 2020). In distant secondary organs, metastatic cells leak out of blood vessels, but the overwhelming bulk of these disseminated cancer cells are either eliminated via cytotoxic lymphocytes or undergo dormancy because of their failure to exploit niche assets (Massagué and Obenauf 2016, Lambert *et al* 2017). The small subset of disseminated cancer cells that evade immune attack and adjust to the local microenvironment can ultimately give rise to lethal metastatic lesions (Fidler 1995, Vanharanta and Massagué 2013).

At each step of this metastatic cascade, cancer cells must adjust to the biophysical characteristics of their microenvironment and move through the intricate network of forces created by the ECM, fluid flow, and passive or active engagement with neighboring cells. During the last years, it emerged that intrinsic cellular mechanical properties, like cytoskeletal stiffness and membrane tension, are central to this process. In this context, current research is aimed at elucidating how cellular mechanics influence the behavior of metastatic cells and their crosstalk with immune and stromal cells in the tumor microenvironment.

6.7.1 Cell mechanics during the growth of the primary tumor and local invasion

The biophysical characteristics of epithelial cells begin to alter in the first precancerous condition. These alterations were associated with transcriptional signatures characteristic for wound healing processes and rearrangement of tissues. Epithelial cells located at the wound margin exhibit a lower mechanical compliance (enhanced stiffness) compared to the other epithelial cells that are not actively involved in the dynamics of wound margin sealing (Wagh *et al* 2008). A comparable stiffening phenotype is triggered via TGFβ (Buckley *et al* 2012), a well-known promoter of both fibrosis and malignancy. Oncogenes can also enhance the stiffness in epithelial cells preceding complete malignant transformation of these cells. For instance, HER2 and KrasG12D expression can stiffen epithelial cells when cultured in biomimetic 3D scaffolds (Panciera *et al* 2020). In addition, mammary tumorigenesis caused by the polyomavirus middle T antigen (PyMT) is linked to a temporary stiffening of the mammary epithelium in genetically engineered mouse models (GEMMs) (Plodinec *et al* 2012). Overall, these results suggest that mechanical alterations are an initial event during epithelial transformation. *In situ* carcinomas are characterized by hypercellularity of the primary tumor, which causes compression of the adjacent tissue. This firm pressure modifies the physiologic state and destiny of multiple cellular components within the TME (Fernández-Sánchez *et al* 2015). In a mouse model of colon tumorigenesis, for instance, it was determined that magnetic replication of solid-state loading through the utilization of captured beads causes anomalous Wnt-β-catenin signal transduction route activation within the colon epithelium, albeit no *de novo* tumor is present. This type of mechanically driven hyperproliferation may be a potential driving force for tumor advancement and genetic heterogeneity. Solid stress and compression, and cytokines produced by cancers have also been associated with the conversion of fibroblasts to a chronically activated and differentiated state referred to as cancer-associated fibroblast (CAF). After their generation, CAFs produce abundant ECM and then promote pathological ECM rearrangement through cellular contractility and crosslinking of collagen fibrils (Cox and Erler 2011). In this way, the ECM stiffens and increases the solid stress. At the same time, CAFs sequester hyaluronan within the tumor microenvironment, creating repulsive electrostatic forces and trapping water, which adds to the swelling (Stylianopoulos *et al* 2012).

As the tumor progresses, the external forces generated by solid stress, physical constriction and swelling lead to cell jamming inside the tumor volume (Han *et al* 2020). Cancer cells evade these physical and topographical restrictions and infiltrate into the surrounding tissue as they switch to a more deformable and more viscous cellular phenotype (Han *et al* 2020, Ilina *et al* 2020). In MMTV-PyMT-facilitated GEMMs of breast cancer, these compliant cells serve as leading cells to establish the invasive frontier of the primary tumor that subsequently yields metastases in secondary targeted organs (Cheung *et al* 2013). The increased deformability of these cells is triggered by the transfer of fluid through gap junctions from the core of the tumor mass toward the cells at the leading edge of the invasion front (Han *et al* 2020). This observation leads to the assumption that, in principle, all cells within a tumor mass can become more deformable when they are at the tumor border. It is probable that increased potassium ion levels within the tumor interstitial fluid of melanomas and glioblastomas (GBM) also enhance mechanical compliance, because extracellular potassium reduces stiffness across multiple cell types (Oberleithner *et al* 2009, Eil *et al* 2016, Jackson 2017). Propagated cancer cells harvested from the breast and lung of patients exhibit higher compliance compared to their corresponding normal counterparts, and there is a causal association between mechanical compliance and invasive primary tumor outgrowth (Cross *et al* 2007). For instance, stiffening of breast cancer organoids in 3D culture through pharmacological stabilized actin filaments or stiffening of pancreatic ductal adenocarcinoma and colon cancer cells in 2D culture through excessive actin-myosin contractility, and that these *in vitro* phenotypes lead to a diminished metastatic spread in xenograft models (Surcel *et al* 2019, Bryan *et al* 2020, Han *et al* 2020). Enhanced cell stiffness probably impairs cellular deformation, which is necessary for cancer cells to maneuver through the restrictive ECM scaffold. Therefore, the ability of cancer cells to flee the primary TME is directly related to their pliability (Surcel *et al* 2019, Bryan *et al* 2020).

The invasive phenotype of leading cells is reinforced through the deposition and reorganization of the ECM through CAFs (Hanley *et al* 2020). Most strikingly, in early-stage breast cancer, cancer cells in 3D spheroid models apply physical force to orient ECM fibrils, which provides access and activation cues for tumor-derived factors toward normal fibroblasts (Jung *et al* 2020). Activated fibroblasts then exert contractile forces to reshape and stiffen the ECM scaffold, causing positive feedback activation of mechanosensitive Yes-associated protein (YAP) regulated gene expression (Calvo *et al* 2013). YAP activity within fibroblasts amplifies the pathological CAF phenotype, which is marked by exaggerated contractility, ECM laydown, and matrix reorganization. This back coupling between CAF mechanics and ECM is a demonstration of mechanoreciprocity, which refers to a cyclic mechanism whereby ECM reorganization through cells fuels their own biophysical alteration. Adipocytes and myofibroblasts in breast cancer, hepatic stellate cells in liver cancer and cancer cells themselves perpetuate this pathological biophysical scenery as they enhance ECM release (Iyengar *et al* 2005, Naba *et al* 2014, Wishart *et al* 2020, Deligne and Midwood 2021). In invasive breast cancer, ECM reorganization through filamentous bundling of higher-order collagen fibers through lysyl oxidase, together with an accompanying increase in matrix stiffness, can trigger a multitude of mechanotransduction events in cancer cells (Payne *et al* 2006, Levental *et al* 2009, Acerbi *et al* 2015). Consecutive activation of focal adhesion kinase (FAK), Rho/Rac/Cdc42 GTPases, and extracellular signal-regulated kinase (ERK) can culminate in the transcriptional activation of myocardin-related transcription factor (MRTF), YAP, and TWIST1, each of which promotes migratory and invasive gene expression for ensuing cancer cell spreading (Medjkane *et al* 2009, Lamar *et al* 2012, Wei *et al* 2015). CAFs also exert mechanical forces that drive vessel rearrangement and open new avenues for hematogenous spread of cancer cells (figure 6.7) (Sewell-Loftin *et al* 2017).

6.7.2 Entry mechanisms of cancer cells for the vascular system

The penetration of cancer cells into the vascular system encompasses both the breaking of the vessel walls and the pathological reorganization of the vascular network. The latter is determined through several physical parameters inside the microenvironment of the tumor. In the first place, ECM stiffening, hyaluronan-driven water uptake and hypercellularity result in vessel compression, which is referred to as vasoconstriction, in tumors of mice and humans (Sewell-Loftin *et al* 2017). In brain tumors, this compression can be fatal because it disrupts blood flow and leads to localized hypoxia and ischemia. In tumors outside the skull, by contrast, compressive hypoxia can be a powerful stimulant of distant metastases. Prolyl hydroxylation and the subsequent breakdown of HIFs is impeded when there is a decline in the oxygen delivery to the cells (Semenza 2013). Upon stabilization, HIF promotes endothelial proliferation and vascular restructuring, in which blood vessels temporarily forfeit their architectural organization because of inadequate pericyte coatings (Jain 2001, Nobre *et al* 2018).

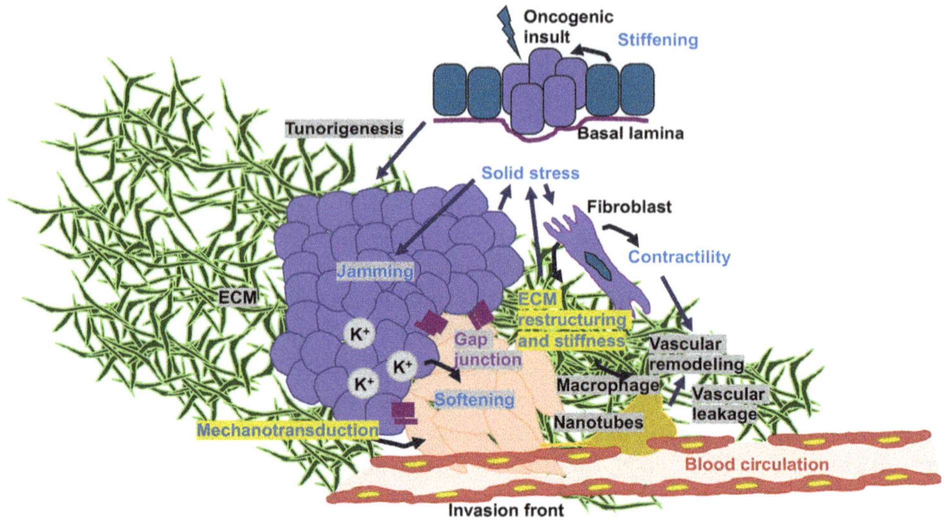

Figure 6.7. Alterations in the mechanical and biophysical characteristics of epithelial cells (in blue) begin at the precancerous phase and proceed dynamically during tumor progression and metastasis. Multiple components of the tumor microenvironment, like fibroblasts, macrophages, and the ECM, participate in these alterations.

This unstable condition offers cancer cells an avenue to enter the bloodstream. Tumor hypoxia can also lead to activation of HIF in cancer cells, which in turn leads to the expression of EMT genes that drive migration and cancer metastasis (Nobre *et al* 2018). In mouse models of breast cancer, the EMT phenotype is reinforced due to the canonical EMT transcription factor TWIST1, whose activation is triggered in response to an increased ECM scaffold stiffening. At the mechanistic level, TWIST1 is localized in the cytoplasm by the anchor protein G3BP2; ECM-controlled mechanotransduction breaks this connection and permits TWIST1 to pass into the cell nucleus (Wei *et al* 2015).

Vascular restructuring due to ECM stiffening correlates with macrophage and monocyte recruitment into breast tumors (Acerbi *et al* 2015). Under physiological conditions, macrophages clear tissue injuries, whereas in the case of tumorigenesis, macrophages adopt phenotypes that support tumor growth and invasiveness (Cassetta and Pollard 2018). Macrophages, for instance, produce nanotubes to connect with adjacent breast cancer cells, which physically tug cancer cells toward the endothelium alongside the ECM scaffold (Hanna *et al* 2019). At the same time, these cell-to-cell contacts trigger RhoA-GTPase signal transduction events in cancer cells, which leads to increased invasiveness across endothelial barriers *in vitro* and promotes spreading in zebrafish models, indicating that macrophage-facilitated nanotubes augment cancer cell intravascular dissemination (Roh-Johnson *et al* 2014). In fact, intravital microscopy of locally infiltrating breast cancer cells reveals that macrophages can generate sites of transient vascular leakiness and guide cancer cells toward these regions (Harney *et al* 2015).

While intravital image techniques have been crucial to advancing the understanding of the physical interplay between migrating cancer cells and their stroma, a series of *in vitro* modeling efforts using transwell assays and microfluidic channels with geometrically defined spatial restrictions have revealed the interaction between cancer cell mechanics and motility. It has been shown that cancer cells reorganize the cytoskeleton and membrane to reshape their cytoplasmic compliance, which is required for them to travel through densely packed areas (Cao *et al* 2015). In contrast to the cytoplasm, nuclei cannot be deformed easily (Denais *et al* 2016). Correspondingly, it was determined that extrusion of cancer cells via narrow openings tears the nuclear envelope, causing chromosome fractures and DNA fission due to cytoplasmic nucleases (Cao *et al* 2015). A nuclear fracture probably constitutes a weak point for cancer cells that enter the bloodstream through intravasation. The resulting DNA defects may, for instance, exacerbate the consequences of oxidative stress and promote cancer cell deterioration during the spread of melanoma and breast cancer cells as observed in experimental metastasis models in mice (Piskounova *et al* 2015, Basnet *et al* 2019). Nevertheless, nuclear breaks in physically limited primary breast tumors have also been linked to increased cancer cell invasiveness at the local site in humans and mice, indicating that mechanically triggered DNA damage could drive metastasis formation either through damage-related signal transduction routes or by giving rise to metastatic private oncogenic mutations (Nader *et al* 2021). A private oncogenic mutation refers to a rarely encountered genetic mutation that generally appears solely in a single family or a limited small population. Both hypotheses have not yet been verified, so the exact link between nuclear breakage and the formation of metastases is uncertain (Pfeifer *et al* 2017).

6.7.3 Cell mechanics foster cancer cell survival following transmigration via the vascular system

When cancer cells enter the bloodstream, they lose their attachment to the ECM scaffold and are suddenly no longer exposed to the mechanical and biochemical signals that normally ensure their survival (figure 6.8).

In normal epithelium, disruption of ECM anchorage results in anoikis, which is a physiological mechanism to preserve epithelial homeostasis. Cell proliferation during epithelial renewal and regenerative events results in overcrowding and mechanical compression, which displaces older epithelial cells and induces apoptosis within the epithelial lumen. This process is well characterized in the colon epithelium (Gudipaty and Rosenblatt 2017). When cells undergo extrusion, the cell surface integrins disengage from the basal lamina and cease signaling for survival through PI3K, AKT, ERK, and YAP (Reginato *et al* 2003, Zhao *et al* 2012, Paoli *et al* 2013). While ECM detachment leads to anoikis and apoptosis in non-transformed cells, oncogenic activation of PI3K/AKT, ERK, and YAP allows cells to avoid this event (Grassian *et al* 2011, Zhao *et al* 2012). Notably, YAP can also be triggered through calcium entry via mechanically regulated piezo channels (Pardo-Pastor *et al* 2018). It is well established that mechanical compression inside an epithelium triggers Piezo1/2 activation (Eisenhoffer *et al* 2012). It is therefore enticing to propose that the mechanical compression encountered by cancer cells passing through a vascular

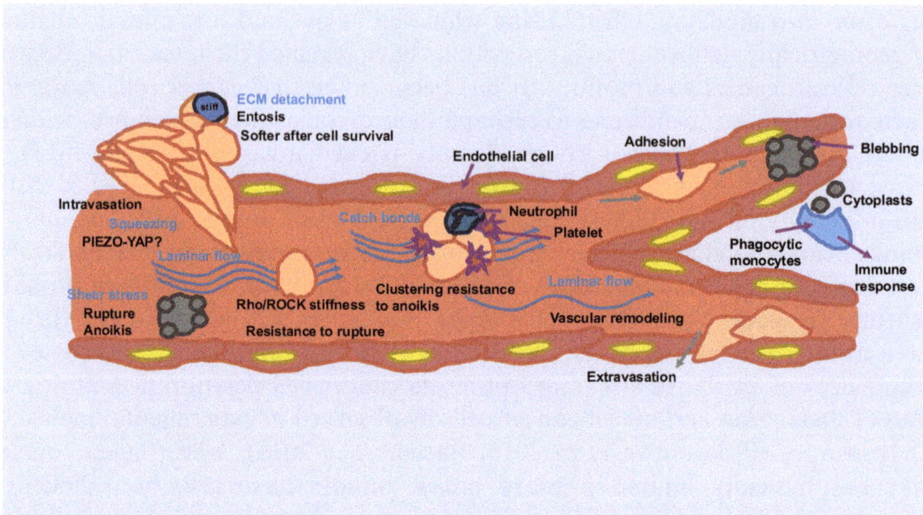

Figure 6.8. Outside the matrix, biophysical forces produced through laminar flow and detaching from the substrate dictate metastatic behavior through multiple avenues including anoikis evasion, entosis, adhesions formation, such as catch-bonds, cluster formation, extravasation, and immune reaction.

endothelial layer might impair anoikis through the Piezo1/2-YAP axis. Nevertheless, it has been reported that cancer cells, comprising breast cancer cells like MDA-MB-231 and 1205Lu, bladder cancer cells like EJ-28, prostate cancer cells like DU145 and melanoma cells like A375 and T24, which transmigrate through the endothelial cell layer and thereafter exhibit increased their invasiveness (Mierke *et al* 2008). These results indicate an increased survival rate of the cancer cells and a reprogramming of the cancer cells' invasive capacity. The absence of ECM attachment results in cancer cells becoming highly dependent on cancer cell–cancer cell interaction for their survival (Li *et al* 2001). In certain cases, this leads to a cannibalistic phenomenon referred to as entosis, whereby a 'winner' cancer cell swallows its neighboring 'loser' cancer cell. Through entosis, cancer cells can withstand the cell death and lack of nutrients that result from cells being detached (Overholtzer *et al* 2007, Hamann *et al* 2017). This mechanism is propelled through mechanical differences; winner cells are softer than loser cells and winner cells use actomyosin-driven contractility for entrapment (Sun *et al* 2014). Entosis has been found in primary and metastatic tumors, and especially in pleural fluid from the lungs of breast cancer patients and in bladder cancer patients' urine samples (Gupta and Dey 2003, Overholtzer *et al* 2007, Hayashi *et al* 2020). Therefore, it appears plausible that entosis functions as a survival mechanism for cancer cells that come out as winners in fluid environments such as the bloodstream. Another outcome of enhanced interaction between cancer cells is the generation of clusters. Breast cancer cells can collectively invade the bloodstream and even take CAFs and monocytes with them when they intravasate in the vessels (Hwang *et al* 2019, Aceto 2020). Even though CSC clusters are considerably less common in breast cancer patients compared to isolated cancer cells, the presence of CSC clusters strongly corresponds with the development of distant metastases and a reduced survival rate (Aceto *et al* 2014). In line with these clinical findings, CSC clusters lead to more intense colonization of secondary organs than single cells in experimental models (Aceto *et al* 2014). Besides promoting favorable cell–cell interactions, breast cancer cell clusters constitute nanolumens that can entrap growth factors and thereby enhance juxtracellular survival signal transduction (Wrenn *et al* 2020).

Even though ECM detachment and intravasation show certain mechanical similarities, a key distinction is that intravasating cells are subjected to strong shear stress that is created by the laminar fluid flow in the bloodstream and which they must withstand to prevent disruption. Most early events in the metastatic cascade, including tumor growth and localized invasion, are enhanced through enhanced mechanical compliance. Cancer cells that gain access to the blood circulation, however, must display increased cell stiffness to survive the damaging effects of shear forces (Moose *et al* 2020). This is achieved through several mechanisms: for instance, nuclear laminins enhance nuclear stiffness, and their exhaustion renders cancer cells sensitive to breakage as shear forces imitate circulation *in vitro* (Mitchell *et al* 2015). Shear stress also induces mechanotransduction in cancer cells, which ultimately results in F-actin polymerization and enhanced cortical stiffness that is triggered through Rho-GTPase/Rho kinase (ROCK). Blocking this signaling route impairs the survival of cancer cells in mouse models for experimental prostate cancer and breast cancer metastasis (Liu *et al* 2009, Schackmann *et al* 2011, Chivukula *et al* 2015, Moose *et al* 2020).

Even though laminar flow produces harmful shear forces, it also delivers tensile forces that are necessary to strengthen the intercellular 'catching bonds' that exist between circulating cancer cells and platelets. While most molecular interactions function as slip bonds and become less stable under tension, the strength of catch-bonds improves up to an optimal imposed force (Dembo *et al* 1988, Thomas *et al* 2008), which is generally in the magnitude of 10–20 pN. It has been established that adhesion molecules like selectins and integrins with their respective ligands create catch-bond adhesion bridges (Marshall *et al* 2003, Yago *et al* 2004, Kong *et al* 2009, Chen *et al* 2010, Rosetti *et al* 2015), and it is commonly assumed that the laminar flow in the blood vessels supplies the necessary force to induce the catch-bond bridging, thereby causing the aggregation of cancer cells and platelets (Felding-Habermann *et al* 1996, Slattery and Dong 2003). Such heterotypic clusters are more resilient to fluid shear stress, anoikis and NK cell-mediated cytotoxicity compared to single cells (Egan *et al* 2014, Haemmerle *et al* 2017, Schmied *et al* 2021). Cancer cells also aggregate with neutrophils via integrin-mediated adhesions to create clusters, and it could be imagined that these adhesive catch-bonds are also reinforced through laminar flow (Szczerba *et al* 2019). Ultimately, vascular endothelial cells also attach more tightly to CSCs circulating under laminar flow through integrins, which has been proven to facilitate the anchoring of cancer cells within capillaries (Follain *et al* 2020).

6.7.4 Cell mechanisms in extravasation and metastasis

Intravital imaging in the brain, lung and liver of mice demonstrates that cancer cells capable of surviving circulation arrest in vascular branches or zones of narrowing in these secondary organs (Kienast *et al* 2010, Ritsma *et al* 2012, Headley *et al* 2016, Entenberg *et al* 2018). A prevailing flow rate is needed for the retention of cancer cells within the circulary system (Follain *et al* 2018). After forming robust adhesions between the circulating cancer cells and the luminal side of the capillary endothelium, the steady flow of fluid activates mechanotransduction within the vascular endothelium, leading to reorganization of the vessel, as reviewed in Mierke (2024b). Flow-induced vascular rearrangement is crucial for the extravasation of cancer cells, since its pharmacological blockade impedes their transendothelial migration (Follain *et al* 2021). In lung metastasis mouse models of melanoma, cancer cells that are unable to extravasate are subjected to sustained shear stress, leading to plasma membrane detachment, which is referred to as blebbing and liberation of the blebs into the microenvironment (Lawler *et al* 2006, Headley *et al* 2016). The phagocytosis of these cancer cell-derived blebs carried out by tissue-resident normal dendritic cells and non-tissue-resident macrophages promotes immune programming inside the metastatic microenvironment, which impacts the destiny of cancer cells that successfully propagate (Headley *et al* 2016).

6.7.5 Mechanotransduction and the transition from metastatic quiescence to outgrowth

Disseminated cancer cells that have finished extravasation can remain non-detectable clinically over months and sometimes decades, and this state of metastatic quiescence engages the orchestration of multiple metabolic, cell cycle and immune surveillance pathways, which is reviewed in Sosa *et al* (2014) and Goddard *et al* (2018). In the metastatic quiescent phase, the micrometastatic proliferation of spread cancer cells is counterbalanced through the death of cancer cells triggered via the immune system (Fidler 1995, Pommier *et al* 2018, Parida *et al* 2022). When immunity is impaired, the uncontrolled proliferation of scattered cancer cells can result in their colonization and deadly spread. The importance of immune surveillance in preventing metastasis is most evident during organ transplantation, where immunosuppressive medications designed to suppress tissue rejection can also stimulate the growth of metastatic tumors (MacKie *et al* 2003, Armanios *et al* 2004). Several notable examples are the occurrence of melanoma in kidneys and extracranial GBM metastases in kidneys, livers and lungs of recipients of transplants. GBM cases are particularly intriguing because brain tumors are conventionally regarded as non-metastatic. These findings illustrate that cancer cells that have mastered all prior phases of the metastatic cascade, including migration, invasion and survival in the bloodstream, cannot establish clinically significant metastatic colonies without being able to avoid immune surveillance in secondary organs throughout the metastatic phase. They also suggest that the proliferative quiescence of spreading cancer cells may be associated with immune bypass. In fact, quiescent lung, breast and pancreatic cancer cells with slow cell cycle dynamics downregulate

T and NK cell activating ligands, like class I MHC I and MICA, and thus escape immune defense through cytotoxic lymphocytes in mice (Pommier *et al* 2018, Parida *et al* 2022).

The resurgence of dispersed cancer cells from a quiescent state and their productive proliferation for colonization necessitate the optimal exploitation of available local resources. The perivascular niche, situated on the abluminal surface of blood vessels, is one of the most important microenvironments for the growth of metastases (Kienast *et al* 2010, Ritsma *et al* 2012, Ghajar *et al* 2013, Entenberg *et al* 2018). Breast, lung, melanoma and colon cancer cells have disseminated and migrated on this surface into the bone marrow, lung and liver in mouse models for secondary niche colonization. Considering that ECM stiffness enhances mechanotransduction and advancement of the cancer, it is very probable that spreading cancer cells aim for a mechanically favorable environment using durotaxis, which is a type of cellular migration that guides cells to stiff substrates (Lo *et al* 2000). Most interestingly, factors sequestered from primary tumors can precondition secondary perivascular cavities to create mechanochemically tolerant environments even prior to the cancer cells entering remote organs. For instance, exosomes originating from orthotopic melanoma and breast tumors can trigger a phenotypic alteration in remote pericytes and vascular smooth muscle cells within the mouse model (Murgai *et al* 2017). Likewise, exosomes released from pancreatic cancer cells were observed to stimulate Kupffer cells, which reside in the liver, to release TGFβ (Costa-Silva *et al* 2015). In both instances, cancer-related exosomes direct the stromal cells to release fibronectin, which acts as a forerunner to tissue fibrosis that stiffens the tissue and is thus a potent activator of integrin-facilitated mechanotransduction, which all translates to better metastatic spread (Cox and Erler 2011, Cooper and Giancotti 2019). The mechanotransduction of scattered breast cancer cells is enhanced through the action of neutrophils that release metalloprotease-containing extracellular DNA trapping agents to digest laminin and uncover cryptic integrin tethering sites (Albrengues *et al* 2018).

Remarkably, the metastatic colonization of secondary targeted organs requires not only the presence of spread cancer cells but also the expression of the respective mechanotransduction pathways. For instance, expression of TM4SF1, a membrane receptor involved in promoting signal transduction that occurs downstream of the collagen receptor discoidin domain receptor 1, is adequate to encourage the outgrowth of quiescent breast cancer cells within secondary organs (Gao *et al* 2016). Propagated cancer cells also express urokinase-type plasminogen activator (uPA), which augments integrin signal transduction. Activation of integrin through uPA shifts the balance of ERK/p38 growth/stress signal transduction in the direction of the former to trigger disseminated cancer cell proliferation (Ghiso *et al* 1999). Using experimental metastasis models, mouse breast cancer cells require integrin-linked kinase (ILK) and beta-parvin activity for appropriate reorganization of the cytoskeleton during lung metastasis (Shibue and Weinberg 2009, Shibue *et al* 2013). Likewise, the neural cell adhesion molecule L1 (L1CAM) that is expressed on disseminated cancer cells adheres to the vascular basal lamina and amplifies integrin-ILK signal transduction to activate mechanosensitive transcription factors such as YAP and MRTFA/B in breast and lung cancer metastasis into the brain and the lung (Valiente *et al* 2014, Er *et al* 2018). Vice versa, astrocytes in the brain cause metastatic quiescence in disseminated cancer cells through the expression of laminin-211, thereby repressing YAP-driven metastatic tumor growth (Dai *et al* 2021). These findings indicate that mechanotransduction is a force underlying the wake-up of metastases after quiescence.

6.7.6 Elementary mechanisms govern mechanosensitive transcription factors required for colonization

Metastases utilize specialized gene expression modules to establish themselves in their newly found habitats and to proliferate. Notably, several transcription factors that are crucial for metastatic colonization, including β-catenin, YAP, and MRTFA/B, sense mechanical cues (Cui *et al* 2015, Röper *et al* 2018). While the mechanical forces that promote β-catenin activity specifically in metastasis are not yet known, YAP and MRTF are activated upon cancer cell dissemination on the abluminal surface of the vasculature in secondary organs (Er *et al* 2018). This process relies on integrin-based cell adhesion, cell polarization, and the formation of F-actin-based architectures. YAP and MRTFA/B cooperate physically to drive transcription of a group of genes that are partially redundant for metastatic growth (Lamar *et al* 2012, Kim *et al* 2017).

YAP and the associated TAZ are recognized as targets of the Hippo signal transduction route, which is mainly accountable for the downregulation of cell growth and proliferation observed during contact inhibited growth (Zhao *et al* 2007). For instance, cell–cell contacts among confluent cells at the injury site trigger Hippo kinases, that phosphorylate YAP/TAZ upon epithelial wound sealing. These repressive phosphorylation incidents result in cytoplasmic sequestration of YAP/TAZ and thus suppression of their transcriptional function in conjunction with the TEAD family of transcription factors. Besides Hippo pathway control during development and disease, activation of YAP/TAZ is also linked to adhesion signaling and the subsequent reorganization of the cytoskeleton. Proteins that are activated upon interaction with integrin, like Rho/Rac family GTPases, FAK, p21-activated kinase (PAK), ILK, and Src, are found to facilitate nuclear translocation of YAP/TAZ (Serrano *et al* 2013, Kim and Gumbiner 2015, Lamar *et al* 2019). The polymerization of F-actin is a crucial part of this mechanism because high levels of cytoplasmic monomeric G-actin enhance the interaction of YAP/TAZ with cytoplasmic effectors of sequestration, like ARID1a, which is a member of the SWI/SNF chromatin reorganization complex, and the angiomotin protein family (Dupont *et al* 2011, Mana-Capelli *et al* 2014, Chang *et al* 2018). The importance of the F-actin/G-actin ratio in the cell is further emphasized by the fact that the transcriptional activity of YAP/TAZ can be repressed through actin capping and severing proteins (Aragona *et al* 2013). Ultimately, mechanotransduction from the cell surface into the nucleus through the linker of the nucleoskeleton and cytoskeleton complex (LINC) is essential for YAP/TAZ nuclear entry via its ability to trigger the opening of nuclear pore complexes (Elosegui-Artola *et al* 2017). When YAP/TAZ translocate to the nucleus, YAP/TAZ upregulate the expression of cell cycle progression cyclins, enhance metabolic programming for cell proliferation, remodel the cytoskeleton and extracellular matrix, and activate a series of target genes that partially intersect with those of MRTFA/B (Zanconato *et al* 2015, Kim *et al* 2017, Lee *et al* 2019a).

MRTFA and B represent two related transcription factors that are required for myoepithelial differentiation throughout the lactation/involution cycles, for cardiovascular organogenesis, and for the migration and incorporation of smooth muscle cells (Oh *et al* 2005, Li *et al* 2006). Like YAP/TAZ, import of MRTF-A/B nucleotides is governed by the LINC complex (Ho *et al* 2013, Plessner *et al* 2015) and reacts strongly to the F-actin cytoskeleton. In this instance, the connection is obvious: monomeric G-actin physically attaches to and deposits MRTF isoforms in the cytoplasm (Miralles *et al* 2003, Vartiainen *et al* 2007, Plessner *et al* 2015). Depletion of G-actin as the cytoskeleton grows disrupts this inhibitory interface, thereby allowing MRTFA/B to translocate into the nucleus. Therefore, several of the structural and signaling molecules that cause G-actin clearance by integrating it into F-actin, like Rho/Rac GTPases, ROCK, and formins, trigger the activation of MRTFA/B. When entering the cell nucleus, MRTF isoforms associate with serum response factor (SRF), which is a ubiquitously expressed DNA-binding protein implicated in both proliferative and morphological reactions in response to growth factor signal transduction. MRTF-SRF complexes especially drive transcription of actin, actin-bundling proteins, and additional genes implicated in cellular contractility. In this way, the MRTF signal transmission precisely controls the cytoskeletal reaction that propels it. The tight connection observed between the activation of mechanosensitive transcription factors like YAP and MRTF, the proliferation of disseminated cancer cells, and cytoskeletal rearrangement indicates that the biophysical condition of disseminated cancer cells undergoes a transformation upon metastatic outgrowth (Albrengues *et al* 2018, Er *et al* 2018, Dai *et al* 2021). A transformation to a malignant cancer is generally accompanied by an enhancement of the mechanical compliance. Conversely, in the perivascular niche, colonization and outgrowth necessitate stiffening of the cancer cells through F-actin polymerization upon adhesion to blood vessel abluminal surfaces. As discussed below, this microenvironment-based transition seems to be of relevance for the detection of incipient metastases through the immune system (Lei *et al* 2021, Liu *et al* 2021c; Tello-Lafoz *et al* 2021).

6.7.7 Antitumor immunosurveillance is controlled via mechanoregulation

Structural plasticity and migration are among the lifestyle aspects of immune cells, involving intercellular interactions, motility, and phagocytosis. All these biological processes are characterized by strong physical effects, and in recent years they have been shown to be governed by mechanical regulation (Huse 2017). In terms of antitumor immunity, the mechanoregulation of cytotoxic lymphocytes, that comprise cytotoxic T cells and natural killer (NK) cells, is critical. T cells and NK cells recognize cancer cells through the detection of cell surface molecules that are upregulated upon stress and transformation. This leads to the generation of a highly dynamic cell–cell communication, referred to as an immune synapse (Stinchcombe and Griffiths 2007, Dustin and Long 2010). The lymphocytes then release toxic granzyme proteases and the pore-forming protein perforin at the interface, which results in the death of the cancer cells. This killing reaction forms a central part of normal immunosurveillance and an essential feature of anti-tumor immunotherapy (Finn 2018).

Immune synapses apply forces in the nanonewton regime that govern lymphocyte cytotoxicity via at least two mechanisms. First, they increase the pore-forming activity of perforin through the physical stress and distortion of the targeted cell membrane (Basu *et al* 2016). To accomplish this synergy between secretory and mechanical performance, lymphocytes employ membrane-tethered actin-based protrusions that temporally and spatially synchronize force delivery and perforin inside the synapse. The Wiskott–Aldrich syndrome protein (WASP), which is a nucleating factor for actin assembly, appears to fulfill a particularly important function in this process (Tamzalit *et al* 2019). Secondly, synaptic forces affect cytotoxicity through modulating mechanotransduction. Several crucial lymphocyte immune receptors, among them the T-cell antigen receptor and the $\alpha L\beta 2$ integrin LFA1, establish high-affinity catch-bonds with their specific ligands only provide high affinity when set in a tensile state (Kong *et al* 2009, Chen *et al* 2010, Liu *et al* 2014, Rosetti *et al* 2015, González *et al* 2019). The signaling potential of these receptors also seems to rely on mechanically regulated conformational shifts (del Rio *et al* 2009, Friedland *et al* 2009, Lee *et al* 2015). These two mechanoregulatory circuits are regulated through the physical characteristics of the target cell. The effectiveness of perforin pore creation relies on the tension of the targeted membrane, that is in turn provided through the subjacent cortical cytoskeleton. Likewise, the mechanical demands of immune receptor activation impose physical requirements on the juxtaposed cell surface, which needs to be sufficiently stiff to withstand synaptic pulling forces. In fact, several studies have demonstrated that the stiffness of a target substrate or cell defines the extent of lymphocyte activation via the synapse, whereby stiffer surfaces elicit a larger response (Judokusumo *et al* 2012, Blumenthal *et al* 2020, Friedman *et al* 2021).

Therefore, the biophysical characteristics of cancer cells appear to be a crucial determinant of their susceptibility to the surveillance of patrolling cytotoxic lymphocytes. In fact, the increased pliability generally linked to malignancy would be anticipated to shield cancer cells during metastatic dissemination from immune-mediated destruction. In specific circumstances, mechanoregulation might also have the contrary impact in that lymphocytes would be able to recognize metastases on the strength of their specific biophysical characteristics. This concept has lately been investigated in the setting of metastatic sprouting within the perivascular compartment. This process, which is described in the above section, involves the polymerization of actin, which allows cells to propagate and migrate over the abluminal vascular surface. While this architectural rearrangement is required for metastasis, it results in enhanced stiffness of the cancer cells, which causes increased mechanotransduction and activation of cooperating cytotoxic T lymphocytes and NK cells. Similarly, it has been established that overexpression of MRTF in cancer cells decreases their ability to colonize within immunocompetent mice, whereas it enhances their colonization in mice deficient in $CD8^+$ T cells or NK cells (Tello-Lafoz *et al* 2021). Remarkably, the MRTF-driven immunosensitization could be rescued through F-actin removal in cancer cells, confirming the biophysical nature of these phenotypes. These findings have uncovered a biophysical mode of immunosurveillance, referred to as mechanosurveillance, that empowers cytotoxic lymphocytes to combat metastatic outgrowth.

In other phases of metastasis, are cancer cells biophysically weak in a way that could be utilized by the immune system? The enhanced osmotic swelling accompanying local invasion of primary tumors may render them possibly more sensitive to cellular cytotoxicity via elevated membrane tension. In fact, it has been established that osmotic swelling increases the lysis caused through purified perforin *in vitro* (Basu *et al* 2016). Likewise, stiffness of the primary tumor, which relies on ECM and is recognized to promote proliferative reactions via YAP (Dupont *et al* 2011, Montagner and Dupont 2020), could also enhance mechanical surveillance through the MRTF signal transduction pathway. The impact of micromilieu mechanics on cytotoxic lymphocyte functioning is a productive field for future investigations. It remains to be seen whether immune synapses present in the bloodstream, in pleural or peritoneal fluids, or in highly compliant or very stiff tissues display the same biophysical characteristics and fulfill the same requirements. The degree to how biophysical immunodeficiencies impact anti-tumor immunotherapy is also an intriguing and vastly understudied area. In this context, it is intriguing to realize that increased MRTF signaling levels are associated with enhanced reactivity to immune checkpoint inhibition therapy for melanoma (Tello-Lafoz *et al* 2021). Future research in this direction should identify even more prognostic biomarkers and new treatment options.

6.7.8 Mechanoregulation of metastasis: a potential target

Medications that specifically target several signaling mediators of the mechanotransduction chain, including integrins, FAK, and PAK, have been the focus of several recent clinical investigations (Lampi and Reinhart-King 2018, Liu *et al* 2021b; Slack *et al* 2022). Molecular redundant nature of mechanotransduction, nevertheless, constitutes a potential hurdle for this strategy. It is therefore hardly astonishing that several clinical trials involving integrin-blocking therapeutics, for instance, have so far yielded no results in terms of progression-free or overall survival (OS) in cancer patients, which has been discussed in the review article (Bergonzini *et al* 2022). Alternatively, it could be envisaged to block the upstream regulatory elements of mechanotransduction in the TME. Fibroblasts, for instance, express angiotensin receptors, which activate Rho-GTPase and subsequent adhesion signal transduction events. This leads to the build-up of fibrotic tissue, which dramatically enhances mechanotransduction in the TME (Shen *et al* 2020). It has been determined that blocking this angiotensin signaling pathway decreases ECM stiffness within the liver and in experimental liver metastases. Moreover, retrospective analysis of colorectal cancer patients indicates that angiotensin inhibitors offer survival advantage for those with liver metastases. Besides halting proliferative mechanotransduction reactions, the decrease in ECM stiffness could also enhance drug permeation and anti-tumor immune influx into the TME by reducing the solid pressure on the vascular system (Jain 2005). Therapies targeting ECM and its mechanical features, nevertheless, can also abolish the physical restraints on tumor growth and allow cancer cells to spread from the primary tumor. For instance, clinical studies showed that sonic hedgehog blockade to decrease CAF-derived desmoplastic ECM resulted in poorer overall survival in pancreatic ductal adenocarcinoma patients, and that CAF- and ECM-targeting treatment is associated with enhanced metastasis in animal models (Rhim *et al* 2014, Catenacci *et al* 2015, Ko *et al* 2016). These studies emphasize the necessity to generate specific ECM-targeting or softening strategies for various pathological circumstances.

Focusing on mechanosensitive transcription factors like YAP, MRTF, and TWIST1 is also a possible way to avoid molecular redundancy at the intermittently transmitted cues. There are complicating issues with this strategy. The first is that transcription factors are dynamically controlled, and transcription factors that drive metastasis may not be functional in the later phases of metastatic outgrowth. For instance, TWIST1 propels EMT in reaction to mechanical cues encountered in early metastatic spread but is scaled down in colonization to fuel metastatic growth in distant secondary organs (Ocaña *et al* 2012, Tsai *et al* 2012, Wei *et al* 2015). In contrast, YAP is found to be expressed and functional in well-established micrometastases (Er *et al* 2018). The development of a specific inhibitor may be hampered by the lack of clear binding sites for small molecules, although a suitable candidate transcription factor is identified (Bushweller 2019). An appealing alternative is to interfere with the regulation of these transcription factors. For instance, blocking TEAD's autopalmitoylation disrupts its association with YAP/TAZ and blocks the proliferation of cancer cells (Tang *et al* 2021). It might also make more sense to specifically target the weaknesses linked to the activation of the transcription factor, rather than the transcription factor itself. In fact, the stiffness of cancer cells is enhanced through the activity of MRTF, which can be further increased through the blocking of the immune checkpoint (Tello-Lafoz *et al* 2021).

Influencing the mechanical characteristics of the cytoskeleton of cancer cells is another interesting approach that has in fact proved effective in several preclinical studies. For instance, blocking the actomyosin contraction machinery in an activated conformation with the small molecule 4-hydroxyacetophenone (4-HAP) abrogates the invasiveness of colon cancer cells and decreases their capacity to metastasize into the liver in mouse models (Surcel *et al* 2019, Bryan *et al* 2020). Stiffening of cancer cells, nevertheless, has not always yielded the expected anti-invasion benefit, as several other treatments that decrease cancer cell compliance can foster invasion *in vitro*. Stiffening of cancer cells, nevertheless, has not always yielded the expected anti-invasion benefit, as several other treatments that decrease cancer cell compliance can foster invasion *in vitro* (Kim *et al* 2016b, Nguyen *et al* 2016). For instance, when prostate cancer cells are excited through β-adrenergic receptors, they become stiffer and more invasive at the same time. In addition, ROCK-controlled F-actin polymerization augments the stiffness of individual breast cancer cells when they invade into collagen gels (Staunton *et al* 2016). The biophysical determinants of metastases are stage-dependent, as described before. Therefore, biophysical impediments to a specific step, such as localized invasion, may encourage another step, like secondary seed dispersal.

While the effects of enhanced stiffness on cancer cell invasion appear to be impacted on by the specific cell type and microenvironmental contextual cues, the impact on immune surveillance (i.e. mechanosurveillance) seems to be shared across many diverse models. The increase in stiffness of breast cancer, melanoma, and hepatocellular carcinoma stimulates the activation of mechanosurveillance mechanisms through NK cells, $CD8^+$ T cells, and adoptively transferred T cells (Lei *et al* 2021, Lv *et al* 2021, Tello-Lafoz *et al* 2021). Theoretically, an enhancement of the stiffness in cancer cells could also facilitate their ingestion through macrophages (figure 6.9) (Beningo and Wang 2002, Sosale *et al* 2015).

At this point in time, efforts to exert mechanical control over metastatic cells through enhancing cancer cell stiffness pose numerous hurdles for clinical implementation. First, the treatments for stiffening cells, like myosin-activating 4-HAP, the actin crosslinking agent jasplakinolide, and the cholesterol-degrading agent methyl-β-cyclodextrin, that have been applied to mice are not specific enough to be effective in humans (Lei *et al* 2021, Liu *et al* 2021c). Second, agents that enhance the stiffness of cancer cells could also possibly activate mechanotransduction paths that favor metastatic colonization. The ultimate translational hurdle, therefore, is to elucidate signaling intermediates and molecular events that establish cortical cell stiffness but are decoupled when pro-metastatic mechanotransduction is activated.

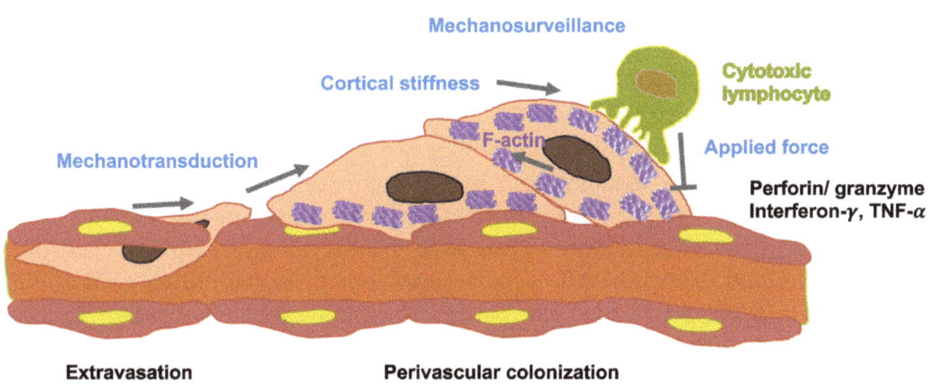

Figure 6.9. Upon arrival in secondary organs, the disseminated cancer cells activate transcription factors like YAP and MRTF through signaling and mechanotransduction to proliferate. Through mechanotransduction, reorganization of the F-actin cytoskeleton, and the accompanying changes in cortical stiffness of the scattered cancer cells, nevertheless, renders them susceptible to mechanical surveillance through cytotoxic lymphocytes. Cytotoxic lymphocyte detection of the cortical stiffness of cancer cells elicits force application and augments their intrinsic mechanotransduction for enhanced lytic granule cytotoxicity and cytokine release. This apparent paradox has important implications for metastatic progression and helps to account for the extreme inefficiency of the metastatic process, since less than 1% of spreading cancer cells succeed in colonizing remote organs.

6.8 Possible connection between physical microenvironmental stresses and increased aggressiveness of metastasizing cancer cells

Cancer cells in secondary tumors metastasize more effectively than their equivalents in primary tumors. This is partly because of the adverse microenvironments that metastatic cancer cells face, leading to the survival of a metastatic phenotype out of the original population. Whether or not mechanical stress contributes to this shift in metastatic capacity is uncertain. Mechanical deformation is able to select a cancer cell subpopulation that is resistant to mechanically triggered cell death when cancer cells are forced to pass through small capillary-sized restrictions (Jiang *et al* 2023). Transcriptomic profiling shows that in this subpopulation, proliferative and DNA injury response signal transduction routes are enhanced, leading to a more proliferative and chemotherapy-resistant cell phenotype. These findings emphasize a possible association between physical stresses of the microenvironment and increased malignant behavior of metastatic cancer cells, that can be exploited as a therapeutic approach to inhibit metastatic dissemination of cancer cells.

Metastasis is an infrequent and rarely observed event, although CTCs are continuously present within the peripheral blood of patients with cancer (Koop *et al* 1995, Riethdorf *et al* 2007). The shear stress of fluids (Regmi *et al* 2017, Moose *et al* 2020), immune surveillance (Mohme *et al* 2017), and small capillaries (Weiss *et al* 1985, Strilic and Offermanns 2017) are able to kill the majority of metastasizing cancer cells that gain entry into the bloodstream. Secondary cancers have distinct biophysical (Swaminathan *et al* 2011) and molecular characteristics (Shah *et al* 2009) in comparison to their primary equivalents, are frequently more deadly, and pose significant hurdles for cancer therapeutics. Clarifying the mechanism of metastasis and identifying possible strategies to inhibit metastatic dissemination may provide a more powerful approach to cancer treatment.

Research has demonstrated that the physical microenvironment of cancer cells has an essential function in the development of cancer (Follain *et al* 2020, Beeghly *et al* 2022). For instance, mechanical stress caused when tissue space is constricted can break open the nuclei of metastasizing cancer cells and exacerbate genomic instability (Denais *et al* 2016, Raab *et al* 2016, Irianto *et al* 2017). In the meantime, the mechanical cues from blood flow shear stress can enhance the survival of cancer cells in the vascular bed, promote extravasation, and enhance their resistance to anti-cancer therapeutics (Fan *et al* 2016, Xin *et al* 2019). These findings indicate that the physical forces experienced by cancer cells as they metastasize can substantially change the metastasizing cancer cells. In contrast, metastasis can be viewed as an evolutionary selection event that favors cancer cells with certain characteristics that enable them to survive various detrimental stresses (Fidler 1973, Minn *et al* 2005, Furlow *et al* 2015). It is well established that metastatic cancer cells are usually more invasive and resistant to anti-cancer treatments when compared to those localized in primary tumors. Nevertheless, it is uncertain whether mechanical stresses directly contribute to this malignant transformation. Specifically, it is essential to determine whether there is a certain cancer subpopulation that is resistant to mechanical stress and what characteristics such a subpopulation might display.

To address this concern, it has been proposed that a subpopulation of cancer cells resistant to mechanical stress could be isolated. The comparison of such a selected subpopulation with its non-selected counterpart offers a novel cellular model to investigate the factors that affect cell survival in response to mechanical stress and to examine how the subpopulation varies from the original population following such selection. This model is devoid of any artificial genetic or protein disturbances and can therefore mirror the molecular hallmarks at their endogenous scale with only the impact of temporal mechanical stress. The observations made in the context of the resilient subpopulation offer new insights into how mechanical stress can modify metastasizing cancer cells, and the procedure and experimental model can simply be adapted to conduct more comprehensive research into metastasis in the context of physical factors.

6.8.1 Resistance to deformation is an optional characteristic

First, a workflow was developed to isolate cancer cells that can persist after undergoing deformation due to bottlenecks with dimensions close to those of the capillaries (Doerschuk *et al* 1993). To carry out high-throughput deformation, a microfluidic deformation assay was established and refined, which can carry out such mechanical selection on 30 000 individual suspended cancer cells per minute. For practical reasons, the driving pressure applied in this experiment lay outside the physiological limits. To ascertain whether the surviving cancer cells are inherently resistant to such deformation after intense mechanical deformation, a two-step deformation experiment was conducted utilizing the microfluidic deformation testing device for H1650 (lung cancer cell line), MCF-7 (breast cancer cell line) and MDA-MB-231-Luc (luciferase-expressing metastatic breast cancer cell line MDA-MB-231, henceforth referred to as MDA-Luc). The results indicated that the majority of cells that survived the first round of deformation (the survivors) also resisted the second deformation, indicating the existence of a cancer subpopulation that is resistant to deformation, which is referred to as 'mechanoresilient cells' (Jiang *et al* 2023). Subsequently, after subjecting the original cancer cell population (ORI) to the microfluidic deformation assay one time, which is referred to as 'mechanical selection', the surviving cells were harvested and amplified. In various cancer cell lines, the mechanoresilient phenotype was found to be inducible in MCF-7, Mia-PaCa2, MDA-Luc, and BT549-Luc, which is a luciferase-expressing BT549 breast cancer cell line, cell lines. Even though the surviving H1650 cells exhibited resilience to deformation in the two-round deformation assay, this phenotype vanished after cell expansion, whereas the expanded CaCo2 surviving cells also failed to exhibit resilience to deformation. Since MCF-7 cells are non-metastatic epithelial breast cancer cells that are at the very onset of the metastatic cascade, it is of special interest to explore whether and how the mechanoresistant subpopulation emerges in these cancer cells during their early evolution. Therefore, MCF-7 cells were subjected to a more detailed examination. Initially, several rounds of mechanical selection were carried out on the cells, whereby it was determined that a single selection round (the selected cells are referred to below as S1 cells) was sufficient to maximize this type of selection effect (22.87% ± 7.31% in ORI versus 52.30% ± 8.12% in S1 MCF-7, mean ± standard deviation) (Jiang *et al* 2023). It is remarkable that at the exact same time, the population gradually became resistant to cell death induced through deformation. The time frame of the mechanoresilient phenotype was, nevertheless, sufficiently long for downstream cellular and molecular characterizations. Together, these findings demonstrated the potential to isolate and extend a novel cancer subtype that exhibits resistance to cell death induced by mechanical deformation. Together with the original, non-selected population, these selected cancer cells can provide excellent cell models for examining the determinants of cell survival under extreme mechanical stress.

6.8.2 Cellular characteristics are linked to mechanoresilience

After examining the size distribution of MCF-7 cells that were selected for various rounds and comparing them to the original population, the cell size turned out not to be a crucial factor for survival following deformation. To confirm this additionally, when MCF-7 cells are grown at various densities, the disparity in size is utilized, and MCF-7 cells of various sizes are observed to have a similar survival rate. As cell deformability is recognized as an essential factor in how cancer cells handle mechanical stresses, a microfluidic micropipette assay was employed to directly assess the deformability of ORI and S1-MCF-7 cells (figure 6.10). Nevertheless, no significant difference in deformability at the cellular level was detected in this

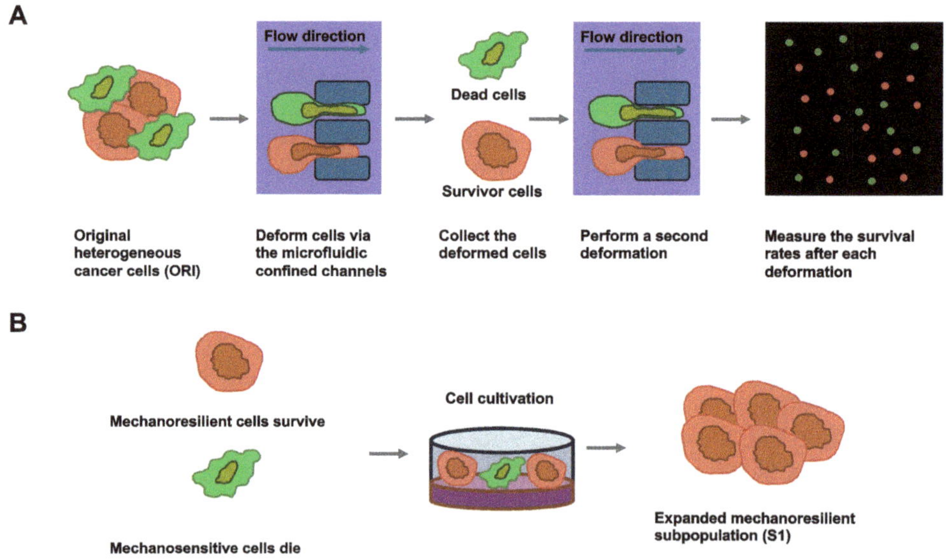

Figure 6.10. Physical deformation can select mechanically resilient cancer cells. (A) Schematic representation of the utilization of microfluidic deformation assay to conduct two rounds of mechanical deformation and measure the survival rate following each deformation round. (B) Schematic representation of the proliferation of surviving cells following deformation of the ORI cells using the microfluidic deformation test.

experiment either. When conducting the selection experiment on various cell lines, it became clear that there are two different types of cellular response following deformation. CaCo2 and Mia PaCa2 cells suffered considerable cell disruption when forced through the bottlenecks, which was not observed for H1650, MCF-7, MDA-Luc, and BT549-Luc cells. This led to substantial cell death in the deformation experiments for Mia-PaCa2 and CaCo2 of 78.2%, whereas the cell loss for the non-rupturing cell lines was only about 3%. The cell death in the second round of Mia PaCa2 cell deformation amounted to about 54.9%, which was close to the expanded S1 population, which indicated a selection impact on the cellular capacity to withstand cell tearing in this cell line. When studying videos of the deformation of a non-rupturing cancer cell line, MCF-7, it was observed that cell disruption occurred in just a small percentage of cancer cells below 1% during cell deformation (Weiss 1990). This implied that the capacity to retain cellular structural integrity during deformation was an effective survival mechanism but was not the primary source of the measured mechanical toughness in MCF-7 cells.

It has already been comprehensively documented that an intact cell nucleus is indispensable for the survival of the cell following severe nuclear deformation (Harada *et al* 2014, Mitchell *et al* 2015, Denais *et al* 2016, Raab *et al* 2016). In fact, a large proportion of the dead cells showed broken cell nuclei following deformation, from which DNA was pushed out of the nuclear lamina. Nuclear laceration was seldom in the surviving cells, although locally deposited lamin A/C or nuclear blebbing was detected in the surviving cells as early as 30 min post-deformation, which is in line with the previously reported nuclear lamina disruption and subsequent healing after directed cell migration (Shah *et al* 2017). These results also imply that nuclear rupture can still arise in surviving cells, albeit to an amount that can be fixed (Harada *et al* 2014, Irianto *et al* 2017). The nuclear deformation was comparatively slow in the limited migration case described above, and cells may have had time to seal a cracked nucleus. In this case, due to the fluid-driven deformation process, the deformation was much quicker and cells with a substantial nuclear crack were improbable to self-repair. Therefore, restricting nuclear lamina breakage during deformation is intrinsically a critical factor for cancer cell survival in the experimental setting. Therefore, to test this, the expression of lamin A/C and lamin B1, two essential structural proteins that underpin the nuclear lamina (Harada *et al* 2014, Gruenbaum and Foisner 2015), was investigated in MCF-7 cells prior to and after mechanical sorting. When compared to the ORI group, S1 MCF-7 cells showed significantly increased lamin B1 levels and slightly elevated lamin A/C levels. To examine the relationship between the expression of lamin and cell survival following deformation, ORI and S1 cells were analyzed immediately upon deformation using a live/dead marker, propidium iodide (PI). This approach identified surviving cells directly after selection. Upon immunostaining, the survival cells exhibited characteristically significantly increased lamin B1 levels, whereas lamin A/C levels failed to correlate consistently with survival. The lamin A/C content also rules out the option that the observed low lamin B1 level in dead cells is due to protein breakdown after cell death. This high lamin B1 phenotype was also identified in the surviving cells of MCF-7-Luc, MDA-Luc, BT549-Luc, H1650, and Mia-PaCa2 following deformation. In addition, in MDA-Luc cells, only 11.3% of the subpopulation with a lamin B1 intensity in the lower 25th percentile survived, whereas 62.4% of the subpopulation with a lamin B1 intensity in the upper 75th percentile survived. In the meantime, the lamin B1 intensity in all examined cell lines is higher in the surviving cells than that of the 70th percentile. Knocking down lamin B1 in ORI- and S1-MCF-7 cells significantly decreased survival following deformation. It is noteworthy that in LMNB1-KD MCF-7 cells, no apparent nuclear bleeding was detected, which had formerly been seen in some cell lines upon the silencing of lamin B1 (Coffinier *et al* 2011, Funkhouser *et al* 2013). Subsequently, overexpression of lamin B1 was found to improve the survival rate of MDA-Luc and BT549-Luc cells post-deformation. These findings indicate that lamin B1 is crucial for determining cell survival following deformation. To obtain a deeper glimpse into the differential behavior of lamin B1 and lamin A/C upon deformation, MCF-7 cells were co-transfected with mCerulean-lamin B1 and mCherry-lamin A/C. Unexpectedly, lamin A/C was temporarily dissociated from a compressed nucleus and the intensity of lamin A/C declined along the nuclear periphery throughout deformation. Lamin B1 remained undisturbed throughout the complete deformation process prior to being constrained to crack at a 2 μm narrowing. These different dynamics of lamin A/C and lamin B1 indicated that lamin B1 mainly functioned to preserve the integrity of the nuclear lamina upon fast nuclear deformation events in MCF-7 breast cancer cells.

To determine whether there were additional alterations in the nuclei of selected cells, the nuclear structures of ORI and S1 MCF-7 cells were examined at various cell stages. The nuclear sizes of the surviving cells were somewhat lower than those of the dead cells. Subsequently, the nuclear lamina was originally invaginated and undulated in the suspended unperturbed state (Malhas *et al* 2011, Kim *et al* 2016a), but became fully extended as the nucleus deformed. These findings are in line with a recent demonstration that the nuclear lamina straightens when it is in a state of compression, which has been described as an important mechanosensory process (Lomakin *et al* 2020). Besides, although the S1 MCF-7 cells had comparable nuclear projected area and perimeter in the suspended state, they had a larger nuclear lamina surface area and larger nuclear projected area and perimeter in the adherent state compared with the ORI cells. Since the nuclear lamina is found to be stretched through cytoskeletal forces generated during cell spreading (Li *et al* 2014), these results suggest that the nuclear lamina surface in S1 cells was more compliant, which could cushion the stress experienced during nuclear stretching as the cells traverse the narrowings. Collectively, these lines of evidence indicated that higher expression of lamin B1 and larger surface areas of nuclear lamina could restrict cell death in the face of severe cellular deformation, thereby contributing to a mechanoresistant breast cancer subset following mechanical selection.

To understand how the expression of lamin genes is linked to carcinogenesis, a cross-cancer meta-analysis was conducted using the TCGA cancer genome database (Vashisth *et al* 2021). First, the expression of LMNA, LMNB1 and LMNB2 in normal and primary tumor tissue specimens was evaluated. Out of the 24 types of cancer examined, both types of B-type lamin were identified: lamin B1 and lamin B2 accumulated considerably in tumor tissues, such 22 out of 24 and 23 out of 24 types of tumors examined, respectively, whereas LMNA accumulated solely in 9 out of 24 types of cancer. Slight downregulation of LMNA was found in breast tumor tissues, whereas LMNB1 and LMNB2 were markedly elevated. These findings pointed to a strong association between B-type laminin and carcinogenesis. Prognostic analyses also indicated that the expression of LMNA, LMNB1, and LMNB2 all conferred a prognostic impact on distant metastasis-free survival (DMFS) of breast cancer patients.

6.8.3 Improved proliferation and resilience to DNA injury upon mechanical selection

The next step was to try to gain a more complete picture of the transcriptomic alterations in the mechanically selected cancer cells. ORI-, S1- and a flow control group (CTRL) MCF-7 cells were analyzed by RNA sequencing. There were only six down-regulated and 19 upregulated genes detected to be differentially expressed between ORI and CTRL cells, suggesting that transient high shearing exposure does not elicit significant transcriptomic alterations in MCF-7 cells. When comparing the groups S1 and ORI or S1 and CTRL, a total of 479 or 280 differently expressed genes were detected, which verified that the mechanical selection caused a significantly modified gene expression pattern in the S1 cells. To concentrate on the impact of mechanical deformation, functional enrichment experiments were conducted on the list of S1 genes in comparison to the CTRL ones. Cell proliferation and the DNA damage response (DDR) are the most affected cellular mechanisms upon mechanical selection. Characteristic gene set enrichment analysis (GSEA) of variably expressed genes indicated that the target genes of two proto-oncogenes, E2F and MYC are pronouncedly upregulated in the selected MCF-7 cells, indicating an enhanced tumorigenic potential of mechanoresistant cancer cells (Dang 1999, Bracken *et al* 2004).

Subsequently, two proliferation assays were conducted to determine whether the modified proliferation genes led to a more proliferative phenotype. In fact, while the selected cancer cells exhibited an initial cell cycle retardation, as reported previously (Pfeifer *et al* 2018, Xia *et al* 2018), when the cells were grown for more than 7 days post-selection, S1 MCF-7 cells proliferated at a much higher level compared to ORI cells. Subsequently, it has been explored whether the modified DDR gene expression had an impact on the function of the cells. To first examine whether aberrant DNA reparation in S1-MCF-7 cells confers resistance to DNA injury, S1-MCF-7 cells were exposed to the chemotherapeutic agent doxorubicin (Dox), which is recognized to elicit DNA double-strand breaks (DSBs) (Yang *et al* 2014). While the IC50 value was not meaningfully altered, the selected cells showed much higher viability at a high concentration of Dox as detected by MTT assays. To assess whether the cells display a resistance to Dox-induced DNA injury, DSB levels in ORI and S1 MCF-7 cells were assessed after exposure to 1 μm Dox, a concentration that has been demonstrated to cause high cell death. S1 cells exposed to 1 μm Dox for 12 h exhibited decreased γH2Ax intensity and a smaller number of γH2Ax foci, reflecting resistance to chemotherapy under high drug concentrations.

6.8.4 Differentially expressed genes upon mechanical selection point to a worse disease outcome

The genes that were variably expressed in the mechanically selected cells yielded several genes that may possibly bridge the gap between intravascular mechanical selection and the malignant metastatic behavior. In the future, the differentially expressed genes that were unique to S1 relative to the ORI and CTRL groups were utilized for a meta-analysis in a large, pooled breast cancer patient cohort. About 40% of the identified genes that were found to be differentially expressed exhibited predictive power in at least one of the analyzed clinical endpoints, such as recurrence-free survival (RFS), overall survival (OS) and DMFS. Genes related to prognosis were most strongly linked to proliferation, indicating that the altered proliferation routes after mechanical selection are related to a worse breast cancer outcome. By performing paired RNA-seq analysis of primary tumors and their matched metastases in the TCGA-BRCA dataset, 7 genes were identified as differentially expressed in metastases and S1-MCF-7 cells, and 31 genes were identified as altered genes in the differentially expressed genes list after removal of the log2-fold alteration cutoff. Moreover, a weighted score of these 31 genes emerged as a strong prognostic marker for breast cancer, which indicated that the breast cancer cells underwent a shift towards a more malignant subtype following mechanical selection. Ultimately, the mechanoresistant gene profiles were analyzed in an *in vivo* selection of metastatic MDA-MB-231 breast cancer cells for metastases within the lungs (MDA_LM2 in GSE2603) (Minn *et al* 2005). There were a remarkable 61 genes in common between the differentially expressed genes of MDA_LM2 and MCF-7_S1. Weighted expression profiling of these 61 genes revealed highly prognostic significance for 3-year lung-metastasis-free survival (LMFS) and low prognostic significance for 5-year LMFS in the identical study cohort of patients with breast cancer. The score also appeared to correlate strongly with the overall survival of patients diagnosed with lymph node metastases. The 61-gene panel was also identified as highly predictive in a larger breast cancer sample set of 1764 patients (Jiang *et al* 2023). Together, these 61 genes constituted a putative set of genes that reflected the mechanical side of *in vivo* metastasis development and coincided with a bad prognosis for breast cancer.

The physical obstacle to metastasis is an essential innate defense of the human organism to tumor invasion and a key to the development of strategies to render malignant tumors locally limited is comprehension of how such impediments are compromised during cancer progression. While it is assumed that mechanical stresses would kill most metastasizing cancer cells (Albertsson *et al* 1995), the surviving cells are of great importance to elucidate the factors that affect cell survival in conditions of extreme mechanical stress. In this case, a mechanical selection workflow with microfluidic assays has been set up that can be easily utilized for the selection of cancer cells that could be viable after a damaging mechanical deformation. Based on this study, the existence of a novel cancer cell subtype, mechanoresistant cancer cells, which can resist mechanical stress-induced cell death, has been demonstrated. This mechanoresilient subpopulation is devoid of any laboratory, genetic or protein disturbances and can be utilized specifically to examine the impact of mechanical stress on cancer cells. The cell model created by a mechanical selection procedure like this can be employed to examine further which type of cancer cell is probably to remain alive under detrimental mechanical stress. While the channel sizes used are mimicking human capillaries, a major caveat of the work is that the pressure used to select the cancer cells was about 50–100 times stronger than what would normally be encountered in capillaries. This is mainly due to the need for such pressure to rinse cancer cells out of the microfluidic system at an adequate rate for subsequent amplification and assessment. While confirmation of the results in a physiologically relevant pressure regime and *in vivo* would provide additional support, both are technically difficult at this time. A compromise like this does not detract from the results, because: (i) such deformation is still transiently attainable in lung capillaries and during muscle contraction; (ii) the conclusions of the subsequent molecular characterization are mostly valid in a lower pressure setting; (iii) the methodology yielded a more comfortable and efficacious *in vitro* operation sequence for examining how mechanical stresses affect the destiny of metastasizing cancer cells. The possible coupled effect of mechanical selection and mechanical perturbation is a further constraint of this methodology. The changed molecules in the original population and the selected subpopulation could be the result of the subpopulation's survival advantage or the mechanical stress that changed the relevant molecular signal transduction cascades in the surviving cells. The live/dead staining approach adopted can largely overcome this problem since the short period of the mechanical selection procedure is usually too short for activation of protein expression, bearing in mind that cells are being fixed right after deformation. Concomitant labeling of proteins with similar structures, such as lamin A/C and lamin B1 in this study, ruled out the influence of protein breakdown resulting from cell death. Nevertheless, this approach cannot rule out the existence of additional up- or downregulation of the protein of interest in the amplified cell population. A combined selection effect can be disentangled using single-cell tracking following mechanical selection, which is a promising avenue for further research.

With this experimental model, the correlation between cell and nucleus size and survival after deformation was investigated. The nuclear size in the surviving fraction is smaller than that in the dead fraction, which is intuitively reasonable as smaller nuclei should suffer less harm during deformation. This is consistent with the similar cell size observed in the MCF7 cell population selected for various rounds or the absence of a difference in survival of MCF-7 cells when grown to various sizes. A plausible interpretation is that the survival benefit of MCF-7 cells with smaller cell sizes or smaller nuclei was negligible relative to the more dominant determinants uncovered in this investigation. Another research study also revealed that the aspect ratio of cells after deformation could be indicative of the metastatic capacity of cells (Yan *et al* 2022). Consistent with this, the more metastatic MDA-MB-231 cells leave the channel with a higher and longer-lasting aspect ratio. In MCF7 cells, the surviving cells after selection had no different nuclear aspect ratios, but it remains to be determined whether the cellular aspect ratio influences survival.

The nuclear lamina was observed to have a more extended area in the selected cells following spreading. In view of the finding that the nuclei in a suspended condition exhibited a larger reserved area, it is assumed that their larger spreading area was mainly caused by such a reserved area prior to spreading It is important to note, nevertheless, that this does not rule out the chance that a higher contractile force of the cytoskeleton leads to a larger lamina area after spreading. This finding indicates that high expression of lamin B1 is critically associated with cell survival after deformation, which is in line with a recent study demonstrating that lamin B1 could limit nuclear breakage in migrating neurons devoid of lamin A/C (Chen *et al* 2019). A consequential question is how Lamin B1 mechanistically relates to the deformed resistance noted. In the first place, according to these results, the surviving cells should naturally exhibit a higher expression pattern following deformation. It is improbable that the detected correlation is due to a mechanically induced upregulation of lamin B1, considering that the cancer cells were directly fixed upon deformation. Therefore, the high amount of lamin B1 in the extended S1 population of MCF-7 cells is also thought to be more attributable to the originally increased lamin B1 expression in the surviving cells. It is worth pointing out that the level of lamin B1 in the selected cells would decline gradually over the prolonged period of cell culture. It is possible that high lamin B1 levels under normal conditions are not a universally beneficial phenotype and may only exist transiently in the MCF-7 population, leading to the dynamic lamin B1 levels (Izdebska *et al* 2017). As mentioned above, it could not be ruled out at this point that the high lamin B1 phenotype in the amplified S1 MCF-7 cells could have been caused through a short-term mechanical activation of lamin B1 expression. Even though lamin B1 was highly related to cell survival following deformation of the various cell lines, it must also be considered that the basic expression of lamin B1 varied greatly in the cell lines tested. For instance, the BT549-Luc and MDA-Luc cell lines employed in this investigation show lower lamin B1 expression relative to MCF7 cells. Nevertheless, MDA-Luc displayed a higher survival rate following deformation in comparison to MCF-7, whereas BT549-Luc displayed the opposite behavior. An accompanying finding is that the assayed cell lines exhibited distinct phenotypes of cellular body integrity following deformation. Importantly, all cell populations universally indicated that their subpopulation with increased lamin B1 expression had a significantly higher survival rate post-deformation. In summary, these results indicate that lamin B1 appears to be a universal factor related to resilience to damaging deformation in all cell lines examined, and that resilience is also a combinatorial function of other, as yet uncharacterized, factors.

Another finding is the dissimilar deformation dynamics of lamin A/C and lamin B1 under deformation. It appears that lamin A/C is more mobilized in the nuclear lamina and can be sequestered from the nucleus during large nuclear stresses in MCF-7 cells, which may be due to its distinct structural architecture (Nmezi *et al* 2019, Patil and Sengupta 2021) in comparison to lamin B1 in the lamina. Even though the results primarily emphasize lamin B1 rather than lamin A/C, this is not inconsistent with previous evidence on the function of lamin A/C in restricted migration (Harada *et al* 2014) or the loss of lamin B1 in enhancing lung cancer metastasis (Jia *et al* 2019). These results show that the functioning of various lamin subtypes at distinct steps of cancer evolution is reliant on specific contexts. Whereas lamin A/C has long been implicated in the control of nuclear mechanics (Lammerding *et al* 2006, Earle *et al* 2020), the findings on the connection between lamin B1 and mechanoresilience reveal a hitherto underappreciated role of B-type lamins in the control of nuclear lamina integrity in extreme deformation and cancer advancement (Vashisth *et al* 2021).

Using RNA sequencing and functional characterization, there was an unsuspected connection between the mechanical selection mechanism and the increased proliferation and activated DNA injury reaction. While the MTT-determined viability distinction between ORI and S1-MCF-7 cells in the doxorubicin concentration assay seems to be marginal, the S7-MCF-7 cells exposed to high mechanical stress for several rounds exhibited considerably higher viability when exposed to high drug concentration. Moreover, S1 MCF-7, although the difference in viability is negligible in comparison to ORI cells, demonstrates a strong decrease in DNA injury following exposure to a high concentration of doxorubicin, which has been proven to cause a strong decline in cell viability. This adds to recent insights into the migration of cancer cells under constraint (Denais *et al* 2016). CTCs and cancer cells in secondary tumors prove to be more cancerogenic and tolerant to DNA-attacking chemotherapeutics. The results indicate that the management of mechanical stress in the metastatic process may enhance proliferative capacity and improve drug resistance. In contrast to previous accounts of mechanically induced proliferation via mechanosensitive channels (Gudipaty *et al* 2017), the modified proliferation and reaction to DNA injury seen in this case is the outcome of a temporary deformation. The expression of cellular lamin B1 has earlier been linked to telomere and chromosome instability (Butin-Israeli *et al* 2015, Pennarun *et al* 2021), the attraction of DNA repair enzymes (Etourneaud *et al* 2021) and senescence of cells (Shimi *et al* 2011). Similar to resistance to deformation, nevertheless, the elevated lamin B1 level was reversed over time, whereas the enhanced proliferation in the mechanically selected cancer cells could persist for more than two months, indicating that the proliferative alteration might be separate from the elevated lamin B1 characteristics. There is growing evidence that both temporary and persistent cellular deformations substantially alter cellular epigenotypes, which compromise phenotypes of cells that rely on epigenetic alterations (Hsia *et al* 2022, Song *et al* 2022). Even though not directly addressed in this approach, the temporary character of several of the effects suggests an epigenetic mechanism as a potential driver for these observations, which deserves future investigation. It is also unclear whether this effect is common to all cells or restricted to cancer cells. While monitoring and summarizing different phenotypes of cancer cells following deformation, non-cancerous cells could exhibit unique characteristics upon selection. This is an important next stage in unravelling the mechanism underpinning the observed mechanoresilience.

Ultimately, the analysis also revealed that the transcriptomic alterations after the selection process are linked to a worse outcome in breast cancer patients. The results of the study indicate that mechanical stress in the metastatic cancer cascade can act to select and alter metastasizing cancer cells, thereby driving a critical step in the transformation of cancer cells into the more difficult-to-treat metastatic phenotype. Metastasis occurs over time, propelling primary cancer cells to become more deadly (Turajlic and Swanton 2016, McGranahan and Swanton 2017). The new aspect is that harmful mechanical deformation can act as a driving force for such metastatic development, whereby the malignant cancer cells become mechanoresistant. The precise mechanisms behind the selection process and how the lamina structure affects cell survival after deformation require additional investigation. The findings should be used to prevent the development of metastases with the assistance of the natural physical microenvironment. Based on these findings, additional work is needed to further elucidate the mechanisms of the selection process. Such studies may lead to new therapeutic strategies that can turn cancer into a manageable localized disease. In addition to mechanoresilience mechanisms, cancer cells' mechanical characteristics are critical for their metastatic capacity. In the following the specific role of stiffer cancer cells in prostate cancer is highlighted and discussed.

6.9 Stiffer prostate cancer cells are aggressive and metastatic

Prostate cancer, nonetheless, seems to diverge from these general oncobiophysical patterns; aggressive metastatic prostate cancer cells seem to be stiffer compared to their weakly metastatic prostate cancer counterparts (Molter *et al* 2022). All cancer types require migration through crowded multicellular microenvironments as they metastasize. At the same time, cancers tend to alter their biophysical characteristics. In fact, cell softening and enhanced contractility turn out to be apparently ubiquitous biomarkers of metastasis formation that can favor metastasis. The stiffness and contractility of cells are also affected through the microenvironment. Stiffer matrices, which mimic the tumor microenvironment, lead to greater contractility of metastatic cells, which in turn enhances contractile cancerogenic phenotypes. Prostate cancer, nevertheless, seems to diverge from these general biophysical patterns in cancer; aggressive metastatic prostate cancer cells tend to be stiffer compared to their weakly metastatic prostate cancer relatives. While it has been suggested that metastatic prostate cancer cells are inherently more contractile compared to normal cells, it is not well understood how cell contractility modifies with rising metastatic capacity of prostate cancer. The biophysical alterations of prostate cancer cells with differing metastatic capabilities were analyzed as a function of the stiffness of the microenvironment. Utilizing a series of cell lines with progressively enhanced metastatic capacity (22RV1, LNCaP, DU145, and PC3), their contractility was quantified with traction force microscopy (TFM), cortical stiffness with optical magnetic twist cytometry (OMTC), and their motile behavior with time-lapse microscopy. Contractility, stiffness and motility of prostate cancer cells were found to not universally increase in line with their metastatic capacity. On the contrary, prostate cancer cells exhibiting varying metastatic capabilities demonstrate unique biophysical reactions that are affected to differing extents by the stiffness of the substrate. Notwithstanding this biophysical heterogeneity, this work suggests that the mechanical microenvironment plays a critical role in dictating the biophysical reaction of prostate cancers of differing metastatic capacity. The study's mechanical emphasis and methodology are unique and provide a complementary strategy to the traditional biochemical and genetic strategies typically employed to elucidate this disease and may yield new perspectives and avenues of approach. Metastasis is characterized as the spread of cancer cells from a primary tumor to a new secondary location, and this phenomenon contributes to at least 66.7% of cancer deaths in solid tumors (Dillekås *et al* 2019). While cancer advancement and metastasis are linked to a variety of mutations, cells in a wide range of diverse cancers tend to alter their biophysical characteristics (Northcott *et al* 2018). More specifically, actin cortex softening (Cross *et al* 2007, Luo *et al* 2016) and augmented contractility (Kraning-Rush *et al* 2012) are emergent conserved hallmarks of metastatic adaptation that can underlie migration in crowded multicellular milieus. These mechanical properties are also modified through the stiffness of the microenvironment. For instance, matrix stiffening, that mimics alterations in the TME, causes metastatic cells to contract more strongly (Kraning-Rush *et al* 2012), thereby favoring cancer phenotypes (Ishihara *et al* 2017). Similarly, invasive cellular transformations accompanying metastatic and resistant prostate cancer, like EMT, experience strong alterations in their contractility (Yoshie *et al* 2018), cortical stiffness (Schneider *et al* 2013), bulk monolayer mechanics (Sutton *et al* 2021), and cell motility (Leggett *et al* 2021).

These commonly seen biophysical patterns in cancer cells are not entirely of universal nature. Prostate cancer, a major silent health threat for men, responsible for 10% of cancer deaths in men (LeBlanc *et al* 2019), seems to diverge from the cell softening tendencies observed in other cancer models, since prostate cancer cells become stiffer as their metastatic potential rises (Faria *et al* 2008, Bastatas *et al* 2012, Lekka *et al* 2012, Khan *et al* 2018, Liu *et al* 2019). The mechanisms and functional implications linked to these aberrations are not yet fully elucidated. The mechanical transitions of prostate cancer in metastatic spread are, nevertheless, an area of increasing clinical relevance. Aggressive and invasive metastatic prostate cancer is accompanied by the emergence of resistance against standard therapeutic approaches (Wang *et al* 2018a; Formaggio *et al* 2021), especially against androgen deprivation therapy (ADT) (Zhu and Kyprianou 2010, Sun *et al* 2012, Huo *et al* 2015, Miao *et al* 2017), that offer valuable mechanical insights under challenging circumstances.

Previous work has typically addressed individual facets of prostate cancer biophysics, like stiffness (Bastatas *et al* 2012, Lekka *et al* 2012) or contractility (Kraning-Rush *et al* 2012) but have not analyzed them together nor assessed them as a function of metastatic capacity or microenvironmental mechanical characteristics. Prostate cancer research typically utilizes a panel of established metastatic cell lines. The best recognized are PC3, DU145 and LNCaP, which metastasize at high, moderate and low rates, respectively (Wu *et al* 2013). These three model cell lines also originate from distinct metastatic niches, where each has a unique mechanical landscape: stiff bone (PC3) (Kaighn *et al* 1979), relatively soft brain (DU145) (Stone *et al* 1978), and soft lymph node metastases (LNCaP) (Horoszewicz *et al* 1980). These can impart unique intrinsic mechanosensitivities and mechanical adjustments. In fact, the mechanical microenvironment plays an important part in the biological and biophysical behavior of prostate cancer cells. For instance, solely PC3 cells showed a substrate stiffness-dependent phenotypic transition (Aw Yong *et al* 2017), comprising cell morphology (Prauzner-Bechcicki *et al* 2015), cytoplasmic compliance (Baker *et al* 2009), proliferation (Liu *et al* 2020c), and migration (Yeoman *et al* 2021). Similar results were also obtained independently by other scientists for LNCaP (Sieh *et al* 2012) and DU145 (Prauzner-Bechcicki *et al* 2015, Aw Yong *et al* 2017). In addition, highly metastatic PC3 cells were found to apply higher forces compared to primary epithelial cells, as a function of stiffness (Kraning-Rush *et al* 2012). So far, there has been no trial to quantify how cell contractility alters with growing metastatic capacity. This is possibly a key characteristic to be addressed, as a recent work demonstrated that highly metastatic PC3 and lowly metastatic LNCaP cell lines exhibit increased *in vivo* metastatic capacity after culturing on substrates mimicking the microenvironmental stiffness of the original site of origin at cell isolation (Liu *et al* 2020c). It should be emphasized that this is irrespective of the absolute value of the substrate stiffness, in contrast to conventional knowledge about the contribution of microenvironmental stiffness in promoting carcinogenesis. These confusing findings show that the mechanisms of the microenvironment are an important but nuanced contributor to prostate cancer advancement. Quantitative biophysical profiling can yield important insights into how these features drive directional cell migration towards metastasis. The interaction between cell mechanics, microenvironmental mechanics and metastatic prostate cancer advancement was elucidated through the quantification of cell contractility, cell stiffness and cell migration (Molter *et al* 2022). The biophysics of various cells of diverse metastatic capacity were found to be dissimilar and scale non-uniformly as a function of metastatic capacity but rather display unique biophysical reactions over a spectrum of microenvironmental stiffening that resembles that *in vivo* (Molter *et al* 2022). The findings underline the necessity to consider the role of the microenvironment when establishing mechanical biomarkers for cancer progression, since it influences the baseline mechanical characteristics, and the mechanical trends measured in several model cell types.

6.9.1 The contractile force of a monolayer of prostate cancer cells exhibits non-monotonic variations with the stiffness of the substrate

The model cell lines display differing morphologies when creating monolayers. 22RV1 and LNCaP are inclined to create continuous monolayers, which hinders the differentiation between individual cells. DU145 also produces continuous monolayers, even though its independent, paving-stone-like shape is simpler to discern. Regarding cell morphology, PC3 cells are the most heterogeneous, displaying a wide range of spreading patterns and diverse geometries. Therefore, PC3 monolayers often develop transient cavities during migration enabling other cells to pass through, while others move collectively as clusters or by gliding alongside their neighboring cells. Cells vary in biophysics and contractility in response to substrate stiffness and metastatic capacity. By linking substrate stiffness and metastatic capacity, it can be explored how increasingly metastatic cells adjust their contractile response in accordance with the varying mechanical microenvironments. For quantifying the contractility of cells, two metrics are employed that characterize their contractility: the root mean square tensile stress (RMST, Pascal) and the strain energy (contractile work, in Joule). The tensile stress refers to the stress applied by the monolayer to the substrate, and the deformation energy refers to the overall energy expended when the monolayer distorts the substrate below it, which is expressed as the work of contractile work (Butler *et al* 2002, Mierke *et al* 2008, 2011).

All cell lines exhibited less deformation on stiffer substrates and converged to similar deformation levels, indicative of noise levels, at 50 kPa substrates. This implies that the 50 kPa substrate is overly stiff for dissipating cell deformations. Thus, the discussion is limited to the contractility trends seen between 1 and 12 kPa. All cells also experience an increase in their traction stress as substrate stiffness rises, which is a commonly anticipated cell response (Califano and Reinhart-King 2010). Nonetheless, strain energy variations with substrate stiffness are observed to differ between cell lines; 22RV1 (tumorigenic) reduces its strain energy with rising substrate stiffness, and this opposite tendency is reflected in the decrease of the peak deformation energies of DU145, although DU145 maintained a wide dispersion of deformation energies with comparable medians over the range of stiffnesses from 3–12 kPa. In contrast, PC3 monolayers displayed a maximum in their mean deformation energy at 3 kPa, which rose from 1 kPa and subsequently dropped back down. In a similar fashion, LNCaP displayed a maximum in their strain energy at 3 kPa. These decoupled correlations between tension and strain energy only emerge when contractility is compared between various substrates, and not in the case of identical stiffnesses, which usually show universally positive relationships for both monolayers and individual cells. A decrease in contractile activity along with increased traction forces in the case of substrate stiffening has already been demonstrated in monolayers (Pasqualini *et al* 2018) and single-cell investigations (Mierke *et al* 2010, 2011, Müller and Pompe 2016). These findings underscore the importance of quantifying contractile forces when dissecting the substrate stiffness reaction of cells.

When various cell types grown on 1-kPa substrates were directly compared, the magnitude of traction stress and deformation energy did not appear to correlate with their relative metastatic capacity. At stiffnesses like prostatic epithelium of 3 kPa and cancer tissue of 12 kPa, LNCaP displays the smallest substrate displacements, traction stresses and strain energies between metastatic cell lines, with median strain energy similar to non-metastatic 22RV1. The aggressive cell lines PC3 and DU145, compared to LNCaP and 22RV1, exhibit significantly elevated strain energy and traction stress on 3-kPa and 12-kPa substrates. This finding is in line with the previously published positive correlation between contractility, metastatic potential and substrate stiffness (Kraning-Rush *et al* 2012). Nevertheless, the strain energy peaks of DU145 are the highest among all cell types for 1 and 3 kPa substrates. Moreover, the median strain energy of DU145 is higher or similar to that of PC3, which is the highest aggressive cell line, over all stiffnesses. These results were unexpected, as traditional biophysical knowledge implies that a more metastatic cell line should be more contractile (Kraning-Rush *et al* 2012).

6.9.2 Primary tumor mechanical microenvironment fails to provide information on stiffness-dependent strain energy trends in metastatic prostate cancer

These metastatic prostate cancer cell lines were derived from different metastatic locations with varying stiffnesses. Consequently, the cell lines may exhibit mechanical adjustments to microenvironmental stiffnesses that are similar to their metastatic site of origin, which may impact their metastatic potential (Liu *et al* 2020c). LNCaP and DU145 were taken from metastatic sites with reasonably low tissue stiffness of about 1 kPa, but they differ in contractility and exhibit different trends in strain energy with rising stiffness. Moreover, LNCaP and PC3, the latter originating from a relatively stiff bone metastasis location, exhibit a similar strain energy at ultralow stiffness of 1 kPa compared to DU145. The discrepancy between DU145 and LNCaP and the occasional similarity between DU145 and PC3 over all stiffnesses indicate that the potential factor of source tissue stiffness alone cannot be brought into line with traction force microscopy.

6.9.3 Metastatic prostate cancer cells with distinct metastatic capabilities exhibit unique thresholds, at which they alter their cortical stiffness distinctively

The apparent moduli (cell stiffness) of the various prostate cancer cell lines are found to change with substrate stiffness. 22RV1 visualized two substrate stiffness thresholds above which a marked increase in apparent cell stiffness, such as the apparent complex shear modulus G*, was evident. LNCaP showed a solitary peak in the apparent complex shear modulus at 12 kPa, though it otherwise displayed comparable median cell moduli. This unique peak at tumor-like substrate stiffness indicates a narrow interval of substrate stiffness across which LNCaP significantly alters its biophysical characteristics. The median stiffness of DU145 is resistant to variations in substrate stiffness, even though there is a slight rise in the apparent modulus when the 1-kPa and 50-kPa groups are compared. Finally, PC3 exhibits a significant rise of the apparent cell modulus beyond a threshold value between 12 and 25 kPa, below and above which the median stiffnesses were similar. For all cell types and stiffnesses, both high and low apparent moduli could be detected, which is probably due to the heterogeneous cell shapes and unrelated mechanical situations present in the monolayer. These measurements were performed in conjunction with TFM to check the contractility of the cells, as actomyosin-driven contractile prestress can influence the cortical stiffness of the cells (Schierbaum *et al* 2019, Chowdhury *et al* 2021). Monolayers taken for cell stiffness measurements produced strain energies and traction stresses that resembled the ranges of the time-course experiments and showed qualitatively analogous stiffness-dependent tendencies. There are several trends of apparent cell modulus as a function of metastatic capacity, while at 3 kPa there is a mild decline in mean and/or median cell moduli as a function of metastatic capacity. In contrast, this trend is inverted, causing stiffening of the cells for the metastatic cell lines grown at 50 kPa. According to established biophysical understanding, cell stiffness should be inversely correlated with increased metastatic behavior (Luo *et al* 2016, Gensbittel *et al* 2021). Contrary to this trend, it has already been observed that the less metastatic LNCaP cell line is softer compared to its more metastatic equivalents (Bastatas *et al* 2012, Liu *et al* 2019). Only LNCaP exhibited the lowest mean apparent modulus on 50-kPa substrates (Molter *et al* 2022). When attached to substrates of medium stiffness, especially those that correspond to the stiffness of the TME, LNCaP exhibited a significant rise in apparent modulus. In this case, the median cell stiffness of LNCaP was higher than that of DU145 and PC3, thereby re-establishing the expected phenotype and indicating that substrate characteristics are a crucial factor for cell mechanics. It is suspected that previous characterizations of LNCaP as the softest metastatic cell line may have been affected by its mechanical characteristics being measured on substrates of extremely high, above-physiological stiffness, like glass.

6.9.4 Velocity and directional persistence of various metastatic prostate cancer cells with distinct potential are regulated differently due to substrate stiffness

Monitoring the motility of prostate cancer cell lines over a prolonged period of time yielded unique stiffness-dependent features. The most significant alterations over the stiffness interval arise in PC3 monolayers, which exhibited an elevation in the 1–12 kPa trajectory regions. When graphing the mean squared displacement (MSD) of the cell as a function of time delay, PC3 curves were seen to shift upwards when compared to less metastatic cells, which indicates more rapid cell motility. This was corroborated when calculating the individual cell velocities, which indicated that PC3 cells raised their mean velocity from 1 to 12 kPa before flattening out from 12 to 50 kPa. The slope of the MSD graphs reflects the character of the cell movement, whereby a slope of larger than 1, 1 and smaller than 1 indicates a directed (persistent) random or confined movement, respectively (Panzetta *et al* 2020). Increasing numbers of MSDs with slopes above 1 are identified as the microenvironment stiffens preferentially for PC3, indicating an increasing fraction of cells with more ballistic, directed migration characteristics that are linked to increased metastatic potential and directed cell migration (Huda *et al* 2018, Liu *et al* 2020b). In addition, the directed persistence of PC3 enhances on higher substrates, which is indicated by the enhancement of the inverse tortuosity of PC3, demonstrating that these cells adopt relatively straight routes. Overall, PC3 shows the most consistent relatively high directionality (MSD slope), directed persistence (tortuosity) and velocity among the three cell types for all stiffnesses. It should be emphasized that despite the observed variations in cell density among the various cell types in this investigation, there was no significant tendency for cells to display increased migration (as indicated by cell velocity and directionality) or contractility (as indicated by traction stress) as a function of cell density.

The stiffness-dependent behavior of PC3 is unparalleled across the measured cell types. Interestingly, 22RV1 showed an opposite relationship between substrate stiffness and inverse tortuosity, with median values falling as a function of substrate stiffness, with the most linear trajectories occurring at lower stiffnesses. The data also revealed that 22RV1 on stiffer substrates generated more constrained motion, as evidenced by the slight leftward shift in the proportion of cells with MSD slopes less than unity. This progressive reverse tendency was not evident in the velocity measurements. Unexpectedly, DU145 seemed to be fairly unaffected by substrate stiffness in both velocity and tortuosity. Importantly, the fraction of cells undergoing persistent directed movement (MSD slope above 1) was generally higher for DU145 compared to 22RV1 and like PC3 although the velocity was quite low under all circumstances. This finding indicates that the collective migration of DU145 might be comparatively slow but still persistent. When these various migratory behaviors are compared with the contractility analyses, it appears that the magnitude of contraction is not associated with a particular velocity or a specific persistent motility phenotype. This indicates that contractility by itself does not allow the motile behavior of prostate cancer in full monolayers to be accurately predicted.

In this research, the biophysical disparities of progressively metastatic prostate cancer cell monolayers are quantitatively characterized as a function of substrate stiffness in relation to their contractility, stiffness and motility. The data reveal that monolayers formed by prostate cancer cell lines of escalating metastatic potential display divergent and unique mechanical signatures in reaction to variable substrate stiffnesses, with their magnitudes and sensitivities failing to scale proportionally or monotonically with metastatic capacity. Instead, these findings suggest that the mechanical characteristics of prostate cancer cells are highly sensitive to their local microenvironment, which may confer distinct migratory benefits and drawbacks during the process of metastasis. In favor of this assumption, it was demonstrated that PC3 and LNCaP exhibit completely distinct metastatic behavior, which is clearly driven by stiff and soft microenvironments, respectively (Liu *et al* 2020c). Such results support the hypothesis that the mechanical uniqueness of the cell lines studied in this work reflects the mechanical adaptations of the cell line source, in contrast to universal adaptations that may accompany the advancement of prostate cancer and the enhancement of its metastatic potential.

Relatively aggressive cells, such as DU145 and PC3 exert higher contractile stresses and perform more work at physiological and cancer-like stiffnesses compared to tumorigenic and weakly metastatic lines, such as 22RV1 and LNCaP, respectively. Nonetheless, TFM-measured contractility cannot absolutely forecast metastatic capacity. Alterations in contractile forces in stiffer microenvironments may impact both cell migration and the global phenotype. Thus, it is possible that cell lines can acquire mechanosensitivity from the mechanical characteristics of their tissue of origin (Liu *et al* 2020c). Moreover, their phenomenon seems to be similar to mechanical memory, as the cells can only react to environmental stiffnesses they had been exposed before, such as the mechanical characteristics at their original tissue site. Therefore, there may be stiffness thresholds above which cells are no longer able to perform the contractile work needed to distort their surroundings. Likewise, it was found that multicellular arrangements of cardiomyocytes exhibit ideal coupling between cell metabolism and contractile work when they are exposed to a narrow spectrum of physiological intermediate stiffnesses (Pasqualini *et al* 2018). Although this is a completely distinct cellular model, this concept could also be relevant for other contractile cells, such as those found in prostate cancer. In fact, the energy expense related to cell-induced substrate translocation has been recognized as a major factor dictating the migratory route of constrained breast cancer cells, which is controlled by both cell stiffness and substrate stiffness (Zanotelli *et al* 2021). A similar rule might exist for prostate cancer, in which resident cells need to interact not merely with the substrate but also with their adjacent cells mechanically. In fact, cooperative cell forces have been demonstrated to influence collective cell migration for various cell types, with cells migrating in the direction of minimal cell–cell shear stress (Tambe *et al* 2011).

In terms of cell stiffness, prostate cancer exhibits both typical trends, that is, decreasing stiffness with metastatic potential, and atypical trends not seen in conventional cancer biophysics. Notably, the atypical trend of prostate cancer cells becoming increasingly stiff as their metastatic potential increases is consistent with earlier findings (Bastatas *et al* 2012, Liu *et al* 2019). These results are in line with a recent study on prostate cancer (Feng *et al* 2023). This indicates that there might be additional mechanisms in prostate cancer that can control its mechanical characteristics. The factor driving this trend, namely environmental substrate stiffness, underlines the relevance of the mechanical microenvironment for the mechanical characteristics of prostate cancer cells. In addition, the stiffness measurements obtained for LNCaP exemplify the potential interaction between cell stiffness and contractility. LNCaP was shown to have both higher contractility and higher cell stiffness ranging from 3 to 12 kPa, which represents a unique range of stiffness over which LNCaP can fine-tune its biophysical characteristics. This dynamic alteration of apparent modulus of LNCaP cells is proposed to arise from the contribution of contraction-mediated cytoskeletal prestress, which is recognized to enhance the dynamic cell stiffness *in situ* (Chowdhury *et al* 2021). At higher substrate stiffnesses, it is hypothesized that LNCaP could relax its cytoskeletal tension, resulting in a reduction in cell stiffness. Former investigations have indicated that LNCaP cells possess a diminished capability to integrate mechanical cues on stiff substrates (Liu *et al* 2020c). This mechanosensitivity could be part of the reason that LNCaP cells exhibited enhanced metastatic potential and expression of EMT biomarkers on soft substrates rather than stiff substrates like PC3 in a mouse model (Liu *et al* 2020c). The study, nevertheless, only examined LNCaP and PC3 cells cultured at 0.7–50 kPa, whereas the differences in contractile and cortical biophysics are revealed in intermediate stiffness regimes that better reflect the transformation from a healthy tissue to a stiffened TME, such as 3–12 kPa (Hoyt *et al* 2008). A separate study in which parental LNCaP cells were analyzed in comparison with their more metastatic-derived cell lines. such as CL-1 and CL-2, indicated that the apparent enhancement of cell stiffness with metastatic capacity could be due to improved adhesion and tension in the cytoskeleton facilitated by focal adhesions (Bastatas *et al* 2012), which the LNCaP monolayer might have acquired through plating on a medium stiffness substrate.

Increased heterogeneity in cell shape was seen in PC3 cells compared to the more cohesive 22RV1, LNCaP, and DU145 Increased heterogeneity mechanical characteristics in reaction to matrix stiffening could be a crucial feature that enables differential migratory capabilities during prostate cancer advancement. On stiffer substrates, this cell shape was also associated with more directed motility and low monolayer tortuosity. Specifically, directed motility may represent a beneficial phenotype during tissue invasion (Huda *et al* 2018, Liu *et al* 2020b). Metastatic cells, such as PC3, have been demonstrated to possess significant heterogeneity in their adhesiveness (adhesion strength), with highly adherent metastatic cells displaying decreased migratory potential (Fuhrmann *et al* 2017). Moreover, loosely adherent PC3 cells paradoxically apply higher traction forces, and the two populations with distinct adhesion strengths exhibit opposite durotactic behavior when exposed to a stiffness gradient (Yeoman *et al* 2021). Even though the differences in motility seen are not attributed to a certain mechanism, the conclusion is that the stiffness-sensitivity does not entirely rely on the metastatic capacity, as the velocity and directionality of DU145 do not seem to surpass that of noninvasive 22RV1 at all stiffnesses (Molter *et al* 2022).

By studying full prostate cancer monolayers, the work by Molter and colleagues differs from other 2D TFM-based characterizations of cancer cell motility and contractility, which have been performed primarily on single cells rather than multicellular assemblies (Lintz *et al* 2017, Lekka *et al* 2021). The increased complexity of mechanical cell–cell interactions and mechanical heterogeneity inside the monolayer poses a hurdle when it comes to assigning contractile measurements to specific cells and their stiffness or migratory tendencies. Nevertheless, the multicellular characterization offers valuable information in a system closer to physiological circumstances. It is known that cell–cell interaction affects the migratory behavior of PC3 cells within a 3D matrix (Cui and Yamada 2013), and both cell–cell adhesion and cell grouping were shown to be crucial for LNCaP metastasis and the accompanying increase in EMT biomarkers (Liu *et al* 2021a). This indicates that mechanical crosstalk between adjacent cells can have a significant impact on the progression of prostate cancer.

Moreover, in addition to the mechanisms discussed here, environmental architecture, such as 3D versus 2D, and the adhesive ligand are also key modulators of prostate cancer mechanical characteristics, which may diverge from the general trends presented. In terms of architecture, matrix stiffening has variable effects on the cytoplasmic stiffness of PC3 in 2D or 3D collagen microenvironments (Baker *et al* 2009), and LNCaP displays strong alterations in shape and cytoskeletal arrangement between 2D and 3D culture (Sieh *et al* 2012). In fact, cell geometry and orientation are accepted predictors of cell mechanics (Oakes *et al* 2014). When comparing the locomotion of metastatic and non-tumorous breast cancer cells, the trends in migration velocity and persistence varied according to the dimensionality of the motility assessment (The Physical Sciences—Oncology Centers Network 2013). Thus, the link between 2D biophysical characterizations and 3D *in vitro* or *in vivo* migratory patterns must be considered with prudence and in the context of the various extrinsic factors that can impact prostate cancer mechanical characteristics, mechanosensing, spatial restriction, and migratory mode (Mierke 2019). In terms of the function of the adhesive ligand, prostate cancer cells can express various integrins, which are the receptors that specifically sense and adhere to the ECM proteins, throughout the progression of prostate cancer, which can consequently weaken or strengthen the mechanical linkage between the cell and the substrate and the resulting biophysical reaction (Goel *et al* 2008, Sutherland *et al* 2012). For instance, elevated expression of collagen-I-binding integrin α2β1 is correlated with bone metastases *in vivo* (Sottnik *et al* 2013). Most interestingly, in an experimental setting, selective culturing on collagen I led to the development of a collagen I-binding LNCaP cells, whereas the parent line is non-binding, which caused enhanced α2β1 expression, augmented invasiveness, and the gained capacity to metastasize to the bone (Hall *et al* 2006, 2008). This increased invasive phenotype was facilitated through collagen/α2β1-facilitated activation of the preceding acto-myosin contractility regulator RhoC (Hall *et al* 2006, 2008). Similarly, it was observed that PC3 exhibits enhanced adhesion and cell stiffness when seeded on collagen I, whereas parental LNCaP cells show no such effect (Docheva *et al* 2010), which is partly attributed to their inherently low adhesion (Liberio *et al* 2014) and absence of the necessary integrin receptor (Hall *et al* 2006). This ligand-dependent spreading and cell stiffening is in line with the contractility causing stiffening tendencies in prostate cancer as a function of metastatic capacity on stiff substrates with high tensile stresses, whereas they soften on soft substrates under lower stresses. Apparently, the varying extracellular environments of metastasizing cells and their specific substrate-perceiving proteome can be an extremely crucial factor in the mechanical phenotypes that develop during the advancement of prostate cancer and can be more varied *in vivo* than those observed in monolayers on fibronectin. Since the biophysics of cells is affected by both biochemical and mechanical alterations in metastasis, a key question is how these two cell-regulatory aspects work together to establish cellular characteristics. Nonetheless, additional substrate-independent effects may account for this response, since enhanced metastatic capacity was also seen with stiffened prostate cancer cells within suspension culture (Liu *et al* 2019).

The implications of these findings are practical for the characterization of cells *in vitro*, as a large number of readouts are performed on cells grown on glass substrates with super-physiological stiffnesses (Faria *et al* 2008, Bastatas *et al* 2012, Lekka *et al* 2021), or in mechanophenotyping platforms, like microfluidic devices, that use suspended cells and apply non-physiological strain rates (Liu *et al* 2019). While these established techniques can detect cancerous cell fractions, it is worth highlighting that contractility and cell stiffness are not absolute predictors of prostate cancer metastasis potential, as these biophysical reactions of these cells have no correlation with each other. In addition, this work shows that distinct mechanical microenvironments elicit divergent alterations in their biophysics and need to be addressed when utilizing biophysical markers to establish a holistic physical 'finger-print' of the cell. In the future, it would be ideal if characterizations were to be carried out within an isogenic cell line with reproducible alterations in its biology and its capacity for metastasis. While bone is the most frequent target site for prostate cancer metastasis (Gandaglia *et al* 2014), and PC3 is the most extensively mechanically characterized prostate cancer cell line in the scientific literature, it is hypothesized that PC3 is perhaps not fully representative of the biophysics of all types of metastatic prostate cancer. In contrast, the observed independent mechanical alterations are proposed to be indicative not merely of their metastatic capacity and the substrate stiffness at which they were assessed, but also of the biophysics that specifically allowed their colonization at their metastatic niche.

6.10 Conclusions and future directions

Due to the heterogeneity of tumors, it is challenging to determine whether the bulk tumor has become stiffer or softer. When individual cells are analyzed for their mechanical cues, such as stiffness, viscoelasticity or force generation it is often just a single timepoint that is analyzed per cell. The dynamical changes are not considered. In summary, the cell lines commonly employed to model prostate cancer metastasis and cancer advancement are mechanically dissimilar. Perhaps most strikingly, migration behavior, contractility, and cell stiffness alteration of prostate cancer are all affected in different ways according to substrate stiffness. It is speculated that advancing prostate cancer may display unique dynamic ranges of stiffness and/or contractility, to which it fine-tunes its biophysical characteristics. These results highlight the need for a proper characterization of cancer cell mechanics in their physiologically relevant mechanical milieus for the proper identification of a more robust repertoire of contractile force-dependent biomarkers. In future work seeking to elucidate the biophysics of prostate cancer, the mechanical environment of the prostate cancer cells *in vivo* during metastasis should be considered, and observations of cells within their physiological mechanical environment may yield more precise interpretations of the biophysical mechanisms that favor prostate cancer advancement.

The mechanoresponsive mechanisms are involved in a wide range of cell shape alterations and are also strongly linked to cancer advancement and the formation of metastases. In addition, expression levels of several of these proteins are dysregulated in multiple cancers in distinct manners, likely resulting in the altered cellular mechanics and mechanoresponsivity linked to cancer advancement, such as myosin IIA, myosin IIB, α-actinin 1, α-actinin 4, filamin A, filamin B and 14-3-3 (figure 6.11) (Cross *et al* 2007, Weins *et al* 2007, Xu *et al* 2010, Gao *et al* 2015, Surcel *et al* 2019).

Notwithstanding their various degrees of cancer severity, mechanobiome proteins have been widely neglected in the development of medications and in studies. This failure can be traced back to four main statements. The first issue is that the classical view of these proteins often assumes a single key role for each of them in the cell, failing to take into consideration the diversity of functions that can be achieved by varying the ratios between the paralog groups. For instance, the conventional definition of non-muscle myosin II is that of the contractility necessary for cytokinesis and motility. The functions of myosin II are far more comprehensive, spanning roles in mechanosensation, elasticity and viscoelasticity, cortical tension and fluidity, regulation of cell adhesion to substrates as well as to other cells, the integration of signaling and mechanical cues, and the effects of global cell mechanics on a plethora of other cellular behaviors (De Lozanne and Spudich 1987, Bai *et al* 2016, Schiffhauer *et al* 2016, Halder *et al* 2019). It is the interaction between the three myosin isoforms that distorts cellular responses in disease settings (Picariello *et al* 2019, Surcel *et al* 2019); thus, comprehension of how the isoforms are fully involved in all NMII functions will elucidate how alterations in their expression lead to transformative cancer cells. Likewise, most research on 14-3-3 in the area of

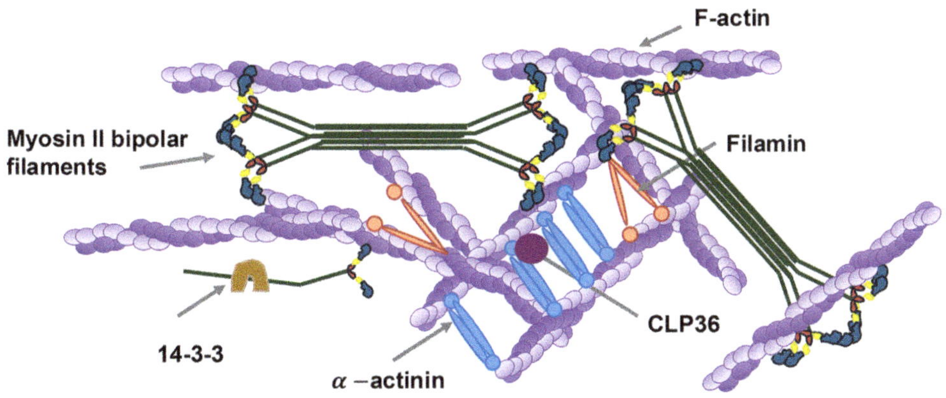

Figure 6.11. Mechanosensitive proteins control cell structure and the ability to adjust to mechanical stress, and their expression levels are often increased during cancer development. Mechanosensitive proteins comprise non-muscle myosin II, α-actinin, and filamin. Myosin II forms bipolar filaments that are then assembled into actin networks and stress fibers within the cells. α-actinins are antiparallel dimers that bundle actin filaments. Filamins, for their part, are V-shaped dimers which also crosslink actin filaments. 14-3-3 proteins can attach to the myosin II tail region, where they perform a number of biochemical functions, including regulating the bipolar alignment of myosin II filaments. CLP36 associates with α-actinin and actin-rich features such as stress fibers and the cell cortex.

cancer has concentrated on its involvement in processes such as DNA injury response (Ferguson *et al* 2000), with lesser emphasis on its crosstalk with cytoskeletal components (Gohla and Bokoch 2002, West-Foyle *et al* 2018). Since seven isoforms have partially redundant function and differential expression in various cancers, a complete functional characterization will enable more selective modulation of individual 14-3-3 isoforms in cancer.

The second issue is that the ubiquitous expression of mechanosensitive proteins throughout multiple cell types has led to the hypothesis that targeting them will be detrimental to human patients. This is true even though the proteins Kras, Rho and Aurora kinase, which are likewise abundantly expressed, are the focus of several pharmaceutical trials for cancer, which is reviewed in (Cicenas 2016, Feng *et al* 2016). Moreover, the families of proteins that are mechanoresponsive are frequently regarded as the aggregate of all their isoforms or the isoform that is most frequently expressed. The most obvious case is that of myosins, where research on non-muscle myosin IIs is concentrated on IIA and IIB, and IIC is rarely investigated because of its low abundance relative to IIA and IIB. Despite its lower levels, such as 18 nM myosin IIC versus 565 nM IIA in pancreatic adenocarcinoma cancer cells, myosin IIC aids in easing actin reorganization and retrograde flux and cooperates with myosin IIA to enhance dissemination and metastasis (Surcel *et al* 2019, Bryan *et al* 2020). These findings are an example of how low-frequency proteins, often neglected in big data analysis, may in fact be suitable candidates for targeted drug delivery. Even though protein abundance must be considered in a new perspective when determining the significance of a particular protein for cancer advancement, it is also noteworthy that mechanoresponsiveness, along with almost all cellular processes, can also be adjusted by post-translational modifications. An important finding is that NMIIB exhibits highly cell-type-specific and even cell-cycle-specific mechanoresponsiveness (Schiffhauer *et al* 2019). Phosphorylation of the myosin II heavy chain, which is performed by PKCζ, confers this variable degree of mechanoresponsivity. This instance certainly shows the importance of bearing in mind the particular setting of each disease, since this is probably what influences the most appropriate strategy for combating it.

Thirdly, and possibly most significantly, proteins that are found to be highly expressed in cancers are frequently used for pharmacological intervention. In fact, in cancer systems, disabling or inhibiting mechanosensitive proteins can lead to increased dissemination and animals with more metastatic burden (Hotulainen and Lappalainen 2006, Conti *et al* 2015, Picariello *et al* 2019, Surcel *et al* 2019). Rather than discarding mechanosensitive proteins as non-targeted, evidence from several research studies indicates that the pharmaceutical paradigm must move beyond mere inhibition. Mechanical adaptability is on a continuous spectrum, with cells in a particular tissue environment having an optimal balance between adaptability/plasticity and stability. Cancer cells strive for all kinds of adaptability, but have developed several ways to assure their survival, so that inhibiting them alone is not adequate to curtail cancer cell behavior and, in several cases, intensify their invasive nature. Alternatively, a feasible experimental strategy is to design small molecules that can selectively activate mechanosensitive proteins and drive them into a regime where they function more robustly than their non-mechanosensitive sister paralogs. A possibility to create this hyperactivation is to enhance the binding affinity of proteins to cytoskeletal binding partners like actin filaments, thereby inhibiting their breakdown and diminishing other abnormal morphologies (figure 6.12). As a matter of fact, a similar approach was employed effectively for omecamtiv mecarbil, which is a highly selective activator of cardiac myosin and is in phase 3 clinical trials for the treatment of hypertrophic cardiomyopathy (Morgan *et al* 2010). In two separate investigations, 4-hydroxyacetophenone (4-HAP) demonstrates encouraging results, specifically activating myosin IIC by attaching it to actin filaments in pancreatic duct adenocarcinoma and colon cancer models. It seems that shifting the activation/inhibition balance in favor of activation can curtail cancer progression in mouse models (Surcel *et al* 2019, Bryan *et al* 2020). Considering mechanoresponsive proteins as focused drug targets (figure 6.12) can go a long way towards transforming the landscape of the battle against cancer.

A final hurdle is that the evolution of anti-metastatic cancer therapies usually relies on the assessment of primary tumor size, focusing on tumor size decrease. The rationale is that the primary tumor is simpler to visualize due to the larger tumor mass, and that tumor shrinkage may be a quicker and more appropriate proxy for the long-term efficacy of the medication for patients. This hurdle, nevertheless, increases the risk of underestimating the true potential of anti-metastatic therapeutic approaches (Welch and Hurst 2019). Interestingly, at least two conventional anti-cancer therapeutics, the DNA-damaging cisplatin and the microtubule-stabilizing docetaxel, also alter the mechanical characteristics of cancer cells, contributing to a decrease in invasiveness (Raudenska *et al* 2019). Improved approaches that more

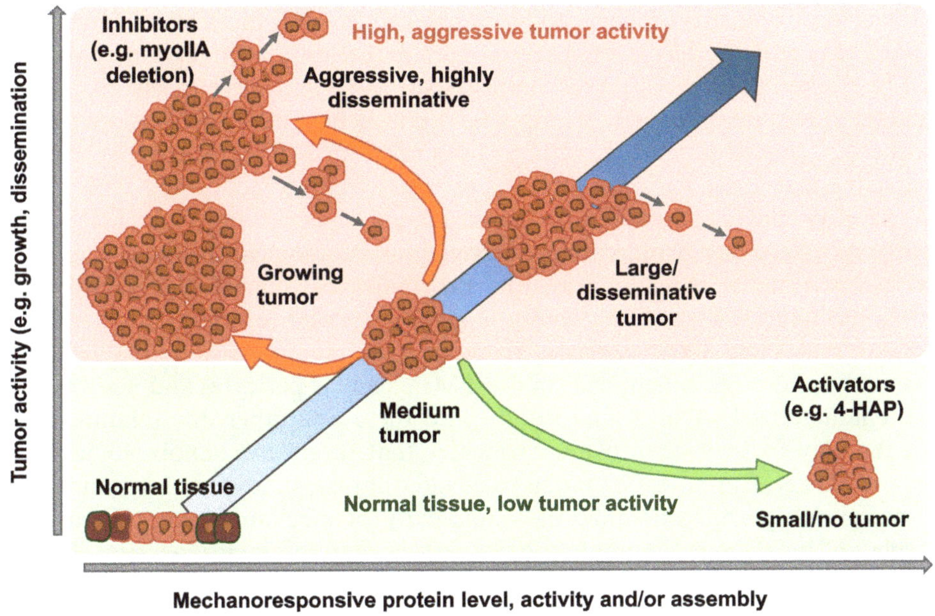

Figure 6.12. The shift of the activation curve of mechanosensitive proteins and their effectors is a feasible approach for the design of anti-cancer medications. Cancer advancement and the formation of tumors are characterized by an aberrant expression of the mechanosensitive proteins, comprising α-actinin, filamin, and non-muscle myosin II (NMII) as well as partner proteins such as 14-3-3 and CLP36. When not counteracted, the modified gene expression levels coincide with enhanced cell activity, especially with metastasis generation. Since mechanoresponsive proteins like NMIIs also exert tumor-suppressing activity through repression of signaling cascades like the ERK pathway, pharmacological intervention may result in enhanced tumor growth. By contrast, activators can trap a cell, resulting in anti-cancer actions. An example of this is the activator 4-HAP, which enhances non-muscular myosin IIB and IIC organization, causing elevated cortical tension and reduced metastatic activity of the tumor.

fully exploit underappreciated mechanosensitive proteins may provide the strength needed to generate new anti-metastatic cancer drugs. Alternatively, upregulation of mechanosensitive proteins could provide a molecular screening panel for metastases. α-actinin 4 is one such possibility for cervical cancer (Ma *et al* 2020). Overall, the mechanobiome, specifically the mechanoresponsive proteins and their regulatory circuitries, offers tremendous possibilities for cancer therapy. By targeting this mechanism, a broader spectrum of cancers can potentially be blocked at their deadliest point of attack, such as the generation of metastases, whereby the side effects for patients are reduced to a minimum.

Biophysical forces clearly regulate several phases of metastasis, and this regulation seems to comprise more than a mere bi-directional crosstalk between the ECM and cancer cells. Nonetheless, the current state of scientific knowledge is still in its infancy. The molecular and cellular intricacy of cancers is integrated in their mechanobiology. Fibroblasts utilize contractility to reshape ECM, endothelial cells react to shear forces and tensile stresses, cancer cells adjust to the biophysical characteristics of distinct niches, and immune cells exploit mechanical forces to lyse targeted cancer cells. The complexity of tumor disease and malignant progression must be studied in terms of its mechanical aspects so that these mechanisms can be elucidated and modified by therapeutic approaches to achieve more effective treatment of cancer patients that can be used in many different stages of the disease. Unraveling how these various pathways progress and interact during tumor growth and development demands more sophisticated experimental systems that are capable of effectively demarcating different microecological, cellular, and subcellular compartments while offering the possibility of quantifying and disrupting distinct mechanical signatures in physiologically relevant settings (Amoh *et al* 2005, Hoffman and Yang 2006b, 2006a). Moreover, the biophysical interaction between cancer cells and other salient cells in the TME, such as pericytes and neurons, is yet to be characterized. A profound insight into these and other interrelations should pave the way for new therapeutic strategies that target mechanobiological weaknesses in cancer. Care must be taken to ensure that hypotheses and assumptions do not oversimplify and generalize the complexity of mechanobiology, ignoring the interplay of biochemical, immunological and mechanobiological aspects. The same mistake must not be made here as with the sequencing of the human genome, namely assuming that it is now possible to identify all tumor-relevant genes. This is an important step in understanding tumors, but it is also no more than one component that is dynamically modified and therefore still represents a major challenge.

The impact of overall health on the biophysical terrain in cancer is also an emerging focus of attention. For instance, obesity is widespread in countries that follow a Western diet and has far-reaching implications for the mechanical scenery of the TME. Obesity can cause profound alterations in the composition and architecture of the ECM scaffold, thereby affecting mechanotransduction in cancer cells, immunological cells, endothelial cells and stromal cells (Druso and Fischbach 2018, Wishart *et al* 2020, Mierke 2024b, 2024a). Obesity is also linked to high cholesterol values, which can impact the phenotypes of immune cells, cancer cells, stroma cells and vascular cells via mechanoregulation of membrane fluidity and stiffness (Levitan 2016, Riscal *et al* 2019, Liu *et al* 2021c). With aging, the arterial wall loses its elasticity and becomes stiffer, which impacts the perivascular metastatic niche (Lee and Oh 2010). Ultimately, physical activity that enhances hydrodynamic shear stress within the circulation may modulate the survival of CTCs (Fan *et al* 2016, Brown *et al* 2018). Investigating these systemic mechanisms and their mechanobiology in the setting of metastases has the power to direct the clinical management of metastatic disease and offer new therapeutic strategies. In summary, a fundamental grasp of the mechanobiology powering growth and suppression of metastases can shed valuable light on the regulatory circuitry of the metastatic cascade and unveil new targets for therapeutic interference. By combining these biophysical findings with data from multiomics, intravital imaging and clinical trials, researchers can investigate mechanoregulatory signaling pathways in their proper physiological setting, thus creating the basis for translationally meaningful advancements in the research field.

References and further reading

Acerbi I, Cassereau L, Dean I, Shi Q, Au A, Park C *et al* 2015 Human breast cancer invasion and aggression correlates with ECM stiffening and immune cell infiltration *Integr. Biol.* **7** 1120–34

Aceto N 2020 Bring along your friends: homotypic and heterotypic circulating tumor cell clustering to accelerate metastasis *Biomed. J.* **43** 18–23

Aceto N, Bardia A, Miyamoto D T, Donaldson M C, Wittner B S, Spencer J A *et al* 2014 Circulating tumor cell clusters are oligoclonal precursors of breast cancer metastasis *Cell* **158** 1110–22

Albertsson P A, Nannmark U and Johansson B R 1995 Melanoma cell destruction in the microvasculature of perfused hearts is reduced by pretreatment with vitamin E *Clin. Exp. Metast.* **13** 269–76

Albrengues J, Shields M A, Ng D, Park C G, Ambrico A, Poindexter M E *et al* 2018 Neutrophil extracellular traps produced during inflammation awaken dormant cancer cells in mice *Science* **361** eaao4227

Alexandrov L B, Nik-Zainal SAustralian Pancreatic Cancer Genome Initiative, ICGC Breast Cancer Consortium, ICGC MMML-Seq Consortium, ICGC PedBrain *et al* 2013 Signatures of mutational processes in human cancer *Nature* **500** 415–21

Al-Hajj M, Wicha M S, Benito-Hernandez A, Morrison S J and Clarke M F 2003 Prospective identification of tumorigenic breast cancer cells *Proc. Natl Acad. Sci. USA* **100** 3983–8

Amendola P G, Reuten R and Erler J T 2019 Interplay between LOX enzymes and integrins in the tumor microenvironment *Cancers* **11** 729

Amoh Y, Yang M, Li L, Reynoso J, Bouvet M, Moossa A R *et al* 2005 Nestin-linked green fluorescent protein transgenic nude mouse for imaging human tumor angiogenesis *Cancer Res.* **65** 5352–7

Amos S E and Choi Y S 2021 The cancer microenvironment: mechanical challenges of the metastatic cascade *Front. Bioeng. Biotechnol.* **9** 625859

Anandi L, Garcia J, Ros M, Janská L, Liu J and Carmona-Fontaine C 2025 Direct visualization of emergent metastatic features within an *ex vivo* model of the tumor microenvironment *Life Sci. Alliance* **8** e202403053

Ansari N A and Rasheed Z 2010 Non-enzymatic glycation of proteins: from diabetes to cancer *Biomed KHIM* **56** 168–78

Aragona M, Panciera T, Manfrin A, Giulitti S, Michielin F, Elvassore N *et al* 2013 A mechanical checkpoint controls multicellular growth through yap/taz regulation by actin-processing factors *Cell* **154** 1047–59

Armanios M Y, Grossman S A, Yang S C, White B, Perry A, Burger P C *et al* 2004 Transmission of glioblastoma multiforme following bilateral lungtransplantation from an affected donor: case study and review of theliterature *Neuro-Oncol.* **6** 259–63

Aw Yong K M, Sun Y, Merajver S D and Fu J 2017 Mechanotransduction-induced reversible phenotypic switching in prostate cancer cells *Biophys. J.* **112** 1236–45

Bai H, Zhu Q, Surcel A, Luo T, Ren Y, Guan B *et al* 2016 Yes-associated protein impacts adherens junction assembly through regulating actin cytoskeleton organization *Am. J. Physiol.-Gastrointest. Liver Physiol.* **311** G396–411

Baker A-M, Bird D, Lang G, Cox T R and Erler J T 2013 Lysyl oxidase enzymatic function increases stiffness to drive colorectal cancer progression through FAK *Oncogene* **32** 1863–8

Baker E L, Bonnecaze R T and Zaman M H 2009 Extracellular matrix stiffness and architecture govern intracellular rheology in cancer *Biophys. J.* **97** 1013–21

Basnet H, Tian L, Ganesh K, Huang Y-H, Macalinao D G, Brogi E *et al* 2019 Flura-seq identifies organ-specific metabolic adaptations during early metastatic colonization *eLife* **8** e43627

Bastatas L, Martinez-Marin D, Matthews J, Hashem J, Lee Y J, Sennoune S *et al* 2012 AFM nano-mechanics and calcium dynamics of prostate cancer cells with distinct metastatic potential *Biochim. Biophys. Acta (BBA)—Gen. Subj.* **1820** 1111–20

Basu R, Whitlock B M, Husson J, Le Floc'h A, Jin W, Oyler-Yaniv A *et al* 2016 Cytotoxic T cells use mechanical force to potentiate target cell killing *Cell* **165** 100–10

Baumgart E 2000 Stiffness—an unknown world of mechanical science? *Injury* **31** 14–84

Beeghly G F, Amofa K Y, Fischbach C and Kumar S 2022 Regulation of tumor invasion by the physical microenvironment: lessons from breast and brain cancer *Annu. Rev. Biomed. Eng.* **24** 29–59

Beier D, Hau P, Proescholdt M, Lohmeier A, Wischhusen J, Oefner P J *et al* 2007 CD133+ and CD133− glioblastoma-derived cancer stem cells show differential growth characteristics and molecular profiles *Cancer Res.* **67** 4010–5

Beningo K A and Wang Y 2002 Fc-receptor-mediated phagocytosis is regulated by mechanical properties of the target *J. Cell Sci.* **115** 849–56

Bergers G and Fendt S-M 2021 The metabolism of cancer cells during metastasis *Nat. Rev. Cancer* **21** 162–80

Bergonzini C, Kroese K, Zweemer A J M and Danen E H J 2022 Targeting integrins for cancer therapy—disappointments and opportunities *Front. Cell Dev. Biol.* **10** 863850

Blumenthal D, Chandra V, Avery L and Burkhardt J K 2020 Mouse T cell priming is enhanced by maturation-dependent stiffening of the dendritic cell cortex *eLife* **9** e55995

Bos R, Van Der Groep P, Greijer A E, Shvarts A, Meijer S, Pinedo H M *et al* 2003 Levels of hypoxia-inducible factor-1α independently predict prognosis in patients with lymph node negative breast carcinoma *Cancer* **97** 1573–81

Boyd N F, Li Q, Melnichouk O, Huszti E, Martin L J, Gunasekara A *et al* 2014 Evidence that breast tissue stiffness is associated with risk of breast cancer *PLoS One* **9** e100937

Boyle S T, Poltavets V, Kular J, Pyne N T, Sandow J J, Lewis A C *et al* 2020 ROCK-mediated selective activation of PERK signalling causes fibroblast reprogramming and tumour progression through a CRELD2-dependent mechanism *Nat. Cell Biol.* **22** 882–95

Bracken A P, Ciro M, Cocito A and Helin K 2004 E2F target genes: unraveling the biology *Trends Biochem. Sci.* **29** 409–17

Brahimi-Horn M C, Chiche J and Pouysségur J 2007 Hypoxia and cancer *J. Mol. Med.* **85** 1301–7

Brooks D L P, Schwab L P, Krutilina R, Parke D N, Sethuraman A, Hoogewijs D *et al* 2016 ITGA6 is directly regulated by hypoxia-inducible factors and enriches for cancer stem cell activity and invasion in metastatic breast cancer models *Mol Cancer* **15** 26

Brown J C, Rhim A D, Manning S L, Brennan L, Mansour A I, Rustgi A K *et al* 2018 Effects of exercise on circulating tumor cells among patients with resected stage I–III colon cancer *PLoS One* **13** e0204875

Bryan D S, Stack M, Krysztofiak K, Cichoń U, Thomas D G, Surcel A *et al* 2020 4-Hydroxyacetophenone modulates the actomyosin cytoskeleton to reduce metastasis *Proc. Natl Acad. Sci. USA* **117** 22423–9

Buckley S T, Medina C, Davies A M and Ehrhardt C 2012 Cytoskeletal re-arrangement in TGF-β1-induced alveolar epithelial-mesenchymal transition studied by atomic force microscopy and high-content analysis *Nanomed. Nanotechnol. Biol. Med.* **8** 355–64

Bushweller J H 2019 Targeting transcription factors in cancer—from undruggable to reality *Nat. Rev. Cancer* **19** 611–24

Butcher D T, Alliston T and Weaver V M 2009 A tense situation: forcing tumour progression *Nat. Rev. Cancer* **9** 108–22

Butin-Israeli V, Adam S A, Jain N, Otte G L, Neems D, Wiesmüller L *et al* 2015 Role of lamin B1 in chromatin instability *Mol. Cell. Biol.* **35** 884–98

Butler J P, Tolić-Nørrelykke I M, Fabry B and Fredberg J J 2002 Traction fields, moments, and strain energy that cells exert on their surroundings *Am. J. Physiol.-Cell Physiol.* **282** C595–605

Califano J P and Reinhart-King C A 2010 Substrate stiffness and cell area predict cellular traction stresses in single cells and cells in contact *Cel. Mol. Bioeng.* **3** 68–75

Calvo F, Ege N, Grande-Garcia A, Hooper S, Jenkins R P, Chaudhry S I *et al* 2013 Mechanotransduction and YAP-dependent matrix remodelling is required for the generation and maintenance of cancer-associated fibroblasts *Nat. Cell Biol.* **15** 637–46

Cao X, Lin Y, Driscoll T P, Franco-Barraza J, Cukierman E, Mauck R L *et al* 2015 A chemomechanical model of matrix and nuclear rigidity regulation of focal adhesion Size *Biophys. J.* **109** 1807–17

Carracedo A, Cantley L C and Pandolfi P P 2013 Cancer metabolism: fatty acid oxidation in the limelight *Nat. Rev. Cancer* **13** 227–32

Carracedo A, Weiss D, Leliaert A K, Bhasin M, De Boer V C J, Laurent G *et al* 2012 A metabolic prosurvival role for PML in breast cancer *J. Clin. Invest.* **122** 3088–100

Cassetta L and Pollard J W 2018 Targeting macrophages: therapeutic approaches in cancer *Nat. Rev. Drug Discov.* **17** 887–904

Catenacci D V T, Junttila M R, Karrison T, Bahary N, Horiba M N, Nattam S R *et al* 2015 Randomized phase Ib/II study of gemcitabine plus placebo or vismodegib, a hedgehog pathway inhibitor, in patients with metastatic pancreatic cancer *J. Clin. Oncol.* **33** 4284–92

Chang L, Azzolin L, Di Biagio D, Zanconato F, Battilana G, Lucon Xiccato R *et al* 2018 The SWI/SNF complex is a mechanoregulated inhibitor of YAP and TAZ *Nature* **563** 265–9

Chaudhuri O, Koshy S T, Branco da Cunha C, Shin J-W, Verbeke C S, Allison K H *et al* 2014 Extracellular matrix stiffness and composition jointly regulate the induction of malignant phenotypes in mammary epithelium *Nat. Mater* **13** 970–8

Chen M B, Whisler J A, Jeon J S and Kamm R D 2013 Mechanisms of tumor cell extravasation in an *in vitro* microvascular network platform *Integr. Biol.* **5** 1262

Chen N Y, Yang Y, Weston T A, Belling J N, Heizer P, Tu Y *et al* 2019 An absence of lamin B1 in migrating neurons causes nuclear membrane ruptures and cell death *Proc. Natl Acad. Sci. USA* **116** 25870–9

Chen W, Lou J and Zhu C 2010 Forcing switch from short- to intermediate- and long-lived states of the Aa domain generates LFA-1/ICAM-1 catch bonds *J. Biol. Chem.* **285** 35967–78

Chen X, Bahrami A, Pappo A, Easton J, Dalton J, Hedlund E *et al* 2014 Recurrent somatic structural variations contribute to tumorigenesis in pediatric osteosarcoma *Cell Rep.* **7** 104–12

Cheung K J, Gabrielson E, Werb Z and Ewald A J 2013 Collective invasion in breast cancer requires a conserved basal epithelial program *Cell* **155** 1639–51

Chivukula V K, Krog B L, Nauseef J T, Henry M D and Vigmostad S C 2015 Alterations in cancer cell mechanical properties after fluid shear stress exposure: a micropipette aspiration study *Cell Health Cytoskelet* **7** 25–35

Choi J, Kim D H, Jung W H and Koo J S 2013 Metabolic interaction between cancer cells and stromal cells according to breast cancer molecular subtype *Breast Cancer Res.* **15** R78
Chowdhury F, Huang B and Wang N 2021 Cytoskeletal prestress: the cellular hallmark in mechanobiology and mechanomedicine *Cytoskeleton* **78** 249–76
Chowdhury F, Na S, Li D, Poh Y-C, Tanaka T S, Wang F *et al* 2010 Material properties of the cell dictate stress-induced spreading and differentiation in embryonic stem cells *Nat. Mater* **9** 82–8
Cicenas J 2016 The Aurora kinase inhibitors in cancer research and therapy *J. Cancer Res. Clin. Oncol.* **142** 1995–2012
Clevers H 2011 The cancer stem cell: premises, promises and challenges *Nat. Med.* **17** 313–9
Coffinier C, Jung H-J, Nobumori C, Chang S, Tu Y, Barnes R H *et al* 2011 Deficiencies in lamin B1 and lamin B2 cause neurodevelopmental defects and distinct nuclear shape abnormalities in neurons *Mol. Biol. Cell* **22** 4683–93
Condeelis J and Pollard J W 2006 Macrophages: obligate partners for tumor cell migration, invasion, and metastasis *Cell* **124** 263–6
Conklin M W, Eickhoff J C, Riching K M, Pehlke C A, Eliceiri K W, Provenzano P P *et al* 2011 Aligned collagen is a prognostic signature for survival in human breast carcinoma *Am. J. Pathol.* **178** 1221–32
Conti A M, Saleh A D, Brinster L R, Cheng H, Chen Z, Cornelius S *et al* 2015 Conditional deletion of nonmuscle myosin II-A in mouse tongue epithelium results in squamous cell carcinoma *Sci. Rep.* **5** 14068
Cooper J and Giancotti F G 2019 Integrin signaling in cancer: mechanotransduction, stemness, epithelial plasticity, and therapeutic resistance *Cancer Cell* **35** 347–67
Cornelison R C, Brennan C E, Kingsmore K M and Munson J M 2018 Convective forces increase CXCR4-dependent glioblastoma cell invasion in GL261 murine model *Sci. Rep.* **8** 17057
Costa-Silva B, Aiello N M, Ocean A J, Singh S, Zhang H, Thakur B K *et al* 2015 Pancreatic cancer exosomes initiate pre-metastatic niche formation in the liver *Nat. Cell Biol.* **17** 816–26
Cox T R 2021 The matrix in cancer *Nat. Rev. Cancer* **21** 217–38
Cox T R and Erler J T 2011 Remodeling and homeostasis of the extracellular matrix: implications for fibrotic diseases and cancer *Dis. Models Mech.* **4** 165–78
Creed S J, Le C P, Hassan M, Pon C K, Albold S, Chan K T *et al* 2015 β2-adrenoceptor signaling regulates invadopodia formation to enhance tumor cell invasion *Breast Cancer Res* **17** 145
Cross S E, Jin Y-S, Rao J and Gimzewski J K 2007 Nanomechanical analysis of cells from cancer patients *Nat. Nanotech.* **2** 780–3
Cui Y, Hameed F M, Yang B, Lee K, Pan C Q, Park S *et al* 2015 Cyclic stretching of soft substrates induces spreading and growth *Nat. Commun.* **6** 6333
Cui Y and Yamada S 2013 N-cadherin dependent collective cell invasion of prostate cancer cells is regulated by the n-terminus of α-catenin *PLoS One* **8** e55069
Dai J, Cimino P J, Gouin K H, Grzelak C A, Barrett A, Lim A R *et al* 2021 Astrocytic laminin-211 drives disseminated breast tumor cell dormancy in brain *Nat. Cancer* **3** 25–42
Dang C V 1999 c-Myc target genes involved in cell growth, apoptosis, and metabolism *Mol. Cell. Biol.* **19** 1–11
Das J, Chakraborty S and Maiti T K 2020 Mechanical stress-induced autophagic response: a cancer-enabling characteristic? *Semin. Cancer Biol.* **66** 101–9

Davis R T, Blake K, Ma D, Gabra M B I, Hernandez G A, Phung A T *et al* 2020 Transcriptional diversity and bioenergetic shift in human breast cancer metastasis revealed by single-cell RNA sequencing *Nat. Cell Biol.* **22** 310–20

De Lozanne A and Spudich J A 1987 Disruption of the *dictyostelium* myosin heavy chain gene by homologous recombination *Science* **236** 1086–91

del Rio A, Perez-Jimenez R, Liu R, Roca-Cusachs P, Fernandez J M and Sheetz M P 2009 Stretching single talin rod molecules activates vinculin binding *Science* **323** 638–41

Deligne C and Midwood K S 2021 Macrophages and extracellular matrix in breast cancer: partners in crime or protective allies? *Front. Oncol.* **11** 620773

Dembo M, Torney D C, Saxman K and Hammer D 1988 The reaction-limited kinetics of membrane-to-surface adhesion and detachment *Proc. R. Soc. Lond. B.* **234** 55–83

Denais C M, Gilbert R M, Isermann P, McGregor A L, te Lindert M, Weigelin B *et al* 2016 Nuclear envelope rupture and repair during cancer cell migration *Science* **352** 353–8

Dieter S M, Ball C R, Hoffmann C M, Nowrouzi A, Herbst F, Zavidij O *et al* 2011 Distinct types of tumor-initiating cells form human colon cancer tumors and metastases *Cell Stem Cell* **9** 357–65

Dillekås H, Rogers M S and Straume O 2019 Are 90% of deaths from cancer caused by metastases? *Cancer Med.* **8** 5574–6

Discher D E, Janmey P and Wang Y 2005 Tissue cells feel and respond to the stiffness of their substrate *Science* **310** 1139–43

Discher D E, Mooney D J and Zandstra P W 2009 Growth factors, matrices, and forces combine and control stem cells *Science* **324** 1673–7

Docheva D, Padula D, Schieker M and Clausen-Schaumann H 2010 Effect of collagen I and fibronectin on the adhesion, elasticity and cytoskeletal organization of prostate cancer cells *Biochem. Biophys. Res. Commun.* **402** 361–6

Doerschuk C M, Beyers N, Coxson H O, Wiggs B and Hogg J C 1993 Comparison of neutrophil and capillary diameters and their relation to neutrophil sequestration in the lung *J. Appl. Physiol.* **74** 3040–5

Douville J, Beaulieu R and Balicki D 2009 ALDH1 as a functional marker of cancer stem and progenitor cells *Stem Cells Dev.* **18** 17–26

Druso J E and Fischbach C 2018 Biophysical properties of extracellular matrix: linking obesity and cancer *Trends Cancer* **4** 271–3

Dupont S, Morsut L, Aragona M, Enzo E, Giulitti S, Cordenonsi M *et al* 2011 Role of YAP/TAZ in mechanotransduction *Nature* **474** 179–83

Dustin M L and Long E O 2010 Cytotoxic immunological synapses *Immunol. Rev.* **235** 24–34

Earle A J, Kirby T J, Fedorchak G R, Isermann P, Patel J, Iruvanti S *et al* 2020 Mutant lamins cause nuclear envelope rupture and DNA damage in skeletal muscle cells *Nat. Mater.* **19** 464–73

Egan K, Cooke N and Kenny D 2014 Living in shear: platelets protect cancer cells from shear induced damage *Clin. Exp. Metastasis* **31** 697–704

Eil R, Vodnala S K, Clever D, Klebanoff C A, Sukumar M, Pan J H *et al* 2016 Ionic immune suppression within the tumour microenvironment limits T cell effector function *Nature* **537** 539–43

Eisenhoffer G T, Loftus P D, Yoshigi M, Otsuna H, Chien C-B, Morcos P A *et al* 2012 Crowding induces live cell extrusion to maintain homeostatic cell numbers in epithelia *Nature* **484** 546–9

Elia I, Doglioni G and Fendt S-M 2018 Metabolic hallmarks of metastasis formation *Trends Cell Biol.* **28** 673–84

Elosegui-Artola A, Andreu I, Beedle A E M, Lezamiz A, Uroz M, Kosmalska A J *et al* 2017 Force triggers YAP nuclear entry by regulating transport across nuclear pores *Cell* **171** 1397–1410.e14

Elosegui-Artola A, Bazellières E, Allen M D, Andreu I, Oria R, Sunyer R *et al* 2014 Rigidity sensing and adaptation through regulation of integrin types *Nat. Mater.* **13** 631–7

Engler A J, Sen S, Sweeney H L and Discher D E 2006 Matrix elasticity directs stem cell lineage specification *Cell* **126** 677–89

Entenberg D, Voiculescu S, Guo P, Borriello L, Wang Y, Karagiannis G S *et al* 2018 A permanent window for the murine lung enables high-resolution imaging of cancer metastasis *Nat. Methods* **15** 73–80

Er E E, Tello-Lafoz M and Huse M 2022 Mechanoregulation of metastasis beyond the matrix *Cancer Res.* **82** 3409–19

Er E E, Valiente M, Ganesh K, Zou Y, Agrawal S, Hu J *et al* 2018 Pericyte-like spreading by disseminated cancer cells activates YAP and MRTF for metastatic colonization *Nat. Cell Biol.* **20** 966–78

Etourneaud L, Moussa A, Rass E, Genet D, Willaume S, Chabance-Okumura C *et al* 2021 Lamin B1 sequesters 53BP1 to control its recruitment to DNA damage *Sci. Adv.* **7** eabb3799

Fan R, Emery T, Zhang Y, Xia Y, Sun J and Wan J 2016 Circulatory shear flow alters the viability and proliferation of circulating colon cancer cells *Sci. Rep.* **6** 27073

Faria E C, Ma N, Gazi E, Gardner P, Brown M, Clarke N W *et al* 2008 Measurement of elastic properties of prostate cancer cells using AFM *Analyst* **133** 1498

Faubert B, Solmonson A and DeBerardinis R J 2020 Metabolic reprogramming and cancer progression *Science* **368** eaaw5473

Felding-Habermann B, Habermann R, Saldívar E and Ruggeri Z M 1996 Role of β3 integrins in melanoma cell adhesion to activated platelets under flow *J. Biol. Chem.* **271** 5892–900

Feng D, Wang J, Shi X, Li D, Wei W and Han P 2023 Membrane tension-mediated stiff and soft tumor subtypes closely associated with prognosis for prostate cancer patients *Eur. J. Med. Res.* **28** 172

Feng Y, LoGrasso P V, Defert O and Li R 2016 Rho kinase (ROCK) inhibitors and their therapeutic potential *J. Med. Chem.* **59** 2269–300

Fenner J, Stacer A C, Winterroth F, Johnson T D, Luker K E and Luker G D 2014 Macroscopic stiffness of breast tumors predicts metastasis *Sci. Rep.* **4** 5512

Ferguson A T, Evron E, Umbricht C B, Pandita T K, Chan T A, Hermeking H *et al* 2000 High frequency of hypermethylation at the 14-3-3 σ locus leads to gene silencing in breast cancer *Proc. Natl Acad. Sci. USA* **97** 6049–54

Fernández-Sánchez M E, Barbier S, Whitehead J, Béalle G, Michel A, Latorre-Ossa H *et al* 2015 Mechanical induction of the tumorigenic β-catenin pathway by tumour growth pressure *Nature* **523** 92–5

Fidler I J 1973 Selection of successive tumour lines for metastasis *Nat. New Biol.* **242** 148–9

Fidler I J 1995 Melanoma metastasis *Cancer Control* **2** 398–404

Filipe E C, Velayuthar S, Philp A, Nobis M, Latham S L, Parker A L *et al* 2024 Tumor biomechanics alters metastatic dissemination of triple negative breast cancer via rewiring fatty acid metabolism *Adv. Sci.* **11** 2307963

Finn O J 2018 A believer's overview of cancer immunosurveillance and immunotherapy *J. Immunol.* **200** 385–91

Follain G, Herrmann D, Harlepp S, Hyenne V, Osmani N, Warren S C *et al* 2020 Fluids and their mechanics in tumour transit: shaping metastasis *Nat. Rev. Cancer* **20** 107–24

Follain G, Osmani N, Azevedo A S, Allio G, Mercier L, Karreman M A *et al* 2018 Hemodynamic forces tune the arrest, adhesion, and extravasation of circulating tumor cells *Dev. Cell* **45** 33–52.e12

Follain G, Osmani N, Gensbittel V, Asokan N, Larnicol A, Mercier L *et al* 2021 Impairing flow-mediated endothelial remodeling reduces extravasation of tumor cells *Sci. Rep.* **11** 13144

Formaggio N, Rubin M A and Theurillat J-P 2021 Loss and revival of androgen receptor signaling in advanced prostate cancer *Oncogene* **40** 1205–16

Friedland J C, Lee M H and Boettiger D 2009 Mechanically activated integrin switch controls α_5 β_1 function *Science* **323** 642–4

Friedman D, Simmonds P, Hale A, Bere L, Hodson N W, White M R H *et al* 2021 Natural killer cell immune synapse formation and cytotoxicity are controlled by tension of the target interface *J. Cell Sci.* **134** jcs258570

Fritz G, Just I and Kaina B 1999 Rho GTPases are over-expressed in human tumors *Int. J. Cancer* **81** 682–7

Fuhrmann A, Banisadr A, Beri P, Tlsty T D and Engler A J 2017 Metastatic state of cancer cells may be indicated by adhesion strength *Biophys. J.* **112** 736–45

Funkhouser C M, Sknepnek R, Shimi T, Goldman A E, Goldman R D and Olvera De La Cruz M 2013 Mechanical model of blebbing in nuclear lamin meshworks *Proc. Natl Acad. Sci. USA* **110** 3248–53

Furlow P W, Zhang S, Soong T D, Halberg N, Goodarzi H, Mangrum C *et al* 2015 Mechanosensitive pannexin-1 channels mediate microvascular metastatic cell survival *Nat. Cell Biol.* **17** 943–52

Gaggioli C, Hooper S, Hidalgo-Carcedo C, Grosse R, Marshall J F, Harrington K *et al* 2007 Fibroblast-led collective invasion of carcinoma cells with differing roles for RhoGTPases in leading and following cells *Nat. Cell Biol.* **9** 1392–400

Gandaglia G, Abdollah F, Schiffmann J, Trudeau V, Shariat S F, Kim S P *et al* 2014 Distribution of metastatic sites in patients with prostate cancer: A population-based analysis *Prostate* **74** 210–6

Gao H, Chakraborty G, Zhang Z, Akalay I, Gadiya M, Gao Y *et al* 2016 Multi-organ site metastatic reactivation mediated by non-canonical discoidin domain receptor 1 signaling *Cell* **166** 47–62

Gao Y, Li G, Sun L, He Y, Li X, Sun Z *et al* 2015 ACTN4 and the pathways associated with cell motility and adhesion contribute to the process of lung cancer metastasis to the brain *BMC Cancer* **15** 277

Gensbittel V, Kräter M, Harlepp S, Busnelli I, Guck J and Goetz J G 2021 Mechanical adaptability of tumor cells in metastasis *Dev. Cell* **56** 164–79

Ghajar C M, Peinado H, Mori H, Matei I R, Evason K J, Brazier H *et al* 2013 The perivascular niche regulates breast tumour dormancy *Nat. Cell Biol.* **15** 807–17

Ghiso J A A, Kovalski K and Ossowski L 1999 Tumor dormancy induced by downregulation of urokinase receptor in human carcinoma involves integrin and MAPK signaling *J. Cell Biol.* **147** 89–104

Giannone G, Dubin-Thaler B J, Rossier O, Cai Y, Chaga O, Jiang G *et al* 2007 Lamellipodial actin mechanically links myosin activity with adhesion-site formation *Cell* **128** 561–75
Gilkes D M, Semenza G L and Wirtz D 2014 Hypoxia and the extracellular matrix: drivers of tumour metastasis *Nat. Rev. Cancer* **14** 430–9
Ginestier C, Hur M H, Charafe-Jauffret E, Monville F, Dutcher J, Brown M *et al* 2007 ALDH1 is a marker of normal and malignant human mammary stem cells and a predictor of poor clinical outcome *Cell Stem Cell* **1** 555–67
Goddard E T, Bozic I, Riddell S R and Ghajar C M 2018 Dormant tumour cells, their niches and the influence of immunity *Nat. Cell Biol.* **20** 1240–9
Goel H L, Li J, Kogan S and Languino L R 2008 Integrins in prostate cancer progression *Endocr. Relat. Cancer* **15** 657–64
Gohla A and Bokoch G M 2002 14-3-3 Regulates actin dynamics by stabilizing phosphorylated cofilin *Curr. Biol.* **12** 1704–10
Golebiewska A, Brons N H C, Bjerkvig R and Niclou S P 2011 Critical appraisal of the side population assay in stem cell and cancer stem cell research *Cell Stem Cell* **8** 136–47
Gong Y, Ji P, Yang Y-S, Xie S, Yu T-J, Xiao Y *et al* 2021 Metabolic-pathway-based subtyping of triple-negative breast cancer reveals potential therapeutic targets *Cell Metab.* **33** 51–64.e9
González C, Chames P, Kerfelec B, Baty D, Robert P and Limozin L 2019 Nanobody-CD16 catch bond reveals NK cell mechanosensitivity *Biophys. J.* **116** 1516–26
Grassian A R, Schafer Z T and Brugge J S 2011 ErbB2 stabilizes epidermal growth factor receptor (EGFR) expression via erk and sprouty2 in extracellular matrix-detached cells *J. Biol. Chem.* **286** 79–90
Gregorieff A, Liu Y, Inanlou M R, Khomchuk Y and Wrana J L 2015 Yap-dependent reprogramming of Lgr5+ stem cells drives intestinal regeneration and cancer *Nature* **526** 715–8
Gruenbaum Y and Foisner R 2015 Lamins: nuclear intermediate filament proteins with fundamental functions in nuclear mechanics and genome regulation *Annu. Rev. Biochem.* **84** 131–64
Guck J, Schinkinger S, Lincoln B, Wottawah F, Ebert S, Romeyke M *et al* 2005 Optical deformability as an inherent cell marker for testing malignant transformation and metastatic competence *Biophys. J.* **88** 3689–98
Gudipaty S A, Lindblom J, Loftus P D, Redd M J, Edes K, Davey C F *et al* 2017 Mechanical stretch triggers rapid epithelial cell division through Piezo1 *Nature* **543** 118–21
Gudipaty S A and Rosenblatt J 2017 Epithelial cell extrusion: pathways and pathologies *Semin. Cell Dev. Biol.* **67** 132–40
Guimarães C F, Gasperini L, Marques A P and Reis R L 2020 The stiffness of living tissues and its implications for tissue engineering *Nat. Rev. Mater.* **5** 351–70
Guo L, Ye C, Hao X, Zheng R, Ju R, Wu D *et al* 2012 Carboxyamidotriazole ameliorates experimental colitis by inhibition of cytokine production, nuclear factor-κB activation, and colonic fibrosis *J. Pharmacol. Exp. Ther.* **342** 356–65
Gupta K and Dey P 2003 Cell cannibalism: diagnostic marker of malignancy *Diagn. Cytopathol.* **28** 86–7
Haage A and Schneider I C 2014 Cellular contractility and extracellular matrix stiffness regulate matrix metalloproteinase activity in pancreatic cancer cells *FASEB J.* **28** 3589–99
Haemmerle M, Taylor M L, Gutschner T, Pradeep S, Cho M S, Sheng J *et al* 2017 Platelets reduce anoikis and promote metastasis by activating YAP1 signaling *Nat. Commun.* **8** 310

Halder D, Saha S, Singh R K, Ghosh I, Mallick D, Dey S K *et al* 2019 Nonmuscle myosin IIA and IIB differentially modulate migration and alter gene expression in primary mouse tumorigenic cells *Mol. Biol. Cell* **30** 1463–76

Hall C L, Dai J, Van Golen K L, Keller E T and Long M W 2006 Type I collagen receptor ($\alpha 2\beta 1$) signaling promotes the growth of human prostate cancer cells within the bone *Cancer Res.* **66** 8648–54

Hall C L, Dubyk C W, Riesenberger T A, Shein D, Keller E T and Van Golen K L 2008 Type I collagen receptor ($\alpha 2\beta 1$) signaling promotes prostate cancer invasion through RhoC GTPase *Neoplasia* **10** 797–803

Hamann J C, Surcel A, Chen R, Teragawa C, Albeck J G, Robinson D N *et al* 2017 ENTOSIS is induced by glucose starvation *Cell Rep.* **20** 201–10

Han Y L, Pegoraro A F, Li H, Li K, Yuan Y, Xu G *et al* 2020 Cell swelling, softening and invasion in a three-dimensional breast cancer model *Nat. Phys.* **16** 101–8

Hanahan D 2022 Hallmarks of cancer: new dimensions *Cancer Discov.* **12** 31–46

Hanley C J, Henriet E, Sirka O K, Thomas G J and Ewald A J 2020 Tumor-resident stromal cells promote breast cancer invasion through regulation of the basal phenotype *Mol. Cancer Res.* **18** 1615–22

Hanna S J, McCoy-Simandle K, Leung E, Genna A, Condeelis J and Cox D 2019 Tunneling nanotubes, a novel mode of tumor cell-macrophage communication in tumor cell invasion *J. Cell Sci.* **223321**

Hanselmann R G and Welter C 2022 Origin of cancer: cell work is the key to understanding cancer initiation and progression *Front. Cell Dev. Biol.* **10** 787995

Harada T, Swift J, Irianto J, Shin J-W, Spinler K R, Athirasala A *et al* 2014 Nuclear lamin stiffness is a barrier to 3D migration, but softness can limit survival *J. Cell Biol.* **204** 669–82

Harney A S, Arwert E N, Entenberg D, Wang Y, Guo P, Qian B-Z *et al* 2015 Real-time imaging reveals local, transient vascular permeability, and tumor cell intravasation stimulated by TIE2hi macrophage–derived VEGFA *Cancer Discov.* **5** 932–43

Hayashi A, Yavas A, McIntyre C A, Ho Y, Erakky A, Wong W *et al* 2020 Genetic and clinical correlates of entosis in pancreatic ductal adenocarcinoma *Mod. Pathol.* **33** 1822–31

Headley M B, Bins A, Nip A, Roberts E W, Looney M R, Gerard A *et al* 2016 Visualization of immediate immune responses to pioneer metastatic cells in the lung *Nature* **531** 513–7

Ho C Y, Jaalouk D E, Vartiainen M K and Lammerding J 2013 Lamin A/C and emerin regulate MKL1–SRF activity by modulating actin dynamics *Nature* **497** 507–11

Hoffman R M and Yang M 2006a Color-coded fluorescence imaging of tumor-host interactions *Nat. Protoc.* **1** 928–35

Hoffman R M and Yang M 2006b Subcellular imaging in the live mouse *Nat. Protoc.* **1** 775–82

Holle A W, Govindan Kutty Devi N, Clar K, Fan A, Saif T, Kemkemer R *et al* 2019 Cancer cells invade confined microchannels via a self-directed mesenchymal-to-amoeboid transition *Nano Lett.* **19** 2280–90

Hope K J, Jin L and Dick J E 2004 Acute myeloid leukemia originates from a hierarchy of leukemic stem cell classes that differ in self-renewal capacity *Nat. Immunol.* **5** 738–43

Horoszewicz J S, Leong S S, Chu T M, Wajsman Z L, Friedman M, Papsidero L *et al* 1980 The LNCaP cell line—a new model for studies on human prostatic carcinoma *Prog. Clin. Biol. Res.* **37** 115–32

Hosokawa M, Hayata T, Fukuda Y, Arakaki A, Yoshino T, Tanaka T *et al* 2010 Size-selective microcavity array for rapid and efficient detection of circulating tumor cells *Anal. Chem.* **82** 6629–35

Hotulainen P and Lappalainen P 2006 Stress fibers are generated by two distinct actin assembly mechanisms in motile cells *J. Cell Biol.* **173** 383–94

Hoyt K, Castaneda B, Zhang M, Nigwekar P, di Sant'Agnese P A, Joseph J V *et al* 2008 Tissue elasticity properties as biomarkers for prostate cancer *Cancer Biomark.* **4** 213–25

Hsia C-R, McAllister J, Hasan O, Judd J, Lee S, Agrawal R *et al* 2022 Confined migration induces heterochromatin formation and alters chromatin accessibility *iScience* **25** 104978

Huda S, Weigelin B, Wolf K, Tretiakov K V, Polev K, Wilk G *et al* 2018 Lévy-like movement patterns of metastatic cancer cells revealed in microfabricated systems and implicated *in vivo* *Nat. Commun.* **9** 4539

Huo C, Kao Y-H and Chuu C-P 2015 Androgen receptor inhibits epithelial–mesenchymal transition, migration, and invasion of PC-3 prostate cancer cells *Cancer Lett.* **369** 103–11

Hupfer A, Brichkina A, Koeniger A, Keber C, Denkert C, Pfefferle P *et al* 2021 Matrix stiffness drives stromal autophagy and promotes formation of a protumorigenic niche *Proc. Natl Acad. Sci. USA* **118** e2105367118

Huse M 2017 Mechanical forces in the immune system *Nat. Rev. Immunol.* **17** 679–90

Hvichia G E, Parveen Z, Wagner C, Janning M, Quidde J, Stein A *et al* 2016 A novel microfluidic platform for size and deformability based separation and the subsequent molecular characterization of viable circulating tumor cells *Int. J. Cancer* **138** 2894–904

Hwang P Y, Brenot A, King A C, Longmore G D and George S C 2019 Randomly distributed K14 + breast tumor cells polarize to the leading edge and guide collective migration in response to chemical and mechanical environmental cues *Cancer Res.* **79** 1899–912

Iinuma H, Watanabe T, Mimori K, Adachi M, Hayashi N, Tamura J *et al* 2011 Clinical significance of circulating tumor cells, including cancer stem-like cells, in peripheral blood for recurrence and prognosis in patients with dukes' stage b and c colorectal cancer *J. Clin. Oncol.* **29** 1547–55

Ilina O, Gritsenko P G, Syga S, Lippoldt J, La Porta C A M, Chepizhko O *et al* 2020 Cell–cell adhesion and 3D matrix confinement determine jamming transitions in breast cancer invasion *Nat. Cell Biol.* **22** 1103–15

Irianto J, Pfeifer C R, Xia Y and Discher D E 2016 SnapShot: mechanosensing matrix *Cell* **165** 1820–0

Irianto J, Xia Y, Pfeifer C R, Athirasala A, Ji J, Alvey C *et al* 2017 DNA damage follows repair factor depletion and portends genome variation in cancer cells after pore migration *Curr. Biol.* **27** 210–23

Ishihara S, Inman D R, Li W-J, Ponik S M and Keely P J 2017 Mechano-signal transduction in mesenchymal stem cells induces prosaposin secretion to drive the proliferation of breast cancer cells *Cancer Res.* **77** 6179–89

Iyengar P, Espina V, Williams T W, Lin Y, Berry D, Jelicks L A *et al* 2005 Adipocyte-derived collagen VI affects early mammary tumor progression *in vivo*, demonstrating a critical interaction in the tumor/stroma microenvironment *J. Clin. Invest.* **115** 1163–76

Izdebska M, Gagat M and Grzanka A 2017 Overexpression of lamin B1 induces mitotic catastrophe in colon cancer LoVo cells and is associated with worse clinical outcomes *Int. J. Oncol.* **52** 89–102

Jackson W F 2017 Potassium channels in regulation of vascular smooth muscle contraction and growth *Adv. Pharmacol.* **78** 89–144

Jain R K 2001 Normalizing tumor vasculature with anti-angiogenic therapy: a new paradigm for combination therapy *Nat. Med.* **7** 987–9

Jain R K 2005 Normalization of tumor vasculature: an emerging concept in antiangiogenic therapy *Science* **307** 58–62

Jain S, Cachoux V M L, Narayana G H N S, de Beco S, D'Alessandro J, Cellerin V *et al* 2020 The role of single-cell mechanical behaviour and polarity in driving collective cell migration *Nat. Phys.* **16** 802–9

Jia Y, Vong J S-L, Asafova A, Garvalov B K, Caputo L, Cordero J *et al* 2019 Lamin B1 loss promotes lung cancer development and metastasis by epigenetic derepression of RET *J. Exp. Med.* **216** 1377–95

Jiang K, Lim S B, Xiao J, Jokhun D S, Shang M, Song X *et al* 2023 Deleterious mechanical deformation selects mechanoresilient cancer cells with enhanced proliferation and chemo-resistance *Adv. Sci.* **10** 2201663

Judokusumo E, Tabdanov E, Kumari S, Dustin M L and Kam L C 2012 Mechanosensing in T lymphocyte activation *Biophys. J.* **102** L5–7

Jung W-H, Yam N, Chen C-C, Elawad K, Hu B and Chen Y 2020 Force-dependent extracellular matrix remodeling by early-stage cancer cells alters diffusion and induces carcinoma-associated fibroblasts *Biomaterials* **234** 119756

Kaighn M E, Narayan K S, Ohnuki Y, Lechner J F and Jones L W 1979 Establishment and characterization of a human prostatic carcinoma cell line (PC-3) *Invest. Urol.* **17** 16–23

Kaiser J 2015 The cancer stem cell gamble *Science* **347** 226–9

Kalli M, Li R, Mills G B, Stylianopoulos T and Zervantonakis I K 2022 Mechanical stress signaling in pancreatic cancer cells triggers p38 MAPK- and JNK-dependent cytoskeleton remodeling and promotes cell migration via Rac1/cdc42/myosin II *Mol. Cancer Res.* **20** 485–97

Kapucuoğlu , Bozkurt N, Başpınar K K, Koçer M, Eroğlu H E, Akdeniz R *et al* 2015 The clinicopathological and prognostic significance of CD24, CD44, CD133, ALDH1 expressions in invasive ductal carcinoma of the breast *Pathol.—Res. Pract.* **211** 740–7

Keysar S B and Jimeno A 2010 More than markers: biological significance of cancer stem cell-defining molecules *Mol. Cancer Therap.* **9** 2450–7

Khan Z S, Santos J M and Hussain F 2018 Aggressive prostate cancer cell nuclei have reduced stiffness *Biomicrofluidics* **12** 014102

Kienast Y, Von Baumgarten L, Fuhrmann M, Klinkert W E F, Goldbrunner R, Herms J *et al* 2010 Real-time imaging reveals the single steps of brain metastasis formation *Nat. Med.* **16** 116–22

Kim D-H 2009 Differential expression of the LOX family genes in human colorectal adenocarcinomas *Oncol. Rep.* **22**

Kim D-H, Li B, Si F, Phillip J M, Wirtz D and Sun S X 2016a Volume regulation and shape bifurcation in the cell nucleus *J. Cell Sci.* **129** 457–7

Kim J and DeBerardinis R J 2019 Mechanisms and implications of metabolic heterogeneity in cancer *Cell Metab.* **30** 434–46

Kim N-G and Gumbiner B M 2015 Adhesion to fibronectin regulates Hippo signaling via the FAK–Src–PI3K pathway *J. Cell Biol.* **210** 503–15

Kim S, Kim D H, Jung W-H and Koo J S 2013 Metabolic phenotypes in triple-negative breast cancer *Tumor Biol.* **34** 1699–712

Kim T, Hwang D, Lee D, Kim J, Kim S and Lim D 2017 MRTF potentiates TEAD-YAP transcriptional activity causing metastasis *EMBO J.* **36** 520–35

Kim T-H, Gill N K, Nyberg K D, Nguyen A V, Hohlbauch S V, Geisse N A *et al* 2016b Cancer cells become less deformable and more invasive with activation of β-adrenergic signaling *J. Cell Sci.* **129** 4563–75

Kim T-H, Ly C, Christodoulides A, Nowell C J, Gunning P W, Sloan E K *et al* 2019 Stress hormone signaling through β-adrenergic receptors regulates macrophage mechanotype and function *FASEB J.* **33** 3997–4006

Kitazawa T, Eto M, Woodsome T P and Khalequzzaman M 2003 Phosphorylation of the myosin phosphatase targeting subunit and CPI-17 during Ca^{2+} sensitization in rabbit smooth muscle *J. Physiol.* **546** 879–89

Ko A H, LoConte N, Tempero M A, Walker E J, Kate Kelley R, Lewis S *et al* 2016 A phase I study of FOLFIRINOX Plus IPI-926, a hedgehog pathway inhibitor, for advanced pancreatic adenocarcinoma *Pancreas* **45** 370–5

Kong F, García A J, Mould A P, Humphries M J and Zhu C 2009 Demonstration of catch bonds between an integrin and its ligand *J. Cell Biol.* **185** 1275–84

Koop S, MacDonald I C, Luzzi K, Schmidt E E, Morris V L, Grattan M *et al* 1995 Fate of melanoma cells entering the microcirculation: over 80% survive and extravasate *Cancer Res.* **55** 2520–3

Koorman T, Jansen K A, Khalil A, Haughton P D, Visser D, Rätze M A K *et al* 2022 Spatial collagen stiffening promotes collective breast cancer cell invasion by reinforcing extracellular matrix alignment *Oncogene* **41** 2458–69

Koundouros N and Poulogiannis G 2020 Reprogramming of fatty acid metabolism in cancer *Br. J. Cancer* **122** 4–22

Kraning-Rush C M, Califano J P and Reinhart-King C A 2012 Cellular traction stresses increase with increasing metastatic potential *PLoS One* **7** e32572

Krndija D, Schmid H, Eismann J-L, Lother U, Adler G, Oswald F *et al* 2010 Substrate stiffness and the receptor-type tyrosine-protein phosphatase alpha regulate spreading of colon cancer cells through cytoskeletal contractility *Oncogene* **29** 2724–38

Kumar S and Weaver V M 2009 Mechanics, malignancy, and metastasis: the force journey of a tumor cell *Cancer Metastasis Rev.* **28** 113–27

Laakkonen P, Waltari M, Holopainen T, Takahashi T, Pytowski B, Steiner P *et al* 2007 Vascular endothelial growth factor receptor 3 is involved in tumor angiogenesis and growth *Cancer Res.* **67** 593–9

Lam W A, Cao L, Umesh V, Keung A J, Sen S and Kumar S 2010 Extracellular matrix rigidity modulates neuroblastoma cell differentiation and N-myc expression *Mol. Cancer* **9** 35

Lamar J M, Stern P, Liu H, Schindler J W, Jiang Z-G and Hynes R O 2012 The Hippo pathway target, YAP, promotes metastasis through its TEAD-interaction domain *Proc. Natl Acad. Sci. USA* **109** E2441–50

Lamar J M, Xiao Y, Norton E, Jiang Z-G, Gerhard G M, Kooner S *et al* 2019 SRC tyrosine kinase activates the YAP/TAZ axis and thereby drives tumor growth and metastasis *J. Biol. Chem.* **294** 2302–17

Lambert A W, Pattabiraman D R and Weinberg R A 2017 Emerging biological principles of metastasis *Cell* **168** 670–91

Lammerding J, Fong L G, Ji J Y, Reue K, Stewart C L, Young S G *et al* 2006 Lamins A and C but not lamin B1 regulate nuclear mechanics *J. Biol. Chem.* **281** 25768–80

Lampi M C and Reinhart-King C A 2018 Targeting extracellular matrix stiffness to attenuate disease: From molecular mechanisms to clinical trials *Sci. Transl. Med.* **10** eaao0475

Lanning N J, Castle J P, Singh S J, Leon A N, Tovar E A, Sanghera A *et al* 2017 Metabolic profiling of triple-negative breast cancer cells reveals metabolic vulnerabilities *Cancer Metab.* **5** 6

Lapidot T, Sirard C, Vormoor J, Murdoch B, Hoang T, Caceres-Cortes J *et al* 1994 A cell initiating human acute myeloid leukaemia after transplantation into SCID mice *Nature* **367** 645–8

Lawler K, Foran E, O'Sullivan G, Long A and Kenny D 2006 Mobility and invasiveness of metastatic esophageal cancer are potentiated by shear stress in a ROCK- and Ras-dependent manner *Am. J. Physiol.-Cell Physiol.* **291** C668–77

Le C P, Nowell C J, Kim-Fuchs C, Botteri E, Hiller J G, Ismail H *et al* 2016 Chronic stress in mice remodels lymph vasculature to promote tumour cell dissemination *Nat. Commun.* **7** 10634

LeBlanc A G, Demers A and Shaw A 2019 Recent trends in prostate cancer in Canada *Health Rep.* **30** 12–7

LeBleu V S, O'Connell J T, Gonzalez Herrera K N, Wikman H, Pantel K, Haigis M C *et al* 2014 PGC-1α mediates mitochondrial biogenesis and oxidative phosphorylation in cancer cells to promote metastasis *Nat. Cell Biol.* **16** 992–1003

Lee C, Jeong S, Jang C, Bae H, Kim Y H, Park I *et al* 2019a Tumor metastasis to lymph nodes requires YAP-dependent metabolic adaptation *Science* **363** 644–9

Lee H-Y and Oh B-H 2010 Aging and arterial stiffness *Circ. J.* **74** 2257–62

Lee J Y, Chang J K, Dominguez A A, Lee H, Nam S, Chang J *et al* 2019b YAP-independent mechanotransduction drives breast cancer progression *Nat. Commun.* **10** 1848

Lee M S, Glassman C R, Deshpande N R, Badgandi H B, Parrish H L, Uttamapinant C *et al* 2015 A mechanical switch couples T cell receptor triggering to the cytoplasmic juxtamembrane regions of CD3$\zeta\zeta$ *Immunity* **43** 227–39

Leggett S E, Hruska A M, Guo M and Wong I Y 2021 The epithelial-mesenchymal transition and the cytoskeleton in bioengineered systems *Cell Commun. Signal* **19** 32

Lei K, Kurum A, Kaynak M, Bonati L, Han Y, Cencen V *et al* 2021 Cancer-cell stiffening via cholesterol depletion enhances adoptive T-cell immunotherapy *Nat. Biomed. Eng.* **5** 1411–25

Lekka M, Gnanachandran K, Kubiak A, Zieliński T and Zemła J 2021 Traction force microscopy—measuring the forces exerted by cells *Micron* **150** 103138

Lekka M, Pogoda K, Gostek J, Klymenko O, Prauzner-Bechcicki S, Wiltowska-Zuber J *et al* 2012 Cancer cell recognition—mechanical phenotype *Micron* **43** 1259–66

Levental K R, Yu H, Kass L, Lakins J N, Egeblad M, Erler J T *et al* 2009 Matrix crosslinking forces tumor progression by enhancing integrin signaling *Cell* **139** 891–906

Levitan I 2016 Paradoxical impact of cholesterol on lipid packing and cell stiffness *Front. Biosci.* **21** 1245–59

Li G, Satyamoorthy K and Herlyn M 2001 N-cadherin-mediated intercellular interactions promote survival and migration of melanoma cells *Cancer Res.* **61** 3819–25

Li Q, Kumar A, Makhija E and Shivashankar G V 2014 The regulation of dynamic mechanical coupling between actin cytoskeleton and nucleus by matrix geometry *Biomaterials* **35** 961–9

Li S, Chang S, Qi X, Richardson J A and Olson E N 2006 Requirement of a myocardin-related transcription factor for development of mammary myoepithelial cells *Mol. Cell. Biol.* **26** 5797–808

Li Y-J, Fahrmann J F, Aftabizadeh M, Zhao Q, Tripathi S C, Zhang C *et al* 2022 Fatty acid oxidation protects cancer cells from apoptosis by increasing mitochondrial membrane lipids *Cell Rep.* **39** 110870

Liberio M S, Sadowski M C, Soekmadji C, Davis R A and Nelson C C 2014 Differential effects of tissue culture coating substrates on prostate cancer cell adherence, morphology and behavior *PLoS One* **9** e112122

Lin F, Zhang H, Huang J and Xiong C 2018 Substrate stiffness coupling TGF-β1 modulates migration and traction force of MDA-MB-231 human breast cancer cells *in vitro ACS Biomater. Sci. Eng.* **4** 1337–45

Lin H-H, Lin H-K, Lin I-H, Chiou Y-W, Chen H-W, Liu C-Y *et al* 2015 Mechanical phenotype of cancer cells: cell softening and loss of stiffness sensing *Oncotarget* **6** 20946–58

Lintz M, Muñoz A and Reinhart-King C A 2017 The mechanics of single cell and collective migration of tumor cells *J. Biomech. Eng.* **139** 021005

Liu B, Chen W, Evavold B D and Zhu C 2014 Accumulation of dynamic catch bonds between TCR and agonist peptide-MHC triggers T cell signaling *Cell* **157** 357–68

Liu C, Li M, Dong Z-X, Jiang D, Li X, Lin S *et al* 2021a Heterogeneous microenvironmental stiffness regulates pro-metastatic functions of breast cancer cells *Acta Biomater.* **131** 326–40

Liu C, Pei H and Tan F 2020a Matrix stiffness and colorectal cancer *OTT* **13** 2747–55

Liu H, Liu K and Dong Z 2021b The role of p21-activated kinases in cancer and beyond: where are we heading? *Front. Cell Dev. Biol.* **9** 641381

Liu J, Tan Y, Zhang H, Zhang Y, Xu P, Chen J *et al* 2012 Soft fibrin gels promote selection and growth of tumorigenic cells *Nat. Mater.* **11** 734–41

Liu N, Du P, Xiao X, Liu Y, Peng Y, Yang C *et al* 2019 Microfluidic-based mechanical phenotyping of androgen-sensitive and non-sensitive prostate cancer cells lines *Micromachines* **10** 602

Liu S, Goldstein R H, Scepansky E M and Rosenblatt M 2009 Inhibition of rho-associated kinase signaling prevents breast cancer metastasis to human bone *Cancer Res.* **69** 8742–51

Liu Y, Liang X, Dong W, Fang Y, Lv J, Zhang T *et al* 2018 Tumor-repopulating cells induce PD-1 expression in CD8 + T cells by transferring kynurenine and AhR activation *Cancer Cell* **33** 480–94.e7

Liu Y, Zhang T, Zhang H, Li J, Zhou N, Fiskesund R *et al* 2021c Cell softness prevents cytolytic T-cell killing of tumor-repopulating cells *Cancer Res.* **81** 476–88

Liu Z, Lee S J, Park S, Konstantopoulos K, Glunde K, Chen Y *et al* 2020b Cancer cells display increased migration and deformability in pace with metastatic progression *FASEB J.* **34** 9307–15

Liu Z, Wang L, Xu H, Du Q, Li L, Wang L *et al* 2020c Heterogeneous responses to mechanical force of prostate cancer cells inducing different metastasis patterns *Adv. Sci.* **7** 1903583

Lo C-M, Wang H-B, Dembo M and Wang Y 2000 Cell movement is guided by the rigidity of the substrate *Biophys. J.* **79** 144–52

Lomakin A J, Cattin C J, Cuvelier D, Alraies Z, Molina M, Nader G P F *et al* 2020 The nucleus acts as a ruler tailoring cell responses to spatial constraints *Science* **370** eaba2894

Lugli A, Iezzi G, Hostettler I, Muraro M G, Mele V, Tornillo L *et al* 2010 Prognostic impact of the expression of putative cancer stem cell markers CD133, CD166, CD44s, EpCAM, and ALDH1 in colorectal cancer *Br. J. Cancer* **103** 382–90

Lumaquin-Yin D, Montal E, Johns E, Baggiolini A, Huang T-H, Ma Y *et al* 2023 Lipid droplets are a metabolic vulnerability in melanoma *Nat. Commun.* **14** 3192

Luo Q, Kuang D, Zhang B and Song G 2016 Cell stiffness determined by atomic force microscopy and its correlation with cell motility *Biochim. Biophys. Acta (BBA)—Gen. Subj.* **1860** 1953–60

Luo T, Mohan K, Iglesias P A and Robinson D N 2013 Molecular mechanisms of cellular mechanosensing *Nat. Mater* **12** 1064–71

Luo T, Mohan K, Srivastava V, Ren Y, Iglesias P A and Robinson D N 2012 Understanding the cooperative interaction between Myosin II and actin cross-linkers mediated by actin filaments during mechanosensation *Biophys. J.* **102** 238–47

Lv J, Liu Y, Cheng F, Li J, Zhou Y, Zhang T *et al* 2021 Cell softness regulates tumorigenicity and stemness of cancer cells *EMBO J.* **40** e106123

Ma X, Xue H, Zhong J, Feng B and Zuo Y 2020 Serum actinin-4 levels as a potential diagnostic and prognostic marker in cervical cancer *Dis. Markers* **2020** 1–6

MacKie R M, Reid R and Junor B 2003 Fatal melanoma transferred in a donated kidney 16 years after melanoma surgery *N. Engl. J. Med.* **348** 567–8

Maimari N, Pedrigi R M, Russo A, Broda K and Krams R 2016 Integration of flow studies for robust selection of mechanoresponsive genes *Thromb. Haemost.* **115** 474–83

Malhas A, Goulbourne C and Vaux D J 2011 The nucleoplasmic reticulum: form and function *Trends Cell Biol.* **21** 362–73

Mana-Capelli S, Paramasivam M, Dutta S and McCollum D 2014 Angiomotins link F-actin architecture to Hippo pathway signaling *Mol. Biol. Cell* **25** 1676–85

Marshall B T, Long M, Piper J W, Yago T, McEver R P and Zhu C 2003 Direct observation of catch bonds involving cell-adhesion molecules *Nature* **423** 190–3

Martinez-Outschoorn U E, Peiris-Pagés M, Pestell R G, Sotgia F and Lisanti M P 2017 Cancer metabolism: a therapeutic perspective *Nat. Rev. Clin. Oncol.* **14** 11–31

Martínez-Reyes I and Chandel N S 2021 Cancer metabolism: looking forward *Nat. Rev. Cancer* **21** 669–80

Massagué J and Obenauf A C 2016 Metastatic colonization by circulating tumour cells *Nature* **529** 298–306

McGrail D J, Kieu Q M N and Dawson M R 2014 Metastatic ovarian cancer cell malignancy is increased on soft matrices through a mechanosensitive Rho/ROCK pathway *J. Cell Sci.* **127** 2621–6

McGranahan N and Swanton C 2017 Clonal heterogeneity and tumor evolution: past, present, and the future *Cell* **168** 613–28

Medema J P 2013 Cancer stem cells: the challenges ahead *Nat. Cell Biol.* **15** 338–44

Medjkane S, Perez-Sanchez C, Gaggioli C, Sahai E and Treisman R 2009 Myocardin-related transcription factors and SRF are required for cytoskeletal dynamics and experimental metastasis *Nat. Cell Biol.* **11** 257–68

Menard J A, Christianson H C, Kucharzewska P, Bourseau-Guilmain E, Svensson K J, Lindqvist E *et al* 2016 Metastasis stimulation by hypoxia and acidosis-induced extracellular lipid uptake is mediated by proteoglycan-dependent endocytosis *Cancer Res.* **76** 4828–40

Miao L, Yang L, Li R, Rodrigues D N, Crespo M, Hsieh J-T *et al* 2017 Disrupting androgen receptor signaling induces snail-mediated epithelial–mesenchymal plasticity in prostate cancer *Cancer Res.* **77** 3101–12

Mierke C T 2019 The matrix environmental and cell mechanical properties regulate cell migration and contribute to the invasive phenotype of cancer cells *Rep. Prog. Phys.* **82** 064602

Mierke C T 2020 Mechanical cues affect migration and invasion of cells from three different directions *Front. Cell Dev. Biol.* **8** 583226

Mierke C T 2024a Extracellular matrix cues regulate mechanosensing and mechanotransduction of cancer cells *Cells* **13** 96

Mierke C T 2024b Mechanosensory entities and functionality of endothelial cells *Front. Cell Dev. Biol. Sec. Cell Adhes. Migrat.* **12** 1446452

Mierke C T, Frey B, Fellner M, Herrmann M and Fabry B 2011 Integrin α5β1 facilitates cancer cell invasion through enhanced contractile forces *J. Cell Sci.* **124** 369–83

Mierke C T, Kollmannsberger P, Zitterbart D P, Diez G, Koch T M, Marg S *et al* 2010 Vinculin facilitates cell invasion into three-dimensional collagen matrices *J. Biol. Chem.* **285** 13121–30

Mierke C T, Zitterbart D P, Kollmannsberger P, Raupach C, Schlötzer-Schrehardt U, Goecke T W *et al* 2008 Breakdown of the endothelial barrier function in tumor cell transmigration *Biophys. J.* **94** 2832–46

Miller T J, McCoy M J, Hemmings C, Bulsara M K, Iacopetta B and Platell C F 2017 The prognostic value of cancer stem-like cell markers SOX2 and CD133 in stage III colon cancer is modified by expression of the immune-related markers FoxP3, PD-L1 and CD3 *Pathology* **49** 721–30

Minn A J, Gupta G P, Siegel P M, Bos P D, Shu W, Giri D D *et al* 2005 Genes that mediate breast cancer metastasis to lung *Nature* **436** 518–24

Miralles F, Posern G, Zaromytidou A-I and Treisman R 2003 Actin dynamics control SRF activity by regulation of its coactivator MAL *Cell* **113** 329–42

Mitchell M J, Denais C, Chan M F, Wang Z, Lammerding J and King M R 2015 Lamin A/C deficiency reduces circulating tumor cell resistance to fluid shear stress *Am. J. Physiol.-Cell Physiol.* **309** C736–46

Mohamed H, Murray M, Turner J N and Caggana M 2009 Isolation of tumor cells using size and deformation *J. Chromatogr.* A **1216** 8289–95

Mohammadi H and Sahai E 2018 Mechanisms and impact of altered tumour mechanics *Nat. Cell Biol.* **20** 766–74

Mohme M, Riethdorf S and Pantel K 2017 Circulating and disseminated tumour cells—mechanisms of immune surveillance and escape *Nat. Rev. Clin. Oncol.* **14** 155–67

Molter C W, Muszynski E F, Tao Y, Trivedi T, Clouvel A and Ehrlicher A J 2022 Prostate cancer cells of increasing metastatic potential exhibit diverse contractile forces, cell stiffness, and motility in a microenvironment stiffness-dependent manner *Front. Cell Dev. Biol.* **10** 932510

Montagner M and Dupont S 2020 Mechanical forces as determinants of disseminated metastatic cell fate *Cells* **9** 250

Moose D L, Krog B L, Kim T-H, Zhao L, Williams-Perez S, Burke G *et al* 2020 Cancer cells resist mechanical destruction in circulation via RhoA/actomyosin-dependent mechano-adaptation *Cell Rep.* **30** 3864–3874.e6

Morgan B P, Muci A, Lu P-P, Qian X, Tochimoto T, Smith W W *et al* 2010 Discovery of *Omecamtiv Mecarbil* the first, selective, small molecule activator of cardiac myosin *ACS Med. Chem. Lett.* **1** 472–7

Moritz M N O, Merkel A R, Feldman E G, Selistre-de-Araujo H S and Rhoades (Sterling) J A 2021 Biphasic α2β1 integrin expression in breast cancer metastasis to bone *IJMS* **22** 6906

Müller C and Pompe T 2016 Distinct impacts of substrate elasticity and ligand affinity on traction force evolution *Soft Matter* **12** 272–80

Murgai M, Ju W, Eason M, Kline J, Beury D W, Kaczanowska S *et al* 2017 KLF4-dependent perivascular cell plasticity mediates pre-metastatic niche formation and metastasis *Nat. Med.* **23** 1176–90

Naba A, Clauser K R, Lamar J M, Carr S A and Hynes R O 2014 Extracellular matrix signatures of human mammary carcinoma identify novel metastasis promoters *eLife* **3** e01308

Nader G P D F, Agüera-Gonzalez S, Routet F, Gratia M, Maurin M, Cancila V *et al* 2021 Compromised nuclear envelope integrity drives TREX1-dependent DNA damage and tumor cell invasion *Cell* **184** 5230–46.e22

Nguyen A V, Nyberg K D, Scott M B, Welsh A M, Nguyen A H, Wu N *et al* 2016 Stiffness of pancreatic cancer cells is associated with increased invasive potential *Integr. Biol.* **8** 1232–45

Nguyen L T S, Jacob M A C, Parajón E and Robinson D N 2022 Cancer as a biophysical disease: targeting the mechanical-adaptability program *Biophys. J.* **121** 3573–85

Nmezi B, Xu J, Fu R, Armiger T J, Rodriguez-Bey G, Powell J S *et al* 2019 Concentric organization of A- and B-type lamins predicts their distinct roles in the spatial organization and stability of the nuclear lamina *Proc. Natl Acad. Sci. USA* **116** 4307–15

Nobre A R, Entenberg D, Wang Y, Condeelis J and Aguirre-Ghiso J A 2018 The Different routes to metastasis via hypoxia-regulated programs *Trends Cell Biol.* **28** 941–56

Northcott J M, Dean I S, Mouw J K and Weaver V M 2018 Feeling stress: the mechanics of cancer progression and aggression *Front. Cell Dev. Biol.* **6** 17

Nukuda A, Sasaki C, Ishihara S, Mizutani T, Nakamura K, Ayabe T *et al* 2015 Stiff substrates increase YAP-signaling-mediated matrix metalloproteinase-7 expression *Oncogenesis* **4** e165–5

Oakes P W, Banerjee S, Marchetti M C and Gardel M L 2014 Geometry regulates traction stresses in adherent cells *Biophys. J.* **107** 825–33

Oberleithner H, Callies C, Kusche-Vihrog K, Schillers H, Shahin V, Riethmüller C *et al* 2009 Potassium softens vascular endothelium and increases nitric oxide release *Proc. Natl Acad. Sci. USA* **106** 2829–34

O'Brien C A, Pollett A, Gallinger S and Dick J E 2007 A human colon cancer cell capable of initiating tumour growth in immunodeficient mice *Nature* **445** 106–10

Ocaña O H, Córcoles R, Fabra Á, Moreno-Bueno G, Acloque H, Vega S *et al* 2012 Metastatic colonization requires the repression of the epithelial-mesenchymal transition inducer Prrx1 *Cancer Cell* **22** 709–24

Oh J, Richardson J A and Olson E N 2005 Requirement of myocardin-related transcription factor-B for remodeling of branchial arch arteries and smooth muscle differentiation *Proc. Natl Acad. Sci. USA* **102** 15122–7

Ondeck M G, Kumar A, Placone J K, Plunkett C M, Matte B F, Wong K C *et al* 2019 Dynamically stiffened matrix promotes malignant transformation of mammary epithelial cells via collective mechanical signaling *Proc. Natl Acad. Sci. USA* **116** 3502–7

Overholtzer M, Mailleux A A, Mouneimne G, Normand G, Schnitt S J, King R W *et al* 2007 A nonapoptotic cell death process, entosis, that occurs by cell-in-cell invasion *Cell* **131** 966–79

Panciera T, Citron A, Di Biagio D, Battilana G, Gandin A, Giulitti S *et al* 2020 Reprogramming normal cells into tumour precursors requires ECM stiffness and oncogene-mediated changes of cell mechanical properties *Nat. Mater.* **19** 797–806

Pang M, Teng Y, Huang J, Yuan Y, Lin F and Xiong C 2017 Substrate stiffness promotes latent TGF-β1 activation in hepatocellular carcinoma *Biochem. Biophys. Res. Commun.* **483** 553–8

Panzetta V, La Verde G, Pugliese M, Arrichiello C, Muto P, La Commara M *et al* 2020 Investigation of biophysical migration parameters for normal tissue and metastatic cancer cells after radiotherapy treatment *Front. Phys.* **8** 575906

Paoli P, Giannoni E and Chiarugi P 2013 Anoikis molecular pathways and its role in cancer progression *Biochim. Biophys. Acta (BBA)—Mol. Cell Res.* **1833** 3481–98

Parajón E, Surcel A and Robinson D N 2021 The mechanobiome: a goldmine for cancer therapeutics *Am. J. Physiol.-Cell Physiol.* **320** C306–23

Pardo-Pastor C, Rubio-Moscardo F, Vogel-González M, Serra S A, Afthinos A, Mrkonjic S *et al* 2018 Piezo2 channel regulates RhoA and actin cytoskeleton to promote cell mechanobiological responses *Proc. Natl Acad. Sci. USA* **115** 1925–30

Parida P K, Marquez-Palencia M, Nair V, Kaushik A K, Kim K, Sudderth J *et al* 2022 Metabolic diversity within breast cancer brain-tropic cells determines metastatic fitness *Cell Metab.* **34** 90–105.e7

Park C C, Zhang H J, Yao E S, Park C J and Bissell M J 2008 β1 Integrin inhibition dramatically enhances radiotherapy efficacy in human breast cancer xenografts *Cancer Res.* **68** 4398–405

Park C C, Zhang H, Pallavicini M, Gray J W, Baehner F, Park C J *et al* 2006 β1 integrin inhibitory antibody induces apoptosis of breast cancer cells, inhibits growth, and distinguishes malignant from normal phenotype in three dimensional cultures and *in vivo Cancer Res.* **66** 1526–35

Pasqualini F S, Agarwal A, O'Connor B B, Liu Q, Sheehy S P and Parker K K 2018 Traction force microscopy of engineered cardiac tissues *PLoS One* **13** e0194706

Paszek M J, Zahir N, Johnson K R, Lakins J N, Rozenberg G I, Gefen A *et al* 2005 Tensional homeostasis and the malignant phenotype *Cancer Cell* **8** 241–54

Pathak A and Kumar S 2012 Independent regulation of tumor cell migration by matrix stiffness and confinement *Proc. Natl Acad. Sci. USA* **109** 10334–9

Patil S and Sengupta K 2021 Role of A- and B-type lamins in nuclear structure–function relationships *Biol. Cell* **113** 295–310

Payne S L, Hendrix M J C and Kirschmann D A 2006 Lysyl oxidase regulates actin filament formation through the p130^{Cas}/Crk/DOCK180 signaling complex *J. Cell. Biochem.* **98** 827–37

Peela N, Sam F S, Christenson W, Truong D, Watson A W, Mouneimne G *et al* 2016 A three-dimensional micropatterned tumor model for breast cancer cell migration studies *Biomaterials* **81** 72–83

Peinado H, Del Carmen Iglesias-de La Cruz M, Olmeda D, Csiszar K, Fong K S K, Vega S *et al* 2005 A molecular role for lysyl oxidase-like 2 enzyme in Snail regulation and tumor progression *EMBO J.* **24** 3446–58

Pelicano H, Zhang W, Liu J, Hammoudi N, Dai J, Xu R-H *et al* 2014 Mitochondrial dysfunction in some triple-negative breast cancer cell lines: role of mTOR pathway and therapeutic potential *Breast Cancer Res* **16** 434

Peng Y, Chen Z, Chen Y, Li S, Jiang Y, Yang H *et al* 2019 ROCK isoforms differentially modulate cancer cell motility by mechanosensing the substrate stiffness *Acta Biomater.* **88** 86–101

Pennarun G, Picotto J, Etourneaud L, Redavid A-R, Certain A, Gauthier L R *et al* 2021 Increase in lamin B1 promotes telomere instability by disrupting the shelterin complex in human cells *Nucl. Acids Res.* **49** 9886–905

Pfeifer C R, Alvey C M, Irianto J and Discher D E 2017 Genome variation across cancers scales with tissue stiffness—an invasion-mutation mechanism and implications for immune cell infiltration *Curr. Opin. Syst. Biol.* **2** 103–14

Pfeifer C R, Xia Y, Zhu K, Liu D, Irianto J, García V M M *et al* 2018 Constricted migration increases DNA damage and independently represses cell cycle *Mol. Biol. Cell* **29** 1948–62

Picariello H S, Kenchappa R S, Rai V, Crish J F, Dovas A, Pogoda K *et al* 2019 Myosin IIA suppresses glioblastoma development in a mechanically sensitive manner *Proc. Natl Acad. Sci. USA* **116** 15550–9

Pickup M W, Mouw J K and Weaver V M 2014 The extracellular matrix modulates the hallmarks of cancer *EMBO Rep.* **15** 1243–53

Piskounova E, Agathocleous M, Murphy M M, Hu Z, Huddlestun S E, Zhao Z *et al* 2015 Oxidative stress inhibits distant metastasis by human melanoma cells *Nature* **527** 186–91

Plessner M, Melak M, Chinchilla P, Baarlink C and Grosse R 2015 Nuclear F-actin formation and reorganization upon cell spreading *J. Biol. Chem.* **290** 11209–16

Plodinec M, Loparic M, Monnier C A, Obermann E C, Zanetti-Dallenbach R, Oertle P *et al* 2012 The nanomechanical signature of breast cancer *Nat. Nanotech.* **7** 757–65

Pommier A, Anaparthy N, Memos N, Kelley Z L, Gouronnec A, Yan R *et al* 2018 Unresolved endoplasmic reticulum stress engenders immune-resistant, latent pancreatic cancer metastases *Science* **360** eaao4908

Prauzner-Bechcicki S, Raczkowska J, Madej E, Pabijan J, Lukes J, Sepitka J *et al* 2015 PDMS substrate stiffness affects the morphology and growth profiles of cancerous prostate and melanoma cells *J. Mech. Behav. Biomed. Mater.* **41** 13–22

Provenzano P P, Inman D R, Eliceiri K W, Knittel J G, Yan L, Rueden C T *et al* 2008 Collagen density promotes mammary tumor initiation and progression *BMC Med.* **6** 11

Quintana E, Shackleton M, Sabel M S, Fullen D R, Johnson T M and Morrison S J 2008 Efficient tumour formation by single human melanoma cells *Nature* **456** 593–8

Raab M, Gentili M, de Belly H, Thiam H-R, Vargas P, Jimenez A J *et al* 2016 ESCRT III repairs nuclear envelope ruptures during cell migration to limit DNA damage and cell death *Science* **352** 359–62

Rankin E B and Giaccia A J 2016 Hypoxic control of metastasis *Science* **352** 175–80

Raudenska M, Kratochvilova M, Vicar T, Gumulec J, Balvan J, Polanska H *et al* 2019 Cisplatin enhances cell stiffness and decreases invasiveness rate in prostate cancer cells by actin accumulation *Sci. Rep.* **9** 1660

Reginato M J, Mills K R, Paulus J K, Lynch D K, Sgroi D C, Debnath J *et al* 2003 Integrins and EGFR coordinately regulate the pro-apoptotic protein Bim to prevent anoikis *Nat. Cell Biol.* **5** 733–40

Regmi S, Fu A and Luo K Q 2017 High shear stresses under exercise condition destroy circulating tumor cells in a microfluidic system *Sci. Rep.* **7** 39975

Reid S E, Kay E J, Neilson L J, Henze A, Serneels J, McGhee E J *et al* 2017 Tumor matrix stiffness promotes metastatic cancer cell interaction with the endothelium *EMBO J.* **36** 2373–89

Reiterer M, Colaço R, Emrouznejad P, Jensen A, Rundqvist H, Johnson R S *et al* 2019 Acute and chronic hypoxia differentially predispose lungs for metastases *Sci. Rep.* **9** 10246

Rhim A D, Oberstein P E, Thomas D H, Mirek E T, Palermo C F, Sastra S A *et al* 2014 Stromal elements act to restrain, rather than support, pancreatic ductal adenocarcinoma *Cancer Cell* **25** 735–47

Rho J Y, Ashman R B and Turner C H 1993 Young's modulus of trabecular and cortical bone material: ultrasonic and microtensile measurements *J. Biomech.* **26** 111–9

Rice A J, Cortes E, Lachowski D, Cheung B C H, Karim S A, Morton J P *et al* 2017 Matrix stiffness induces epithelial–mesenchymal transition and promotes chemoresistance in pancreatic cancer cells *Oncogenesis* **6** e352–2

Riehl B D, Kim E, Bouzid T and Lim J Y 2021 The role of microenvironmental cues and mechanical loading milieus in breast cancer cell progression and metastasis *Front. Bioeng. Biotechnol.* **8** 608526

Riethdorf S, Fritsche H, Müller V, Rau T, Schindlbeck C, Rack B *et al* 2007 Detection of circulating tumor cells in peripheral blood of patients with metastatic breast cancer: a validation study of the cellsearch system *Clin. Cancer Res.* **13** 920–8

Riscal R, Skuli N and Simon M C 2019 Even cancer cells watch their cholesterol! *Mol. Cell* **76** 220–31

Ritsma L, Steller E J A, Beerling E, Loomans C J M, Zomer A, Gerlach C *et al* 2012 Intravital microscopy through an abdominal imaging window reveals a pre-micrometastasis stage during liver metastasis *Sci. Transl. Med.* **4** 158ra145

Rizvi I, Gurkan U A, Tasoglu S, Alagic N, Celli J P, Mensah L B *et al* 2013 Flow induces epithelial-mesenchymal transition, cellular heterogeneity and biomarker modulation in 3D ovarian cancer nodules *Proc. Natl Acad. Sci. USA* **110** E1974–83

Roh-Johnson M, Bravo-Cordero J J, Patsialou A, Sharma V P, Guo P, Liu H *et al* 2014 Macrophage contact induces RhoA GTPase signaling to trigger tumor cell intravasation *Oncogene* **33** 4203–12

Romani P, Nirchio N, Arboit M, Barbieri V, Tosi A, Michielin F *et al* 2022 Mitochondrial fission links ECM mechanotransduction to metabolic redox homeostasis and metastatic chemotherapy resistance *Nat. Cell Biol.* **24** 168–80

Röper J-C, Mitrossilis D, Stirnemann G, Waharte F, Brito I, Fernandez-Sanchez M-E *et al* 2018 The major β-catenin/E-cadherin junctional binding site is a primary molecular mechanotransductor of differentiation *in vivo eLife* **7** e33381

Rosetti F, Chen Y, Sen M, Thayer E, Azcutia V, Herter J M *et al* 2015 A lupus-associated Mac-1 variant has defects in integrin allostery and interaction with ligands under force *Cell Rep.* **10** 1655–64

Roshanzamir F, Robinson J L, Cook D, Karimi-Jafari M H and Nielsen J 2022 Metastatic triple negative breast cancer adapts its metabolism to destination tissues while retaining key metabolic signatures *Proc. Natl Acad. Sci. USA* **119** e2205456119

Schackmann R C J, Van Amersfoort M, Haarhuis J H I, Vlug E J, Halim V A, Roodhart J M L *et al* 2011 Cytosolic p120-catenin regulates growth of metastatic lobular carcinoma through Rock1-mediated anoikis resistance *J. Clin. Invest.* **121** 3176–88

Schafer Z T, Grassian A R, Song L, Jiang Z, Gerhart-Hines Z, Irie H Y *et al* 2009 Antioxidant and oncogene rescue of metabolic defects caused by loss of matrix attachment *Nature* **461** 109–13

Schatton T, Murphy G F, Frank N Y, Yamaura K, Waaga-Gasser A M, Gasser M *et al* 2008 Identification of cells initiating human melanomas *Nature* **451** 345–9

Schierbaum N, Rheinlaender J and Schäffer T E 2019 Combined atomic force microscopy (AFM) and traction force microscopy (TFM) reveals a correlation between viscoelastic material properties and contractile prestress of living cells *Soft Matter* **15** 1721–9

Schiffhauer E S, Luo T, Mohan K, Srivastava V, Qian X, Griffis E R *et al* 2016 Mechanoaccumulative elements of the mammalian actin cytoskeleton *Curr. Biol.* **26** 1473–9

Schiffhauer E S, Ren Y, Iglesias V A, Kothari P, Iglesias P A and Robinson D N 2019 Myosin IIB assembly state determines its mechanosensitive dynamics *J. Cell Biol.* **218** 895–908

Schmied L, Höglund P and Meinke S 2021 Platelet-mediated protection of cancer cells from immune surveillance—possible implications for cancer immunotherapy *Front. Immunol.* **12** 640578

Schneider D, Baronsky T, Pietuch A, Rother J, Oelkers M, Fichtner D *et al* 2013 Tension monitoring during epithelial-to-mesenchymal transition links the switch of phenotype to expression of moesin and cadherins in NMuMG cells *PLoS One* **8** e80068

Schrader J, Gordon-Walker T T, Aucott R L, Van Deemter M, Quaas A, Walsh S *et al* 2011 Matrix stiffness modulates proliferation, chemotherapeutic response, and dormancy in hepatocellular carcinoma cells *Hepatology* **53** 1192–205

Semenza G L 2010 Defining the role of hypoxia-inducible factor 1 in cancer biology and therapeutics *Oncogene* **29** 625–34

Semenza G L 2013 HIF-1 mediates metabolic responses to intratumoral hypoxia and oncogenic mutations *J. Clin. Invest.* **123** 3664–71

Serrano I, McDonald P C, Lock F, Muller W J and Dedhar S 2013 Inactivation of the Hippo tumour suppressor pathway by integrin-linked kinase *Nat. Commun.* **4** 2976

Sewell-Loftin M K, Bayer S V H, Crist E, Hughes T, Joison S M, Longmore G D *et al* 2017 Cancer-associated fibroblasts support vascular growth through mechanical force *Sci. Rep.* **7** 12574

Shah P, Wolf K and Lammerding J 2017 Bursting the bubble—nuclear envelope rupture as a path to genomic instability? *Trends Cell Biol.* **27** 546–55

Shah S P, Morin R D, Khattra J, Prentice L, Pugh T, Burleigh A *et al* 2009 Mutational evolution in a lobular breast tumour profiled at single nucleotide resolution *Nature* **461** 809–13

Shen Y, Wang X, Lu J, Salfenmoser M, Wirsik N M, Schleussner N *et al* 2020 Reduction of liver metastasis stiffness improves response to bevacizumab in metastatic colorectal cancer *Cancer Cell* **37** 800–817.e7

Shibue T, Brooks M W and Weinberg R A 2013 An integrin-linked machinery of cytoskeletal regulation that enables experimental tumor initiation and metastatic colonization *Cancer Cell* **24** 481–98

Shibue T and Weinberg R A 2009 Integrin β_1 -focal adhesion kinase signaling directs the proliferation of metastatic cancer cells disseminated in the lungs *Proc. Natl Acad. Sci. USA* **106** 10290–5

Shimi T, Butin-Israeli V, Adam S A, Hamanaka R B, Goldman A E, Lucas C A *et al* 2011 The role of nuclear lamin B1 in cell proliferation and senescence *Genes Dev.* **25** 2579–93

Shin J-W and Mooney D J 2016 Extracellular matrix stiffness causes systematic variations in proliferation and chemosensitivity in myeloid leukemias *Proc. Natl Acad. Sci. USA* **113** 12126–31

Shmelkov S V, Butler J M, Hooper A T, Hormigo A, Kushner J, Milde T *et al* 2008 CD133 expression is not restricted to stem cells, and both CD133+ and CD133− metastatic colon cancer cells initiate tumors *J. Clin. Invest.* **118** 2111–20

Sieh S, Taubenberger A V, Rizzi S C, Sadowski M, Lehman M L, Rockstroh A *et al* 2012 Phenotypic characterization of prostate cancer LNCaP cells cultured within a bioengineered microenvironment *PLoS One* **7** e40217

Simões R V, Serganova I S, Kruchevsky N, Leftin A, Shestov A A, Thaler H T *et al* 2015 Metabolic plasticity of metastatic breast cancer cells: adaptation to changes in the microenvironment *Neoplasia* **17** 671–84

Singh R, Davies P and Bajaj A K 2003 Identification of nonlinear and viscoelastic properties of flexible polyurethane foam *Nonlinear Dyn.* **34** 319–46

Slack R J, Macdonald S J F, Roper J A, Jenkins R G and Hatley R J D 2022 Emerging therapeutic opportunities for integrin inhibitors *Nat. Rev. Drug Discov.* **21** 60–78

Slattery M J and Dong C 2003 Neutrophils influence melanoma adhesion and migration under flow conditions *Int. J. Cancer* **106** 713–22

Sloan E K, Priceman S J, Cox B F, Yu S, Pimentel M A, Tangkanangnukul V *et al* 2010 The sympathetic nervous system induces a metastatic switch in primary breast cancer *Cancer Res.* **70** 7042–52

Song Y, Soto J, Chen B, Hoffman T, Zhao W, Zhu N *et al* 2022 Transient nuclear deformation primes epigenetic state and promotes cell reprogramming *Nat. Mater.* **21** 1191–9

Sosa M S, Bragado P and Aguirre-Ghiso J A 2014 Mechanisms of disseminated cancer cell dormancy: an awakening field *Nat. Rev. Cancer* **14** 611–22

Sosale N G, Rouhiparkouhi T, Bradshaw A M, Dimova R, Lipowsky R and Discher D E 2015 Cell rigidity and shape override CD47's 'self'-signaling in phagocytosis by hyperactivating myosin-II *Blood* **125** 542–52

Sottnik J L, Daignault-Newton S, Zhang X, Morrissey C, Hussain M H, Keller E T *et al* 2013 Integrin alpha2beta1 ($\alpha 2\beta 1$) promotes prostate cancer skeletal metastasis *Clin. Exp. Metastasis* **30** 569–78

Staunton J R, Doss B L, Lindsay S and Ros R 2016 Correlating confocal microscopy and atomic force indentation reveals metastatic cancer cells stiffen during invasion into collagen I matrices *Sci. Rep.* **6** 19686

Stavropoulou V, Kaspar S, Brault L, Sanders M A, Juge S, Morettini S *et al* 2016 MLL-AF9 expression in hematopoietic stem cells drives a highly invasive AML expressing emt-related genes linked to poor outcome *Cancer Cell* **30** 43–58

Stinchcombe J C and Griffiths G M 2007 Secretory mechanisms in cell-mediated cytotoxicity *Annu. Rev. Cell Dev. Biol.* **23** 495–517

Stone K R, Mickey D D, Wunderli H, Mickey G H and Paulson D F 1978 Isolation of a human prostate carcinoma cell line (DU 145) *Int. J. Cancer* **21** 274–81

Streitberger K-J, Lilaj L, Schrank F, Braun J, Hoffmann K-T, Reiss-Zimmermann M *et al* 2020 How tissue fluidity influences brain tumor progression *Proc. Natl Acad. Sci. USA* **117** 128–34

Strilic B and Offermanns S 2017 Intravascular survival and extravasation of tumor cells *Cancer Cell* **32** 282–93

Stylianopoulos T, Martin J D, Chauhan V P, Jain S R, Diop-Frimpong B, Bardeesy N *et al* 2012 Causes, consequences, and remedies for growth-induced solid stress in murine and human tumors *Proc. Natl Acad. Sci. USA* **109** 15101–8

Sun Q, Luo T, Ren Y, Florey O, Shirasawa S, Sasazuki T *et al* 2014 Competition between human cells by entosis *Cell Res.* **24** 1299–310

Sun Y, Wang B-E, Leong K G, Yue P, Li L, Jhunjhunwala S *et al* 2012 Androgen deprivation causes epithelial–mesenchymal transition in the prostate: implications for androgen-deprivation therapy *Cancer Res.* **72** 527–36

Surcel A, Schiffhauer E S, Thomas D G, Zhu Q, DiNapoli K T, Herbig M *et al* 2019 Targeting mechanoresponsive proteins in pancreatic cancer: 4-hydroxyacetophenone blocks dissemination and invasion by activating MYH14 *Cancer Res.* **79** 4665–78

Sutherland M, Gordon A, Shnyder S, Patterson L and Sheldrake H 2012 RGD-binding integrins in prostate cancer: expression patterns and therapeutic prospects against bone metastasis *Cancers* **4** 1106–45

Sutton A A, Ehrlicher A J, Molter C W, Amini A, Idicula J, Furman M *et al* 2021 Cell monolayer deformation microscopy—a new method to measure the rheology of cell monolayers reveals mechanical fragility of the cell network in the epithelial to mesenchymal transition *Biophys. J.* **120** 63a–4a

Swaminathan V, Mythreye K, O'Brien E T, Berchuck A, Blobe G C and Superfine R 2011 Mechanical stiffness grades metastatic potential in patient tumor cells and in cancer cell lines *Cancer Res.* **71** 5075–80

Swift J, Ivanovska I L, Buxboim A, Harada T, Dingal P C D P, Pinter J *et al* 2013 Nuclear lamin-A scales with tissue stiffness and enhances matrix-directed differentiation *Science* **341** 1240104

Szczerba B M, Castro-Giner F, Vetter M, Krol I, Gkountela S, Landin J *et al* 2019 Neutrophils escort circulating tumour cells to enable cell cycle progression *Nature* **566** 553–7

Tambe D T, Corey Hardin C, Angelini T E, Rajendran K, Park C Y, Serra-Picamal X *et al* 2011 Collective cell guidance by cooperative intercellular forces *Nat. Mater* **10** 469–75

Tamzalit F, Wang M S, Jin W, Tello-Lafoz M, Boyko V, Heddleston J M *et al* 2019 Interfacial actin protrusions mechanically enhance killing by cytotoxic T cells *Sci. Immunol.* **4** eaav5445

Tan Y, Tajik A, Chen J, Jia Q, Chowdhury F, Wang L *et al* 2014 Matrix softness regulates plasticity of tumour-repopulating cells via H3K9 demethylation and Sox2 expression *Nat. Commun.* **5** 4619

Tang K, Li S, Li P, Xia Q, Yang R, Li T *et al* 2020 Shear stress stimulates integrin β1 trafficking and increases directional migration of cancer cells via promoting deacetylation of microtubules *Biochim. Biophys. Acta (BBA)—Mol. Cell Res.* **1867** 118676

Tang K, Yu Y, Zhu L, Xu P, Chen J, Ma J *et al* 2019 Hypoxia-reprogrammed tricarboxylic acid cycle promotes the growth of human breast tumorigenic cells *Oncogene* **38** 6970–84

Tang T T, Konradi A W, Feng Y, Peng X, Ma M, Li J *et al* 2021 Small molecule inhibitors of TEAD auto-palmitoylation selectively inhibit proliferation and tumor growth of *NF2*-deficient mesothelioma *Mol. Cancer Therap.* **20** 986–98

Taussig D C, Miraki-Moud F, Anjos-Afonso F, Pearce D J, Allen K, Ridler C *et al* 2008 Anti-CD38 antibody–mediated clearance of human repopulating cells masks the heterogeneity of leukemia-initiating cells *Blood* **112** 568–75

Taussig D C, Vargaftig J, Miraki-Moud F, Griessinger E, Sharrock K, Luke T *et al* 2010 Leukemia-initiating cells from some acute myeloid leukemia patients with mutated nucleophosmin reside in the CD34− fraction *Blood* **115** 1976–84

Tello-Lafoz M, Srpan K, Sanchez E E, Hu J, Remsik J, Romin Y *et al* 2021 Cytotoxic lymphocytes target characteristic biophysical vulnerabilities in cancer *Immunity* **54** 1037–54.e7

The Physical Sciences—Oncology Centers Network 2013 A physical sciences network characterization of non-tumorigenic and metastatic cells *Sci. Rep.* **3** 1449

Thomas W E, Vogel V and Sokurenko E 2008 Biophysics of catch bonds *Annu. Rev. Biophys.* **37** 399–416

Truong H H, Xiong J, Ghotra V P S, Nirmala E, Haazen L, Le Dévédec S E *et al* 2014 β_1 Integrin Inhibition elicits a prometastatic switch through the tGFβ–miR-200–ZEB network in E-cadherin–positive triple-negative breast cancer *Sci. Signal.* **7** ra15

Tsai J H, Donaher J L, Murphy D A, Chau S and Yang J 2012 Spatiotemporal regulation of epithelial-mesenchymal transition is essential for squamous cell carcinoma metastasis *Cancer Cell* **22** 725–36

Turajlic S and Swanton C 2016 Metastasis as an evolutionary process *Science* **352** 169–75

Ulrich T A, de Juan Pardo E M and Kumar S 2009 The mechanical rigidity of the extracellular matrix regulates the structure, motility, and proliferation of glioma cells *Cancer Res.* **69** 4167–74

Urbanska M, Winzi M, Neumann K, Abuhattum S, Rosendahl P, Müller P *et al* 2017 Single-cell mechanical phenotype is an intrinsic marker of reprogramming and differentiation along the mouse neural lineage *Development* **144** 4313–21

Valiente M, Obenauf A C, Jin X, Chen Q, Zhang X H-F, Lee D J *et al* 2014 Serpins promote cancer cell survival and vascular co-option in brain metastasis *Cell* **156** 1002–16

Van Tienen L M, Mieszczanek J, Fiedler M, Rutherford T J and Bienz M 2017 Constitutive scaffolding of multiple WNT enhanceosome components by Legless/BCL9 *eLife* **6** e20882

Vander Heiden M G, Cantley L C and Thompson C B 2009 Understanding the Warburg effect: the metabolic requirements of cell proliferation *Science* **324** 1029–33

Vanharanta S and Massagué J 2013 Origins of metastatic traits *Cancer Cell* **24** 410–21

Vartiainen M K, Guettler S, Larijani B and Treisman R 2007 Nuclear actin regulates dynamic subcellular localization and activity of the SRF cofactor MAL *Science* **316** 1749–52

Vashisth M, Cho S, Irianto J, Xia Y, Wang M, Hayes B *et al* 2021 Scaling concepts in 'omics: nuclear lamin-B scales with tumor growth and often predicts poor prognosis, unlike fibrosis *Proc. Natl Acad. Sci. USA* **118** e2112940118

Vaupel P, Höckel M and Mayer A 2007 Detection and characterization of tumor hypoxia using pO_2 histography *Antioxid. Redox Signal.* **9** 1221–36

Wagh A A, Roan E, Chapman K E, Desai L P, Rendon D A, Eckstein E C *et al* 2008 Localized elasticity measured in epithelial cells migrating at a wound edge using atomic force microscopy *Am. J. Physiol. Lung Cell. Mol. Physiol.* **295** L54–60

Wakamatsu Y, Sakamoto N, Oo H Z, Naito Y, Uraoka N, Anami K *et al* 2012 Expression of cancer stem cell markers ALDH1, CD44 and CD133 in primary tumor and lymph node metastasis of gastric cancer *Pathol. Int.* **62** 112–9

Wang G, Zhao D, Spring D J and DePinho R A 2018a Genetics and biology of prostate cancer *Genes Dev.* **32** 1105–40

Wang N, Butler J P and Ingber D E 1993 Mechanotransduction across the cell surface and through the cytoskeleton *Science* **260** 1124–7

Wang T, Shigdar S, Gantier M P, Hou Y, Wang L, Li Y *et al* 2015 Cancer stem cell targeted therapy: progress amid controversies *Oncotarget* **6** 44191–206

Wang Y, Zeng Z, Lu J, Wang Y, Liu Z, He M *et al* 2018b CPT1A-mediated fatty acid oxidation promotes colorectal cancer cell metastasis by inhibiting anoikis *Oncogene* **37** 6025–40

Weaver V M, Petersen O W, Wang F, Larabell C A, Briand P, Damsky C *et al* 1997 Reversion of the malignant phenotype of human breast cells in three-dimensional culture and *in vivo* by integrin blocking antibodies *J. Cell Biol.* **137** 231–45

Wei S C, Fattet L, Tsai J H, Guo Y, Pai V H, Majeski H E *et al* 2015 Matrix stiffness drives epithelial–mesenchymal transition and tumour metastasis through a TWIST1–G3BP2 mechanotransduction pathway *Nat. Cell Biol.* **17** 678–88

Weinberg R A 2013 *The Biology of Cancer* (New York: W.W. Norton & Company)

Weins A, Schlondorff J S, Nakamura F, Denker B M, Hartwig J H, Stossel T P *et al* 2007 Disease-associated mutant α-actinin-4 reveals a mechanism for regulating its F-actin-binding affinity *Proc. Natl Acad. Sci. USA* **104** 16080–5

Weiss L 1990 Metastatic inefficiency *Advances in Cancer Research* (Amsterdam: Elsevier) pp 159–211

Weiss L, Dimitrov D S and Angelova M 1985 The hemodynamic destruction of intravascular cancer cells in relation to myocardial metastasis *Proc. Natl Acad. Sci. USA* **82** 5737–41

Welch D R and Hurst D R 2019 Defining the Hallmarks of metastasis *Cancer Res.* **79** 3011–27

West-Foyle H, Kothari P, Osborne J and Robinson D N 2018 14-3-3 Proteins tune non-muscle myosin II assembly *J. Biol. Chem.* **293** 6751–61

Wishart A L, Conner S J, Guarin J R, Fatherree J P, Peng Y, McGinn R A *et al* 2020 Decellularized extracellular matrix scaffolds identify full-length collagen VI as a driver of breast cancer cell invasion in obesity and metastasis *Sci. Adv.* **6** eabc3175

Wozniak M A, Desai R, Solski P A, Der C J and Keely P J 2003 ROCK-generated contractility regulates breast epithelial cell differentiation in response to the physical properties of a three-dimensional collagen matrix *J. Cell Biol.* **163** 583–95

Wrenn E D, Yamamoto A, Moore B M, Huang Y, McBirney M, Thomas A J *et al* 2020 Regulation of collective metastasis by nanolumenal signaling *Cell* **183** 395–410.e19

Wu B, Liu D-A, Guan L, Myint P K, Chin L, Dang H *et al* 2023 Stiff matrix induces exosome secretion to promote tumour growth *Nat. Cell Biol.* **25** 415–24

Wu P-H, Aroush D R-B, Asnacios A, Chen W-C, Dokukin M E, Doss B L *et al* 2018 A comparison of methods to assess cell mechanical properties *Nat. Methods* **15** 491–8

Wu X, Gong S, Roy-Burman P, Lee P and Culig Z 2013 Current mouse and cell models in prostate cancer research *Endocr.-Relat. Cancer* **20** R155–70

Wyckoff J B, Jones J G, Condeelis J S and Segall J E 2000 A critical step in metastasis: *in vivo* analysis of intravasation at the primary tumor *Cancer Res.* **60** 2504–11

Xia Y, Ivanovska I L, Zhu K, Smith L, Irianto J, Pfeifer C R *et al* 2018 Nuclear rupture at sites of high curvature compromises retention of DNA repair factors *J. Cell Biol.* **217** 3796–808

Xin Y, Chen X, Tang X, Li K, Yang M, Tai W C-S *et al* 2019 Mechanics and actomyosin-dependent survival/chemoresistance of suspended tumor cells in shear flow *Biophys. J.* **116** 1803–14

Xu W, Mezencev R, Kim B, Wang L, McDonald J and Sulchek T 2012 Cell stiffness is a biomarker of the metastatic potential of ovarian cancer cells *PLoS One* **7** e46609

Xu Y, Bismar T A, Su J, Xu B, Kristiansen G, Varga Z *et al* 2010 Filamin A regulates focal adhesion disassembly and suppresses breast cancer cell migration and invasion *J. Exp. Med.* **207** 2421–37

Yago T, Wu J, Wey C D, Klopocki A G, Zhu C and McEver R P 2004 Catch bonds govern adhesion through L-selectin at threshold shear *J. Cell Biol.* **166** 913–23

Yan Z, Xia X, Cho W C, Au D W, Shao X, Fang C *et al* 2022 Rapid plastic deformation of cancer cells correlates with high metastatic potential *Adv Healthc. Mater.* **11** 2101657

Yang F, Teves S S, Kemp C J and Henikoff S 2014 Doxorubicin, DNA torsion, and chromatin dynamics *Biochim. Biophys. Acta (BBA)—Rev. Cancer* **1845** 84–9

Yang J-H, Sakamoto H, Xu E C and Lee R T 2000 Biomechanical regulation of human monocyte/macrophage molecular function *Am. J. Pathol.* **156** 1797–804

Yeoman B, Shatkin G, Beri P, Banisadr A, Katira P and Engler A J 2021 Adhesion strength and contractility enable metastatic cells to become adurotactic *Cell Rep.* **34** 108816

Yoshie H, Koushki N, Kaviani R, Tabatabaei M, Rajendran K, Dang Q *et al* 2018 Traction force screening enabled by compliant PDMS elastomers *Biophys. J.* **114** 2194–9

Yu H, Mouw J K and Weaver V M 2011 Forcing form and function: biomechanical regulation of tumor evolution *Trends Cell Biol.* **21** 47–56

Zanconato F, Forcato M, Battilana G, Azzolin L, Quaranta E, Bodega B *et al* 2015 Genome-wide association between YAP/TAZ/TEAD and AP-1 at enhancers drives oncogenic growth *Nat. Cell Biol.* **17** 1218–27

Zanotelli M R, Zhang J and Reinhart-King C A 2021 Mechanoresponsive metabolism in cancer cell migration and metastasis *Cell Metab.* **33** 1307–21

Zhang W, Kai K, Choi D S, Iwamoto T, Nguyen Y H, Wong H *et al* 2012 Microfluidics separation reveals the stem-cell–like deformability of tumor-initiating cells *Proc. Natl Acad. Sci. USA* **109** 18707–12

Zhao B, Li L, Wang L, Wang C-Y, Yu J and Guan K-L 2012 Cell detachment activates the Hippo pathway via cytoskeleton reorganization to induce anoikis *Genes Dev.* **26** 54–68

Zhao B, Wei X, Li W, Udan R S, Yang Q, Kim J *et al* 2007 Inactivation of YAP oncoprotein by the Hippo pathway is involved in cell contact inhibition and tissue growth control *Genes Dev.* **21** 2747–61

Zhu M and Kyprianou N 2010 Role of androgens and the androgen receptor in epithelial-mesenchymal transition and invasion of prostate cancer cells *FASEB J.* **24** 769–77

IOP Publishing

Physics of Cancer, Volume 6 (Second Edition)
Cellular mechanisms to foster or fight cancer
Claudia Tanja Mierke

Chapter 7

The dynamic and adaptive mechanical signature of cancer cells and their molecular biological features regulate their metastatic journey

7.1 Summary

Most mechanical analyses of cancer cells are analyzed at a single time point, and dynamic or repeated analysis is not routinely performed. Moreover, the mechanical and molecular, such as multiomics, were examined separately. The combination of mechanical and molecular biological examination, referred to as ELAStomics, appears to be necessary to fully comprehend the mechanobiological features of cancer cells. Since the mechanical properties of cancer cells are not a stationary phenotype, single-cell fluid force spectroscopy appears to be necessary to decipher the dynamic mechanical signatures of malignant cancers such as breast cancer. Based on the dynamic mechanical properties of cancer cells, it is hypothesized that the dynamic adaptation of the mechanical properties of cancer cells is crucial for cancer cell metastasis. Moreover, it can be hypothesized that the biophysical properties of subpopulations of breast cancer cells determine their metastatic preference, such as organotropism. In addition to cellular mechanical characteristics, the mechanical cues of the metastatic pathway also play a role in cancer metastasis. In this chapter, the mechanical adaptability and dynamic mechanical nature of cancer cells during malignant progression of cancer are emphasized. The necessity of combinatorial analysis of molecular biological and mechanobiological features is discussed. Finally, it can be concluded that multiomics analysis may hold a potential for cancer therapy, as it allows dynamic and repeated analysis of the cancer cells at a single-cell level.

7.2 Single-cell fluidic force spectroscopy unveils dynamic mechanical signatures of malignant cancers in breast cancer

The crosstalk between the physical characteristics of cancer cells and metastatic potential emphasizes the importance of cancer cell mechanobiology. Leveraging

doi:10.1088/978-0-7503-4007-6ch7

fluid-based single-cell force spectroscopy (SCFS), dissipation quartz crystal microbalance (QCM-D), and a model of cells with a range of metastatic potentials, the evolution of biomechanics throughout the metastatic conditions can be explored. Therefore, the adhesive forces between the cell and the substrate and between the cells, the spring constant of the cells, the cell height and the viscoelasticity of the cells can be measured. The general observation is that cancer cells (excluding prostate cancer cells) at the lower end of the metastasizing spectrum are systematically stiffer, less viscoelastic, and larger compared to highly metastasizing cells. These mechanical changes in cells within a cluster are significantly associated with the cells' metastatic potential but are significantly lacking in individual cells. In addition, the responsiveness to chemotherapy in terms of response time and extent relies strongly on the viscoelastic properties of the cells. Changes in the softness and elasticity of cells could act as mechanoadaptive mechanisms involved in cancer cell metastasis and provide insights into metastasis and the efficacy of prospective treatment interventions.

Solid tumors are frequently recognized based on their increased stiffness and strength relative to normal tissues, such as on palpation, but the mechanical characteristics of solid tumors can take on more complex forms (Lee *et al* 2012, Xu *et al* 2012, Alibert *et al* 2017, Nguyen *et al* 2022). During cancer invasion and metastasis, the characteristics of cancer cells undergo changes that affect their proliferation, differentiation, migration, contractility, and apoptosis, leading to alterations in their biomechanical features. Divergences from the mechanical homeostasis are linked to marked cell invasiveness and are accompanied by a rearrangement of the extracellular matrix (ECM), diminished cell–cell and cell–matrix adhesion, enhanced cell elasticity and deformability, and augmented cell motility (Paszek *et al* 2005, Mierke *et al* 2011, Mierke 2013, Chaudhuri *et al* 2014, Fischer *et al* 2017). These changes are linked to variations in cytoskeletal structure and membrane viscosity, which impact on cell softness and the likelihood of malignancy and metastasis (Swaminathan *et al* 2011, Xu *et al* 2012, Nematbakhsh *et al* 2017, Aseervatham 2020, Kashani and Packirisamy 2020). This malignant transformation and mechanical rearrangement are facilitated through epithelial–mesenchymal transition (EMT) under the impact of upregulation, downregulation or even silencing of genes and cancer marker expressions like integrins, actin, fibronectin, Rho-GTPase, gap junction proteins and EMT markers (Gerashchenko *et al* 2019).

Gap junction intercellular communication (GJIC) is constructed from connexin (Cx) proteins, which act as transmembrane proteins to facilitate cell–cell signaling and the passage of ions and metabolites across adjacent cells (Fostok *et al* 2019). Connexin 43 (Cx43), a component of channel-forming gap junctions (GJs), displays spatio-temporal expression profiles and is essential for development, differentiation and tumor suppression in mammary tissues (Zhou *et al* 2023). Disrupted expression and localization of Cx43 can modify mammary gland functioning and promote cancer initiation and advancement, where cancer cells physically separate from their tumor microenvironment (TME), invade tissue, and eventually colonize remote organ locations (Kanczuga-Koda *et al* 2006, Banerjee 2016). Conversely, upon re-expression of Cx43, it re-establishes active GJIC between adjacent cells and aids cancer cells in their interaction with the endothelium, thus improving their Intravasation and extravasation (Pollmann *et al* 2005, Naser Al Deen *et al* 2019). Recently, upregulation of Cx43 in metastatic human breast epithelial cells of the cancer cell line MDA-MB-231 was shown to cause elevated expression of epithelial markers like E-cadherin and ZO-1. Subsequently, it leads to the sequestration of β-catenin at the cell membrane *in vitro*, in addition to weakening the growth of primary tumors and reducing the malignancy level of these cells *in vivo*. In contrast, silencing of Cx43 caused these breast cancer cells to acquire a mesenchymal phenotype with enhanced N-cadherin expression *in vitro* and a more aggressive metastatic phenotype *in vivo* (Zibara *et al* 2015, Kazan *et al* 2019).

There has been much recent excitement about recent efforts in breast cancer therapy that target inhibitory signal transduction routes implicated in the regulation of cell mechanical characteristics like stiffness, elasticity, and adhesiveness (Lei *et al* 2021, Luo *et al* 2022). Docetaxel (DTX), a stabilizer of microtubules, has been found to interfere with intercellular adhesion forces through interference with cytoskeletal adhesion proteins like actin and integrins, thereby altering cellular shape, division, and motile behavior (Raudenska *et al* 2019). Determining the mechanical characteristics of cancer cells may offer greater mechanistic understanding of tumor development and metastasis, act as cancer biomarkers for early diagnosis, and provide a new foundation for designing new therapies that can modify the mechanical signatures of cells for enhanced intervention. This chapter discusses in the first part the relationship between the mechanical characteristics of MDA-MB-231 breast epithelial cancer and its metastatic capacity using SCFS. These results show that downregulation of Cx43 caused softening of cancer cells, which was linked to diminished cell–cell adhesion, elevated elasticity and deformability, and heightened their potential for malignant behavior and increased aggressiveness. Conversely, upregulation of Cx43 was linked to tight cell–cell adhesion, stiffness, whereby the malignant potential is decreased. Ultimately, the temporal progression of biomechanical alterations in reaction to DTX pointed to heightened cell stiffness in aggressive and more malignant cell subtypes.

In the second part of this chapter, it is described how the biophysical technique ELASTomics can be incorporated into established single-cell sequencing setups, thereby facilitating the simultaneous investigation of cell surface mechanics and their transcriptional control at an unparalleled level of resolution. In the third part, the various mechanical demands that cancer cells must satisfy to perform their tasks on the metastatic journey are explained, and the option that they dynamically adjust their characteristics to suit these demands is proposed. In the fourth part, the difference in cell stiffness is equivalent to the difference in the stiffness of their metastasized organs. Moreover, it is discussed how the cell cytoskeleton and mechanics are related with organotropism. The fifth part discusses the lack of agreement on whether there is a positive or negative correlation between the stiffness and invasiveness of cancer cells. Finally, it is discussed whether the relationship between the stiffness of cancer tissue and cancer cells and the invasive behavior depends on external circumstances.

7.2.1 Control of Cx43 expression triggers shape alterations related with the metastatic potential

Downregulation of Cx43 expression resulted in significant cellular shape alterations in MDA-MB-231 cells, with shCx43 cells acquiring a mesenchymal phenotype and Cx43D cells, which overexpress a fusion protein of Dendra-2 fused at the N-terminus of Cx43, adopting a more epithelial appearance. The Cx43Ds expressing the Dendra-2-Cx43 fusion protein can still create functional GJs (Al-Ghadban *et al* 2016). Quantitative evaluation of this alteration was performed in individual cells in the center of the epithelium utilizing digital holographic microscopy to determine the cross-sectional diameter and the cell height. In all three cell lines, the cells had nearly the same diameter of about 40 μm, but the mean cross-sectional height differed among the cells, with Cx43D cells having the largest height (21.6 ± 1.4 μm), then WT (15.8 ± 0.9 μm), and finally shCx43 (12.9 ± 0.6 μm) (Habli *et al* 2024). It should be noted that the three subpopulations of MDA-MB-231 cells with down-regulated or upregulated Cx43 expression were established and characterized earlier, whereby a triple-negative breast cancer *in vitro* model with different metastatic potential was generated from the same cell line (Zibara *et al* 2015, Kazan *et al* 2019). Expression of Cx43 was assessed using qPCR in sorted cells as a matter of routine and analyzed in comparison to Cx43 expression in MDA-MB-231 cells and wild-type cells (control).

7.2.2 Cellular metastatic potential modifies cell–matrix and cell–cell adhesions, and cellular elasticity

The mechanical and biophysical characteristics of individual cells cultured as single cells in a non-cell–cell contact environment and within a cluster of cells in a fully enclosed environment, under equivalent conditions of confluence, were determined using fluidic force microscopy. This SCFS configuration was utilized to detect and gauge the effects of cell–cell interaction in the presence of Cx43 expression on the total adhesive forces in cells on a metric of metastatic potential. Brightfield images of SCFS-treated cells prior to and after detachment were taken with the respective representative force-distance curves (*F–D*). The obtained *F–D* curves of the SCFS-treated cells before and after detachment reveal a distinct evolution from a single-cell status to cells in a cluster, indicating the influence of Cx43 on the total binding. MDA-MB-231 cells at the single-cell level displayed the identical adhesion strength with an average adhesion force of about 70 nN, independent of Cx43 expression. Unexpectedly, the adhesion of MDA-MB-231 cells in a cluster differed significantly between the different subgroups and in comparison, to the single-cell state within the same subgroup, indicating a remarkable disparity in adhesion between the cells examined and a prominent involvement of Cx43 in cell–cell adhesion. Cx43D cells displayed a significant rise of F_{adh} to over 380 nN; a similar trend, albeit to a smaller extent, was seen in WT cells, which displayed an average F_{adh} of 300 nN. Both WT and Cx43D cells displayed a significantly about four-fold higher F_{adh} compared to individual cells, which displayed 70 nN. In contrast, the highly metastatic shCx43 cells in clusters showed just a two-fold rise in F_{adh} relative to a single-cell condition with an average of about 145 nN. Whereas the force determined for an individual cell only accounts for the adhesion force between the cell and the supporting substrate ($F_{cell-substrate}$), the force determined for a cell within a cluster comprises the intercellular adhesion forces in addition ($F_{cell-substrate} + F_{cell-cell}$). For the purpose of this investigation, the intercellular adhesion forces, which are represented as the relative force alteration between cells in a cluster and cells in a solitary state, can be obtained according to the equation (7.1) for relative force (Sancho *et al* 2017):

$$\text{Percentage relative force change} = \frac{\text{force}_{\text{clustered cell}} - \text{force}_{\text{individual cell}}}{\text{force}_{\text{individual cell}}} \cdot 100 \quad (7.1)$$

Intercellular forces amounted to 230 nN for control WT cells, 74 nN for highly metastatic shCx43 cells, and 290 nN for non-metastatic Cx43D cells. When transitioning from a single-cell state to a group cell state, WT cells exhibited a 320% enhancement in force alteration, shCx43 cells exhibited a 100% enhancement in force alteration, and Cx43D cells displayed an ≅400% enhancement in force alteration. The displacement of the cantilever just before the final break-off event provides information about the cell's stretching capacity before the maximum break-off force, such as F_{adh} is achieved. Single MDA-MB-231 cells, irrespective of Cx43 expression, showed similar D_{max} values (≅5.5 μm), in contrast to cells inside a cluster, where the values substantially differed between the different cell subtypes. The lowest D_{max} scores were observed in WT and shCx43 cells, averaging approximately 6.6 and 8.8 μm, respectively, which represents a significant reduction in the D_{max} values observed for Cx43D cells, which averaged approximately 10.2 μm. In fact, cells that need a higher D_{max} to completely dislodge typically demonstrate greater adhesion strength, which is in line with the F_{adh} data obtained. To assess the stiffness of the cells in this model, the spring coefficient (Sc) deduced from F_{adh}/D_{max} is utilized to estimate the elastic deformation potential of the cells assessed (figure 7.1) (Nagy *et al* 2022). The mean Sc values of individual MDA-MB-231 cells varied between 9.2 and 11.9 nN μm^{-1}, whereby no statistical significance

Figure 7.1. Concept of Sc and a schematic (out-of-scale) illustration demonstrating the elastic deformation capability of the cell during pull-off from a surface with a fluid-filled hollow FM cantilever. The relationship between F_{adh} (maximum force) and D_{max} (maximum elongation prior to detachment) when detaching the cell provides information about the cell's elastic properties and can be evaluated based on the *F–D* curves.

existed among the various subgroups (Habli *et al* 2024). When Cx43 is overexpressed, nevertheless, the mean Sc value in clusters rises considerably compared to other subpopulations and compared to individual cells, whereas there is no noticeable alteration upon downregulation of Cx43 expression.

7.2.3 Cell viscoelasticity is governed by the cellular metastatic potential

The quartz crystal microbalance with dissipation (QCM-D) was utilized to relate cellular elasticity and viscosity to their metastatic capacity and to establish a time course of potential alterations as a reaction to treatments. A typical scheme of time-dependent cell unbinding observed with QCM-D is shown in figure 7.2, with the respective shifts in frequency and dissipation caused by the viscoelastic nature of the cells under examination.

When studying the dynamics of cell dissociation with trypsin utilizing QCM-D, a single-phase QCM-D response can be detected for all cell subtypes, showing a steep rise in ΔF and a strong decline in ΔD. After that, a plateau was achieved within 20 min, which stayed almost unchanged also after injection of 1× phosphate buffered saline (PBS) to guarantee washing away of loosely attached cells. Figure 7.3 displays a schematic sketch of the Df plots (revealed by the $\Delta F/\Delta D$ ratio) of the three MDA-MB-231 subtypes along with their corresponding linearization.

The resulting diagrams correspond to the typical frequency and dissipation reactions that occur during cell rounding and subsequent cell delamination. The measured frequency and dissipation cues and the derived Df diagram exhibited nearly the same patterns for the three subtypes and cell concentrations, albeit with

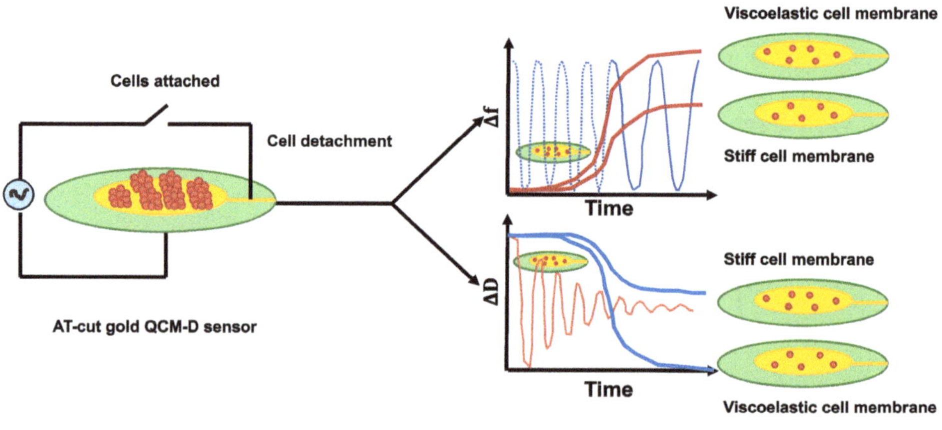

Figure 7.2. Representation of cell-based QCM-D experiments under liquid regime conditions and the corresponding diagrams showing alterations in frequency (ΔF) and dissipation energy (ΔD) with respect to time, reflecting the stiffness or softness properties of the cell layer. QCM-D exploits an inverse piezoelectric phenomenon: the AT-quartz crystal starts to oscillate in response to an imposed electrical voltage. The mechanical vibrations can be translated into an equivalent electrical circuit model that provides a full characterization of the vibration in the event of a mass and viscous loading. For soft and viscoelastic cell monolayers, the amplitude of the crystal oscillation decays quickly due to high energy dissipation as contrasted with stiff cell monolayers.

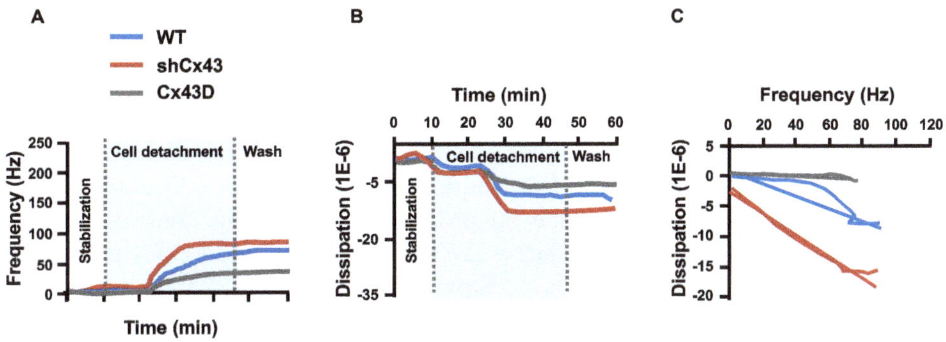

Figure 7.3. The metastatic state of MDA-MB-231 cells affects cellular viscoelasticity as detected by QCM-D. Real-time QCM-D assessments. (A) ΔF responses and (B) ΔD responses and equivalent (C) Df plots with linear regression of low-cell-number MDA-MB-231 cells with differential expression of Cx43 which undergo detachment.

variations in extent. At low cell numbers, the increase in ΔF was nearly the same for all subtypes, in contrast to conditions with high cell numbers, where the observed ΔF values across subtypes varied substantially, with shCx43 showing the largest frequency displacement at about 150 Hz and Cx43-cells and WT cells exhibiting similar displacements in the area between 80 and 90 Hz. Comparing ΔF between cell number densities, shCx43 and WT cells displayed a substantial augmentation in frequency, whereas no significance was detected in Cx43D cells. In contrast, ΔD varied significantly between the three subgroups, including both low and high cell densities, thereby confirming a unique viscoelastic fingerprint for every cell type. The highly metastatic shCx43 cells exhibited the greatest ΔD reaction with about 40×10^{-6}, whereas non-metastatic Cx43D cells showed a ΔD reaction that was three-fold smaller with an average of about 12×10^{-6}, and WT cells exhibited an intermediate ΔD reaction with an average of about 20×10^{-6} (Habli *et al* 2024). When examining the Df Plots slopes in the low cell number condition, shCx43 cells exhibited the lowest rise, with an average of −0.252, trailed by WT and Cx43D, displaying intermediate rises, with averages of about −0.406. At the same time, in experiments with a large number of cells, the slopes of the Df diagrams were found to be similar in all three subtypes, with rather small values lying within the range of −0.145 to −0.236. Remarkably, the slopes differed among the various cell confluences, with Cx43D and WT showing a steeper slope in their low-density conditions, whereas shCx43 hardly exhibited any alterations between the two cell densities.

7.2.4 The sensitivity of chemotherapy is impacted by the metastatic potential of the cells

In contrast to trypsin-based cell dissociation, treatment of cells with DTX at two different doses, such as 20 and 50 nM led to a biphasic QCM-D reaction with a steep decline followed by a gradual rise of ΔF and two very distinct ΔD reactions between the different doses, with distinct magnitudes and reaction rates. The highly metastatic shCx43 cells showed a fast, negative and sharp ΔF reaction just a few minutes following the start of treatment, in contrast to the non-metastatic Cx43D cells and the WT parent cells, which only started to react 30 min following the beginning of treatment. ShCx43 cells displayed the highest negative ΔF displacement upon treatment with 20 and 50 nM DTX, which averaged around 35 and 355 Hz, respectively. Thereafter, the frequency increased back to zero, with the reaction being much quicker at the 50 nM concentration. In contrast, WT and Cx43D cells showed the similar negative ΔF shift reactions upon DTX treatment: a gradual decline of ΔF and then a slow increase to the baseline, with WT demonstrating higher frequency responses at the 50 nM treatment concentration. No significance was detected between the three cell subtypes at comparable treatment concentrations, but when the two treatment regimens were juxtaposed, WT and shCx43 cells displayed a significant disparity in ΔF reaction. It is noteworthy that the decline of ΔF below equilibrated values, which adds to a negative ΔF, could point to cellular stiffening.

In contrast to frequency reactions, dissipation reactions reveal the viscoelastic dynamics of the cell membrane in DTX applications. Treatment with 20 nM DTX resulted in a positive ΔD reaction of the three MDA-MB-231 subtypes, suggesting potential cellular stiffening because of chemotherapy. Analogous to the ΔF displacement, the highly metastatic shCx43 cells demonstrated a fast, strong rise of ΔD, in contrast to the WT and Cx43D subtypes, which reached a plateau at zero 30 min after the onset of treatment and then exhibited a modest increase before stabilizing. shCx43 cells displayed the largest ΔD reaction with an average value of about 21×10^{-6}, then WT, and finally Cx43D with ΔD reactions with average scores of about 18×10^{-6} and 11×10^{-6}, respectively (Habli *et al* 2024). In contrast, 50 nM DTX led to negative ΔD reactions with much higher values than for 20 nM DTX exposure. As anticipated, shCx43 cells showed the strongest ΔD reaction with an average value of about -240×10^{-6}, which represents a six-fold disparity in comparison to WT cells and Cx43D cells. Notably, Cx43D cells displayed a small increment in dissipation then followed by a nearly equal decline with an average of 25×10^{-6} in reaction to DTX exposure. There was no observed significance within each treated group, but when WT and shCx43 cells were assessed together at the two different treatment dose levels, the recorded displacements were significantly dissimilar. Notably, when imaging cell disengagement after treatment with bright-field pictures, almost 90 min into the experiment, similar disengagement signatures were seen in each cell subtype, with much earlier reactions seen in the 50 nM DTX treatment condition compared to the 20 nM treatment condition.

20 nM DTX-QCM-D readings are combined with actin fluorescent staining and confocal time-lapse recording to trace biomechanical alterations after treatment at various time points in terms of actin remodeling and cellular roundedness. Upon analysis of non-treated control cells, differences in actin reorganization and cell area between the three subtypes are immediately visualized and in line with each cell's metastatic capacity. The less metastatic Cx43D cells showed elongated and well-organized actin filament meshworks that extended almost homogeneously throughout the cell. With growing metastatic activity of the cells, actin filaments become less

ordered and more fragmented, as seen in shCx43 cells and to a smaller degree in WT cells. After exposure, cells exhibited a strong actin reorganization in both dynamics and local distribution within treated cells after a period of time. In all cell subtypes, with variations in rates and strengths of reaction, DTX induced a pronounced rearrangement of actin filaments towards the nucleus and away from the cytoplasmic periphery, causing alterations in cell shape, marked rounding of cells and a reduction in cell area, indicative of cell stiffening and delamination (Khraiche *et al* 2019, Hou *et al* 2021). The last of these was measured using a QCM-D, with frequency drops noted in all cell subtypes, which can be attributed to stiffening of the cells. Remarkably, shCx43 cells displayed the most rapid decrease in cell area as early as 5 min and lost close to 30% of their overall area because of actin stabilization and reorganization, coinciding with the strong 60-Hz dip in their frequency-detachment profile. In contrast, the parental WT and Cx43D cells showed hardly any decrease in their cell surface area 5 min following treatment. These cells exhibited a marked shrinkage of the cell area at about 50% of the entire area, which was accompanied by a marked spheroidization of the cells at around 15–30 min of exposure, which was at a later time point in comparison to shCx43 cells. The actin filaments, which were comparatively highly organized relative to those in shCx43, needed additional time to completely reorganize to the central part of the cells; hence the retarded small frequency declines (nearly 20 Hz at approximately 40 min of exposure). After mapping the QCM-D cell detachment reaction using time-lapse videos, the restructuring of actin filaments was explored. Even though the highly metastatic shCx43 cells displayed the most rapid reaction in the QCM-D data and actin reorganization, their dissociation and morphological alterations seemed to be gradual and individualized. It took about 60 min before the monolayer of cells started to become rounded and then disintegrated into a single-cell status. Interestingly, shCx43 cells, which displayed a mesenchymal-like phenotype, were the most robust following a 30-minute treatment and barely revealed any evidence of cell blebs. When the level of metastasis was relatively low, detachment reaction of Cx43D cells was unique in that cells in monolayers formed rounded structures and bulging membranes prior to becoming detached. Conversely, the WT parental control cells required a mid-range time of 30 min to become circularized and begin detaching from the substrate, with reactions that were like shCx43, but to a lower degree and more slowly. It should be noted that the DTX treatment had no effect on cell viability even after 90 min, with just 3%–5% of cells dying in any group.

This chapter examines the progressive alteration of cancer cell biomechanical characteristics over the full spectrum of metastatic potential deduced from the identical cell line through modulation of Cx43 expression. Several investigations have demonstrated that aggressive cancer cells may undergo adaptive softening, spread out of their microenvironment, lengthen to squeeze themselves through capillaries, and metastasize to far-flung sites, creating life-threatening metastatic colonies (El-Sabban *et al* 2003, Marx *et al* 2005, 2007, El-Saghir *et al* 2011, Weber *et al* 2011, Chen *et al* 2012, Tymchenko *et al* 2012, Rother *et al* 2014, Nowacki *et al* 2015, Chang *et al* 2016, Lekka 2016, Nguyen *et al* 2016b, Fuhrmann *et al* 2017, Gandalovičová *et al* 2017, Abotaleb *et al* 2018, Osmani *et al* 2019, Habli *et al* 2020,

Han *et al* 2020, Rianna *et al* 2020, Tishchenko *et al* 2020, Gensbittel *et al* 2021, Lv *et al* 2021, Easley *et al* 2022, Habli *et al* 2023, Liboz *et al* 2023). A detailed picture of the cell mechanics of different cancer cell types is provided in table 7.1.

Earlier endeavors to answer this question often concentrated on mechanical analogies between cancer and normal cells or two cell lines with different metastatic potential derived from the same parental cell line (table 7.1) (Lekka *et al* 1999, Cross *et al* 2008, Faria *et al* 2008, Li *et al* 2008, Rebelo *et al* 2013). These investigations have provided enormous insight into the importance of cellular mechanics in invasion and metastasis, and have spawned the idea that cancer cells must be highly deformable to pass through tight spaces in tissues (Guck *et al* 2005, Xu *et al* 2012). Although many investigations have in fact demonstrated that cancer cells are softer compared to normal cells (Lekka *et al* 1999, Cross *et al* 2008, Faria *et al* 2008, Li *et al* 2008, Rebelo *et al* 2013), the picture is far from conclusive, with several studies indicating that cancer cells are stiffer than normal cells (Zhang *et al* 2002, Lam *et al* 2008, 2008, Pogoda *et al* 2021), particularly when they encounter an ECM (Pickup *et al* 2014, Acerbi *et al* 2015, Rianna and Radmacher 2017a, 2017b, Abidine *et al* 2018).

In addition, the difficulties in deciphering such studies are exacerbated because of the differences in genetic background between normal and tumorigenic culture models, so it is challenging to determine whether observed differences in cell stiffness are truly related to variations in tumorigenic capacity. Actually, in mechanical measurements within an isogenic cancer progression sequence, it was found that cell stiffness rises with tumorigenic capacity (Baker *et al* 2010). Additional insights can be gained by characterizing tumor cells at different stages of restriction and invasion. As cancer cells become increasingly constrained, they undergo a significant softening, which is concordant with the idea that deformability supports invasion and metastasis.

Experimental techniques to assess the mechanical characteristics of cells are limited, but they all suggest a relationship between cell stiffness and cell behavior. In the past, micropipette aspiration was the predominant experimental technique (Hochmuth 2000). With this technique, a 50% decrease in the elasticity of malignant fibroblasts in comparison to their normal counterparts was found (Ward *et al* 1991). More lately, AFM has been employed to assess cell stiffness (Rotsch *et al* 1999, Mahaffy *et al* 2000). AFM has been used to study normal human bladder cell lines and complementary cancer cell lines and it was revealed that the stiffness of the cells had a difference of an order of magnitude, whereby the cancer cells possessed stiffer mechano-phenotypes (Lekka *et al* 1999). In addition, magnetic bead rheology (Wang *et al* 1993), microneedle probes, such as cell poking (Zahalak *et al* 1990), microplate manipulation (Thoumine and Ott 1997), acoustic microscopes (Kundu *et al* 2000), sorting in microfabricated sieves (Carlson *et al* 1997), and the manipulation of beads bound to cells using optical tweezers (Sleep *et al* 1999) have been employed to explore the mechanical characteristics of cells or cell membanes. In most cases, malignant cells responded either less elastically (softer) or less viscously (less resistant to flow) to the exerted stresses, which varied depending on the measurement technique and cancer model being used.

Table 7.1. Comparison of different investigations on cancer cell mechanics, migration and metastasis.

Cell mechanics	Cell type	Biophysical technique	Description	Reference
Cancer cells soften to migrate through narrow spaces.	U2OS osteosarcoma cell line.	AFM	Using a polydimethylsiloxane open-roof microdevice featuring tapered, fibronectin-coated channels. Cell softening as a mechanoadaptive mechanism during invasion.	Rianna *et al* (2020)
Cancer cells are softer than normal cells.	Cancerous cells (bladder transitional cell carcinoma Hu456, bladder transitional cell carcinoma T24, BC3726, which originates from non-malignant bladder urothelium HCV29 transformed by the v-ras oncogene) and normal cells (non-malignant ureter Hu609 and HCV29).	Scanning force microscopy (SFM) (AFM derived).	The Young's modulus (cell stiffness) has been measured in cell culture medium.	Lekka *et al* (1999)
Metastatic cancer cells (patient-derived cells, such as pleural effusion samples) are softer compared of benign cells (normal mesothelial cells).	Metastatic cancer cells (patient-derived cells, such as pleural effusion samples of non-small carcinoma of the lung, breast ductal adenocarcinoma and pancreatic adenocarcinoma) and normal mesothelial cells.	AFM	Metastatic cancer cells tend to be more than 80% softer compared to benign cells, and their scattering is six times tighter than that of normal cells.	Cross *et al* (2008)

(*Continued*)

Table 7.1. (*Continued*)

Cell mechanics	Cell type	Biophysical technique	Description	Reference
Prostate cancer cells are softer compared to normal cells.	LNCaP, PC-3 and benign prostate hyperplasia (BPH) cells and normal prostate tissue, such as PNT2.	AFM	Young's moduli.	Faria *et al* (2008)
PC-3 cells are softer than PZ-HPV-7 cells.	Metastatic prostate cells (PC-3) and normal human prostate cells (PZ-HPV-7).	AFM	The modulus of elasticity was reduced in the advancement of prostate cancer, and the elasticity of the cellular element in prostate cancer tissue was lower than that of normal prostate tissue.	Zeng *et al* (2023)
Non-malignant prostate cells are stiffer than cancer cells while the metastatic cells are much softer than malignant cells from the primary tumor site.	Normal human prostate (RWPE-1) cells and a range of malignant (22Rv1) and metastatic prostate cells (LNCaP, Du145 and PC3).	AFM	The Young's modulus is being proposed as a new diagnostic biomarker for detecting cancer in its early stages.	Pogoda *et al* (2021)
Benign (MCF-10A) are stiffer comapred to cancerous MCF-7 human breast epithelial cells.	Elasticity of benign (MCF-10A) and cancerous (MCF-7) human breast epithelial cells.	AFM	Malignant MCF-7 breast cells exhibit an apparent Young's modulus that is significantly lower (1.4–1.8 times) compared to that of their non-malignant MCF-10A counterparts at a physiological temperature of 37 °C.	Li *et al* (2008)

Different kidney tumor types, such as carcinoma A-498 cells and adenocarcinoma ACHN cells are compared to a non-tumorigenic cell line RC-124.	Non-tumorigenic cells are less deformable and more viscous compared to cancerous cells. Cancer cell lines exhibit distinctive viscoelastic cues.	AFM	Viscoelastic properties, such as elasticity modulus E and apparent viscosity η $E(RC\text{-}124) > E(A\text{-}498) > E(ACHN)$ and $\eta(RC\text{-}124) > \eta(A\text{-}498) > \eta(ACHN)$.	Rebelo *et al* (2013)
Malignantly transformed SV-T2 are more deformable (softer) compared to normal BALB/3T3. Malignantly transformed modMCF-7 are more deformable compared to MCF-7. MCF-7 are more deformable compared to MCF-10A. modMDA-MB-231 are more deformable compared to MDA-MB-231 and	Chemically modified MCF-7 cells, referred to as modMCF-7 were produced by exposing MCF-7 cells to 100 nM 12-O-tetradecanoyl-phorbol-13-acetate (TPA) for 18 h. MDA-MB-231 cells were exposed to all-trans retinoic acid to produce modMDA-MB-231 cells, using the procedures described in Wang *et al* (2002).	Microfluidic optical cell stretcher.	Malignant and less malignant cell lines from one original cell line are compared.	Guck *et al* (2005)
Cloned rat embryo fibroblast (CREF, which are Wt cells generated by infecting the CREF cells with a wild-type adenovirus type 5) are stiffer compared to ras-transformed cells CREF-T24. Primary rat fibroblasts (WT) and WT-T24.	Non-transformed and transformed rat fibroblasts derived from the same normal cell line.	Micropipette aspiration.	Elevated cell deformability represents an important characteristics of cancer cell progression.	Ward *et al* (1991)

(*Continued*)

Table 7.1. (*Continued*)

Cell mechanics	Cell type	Biophysical technique	Description	Reference
Apparent viscosity is 30% lower for transformed compared to normal fibroblasts. Transformed fibroblasts displayed a more fragile surface compared to their normal counterparts.	Normal human dermal fibroblasts and their SV40-transformed counterparts.	Micropipette aspiration.	Suspended cells are measured. Normal fibroblasts exhibited much more contractility than their transformed counterparts.	Thoumine and Ott (1997)
Aggressive HEY A8 are softer compared to less invasive HEY cells.	Highly invasive ovarian cancer cells (HEY A8) and their less invasive parental cells (HEY).	AFM	Decreased stiffness of highly metastatic HEY A8 cells is coupled to the actin cytoskeleton rearrangement.	Xu *et al* (2012)
Hepatocytes are stiffer compared to hepatocellular carcinoma (HCC) cells.	Viscoelastic characteristics of human hepatocytes and hepatocellular carcinoma (HCC) cells under cytoskeletal perturbation.	Micropipette aspiration.	Colchicines (Col), cytochalasin D (CD) and vinblastine (VBL) are employed to treat HCC cells HCC displayed a litte decrease in the viscous coefficient.	Zhang *et al* (2002)
Normal thyroid cells and cancer cells grown on differently stiff substrates, such as 3–5 k Pa (soft) and 30–40 kPa (stiff).	Human primary thyroid cells (S747) and anaplastic thyroid carcinoma cells (S277).	AFM	While normal thyroid cells adjusted their mechanical characteristics to various stiff substrates, cancer cells were less impacted by the ambient stiffness In normal cells, the modulus of elasticity varied from 1.2 to 1.6 and 2.6 kPa with rising substrate stiffness; the dynamic viscosity values changed accordingly from 230 to 515 and 470 Pa·s.	Rianna and Radmacher (2017a)

			In opposition to this, the values for cancer cells were fairly constant independent of the substrate stiffness (on average, the Young's modulus was 1.3 kPa and the dynamic viscosity was 300 Pa·s).	
	Three bladder epithelial cancer cell lines were used that reflect increasing states of malignancy in the progression of metastasis: RT112, T24, and J82.		The impact of an elastic substrate on the mechanical characteristics of adherent cancer cells and the assessment of their mechanosensitivity was investigated using three distinct elastic gels, namely 5, 8 and 28 kPa and an endothelial substrate. A global trend is that the cell elasticity increases with substrate stiffness, and the stiffening is linked to the development of actin fibers on stiffer substrates.	Abidine *et al* (2018)
Breast MCF-7 and MDA-MB-231 cancer cells are softer than benign mammary gland non-tumorigenic MCF-10A cells.	MDCKII cells and CaKi-1 cells originate from the kidney; SW13 are derived from the adrenal gland; NMuMG, MCF-10A, MCF-7 and MDA-MB-231 are cells from the mammary gland, and A549 cells are derived from the lung epithelial layer MDCKII and NMuMG cells show epithelial morphology.	AFM	There exist tissue-specific differences between the cell lines regarding their elastic and viscose features.	Rother *et al* (2014)

(*Continued*)

Table 7.1. (*Continued*)

Cell mechanics	Cell type	Biophysical technique	Description	Reference
ALL-patient-derived cells ($n = 15$).	Primary cell samples.	AFM	The difference in cell stiffness was more pronounced in symptomatic patients, suggesting that a small fraction of stiff cells in the population is adequate to cause leucostasis in paediatric ALL.	Lam *et al* (2008)
Soft cells are highly tumorigenic and metastatic compared to stiff cells sorted from the same parental cell line.	Cancer cells derived from mice with breast cancer (4T1), from humans with breast cancer (MCF-7), from mice with B16 melanoma and from humans with primary melanoma (MP-1).	AFM and microfluidic device for cell sorting.	Intrinsic softness represents a unique biomarker of highly tumorigenic and metastatic cancer cells.	Lv *et al* (2021)
Increase in deformability in more invasive cells such as transformed MCF7 breast cancer cells when compared to non-metastatic MCF10 and non-transformed MCF7 cells.	Non-tumorigenic breast epithelial MCF10 cells; a non-motile, non-metastatic breast epithelial cancer MCF7 cells; and MCF7 cells transformed with phorbol ester.	AFM	Deformability depends on the modulus of elasticity, which provides a quantitative measure of the elastic behavior of cells.	Lekka (2016)
Human colorectal adenocarcinoma HCT116 cells and TGF-β stimulated HCT116 cells (mimic CTCs).	HCT116 and CTC-mimicking HCT116 cells by treating HCT116 cells with TGF-β.	Digital acousto-holographic microscope.	Acoustic volume wave compression of cells permits the measurement of the distribution of cell stiffness with a resolution in the sub-micrometer scale, with no adverse effect on cell viability.	Varol *et al* (2022)
Cells from the tumor displayed a larger cell size and higher deformation compared to healthy equivalents.	Primary single cells of biopsies of colon, intestine, kidney, liver lung, MLN, pancreas, spleen, stomach, thymus.	Real-time fluorescence and deformability cytometry (RT-FDC).	The physical phenotype of cells from tumor tissue is significantly altered compared to control cells.	Soteriou *et al* (2023)

Cell biomechanics is implicated in cancer formation and progression. Therefore, comprehension of the mechanical fingerprints of cancer cells with varying malignant potential may yield insights into invasion and metastasis and enhance the possibility to physically track the generation of circulating tumor cells (CTCs) and forecast their invasive potential. In human TNBC breast cancer cells, overexpression of Cx43 favors the mesenchymal-to-epithelial transition (MET), thereby inhibiting cell proliferation, invasiveness, xenograft tumor incidence and growth, and metastasis, and reestablishing cellular capacity for differentiation, indicating that Cx43 exerts a tumor-suppressive function (Habli *et al* 2023). These non-metastatic cells tend to be taller, have a larger cell surface area and, subsequently, a larger cell volume. Conversely, repression of Cx43 has been associated with the suppression of these epithelial properties, involving the dissolution of cell–cell junctions, adherens junctions and the breakdown of epithelial polarity, a collection of events that are traits for pro-metastatic characteristics. Consequently, the cells adopt a mesenchymal phenotype marked by actin restructuring and stress fiber production, and the cells acquire the capacity for migration and invasion. These cells are less voluminous, with less height and a smaller cell surface area (El-Saghir *et al* 2011, Kazan *et al* 2019, Tishchenko *et al* 2020, Habli *et al* 2023). Prior work has demonstrated that metastatic cells are highly deformable relative to their non-metastatic equivalents, although AFM-based SCFS and a variety of cell lines have frequently been utilized. Interpretation of the majority of these investigations is confounded by many shortcomings associated with AFM. These range from low-precision measurements to the utilization of coated cantilever tips that interfere with the cells under detachment and give rise to measurement inaccuracies. Moreover, it is not possible to discern whether the variations are due to intrinsic genetic characteristics associated with the adapted culture models or to the actual effect of the cells' metastatic capacity on the mechanical characteristics of the cells.

The work of Habli and colleagues utilizes the innovative high throughput and high fidelity fluid-based SCFS of a uniquely derived *in vitro* progression model for breast cancer metastases originating from the identical cell line lineage (El-Sabban *et al* 2003, Han *et al* 2020, Rianna *et al* 2020, Lv *et al* 2021). This strategy provided comprehensive mechanostructural profiles of breast cancer cells across a spectrum of malignant potential. By directly comparing the adhesion forces of single cells with cells within clusters, the intercellular adhesion forces of well-established mature tight and gap junctions could be determined (Lekka 2016, Sancho *et al* 2017). Unexpectedly, SCFS adhesion data revealed that variations in cellular mechanical characteristics are obscured when MDA-MB-231 cells are out of physical engagement with neighboring cells, such as in the single-cell status, regardless of Cx43 expression and metastatic capacity. This mechanical homogeneity is lost, however, when the same cells adhere to neighboring cells, pointing to a complex crosstalk of cell–cell contacts in determining the mechanical characteristics of cells and a progressive relationship between the strength of cell adhesion and stiffness with the degree of malignant behavior of the cells when the cells are clustered together. The increase in $D_{\max}$ in non-metastatic cells inside a cluster relative to both parental and highly invasive cells is ascribed to the existence of tight and GJ proteins, as well as the greater cell height and volume, so the retracting cantilever needs more distance to completely detach the cells. When assessing the apparent cell elasticity, which relates cell stretching to its adhesion strength, non-metastatic cells in clusters were found to have low deformation capacity, which is concordant with a stiffer/rigid phenotype and in line with their non-invasive nature (Weber *et al* 2011, Chang *et al* 2016, Liboz *et al* 2023), a prominent characteristic lacking when cells are out of physical contact with neighboring cells. When considering the steepness of the *F–D* curves, cells in individual cell states show nearly the same steepness, in contrast to cells in clusters. In the case of the latter, the sharpest slope was observed for non-metastatic cells inside a cluster, indicating an enhancement in stiffness, in opposition to the elastic aggressive and strongly metastatic cells. Actually, normal/non-tumorigenic cells typically form highly organized, polymerized, and properly dispersed actin filaments that provide cell stiffness and strength, in opposition to their tumorigenic equivalents (Aseervatham 2020, Kashani and Packirisamy 2020). The latter was verified using actin filament staining to examine localization and intensity within the three cell subsets. Whereas these biomechanical findings are new with respect to Cx43-modulated cells, the link between the mechanical characteristics of cells and their utilization as biomarkers for predicting the malignant behavior of cells has been previously described in several metastatic cell models (Swaminathan *et al* 2011, Nematbakhsh *et al* 2017, Han *et al* 2020, Kashani and Packirisamy 2020, Varol *et al* 2022).

The fluid-based SCFS data using QCM-D measurements were supplemented where frequency displacements point to differences in the convoluted viscoelastic performance of the cell membrane and the subjacent ECM and substrate (the layer that establishes direct mechanical interconnection between the matrix/sensor surface and the cell membrane (Fuhrmann *et al* 2017), and dissipation shifts provide information on the viscoelasticity and stiffness of the cell membrane of peeling cells (Tymchenko *et al* 2012). Notably, cells provide an acoustically thick layer of between 10 and 20 μm, which is much greater than the characteristic penetration depth (δ) (>250 nm in water at $f_0 = 5$ MHz), so the layer directly over the sensor such as the cell membrane and cytoskeleton affects the QCM-D measurements. Upon trypsin application, characteristic ΔF and ΔD reactions were seen when cells detached (Chen *et al* 2012, Nowacki *et al* 2015, Easley *et al* 2022). An enhancement of ΔF suggested that a lower mass was found at the sensor surface, thereby validating cell detachment and cytoskeleton perturbation. The magnitude of the frequency shifts differed significantly between the three subtypes when tight cell–cell contacts were evident. The same trend of ΔF reaction was seen at low cell number albeit to a lower degree, which is due to a lower mass of cells grown on the sensor area. The data demonstrated a close correlation between ΔD reactions and the degree of malignancy of the cells, which was affected by cellular mechanical characteristics, cytoskeletal restructuring, and the number of focal adhesion sites, and not by cell–cell junctions. Highly invasive cells gave off more energy at both cell densities, which points to a viscoelastic cell membrane, which is an intrinsic cell characteristic. Moreover, when the magnitude of ΔD shift was analyzed, non-metastatic cells exhibited a smaller reaction at both cell concentrations, which indicates that the cell membrane of non-metastatic cells has a lower viscous nature and higher stiffness compared to metastatic cells. Moreover, a higher Df slope indicates that the cell layer is more viscoelastic (softer and higher dissipative) and has a softer phenotype. These findings demonstrated that non-invasive cells in the spectrum of metastasis potential possess a viscous cell membrane and a stiff cytoskeleton with elevated rigidity, while metastatic cells have a less viscous membrane although a softer cytoskeleton, which is in line with other investigations (Marx *et al* 2005, Rother *et al* 2014, Nematbakhsh *et al* 2017). Confocal fluorescence micrographs revealed that cells with low metastatic capacity form relatively better-organized filamentous actin networks with elongated filaments homogeneously dispersed from the periphery toward the center, which accounts for the stiff cellular phenotype. In marked contrast, highly metastatic cells displayed a more elastic and softer phenotype, perhaps attributable to actin distribution at the edge of the cell and not at the midpoint. Displacements in dissipation and Df plots are much more sensitive probes of membrane viscosity and elasticity compared to frequency shifts. This is mirrored in the amount of dissipation processes occurring while the cell is detaching, like viscous sliding between the cell's basal membrane and the fluid medium confined between the cells and the sensor surface, reorganization of the actin cytoskeleton, and frictional sliding of integrins on the ECM (Nowacki *et al* 2015).

Ultimately, QCM-D was utilized in conjunction with fluorescent staining to assess the biomechanical cellular reactions to DTX exposure at various concentrations and time points on the three cell subtypes with a range of metastatic disease states. DTX treatments revealed time-dependent changes in the mechanical characteristics of the basal area of the treated cells. Following DTX injection, metastatic cells at the upper end of the metastatic spectrum showed an immediate and dramatic reaction in ΔF and ΔD, which was almost six times stronger than in cells at the lower limit of the metastatic range. Although treated cells are prone to contract and separate, there was a clear negative displacement of frequency with a positive displacement of dissipation, whereupon both signals returned to ground level. This sequence of reactions indicates that cells first stiffen due to the stabilization of microtubules and then contract before separating. The identical reaction signature was seen in the other two cell subtypes, albeit to a lesser degree and at a far slower speed. This differential reaction could be attributed to the intrinsically disrupted and remodeled cytoskeleton of highly metastatic cells, thereby predisposing them to DTX treatment and sensitizing them to alterations in microtubule dynamics relative to their less invasive analogues. Moreover, sequential imaging of cells treated with DTX indicated that highly metastatic cells mainly show collective cell detachment (in groups), while their single-celled counterparts typically detach individually. These observations are in line with the reported impacts of DTX on cells (Nguyen *et al* 2016b, Gandalovičová *et al* 2017, Hou *et al* 2021). Despite mainly targeting microtubules, DTX can also have secondary impacts on other cytoskeleton components inside the cell, primarily the reallocation of actin filaments out of the cell periphery and into the cellular center (Marx *et al* 2007, Nguyen *et al* 2016b, Hou *et al* 2021), which has been monitored through fluorescent staining. The consequential impact on actin dynamics leads to the generation of thicker, more stabilized stress fibers, leading to enhanced cellular stiffness. From a biological point of view, such treatments could reduce the deformability of cancer cells and perhaps impair their capacity for invasion and metastasis in distant tissues. This analysis yields compelling evidence for the evolving biomechanical alterations of breast epithelial cancer cells as their malignant plasticity diversifies at the various stages of EMT and MET. With increasing malignant transformation of the cell from the non-invasive Cx43D cells to the aggressive WT cells and the highly invasive shCx43 cells in the process of metastasis, the cells change from becoming less adherent and increasingly viscoelastic and deformable. The higher the deformability and viscoelasticity of the cells, the larger the interface are between the dislodged soft tissue cells and the adjacent endothelial cells, which provides more possibilities for the cells to establish adhesive bonds, thus favoring tumorigenesis (Abotaleb *et al* 2018).

7.3 Combined mechanical and molecular biological probing of cancer cells

The molecular system that regulates the mechanical characteristics of cells is still uncharted at the single-cell level, mostly due to the restricted capacity to pair mechanophenotyping with unsupervised transcriptional profiling. In the following section, a method is presented that consists of an electroporation-based lipid bilayer assay for cell surface tension and transcriptomics (ELASTomics), in which oligonucleotide-labeled macromolecules are introduced into cells by nanopore electroporation to determine the mechanical properties of the cell surface and counted via sequencing (Shiomi *et al* 2024). ELASTomics can be easily incorporated into established single-cell sequencing workflows, allowing the simultaneous analysis of cell-surface mechanics and the underpinning transcriptional regulatory landscape at unparalleled resolution. ELASTomics was verified through the analysis of cancer cell lines of different malignant tumors and demonstrates that the technology can precisely identify cell types and evaluate cell surface tension. ELASTomics provides a system for exploring the links between cell surface tension, surface proteins and transcripts across the cell lines that developfrom mouse hematopoietic progenitor cells. The cell surface mechanical characteristics of cellular senescence were examined, and it was found that RRAD modulates the cell surface tension of senescent TIG-1 cells. ELASTomics offers a powerful way to characterize the mechanical and molecular cellular phenotypes of single cells and the crosstalk between them in a variety of biological settings. It is common knowledge that the mechanical characteristics of cells are related to the functionalities of cancer cells (Gensbittel *et al* 2021, Lv *et al* 2021, Tsujita *et al* 2021), stem cells (Titushkin and Cho 2006, Pajerowski *et al* 2007, Chowdhury *et al* 2010, Bergert *et al* 2021), and senescent cells (Berdyyeva *et al* 2005, Phillip *et al* 2015, Shiomi *et al* 2020). Technical breakthroughs in single-cell mechanical phenotyping (Gossett *et al* 2012, Otto *et al* 2015, Rosendahl *et al* 2018, Shiomi *et al* 2021) have opened up possibilities for profiling mechanical indicators of cellular functioning and cellular states as pieces of complementary information orthogonal to the traditional profiling of cellular states in terms of molecular abundance, that is, in terms of molecular phenotype (Eling *et al* 2019). Nevertheless, there are limitations to the present techniques for dissecting the relationship between the mechanical and molecular phenotypes as determined through unsupervised transcriptional profiling (Abdelmoez *et al* 2018, Oguchi *et al* 2021). Therefore, the underlying molecular system on the single-cell scale is still largely enigmatic (Diz-Muñoz *et al* 2018). ELASTomics, which is a methodology that is capable of combining cell surface phenotypic profiling with unsupervised transcriptomics for thousands of individual cells, has been launched (Shiomi *et al* 2024). ELASTomics uses nanopore electroporation, whereby molecules are transported into the cells according to their surface tension (Kang *et al* 2018, Cao *et al* 2019, Nathamgari *et al* 2019, Patino *et al* 2022, Pathak *et al* 2023). This technique has been easily implemented in cultured cell lines and primary cells, and the joint analysis reveals the transcriptional control of cell surface mechanics in a variety of biological settings, ranging from malignant cancers to hematopoietic cell differentiation and senescence of cells.

7.3.1 The strategic approach of ELASTomics

The approach to determine the mechanical properties of the cell surface was to generate nanopores by electroporation to render the plasma membrane of living cells permeable (Mukherjee *et al* 2018, Cao *et al* 2019), and subsequently to import electrophoretically DNA-targeted dextran molecules (DTD) with different Stokes radii in the range of 4.1–17.0 nm, and to perform digital counting of the imported DTD molecules by sequencing. Since it has been shown that the size of the pore that develops in a plasma membrane following nanopore electroporation grows with the tension of the plasma membrane (Neu and Krassowska 1999, Mukherjee *et al* 2018), it has been hypothesized that the quantity of imported DTD molecules mirrors the mechanical characteristics of the cell surface. To facilitate concurrent detection of DTD and mRNA employing existing single-cell sequencing methodologies, oligonucleotides were generated for DTD that can be trapped by oligo-polythymidine primers and include a barcode sequence for DTD identification and a primer handle for amplification by polymerase chain reaction (PCR). In ELASTomics, cells were placed on a membrane etched with grooves containing numerous nanopores with a diameter of 100 nm. A DTD-containing buffer solution was deposited on the uncolonized side of the membrane and then pulsed voltages (40 V, 5 ms width, square waves at 20 Hz, 500 cycles, unless indicated differently) were imposed over the track-etched membrane with a pair of electrodes to create a focused electric field in the proximity of the nanopores (Shiomi *et al* 2024). The tightly focused electric field can reversibly electroporate the plasma membrane in the vicinity of the nanopores and thereby facilitate the uptake of DTD through electrophoresis. In contrast to conventional bulk electroporation, the nanopore-focused electric field delivers highly efficient (Mukherjee *et al* 2018, Cao *et al* 2019) and stable electroporation independent of cell size, morphology or orientation of the cell (Kurosawa *et al* 2006, Shintaku *et al* 2014). Thus, the plasma membrane is electroporated in a surface-tension-dependent fashion. Combining ELASTomics and single-cell RNA sequencing (scRNA-seq) using 10x Genomics Single Cell 3′ v3.1, the mechanical phenotype and transcriptomic expression can be analyzed concurrently in thousands of individual cells, utilizing the Cellular Indexing of Transcriptomes and Epitopes by Sequencing (CITE-seq) protocol (Stoeckius *et al* 2017) to enrich DNA products from cDNA and DTD and generate two Illumina sequencing libraries.

7.3.2 Nanopore electroporation is designed for ELASTomics

To compensate for the sensitivity and adverse effects of nanopore electroporation on cells, the number of molecules transferred, and cell viability were assessed after nanopore electroporation with fluorescein isothiocyanate-labeled bovine serum albumin (FITC-BSA; Stokes radius of 3.6 ± 1.4 nm) as a replacement for DTD at several voltage settings, with subsequent evaluation by flow cytometry. The quantity of incorporated FITC-BSA closely correlated with the quantity of DTDs. To avoid cell activation and cell death caused by the influx of calcium ions, calcium-free solutions, PBS (−) and HEPES-based buffer (20 mM HEPES/NaOH, pH 7.0 and 260 mM sucrose) were chosen for the nanopore electroporation. The experimental results indicated that the higher applied voltage enhanced the number of imported molecules and increased the sensitivity of the cell surface tension assessment. Excessive voltage levels, nevertheless, negatively affected cell survival and caused gene expression disturbances. Based on experimental monitoring, ELASTomic settings yielding over 90% viability were determined to be 40 V for PC-3 human prostate cancer cells, MDA-MB-231 human breast cancer cells, MCF7 human breast cancer cells, and MCF10A human breast epithelial cells, 75 V for mouse hematopoietic stem/progenitor cells (mHSPCs), and 50 V for TIG-1 cells. The number of molecules introduced was repeatable using the same settings. Nanopore electroporation can be utilized for a variety of cell lines, comprising HeLa human cervix cancer cells, PC-3, MDA-MB-231, MCF7, MCF10A, TIG-1, OVCAR-3 human ovarian cancer cell line, Chinese hamster ovary cells subclone K1 (CHO-K1), GEM-81 cells (adherent cells, goldfish), primary suspended mHPSCs and suspended K562 human lymphoma cell line. Non-adherent cells such as K562 and mHSPCs need higher applied voltages compared to adherent cells to take up comparable amounts of molecules through nanopore electroporation. Crucially, a two-minute incubation after the electric pulses were delivered was necessary to reseal the pores in the phospholipid bilayers and prevent the cells from calcium activation and subsequent cell death. Following trypsinization, cells were maintained on ice to quench alterations in gene expression and processed for single-cell RNA sequencing within one hour. To minimize the impact of nanopore electroporation and DTD capture on gene expression analysis even further, single-cell data from non-nanopore-captured cells were acquired and incorporated into the ELASTomics data prior to analysis. This process made it possible to perform reliable gene expression analysis.

7.3.3 Cell surface tension is heterogeneously distributed

Based on a theoretical comprehension of the electroporation mechanism (Neu and Krassowska 1999, Kotnik *et al* 2019), surface tension affects the size of a pore in a plasma membrane through free energy alterations upon pore creation by liberating surface energy. For the cases of equal transmembrane tension and line tension, a higher surface tension results in a greater likelihood of a more extended pore formation. The DTD molecules incorporated into cells thus rise with cell surface tension. To determine whether the cell–cell variability mirrored by DTD frequency is consistent with quantitative variations in single-cell surface tension, the association between the number of molecules taken up and the surface tension of single MCF10A cells has been explored using fluorescence microscopy and AFM (Ding *et al* 2018). Even though surface tension measurements via AFM were conducted within 0.5–3 h post nanopore electroporation, a correlation was found between the area-normalized amount of FITC-BSA and the surface tension determined via AFM, which suggests that the fluctuations in the number of molecules absorbed mirror the fluctuations in surface tension from cell to cell. Remarkably, the total number of molecules imported into a cell through nanopore electroporation exhibited a comparable relationship with surface tension, although it is expected to be related to the area of adhesion, which is the quantity of nanopores in the track-etched membrane located below the cell membrane. This is ascribed to the reliance of the adhesion area on the cell surface tension (Xie *et al* 2018), which was also found in the following study (Shiomi *et al* 2024), indicating that the impact of the adhesion area on nanopore electroporation strengthens rather than attenuates the relationship between surface tension and the quantity of molecules internalized.

7.3.4 Impact of Stokes radius of DTD on electrophoretic translocation

In the next step, the impact of the Stokes radius of DTD on electrophoretic relocation via a pore in a lipid bilayer was analyzed. The Stokes radii (4.1 ± 0.0 nm–17.0 ± 12.2 nm) cover the critical radius of 15 nm, which is the predicted maximum radius that can be reclosed in an equilibrium process (Mukherjee *et al* 2018). The relative permeability of DTD was calculated as counts/μDTD cDTD, whereby μDTD and cDTD are the electrophoretic mobility and concentration of DTD within the buffer, respectively. The permeability of different DTD sizes dropped with rising Stokes radius, implying that relocation was decreased due to steric hindrance caused by the DTD size. Remarkably, the degree of permeability was highly conserved between DTDs of varying molecular weights, confirming the robustness of DTD enumeration for cell-type-independent quantification. The permeability ratio for a DTD pair was unaffected by cell type, indicating that the impact of cell size can be adjusted (normalized) through the permeability ratio. Nevertheless, the permeability ratio was determined to be somewhat more noise-sensitive compared to the DTD counting. Therefore, mainly normalized 4 kDa DTD counts using the centered log ratio transformation were utilized to predict cell surface tension due to the relationship between the total amount and surface tension. The anticipated error in measuring cell surface tension when using the total number of incoming molecules was estimated to be approximately 10 ± 0.39 times in the region of 10–1000 pN μm^{-1}, due to the correlation between them.

7.3.5 Impact of cortical and plasma membrane tension on cell surface tension

To gauge the effectiveness of ELASTomics in creating surface mechanics maps of individual cells, two proof-of-principle experiments were conducted in which cells with a variety of surface tensions affected through cortical and plasma membrane tension were examined (Tinevez *et al* 2009, Clark *et al* 2014, Tsujita *et al* 2021). The cell surface tension of MCF10A cells has been disrupted with cytochalasin D, which lowers cell surface tension under the governance of cortical tension through impeding actin polymerization and destroying filamentous actin structures. After subjecting MCF10A cells to treatment with cytochalasin D or vehicle, ELASTomics was carried out on the respective MCF10A cells, generating scRNA-seq libraries and enumerating the number of DTD-derived DNA tags within individual cells. As anticipated, the MCF10A cells exposed to Cytochalasin D exhibited lower DTD levels than the vehicle-treated control cells, as predicted according to the theoretical model. In addition, the ability of nanopore electroporation to detect changes in cell surface tension induced either by blebbistatin and Y-27632, which alter cortical tension, or methyl-β-cyclodextrin, which modifies the plasma membrane tension, was verified, indicating the strength of the technique. In a subsequent proof-of-principle experiment, ELASTomics was performed on four cell lines with various plasma membrane tensions that are linked to malignant cancers and invasiveness (Tsujita *et al* 2021). The mechanical characteristics of cancer cells have an essential impact on metastatic spread (Gensbittel *et al* 2021) and the stemness of cancers (Lv *et al* 2021). In a recent study, it was found that low plasma membrane tension is a requirement for enhanced cancer cell invasiveness (Tsujita *et al* 2021). In this experiment, noninvasive MCF-10A epithelial cells, low-invasive MCF-7 breast cancer cells, highly-metastatic MDA-MB-231 breast cancer cells, and aggressive PC-3 prostate cancer cells were employed, which were characterized according to their plasma membrane tension of 91.89, 82.78, 45.19 pN μm^{-1} (ruffling) (50.45 pN μm^{-1} blebbing) and 38.33 (ruffling) (42.61 pN μm^{-1}, blebbing) (Tsujita *et al* 2021). MCF10A cells had the highest mean DTD scores among the cell types, and MCF7 cells had higher scores compared to MDA-MB-231 and PC-3 cells (Shiomi *et al* 2024). This finding emphasizes the low cell surface tension in highly invasive and aggressive cancer cells and is in line with the findings of plasma membrane tension determination employing optical tweezers (Tsujita *et al* 2021).

7.3.6 Mechanics of the cell surface mechanics is linked to cancer malignancy

The purpose of this study was to determine whether ELASTomics can be used to perform a joint analysis of cell surface tension and gene expression using four different cell types comprising MCF10A, MCF7, MDA-MB-231 and PC-3, whereby simultaneous mapping of the mechanical and molecular phenotypes can be conducted. To ease cell type identification in scRNA-seq data analysis, 62-nucleotide tags without dextran as a cell hashing tag were added to the DTD buffer (Shin *et al* 2019). Thereafter, individual cell types were electroporated with DTD, and the 10x Genomics Single Cell 3′ v3.1 on a pool of DTD-labeled cells was performed. Data from six batches were integrated and two or three different cell types or experimental conditions were combined into a single set of data. 3804 PC-3 cells (median of 10 805 UMI and 3432 genes detected per cell), 2275 MDA-MB-231 cells (median of 4594 UMI and 1924 genes detected per cell), 1575 MCF7 cells (median of 13 308 UMI and 3536 per cell detected genes) and 2083 MCF10A cells (median of 6124 UMI and 2052 per cell detected genes) have been analyzed using nanopore-electroporated cells. In addition, 4018 PC-3 cells (with a median of 10 800 UMI and 3440 genes analyzed per cell), 1691 MDA-MB-231 cells (with a median of 6613 UMI and 2494 genes analyzed per cell), 3087 MCF7 cells (with a median of 13 136 UMI and 3363 genes analyzed per cell), and 3412 MCF10A cells (with a median of 4306 UMI and 1531 genes analyzed per cell) have been revealed with the control cells (Shiomi *et al* 2024). For purposes of comparison, 5072 PC-3 cells treated with 75 V (with a median of 8086 UMI and 2888 genes expressed per cell) were also characterized, along with 180 MCF10A cells exposed to cytochalasin D and subjected to treatment with 40 V (with a median of 32 706 UMI and 5316 genes expressed per cell), and 822 MCF10A cells processed at 40 V (with a median of 14 350 UMI and 3534 genes expressed per cell). Unified Multivariate Approach and Projection (UMAP) and Louvain community detection using Seurat (version 4.0.4) defined four discrete clusters of expressed marker genes and DTD abundances for the individual cell types (Shiomi *et al* 2024). Cell hashing tag resulted in consistent cell numbers within each cluster, which assisted in the characterization of cell types from clusters through gene expression. To determine genes that are likely to be common regulators of cell surface tension in various cell types, the correlation coefficient between DTD frequency and gene expression was determined for MDA-MB-231 and MCF-7. Consistent with predictions, genes with higher expression in MCF7 compared to MDA-MB-231 were positively correlated with DTD values and vice versa, albeit with remarkable outliers. For example, the expression of the ribosomal proteins L11 (RPL11) and L37 (RPL37) exhibited an inverse correlation with the DTD frequency, whereas the expression level variations between the two cell types were not significant. Analysis of the accumulation of gene ontology (GO) in positively (correlation above 0.2) or negatively (correlation below −0.15) related genes revealed an accumulation in genes that control cell adhesion, which comprises cadherin linkage and cell adhesion mediator activity, and essential genes that govern cell surface mechanics, composed of actin tethering and myosin tethering. To clarify the functional augmentation that was clearly revealed by the correlation with DTD frequency, a Genset Enrichment Analysis (GSEA) was conducted, in which the genes were ranked based on the correlation coefficient. Importantly, GSEA revealed an accumulation in several gene sets, including myosin heavy chain class II complex binding (GO:0023026) and symporter activity (GO: 0015293) that remained marginal in a standard GSEA relying on the alteration of gene expression across the two cell types, underlining the co-enrichment of cytoskeletal proteins in MCF7 and MDA-MB-231 cells. It is observed that gene expression in MCF10A cells was mildly disrupted with nanopore electroporation. GSEA revealed an accumulation of cellular reactions toward stress, cell mortality, and the control of cell mortality, which is concurrent with a former study (Mukherjee *et al* 2023). The following investigations, such as hematopoietic cell differentiation and senescence, reveal that the technique can be used for other purposes related to cancer development and malignant progression. It uncovered the enormous potential of the technique and highlights its modifications. All of which clarifies the results obtained with various cancer cell lines.

7.3.7 Cell surface mechanical cues in hematopoietic cell differentiation

Cell differentiation, involving the production of mature blood cells from hematopoietic stem cells, typically proceeds in parallel with the maturation of cell membranes to facilitate their specialized cellular functions. Hence, the question was examined whether ELASTomics can visualize cell surface tension evolution in normal hematopoietic cell differentiation in murine bone marrow (BM). For the analysis, BM cells, comprising HSPCs, were co-isolated using a c-kit^{+} magnetic column from 8-week-old C57BL/6 J mice. To use ELASTomics on primary non-adherent cells, which are fastidious cell types, cells were dropped for 1 h on the 100 μg ml^{-1} fibronectin-precoated membrane that was etched with traces, and then 75 V was employed for nanopore electroporation. To facilitate the identification of cell phenotypes, HSPCs treated with nanopore electroporation were marked with antibody-oligonucleotide conjugates that specifically recognized three surface markers, namely Sca-1, CD48, and CD150, following nanopore electroporation. The CITE-seq experimental procedure was carried out, generating a (long) library for mRNA-seq, a (short) library for DTD, and antibody-derived tag library. A total of 10 011 mHSPCs were captured, with a median of 6893 UMI and 1919 genes expressed on a per-cell basis. The UMAP of the mHSPCs revealed a hierarchical and progressive alteration of cell states throughout differentiation as determined from the expression of lineage-marker genes and surface proteins. Superimposition of DTD quantifications on UMAP revealed that nanopore electroporation facilitated the transport of DTD into non-adherent cells in a variety of different states.

The focus fell on ELASTomics readouts of erythroid maturation from HSPCs because structural alterations in the cell membrane are of fundamental importance throughout the course of erythroid ripening and differentiation, involving denuclearization (Peters *et al* 1992). The ELASTomics data revealed that a substantial fraction of megakaryocyte-erythroid progenitors/erythroid precursors (MEP/EryP) and erythroid cells acquired more DTDs compared to other cell, suggesting that cells that have decided to undergo erythrocyte differentiation temporarily raise their cell surface tension prior to denuclearization in the erythrocyte phase. In addition, using FITC-BSA, nanopore electroporation of mHSPCs was conducted and it was verified that Ter-119 positive cells, being erythroid progenitor cells, exhibited higher cell surface tension compared to Ter-119 negative cells, corroborating the findings of ELASTomics.

To investigate genes linked to alterations in cell surface tension in cell differentiation, the correlation coefficient between DTD frequency and gene expression within the erythrocyte erythrocyte lineage (hematopoietic stem cell (HSC)–multipotential progenitor (MPP)–common myeloid progenitor (CMP)–MEP/EryP–erythroid–erythrocyte) has been determined. To surmount the relatively low DTD values in mHSPCs, during correlation calculation cells with a centered logarithmic ratio of DTD lower than 1 were excluded. The dataset revealed 54 genes with expression patterns positively associated with DTD numbers, highlighting the participation of Spectrin Alpha 1 (Spta1) and Spectrin Beta (Sptb), genes linked to the control of erythroid membrane structure, during the temporary elevation of cell surface tension. GSEA also revealed an increase in spectrin association (GO:0030507) and gene sets associated with the modulation of spectrin through ATP-dependent phospholipid flippase (Manno *et al* 2010). In contrast, only one gene in the lineage from mHSPCs to granulocytes exhibited a significant positive relationship between gene expression and cell surface tension.

7.3.8 Cell surface mechanical cues in senescence of cells

Senescent cells undergo alterations in their mechanical characteristics, with an enhancement of stiffness or traction force in conjunction with a decrease in elasticity and strength (Bajpai *et al* 2021). Changes in the mechanical characteristics of the cells result in a progressive imbalance in mechanosensitive signal transduction (Park *et al* 2000, Phillip *et al* 2017) and the cytoskeletal scaffold (Mu *et al* 2020). Consequently, it has been explored how multimodal data from ELASTomics could improve the comprehension of surface mechanical alterations during cellular senescence (Phillip *et al* 2015). ELASTomics was conducted with human fetal lung fibroblasts (TIG-1), which display replicative senescence characterized with alterations in cholesterol amount in lipid rafts (Nakamura *et al* 2003), chromosomal instability (Ohshima and Seyama 2010), aberrant glycation and Golgi trafficking (Udono *et al* 2015), anomalous lipid build-up (Inoue *et al* 2017), and mitochondrial impairment (Fujita *et al* 2022) as cellular senescence advances using 50 V pulses. 4654 TIG-1 cells (applied voltage: 0 V) with a median of 13 130 UMI and 3648 genes and 4711 TIG-1 cells (applied voltage: 50 V) with a median of 19 207 UMI and 4474 genes identified per cell. Cells with senescence hallmarks were detected more frequently in a comparably senescent TIG-1 population (with a population doubling level (PDL) of 50–60) compared to a young TIG-1 population (PDL = 30–40) at the bulk population level, whereas the expression levels of different senescence-associated genes were found to be heterogeneous in individual TIG-1 cells at the single-cell level. In line with this, ELASTomics demonstrated that at the population level, the amount of the high cell surface tension TIG-1 cells was increased in the aging TIG-1 population as compared to the young TIG-1 population. It was confirmed that the surface tension of senescent TIG-1 cells was on average stronger compared to young TIG-1 cells, which was verified by AFM. Moreover, the surface tension of individual cells exhibited heterogeneity, which is in line with the heterogeneous expression of senescence-associated genes. To examine genes associated with alterations in cell surface tension in the process of aging, the correlation coefficient between DTD frequency and gene expression was calculated. The expression intensities of genes upregulated in senescence correlated positively with the DTD frequency, among them the Fos proto-oncogene (FOS), adrenomedullin (ADM) (Hwang *et al* 2007), and cyclin-dependent kinase inhibitor 1 A (CDKN1A) (Zhang *et al* 2021). Gene expression levels that were repressed during senescence negatively associated with DTD frequency, for instance for deoxythymidylate kinase (DTYMK) and disc-associated protein 5 (DLGAP5) (Lee and Shivashankar 2020). Moreover, the expression levels of genes associated with DNA replication and repair, for example nucleolin (NCL) and clamp-associated factor (PCLAF), exhibited a negative relationship with DTD frequency, consistent with the decrease in cell proliferation that occurs as cells age.

To identify genes that govern cell surface tension regulation in senescence, 210 genes were obtained by thresholding at Pearson's $r > 0.15$ and $P < 10^{-10}$. The genes Ras-related glycolysis inhibitor and calcium channel regulator (RRAD) were identified as a candidate modulator by regression utilizing the generalized linear model of the elastic network, in which RRAD, after two long non-coding RNAs, AC007952.4 and AC091271.1, and the KLF2 gene, displayed the fourth highest impact on cell surface tension (Shiomi *et al* 2024). In addition, bulk RNA-seq and nanopore electrophoresis were carried out for TIG-1 cells at various PDL levels, and the gene expression and the quantity of imported FITC-BSA were determined. Consistent with the results of ELASTomics, the TIG-1 cell population enhanced both the amount of imported FITC-BSA and the expression of RRAD, while PDL enhanced but KLF2 did not, suggesting that cell surface tension and RRAD expression are associated in cellular senescence. RRAD acts as a biomarker that is induced through senescence-related signals and whose expression rises with advancing age in human skin and fatty tissue (Wei *et al* 2019). RRAD suppresses glycolysis primarily by preventing the relocalization of glucose transporter 1 (GLUT1) toward the plasma membrane (Zhang *et al* 2014). The expression of RRAD causes a p53-driven block in lung cancer cell migration (Hsiao *et al* 2011), and hypermethylation is associated with unfavorable outcomes in lung adenocarcinoma (Suzuki *et al* 2007). Therefore, it can be assumed that RRAD is crucial for controlling cell surface tension and the associated functional alterations occurring in TIG-1 cells during senescence through the glycolysis signaling pathway. GSEA revealed an accumulation in glycolysis-related gene signatures like ATP-dependent activity (GO:0140657), ATP-hydrolysis activity (GO:0016887) and aging (GO:0007568). To examine the functional relationship between RRAD expression and cell surface tension, the expression of RRAD was disrupted with a small interfering RNA. Knocking down RRAD expression resulted in a pronounced recovery of cell surface tension in TIG-1 cells at PDL levels of 40 and 56, when measured against cell surface tension of negative controls. Moreover, inhibition of glycolysis with 2-deoxy-D-glucose (2-DG) enhanced surface tension of TIG-1 cells. These findings propose that RRAD aids an enhanced cell surface tension in the course of senescence through glycolysis within TIG-1 cells. This special multimodal approach uniquely revealed the identification of RRAD as a senescence-related gene linked to cell surface mechanical characteristics.

7.3.9 Discussion on the suitability of ELASTomics in mechanobiology

ELASTomics was introduced as a methodology to incorporate cell surface mechanics profiling with unsupervised transcriptomic assessments of thousands of individual cells. ELASTomics measures cell surface tension through the uptake of DTD into cells via nanopore electroporation and links DTD frequency quantification to gene expression signatures derived from single-cell RNA-seq. ELASTomics offers a holistic view of the gene expression underpinning variations in cell surface tension in diverse biological settings and provides us with the opportunity to identify key genes that govern cell surface tension. In this study, it has been demonstrated that ELASTomics can be utilized to analyze cancer cells of different types of malignancies, primary HSPCs isolated from murine BM, and TIG-1 cells at various senescence stages. ELASTomics offers the flexibility to customize sensitivity through regulation of electric field strength when the electric field does not interfere with gene expression in cells, thereby extending its applicability to a broad array of cell types, encompassing suspended cells and non-mammalian cells.

ELASTomics analysis of MCF7 and MDA-MB-231 cells revealed a negative association of RPL11 and RPL37 expression with DTD scores. RPL37 modulates the expression pattern of p53, a tumor suppressor, by interacting with Mdm2, which acts as a RING-type E3 ubiquitin ligase (Llanos and Serrano 2010, Daftuar *et al* 2013). RPL11 also interacts with Mdm2 to prevent p53 from being ubiquitinated and broken down (Zhang *et al* 2003). Collectively, the negative association of RPL11 and RPL37 with DTD scores may imply that p53-axis control of cell surface tension is preserved across cancer cell types in MCF7 and MDA-MB-231. GSEA revealed an accumulation not just for cytoskeleton-related gene sets like MHC class II protein complex tethering, but additionally for symporter activity and sodium ion transmembrane transporter activity. Alterations in intracellular ion concentrations lead to alterations in osmotic pressure and in cell surface tension (Xie *et al* 2018, Chadwick *et al* 2021). Ion channels and transporters can affect cell surface tension as a result of transporting intracellular ions to modify osmolarity (Mukherjee *et al* 2023).

Cell surface epitopes are robust identifiers for defining cellular states in a heterogeneous population. CITE-seq couples cell surface epitope detection and unrestrained transcriptomic profile generation, and ELASTomics can be incorporated into the CITE-seq process. In a Chromium X or equivalent, the individual suspended cells are disaggregated into water droplets containing reagents and barcoded beads. These barcoded gel beads are coated with oligonucleotides, such as short sequences of nucleotides. Each oligo comprises a pair of barcoding elements and an RNA-binding element. The latter portion is a poly(T) extension that anneals to the mRNA. A barcode element serves to label each bead with a unique barcode. This is appended to the mRNA, rendering it traceable to the cell from which it originated. Another portion of the oligo is a distinct barcode for every individual RNA molecule. It is referred to as a Unique Molecular Identifier (UMI) and is essential to properly quantify gene expression. The filled water droplets in oil are referred to as GEMs, and each GEM should include a cell, a barcoded bead, and reagents. In every GEM, the reagents trigger the cells to liberate the RNA, and the mRNA molecules attach to the barcode-labeled gel beads. GEMs are then pooled and barcoded DNA fragment copies are generated in a multi-step process. Every fragment contains both unique barcodes from the gel bead and a piece of the mRNA code, ensuring that each fragment has all the required information to connect an mRNA molecule to its specific gene and cell. Following next-generation sequencing and data processing, a transcriptome profile of the individual cells in the tissue has been created.

In the next step, the integration of ELASTomics and CITE-seq was presented and cell states across mHSPC lineages were profiled using molecular and mechanical phenotyping. The multimodal dataset demonstrated that erythroid lineage progenitor cells temporarily elevate cell surface tension throughout lineage dedication and erythrocyte ripening. In addition, the analysis of the relationship between cell surface tension and gene expression patterns indicated that Spta1 and Sptb, which are components of the spectrin tetramer, are implicated in the transient change of cell surface tension that occurs during cell differentiation. As the spectrin tetramer forms a primary structural element of the membrane skeleton for maintaining erythrocyte structural stability and deformability (Stokke *et al* 1986, Krieg *et al* 2014, Smith *et al* 2018), it is plausible that the expression of spectrin genes is increased in tandem with an upsurge in cell surface tension as erythroid cells mature. The findings indicate that ELASTomics provides unparalleled elucidation of cell-surface mechanics and related gene control in mHSPC differentiation.

ELASTomics provides a high-throughput strategy for identifying regulatory genes that affect single-cell surface tension. The ELASTomics analysis of TIG-1 cells at various stages of senescence indicated that cell surface tension rose during replicative senescence. The analysis discovered genes associated with cellular senescence, among them FOS, CDKN1A, and PCLAF, through positive or negative association between gene expression and DTD frequency. In addition, the regression analysis revealed RRAD as a candidate for a modulator of cell surface tension in senescence in TIG-1 cells. The effects of RRAD expression on cell surface tension were confirmed after silencing RRAD and monitoring the recovery of cell surface tension for senescent TIG-1 cells. RRAD is recognized to suppress glycolysis through blocking GLUT1 trafficking to plasma membrane (Zhang *et al* 2014). Building on these insights, although additional research is required to clarify the complete molecular pathway, it is possible that cell surface tension can intensify in cellular senescence through the suppression of glycolysis through the increase of RRAD (Park *et al* 2020a). The TIG-1 cell was isolated from a human fetal lung to examine cellular senescence (Ohashi *et al* 1980). TIG-1 cells have also been used to investigate age-related phenomena, which involve mitochondrial impairment, overproduction of reactive oxygen species (Fujita *et al* 2022), and aberrant lipid accumulation (Inoue *et al* 2017). Analysis using the TIG-1 cell line revealed that ATP6V0A2, which encodes a gene for autosomal recessive cutis laxa type 2 (ARCL2), causes aberrant glycosylation and Golgi trafficking as a cellular senescence pathway (Udono *et al* 2015). Consequently, the TIG-1 cell line is advantageous for examining certain genes associated with cell surface tension during cellular senescence because of its low genetic diversity and high degree of reproducibility. Even though TIG-1 cells represent an extensively used cell line for investigating replicative senescence (Kim *et al* 2012, Udono *et al* 2015, Fujita *et al* 2022), cultured senescent cells, such as TIG-1 cells, exhibit slightly different characteristics compared to their *in vivo* counterparts (Sherr and DePinho 2000). ELASTomics can also be tailored to primary fibroblasts by incorporating it into fast cell isolation protocols (Hirano *et al* 2023, Soteriou *et al* 2023). Cytoskeletal regulation is also closely associated with glycolysis in cancer cells, which exhibit high levels of glycolysis (Park *et al* 2020a). The expression of RRAD is repressed through DNA methylation within malignant lung and breast tumors (Suzuki *et al* 2007). It has also been proposed that TXNIP, acting in parallel to RRAD to block GLUT1 trafficking to the plasma membrane, changes cell mechanics in certain cancer cells (Sullivan *et al* 2018). In fact, a positive relationship between cell surface tension and TXNIP expression was found in cancer cell lines. Therefore, it can be speculated that blocking glycolysis signaling routes, which involves the upregulation of RRAD, may have a universal function in modulating cell surface tension in various settings. Importantly, while the correlation between RRAD expression and cell surface tension was found in MCF7, MDA-MB-231, and PC-3 cells, the weak sensitivity of scRNA-seq limited efforts to validate this relationship in the current dataset.

While ELASTomics is a multifaceted strategy that can be utilized to study different cell types, even non-adherent cells, the strength of the electric field needs to be tailored to achieve the best signal-to-noise ratio (SNR) and to minimize interference with gene expression. For example, for best results on mHSPCs, that are non-adherent cells, 75 V was utilized to ensure robust DTD enumeration, while 40 or 50 V was utilized for adherent cells. This infers that cell adhesion to the substrate impacts nanopore electroporation and DTD uptake, which in turn influences the SNR in DTD enumeration. Consequently, when examining a highly diverse cell population with different adhesion modes, the DTD frequency also mirrors other cellular conditions along with cell surface tension. The surface area of adhesion of a cell on the track-etched membrane is an additional parameter that impacts the amount of DTD, since the quantity of DTD taken up by a cell varies with the quantity of pores. This variation can be standardized by employing DTD with a series of Stokes radii (Xie *et al* 2018). Collectively, ELASTomics maps the cell surface mechanics repertoire and the underpinning transcriptional control network and allows the elucidation of genes that are key drivers of cell surface tension regulation. It is conceivable that ELASTomics will yield critical mechanistic knowledge about cell surfaces in wide biological settings, ranging from cancer biology to tissue engineering to cellular senescence.

7.4 Dynamical adaption of cancer cells' mechanical characteristics: mechanoadaptive cancer cells metastasize

The most threatening facet of cancer lies in metastasis. Cancer cells can establish life-threatening metastases after passing certain steps on their way from the primary tumor to remote targeted organs. From a biomechanical point of view, growth, invasion, intravasation, circulation, arrest/adhesion and extravasation of cancer cells require certain cell-mechanical characteristics to survive and finalize the metastatic process. Since metastatic cells are generally softer compared to their non-malignant equivalents, it is assumed that a high degree of deformability of both the cell and its nucleus confers a considerable advantage in terms of metastatic capacity. It is still uncertain, nevertheless, as to whether there is a delicately balanced, but stable, mechanical condition that exhibits all the mechanical characteristics needed for survival during the cascade, or whether cancer cells must dynamically fine-tune their characteristics and intracellular elements with every new stage. In this section, the diverse mechanical demands that may be encountered by successful cancer cells on their travels are outlined and the possible requirement for cancer cells to dynamically adjust their characteristics in line with these demands is proposed. The mechanical fingerprint of a successful cancer cell might in fact be its capacity to adjust to the successive restrictions of the microenvironment encountered during the various stages of the travel (Gensbittel *et al* 2021).

7.4.1 During metastatic spread, cancer cells withstand stress and external forces

Metastasis is clearly the most deadly feature of cancer, with the progression towards secondary tumors being accountable for over 66% of total cancer-related deaths (Steeg 2016, Elmehrath *et al* 2021). The development of these deadly secondary sites involves a series of sequential steps, referred to as the metastatic cascade. It begins with a healthy cell experiencing mutations and epigenetic alterations, which lead to out-of-control proliferation and growth of a solid tumor from an original primary neoplasm within a healthy tissue (Grandér 1998, Vicente-Dueñas *et al* 2018). Such initial growth is followed by the development of a supportive microenvironment (Werb and Lu 2015), especially angiogenesis/lymphangiogenesis around the tumor involving newly formed blood/lymphatic vessels (Watnick 2012, Stacker *et al* 2014), inflammatory-like immune reactions (Whiteside 2006), and reorganization of the adjacent ECM (Alexander and Cukierman 2016). Simultaneously, some cancer cells of the primary tumor develop new characteristics that make it easier for them to leave the primary tumor and enter the adjacent stroma or move along close-lying blood vessels (Gritsenko *et al* 2017). The invasion into the local parenchyma can be driven either actively (cell-autonomously) and/or by the interstitial fluid pressure inside the primary tumor, which is generated by the intratumorally proliferating cells and the pressure differences across the leaky blood vessels and lymphatics (Piotrowski-Daspit *et al* 2016, Stuelten *et al* 2018). The formation of intratumoral lymphatic vessels promotes lymph node metastasis (Karaman and Detmar 2014) which, while not fatal, frequently occurs with the metastasis progression. Life-threatening distant metastases, in contrast, arise from spreading in the bloodstream, which is accomplished by direct intravasation into the blood vessels. Importantly, emerging work has established that lymph node metastases can serve as a portal for blood colonization (Brown *et al* 2018, Pereira *et al* 2018). Intravasation can be supported through immune cells (Harney *et al* 2015, Linde *et al* 2018) and can involve individual cells or collective groups of invasive cells (Cheung *et al* 2016). Cancer cells are then termed CTCs because they are found in the bloodstream. When the CTCs succeed in evading the immune system and overcoming the struggle between the forces of blood flow and the forces of adhesion (Fan *et al* 2016), they will finally come to a halt and remain stuck to the vessel walls in an adhesive manner

(Osmani *et al* 2019). Another possibility is that they become trapped in small capillaries, regardless of adhesion forces (Kienast *et al* 2010). In rare cases, particularly with lung metastases, endothelial-bound cancer cells can proliferate and establish intravascular metastases (Al-Mehdi *et al* 2000). Most commonly, nevertheless, cancer cells must undergo extravasation, which refers to the process by which halted cancer cells leave the bloodstream. Several mechanisms of extravasation have been proposed, like diapedesis (Strell and Entschladen 2008) and endothelial reorganization (Lapis *et al* 1988). By leaving the bloodstream, the cancer cells can eventually enter the stroma/parenchyma of adjacent organs, where they can either proliferate and establish metastases when the microenvironment of the newly accessed organ is favorable (Peinado *et al* 2017) or persist in a quiescent state until their surroundings progress and permit them to become proliferative (Nguyen *et al* 2009, Aguirre-Ghiso 2018).

As cancer cells traverse all these stages, they are exposed to distinct and transforming external mechanical stresses that jeopardize their survival and advancement, and demand specialized mechanical attributes. In the early stages of tumor growth, cells experience growing pressure (figure 7.4), due to imbalances in the inflow and outflow of fluids in the interstitial compartment (Mohammadi and Sahai 2018) and the microenvironment pressing on the enlarging tumor bulk (Stylianopoulos *et al* 2012, Jain *et al* 2020). The rise in compressive strain is attributed to the combinatorial effect of increasing tumor size and stiffening its surrounding microenvironment through stromal rearrangement, which is common to various cancer types like breast or pancreatic cancer. Driven mainly by stromal cells like tumor-associated fibroblasts, stroma rearrangement involves mostly ECM topographical disturbances and stiffening and is recognized to promote invasion and metastasis (Goetz *et al* 2011, Pickup *et al* 2014, Clark and Vignjevic 2015), in particular through collagen cross-linking, the clustering of integrins (Paszek *et al* 2005, Levental *et al* 2009) and, above all, the generation of paths formed by radially aligned fibers (Provenzano *et al* 2006). These pathways can then be utilized for collective cancer cell migration guided through frontline fibroblasts (Gaggioli *et al* 2007). Among the impacts of the rising pressure caused by the reorganization of the stroma, nevertheless, is also a decrease in tumor cell size, a proliferation failure, and a halt in the cell cycle, all of which interferes with tumor growth when examined *in vitro* (Delarue *et al* 2014, Taubenberger *et al* 2019). Due to the enlargement of tumor volume, enhanced compressive stress and stiffening of ECM scaffold, cancer tissues are usually stiffer compared to normal tissues. At the clinical scale, malignant tumors have proven to be significantly stiffer compared to benign equivalents (Lorenzen *et al* 2002, Venkatesh *et al* 2008). In addition to enhancing pressure inside the tumor, a restrictive primary tumor microenvironment also promotes collective invasion of cancer cells rather than individual cell invasion (Haeger *et al* 2014). Effects such as these could favor the intravascularization of nascent invasive CTC clusters, which exhibit a 50-fold enhanced metastatic potential in comparison to individual CTCs (Aceto *et al* 2014). Interestingly, cancer cell clusters were detected to express keratin-14 in several models (Cheung *et al* 2016), which may be linked to enhanced stiffness (Seltmann *et al* 2013) relative to cancer cells that have acquired an

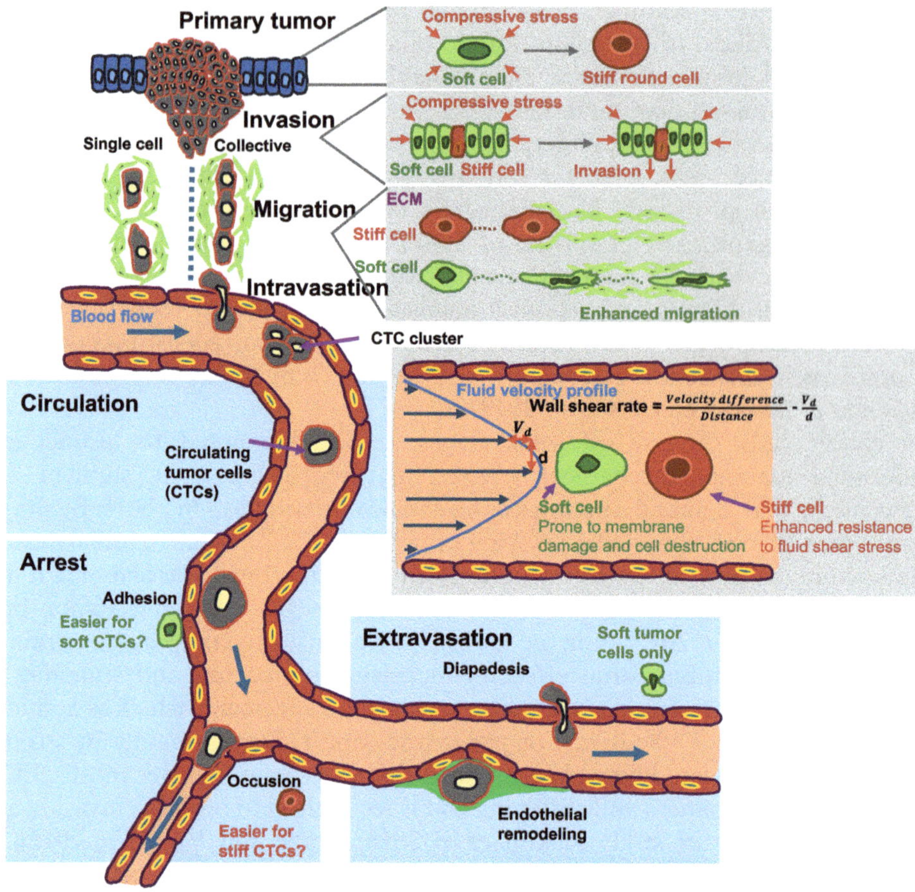

Figure 7.4. In a speculative perspective on the relevance of cancer cell stiffness during the entire metastatic cascade. Invasion, migration, intravasation, circulation, confinement, adhesion and extravasation pose various demands on the mechanical characteristics of cancer cells and favor various mechanical phenotypes. There may be steps in the metastatic cascade where it is better to be a soft cancer cell and a soft CTC or a stiff cancer cell and a stiff CTC.

EMT and demonstrate instead elevated expression of vimentin, for instance. The effects of cancer cell clustering on the mechanical characteristics and mechanical resilience of cancer cells are presently unclear and are also touched on briefly in the section of this chapter.

Cancer cells experience mechanical stress not merely in the vicinity of the primary tumor, but especially after intravasation, when they are moving through the bloodstream and progressing through the intravascular phases of the metastatic cascade (Follain *et al* 2020). When they have entered the bloodstream (or lymph vessels), cancer cells are subject to a totally different set of mechanical stresses (figure 7.4), which arise from hemodynamic forces and bottlenecks in blood vessels.

CTCs must deal with these stresses as they seek to progress through the metastatic cascade (Follain *et al* 2020). For instance, it has been demonstrated that high fluid shear stresses of above 1 Pa trigger cell cycle standstill (Chang *et al* 2008) or even kill cancer cells while they are in circulation (Regmi *et al* 2017). Ultimately, only a limited number of CTCs evade the impact of fluid shear stress and are afforded the opportunity to become fixed and adhere to the endothelium. It is also noteworthy that clusters of CTCs can move through capillary-sized vessels, which could afford additional protection (Au *et al* 2016).

There are two models for CTC capture in the bloodstream: they can either be stopped through active adhesion to the vessel walls (Follain *et al* 2018, Osmani *et al* 2019), or through occlusion when the vessel is topographically disturbed and/or possesses a small diameter (Kienast *et al* 2010, Wirtz *et al* 2011, Headley *et al* 2016, Entenberg *et al* 2018, Paul *et al* 2019) (figure 7.4). The latter often takes place in microcapillaries, where CTCs undergo squeezing as they must be transported through narrow restrictions to arrive at the extravasation spots near their targeted organs. Ultimately, effective macrometastatic outgrowth is instigated by extravasation of locked CTCs within perivascular niches. Along this journey, metastatic cells must overcome obstacles such as the vessel wall or the blood–brain barrier (for brain metastases) to become either dormant or invasive, a hallmark of life-threatening metastases. All the mechanical stresses that cancer cells experience in these sequential phases of the metastatic cascade clearly affect their intrinsic mechanical behavior. Fine-tuning their mechanical characteristics would allow cancer cells to adjust to the various mechanical stresses that pose a threat to metastatic progression in each phase. Specifically, their intrinsic adjustable mechanical phenotype could serve as a pivotal feature in overcoming the stresses under discussion in this section. Although this is likely, there is currently insufficient evidence to confirm that cancer cells modify their mechanical signature as they acquire specific mechanical or other stimuli while progressing along the metastatic cascade. These concepts will be discussed in this section and speculations will be made on how the mechanical characteristics of cancer cells might progress throughout the cascade. While not the focus here, chemical and biochemical cues like pH, temperature, cytokines, adhesive receptors or oxidative stress may also be implicated in the mechanical adjustment of cells (Sunyer *et al* 2009, Sun *et al* 2014, Scholz 2018). In this chapter, the mechanical profiles of cancer cells on their way to metastasis will be briefly outlined, thereby complementing, actualizing and augmenting existing overviews on this theme (Suresh 2007, Kumar and Weaver 2009, Wirtz *et al* 2011, Mierke 2019a). How these mechanical profiles are induced, maintained, and disrupted will also be introduced. This covers the identification of the first mechanosensitive protein in eukaryotic cells. Speculating further on how cancer cells' deformation capabilities might confer tumor metastasis capabilities will help to understand the importance of the adaption of cancer cell mechanics. Thereby it is highlighted why these aspects seem to be crucial research areas for the future. Finally, the concept is promoted that the view of a clearly defined static mechanical phenotype of a cancer cell is at odds with the changing mechanical characteristics needed for effective metastasis. In contrast, the capacity to dynamically adapt to these demands could be the mechanical signature of a functionally effective metastatic cancer cell.

7.4.2 Analysis of cellular mechanics: techniques, outcome and restrictions

It is imperative to consider the mechanics of individual cells separately from tumor mechanics. So far, the mechanics of tumors have been examined primarily by considering the tumor as a single entity or in conjunction with its stroma. It is common knowledge that the tissue of most solid tumors is stiffer and less elastic than healthy tissue, which is actually utilized in manual palpation (Levental *et al* 2009). Nevertheless, the stiffness of the entire tumor cannot be deduced from the stiffness of the individual cancer cells of which it is composed. Firstly, it consists of many accessory stromal elements, such as ECM (Eble and Niland 2019) and stromal cells, (Denton *et al* 2018), all of which are expected to contribute to tumor stiffness. Secondly, numerous original research articles state that cancer cells are in fact softer in comparison to healthy cells (Alibert *et al* 2017), even when they are analyzed in acute tumor sections (Plodinec *et al* 2012). This agreement was achieved by utilizing a wide range of tools and techniques available in many studies investigating single-cell mechanics. This is still based on the assumption that all cancer types behave in a similar manner and the observed phenomenon is of universal nature. While these methods will not be the center of attention in this chapter, as they have been covered extensively in several review articles (Van Vliet *et al* 2003, Darling and Di Carlo 2015, Wu *et al* 2018, Urbanska *et al* 2020), a brief discussion of the pioneering techniques and the latest state-of-the-art instruments (table 7.2) will be presented to highlight the development of the discipline. These techniques have revealed that the stiffness of cancer cells is crucial for tumor growth and therapy. The strengths and weaknesses are listed in table 7.2.

Shortly after it became possible to visualize cells for the first time with the invention of the microscope, the questions arose as to cell mechanics and how to examine them or understand their significance (Pelling and Horton 2008). Since then, many advances have been made toward the creation of the many modern tools available at the present time. Micropipette aspiration (MPA) was first employed in the 1950s (Mitchison and Swann 1954) and provides the first commonly used tool for investigating the mechanical properties of individual cells that is still in regular use. The principle involves applying suction to the tip of the micropipette, drawing part of a cell into a glass capillary, and observing the length of the aspirated cell as time passes. There are a number of various continuum models for extracting the rheological properties of cells (Hochmuth 2000, Mierke *et al* 2010, González-Bermúdez *et al* 2019). The MPA can be utilized to investigate the mechanics of either adherent or suspended cells. AFM eventually became the gold standard in individual cell mechanical analysis (Radmacher *et al* 1996, Thomas *et al* 2013, Dong *et al* 2016). AFM examines the mechanical characteristics of the cells in the reverse way, in that it creates an indentation in the cell with a tip, as opposed to sucking it up. Although typically restricted to measuring cells attached to a surface, it is widely used because of its robustness and sensitivity. A key feature is the use of probes of various sizes, which enable the measurement of both the stiffness of entire cells and subcellular stiffness mappings over the entire cell surface. This area encompasses a wide range of mechanical measurements, from the bending rigidity of the plasma

Table 7.2. Selected techniques frequently employed to measure the mechanical characteristics of cancer cells.

Biophysical technique	Strength	Limitations	References
Micropipette aspiration (MPA)	Very diverse and resilient. Capable of measuring both adherent and suspended cells. Can characterize the mechanical reaction as a function of substrate stiffness. There are models for deriving viscoelastic characteristics.	Low performance, such as 10 cells per hour.	Ward *et al* (1991), Hochmuth (2000), González-Bermúdez *et al* (2019)
Atomic force microscopy (AFM)	Extremely sensitive. High spatial resolution available to visualize the spatial distribution of mechanical characteristics on the cell surface. Can provide quantification of the mechanical reaction toward substrate stiffness. There are models for extracting viscoelastic characteristics.	Low performance, such as 10 cells per hour. Can frequently only analyze cells in contact with a surface under normal conditions. Can measure cells in suspension, when a flat cantilever is employed.	Radmacher *et al* (1996), Lekka *et al* (1999), Cross *et al* (2007, 2008), Darling *et al* (2007), Li *et al* (2008), Plodinec *et al* (2012), Xu *et al* (2012), The Physical Sciences - Oncology Centers Network (2013), Thomas (2015), Dong *et al* (2016), Tavares *et al* (2017), Liu *et al* (2020), Fischer *et al* (2017)
Optical traps, such as OS	Can contactless determine the mechanical characteristics of entire cells. There are models for obtaining viscoelastic characteristics.	Low performance, such as 100 cells per hour. Cannot probe cells that are in attached to a substrate.	Guck *et al* (2001), Wottawah *et al* (2005), Lincoln *et al* (2007b), Remmerbach *et al* (2009), Faigle *et al* (2015), Gullekson *et al* (2017), Mierke (2019b)

(Continued)

Table 7.2. (*Continued*)

Biophysical technique	Strength	Limitations	References
	Can be integrated with microfluidic dispensing for automation and cell sorting. Can directly analyze native circulating cells.	Cannot analyze cells with dark granules in them, such as melanoma cells.	
Microfluidic technique, such as RT-DC	High performance, such as 10–10 000 cells/s. Can characterize circulating cells under native conditions. Stable and user-friendly.	Most approaches lack models to retrieve elasticity or even viscosity. Only investigate mechanics on extremely short time-scales (ms–μs). Cannot probe cells in physical interaction with a substrate	Zhang *et al* (2012), Tse *et al* (2013), Otto *et al* (2015), Guillou *et al* (2016), Ciucci *et al* (2017), Deng *et al* (2017), Ahmmed *et al* (2018), Rosendahl *et al* (2018), Wullkopf *et al* (2018), Holenstein *et al* (2019)
Magnetic Tweezer	Capable of measuring both adherent	Low performance, such as 30–40 cells per hour.	Mierke *et al* (2008b), Mierke (2013), Aermes *et al* (2020, 2021)

membrane and local cytoskeletal structures underneath the plasma membrane to measurements of the entire cytoskeleton or cell nucleus. It is also well adapted for investigating the effect of substrate stiffness on cell mechanics, although the existence of a compliant substrate must be adequately taken into consideration during the analysis (Rheinlaender *et al* 2020). In distinction to the measurement of cells attached to a surface, instruments that use optical forces, like the optical stretcher (OS), can capture and deform cells that are in suspension (Guck *et al* 2001, Mierke 2019b). The key characteristic that differentiates this technique from MPA or AFM is that the probing is fully contactless. Thereby, even naturally adherent cells need to be kept in suspension during their measurement. These naturally adherent cells must be in a rounded shape for their probing and further analysis. This technique was also the initial approach to be combined with microfluidic dispensing (Lincoln *et al* 2007a) and cell sorting (Faigle *et al* 2015) to circumvent the limited throughput of previous techniques. It is only with the latest developments in microfluidic approaches, nevertheless, that cell-mechanical analysis has achieved a high throughput and is now catching up with the performance of conventional fluorescence-based flow cytometers (table 7.2). Real-time deformation cytometry (RT-DC) is a variant of a large family of such microfluidic techniques that have been established in recent years (Otto *et al* 2015, Urbanska *et al* 2020). Cells are guided at high speed (10 cm s^{-1}) in a microfluidic channel, where they undergo contact-free deformation through shear stress and pressure gradients. This technique enables mechanical phenotyping at speeds of >100 cells/s and enables completely new biotechnology and medical applications (Toepfner *et al* 2018). In summary, the tools for measuring the mechanics of individual cells are available in all forms and configurations, enabling the analysis of cells in suspension or on substrates, at the level of an entire cell or a specific area of interest at the cell surface. Both the elasticity and viscosity characteristics can be determined using techniques with varying levels of performance. The main restriction of all these techniques is that they cannot deliver *in situ* information on the mechanical condition of cancer cells, which is also outlined in the following.

These measuring instruments were employed in many investigations to examine the correlation between the mechanical characteristics of individual tumor cells and their metastasis capacity (Suresh 2007). Tumorigenic rat fibroblasts with modified viscoelastic characteristics were detected in MPA studies conducted as early as the 1990s. These characteristics conferred enhanced deformability on the tumorigenic cells in comparison to their non-tumorigenic equivalents (Ward *et al* 1991). AFM measurements revealed the exact same tendency. In the late 1990s, Lekka and his colleagues were the first to examine three cancerous cell lines and two normal human bladder cell lines, and discovered that the cancerous cells were considerably more elastic (deformable) compared to the normal cells (Lekka *et al* 1999). Substantially the identical findings were documented for breast epithelial cell lines (Li *et al* 2008, Fischer *et al* 2017, Liu *et al* 2020), chondrosarcoma cell lines (Darling *et al* 2007), gastrointestinal cancer cells (Suresh *et al* 2005), and ovarian epithelial cancer cells (Xu *et al* 2012). When examining the stiffness patterns of human breast biopsies, AFM indicated that cancer cells are the softest spots in malignant tissues (Plodinec *et al* 2012). Moreover, a nanomechanical AFM analysis of cancer cells and healthy cells taken from patients diagnosed with lung, breast or pancreatic cancer revealed that cancer cells are 70% softer (Cross *et al* 2007). In a follow-up study, this value was raised to 80% (Cross *et al* 2008). At the same time, the analysis of different cell shape features revealed, not surprisingly, several significant geometric distinctions between metastatic and non-metastatic human osteosarcoma cells, particularly in cell volume, cell area, cell circularity, and cell elongation (Lyons *et al* 2016). A recent investigation comparing the deformability of an isogenic set of cancer cells also found that cells that had effectively metastasized to the lung were considerably softer compared to both circulating and primary cancer cells (Liu *et al* 2020). Deformability appears to provide a selective advantage to metastatic cells and could be exploited for prognosis (Swaminathan *et al* 2011).

Conducting all these assays with cells attached to substrates also enabled the mechanical difference between normal and cancerous cells to be perpetuated through the detachment and measurement of cells in suspension. The use of an optical stretcher to assay multiple breast cancer cell lines, such as MCF-7, MCF-10 and MDA-MB-231, upon dissociation in suspension validated the close relationship between cell elasticity and metastatic capacity (Guck *et al* 2005). The result was confirmed in subsequent investigations on Ras-transformed epithelial cells (Gullekson *et al* 2017) and also on primary squamous cell carcinoma cells isolated from the oral mucosa of cancer patients, utilizing the identical optical stretching instrument (Remmerbach *et al* 2009). These are encouraging signs for the latest generation of high-throughput microfluidic techniques that also probe cells in a suspended state. In fact, the vast preponderance of published work utilizing high-throughput techniques has reconfirmed the enhanced deformability of cancer cells (Zhang *et al* 2012, Guillou *et al* 2016, Ciucci *et al* 2017, Deng *et al* 2017, Ahmmed *et al* 2018, Wullkopf *et al* 2018, Holenstein *et al* 2019). Specifically, several osteosarcoma cells assessed by RT-DC were observed to have increasing deformability with enhanced metastatic aggressiveness (Holenstein *et al* 2019). An alternative version of deformability cytometry has already demonstrated the ability to detect malignant cells in the pleural effusions of patients on the basis of their deformability (Tse *et al* 2013), meaning that mechanical phenotyping is on the verge of clinical implementation (Guck and Chilvers 2013). Nonetheless, in these studies, it has not been checked that these analyzed more deformable cells can induce a tumor in a mouse model system.

Although all these investigations demonstrate a positive correlation between the deformability of individual cancer cells and their metastatic capacity, only a small number of investigations have yielded the opposite finding. An example of this is a report that highly metastatic human breast cancer tumor cells acquired increased invasiveness when stiffened by β-adrenergic signal transduction (Kim *et al* 2016). In more recent work, glioblastoma cell invasiveness capacity was also found to positively correlate with cell stiffness, which is conveyed by the formin FMN1 (Monzo *et al* 2021). This mismatch indicates that the well-established correlation between cellular plasticity and metastasis might be much more intricate than it appears at first glance, it could rely on the technique employed (Wu *et al* 2018,

Holenstein *et al* 2019), and it possibly varies depending on the context, such as the specific environment or even on cell memory effects (Ekpenyong *et al* 2012, Toepfner *et al* 2018). This phenomenon is likely because mature cells in the hematopoietic system must be deformable to be able to circulate, whereas cancer cells are not completely differentiated blood cells and are therefore actually stiffer compared to healthy blood cells.

The major drawback of most of the above studies is the lack of representative context. Whereas the research on blood cancers was on the verge of achieving the goal of performing measurements in a close-to-natural *in vivo* environment, using microfluidic techniques, the situation for solid tumors is somewhat divergent: measurements were mostly carried out on cancer cell lines, which are grown in culture and are thus solely representative of a specific mechanical condition that is attained following culture on 2D substrates, which bears little relation to the context in which a cancer cell resides when metastasizing *in vivo*. Without the establishment of novel instruments that enable the non-destructive *in vivo* examination of cell mechanics, insights into the biomechanical characteristics of metastatic cancer cells cannot be reliably linked to successful metastasis. Three-dimensional cell culture is a technique that can be employed to close this knowledge gap and to examine cancer cells *in vitro* under conditions that more closely reflect reality (Lv *et al* 2017). Ultimately, the variety of instruments and methodologies used to investigate cell mechanics has made it possible to address the topic from multiple perspectives. Ultimately, the variety of instruments and methodologies used to investigate cell mechanics has made it possible to address the topic from multiple perspectives (Wu *et al* 2018). In specific, cell elasticity and viscosity measurements are highly sensitive to the frequency or strain rate at which specimens are examined, and the type and dimensions of the probe (Urbanska *et al* 2020), which adds to the difficulty of deriving unambiguous conclusions across various investigations. Ultimately, it cannot be ruled out that the mechanical alterations seen in cancer cells are merely an epiphenomenon and a side effect of other metastatic processes, and not the primary cause, since the studies discussed so far were merely correlative. A causal demonstration is a key next milestone in this endeavor (Guck 2019).

7.4.3 Passive characteristics and active regulatory components of cancer cell mechanics

So far, the passive mechanical characteristics of cells have been the subject of discussion, which can be determined through the exertion of external forces and the subsequent quantifying of the occurring deformation. There is, nevertheless, also an active element in the way cells react to mechanical forces. The mechanical characteristics of tumor cells fall into two distinctive categories: active and passive. Intrinsic properties that govern how a cancer cell or its organelles deform in reaction to applied external forces could be categorized as passive. Active characteristics, in contrast, consist of types of behavior in which mechanosensory mechanisms are involved to adjust the mechanical parameters/behavior of cells in reaction to stimuli from their surrounding environment. To put it another way: the intrinsic mechanical characteristics influence the performance of cells in their surroundings, but simultaneously, the surroundings can cause active reactions in the cells that impact their mechanical characteristics (Mierke 2019a, 2021a). This dynamic mechanical-environmental relationship perspective is the element most lacking in the investigations reviewed in the preceding section. It is therefore imperative to build an accurate picture of the contribution of cell mechanics in the entire metastatic process, starting from primary tumors and leading to secondary tumors. In this section, the structural elements of cancer cells that are implicated in cell mechanics and are crucial for the cells' capacity to deform are discussed. Their implication is assigned to either the passive or active facets of cell mechanics. However, the mechanisms of mechanosensation are not elaborated upon in depth as they are covered extensively elsewhere (Hoffman and Crocker 2009, Janmey and Miller 2011, Chin *et al* 2016, Cheng *et al* 2017, Broders-Bondon *et al* 2018, Mierke 2024a, 2024b).

The cell membrane serves as an interface between the cell and its surroundings and, consequently, is the first cell element to be impacted through external forces (Le Roux *et al* 2019). The cell membrane exhibits elastic characteristics that can be characterized using a bending modulus, which is a passive mechanical property that reflects the membrane's capacity to withstand bending deformation. The bending modulus represents an inherent cellular parameter that differs according to cell type (Pontes *et al* 2013). In addition, newer studies have revealed that cells actively react to mechanical restrictions and buffer them with little, organized indentations of the plasma membrane referred to as caveolae (Nassoy and Lamaze 2012). These influence the surface tension, add to the mechanical characteristics of the membrane and afford a certain level of protection against membrane disruption (Sinha *et al* 2011, Echarri *et al* 2019). In addition, the membrane is recognized to actively react to different mechanical stresses like tension, compression or shear and topographical stimuli, for instance in the guise of membrane flattening, ruffling or fluidization (Le Roux *et al* 2019). In conclusion, the cell membrane is part of both the passive and active aspects of cell mechanics, with its elastic characteristics that enable deformation and its function as a force-sensitive and reaction-inducing boundary between the cell and its surrounding environment. The cell membrane interfaces with the cytoskeleton and, in particular, with the actomyosin cortex that resides just beneath it. This component of the cytoskeleton has long been implicated in membrane deformation (Farsad and Camilli 2003). Most significantly, the actomyosin cortex is required for cortical tension of the cell. In additive to its involvement in cell stiffness, stress levels or fractures in the cortex, cortical tension constrains cell deformation (Chugh and Paluch 2018) and is engaged in membrane protrusions (Paluch *et al* 2005) and contractions at the cellular level (Koenderink and Paluch 2018), thereby becoming an active agent in cellular mechanics that offers the main opposition to cellular deformation when small strains are applied. The actomyosin cortex is just one of the numerous building blocks of the cytoskeleton, which is comprised of microfilaments, microtubules, and intermediate filaments. These various constituents are biopolymers, all of which have distinct mechanical characteristics and can dynamically combine and separate (Janmey 1991). Complementing the low-strain resistance of the actomyosin cortex, intermediate filaments are in charge of high-strain cell deformation resistance, acting as a safety belt to prevent excessive deformation that can disrupt cells (Charrier and Janmey 2016). As a consequence, the cytoskeleton, as the aggregate of its components, is most directly involved in cell compliance, as it controls the overall viscoelastic characteristics of the cell (Wottawah *et al* 2005).

Due to its dynamic character, it is implicated in cellular processes like cell cycle, cell migration and adhesion (Hall 2009), where cells actively reshape themselves via actin rearrangement. Therefore, the cytoskeleton is responsible not just for cell stiffness and passive cell deformation, but also for the general cell shape that they actively assume in various cellular processes (Bhadriraju and Hansen 2002). Moreover, the cytoskeleton locates the organelles within its filament meshwork, thereby establishing a direct connection between cell shape/deformation and the relocalization of key organelles throughout metastasis (figure 7.5). The cytoskeleton is far more implicated in active and passive mechanical events compared to the cell membrane and can be regarded as the main actor in cell mechanics (Fletcher and Mullins 2010).

The nucleus represents one of the numerous organelles associated with the cytoskeleton (Starr and Fridolfsson 2010) and is a hot topic in the area of cancer research (Denais and Lammerding 2014) and physics of cancer research (Fischer *et al* 2020). The nuclear envelope is more than just a simple protective shield. It is responsible for a number of crucial functions in the development of cancer, from preventing genomic alterations and ensuring cell cycle control to managing overall organization of the cytoskeleton and even cell migration (de las Heras *et al* 2013). In particular, the invasion of cancer cells strongly relies on the passive mechanical characteristics of the nucleus and its capacity to squeeze itself through narrow pores

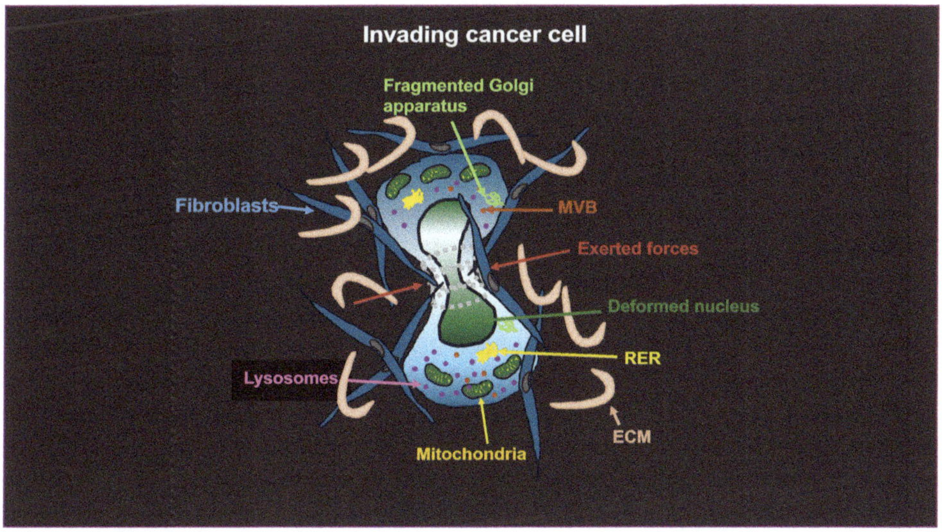

Figure 7.5. A migrating cancer cell experiences forces and deformations that affect the shape of their cell organelles, such as nucleus, mitochondria, lysosomes, rough ER (RER), Golgi apparatus and multivesicular body (MVB). The external forces and confinements are imposed by stromal fibroblasts and the ECM scaffold in a bidirectional interplay.

(Harada *et al* 2014), as it has been evidenced both in *in vitro* microfluidic systems (Fu *et al* 2012) and *in vivo* (Denais *et al* 2016) that the nucleus inside cancer cells is pronouncedly reshaped when the cells perform transmigration. Whereas the cytoskeleton can undergo dynamic rearrangements to allow cells to pass through constrictions, mechanical restraints are conferred on organelles like the nucleus, which is a fairly static and stiff elastic entity. This imposes a severe constraint on vital migration across very narrow bottlenecks (Harada *et al* 2014).

External forces are perceived directly within the nucleus, which acts as a mechanosensor (Shivashankar 2011, Kirby and Lammerding 2018), influencing major cellular processes like chromatin organization and the transcription of genes. Cells also utilize their nucleus as a benchmark to gauge the level of their spatial restriction and adjust their reaction correspondingly (Lomakin *et al* 2020). An important mechanism employed by cells to enhance the forces exerted on the nucleus and favor its deformation as it tries to make its way through narrow constraints involves the fast perinuclear actin polymerization powered by the Arp2/3 protein (Thiam *et al* 2016). In this scenario, the actin filaments that assemble at the nuclear periphery apply lateral forces to the nuclear envelope, which in turn leads to a temporary rupture of the nuclear lamina to allow more extensive deformation of the nucleus and consequently, the passage of the cancer cell. This mechanism relies on the availability of lamin A/C, a type of intermediate filament and the principal structural protein of the nuclear envelope. Lamins are associated with genome safeguard, are mechanosensitive and can affect key transcription modulators of metastasis like SRF and YAP/TAZ (Irianto *et al* 2016). Lamin A/C concentrations are crucial for the viscoelastic characteristics of the nucleus (Lammerding *et al* 2006, Swift and Discher 2014) and thus govern its capacity to contract passively when exposed to forces that enable the cell to move through confined areas (Cao *et al* 2016). Chromatin has also been identified as participating with lamin A/C in nuclear deformability (Stoeckius *et al* 2017, Hobson *et al* 2020). The composition of the lipids that make up the nuclear membrane was recently determined and revealed that it also contributes to deformation (Dazzoni *et al* 2020). Whereas nuclear deformation is associated with cell deformation solely under high loads, it is a severely restricting factor in migration. Nuclear deformation facilitates not only metastasis through transmigration of cancer cells but also causes DNA defects. DNA injury can in turn enhance metastatic potential of cancer cells through increased genomic instability (Lim *et al* 2016), unless the nuclear envelope and DNA repair systems intervene to avoid cell death (Denais *et al* 2016). For instance, DNA injuries resulting from recurrent disruption of the nuclear envelope are emerging as an important determinant of the invasive phenotype in cancer cells (Nader *et al* 2021). Besides the nucleus, other organelles have also been found to change shape and/or location in association with metastases. While the connections between these alterations and the TME and their role in active/passive mechanical phenomena are presently elusive, they are anticipated to be an important area of future investigation, utilizing recent advances in correlative *in vivo* correlative light and electron microscopy (Karreman *et al* 2016) to follow the shapes/localizations of organelles at various steps of the metastatic dissemination cascade (figure 7.5).

7.4.4 Cellular mechanics and cancer metastasis

The functional role of cancer cell mechanics in metastasis is likely to be overseen at each step of the process and is more complex than it appears at first glimpse. In this section, it is proposed to walk through the stages of the metastatic cascade one by one and muse on how the mechanics of the cancer cell come into the picture at every one of them, touching on the numerous alterations of the environment that cancer cells undergo during their metastatic voyage. This involves questioning the simplistic idea that an aggressive metastatic cell is inevitably a soft cancer cell.

Based on the initial invasion stage, it can be postulated that in an epithelium exposed to progressively higher compressive stresses in the vicinity of the primary tumor, stiffer, less deformable cancer cells are the first to be pushed out of the epithelium and passively profit from a selection advantage in comparison to softer cells. In the above-mentioned study, it was determined that the invasive capacity of glioblastoma cells directly relates to cell stiffness. The upregulation of the formin FMN1, that causes cells to become stiffer through the stabilization of microtubules, and at the same time increases motility and shapes the assembly of protrusions rich in actin (Monzo *et al* 2021). Nevertheless, the concept that invading cancer cells are more compliant seems to be more prevalent at present. Recent research demonstrated that invasive cancer cells found at the periphery of primary tumors are more prone to softening and develop invasive spikes as water is pushed from the center to the perimeter of the primary tumor (Han *et al* 2020). Even more importantly, in addition to this passive selection of softer cancer cells, the EMT regimen results in softer cancer cells with modified cell shape leaving the location of the primary tumor (Chen *et al* 2018c). Most evidence suggests that deformability is key to the invasion event, but whether this mechanical phenotype is a prerequisite event for, or a consequential event of, invasion is not completely resolved. Remarkably, recent studies demonstrate that cancer cells soften when migrating in a constrained environment (Rianna *et al* 2020) highlighting the possibility that cancer cells can adjust their mechanical phenotypes to adapt to their environment continuously, as can be seen through the exertion of forces within a ECM scaffold that change over time (Hayn *et al* 2023). The dynamic alterations of induction of displacements of ECM scaffold was elevated in highly invasive MDA-MB-231 cancer cells compared to weakly invasive MCF-7 cancer cells (Hayn *et al* 2023).

The next hurdle for cancer cells after leaving the primary tumor is 3D migration throughout tissues. As stated earlier, it is likely correct that cancer cells would profit from a softer nucleus for migration across narrow interstices, as the nucleus is a passive unit that restricts the passageway (Friedl *et al* 2011). A softer nucleus may also be subject to more mechanical stress and subsequent injury (figure 7.6) (Harada *et al* 2014), which in turn is likely to enhance cancer cell aggressiveness. The process of migration through the ECM is, nevertheless, complicated and influenced by a wide range of mechanical and non-mechanical factors. The mechanism of protein degradation can promote migration and contribute to counteracting the disadvantages of a stiff nucleus (Wolf *et al* 2013). The realignment of fibers during ECM remodeling (Provenzano *et al* 2006) and the interaction of cancer cells with stromal

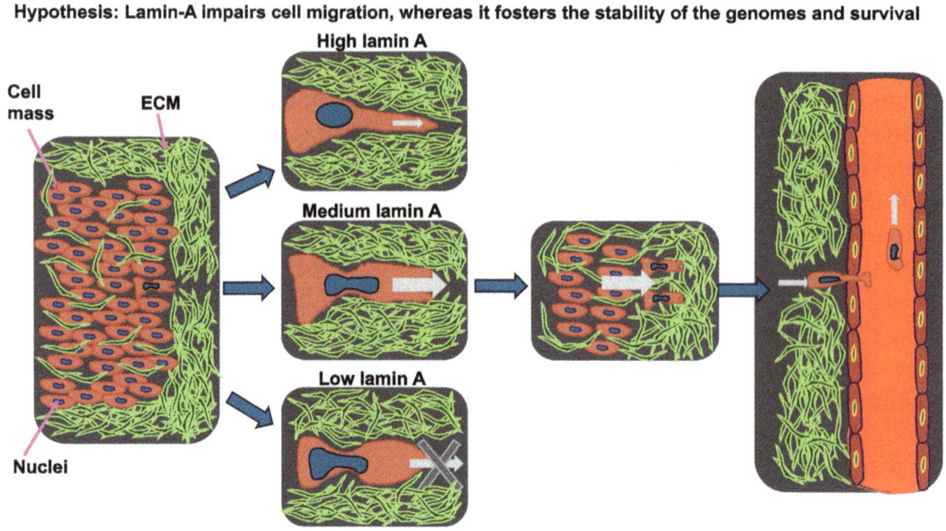

Figure 7.6. 3D migration is extremely sensitive to lamin A abundance, even in the complete lack of major proteomic alterations. A hypothesis is the there is a major effect of lamin-A concentrations on the migration of cells. While low expression levels enable migration, cells with low values cannot survive the stress and high values inhibit migratory behavior.

cells are additional mechanisms that probably mitigate mechanical damage. In addition, cancer cells that migrate cooperatively express keratin-14 (Cheung *et al* 2016), indicating that collective migration decreases the requirement for mechanical characteristics connected to deformability (Seltmann *et al* 2013). Whether this is the case, or not, can only be speculated upon at this point. To complete the mechanical aspect of this step of the metastatic journey, the statement that cancer cells must be 'adaptable' is probably more accurate than 'soft', since the amoeboid migration mechanism is recognized as an active motility process that involves repeated active adjustment of the cell form (Guck *et al* 2010). At the same time, the invasive capacity of cancer cells was found to depend on their capacity to mechanically adjust to the stiffness of the ECM scaffold they migratorily invade (Wullkopf *et al* 2018). Thus, a high activity/adaptivity of the cytoskeleton (possibly leading to an apparently low viscosity) and presumably a soft nucleus seem most appropriate for this migration phase of the metastatic cascade.

Next, the first step of transendothelial migration takes place during intravasation. During the transendothelial migration step, not only the mechanical properties of cancer cells are altered. In addition, also the mechanical characteristics of endothelial cells are altered, as they are softer, more fluid-like and exhibit increased cytoskeletal remodeling dynamics, and hence they possibly adapt to a changed endothelial surrounding microenvironment, such as the presence of transmigrating cancer cells (Mierke 2011). Intravascular and extravascular events can be discussed jointly from a highly conceptual, cell-mechanics point of view, since both processes pose similar mechanical demands: an object must deform to squeeze through a small opening in a thin layer. Naturally, there are also numerous differences in the biological peculiarities of how this occurs. Typically, both are two-step processes that require interactions (adhesion) between cancer cells and endothelial cells before cancer cells are capable of squeezing through endothelial cells to access/exit the circulatory system, in conjunction with crossing the basal lamina. These mechanisms may be both active and passive and may entail severe cell deformation (Chen *et al* 2016). It is uncertain, nonetheless, to what extent cell deformation is involved in these mechanisms, as several mechanisms of endothelial transmigration have been proposed (Mierke 2008, Mierke *et al* 2008b, Strilic and Offermanns 2017). For instance, the intravasation of cancer cells may differ greatly from that of non-cancer cells, as cancer cells can severely impair the structure and integrity of blood vessels, leading to high leakage and at times replacement of endothelial cells inside the endothelium (Chiang *et al* 2016). Importantly, cancer cells can migrate as clusters intravascularly, which especially is enhanced under hypoxia (Donato *et al* 2020). Whether this has an effect on the mechanical demands of cancer cells is unclear. In addition, extravasation of cancer cells can be facilitated through reorganization of the endothelial wall (Lapis *et al* 1988, Allen *et al* 2017, Follain *et al* 2018) and alteration of the mechanical characteristics of endothelial cells (Mierke *et al* 2008b). An alternative way of extravasation is angiopellosis, whereby the cancer cells are passive, only the endothelial cell wall is restructured to enable extravasation (Allen *et al* 2017). During this event, endothelial reorganization facilitates cancer cells to move in and out of the bloodstream without the requirement for cancer cells to deform for effective transmigration. These processes make it difficult to determine the mechanical characteristics of cancer cells necessary for intra-/extravasation, since it is still uncertain what these characteristics are when cells move in and out of the vascular system.

Similarly, this brings an end to the speculative discussion on the metastatic circulation step, for which both high and low stiffness appear to confer significant benefits for CTCs. CTCs are subjected to fluid shear stress (FSS) and would therefore profit from a quick stop, attachment and extravasation to minimize the threat of their destruction. More interestingly, the mechanical characteristics assumed to promote resilience to shear flow stress and rapid cessation are still inconclusive. In terms of FSS survival, two separate pieces of research have identified the actomyosin cytoskeleton as a key player, yet it has entirely opposed functions. The first study, performed entirely *in vitro*, revealed that FSS-surviving CTCs have decreased actomyosin activity, lower cell stiffness, and enhanced chemoresistance (Xin *et al* 2019). The second and more recent work came to the opposite conclusion that actomyosin activity and cell stiffening augmented FSS survival of cancer cells, a finding that was confirmed *in vivo*. Moreover, this research revealed that RhoA and non-muscle myosin II are involved in a mechanosensing mechanism whereby CTCs upregulate actomyosin to confer increased stiffness in the face of FSS (Moose *et al* 2020). Once again, it is worth noting that while it has not been demonstrated, the formation of CTC clusters may influence the mechanical demands of cancer cells throughout these intravascular cascade steps, as clustering may provide mechanical shielding of the cells against shear forces. Shear survival may not be the sole facet of CTC trafficking where the intrinsic mechanical characteristics of cancer cells come into effect, as their ultimate goal of immobilization may also rely on their amount of stiffness. Softer cancer cells were observed to take considerably more time to adhere *in vitro* (Xin *et al* 2019). This appears contrary to intuition, since one could logically assume that deformability increases the area of contact between cancer cells and endothelial cells, and thus has the opposite effect, enabling the creation of more adhesive connections and promoting cell immobilization (Osmani *et al* 2019), but this has not been proven. Because occlusion is the default alternative to adhesion for CTC cessation, it could also be hypothesized that CTC softness decreases the likelihood of them being captured in small capillaries, which in turn helps them to stay suspended in hostile blood flow for a longer time. This ability to evade occlusion within small capillaries has been observed in softer cells, specifically mesenchymal stromal cells (MSCs) (Tietze *et al* 2019) and in circulating blood cells (Ekpenyong *et al* 2012). In contrast, blood cell stiffness has been associated with vascular occlusion in reaction to chemotherapy (Lam *et al* 2008). In conclusion, the ideal mechanical characteristics for CTCs *in vivo* remain challenging to characterize and probe. There is conflicting evidence about the survival of FSS of CTCs, and both stiffness and softness appear to provide benefits for survival and rapid immobilization.

In summary, the importance of cancer cell mechanics for metastasis cannot be boiled down to an elevated metastatic aggressiveness with elevated deformability. Ideal mechanical characteristics for achieving a successful outcome at every step of the metastatic cascade could either have an ultra-sensitive zone or vary substantially between the different steps, although the latter possibility is more plausible. Some pathways to effective metastatic outgrowth are more straightforward than others; for example, shear stress is reduced in veins relative to arteries, and there are benefits to collective migration/dispersal over single-cell progression along the cascade, so cancer cells of varying mechanical competence may arrive at the end of metastasis with an unequal chance of survival. Since cancer cells have no choice but to migrate, they must perceive their environment and dynamically adjust their mechanical characteristics. Achieving the right level of cancer cell stiffness seems to be very difficult in any case, which explains why metastasis is such an extremely inefficient process (Weiss 1990).

7.5 A cancer-intrinsic mechanism mechanically remodels cancer cells

Prior research has found a positive correlation between enhanced tissue rigidity and cancer progression, indicating that cancer advancement relies on the stiffening of cancerous tissues (Paszek *et al* 2005, Ondeck *et al* 2019, Safaei *et al* 2023). Moreover, tissue stiffness can be caused by enhanced matrix protein accumulation, like type I collagen, produced by cancer-associated fibroblasts (Provenzano *et al* 2008, Cox and Erler 2011, Lu *et al* 2012, Wullkopf *et al* 2018, Nissen *et al* 2019). Nevertheless, it has been recently reported that suppression of tumor tissue stiffness in the liver of pancreatic ductal adenocarcinoma (PDAC) metastases increased tumor growth and led to reduced survival in orthotopic PDAC mouse models (Jiang *et al* 2020). This study implies that tumor stroma plays a cancer-protective function in PDAC, which leads to an opposing conclusion to the prediction of a pro-tumorigenic effect on PDAC progression. These reports indicate a general lack of agreement on the correlation between cancer tissue stiffness and cancer progression. This missing consensus also extends to the same tendency in the relationship between cancer cell stiffness and invasiveness. Stiffer pancreatic cancer cells were demonstrated to be more invasive (Nguyen *et al* 2016a). Conversely, cancer cells and metastatic cancer cells have been found to be soft (Guck *et al* 2005, Cross *et al* 2007). Moreover, recent studies have demonstrated that soft cancer cells can escape elimination through T cell-mediated killing by preventing the assembly of perforin pores (Liu *et al* 2021). In addition, cell softness can be utilized as a biomarker for cancer cells with tumorigenic potential and as a stem cell label (Liu *et al* 2021). In the course of cancer, heterogeneity of cancer cells is critical since some cancer cells are cancer stem cells or tumor stem cells, which are inherently soft and have greater resistance to anti-cancer medications. Therefore, the utilization of cell softness as a biomarker for cancer cells having more aggressive characteristics is yet to be elucidated. Research demonstrated that more deformable cancer cells have a predisposition to be more invasive (Swaminathan *et al* 2011, Xu *et al* 2012). Conversely, there are reports that invasive cancer cells exhibit a tendency to be stiffer (Yu *et al* 2015, Kim *et al* 2016, Nguyen *et al* 2016a). Overall, the relationship between cancer tissue, cell stiffness and invasive potential varies depending on the specific context or possibly on the cancer type or cancer stage. Considering the context-dependent way cancer acts, it is essential to consider the external environment in which cancer cells reside and how the intracellular mechanical apparatus needs to react to these influences, instead of concentrating on individual mechanical phenotypes like cell stiffness. Like healthy cells, cancer cells must be capable of integrating chemical and physical cues from their external microenvironment. In contrast to their healthy equivalents, cancer cell survival depends on their capacity to adjust to mechanically diverse and constantly fluctuating microenvironments. This adaptation involves the mechanosensitive molecular machinery, which comprises proteins that are unique in their capacity to perceive and respond to mechanical stresses, and which help set up the mechanical and force-producing operations of the cell. Moreover, mechanosensory elements can be organelles or cellular structures (Mierke 2024b). In humans, the mechanoresponsive cytoskeletal mechanism comprises NMII proteins, like NMIIA, NMIIB and NMIIC, specific paralogs of α-actinins, such as ACTN4, but not ACTN1, and FLN, such as FLNB and to a much lesser degree also FLNA (Schiffhauer *et al* 2016, Kothari *et al* 2019a, Surcel *et al* 2019).

In cancer, NMII proteins are crucial for tumor initiation, growth, and metastasis, which are driven by the involvement of NMII in adhesion, mechanotransduction (Halder *et al* 2019), migration, and contractility. In a comprehensive review, a catalog of various cancers with modified NMII expression and/or regulatory events was generated (Parajón *et al* 2021). Notably, various cancers express distinct subgroups of NMII paralogs, with different patterns between cancer types and different detection methodologies. For instance, PDACs express NMIIA and NMIIC, but not NMIIB (Surcel *et al* 2019). Moreover, myosin assembling decreases spread and metastasis of cancer cells in *in vitro* and *in vivo* models (Surcel *et al* 2019,

Bryan *et al* 2020). The mechanical characteristics of breast cancer tissue were imitated in a recent study to examine the part played by NMIIA and NMIIB in controlling the cell's reactions to substrate stiffness. Specifically, the signaling pathways, such as Rac1-NMIIA and PKCζ-NMIIB, induced in response to substrate stiffness were determined to govern the distribution and activation of NMIIA and NMIIB (Peng *et al* 2022). Notably, the functionality of each NMII isoform does not correlate with its cellular abundance. The isoforms are present in varying amounts that can differ by a factor of 100, for example, NMIIC is in the 5–10 nM range when available, whereas NMIIA is in the 500–750 nM range when present in pancreatic cells. Nevertheless, depletion of either leads to a reduction in cortical tension of a similar extent (Surcel *et al* 2019).

ACTN have a different function compared to NMII, but their involvement is ubiquitous across various cancers. The ACTN family comprises four paralogs that act as actin crosslinkers, namely ACTN1 to ACTN4. ACTN4 has been associated to metastasis through changes in cell performance and differentiation (Parajón *et al* 2021). It has been reported that ACTN4 promotes proliferation, migration, and metastasis in osteosarcoma and enhances invasive capacity via the nuclear factor-κB signaling pathway (Huang *et al* 2020). Moreover, a recent study showed that ACTN4 serum concentrations were increased in cervical cancer patients when compared to healthy controls, indicating that ACTN4 could be a cervical cancer predictive biomarker (Ma *et al* 2020b). In cancer stem cells, blocking ACTN4 resulted in enhanced sensitivity toward anti-cancer agents and diminished spheroid generation, proliferative capacity, and tumorigenesis *in vivo* (Jung *et al* 2020). In addition, ACTN4 enhances breast cancer invasiveness through a mechanism that controls the localization and dispersion of NMIIA and the expression of NMIIB mRNA and protein (Barai *et al* 2021). In PDAC cells, ACTN4 was demonstrated to interface with the GTPase dynamin 2 to modify invasion capacity and ECM rearrangement. The breakdown in the interplay between dynamin 2 and ACTN4 inhibits migration and invasion as a result of the impaired actin-rich organization at the basal surface of the cells (Burton *et al* 2020).

The FLN proteins constitute a family of actin crosslinkers that bind to actin filaments (F-actin) to form a tight gel-like meshwork (Wang and Singer 1977). Human cells express three isoform, termed FLNA, FLNB and FLNC (Stossel *et al* 2001). FLNs are localized in a specific subcompartment of the cell, the cortical F-actin-rich region that underpins the plasma membrane (Glogauer *et al* 1998, Stossel *et al* 2001, Feng and Walsh 2004). FLNB knockout in HeLa cells was recently revealed to modify the expression of genes that are involved in apoptosis, carcinogenesis, and metastasis (Ma *et al* 2020a). In addition, the expression of FLNB is increased particularly in the cancer glands of the pancreas, whereas FLNA is increased in the whole pancreatic tissue and in the stroma (Surcel *et al* 2019). A recently conducted clinical investigation demonstrated that the expression of FLNA is elevated following chemotherapy treatment of colorectal cancer patients (Yeşilkaya 2019). Moreover, elevated FLNA expression resulted in acquired resistance to a tyrosine kinase receptor antagonist, which is applied in the treatment of non-small cell lung cancer (Cheng and Tong 2021). Ultimately, the co-expression of FLNA and clusterin, which is a secreted glycoprotein, has the capability of serving as a biomarker for hepatocellular carcinomas (Patarat *et al* 2021).

A distinguished and emerging family of proteins that may be important for the mechanical program of the cell comprises the dynamins. The dynamin superfamily consists of dynamins 1, 2, and 3. The superfamily includes multi-domain GTPases that exhibit a modular architecture, are larger than 70 kDa, and possess a low affinity to guanine nucleotides (Ramachandran and Schmid 2018). The proteins of the dynamin superfamily are involved in processes including membrane fission and merging as part of endocytosis and organelle formation, and in organizing the cytoskeleton (Jimah and Hinshaw 2019). Dynamins 1, 2, and 3 possess various expression levels within tissues. In a recent investigation, qRT-PCR and immuno-histochemistry were employed to assess the expression of dynamins 1, 2 and 3 levels in human hepatocellular carcinoma (HCC) tissue specimens. In particular, it was observed that the expression of dynamins 1, 2 and 3 was increased in patients suffering from human hepatocellular carcinoma (Tian *et al* 2020). In another set of experiments, dynamin was blocked in cell lines of pediatric acute leukemia and suppressed the proliferation of cells, causing them to undergo caspase-dependent apoptotic cell death (Von Beek *et al* 2021). A further study revealed how dynamin GTPase concentrates F-actin through the formation of a helical structure (Zhang *et al* 2020). They also determined that the build-and-break cycles of dynamin produce actin bundles that are mechanically stiff (Gu *et al* 2010, Zhang *et al* 2020). Collectively, NMII, ACTN, and FLN proteins are strongly involved in cancer evolution and act as biomarkers for cancer initiation and advancement. Taken together, these findings emphasize the importance of mechanosensitive proteins in enabling cancer cells to endure a variety of fluctuating mechanical environments. It is not surprising that upregulation of this mechanical pathway is linked to unfavorable clinical prognosis (Honda 2015, Liu and Chu 2017, Pecci *et al* 2018, Kamil *et al* 2019, Park *et al* 2020b, Tian *et al* 2020, Wang *et al* 2021). Pancreatic cancer appears to be especially susceptible, as several mechanosensitive molecules are found to be elevated in PDAC advancement.

7.6 Mechanical adaptability and its challenges are focused

The mechanical-adaptability mechanism integrated into the mechanical system represents a promising target for cancer therapeutics. Because of its direct involvement in cancer metastasis and proliferation, it is an attractive drug target (Yamaguchi and Condeelis 2007). It is not surprising that proteins within this mechanical program exhibit aberrant expression in a wide range of cancers (Yamaguchi and Condeelis 2007). It is conceivable that interfering with this program through modifying these proteins can result in slow or even reversal of the advancement of cancer. The following section introduces several studies that are aimed at mechanical program elements in the attempt to cure cancer. Among these target molecules are F-actins, α-actinin, NMII, and 14-3-3 (Hsu and Kao 2013, Foerster *et al* 2014, Bryan *et al* 2020, Parajón *et al* 2021). From a cell biology perspective, caution should be exercised when blocking individual mechanical proteins across the system, as disrupting a single constituent of the mechanical system can lead to unanticipated effects. For this reason, extensive deliberations are required before trying to change the mechanical system. The mechanical signaling network is far from being an insulated system, as it is intertwined with a multitude of other cellular processes and cellular features that may also be impacted. These diverse cellular processes involve signal transduction, expression of genes and cellular metabolism (Etienne-Manneville 2004, Blanchoin *et al* 2014, Fife *et al* 2014, Bai *et al* 2016, Picariello *et al* 2019, Ong *et al* 2020, Park *et al* 2020a, DeWane *et al* 2021, Angstadt *et al* 2022). Therefore, fixing errors in one network process can lead to a malfunction in a different cellular process. It is thus essential to comprehend the downstream impact of manipulating mechanical network elements in cancer treatment.

7.6.1 Actin-dependent orientation of the mechanical program

Actin constitutes a globular protein that forms filaments and a polymeric meshwork in the cell cortex and cytoplasm. Actin functions as a major structural constituent of the cell and is involved in a wide range of cellular activities, ranging from cell migration and division to adhesion and signal transduction (Yamaguchi and Condeelis 2007, Bryan *et al* 2020). In the case of cancer, rearrangement of actin is primarily associated with the proliferation and metastasis of tumors (Yamaguchi and Condeelis 2007). Actin has been targeted to serve as a possible therapeutic. Cancer cells were treated with magnetic particles that connect to the actin cytoskeleton of the cell. When a magnetic alternating current is generated, the compounds cluster and disorganize the cell's actin cytoskeleton, which results in a subsequent halt in the cell cycle and ultimately in cell death (Yu *et al* 2020). The main hurdle in this strategy is the delivery of the drugs to the cancer cells. Another potential therapeutic approach is to target the function of several actin-remodeling proteins in the development of cancer. These proteins are N-WASP, cofilin and cortactin, which all show modified expression patterns in various types of cancer, resulting in increased metastatic potential (Yamaguchi and Condeelis 2007). Overall, these alterations in the expression of this protein network indicate the potential of these molecules as targets for cancer therapy. The subsequent studies highlight the difficulties in targeting actin and actin assembly modulators in a tumor context. In breast cancer cells, the elimination of N-WASP or the Rho-subfamily GTPase CDC42 reduced the fast F-actin rearrangement and aggregation in the immune synapse zone in reaction to natural killer cells (NK cells) (Al Absi *et al* 2018). By interfering with actin reorganization, NK cells were enabled to identify cancer cells and tag them for elimination (Al Absi *et al* 2018). Therefore, it is important to acknowledge the importance of actin in other cellular processes when working on therapeutics, to avoid unintended effects like enhanced metastasis and immune escape. Apart from the traditional cancer-related outcomes, the functions of actin in glycolytic activity established a linkage between the cytoskeleton and metabolism (Park *et al* 2020a). Actin assembly binds TRIM21, which is an E3 ubiquitin ligase that targets and recruits the ubiquitin-proteasome system for breakdown of phosphofructokinase proteins, thereby inactivating them. TRIM21 actin polymer sequestration enhances protein levels of phosphofructokinase, thus boosting glycolytic activity in non-small cell lung cancer (Ayad and Weaver 2020, Park *et al* 2020a). This research adds to the picture of aerobic glycolysis, otherwise referred to as the Warburg effect, in cancer cells, in which glycolysis is preferentially utilized even though they have access to the ATP-efficient oxidative phosphorylation pathway (Liberti and Locasale 2016). In summary, the intersection of actin with metabolism highlights the crossroads that must be considered when focusing on the elements of the mechanical program that are connected to actin assembly and arrangement.

7.6.2 The mechanical program aimed at by myosin II

As mentioned before, myosin II serves as a critical component in the program of mechanical adaptability. In the case of cancer, myosin II is involved in a variety of processes, from the onset of the disease to its advancement (Halder *et al* 2019). The three paralogs of myosin II, namely NMIIA, NMIIB and NMIIC fulfill different functions in cell motility and spatially-resolving force generation, which have diverse effects according to cell type (Ouderkirk and Krendel 2014). Depletion of NMII decreased cell migration and cell–cell adhesion capacity in mammalian cells (Vicente-Manzanares *et al* 2009). Although NMII proteins are not specifically defined as genetic drivers of cancer, they are at the interface of multiple signaling routes that permit cancer cells to hijack the myosin II apparatus (Vicente-Manzanares *et al* 2009, Halder *et al* 2019). For instance, it has been demonstrated that the Rho-(Rho-associated protein kinase) ROCK signal transduction route in cancer induces migration and invasion of cancer cells through alterations in ROCK functionality, which enhances the assembly and contraction of actin-myosin (Chin *et al* 2015). In addition, NMIIA has been associated as a tumor suppressor in squamous cell carcinomas where its expression enhanced p53 stability and its nuclear abundance (Ouderkirk and Krendel 2014, Schramek *et al* 2014, Conti *et al* 2015). These findings highlight the promise of directly targeting NMII rather than the cancer-altered signal transduction routes. It is not surprising, nevertheless, that precisely modulating NMII is far from easy.

Myosin II proteins are involved in a number of cellular processes, among them membrane-associated proteins, RNA-interacting proteins, proteins in the nucleus and metabolically active enzymes (Nguyen and Robinson 2020). This idea of a built-in contraction grid was investigated in *Dictyostelium discoideum* and sheds light on the complexity of the mammalian system. For instance, several metabolic proteins were identified, among them adenine nucleotide translocase and methylmalonate semialdehyde dehydrogenase, which function as genetic suppressors of myosin II mutants when overexpressed (Ren *et al* 2014). A subsequent proteome screen revealed interactions between metabolic constituents and the cytoskeleton, and methylmalonic semialdehyde dehydrogenase was re-identified as a biochemical interactor of the contractile filament network (Kothari *et al* 2019b).

Adding further complexity to the therapeutic implications of NMII, several studies have demonstrated the equivocal nature of NMII in various cell types. In glioblastoma, *in vivo* modeling showed that single knockout of NMIIA or NMIIB could not individually inhibit cancer proliferation and invasion (Picariello *et al* 2019). Removal of NMIIA alone resulted in decreased cancer invasion but enhanced cancer mortality due to increased proliferation. Conversely, when both NMIIA and NMIIB were removed in tandem, there was a decrease in the tumor's invasiveness and proliferation. Likewise, the function of NMIIA as an oncogene or tumor suppressor is not yet clear. It is debated that high expression of NMIIA in gastric and esophageal cancer corresponds to worse prognosis and the presence of metastases. In contrast, it was also found that NMIIA levels were decreased in squamous cell carcinomas of the skin, head and neck. Therefore, it was inferred that NMIIA is essential for the post-transcriptional activity and nuclear retainment of p53 (Wang *et al* 2019). Hence, NMIIs can either enhance or suppress cancer according to the environmental circumstances. These diverse effects certainly mirror the cell type and its respective physiological state, along with the tumor cell's continuously transforming surroundings. Therefore, although it is feasible to attack specific parts of the mechanical network, the off-target effects resulting from the crosstalk with other cellular processes need to be accounted for.

7.6.3 Different outcomes in diverse experimental setups

Due to the complexity of various experimental model systems, the outcomes of investigations on the mechanical adaptability response program using cultured cells and investigations utilizing *in vivo* model systems may not consistently agree. For instance, the use of blebbistatin to block NMIIA in an *ex vivo* glioma model reduced cancer invasion (Ivkovic *et al* 2012). As a result of this study, another laboratory found that blebbistatin actually inhibits the invasion of glioma cells in the cerebral cortex of rats (Beadle *et al* 2008). Nevertheless, although the glioma cells are killed faster, another study found that in their glioma rat models, eliminating NMIIA shifts the stiffness of the glioma cells toward optimal proliferation, causing the glioma rat models to die sooner, even though glioma invasion is impeded (Picariello *et al* 2019). This highlights the challenge of targeting a mechanical adaptability component of the program because of its involvement in so many cellular processes. Moreover, cancer advancement increases the degree of complexity in understanding the outcomes of various investigations. In this context, the topic of cancer and its 'dependence' on K-ras was debated, and how K-ras, although the most common mutated oncogenic incident that fosters cancer development, may not be necessary for tumor perpetuation throughout the advancement of malignant cancer (Singh and Settleman 2009). This infers that the biological setting of primary cancer cells may differ considerably from that of malignant cancer cells. Even though the above points represent major hurdles, the mechanical-adaptability program as a whole is certainly worth pursuing, as it leads to promising therapeutic outcomes and the research being conducted will undoubtedly strengthen the pipeline of cancer treatments.

7.6.4 Therapy development highlights targeting the mechanical program

The development of therapies that target these mechanisms is a major objective for researchers in cancer mechanobiology, in view of the significance of the mechanisms of mechanical adaptability in carcinogenesis and their potential for being targeted. Recent trials in this regard have been encouraging (figure 7.7). Metformin, which is used to treat diabetes, has an anti-cancer action through its effect on tumor-associated fibroblasts (Suissa and Azoulay 2014, Wheaton *et al* 2014, Podhorecka *et al* 2017, Song *et al* 2017, Whitburn *et al* 2017, Chen *et al* 2018a). The researchers conducted a proteomic analysis of tumor-associated fibroblasts exposed to metformin and observed that proteins associated with F-actin depolymerization and regulation of the cortical cytoskeleton accounted for a high percentage of the misregulated proteins (Chen *et al* 2018a). It has also been reported that cisplatin and paclitaxel, which are widely used chemotherapeutic agents, influence cell mechanics. These medications have long been considered to delay cancer development due to DNA damage. More recent work, nonetheless, indicates that they also affect the assembly of actin stress fibers and cytoskeletal remodeling in several types of cancer, resulting in enhanced cell stiffness (Köpf-Maier and Mühlhausen 1992, Vassilopoulos *et al* 2014, Kung *et al* 2016, Raudenska *et al* 2019).

An additional prominent class of anti-cancer therapeutics that target the adaptability program acts on the Rho/ROCK signaling pathway, which governs many

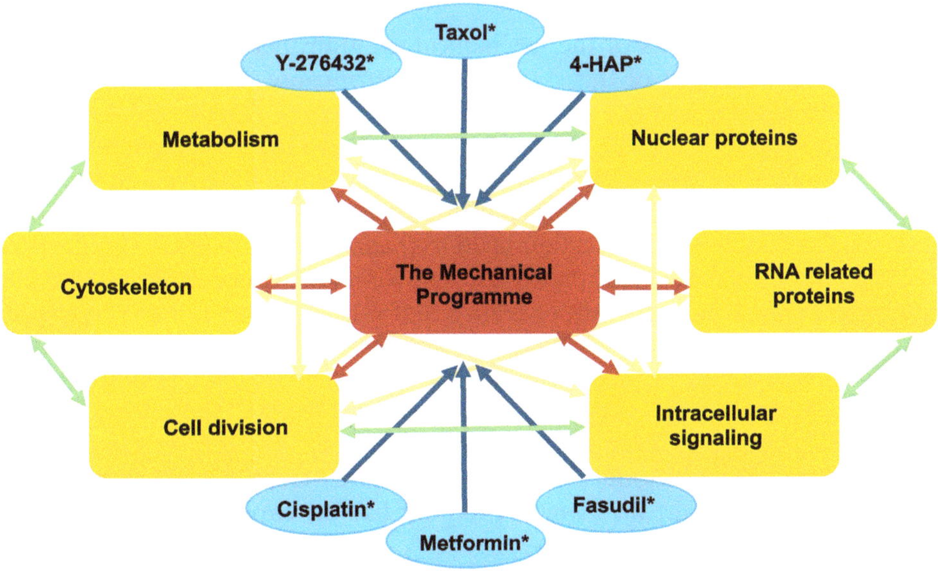

Figure 7.7. The complexities of targeting the mechanical program because of its intricate association with other cellular processes. As there is more excitement to explore the constituents of the mechanical program for cancer therapeutic agents, a more comprehensive systems view is warranted. The mechanical program, as with other cellular processes, influences various processes and is influenced in return. Consequently, while modulating a process that leads to tumor suppressing outcomes, it may cause tumor promoting effects in another process. The asterisked drugs are examples of drugs that modulate the mechanical program.

cellular events, such as actin reorganization and cell migration, cell shape, and proliferation. F-actin and NMII proteins are downstream target proteins of the Rho/ROCK signaling pathway. Recent research indicates that ROCK1 and ROCK2 are frequently elevated in a variety of cancers (Kamai *et al* 2003, Lane *et al* 2008, Kalender *et al* 2009) and are linked to the size of the tumor, the aggressiveness of the cancer and the spread of metastases (Kamai *et al* 2003, Liu *et al* 2009). Several ROCK inhibiting compounds have demonstrated promising anti-cancer effects. For instance, the ROCK inhibiting agent Fasudil inhibits the proliferation and spread of fibrosarcoma, melanoma, breast and urinary bladder cancer cells (Miyamoto *et al* 2012, Xia *et al* 2015, Guerra *et al* 2017, Lee *et al* 2019). This agent also decreases transendothelial migration of PDAC cells and their adhesion to secondary locations (Vennin *et al* 2017). Another ROCK inhibitor compound is Y-27632. This combination is aimed at attenuating the growth and invasive phenotypic characteristics of bladder, ovarian and skin cancers and works in part through the suppression of ROCK-driven phosphorylation of myosin light chain kinase, which directly stimulates the activation of myosin II (Mierke *et al* 2008a, Routhier *et al* 2010, Jiang *et al* 2015, Li *et al* 2015, Cascione *et al* 2017). Intriguingly, Y-27632 has been found to cause an elevation in the stiffness of breast cancer cells (Cascione *et al* 2017). This finding is counterintuitive, as it is initially expected that ROCK blockade would cause inactive myosin II, thereby in turn causing softer cells. Although the magnitude of the effects in this work is relatively low, the finding underscores the need to validate predictions when the mechanical system of a cell is perturbed.

As multiple constituents of the mechanical program perform tumor suppressor activities, the impairment of this system can in turn cause cancer enhancement. 4-hydroxyacetophenone (4-HAP) was found to activate the system and decrease the metastatic potential of PDAC (Surcel *et al* 2019). 4-HAP activates NMIIB and NMIIC through enhancing their assembly, thus enhancing cell stiffness and increasing cortical tension. Enhanced NMII assembly in turn decreases invasive phenotypes and inhibits proliferation of PDAC cancer cells. 4-HAP also decreased metastasis in mouse models for pancreatic and colorectal cancers (Surcel *et al* 2019, Bryan *et al* 2020). It is important to realize that 4-HAP seems to have no effect on NMIIA, which could enable this molecule to circumvent some of the implications of targeting all NMII paralogs concurrently. In addition, by promoting NMIIB and/or NMIIC assembly, 4-HAP is not anticipated to interfere with the tumor-suppressive function of NMII as implicated in a variety of settings such as squamous cell carcinomas and glioblastomas as debated before. Collectively, these findings indicate that targeting the mechanical program has the capacity to generate future therapeutic agents in cancer therapy.

7.7 Cellular cytoskeleton and stiffness are mechanical predictors of breast cancer organotropism

When a tumor metastasizes, cancer cells from the primary lesion spread to other organs, where they establish secondary tumors that are the cause of most cancer-related deaths. The clinical evidence indicates that cancer cell spread is not a random process but instead exhibits an organ preference or organotropism. Whereas intrinsic biochemical determinants of cancer cells have been thoroughly investigated in the framework of organotropy, the contribution of the cell cytoskeleton and mechanics is far less understood. The cytoskeleton and cell mechanics are shown to relate to organotropism. The measurements of cell stiffness indicate that breast cancer cells with bone tropism are significantly stiffer and display more F-actin, while brain-targeted cancer cells are softer and exhibit less F-actin compared to their original parent cells. (Tang *et al* 2021). The disparity in cellular stiffness corresponds to the disparity in the stiffness of their metastatic organs. Moreover, perturbation of the cytoskeleton of bone-targeted breast cancer cells not only leads to increased expression of genes associated with brain metastases, but also to enhanced cell spreading and proliferation on soft substrates that imitate the stiffness of brain tissue. Cytoskeletal stabilization of brain-invading cancer cells results in the induction of genes linked to bone metastasis, whereas their capacity for mechanoadaptation on soft substrates is diminished. Collectively, these results indicate that the cellular cytoskeleton and biophysical characteristics of breast cancer subpopulations coincide with their metastatic advantage relative to the gene expression profile and mechanoadaptation capacity, suggesting a potential function for the cellular cytoskeleton during organotropism. Metastases are the most common reason for cancer-related deaths (Lambert *et al* 2017). Cancer cells metastasize preferentially to certain organs, a phenomenon defined as organotropism, whereby different cancer types and cancer subpopulations metastasize to specific organs. For instance, prostate cancer metastasizes preferentially to the bone, whereas uveal melanoma metastasizes preferentially to the liver (Nguyen *et al* 2009). Breast cancer has a tendency to metastasize to the bones, brain, liver, lungs, and distal lymph nodes (Yates *et al* 2017). The most prevalent breast cancer metastasis is bone metastases, which appear in about 50% of metastatic breast cancer patients (Hess *et al* 2006). About 25% of breast cancer recurrences can be found in the liver, and 20% in the lung and brain (Hess *et al* 2006, Lu and Kang 2007). There is evidence that a variety of endogenous and exogenous elements are involved in the organotropism of cancer cells, which involves the patterns of circulation, endogenous characteristics of cancer cells, organ tropism, environmental niche, and the microenvironment of cancer cells and the host (Wei and Siegal 2017, Gao *et al* 2019). Nevertheless, the molecular mechanisms are still elusive. Thus, elucidating the mechanisms underlying organotropic metastasis will aid in the design of improved diagnostic and/or therapeutic approaches and ultimately lead to improved outcomes for cancer patients.

The impact of endogenous biochemical cues on organotropic metastasis of cancers has been extensively investigated. Metastatic cancer cells originating from certain organs have a marked preference to metastasize to the identical organ of origin (Lu and Kang 2007). The organotropism gene profiles and signaling cascades involved in the intrinsic characteristics of cancer cells contribute to their organ-specific extravasation and colonization (Kang *et al* 2003, Minn *et al* 2005, Bos *et al* 2009, Chang *et al* 2013). Dickkopf-1, which is released by breast cancer cells, had different impacts on lung metastases and bone metastases, enhancing bone metastases via canonical Wnt signal transduction and suppressing lung metastases through non-canonical Wnt signal transduction (Zhuang *et al* 2017). The COX2-MMP1/CCL7 axis improves the capacity of breast cancer cells to form brain metastases through increased blood–brain barrier permeability and tumorigenic cell growth (Xia *et al* 2015). The mechanical characteristics of cancer cells, which are key characteristics intrinsic to the cells, are recognized to be closely associated with their malignant potential (Swaminathan *et al* 2011, Luo *et al* 2016). The malignant transformation caused through genetic mutations is linked to certain alterations in the mechanical characteristics of the cells like stiffness and viscosity (Malandrino *et al* 2018). Numerous earlier investigations have demonstrated that the stiffness of cancer cells is less compared to the equivalent normal cells. For instance, normal mammary epithelial cells soften considerably following transformation (Guck *et al* 2005, Suresh 2007). The mechanical stiffness of cancer cells is extremely heterogeneous (Plodinec *et al* 2012) and shows a significant relationship with their malignant capacity (Swaminathan *et al* 2011). The softness of cells is a distinctive mechanical characteristic of highly cancerous and metastasizing cancer cells (Lv *et al* 2021). The low stiffness of cancer cells can contribute to their transmigration through the endothelial vessel lining, such as the extravasation step in metastasis (Chen *et al* 2016). The softening of the cancer cells causes their self-renewal ability to be enhanced (Ohata *et al* 2012, Castro *et al* 2013). While the link between cell stiffness and malignancy of cancer cells has been established, the connection between the mechanical characteristics of cells and the predisposition for metastasis is still obscure.

The mechanical characteristics of a cell are primarily defined by its cytoskeleton and the proteins associated with it, which comprise three main components: actin filaments, intermediate filaments and microtubules (Fletcher and Mullins 2010). Most prominently, actin filaments are considered to be a major determinant of cell stiffness (Gavara and Chadwick 2016). A tight connection between actin filaments and cell stiffness was verified by the application of substances with a disturbing effect on cell functions, like cytochalasin D (Cyto D) (Wakatsuki *et al* 2001). Moreover, it is common knowledge that the majority of adherent cells are prestressed and generate contractile traction forces on substrates to perceive their characteristics (Wang *et al* 2002). There are investigations that demonstrate that the contractility of the cells has a direct influence on the cell mechanics (Sen and Kumar 2009, Vichare *et al* 2014). Myosin II attaches to actin filaments and produces cellular contractility. Its activity is controlled through myosin light chain kinase (MLCK) and ROCK (Matsumura 2005). Defective myosin bundles in cancer cells result in increased deformability relative to control cells (Vicente-Manzanares *et al* 2009, Hindman *et al* 2015). Thus, the stiffness of cells can be tuned in an effective manner through modulation of the cytoskeleton and contractility. Nevertheless, there are hardly any investigations on the contribution of the actin cytoskeleton and cellular contractility toward organotropism in cancer. Cytoskeleton and stiffness of various breast cancer cell subpopulations with distinct metastatic tropism were assessed using immunofluorescence staining and AFM (Tang *et al* 2021). To examine the contribution of the cytoskeleton in organotropism, the cytoskeleton was either stabilized or destroyed in breast cancer cells exhibiting brain or bone tropism, respectively. Gene expression of genes specifically associated with brain/bone metastases has been analyzed. The mechanosensitivity of these modified cells was assessed on soft substrates that imitated the stiffness of the brain tissue, along with cellular spreading and proliferation (Tang *et al* 2021).

7.7.1 The biophysical features of breast cancer cell subpopulations reflect their preference for metastasis

Cell mechanics are associated with diverse cellular behaviors (Pegoraro *et al* 2017). Nevertheless, the connection between the mechanical characteristics of cancer cells and organotropism is far from elucidated. To answer this question, the stiffness of parental MDA-MB-231 cells and derivatives was determined with a focus on metastasis in bone (MDA-BoM-1833, abbreviated 231-BoM), lung (MDA231-LM2–4175, short 231-LM) and brain (MDA231-BrM2–831, short 231-BrM) using AFM (Kang *et al* 2003, Minn *et al* 2005, Bos *et al* 2009). The data revealed that 231-BrM cells displayed lower cell stiffness compared to 231 parental cells, whereas 231-LM and 231-BoM cells possessed higher stiffness. Moreover, 231-BoM cells manifested the highest cell stiffness within these populations. These findings provide evidence that the stiffness of 231-BrM, 231-LM, and 231-BoM cells displays a progressive increase that coincides with the rising stiffness of their favorite metastatic organs. Actin filament cytoskeleton is a critical determinant of cell stiffness (Blanchoin *et al* 2014, Fischer *et al* 2017, 2020). Subsequently, F-actin was assessed using FITC-phalloidin immunofluorescence labelling in breast cancer cells with varying organotropism. In line with the cell stiffness results, 231-BrM cells showed the lowest F-actin level, whereas 231-BoM cells exhibited the highest F-actin level. F-actin levels were also significantly elevated in an increasing manner from 231-BrM, 231-LM, toward 231-BoM cells. All these data indicate that breast cancer cell subsets with distinct metastatic potential display different biophysical characteristics and that the cytoskeleton and softness could mirror the organotropism of breast cancer cells.

7.7.2 Cell cytoskeleton impacts organotropism-based gene expression profiles

It has been established that there is a connection between the mechanical characteristics of cells and metastatic organotropism. The contribution of biophysical characteristics of cancer cells to metastatic tropism is, nevertheless, still elusive. The hypothesis was put forward that the cytoskeleton could affect the organotropism of breast cancer cells. This idea was explored by investigating the impact of the cytoskeleton on the gene expression pattern associated with brain and bone metastases. The F-actin inhibitor Cyto D, the ROCK inhibitor Y27632 or the myosin II inhibitor blebbistatin were employed to disturb the cytoskeleton of 231-BoM cells. In contrast, the F-actin polymerization activator jasplakinolide (JAS) and the Rho activator narciclasin (NARCI) were utilized to strengthen the cytoskeleton of 231-BrM cells. The expressions of the aforementioned gene signatures for brain and bone metastases were analyzed (Kang *et al* 2003, Minn *et al* 2005, Bos *et al* 2009, Wu *et al* 2015, Pardo-Pastor *et al* 2018). Cytoskeletal perturbation of 231-BoM cells with Cyto D or Y27632 resulted in minimal impact on genes associated with bone metastases, with limited exceptions including ADAMTS1, OPN, and PTHrP. Conversely, most genes associated with brain metastases showed significant increases in expression in 231 BoM cells in reply to these pharmacological interventions, including COX2, ANGPTL4, SERPIN B2, LTBP1, PIEZO2, EREG, HBEGF, and ITGB3. Blebbistatin, by contrast, had no impact on the expression of genes related to bone and brain metastases. Treatment of 231-BrM cells with Jas had no apparent impact on the expression of genes linked to bone and brain metastases. In comparison, the Narci treatment markedly increased five of nine genes linked to bone metastases, including CXCR4, FGF5, ADAMTS1, FST, and PTHrP, and showed minimal impact on the expression of genes linked to brain metastases, aside from the unanticipated increases in COX2 and Serpin B2. After the drug treatment, the F-actin concentration was determined using FITC-phalloidin immunofluorescence staining to evaluate the efficacy of the pharmacological interference. A significant reduction in F-actin was seen in 231 BoM cells when treated with Cyto D. Y27632 and blebbistatin failed to reduce F-actin levels in 231 BoM cells and only had a minor impact on organotropism-linked gene expression, which could be a result of the low concentrations of these compounds. Treatment with both Jas and Narci increased F-actin intensity in 231 BrM cells. These data suggest that compound-based cytoskeletal perturbation/stabilization of breast cancer cell subpopulations displaying bone/brain tropism impacts the expression of organotropism-linked genes.

7.7.3 Cell spreading and proliferation of breast cancer subpopulations of different metastatic tropism react to substrate stiffness via the cytoskeleton

These data demonstrate that the cytoskeleton influences the expression of gene profiles related to organotropism. As a next step, the influence of the cytoskeleton on the reaction of breast cancer subpopulations that vary in their organotropism toward substrate stiffness was investigated. In particular, 231-BrM, 231-LM, and 231-BoM cells each metastasize specifically to the brain of stiffness range between 0.1 and 1.0 kPa, lung of a stiffness range between 0.44 and 7.5 kPa, and bone of a stiffness range of 25–40 kPa (Tse and Engler 2010, Booth *et al* 2012). For this purpose, the cytoskeleton of 231-BrM cells was strengthened using Narci, whereas the cytoskeleton of 231-BoM cells was destroyed using Cyto D. These treated cancer cells are then grown on tissue culture plates (TCPs) or polyacrylamide hydrogels of 0.6 kPa stiffness that resemble the mechanical stiffness of brain tissue. The spread and shape of the cells are positively related to cell proliferation and migration and can therefore be utilized as effective markers to identify whether the microenvironment is conducive to cancer cells (Chen *et al* 1997, Mogilner and Keren 2009). The findings demonstrate that treatment of 231-BrM cells with Narci resulted in an increase/decrease in the roundness-to-aspect ratio on soft substrates, even though the spreading area was unaltered. Conversely, perturbation of the cytoskeleton of 231-BoM cells resulted in increased cell spreading area, improved aspect ratio, but not enhanced roundness on TCPs and soft substrates. Cell proliferation is critical for the evolution of disseminated cancer cells into secondary tumors when they colonize remote organs. It has been seen that 25 nM Narci treatment reduced the rate of cell proliferation on soft substrates, whereas there is no alteration on TCP. Treatment with 50 nM Narci enhanced cell proliferation on TCP, whereas this enhancement was abrogated on soft substrates. Cytoskeletal modulation of 231-BoM cells with Cyto D failed to have any apparent impact on their proliferation on TCPs. Conversely, disruption of the cytoskeleton of 231-BoM cells enhanced their proliferation when cultured on soft substrates. Collectively, these results indicate that the cytoskeleton of cells could be a potential novel modulator of the mechanical favoring of breast cancer cells that display pronounced organotropism on soft substrates.

7.7.4 Knockdown of mDia1 increases the features of brain metastasis in 231-BoM cells

mDia1, which acts downstream of RhoA, is an important driver of F-actin polymerization, which in turn impacts cell mechanics (Hotulainen and Lappalainen 2006, Kostic and Sheetz 2006). Earlier research demonstrates that the Piezo2-RhoA-mDia signaling axis is required for actin cytoskeleton homeostatic control and force generation in 231-BrM cells (Pardo-Pastor *et al* 2018). For further investigation of the involvement of the cytoskeleton and mechanics, 231 BoM cells were transfected with mDia1 siRNAs and their gene expression and proliferation were determined. Knocking down mDia1 in 231 BoM cells altered the expression of bone metastasis genes non-uniformly, with three of nine bone metastasis genes induced, namely IMPG1, OPN, and PTHrP and one, the FST gene, being repressed. Conversely, knockdown of mDia1 impressively inhibited seven of nine brain metastasis genes in 231 BoM cells, namely ANGPTL4, LTBP1, PIEZO2, EREG, ITGAV and ITGB3. Moreover, low dose of 1 nM, inhibition of mDia1 enhanced cell proliferation on soft substrates, which was not observed on TCPs. The data indicate that mDia1 silencing increases the expression of brain metastasis genes and cellular proliferation on soft substrates in 231-BoM cells, suggesting an enhancement of their brain metastatic potential.

Metastasis of many cancers follows an organ-specific distribution pattern. This phenomenon was initially elucidated by the ‘seed and soil’ theory proposed by Steven Paget (Paget 1889), where ‘seed’ relates to cancer cells with the capacity to metastasize, while ‘soil’ relates to the organs with the conducive microenvironment. Once the seeds and soil are matched, organ-specific metastases can arise. Both cancer cells and the TME predetermine the futility of the metastatic spread (Wei and Siegal 2017, Gao *et al* 2019). In cancer cells, the most important factors are the gene signature, stem cell characteristics, the quiescent state of the tumor and the factors secreted by the tumor, which crucially impact the formation of organ-specific metastases (Chen *et al* 2018b). Apart from these biochemical mechanisms, mechanical signals have been found to be critical in cancer metastasis (Mierke 2014, 2018b, 2018a, 2019a, 2021b, 2023), incorporating the mechanics of cancer cells. It is generally acknowledged that the mechanical characteristics of cancer cells are related to the invasive capacity of pancreatic, ovarian and breast cancers (Swaminathan *et al* 2011, Xu *et al* 2012, Nguyen *et al* 2016a). Highly invasive and metastatic cancer cells are significantly more deformable compared to their less invasive peers, which may help them to pass through narrow spaces during the metastasis journey. Previous work demonstrates that decreased cytoskeleton and thus cell mechanics improve survival of CTCs subjected to fluid shear stress in the vasculature and enhances their chemoresistance (Jin *et al* 2018, Xin *et al* 2019). The contribution of cancer cell cytoskeleton and mechanics to organotropism has been established. Subpopulations of breast cancer cells with distinct organotropism display divergent F-actin and cell stiffness, which parallels the stiffness of their target organs. Significantly, perturbation of the cytoskeleton of bone-tropic breast cancer cell subpopulations results in upregulation of genes linked to brain metastases and enhances their spread and proliferation on soft substrates. Conversely, stabilization of the cytoskeleton of subpopulations of brain-invading breast cancer cells upregulates genes associated with bone metastasis and represses genes associated with proliferation on soft substrates. These results indicate that the cytoskeleton does not just relate to organotropism but also affects it, and that a cytoskeleton that conforms to organ mechanics can improve the capacity of cancer cells to match the metastatic organ, which may reveal a new function of the cytoskeleton in cancer metastasis. Notwithstanding, the data demonstrate that Cyto D treatment could upregulate multiple genes related to bone metastasis, among which ADAMTS1, OPN and PTHrP, whereas mDia1 silencing enhanced the expression of OPN and PTHrP. Alternatively, the treatment of Narci and Jas could increase the expression of the brain metastasis-linked gene SERPIN B2 in 231-BrM cells. The unanticipated alterations in gene expression following perturbation of the cytoskeleton could be because pharmacological treatment and genetic alteration not only result in a modification of the cytoskeleton but also cause a cascade of alterations in downstream signaling and cytoskeletal rearrangement.

It is important to note that the cytoskeleton and stiffness of cells can be affected by several extracellular cues (Zanotelli *et al* 2017). Extracellular vesicles (EVs) released from cancer-associated fibroblasts can affect the mechanical phenotype of cancer cells (Hoshino *et al* 2015). Mechanical microenvironments can also deliver a variety of signals that allow a cell to modulate its cytoskeleton and alter its mechanical state (Zanotelli *et al* 2017). Several lines of evidence have indicated that cells adjust their stiffness by remodeling their cytoskeleton to conform to the compliance of their substrates (Solon *et al* 2007, Rianna and Radmacher 2017b). In line with this idea, different levels of stiffness in cancer cells within the identical tumor tissue may be attributed to the heterogeneity of the mechanical microenvironment of the tumor (Plodinec *et al* 2012). The MDA-MB-231 derivatives were obtained by harvesting metastatic cells from the secondary tumors located in the brain, lung, and bone, with diverse tissue mechanics. Therefore, the question emerges as to whether the exceptional cytoskeleton and stiffness of these derivatives are a result of their accommodation to the specialized mechanical microenvironment of the organ of destination or of their intrinsic characteristics uncoupled from extracellular determinants. In addition, after cell extraction, these metastatic cells were grown on TCPs for a prolonged period of time, which may affect the cytoskeleton and stiffness of the respective primary metastatic cells. These important questions must be thoroughly examined in the future. The cytoskeleton is the key structure for transmitting external mechanical signals from the cell membrane into the nucleus, which then governs a variety of cell functions (Stevenson *et al* 2012). Actin polymerization and myosin contractility work synergistically to govern cell shape, motility, division, and secretion of proteins (Carpenter 2000). Previous findings show that manipulation of the cell cytoskeleton and contractility affects the expression of genes associated with survival and resistance to drugs, and governs apoptosis of CTCs in shear flow and their ability to resist chemotherapy (Xin *et al* 2019). When 231-BrM cells were exposed to Narci, the circularity/aspect ratio was enhanced/reduced on soft substrates, despite the spreading area being unperturbed. Conversely, disruption of the cytoskeleton of 231-BoM cells led to an enlargement of the cell spreading area, enhanced aspect ratio, but not enhanced circularity on TCPs and soft substrates. Cytoskeleton-driven mechanical cues have been implicated in cell proliferation (Provenzano and Keely 2011). Targeting the actin cytoskeleton might have different implications for cancer cells on substrates with varying stiffness, as treatment with 50 nM Narci augmented cell proliferation on TCP, while this effect was abrogated on soft substrates. Cyto D increased the proliferation of 231 BoM cells on soft substrates, although not on TCP. It is noteworthy that while brain tissue is mechanically heterogeneous, the stiffness of brain areas that can be invaded by cancer cells remains below 1 kPa, encompassing cortex, cerebellum, and corpus callosum (Koser *et al* 2015, Moeendarbary *et al* 2017, Eberle *et al* 2018). Hence, it is useful to utilize a 0.6 kPa substrate to mimic the mechanical micro-environment of brain tissue and examine the effect on the destiny of cancer cells in the early phase of brain metastasis.

It is established that the cytoskeleton and actomyosin-mediated contractility add to the mechanical stiffness of the cell. Cytoskeletal perturbation or myosin activity inhibition reduces cell stiffness, whereas cytoskeletal strengthening or myosin activity activation enhances mechanical stiffness (Sen and Kumar 2009, Vichare *et al* 2014, Guo *et al* 2020, Liu *et al* 2021). To dissect the contribution of the cytoskeleton to organotropism, several pharmacological inhibitors that modulate actin cytoskeleton dynamics and cellular contractility are employed. Cyto D blocks actin polymerization by blocking the rapidly growing barbed ends of actin filaments. Jas tethers actin filaments and inhibits their break down. Narci induces actin stress fiber formation through activation of the small GTPase RhoA. Blebbistatin inhibits the activity of myosin II directly to decrease cytoskeletal contractility, while Y27632 blocks the myosin II activator ROCK. Hence, these compounds affect the cytoskeleton and cell stiffness via distinct mechanisms. These findings uniformly demonstrate that treatment with Cyto D, but not the low dose of Blebbistatin and Y-27632, significantly decreases F-actin levels, whereas Jas and Narci cause an elevation in F-actin levels. Cyto D appears to significantly modify gene expression, while blebbistatin and Y-27632 exhibit little impact on the gene expression signature in 231-BoM cells. This phenomenon may be attributed to the distinct mechanisms of action of these pharmacological interventions and the low dosage of the medications utilized, since the administered dose of Cyto D, but not of Y-27632/Blebbistatin, resulted in a significant decrease in F-actin abundance. In addition, earlier studies indicate that the cytoskeleton impacts the human pluripotent stem cell differentiation into pancreatic β-cells. However, it was noted that not all cytoskeleton-targeting agents can effectively trigger endocrine differentiation (Hogrebe *et al* 2020). This disparity is likely due to the differing mechanistic actions of these agents on the cytoskeleton (Peng *et al* 2011). Thus, this phenomenon could not be eliminated so far and may account for differential effects of pharmacological interventions on gene expression.

The reaction of cancer cells to mechanical stimuli is critical for their growth in metastatic organs and relies on both intracellular architecture and mechanical characteristics. In addition, the cytoskeleton and stiffness of a cell have to match the mechanics of the microenvironment so that the cell can properly perceive and react to the ambient mechanical stimuli (Discher *et al* 2005, 2009, Wu *et al* 2018), indicating that soft/stiff cancer cells are probably able to survive and grow best in a soft/stiff matrix niche. This concept is reinforced by the observation that cloning and propagating breast cancer cells that tend to metastasize to the bone and lungs can greatly enhance cell proliferation and migration on substrates of matching stiffness (Kostic *et al* 2009). Tang and coworkers explored the functional role of cell cytoskeleton in the reaction of cancer cells with varying metastatic tropism to the soft matrix (Tang *et al* 2021). Stabilization of the cytoskeleton of 231-BrM cells reduces cell spread and proliferation on soft substrates, whereas perturbation of the cytoskeleton of 231-BoM cells increases spread and proliferation. These findings indicate that deregulation of the cytoskeleton can give cells advantages when propagating in soft brain tissue and should be investigated more thoroughly. While gene expression and the capacity for substrate stiffness mechanoadaptation by cancer cells may indicate organotropism to some degree, direct evidence is warranted in the future to prove the impact of cell cytoskeleton and stiffness on organotropism, particularly the use of animal models of organotropism after intracardiac injection of cancer cells with altered cytoskeleton and mechanics.

7.8 The mechanical characteristic of the route controls cancer metastasis

The critical point in the advancement of cancer is metastasis, when cells detach from the primary tumor and migrate through the body to settle and colonize in other targeted tissues. Unlike cells grown in a lab in an aqueous solution, in reality, cancer cells face resistance as they travel through bodily fluids. Contrary to intuition, previous work has shown that cells increase their speed when migrating through thicker solutions, by an increased ruffling of the membrane, which senses mechanical stimuli like viscosity (Pittman *et al* 2022). In the experiments, by contrast, fluids were employed that were much more viscous than those occurring in the body. To metastasize to distant organs, cancer cells have to detach from the primary tumor and gain access to the blood circulation. A recent study indicates that this spread usually takes place while the host is sleeping (Diamantopoulou *et al* 2022). The authors observed that CTCs, that are cells that have detached from the primary tumor and migrated to distant targeted organs, were detected in blood samples taken during quiescence (sleep period) in breast cancer patients and mice at significantly higher numbers than in blood samples taken during active period of breast cancer patients. These new very strong data add further weight to the role of circadian rhythms in tumor spread. In a 2020 review article, it has been speculated that CTCs could be influenced by biological cycles (Cortés-Hernández *et al* 2020). At that time, there were no in-depth investigations comparing CTCs that were produced during the day or at night. It has since been confirmed by another study (Houshyari and Taghizadeh-Hesary 2023).

Moreover, unexpectedly there is a high variation in CTC numbers that rely on the time at which blood samples of cancer patients were obtained (Diamantopoulou *et al* 2022). These daily oscillations in CTC counts have been attributed to the circadian rhythm (Zhu *et al* 2021). To explore this disparity, 30 breast cancer patients were sampled in the hospital at 4:00 a.m. while they were resting and again at 10:00 a.m. that same day, and the CTC frequency of the samples was assessed. A striking difference between day and night collections was observed: 78% of the CTCs detected in all patients were found in the blood samples collected at 4:00 a.m (Diamantopoulou *et al* 2022). Repeating the experiment in four distinct mouse models of breast cancer, blood samples were collected from the animals while they were in the resting phase (during the day) and in the active phase (at night). The result was consistent with findings in humans: 87%–99% of CTCs were found while the mice were in the quiescent phase. In addition, comparing the metastatic capacity of cells sampled at each time point after injection into tumor-free mice, revealed that cancer cells liberated during quiescence had a dramatically enhanced potential to colonize lung tissue (Diamantopoulou *et al* 2022). The half-life of CTCs in humans is very brief, lasting only a few hours or even less. All CTCs detected in the bloodstream must thus have been released recently. Gaining insights into when these circulating cells arise and when they are the most aggressive is essential because this knowledge will aid in the creation of therapeutics that specifically target and destroy this specific subset of cells. The question as to why CTCs shed during the night are better able to metastasize has not yet been fully resolved. Various analyses were carried out to explore this question. The gene expression patterns of CTCs in patients and mice were examined and it was found that those CTCs released during dormancy had higher expression levels of genes related to cell division and mitosis compared to CTCs released during the active phase, which in contrast exhibit higher expression levels of genes related to translation. These results suggest an increased ability to proliferate in the cells of the sleep-shed phase. Glucocorticoid, androgen, and insulin receptors have been found to be highly abundant in CTCs. Since the ligands for these receptors are controlled in a circadian rhythm, this finding implies that CTCs might react to the daily fluctuations of hormones. When mice were treated with dexamethasone or with a continuous liberation of testosterone, a glucocorticoid or an androgen ligand, respectively, throughout their sleep, the quantity of CTCs in their blood was considerably decreased during the resting phase. Insulin treatment during sleep reversed the proliferation cycle, thereby significantly decreasing the number of CTCs identified when the mice were at the resting state and increasing the number of CTCs when the mice became active. It is possible to speculate that there is an immune element in addition to the part played by hormones. Nevertheless, it is difficult to separate the influence of the various factors, and it is not easy to decipher the mechanistic basis of these observations. It is likely that oscillatory expression of circadian clock genes can be identified in the CTCs. In certain breast cancer subtypes, there exist oscillatory expressed circadian clock genes that exhibit rhythmicity at the cellular level. The results on breast cancer should be tested on other cancer types whether there exists such circadian dependency of metastasis formation and whether circadian genes can be identified.

It is well known that as tumors extend in size and crowd together in a confined space, the subsequent compression causes them to metastasize more easily and display enhanced survivability, invasiveness, and aggressiveness. *In vitro* and in mice, it has been shown that a distinct mode of compression, referred to as confined migration, which metastatic cancer cells undergo as they squeeze through narrowed blood vessels, can stimulate various transformations that assist the cells in surviving (Fanfone *et al* 2022). It is a rather unexpected finding, because in non-cancerous cells, compression is typically linked to cell death and apoptosis. These findings show that the direct impact of mechanical stimuli on cancer cells must be taken into account not just at the primary tumor, but also at all stages of metastasis for the purpose of developing better therapeutic approaches. Normally, metastasizing is dangerous for cancer cells. While metastatic cancers are significantly more challenging to target than tumors that remain in place, they also turn more susceptible to a cell death mechanism referred to as anoikis, which is a type of apoptosis that can arise during transmigration and is elicited when integrins rupture and disengage from the ECM. In healthy cells, anoikis serves to stop cells from proliferating in the incorrect location. The mechanical harassment caused by confined migration can exert a highly mutagenic force on cancer cells. In fact, previous research has demonstrated that confined migration modifies gene expression and induces alterations in signal transduction pathways. Occasionally, cancer cells have difficulty passing through a small hole, the nucleus breaks open, and at a later stage the nucleus rebuilds itself. This leads to DNA defects and, consequently, to numerous mutations. It has been hypothesized that it is not the external mechanical compression of the original tumor but rather the process of confined migration of metastasizing cells that causes alterations that prevent anoikis and increase the cell's probability of survival (Fanfone *et al* 2022). The confined space that a migrating cancer cell may experience in the body is simulated by constraining human breast cancer cells to migrate through channels as narrow as 3 μm and subsequently culturing them. The cells were examined for apoptosis biomarkers, and it was found that none of them had been switched on, which is an indication that squeezing though the channel did not affect the viability or proliferative capacity of the cells. Moreover, the capacity for colony formation was examined under conditions where cancer cells are anchored to nothing and left to succumb to the mechanism of cell death. It was observed that the cells that had passed through narrow channels performed better compared to control cells in the first few days following the migration ordeal. This indicates that the compressed cells were actually better able to withstand anoikis, which is consistent with a study on pre-metastatic conditions that bypass cell death (Conod *et al* 2022), although the effect diminished over time. At the same time, cells from compressed tumors that failed to pass through the channels exhibited no enhanced resistance to anoikis. It is possible that the phenotypic alterations caused by compression of the primary tumor were pushed further through confined migration, although this is not a question the experiment was set up to address. The pores are a convenient but somewhat imperfect and oversimplified model, since the actual confined migration is a longer voyage that requires traversing a fiber network that can be forced out of the way. The question now arises as to whether cancer cells constrained by a hydrogel matrix would produce the identical results.

A study indicates that cancer cells sense and react to physiological levels of viscosity (Bera *et al* 2022). In viscous environments, cells undergo architectural rearrangements that enable them to resist external forces and migrate more efficaciously. In fact, the cells seem to have a kind of viscosity memory and keep moving fast when they are reintroduced into a watery medium. These results are promising, and viscosity extends the list of mechanical stimuli perceived by cells and driving their behavioral output. It was found that cancer cells move in confined spaces by taking in water at the front of the cell and squirting it out at the rear, enabling them to move like octopuses through confined spaces (Leijnse *et al* 2022). To investigate how cells move in a viscous medium in the new study, the researchers used a mathematical model that had previously been designed to forecast cellular movements and adapted it to a higher viscosity (Zhang *et al* 2022). The revised model proposed that external resistance causes cells to rearrange actin at the leading edge of the cell. Thereafter, a transport protein referred to as NHE1, which is attracted by actin-binding proteins, accumulates on the membrane and facilitates the uptake of water (Zhang *et al* 2022). The cell swells, causing its membrane to tighten and open TRPV4, an ion channel that reacts to membrane tension. Calcium ions flow into the cell and cause it to contract, thereby exerting a force that surmounts the high viscosity and pushes the cell forward. Super-resolution microscopy of human breast cancer cells verified that actin does in fact assemble at the leading edge of cells that move through thicker substrates. Through systematic examination of every single component, the sequence of events in the signaling path was verified. For example, rapid migration was re-established in NHE1-deficient cells through activation of TRPV4, demonstrating that the TRPV4 channel works subsequent to water transportation. The finding was surprising because it questions the conventional idea that ion channels are the initial responders to external stimuli. Unexpectedly, cells were found to have a memory of whether they were exposed to viscous conditions: human breast cancer cells that had been grown in high-viscosity media for six days and then transferred to aqueous conditions maintained their fast movement compared to cells that had remained in the less viscous solution the entirety of the time. Likewise, breast cancer cells cultured in a sticky medium and subsequently injected into mice metastasized more strongly than those that had been primed in a low-viscosity solution. Cells primed with a high-viscosity medium also migrated faster in zebrafish and rapidly migrated out of the bloodstream in chicken embryos. The establishment of this memory seems to rely on TRPV4. Cells that were cultured for six days in a viscous medium but did not express the ion channel, produced fewer tumor colonies in mice compared to cells with functional channels treated in the same manner. This indicates that the pre-treatment without the channel did not influence the velocity within the animals. It has been hypothesized that the presence of stiff and soft cells in tumors provide tumor growth and metastatic spread at the same time. The results suggest that TRPV4 is a potential drug target for inhibiting cancer metastasis. Animals lacking TRPV4 channels have been found to function normally, indicating that healthy cells may not need to migrate quickly through viscous fluids. In this case, a therapy aimed at TRPV4 would not be expected to produce significant side effects. In fact, the fluid containing cancer cells has a higher viscosity than that surrounding healthy cells because tumors tend to break down nearby tissue and obstruct lymph vessels. However, the question of what role, if any, this signaling pathway serves in normal cells remains to be elucidated. The results should be treated with caution, because cells merely moving more quickly in thicker solutions does not imply that they are more prone to secondary cancers. Tumor metastasis is a highly complex event comprising a long chain of steps, several of which are unrelated to migration. Regardless of whether they can be directly translated into medical practice or not, the results could result in important modifications to cell-based cancer research. The overwhelming bulk of research conducted in cell cultures utilizes media with a viscosity close to that of water, which is a weakness. Utilizing media with a similar viscosity to body fluids could aid in identifying metastasis-blocking druggable targets that would otherwise be difficult to identify. The goal in cancer research is to mimic the extracellular environment as precisely as achievable, and medium viscosity could be one more factor to address. Stiffness-triggered phenotypes facilitate metastasis in softer tissues like the bone marrow (Watson *et al* 2021). Is this in contrast to soft cells that metastasize more potently? Whereas the immediate and transient reaction of breast cancer cells to pathological stiffness in their natural microenvironment is well established, it stays elusive how stiffness-triggered phenotypes are sustained over a prolonged period of time following cancer cell spread *in vivo*. It has been established that fibrotic matrix stiffness drives certain metastatic phenotypes in cancer cells that persist even after the transition to softer microenvironments, like the bone marrow (Watson *et al* 2021). Using a differential gene expression analysis of breast cancer cells that react to stiffness, a multigene-based score of mechanical conditioning was examined and found to be related to bone metastases in breast cancer patients. Sustaining the mechanical stimulation is orchestrated by RUNX2, an osteogenic transcription factor that is known to promote bone metastasis and to act as a mitotic landmarker that maintains chromatin openness at target loci. Genetic and functional approaches revealed that the mechanical conditioning can be sustained, suppressed or prolonged, with commensurate alterations in the bone metastasis capacity.

7.9 Conclusions and future directions

Cancer represents a devastating disease state that is complex to address. For decades, scientists and researchers have tirelessly explored new approaches to developing treatments to combat this group of diseases. To date, most current therapeutic approaches have primarily targeted a limited set of systems that govern cell cycle progression, replication, and evasion of the immune system. To more effectively target cancer, it is necessary to simultaneously attack multiple systems that favor cancer progression. This chapter has discussed the effects of the external mechanical environment on cancer cells, and how external mechanical cues can cause alterations in gene expression that promote tumor progression and favor invasion and metastasis. These challenges are mechanical, forcing the cancer cells to evolve systems that enable them to adapt and survive. The interaction between cancer cells and their mechanical environment functions similarly to a feedback system, which gives cancer the immense capacity to adapt to various environments at different levels of cancer progression. At the core of this feedback mechanism is the mechanical program, which consists of F-actin and mechanosensitive proteins, such as NMII, ACTN, FLN and probably others. This program gives cancer cells the capacity to perceive the countless mechanical signals from their constantly shifting environment, to react to them and to adapt to those signals. Because of its importance in promoting cancer adaptability and progression, this mechanical program represents an optimal target for the design of treatment methodologies. As stated earlier, encouraging results for anti-cancer activity have been achieved with several promising compounds that act on various key components of this system. Nevertheless, there are still a number of challenges to be met. For one thing, some elements of this system have both cancer-promoting and cancer-inhibiting effects. In addition, the mechanical system is intricately linked to a plethora of other cellular processes, such as signal transduction, metabolism, and regulation of RNA. Thus, further investigation is needed to ascertain whether potential therapeutic agents that target this system can do so in an efficient manner without perturbing the global cellular equilibrium. Nonetheless, the mechanical program is an appealing target for cancer therapy, particularly for future combination treatments. It is timely that biophysical mechanisms of cancer are being recognized by the scientific community, as both extrinsic and intrinsic mechanical cues appear to be important in cancer initiation and progression. In view of the deadly character of cancer metastases, it is of crucial importance to focus research endeavors not only on traditional molecular biology, but on all possible fronts, which includes apparently simpler facets of biology like the mechanical characteristics of cancer cells. The biophysical angle is not only crucial for a complete mechanistic understanding of the process by which metastases arise but could also result in the discovery of new therapeutic strategies. Although cancer cells are currently assumed to be softer, which is almost exclusively based on *ex vivo* measurements, many factors must be addressed to fully appreciate how stiffness can influence cancer cell metastasis. For example, the subtleties of the various timescales in which the mechanical characteristics of cancer cells can be expressed are seldom emphasized (Ekpenyong *et al* 2012).

An elastic cell that can deform immediately and reversibly upon colliding with an obstacle has completely distinct mechanical characteristics to a cell that gradually forces its way through a narrowed blood vessel for extended periods of time. Thus, the relative impact of elasticity and viscosity on the individual stages of the metastatic cascade requires further investigation. An additional facet that is still insufficiently appreciated is the dynamics of changes in cancer cell stiffness. Whereas it is more and more evident that cancer cells are capable of sensing their environment and adjusting their stiffness in response, the optimal mechanical characteristics for each step of the metastatic seeding journey and the means by which cancer cells achieve these ideal conditions are yet to be determined for many steps of the metastatic voyage. Interestingly, metastatic cancer cells appear to alter their mechanical properties in reaction to passive and active mechanical cues to a much larger extent compared to normal cells (Baker *et al* 2010, Wullkopf *et al* 2018).

Future efforts will involve a full characterization of the contribution of cancer cell deformability to each step of metastasis using the ever more advanced instruments available individually and in combination. Optical tweezers were applied successfully *in vivo* (Paul *et al* 2019), and a recent technical progress report also gave hope in this regard by investigating the mechanical states of cancer cells or healthy peers *in vivo* (Wu *et al* 2020). Employing particle tracking microrheology in conjunction with intravital imaging, the research reported that cancer cells exhibited reduced compliance while forming tumor-like structures. The evolution of Brillouin microscopy will likely become very useful in the future, as it provides the ability to examine mechanical characteristics *in vivo* (Elsayad *et al* 2019) in a completely particle-free manner. Brillouin microscopy has previously been applied to characterize the mechanical characteristics of 3D ovarian cancer nodules (Conrad *et al* 2019). Quantitative phase imaging also has the capability of being an all-optical tool for characterizing the mechanical features of cells (Eldridge *et al* 2019), and potentially one day *in vivo*. The ultimate goal would be to identify the mechanisms by which cancer cells perceive not only mechanical stress but also chemical and biochemical signals, such as cytokines, pH and temperature in their environment, and the signaling cascades that allow them to adjust their mechanical characteristics as needed. In the future, emphasis in the field should be placed on measuring and reconciling mechanobiological phenotypes of cancer cells throughout the entire metastatic cascade, both in pertinent animal models and in patient specimens. Accurate quantification of the mechanical stress to which cancer cells are exposed during these different steps, utilizing innovative and revealing biophysical and imaging techniques, will be crucial to achieving this aim. In summary, the most important mechanical signature of a victorious metastatic cell may ultimately prove to be the capacity to dynamically adopt the mechanical characteristics needed to traverse each step of the metastatic voyage, regardless of what those characteristics are. As soon as the molecular mechanisms involved have been clearly defined, new therapies that focus on these mechanisms could be effective in the fight against metastases.

In a final note, while it is important to state that gaining an understanding of how mechanical alterations in cancer cells may dictate cancer advancement and reaction to chemotherapy agents, it is important to state some prospective avenues for extension, given that sometimes only one cell line was utilized, this could be extended to additional breast cancer cell lines of various breast cancer types, thereby broadening the comprehension of different cancer behaviors and reactions. In addition, although there is great promise for label-free CTC detection and sensor development, there are still remaining hurdles, such as the genetic and biomechanical heterogeneity of CTCs. The biomechanical role of tumor cells throughout the various steps of metastasis and chemotherapeutic interventions is underappreciated and more complex than it appears. By analyzing the data, the overly simplistic notion that metastatic tumors are uniformly soft can be questioned. In real life, their mechanical characteristics, including softness and stiffness, vary depending on the stage of invasion. Moreover, it was demonstrated that there may be a link between the cytoskeleton/cell stiffness and organotropism. The stiffness of subpopulations of breast cancer cells with varying metastatic preference reflects the mechanical characteristics of the metastasized organs. The cytoskeleton substantially affects the organotropism-related gene expression signature and the mechanical responsiveness to soft substrates mimicking brain tissue stiffness. These results emphasize the pivotal role of the cytoskeleton in organ-specific metastasis, which not just mirrors but may also influence the preference for metastatic organs.

References and futher reading

Abdelmoez M N, Iida K, Oguchi Y, Nishikii H, Yokokawa R, Kotera H *et al* 2018 SINC-seq: correlation of transient gene expressions between nucleus and cytoplasm reflects single-cell physiology *Genome Biol.* **19** 66

Abidine Y, Constantinescu A, Laurent V M, Sundar Rajan V, Michel R, Laplaud V *et al* 2018 Mechanosensitivity of cancer cells in contact with soft substrates using AFM *Biophys. J.* **114** 1165–75

Abotaleb M, Kubatka P, Caprnda M, Varghese E, Zolakova B, Zubor P *et al* 2018 Chemotherapeutic agents for the treatment of metastatic breast cancer: an update *Biomed. Pharmacother.* **101** 458–77

Acerbi I, Cassereau L, Dean I, Shi Q, Au A, Park C *et al* 2015 Human breast cancer invasion and aggression correlates with ECM stiffening and immune cell infiltration *Integr. Biol.* **7** 1120–34

Aceto N, Bardia A, Miyamoto D T, Donaldson M C, Wittner B S, Spencer J A *et al* 2014 Circulating tumor cell clusters are oligoclonal precursors of breast cancer metastasis *Cell* **158** 1110–22

Aermes C, Hayn A, Fischer T and Mierke C T 2020 Environmentally controlled magnetic nano-tweezer for living cells and extracellular matrices *Sci. Rep.* **10** 13453

Aermes C, Hayn A, Fischer T and Mierke C T 2021 Cell mechanical properties of human breast carcinoma cells depend on temperature *Sci. Rep.* **11** 10771

Aguirre-Ghiso J A 2018 How dormant cancer persists and reawakens *Science* **361** 1314–5

Ahmmed S M, Bithi S S, Pore A A, Mubtasim N, Schuster C, Gollahon L S *et al* 2018 Multi-sample deformability cytometry of cancer cells *APL Bioeng.* **2** 032002

Al Absi A, Wurzer H, Guerin C, Hoffmann C, Moreau F, Mao X *et al* 2018 Actin cytoskeleton remodeling drives breast cancer cell escape from natural killer–mediated cytotoxicity *Cancer Res.* **78** 5631–43

Alexander J and Cukierman E 2016 Stromal dynamic reciprocity in cancer: intricacies of fibroblastic-ECM interactions *Curr. Opin. Cell Biol.* **42** 80–93

Al-Ghadban S, Kaissi S, Homaidan F R, Naim H Y and El-Sabban M E 2016 Cross-talk between intestinal epithelial cells and immune cells in inflammatory bowel disease *Sci. Rep.* **6** 29783

Alibert C, Goud B and Manneville J-B 2017 Are cancer cells really softer than normal cells?: Mechanics of cancer cells *Biol. Cell* **109** 167–89

Allen T A, Gracieux D, Talib M, Tokarz D A, Hensley M T, Cores J *et al* 2017 Angiopellosis as an alternative mechanism of cell extravasation *Stem Cells* **35** 170–80

Al-Mehdi A B, Tozawa K, Fisher A B, Shientag L, Lee A and Muschel R J 2000 Intravascular origin of metastasis from the proliferation of endothelium-attached tumor cells: a new model for metastasis *Nat. Med.* **6** 100–2

Angstadt S, Zhu Q, Jaffee E M, Robinson D N and Anders R A 2022 Pancreatic ductal adenocarcinoma cortical mechanics and clinical implications *Front. Oncol.* **12** 809179

Aseervatham J 2020 Cytoskeletal remodeling in cancer *Biology* **9** 385

Au S H, Storey B D, Moore J C, Tang Q, Chen Y-L, Javaid S *et al* 2016 Clusters of circulating tumor cells traverse capillary-sized vessels *Proc. Natl Acad. Sci. USA* **113** 4947–52

Ayad N M E and Weaver V M 2020 Tension in tumour cells keeps metabolism high *Nature* **578** 517–8

Bai H, Zhu Q, Surcel A, Luo T, Ren Y, Guan B *et al* 2016 Yes-associated protein impacts adherens junction assembly through regulating actin cytoskeleton organization *Am. J. Physiol.-Gastrointest. Liver Physiol.* **311** G396–411

Bajpai A, Li R and Chen W 2021 The cellular mechanobiology of aging: from biology to mechanics *Ann. N.Y. Acad. Sci.* **1491** 3–24

Baker E L, Lu J, Yu D, Bonnecaze R T and Zaman M H 2010 Cancer cell stiffness: integrated roles of three-dimensional matrix stiffness and transforming potential *Biophys. J.* **99** 2048–57

Banerjee D 2016 Connexin's connection in breast cancer growth and progression *Int. J. Cell Biol.* **2016** 1–11

Barai A, Mukherjee A, Das A, Saxena N and Sen S 2021 α-Actinin-4 drives invasiveness by regulating myosin IIB expression and myosin IIA localization *J. Cell Sci.* **134** jcs258581

Beadle C, Assanah M C, Monzo P, Vallee R, Rosenfeld S S and Canoll P 2008 The role of myosin II in glioma invasion of the brain *Mol. Biol. Cell* **19** 3357–68

Bera K, Kiepas A, Godet I, Li Y, Mehta P, Ifemembi B *et al* 2022 Extracellular fluid viscosity enhances cell migration and cancer dissemination *Nature* **611** 365–73

Berdyyeva T K, Woodworth C D and Sokolov I 2005 Human epithelial cells increase their rigidity with ageing *in vitro*: direct measurements *Phys. Med. Biol.* **50** 81–92

Bergert M, Lembo S, Sharma S, Russo L, Milovanović D, Gretarsson K H *et al* 2021 Cell surface mechanics gate embryonic stem cell differentiation *Cell Stem Cell* **28** 209–216.e4

Bhadriraju K and Hansen L K 2002 Extracellular matrix- and cytoskeleton-dependent changes in cell shape and stiffness *Exp. Cell. Res.* **278** 92–100

Blanchoin L, Boujemaa-Paterski R, Sykes C and Plastino J 2014 Actin dynamics, architecture, and mechanics in cell motility *Physiol. Rev.* **94** 235–63

Booth A J, Hadley R, Cornett A M, Dreffs A A, Matthes S A, Tsui J L *et al* 2012 Acellular normal and fibrotic human lung matrices as a culture system for *In Vitro* investigation *Am. J. Respir. Crit. Care Med.* **186** 866–76

Bos P D, Zhang X H-F, Nadal C, Shu W, Gomis R R, Nguyen D X *et al* 2009 Genes that mediate breast cancer metastasis to the brain *Nature* **459** 1005–9

Broders-Bondon F, Nguyen Ho-Bouldoires T H, Fernandez-Sanchez M-E and Farge E 2018 Mechanotransduction in tumor progression: the dark side of the force *J. Cell Biol.* **217** 1571–87

Brown M, Assen F P, Leithner A, Abe J, Schachner H, Asfour G *et al* 2018 Lymph node blood vessels provide exit routes for metastatic tumor cell dissemination in mice *Science* **359** 1408–11

Bryan D S, Stack M, Krysztofiak K, Cichoń U, Thomas D G, Surcel A *et al* 2020 4-Hydroxyacetophenone modulates the actomyosin cytoskeleton to reduce metastasis *Proc. Natl Acad. Sci. USA* **117** 22423–9

Burton K M, Cao H, Chen J, Qiang L, Krueger E W, Johnson K M *et al* 2020 Dynamin 2 interacts with α-actinin 4 to drive tumor cell invasion *Mol. Biol. Cell* **31** 439–51

Cao X, Moeendarbary E, Isermann P, Davidson P M, Wang X, Chen M B *et al* 2016 A chemomechanical model for nuclear morphology and stresses during cell transendothelial migration *Biophys. J.* **111** 1541–52

Cao Y, Ma E, Cestellos-Blanco S, Zhang B, Qiu R, Su Y *et al* 2019 Nontoxic nanopore electroporation for effective intracellular delivery of biological macromolecules *Proc. Natl Acad. Sci. USA* **116** 7899–904

Carlson R H, Gabel C V, Chan S S, Austin R H, Brody J P and Winkelman J W 1997 Self-sorting of white blood cells in a lattice *Phys. Rev. Lett.* **79** 2149–52

Carpenter C L 2000 Actin cytoskeleton and cell signaling *Crit. Care Med.* **28** N94–9

Cascione M, De Matteis V, Toma C C, Pellegrino P, Leporatti S and Rinaldi R 2017 Morphomechanical and structural changes induced by ROCK inhibitor in breast cancer cells *Exp. Cell. Res.* **360** 303–9

Castro D J, Maurer J, Hebbard L and Oshima R G 2013 ROCK1 inhibition promotes the self-renewal of a novel mouse mammary cancer stem cell *Stem Cells* **31** 12–22

Chadwick S R, Wu J-Z and Freeman S A 2021 Solute transport controls membrane tension and organellar volume *Cell. Physiol. Biochem.* **55** 1–24

Chang C-W, Yu J-C, Hsieh Y-H, Yao C-C, Chao J-I, Chen P-M *et al* 2016 MicroRNA-30a increases tight junction protein expression to suppress the epithelial–mesenchymal transition and metastasis by targeting Slug in breast cancer *Oncotarget* **7** 16462–78

Chang Q, Bournazou E, Sansone P, Berishaj M, Gao S P, Daly L *et al* 2013 The IL-6/JAK/Stat3 feed-forward loop drives tumorigenesis and metastasis *Neoplasia* **15** 848–IN45

Chang S-F, Chang C A, Lee D-Y, Lee P-L, Yeh Y-M, Yeh C-R *et al* 2008 Tumor cell cycle arrest induced by shear stress: roles of integrins and Smad *Proc. Natl Acad. Sci. USA* **105** 3927–32

Charrier E E and Janmey P A 2016 Mechanical properties of intermediate filament proteins *Methods Enzymol.* **568** 35–57

Chaudhuri O, Koshy S T, Branco da Cunha C, Shin J-W, Verbeke C S, Allison K H *et al* 2014 Extracellular matrix stiffness and composition jointly regulate the induction of malignant phenotypes in mammary epithelium *Nat. Mater* **13** 970–8

Chen C S, Mrksich M, Huang S, Whitesides G M and Ingber D E 1997 Geometric control of cell life and death *Science* **276** 1425–8

Chen G, Yu C, Tang Z, Liu S, An F, Zhu J *et al* 2018a Metformin suppresses gastric cancer progression through calmodulin-like protein 3 secreted from tumor-associated fibroblasts *Oncol. Rep.* **41** 405–14

Chen J Y, Shahid A, Garcia M P, Penn L S and Xi J 2012 Dissipation monitoring for assessing EGF-induced changes of cell adhesion *Biosens. Bioelectron.* **38** 375–81

Chen J, Zhou W, Jia Q, Chen J, Zhang S, Yao W *et al* 2016 Efficient extravasation of tumor-repopulating cells depends on cell deformability *Sci. Rep.* **6** 19304
Chen W, Hoffmann A D, Liu H and Liu X 2018b Organotropism: new insights into molecular mechanisms of breast cancer metastasis *npj Precis. Oncol.* **2** 4
Chen Y, Lan H, Wu Y, Yang W, Chiou A and Yang M 2018c Epithelial-mesenchymal transition softens head and neck cancer cells to facilitate migration in 3D environments *J. Cell. Mol. Med.* **22** 3837–46
Cheng B, Lin M, Huang G, Li Y, Ji B, Genin G M *et al* 2017 Cellular mechanosensing of the biophysical microenvironment: a review of mathematical models of biophysical regulation of cell responses *Phys. Life Rev.* **22–23** 88–119
Cheng L and Tong Q 2021 Interaction of FLNA and ANXA2 promotes gefitinib resistance by activating the Wnt pathway in non-small-cell lung cancer *Mol. Cell. Biochem.* **476** 3563–75
Cheung K J, Padmanaban V, Silvestri V, Schipper K, Cohen J D, Fairchild A N *et al* 2016 Polyclonal breast cancer metastases arise from collective dissemination of keratin 14-expressing tumor cell clusters *Proc. Natl Acad. Sci. USA* **113** E854–63
Chiang S P H, Cabrera R M and Segall J E 2016 Tumor cell intravasation *Am. J. Physiol. Cell Physiol.* **311** C1–C14
Chin L, Xia Y, Discher D E and Janmey P A 2016 Mechanotransduction in cancer *Curr. Opin. Chem. Eng.* **11** 77–84
Chin V T, Nagrial A M, Chou A, Biankin A V, Gill A J, Timpson P *et al* 2015 Rho-associated kinase signalling and the cancer microenvironment: novel biological implications and therapeutic opportunities *Expert Rev. Mol. Med.* **17** e17
Chowdhury F, Na S, Li D, Poh Y-C, Tanaka T S, Wang F *et al* 2010 Material properties of the cell dictate stress-induced spreading and differentiation in embryonic stem cells *Nat. Mater.* **9** 82–8
Chugh P and Paluch E K 2018 The actin cortex at a glance *J. Cell Sci.* **131** jcs186254
Ciucci S, Ge Y, Durán C, Palladini A, Jiménez-Jiménez V, Martínez-Sánchez L M *et al* 2017 Enlightening discriminative network functional modules behind Principal component analysis separation in differential-omic science studies *Sci. Rep.* **7** 43946
Clark A G and Vignjevic D M 2015 Modes of cancer cell invasion and the role of the microenvironment *Curr. Opin. Cell Biol.* **36** 13–22
Clark A G, Wartlick O, Salbreux G and Paluch E K 2014 Stresses at the cell surface during animal cell morphogenesis *Curr. Biol.* **24** R484–94
Conod A, Silvano M and Ruiz I Altaba A 2022 On the origin of metastases: induction of pro-metastatic states after impending cell death via ER stress, reprogramming, and a cytokine storm *Cell Rep.* **38** 110490
Conrad C, Gray K M, Stroka K M, Rizvi I and Scarcelli G 2019 Mechanical characterization of 3D ovarian cancer nodules using brillouin confocal microscopy *Cel. Mol. Bioeng.* **12** 215–26
Conti A M, Saleh A D, Brinster L R, Cheng H, Chen Z, Cornelius S *et al* 2015 Conditional deletion of nonmuscle myosin II-A in mouse tongue epithelium results in squamous cell carcinoma *Sci. Rep.* **5** 14068
Cortés-Hernández L E, Eslami-S Z, Dujon A M, Giraudeau M, Ujvari B, Thomas F *et al* 2020 Do malignant cells sleep at night? *Genome Biol.* **21** 276
Cox T R and Erler J T 2011 Remodeling and homeostasis of the extracellular matrix: implications for fibrotic diseases and cancer *Dis. Model. Mech.* **4** 165–78

Cross S E, Jin Y-S, Rao J and Gimzewski J K 2007 Nanomechanical analysis of cells from cancer patients *Nat. Nanotechnol.* **2** 780–3

Cross S E, Jin Y-S, Tondre J, Wong R, Rao J and Gimzewski J K 2008 AFM-based analysis of human metastatic cancer cells *Nanotechnology* **19** 384003

Daftuar L, Zhu Y, Jacq X and Prives C 2013 Ribosomal proteins RPL37, RPS15 and RPS20 regulate the Mdm2-p53-MdmX network *PLoS One* **8** e68667

Darling E M and Di Carlo D 2015 High-throughput assessment of cellular mechanical properties *Annu. Rev. Biomed. Eng.* **17** 35–62

Darling E M, Zauscher S, Block J A and Guilak F 2007 A thin-layer model for viscoelastic, stress-relaxation testing of cells using atomic force microscopy: do cell properties reflect metastatic potential? *Biophys. J.* **92** 1784–91

Dazzoni R, Grélard A, Morvan E, Bouter A, Applebee C J, Loquet A *et al* 2020 The unprecedented membrane deformation of the human nuclear envelope, in a magnetic field, indicates formation of nuclear membrane invaginations *Sci. Rep.* **10** 5147

de las Heras J I, Batrakou D G and Schirmer E C 2013 Cancer biology and the nuclear envelope: a convoluted relationship *Semin. Cancer Biol.* **23** 125–37

Delarue M, Montel F, Vignjevic D, Prost J, Joanny J-F and Cappello G 2014 Compressive stress inhibits proliferation in tumor spheroids through a volume limitation *Biophys. J.* **107** 1821–8

Denais C and Lammerding J 2014 Nuclear mechanics in cancer *Cancer Biology and the Nuclear Envelope* ed E C Schirmer and J I de las Heras (New York: Springer) pp 435–70

Denais C M, Gilbert R M, Isermann P, McGregor A L, te Lindert M, Weigelin B *et al* 2016 Nuclear envelope rupture and repair during cancer cell migration *Science* **352** 353–8

Deng Y, Davis S P, Yang F, Paulsen K S, Kumar M, Sinnott DeVaux R *et al* 2017 Inertial Microfluidic Cell Stretcher (iMCS): fully automated, high-throughput, and near real-time cell mechanotyping *Small* **13** 1700705

Denton A E, Roberts E W and Fearon D T 2018 Stromal cells in the tumor microenvironment *Stromal Immunology* ed B M J Owens and M A Lakins (Cham: Springer International Publishing) pp 99–114

DeWane G, Salvi A M and DeMali K A 2021 Fueling the cytoskeleton—links between cell metabolism and actin remodeling *J. Cell Sci.* **134** jcs248385

Diamantopoulou Z, Castro-Giner F, Schwab F D, Foerster C, Saini M, Budinjas S *et al* 2022 The metastatic spread of breast cancer accelerates during sleep *Nature* **607** 156–62

Ding Y, Wang J, Xu G-K and Wang G-F 2018 Are elastic moduli of biological cells depth dependent or not? Another explanation using a contact mechanics model with surface tension *Soft Matter* **14** 7534–41

Discher D E, Janmey P and Wang Y 2005 Tissue cells feel and respond to the stiffness of their substrate *Science* **310** 1139–43

Discher D E, Mooney D J and Zandstra P W 2009 Growth factors, matrices, and forces combine and control stem cells *Science* **324** 1673–7

Diz-Muñoz A, Weiner O D and Fletcher D A 2018 In pursuit of the mechanics that shape cell surfaces *Nat. Phys.* **14** 648–52

Donato C, Kunz L, Castro-Giner F, Paasinen-Sohns A, Strittmatter K, Szczerba B M *et al* 2020 Hypoxia triggers the intravasation of clustered circulating tumor cells *Cell Rep.* **32** 108105

Dong C, Dinu C Z and Hu X 2016 Current status and perspectives in atomic force microscopy-based identification of cellular transformation *Int. J. Nanomedicine* **2016** 2107–118

Easley A D, Ma T, Eneh C I, Yun J, Thakur R M and Lutkenhaus J L 2022 A practical guide to quartz crystal microbalance with dissipation monitoring of thin polymer films *J. Polym. Sci.* **60** 1090–107

Eberle D, Fodelianaki G, Kurth T, Jagielska A, Möllmert S, Ulbricht E *et al* 2018 Acute but not inherited demyelination in mouse models leads to brain tissue stiffness changes bioXriv: https://doi.org/10.1101/449603

Eble J A and Niland S 2019 The extracellular matrix in tumor progression and metastasis *Clin. Exp. Metastasis* **36** 171–98

Echarri A, Pavón D M, Sánchez S, García-García M, Calvo E, Huerta-López C *et al* 2019 An Abl-FBP17 mechanosensing system couples local plasma membrane curvature and stress fiber remodeling during mechanoadaptation *Nat. Commun.* **10** 5828

Ekpenyong A E, Whyte G, Chalut K, Pagliara S, Lautenschläger F, Fiddler C *et al* 2012 Viscoelastic properties of differentiating blood cells are fate- and function-dependent *PLoS One* **7** e45237

Eldridge W J, Ceballos S, Shah T, Park H S, Steelman Z A, Zauscher S *et al* 2019 Shear modulus measurement by quantitative phase imaging and correlation with atomic force microscopy *Biophys. J.* **117** 696–705

Eling N, Morgan M D and Marioni J C 2019 Challenges in measuring and understanding biological noise *Nat. Rev. Genet.* **20** 536–48

Elmehrath A O, Afifi A M, Al-Husseini M J, Saad A M, Wilson N, Shohdy K S *et al* 2021 Causes of death among patients with metastatic prostate cancer in the US from 2000 to 2016 *JAMA Netw. Open* **4** e2119568

El-Sabban M E, Sfeir A J, Daher M H, Kalaany N Y, Bassam R A and Talhouk R S 2003 ECM-induced gap junctional communication enhances mammary epithelial cell differentiation *J. Cell Sci.* **116** 3531–41

El-Saghir J A, El-Habre E T, El-Sabban M E and Talhouk R S 2011 Connexins: a junctional crossroad to breast cancer *Int. J. Dev. Biol.* **55** 773–80

Elsayad K, Polakova S and Gregan J 2019 Probing mechanical properties in biology using brillouin microscopy *Trends Cell Biol.* **29** 608–11

Entenberg D, Voiculescu S, Guo P, Borriello L, Wang Y, Karagiannis G S *et al* 2018 A permanent window for the murine lung enables high-resolution imaging of cancer metastasis *Nat. Methods* **15** 73–80

Etienne-Manneville S 2004 Actin and microtubules in cell motility: which one is in control? *Traffic* **5** 470–7

Faigle C, Lautenschläger F, Whyte G, Homewood P, Martín-Badosa E and Guck J 2015 A monolithic glass chip for active single-cell sorting based on mechanical phenotyping *Lab Chip* **15** 1267–75

Fan R, Emery T, Zhang Y, Xia Y, Sun J and Wan J 2016 Circulatory shear flow alters the viability and proliferation of circulating colon cancer cells *Sci. Rep.* **6** 27073

Fanfone D, Wu Z, Mammi J, Berthenet K, Neves D, Weber K *et al* 2022 Confined migration promotes cancer metastasis through resistance to anoikis and increased invasiveness *eLife* **11** e73150

Faria E C, Ma N, Gazi E, Gardner P, Brown M, Clarke N W *et al* 2008 Measurement of elastic properties of prostate cancer cells using AFM *Analyst* **133** 1498

Farsad K and Camilli P D 2003 Mechanisms of membrane deformation *Curr. Opin. Cell Biol.* **15** 372–81

Feng Y and Walsh C A 2004 The many faces of filamin: a versatile molecular scaffold for cell motility and signalling *Nat. Cell Biol.* **6** 1034–8

Fife C M, McCarroll J A and Kavallaris M 2014 Movers and shakers: cell cytoskeleton in cancer metastasis *Br. J. Pharmacol.* **171** 5507–23

Fischer T, Hayn A and Mierke C T 2020 Effect of nuclear stiffness on cell mechanics and migration of human breast cancer cells *Front. Cell Dev. Biol.* **8** 393

Fischer T, Wilharm N, Hayn A and Mierke C T 2017 Matrix and cellular mechanical properties are the driving factors for facilitating human cancer cell motility into 3D engineered matrices *Converg. Sci. Phys. Oncol.* **3** 044003

Fletcher D A and Mullins R D 2010 Cell mechanics and the cytoskeleton *Nature* **463** 485–92

Foerster F, Braig S, Moser C, Kubisch R, Busse J, Wagner E *et al* 2014 Targeting the actin cytoskeleton: selective antitumor action via trapping PKCε *Cell Death Dis.* **5** e1398–8

Follain G, Herrmann D, Harlepp S, Hyenne V, Osmani N, Warren S C *et al* 2020 Fluids and their mechanics in tumour transit: shaping metastasis *Nat. Rev. Cancer* **20** 107–24

Follain G, Osmani N, Azevedo A S, Allio G, Mercier L, Karreman M A *et al* 2018 Hemodynamic forces tune the arrest, adhesion, and extravasation of circulating tumor cells *Dev. Cell* **45** 33–52.e12

Fostok S F, El-Sibai M, El-Sabban M and Talhouk R S 2019 Gap junctions and Wnt signaling in the mammary gland: a cross-talk? *J. Mammary Gland Biol. Neoplasia* **24** 17–38

Friedl P, Wolf K and Lammerding J 2011 Nuclear mechanics during cell migration *Curr. Opin. Cell Biol.* **23** 55–64

Fu Y, Chin L K, Bourouina T, Liu A Q and VanDongen A M J 2012 Nuclear deformation during breast cancer cell transmigration *Lab Chip* **12** 3774

Fuhrmann A, Banisadr A, Beri P, Tlsty T D and Engler A J 2017 Metastatic state of cancer cells may be indicated by adhesion strength *Biophys. J.* **112** 736–45

Fujita Y, Iketani M, Ito M and Ohsawa I 2022 Temporal changes in mitochondrial function and reactive oxygen species generation during the development of replicative senescence in human fibroblasts *Exp. Gerontol.* **165** 111866

Gaggioli C, Hooper S, Hidalgo-Carcedo C, Grosse R, Marshall J F, Harrington K *et al* 2007 Fibroblast-led collective invasion of carcinoma cells with differing roles for RhoGTPases in leading and following cells *Nat. Cell Biol.* **9** 1392–400

Gandalovičová A, Rosel D, Fernandes M, Veselý P, Heneberg P, Čermák V *et al* 2017 Migrastatics—anti-metastatic and anti-invasion drugs: promises and challenges *Trends Cancer* **3** 391–406

Gao Y, Bado I, Wang H, Zhang W, Rosen J M and Zhang X H-F 2019 Metastasis organotropism: redefining the congenial soil *Dev. Cell* **49** 375–91

Gavara N and Chadwick R S 2016 Relationship between cell stiffness and stress fiber amount, assessed by simultaneous atomic force microscopy and live-cell fluorescence imaging *Biomech. Model. Mechanobiol.* **15** 511–23

Gensbittel V, Kräter M, Harlepp S, Busnelli I, Guck J and Goetz J G 2021 Mechanical adaptability of tumor cells in metastasis *Dev. Cell* **56** 164–79

Gerashchenko T S, Novikov N M, Krakhmal N V, Zolotaryova S Y, Zavyalova M V, Cherdyntseva N V *et al* 2019 Markers of cancer cell invasion: are they good enough? *J. Clin. Med.* **8** 1092

Glogauer M, Arora P, Chou D, Janmey P A, Downey G P and McCulloch C A G 1998 The role of actin-binding protein 280 in integrin-dependent mechanoprotection *J. Biol. Chem.* **273** 1689–98

Goetz J G, Minguet S, Navarro-Lérida I, Lazcano J J, Samaniego R, Calvo E *et al* 2011 Biomechanical remodeling of the microenvironment by stromal caveolin-1 favors tumor invasion and metastasis *Cell* **146** 148–63

González-Bermúdez B, Guinea G V and Plaza G R 2019 Advances in micropipette aspiration: applications in cell biomechanics, models, and extended studies *Biophys. J.* **116** 587–94

Gossett D R, Tse H T K, Lee S A, Ying Y, Lindgren A G, Yang O O *et al* 2012 Hydrodynamic stretching of single cells for large population mechanical phenotyping *Proc. Natl Acad. Sci.* **109** 7630–5

Grandér D 1998 How do mutated oncogenes and tumor suppressor genes cause cancer? *Med. Oncol.* **15** 20–6

Gritsenko P, Leenders W and Friedl P 2017 Recapitulating *in vivo*-like plasticity of glioma cell invasion along blood vessels and in astrocyte-rich stroma *Histochem Cell Biol.* **148** 395–406

Gu C, Yaddanapudi S, Weins A, Osborn T, Reiser J, Pollak M *et al* 2010 Direct dynamin–actin interactions regulate the actin cytoskeleton *EMBO J.* **29** 3593–606

Guck J 2019 Some thoughts on the future of cell mechanics *Biophys. Rev.* **11** 667–70

Guck J, Ananthakrishnan R, Mahmood H, Moon T J, Cunningham C C and Käs J 2001 The optical stretcher: a novel laser tool to micromanipulate cells *Biophys. J.* **81** 767–84

Guck J and Chilvers E R 2013 Mechanics meets medicine *Sci. Transl. Med.* **5** 212fa41

Guck J, Lautenschläger F, Paschke S and Beil M 2010 Critical review: cellular mechanobiology and amoeboid migration *Integr. Biol.* **2** 575–83

Guck J, Schinkinger S, Lincoln B, Wottawah F, Ebert S, Romeyke M *et al* 2005 Optical deformability as an inherent cell marker for testing malignant transformation and metastatic competence *Biophys. J.* **88** 3689–98

Guerra F S, Oliveira R G D, Fraga C A M, Mermelstein C D S and Fernandes P D 2017 ROCK inhibition with Fasudil induces beta-catenin nuclear translocation and inhibits cell migration of MDA-MB 231 human breast cancer cells *Sci. Rep.* **7** 13723

Guillou L, Dahl J B, Lin J-M G, Barakat A I, Husson J, Muller S J *et al* 2016 Measuring cell viscoelastic properties using a microfluidic extensional flow device *Biophys. J.* **111** 2039–50

Gullekson C, Cojoc G, Schürmann M, Guck J and Pelling A 2017 Mechanical mismatch between Ras transformed and untransformed epithelial cells *Soft Matter* **13** 8483–91

Guo A, Wang B, Lyu C, Li W, Wu Y, Zhu L *et al* 2020 Consistent apparent Young's modulus of human embryonic stem cells and derived cell types stabilized by substrate stiffness regulation promotes lineage specificity maintenance *Cell Regen.* **9** 15

Habli Z, AlChamaa W, Saab R, Kadara H and Khraiche M L 2020 Circulating tumor cell detection technologies and clinical utility: challenges and opportunities *Cancers* **12** 1930

Habli Z, Zantout A, Al-Haj N, Saab R, El-Sabban M and Khraiche M L 2024 Single-cell fluidic force spectroscopy reveals dynamic mechanical fingerprints of malignancy in breast cancer *ACS Appl. Mater. Interfaces* **16** 50147–59

Habli Z, Zantout A, El-Sabban M and Khraiche M L 2023 Investigating malignancy-dependent mechanical properties of breast cancer cells *2023 45th Annual Int. Conf. of the IEEE Engineering in Medicine & Biology Society (EMBC)* (Sydney: IEEE) pp 1–4

Haeger A, Krause M, Wolf K and Friedl P 2014 Cell jamming: collective invasion of mesenchymal tumor cells imposed by tissue confinement *Biochim. Biophys. Acta (BBA)—Gen. Subj.* **1840** 2386–95

Halder D, Saha S, Singh R K, Ghosh I, Mallick D, Dey S K *et al* 2019 Nonmuscle myosin IIA and IIB differentially modulate migration and alter gene expression in primary mouse tumorigenic cells *Mol. Biol. Cell* **30** 1463–76

Hall A 2009 The cytoskeleton and cancer *Cancer Metastasis Rev.* **28** 5–14

Han Y L, Pegoraro A F, Li H, Li K, Yuan Y, Xu G *et al* 2020 Cell swelling, softening and invasion in a three-dimensional breast cancer model *Nat. Phys.* **16** 101–8

Harada T, Swift J, Irianto J, Shin J-W, Spinler K R, Athirasala A *et al* 2014 Nuclear lamin stiffness is a barrier to 3D migration, but softness can limit survival *J. Cell Biol.* **204** 669–82

Harney A S, Arwert E N, Entenberg D, Wang Y, Guo P, Qian B-Z *et al* 2015 Real-time imaging reveals local, transient vascular permeability, and tumor cell intravasation stimulated by TIE2hi macrophage-derived VEGFA *Cancer Discov.* **5** 932–43

Hayn A, Fischer T and Mierke C T 2023 The role of ADAM8 in mechanophenotype of cancer cells in 3D extracellular matrices *Front. Cell Dev. Biol* **11** 1148162

Headley M B, Bins A, Nip A, Roberts E W, Looney M R, Gerard A *et al* 2016 Visualization of immediate immune responses to pioneer metastatic cells in the lung *Nature* **531** 513–7

Hess K R, Varadhachary G R, Taylor S H, Wei W, Raber M N, Lenzi R *et al* 2006 Metastatic patterns in adenocarcinoma *Cancer* **106** 1624–33

Hindman B, Goeckeler Z, Sierros K and Wysolmerski R 2015 Non-muscle myosin II isoforms have different functions in matrix rearrangement by MDA-MB-231 cells *PLoS One* **10** e0131920

Hirano K, Tsuchiya M, Shiomi A, Takabayashi S, Suzuki M, Ishikawa Y *et al* 2023 The mechanosensitive ion channel PIEZO1 promotes satellite cell function in muscle regeneration *Life Sci. Alliance* **6** e202201783

Hobson C M, Kern M, O'Brien E T, Stephens A D, Falvo M R and Superfine R 2020 Correlating nuclear morphology and external force with combined atomic force microscopy and light sheet imaging separates roles of chromatin and lamin A/C in nuclear mechanics *Mol. Biol. Cell* **31** 1788–801

Hochmuth R M 2000 Micropipette aspiration of living cells *J. Biomech.* **33** 15–22

Hoffman B D and Crocker J C 2009 Cell mechanics: dissecting the physical responses of cells to force *Annu. Rev. Biomed. Eng.* **11** 259–88

Hogrebe N J, Augsornworawat P, Maxwell K G, Velazco-Cruz L and Millman J R 2020 Targeting the cytoskeleton to direct pancreatic differentiation of human pluripotent stem cells *Nat. Biotechnol.* **38** 460–70

Holenstein C N, Horvath A, Schär B, Schoenenberger A D, Bollhalder M, Goedecke N *et al* 2019 The relationship between metastatic potential and *in vitro* mechanical properties of osteosarcoma cells *Mol. Biol. Cell* **30** 887–98

Honda K 2015 The biological role of actinin-4 (ACTN4) in malignant phenotypes of cancer *Cell Biosci.* **5** 41

Hoshino A, Costa-Silva B, Shen T-L, Rodrigues G, Hashimoto A, Tesic Mark M *et al* 2015 Tumour exosome integrins determine organotropic metastasis *Nature* **527** 329–35

Hotulainen P and Lappalainen P 2006 Stress fibers are generated by two distinct actin assembly mechanisms in motile cells *J. Cell Biol.* **173** 383–94

Hou Y, Zhao C, Xu B, Huang Y and Liu C 2021 Effect of docetaxel on mechanical properties of ovarian cancer cells *Exp. Cell. Res.* **408** 112853

Houshyari M and Taghizadeh-Hesary F 2023 The metastatic spread of breast cancer accelerates during sleep: how the study design can affect the results *Asian Pac. J. Cancer Prev.* **24** 353–5

Hsiao B-Y, Chen C-C, Hsieh P-C, Chang T-K, Yeh Y-C, Wu Y-C *et al* 2011 Rad is a p53 direct transcriptional target that inhibits cell migration and is frequently silenced in lung carcinoma cells *J. Mol. Med.* **89** 481–92

Hsu K-S and Kao H-Y 2013 Alpha-actinin 4 and tumorigenesis of breast Cancer *Vitamins & Hormones* (Amsterdam: Elsevier) pp 323–51

Huang Q, Li X, Huang Z, Yu F, Wang X, Wang S *et al* 2020 ACTN4 promotes the proliferation, migration, metastasis of osteosarcoma and enhances its invasive ability through the NF-κB pathway *Pathol. Oncol. Res.* **26** 893–904

Hwang I S S, Fung M L, Liong E C, Tipoe G L and Tang F 2007 Age-related changes in adrenomedullin expression and hypoxia-inducible factor-1 activity in the rat lung and their responses to hypoxia *J. Gerontol. A Biol. Sci. Med. Sci.* **62** 41–9

Inoue C, Zhao C, Tsuduki Y, Udono M, Wang L, Nomura M *et al* 2017 SMARCD1 regulates senescence-associated lipid accumulation in hepatocytes *npj Aging Mech. Dis.* **3** 11

Irianto J, Pfeifer C R, Ivanovska I L, Swift J and Discher D E 2016 Nuclear lamins in cancer *Cel. Mol. Bioeng.* **9** 258–67

Ivkovic S, Beadle C, Noticewala S, Massey S C, Swanson K R, Toro L N *et al* 2012 Direct inhibition of myosin II effectively blocks glioma invasion in the presence of multiple motogens *Mol. Biol. Cell* **23** 533–42

Jain S, Cachoux V M L, Narayana G H N S, de Beco S, D'Alessandro J, Cellerin V *et al* 2020 The role of single-cell mechanical behaviour and polarity in driving collective cell migration *Nat. Phys.* **16** 802–9

Janmey P A 1991 Mechanical properties of cytoskeletal polymers *Curr. Opin. Cell Biol.* **3** 4–11

Janmey P A and Miller R T 2011 Mechanisms of mechanical signaling in development and disease *J. Cell Sci.* **124** 9–18

Jiang H, Torphy R J, Steiger K, Hongo H, Ritchie A J, Kriegsmann M *et al* 2020 Pancreatic ductal adenocarcinoma progression is restrained by stromal matrix *J. Clin. Invest.* **130** 4704–9

Jiang L, Wen J and Luo W 2015 Rho-associated kinase inhibitor, Y-27632, inhibits the invasion and proliferation of T24 and 5367 bladder cancer cells *Mol. Med. Rep.* **12** 7526–30

Jimah J R and Hinshaw J E 2019 Structural insights into the mechanism of dynamin superfamily proteins *Trends Cell Biol.* **29** 257–73

Jin J, Tang K, Xin Y, Zhang T and Tan Y 2018 Hemodynamic shear flow regulates biophysical characteristics and functions of circulating breast tumor cells reminiscent of brain metastasis *Soft Matter* **14** 9528–33

Jung J, Kim S, An H-T and Ko J 2020 α-Actinin-4 regulates cancer stem cell properties and chemoresistance in cervical cancer *Carcinogenesis* **41** 940–9

Kalender M E, Demiryürek S, Oztuzcu S, Kizilyer A, Demiryürek A T, Sevinc A *et al* 2009 Association between the Thr431Asn polymorphism of the ROCK2 gene and risk of developing metastases of breast cancer *Oncol. Res.* **18** 583–91

Kamai T, Tsujii T, Arai K, Takagi K, Asami H, Ito Y *et al* 2003 Significant association of Rho/ROCK pathway with invasion and metastasis of bladder cancer *Clin. Cancer Res.* **9** 2632–41

Kamil M, Shinsato Y, Higa N, Hirano T, Idogawa M, Takajo T *et al* 2019 High filamin-C expression predicts enhanced invasiveness and poor outcome in glioblastoma multiforme *Br. J. Cancer* **120** 819–26

Kanczuga-Koda L, Sulkowski S, Lenczewski A, Koda M, Wincewicz A, Baltaziak M *et al* 2006 Increased expression of connexins 26 and 43 in lymph node metastases of breast cancer *J. Clin. Pathol.* **59** 429–33

Kang N, Guo Q, Islamzada E, Ma H and Scott M D 2018 Microfluidic determination of lymphocyte vascular deformability: effects of intracellular complexity and early immune activation *Integr. Biol.* **10** 207–17

Kang Y, Siegel P M, Shu W, Drobnjak M, Kakonen S M, Cordón-Cardo C *et al* 2003 A multigenic program mediating breast cancer metastasis to bone *Cancer Cell* **3** 537–49

Karaman S and Detmar M 2014 Mechanisms of lymphatic metastasis *J. Clin. Invest.* **124** 922–8

Karreman M A, Hyenne V, Schwab Y and Goetz J G 2016 Intravital correlative microscopy: imaging life at the nanoscale *Trends Cell Biol.* **26** 848–63

Kashani A S and Packirisamy M 2020 Cancer cells optimize elasticity for efficient migration *R. Soc. Open Sci.* **7** 200747

Kazan J M, El-Saghir J, Saliba J, Shaito A, Jalaleddine N, El-Hajjar L *et al* 2019 Cx43 expression correlates with breast cancer metastasis in MDA-MB-231 cells *in vitro*, in a mouse xenograft model and in human breast cancer tissues *Cancers* **11** 460

Khraiche M L, Rogul J and Muthuswamy J 2019 Design and development of microscale Thickness Shear Mode (TSM) resonators for sensing neuronal adhesion *Front. Neurosci.* **13** 518

Kienast Y, Von Baumgarten L, Fuhrmann M, Klinkert W E F, Goldbrunner R, Herms J *et al* 2010 Real-time imaging reveals the single steps of brain metastasis formation *Nat. Med.* **16** 116–22

Kim G, Meriin A B, Gabai V L, Christians E, Benjamin I, Wilson A *et al* 2012 The heat shock transcription factor Hsf1 is downregulated in DNA damage–associated senescence, contributing to the maintenance of senescence phenotype *Aging Cell* **11** 617–27

Kim T-H, Gill N K, Nyberg K D, Nguyen A V, Hohlbauch S V, Geisse N A *et al* 2016 Cancer cells become less deformable and more invasive with activation of β-adrenergic signaling *J. Cell Sci.* **129** 4563–75

Kirby T J and Lammerding J 2018 Emerging views of the nucleus as a cellular mechanosensor *Nat. Cell Biol.* **20** 373–81

Koenderink G H and Paluch E K 2018 Architecture shapes contractility in actomyosin networks *Curr. Opin. Cell Biol.* **50** 79–85

Köpf-Maier P and Mühlhausen S K 1992 Changes in the cytoskeleton pattern of tumor cells by cisplatin *in vitro Chem. Biol. Interact.* **82** 295–316

Koser D E, Moeendarbary E, Hanne J, Kuerten S and Franze K 2015 CNS cell distribution and axon orientation determine local spinal cord mechanical properties *Biophys. J.* **108** 2137–47

Kostic A, Lynch C D and Sheetz M P 2009 Differential matrix rigidity response in breast cancer cell lines correlates with the tissue tropism *PLoS One* **4** e6361

Kostic A and Sheetz M P 2006 Fibronectin rigidity response through Fyn and p130Cas recruitment to the leading edge *Mol. Biol. Cell* **17** 2684–95

Kothari P, Johnson C, Sandone C, Iglesias P A and Robinson D N 2019a How the mechanobiome drives cell behavior, viewed through the lens of control theory *J. Cell Sci.* **132** jcs234476

Kothari P, Srivastava V, Aggarwal V, Tchernyshyov I, Van Eyk J E, Ha T *et al* 2019b Contractility kits promote assembly of the mechanoresponsive cytoskeletal network *J. Cell Sci.* **132** jcs226704

Kotnik T, Rems L, Tarek M and Miklavčič D 2019 Membrane electroporation and electropermeabilization: mechanisms and models *Annu. Rev. Biophys.* **48** 63–91

Krieg M, Dunn A R and Goodman M B 2014 Mechanical control of the sense of touch by β-spectrin *Nat. Cell Biol.* **16** 224–33

Kumar S and Weaver V M 2009 Mechanics, malignancy, and metastasis: the force journey of a tumor cell *Cancer Metastasis Rev.* **28** 113–27

Kundu T, Bereiter-Hahn J and Karl I 2000 Cell property determination from the acoustic microscope generated voltage versus frequency curves *Biophys. J.* **78** 2270–9

Kung M-L, Hsieh C-W, Tai M-H, Weng C-H, Wu D-C, Wu W-J *et al* 2016 Nanoscale characterization illustrates the cisplatin-mediated biomechanical changes of B16-F10 melanoma cells *Phys. Chem. Chem. Phys.* **18** 7124–31

Kurosawa O, Oana H, Matsuoka S, Noma A, Kotera H and Washizu M 2006 Electroporation through a micro-fabricated orifice and its application to the measurement of cell response to external stimuli *Meas. Sci. Technol.* **17** 3127–33

Lam W A, Rosenbluth M J and Fletcher D A 2008 Increased leukaemia cell stiffness is associated with symptoms of leucostasis in paediatric acute lymphoblastic leukaemia *Br. J. Haematol.* **142** 497–501

Lambert A W, Pattabiraman D R and Weinberg R A 2017 Emerging biological principles of metastasis *Cell* **168** 670–91

Lammerding J, Fong L G, Ji J Y, Reue K, Stewart C L, Young S G *et al* 2006 Lamins A and C but not lamin B1 regulate nuclear mechanics *J. Biol. Chem.* **281** 25768–80

Lane J, Martin T A, Watkins G, Mansel R E and Jiang W G 2008 The expression and prognostic value of ROCK I and ROCK II and their role in human breast cancer *Int. J. Oncol.* **33** 585–93

Lapis K, Paku S and Liotta L A 1988 Endothelialization of embolized tumor cells during metastasis formation *Clin. Exp. Metast.* **6** 73–89

Le Roux A-L, Quiroga X, Walani N, Arroyo M and Roca-Cusachs P 2019 The plasma membrane as a mechanochemical transducer *Phil. Trans. R. Soc.* **B 374** 20180221

Lee M-H, Kundu J K, Chae J-I and Shim J-H 2019 Targeting ROCK/LIMK/cofilin signaling pathway in cancer *Arch. Pharm. Res.* **42** 481–91

Lee M-H, Wu P-H, Staunton J R, Ros R, Longmore G D and Wirtz D 2012 Mismatch in mechanical and adhesive properties induces pulsating cancer cell migration in epithelial monolayer *Biophys. J.* **102** 2731–41

Lee Y and Shivashankar G V 2020 Analysis of transcriptional modules during human fibroblast ageing *Sci. Rep.* **10** 19086

Lei K, Kurum A, Kaynak M, Bonati L, Han Y, Cencen V *et al* 2021 Cancer-cell stiffening via cholesterol depletion enhances adoptive T-cell immunotherapy *Nat. Biomed. Eng.* **5** 1411–25

Leijnse N, Barooji Y F, Arastoo M R, Sønder S L, Verhagen B, Wullkopf L *et al* 2022 Filopodia rotate and coil by actively generating twist in their actin shaft *Nat. Commun.* **13** 1636

Lekka M 2016 Discrimination between normal and cancerous cells using AFM *Bio. Nano. Sci.* **6** 65–80

Lekka M, Laidler P, Gil D, Lekki J, Stachura Z and Hrynkiewicz A Z 1999 Elasticity of normal and cancerous human bladder cells studied by scanning force microscopy *Eur. Biophys. J.* **28** 312–6

Levental K R, Yu H, Kass L, Lakins J N, Egeblad M, Erler J T *et al* 2009 Matrix crosslinking forces tumor progression by enhancing integrin signaling *Cell* **139** 891–906

Li Q S, Lee G Y H, Ong C N and Lim C T 2008 AFM indentation study of breast cancer cells *Biochem. Biophys. Res. Commun.* **374** 609–13

Li Y, Li X, Liu K-R, Zhang J-N, Liu Y and Zhu Y 2015 Visfatin derived from ascites promotes ovarian cancer cell migration through Rho/ROCK signaling-mediated actin polymerization *Eur. J.Cancer Prev.* **24** 231–9

Liberti M V and Locasale J W 2016 The warburg effect: how does it benefit cancer cells? *Trends Biochem. Sci.* **41** 211–8

Liboz M, Allard A, Malo M, Lamour G, Letort G, Thiébot B *et al* 2023 Using adhesive micropatterns and AFM to assess cancer cell morphology and mechanics *ACS Appl. Mater. Interfaces* **15** 43403–13

Lim S, Quinton R J and Ganem N J 2016 Nuclear envelope rupture drives genome instability in cancer *Mol. Biol. Cell* **27** 3210–3

Lincoln B, Schinkinger S, Travis K, Wottawah F, Ebert S, Sauer F *et al* 2007a Reconfigurable microfluidic integration of a dual-beam laser trap with biomedical applications *Biomed. Microdevices* **9** 703–10

Lincoln B, Wottawah F, Schinkinger S, Ebert S and Guck J 2007b High-throughput rheological measurements with an optical stretcher *Methods Cell. Biol.* **83** 397–423

Linde N, Casanova-Acebes M, Sosa M S, Mortha A, Rahman A, Farias E *et al* 2018 Macrophages orchestrate breast cancer early dissemination and metastasis *Nat. Commun.* **9** 21

Liu S, Goldstein R H, Scepansky E M and Rosenblatt M 2009 Inhibition of Rho-associated kinase signaling prevents breast cancer metastasis to human bone *Cancer Res.* **69** 8742–51

Liu X and Chu K-M 2017 α-Actinin-4 promotes metastasis in gastric cancer *Lab. Invest.* **97** 1084–94

Liu Y, Zhang T, Zhang H, Li J, Zhou N, Fiskesund R *et al* 2021 Cell softness prevents cytolytic T-cell killing of tumor-repopulating cells *Cancer Res.* **81** 476–88

Liu Z, Lee S J, Park S, Konstantopoulos K, Glunde K, Chen Y *et al* 2020 Cancer cells display increased migration and deformability in pace with metastatic progression *FASEB J.* **34** 9307–15

Llanos S and Serrano M 2010 Depletion of ribosomal protein L37 occurs in response to DNA damage and activates p53 through the L11/MDM2 pathway *Cell Cycle* **9** 4005–12

Lomakin A J, Cattin C J, Cuvelier D, Alraies Z, Molina M, Nader G P F *et al* 2020 The nucleus acts as a ruler tailoring cell responses to spatial constraints *Science* **370** eaba2894

Lorenzen J, Sinkus R, Lorenzen M, Dargatz M, Leussler C, Röschmann P *et al* 2002 MR elastography of the breast:preliminary clinical results *Rofo Fortschr. Geb. Rontgenstr. Neuen. Bildgeb. Verfahr.* **174** 830–4

Lu P, Weaver V M and Werb Z 2012 The extracellular matrix: a dynamic niche in cancer progression *J. Cell Biol.* **196** 395–406

Lu X and Kang Y 2007 Organotropism of breast cancer metastasis *J. Mammary Gland Biol. Neoplasia* **12** 153–62

Luo Q, Kuang D, Zhang B and Song G 2016 Cell stiffness determined by atomic force microscopy and its correlation with cell motility *Biochim. Biophys. Acta (BBA)—Gen. Subj.* **1860** 1953–60

Luo Z, Yao X, Li M, Fang D, Fei Y, Cheng Z *et al* 2022 Modulating tumor physical microenvironment for fueling CAR-T cell therapy *Adv. Drug Deliv. Rev.* **185** 114301

Lv D, Hu Z, Lu L, Lu H and Xu X 2017 Three-dimensional cell culture: a powerful tool in tumor research and drug discovery (review) *Oncol Lett.*

Lv J, Liu Y, Cheng F, Li J, Zhou Y, Zhang T *et al* 2021 Cell softness regulates tumorigenicity and stemness of cancer cells *EMBO J.* **40**

Lyons S M, Alizadeh E, Mannheimer J, Schuamberg K, Castle J, Schroder B *et al* 2016 Changes in cell shape are correlated with metastatic potential in murine and human osteosarcomas *Biol. Open* **5** 289–99

Ma H, Cao L, Wang F, Cheng C, Jiang R, Zhou H *et al* 2020a Filamin B extensively regulates transcription and alternative splicing, and is associated with apoptosis in HeLa cells *Oncol Rep.*

Ma X, Xue H, Zhong J, Feng B and Zuo Y 2020b Serum actinin-4 levels as a potential diagnostic and prognostic marker in cervical cancer *Dis. Markers* **2020** 1–6

Mahaffy R E, Shih C K, MacKintosh F C and Käs J 2000 Scanning probe-based frequencydependent microrheology of polymer gels and biological cells *Phys. Rev. Lett.* **85** 880–3

Malandrino A, Kamm R D and Moeendarbary E 2018 *In vitro* modeling of mechanics in cancer metastasis *ACS Biomater. Sci. Eng.* **4** 294–301

Manno S, Mohandas N and Takakuwa Y 2010 ATP-dependent mechanism protects spectrin against glycation in human erythrocytes *J. Biol. Chem.* **285** 33923–9

Marx K A, Zhou T, Montrone A, McIntosh D and Braunhut S J 2005 Quartz crystal microbalance biosensor study of endothelial cells and their extracellular matrix following cell removal: evidence for transient cellular stress and viscoelastic changes during detachment and the elastic behavior of the pure matrix *Anal. Biochem.* **343** 23–34

Marx K A, Zhou T, Montrone A, McIntosh D and Braunhut S J 2007 A comparative study of the cytoskeleton binding drugs nocodazole and taxol with a mammalian cell quartz crystal microbalance biosensor: different dynamic responses and energy dissipation effects *Anal. Biochem.* **361** 77–92

Matsumura F 2005 Regulation of myosin II during cytokinesis in higher eukaryotes *Trends Cell Biol.* **15** 371–7

Mierke C 2018a *Physics of Cancer. Volume 1: Interplay Between Tumor Biology, Inflammation and Cell Mechanics/Claudia Tanja Mierke (University of Leipzig)* 2nd edn (Bristol: IOP Publishing)

Mierke C 2023 *Physics of Cancer. Volume 4: Mechanical Characterization of Cells/Claudia Tanja Mierke (Biological Physicy Division, University of Leipzig, Leipzig, Germany)* 2nd edn (Bristol: IOP Publishing)

Mierke C T 2008 Role of the endothelium during tumor cell metastasis: is the endothelium a barrier or a promoter for cell invasion and metastasis? *J. Biophys.* **2008** 1–13

Mierke C T 2011 Cancer cells regulate biomechanical properties of human microvascular endothelial cells *J. Biol. Chem.* **286** 40025–37

Mierke C T 2013 The integrin alphav beta3 increases cellular stiffness and cytoskeletal remodeling dynamics to facilitate cancer cell invasion *New J. Phys.* **15** 015003

Mierke C T 2014 The fundamental role of mechanical properties in the progression of cancer disease and inflammation *Rep. Prog. Phys.* **77** 076602

Mierke C T 2018b *Physics of Cancer, Volume 2 (Second Edition): Cellular and Microenvironmental Effects* (Bristol: IOP Publishing)

Mierke C T 2019a The matrix environmental and cell mechanical properties regulate cell migration and contribute to the invasive phenotype of cancer cells *Rep. Prog. Phys.* **82** 064602

Mierke C T 2019b The role of the optical stretcher is crucial in the investigation of cell mechanics regulating cell adhesion and motility *Front. Cell Dev. Biol.* **7** 184

Mierke C T 2021a Bidirectional mechanical response between cells and their microenvironment *Front. Phys.* **9** 749830

Mierke C T 2021b *Physics of Cancer, Volume 3 (Second Edition): Experimental Biophysical Techniques in Cancer Research* (Bristol: IOP Publishing)

Mierke C T 2024a Extracellular matrix cues regulate mechanosensing and mechanotransduction of cancer cells *Cells* **13** 96

Mierke C T 2024b Mechanosensory entities and functionality of endothelial cells *Front. Cell Dev. Biol. Sec. Cell Adhes. Migr.* **12**

Mierke C T, Bretz N and Altevogt P 2011 Contractile forces contribute to increased glycosylphosphatidylinositol-anchored receptor CD24-facilitated cancer cell invasion *J. Biol. Chem.* **286** 34858–71

Mierke C T, Kollmannsberger P, Zitterbart D P, Diez G, Koch T M, Marg S *et al* 2010 Vinculin facilitates cell invasion into three-dimensional collagen matrices *J. Biol. Chem.* **285** 13121–30

Mierke C T, Rösel D, Fabry B and Brábek J 2008a Contractile forces in tumor cell migration *Eur. J. Cell Biol.* **87** 669–76

Mierke C T, Zitterbart D P, Kollmannsberger P, Raupach C, Schlötzer-Schrehardt U, Goecke T W *et al* 2008b Breakdown of the endothelial barrier function in tumor cell transmigration *Biophys. J.* **94** 2832–46

Minn A J, Gupta G P, Siegel P M, Bos P D, Shu W, Giri D D *et al* 2005 Genes that mediate breast cancer metastasis to lung *Nature* **436** 518–24

Mitchison J M and Swann M M 1954 The mechanical properties of the cell surface *J. Exp. Biol.* **31** 443–60

Miyamoto C, Maehata Y, Ozawa S, Ikoma T, Kubota E, Izukuri K *et al* 2012 Fasudil suppresses fibrosarcoma growth by stimulating secretion of the chemokine CXCL14/BRAK *J. Pharmacol. Sci.* **120** 241–9

Moeendarbary E, Weber I P, Sheridan G K, Koser D E, Soleman S, Haenzi B *et al* 2017 The soft mechanical signature of glial scars in the central nervous system *Nat. Commun.* **8** 14787

Mogilner A and Keren K 2009 The shape of motile cells *Curr. Biol.* **19** R762–71

Mohammadi H and Sahai E 2018 Mechanisms and impact of altered tumour mechanics *Nat. Cell Biol.* **20** 766–74

Monzo P, Crestani M, Chong Y K, Ghisleni A, Hennig K, Li Q *et al* 2021 Adaptive mechanoproperties mediated by the formin FMN1 characterize glioblastoma fitness for invasion *Dev. Cell* **56** 2841–55.e8

Moose D L, Krog B L, Kim T-H, Zhao L, Williams-Perez S, Burke G *et al* 2020 Cancer cells resist mechanical destruction in circulation via RhoA/actomyosin-dependent mechano-adaptation *Cell Rep.* **30** 3864–74.e6

Mu X, Tseng C, Hambright W S, Matre P, Lin C, Chanda P *et al* 2020 Cytoskeleton stiffness regulates cellular senescence and innate immune response in Hutchinson–Gilford Progeria Syndrome *Aging Cell* **19** e13152

Mukherjee P, Nathamgari S S P, Kessler J A and Espinosa H D 2018 Combined numerical and experimental investigation of localized electroporation-based cell transfection and sampling *ACS Nano* **12** 12118–28

Mukherjee P, Peng C-Y, McGuire T, Hwang J W, Puritz C H, Pathak N *et al* 2023 Single cell transcriptomics reveals reduced stress response in stem cells manipulated using localized electric fields *Mater. Today Bio.* **19** 100601

Nader G P D F, Agüera-Gonzalez S, Routet F, Gratia M, Maurin M, Cancila V *et al* 2021 Compromised nuclear envelope integrity drives TREX1-dependent DNA damage and tumor cell invasion *Cell* **184** 5230–46.e22

Nagy Á G, Kanyó N, Vörös A, Székács I, Bonyár A and Horvath R 2022 Population distributions of single-cell adhesion parameters during the cell cycle from high-throughput robotic fluidic force microscopy *Sci. Rep.* **12** 7747

Nakamura M, Kondo H, Shimada Y, Waheed A A and Ohno-Iwashita Y 2003 Cellular aging-dependent decrease in cholesterol in membrane microdomains of human diploid fibroblasts *Exp. Cell. Res.* **290** 381–90

Naser Al Deen N, AbouHaidar M and Talhouk R 2019 Connexin43 as a tumor suppressor: proposed connexin43 mRNA-circularRNAs-microRNAs axis towards prevention and early detection in breast cancer *Front. Med.* **6** 192

Nassoy P and Lamaze C 2012 Stressing caveolae new role in cell mechanics *Trends Cell Biol.* **22** 381–9

Nathamgari S S P, Mukherjee P, Kessler J A and Espinosa H D 2019 Localized electroporation with track-etched membranes *Proc. Natl Acad. Sci. USA* **116** 22909–10

Nematbakhsh Y, Pang K T and Lim C T 2017 Correlating the viscoelasticity of breast cancer cells with their malignancy *Converg. Sci. Phys. Oncol.* **3** 034003

Neu J C and Krassowska W 1999 Asymptotic model of electroporation *Phys. Rev.* E **59** 3471–82

Nguyen A V, Nyberg K D, Scott M B, Welsh A M, Nguyen A H, Wu N *et al* 2016a Stiffness of pancreatic cancer cells is associated with increased invasive potential *Integr. Biol.* **8** 1232–45

Nguyen D X, Bos P D and Massagué J 2009 Metastasis: from dissemination to organ-specific colonization *Nat. Rev. Cancer* **9** 274–84

Nguyen L T S, Jacob M A C, Parajón E and Robinson D N 2022 Cancer as a biophysical disease: targeting the mechanical-adaptability program *Biophys. J.* **121** 3573–85

Nguyen L T S and Robinson D N 2020 The unusual suspects in cytokinesis: fitting the pieces together *Front. Cell Dev. Biol.* **8** 441

Nguyen N, Shao Y, Wineman A, Fu J and Waas A 2016b Atomic force microscopy indentation and inverse analysis for non-linear viscoelastic identification of breast cancer cells *Math. Biosci.* **277** 77–88

Nissen N I, Karsdal M and Willumsen N 2019 Collagens and Cancer associated fibroblasts in the reactive stroma and its relation to cancer biology *J. Exp. Clin. Cancer Res.* **38** 115

Nowacki L, Follet J, Vayssade M, Vigneron P, Rotellini L, Cambay F *et al* 2015 Real-time QCM-D monitoring of cancer cell death early events in a dynamic context *Biosens. Bioelectron.* **64** 469–76

Oguchi Y, Ozaki Y, Abdelmoez M N and Shintaku H 2021 NanoSINC-seq dissects the isoform diversity in subcellular compartments of single cells *Sci. Adv.* **7** eabe0317

Ohashi M, Aizawa S, Ooka H, Ohsawa T, Kaji K, Kondo H *et al* 1980 A new human diploid cell strain, TIG-1, for the research on cellular aging *Exp. Gerontol.* **15** 121–33

Ohata H, Ishiguro T, Aihara Y, Sato A, Sakai H, Sekine S *et al* 2012 Induction of the stem-like cell regulator CD44 by Rho kinase inhibition contributes to the maintenance of colon cancer–initiating cells *Cancer Res.* **72** 5101–10

Ohshima S and Seyama A 2010 Cellular aging and centrosome aberrations *Ann. N. Y. Acad. Sci.* **1197** 108–17

Ondeck M G, Kumar A, Placone J K, Plunkett C M, Matte B F, Wong K C *et al* 2019 Dynamically stiffened matrix promotes malignant transformation of mammary epithelial cells via collective mechanical signaling *Proc. Natl Acad. Sci. USA* **116** 3502–7

Ong M S, Deng S, Halim C E, Cai W, Tan T Z, Huang R Y-J *et al* 2020 Cytoskeletal proteins in cancer and intracellular stress: a therapeutic perspective *Cancers* **12** 238

Osmani N, Follain G, García León M J, Lefebvre O, Busnelli I, Larnicol A *et al* 2019 Metastatic tumor cells exploit their adhesion repertoire to counteract shear forces during intravascular arrest *Cell Rep.* **28** 2491–2500.e5

Otto O, Rosendahl P, Mietke A, Golfier S, Herold C, Klaue D *et al* 2015 Real-time deformability cytometry: on-the-fly cell mechanical phenotyping *Nat. Methods* **12** 199–202

Ouderkirk J L and Krendel M 2014 Non-muscle myosins in tumor progression, cancer cell invasion, and metastasis *Cytoskeleton* **71** 447–63

Paget S 1889 The distribution of secondary growths in cancer of the breast *Lancet* **133** 571–3

Pajerowski J D, Dahl K N, Zhong F L, Sammak P J and Discher D E 2007 Physical plasticity of the nucleus in stem cell differentiation *Proc. Natl Acad. Sci.* **104** 15619–24

Paluch E, Piel M, Prost J, Bornens M and Sykes C 2005 Cortical actomyosin breakage triggers shape oscillations in cells and cell fragments *Biophys. J.* **89** 724–33

Parajón E, Surcel A and Robinson D N 2021 The mechanobiome: a goldmine for cancer therapeutics *Am. J. Physiol. Cell Physiol.* **320** C306–23

Pardo-Pastor C, Rubio-Moscardo F, Vogel-González M, Serra S A, Afthinos A, Mrkonjic S *et al* 2018 Piezo2 channel regulates RhoA and actin cytoskeleton to promote cell mechanobiological responses *Proc. Natl Acad. Sci. USA* **115** 1925–30

Park J S, Burckhardt C J, Lazcano R, Solis L M, Isogai T, Li L *et al* 2020a Mechanical regulation of glycolysis via cytoskeleton architecture *Nature* **578** 621–6

Park S, Kang M, Kim S, An H-T, Gettemans J and Ko J 2020b α-Actinin-4 promotes the progression of prostate cancer through the Akt/GSK-3β/β-catenin signaling pathway *Front. Cell Dev. Biol.* **8** 588544

Park W-Y, Park J-S, Cho K-A, Kim D-I, Ko Y-G, Seo J-S *et al* 2000 Up-regulation of caveolin attenuates epidermal growth factor signaling in senescent cells *J. Biol. Chem.* **275** 20847–52

Paszek M J, Zahir N, Johnson K R, Lakins J N, Rozenberg G I, Gefen A *et al* 2005 Tensional homeostasis and the malignant phenotype *Cancer Cell* **8** 241–54

Patarat R, Riku S, Kunadirek P, Chuaypen N, Tangkijvanich P, Mutirangura A *et al* 2021 The expression of FLNA and CLU in PBMCs as a novel screening marker for hepatocellular carcinoma *Sci. Rep.* **11** 14838

Pathak N, Patino C A, Ramani N, Mukherjee P, Samanta D, Ebrahimi S B *et al* 2023 Cellular delivery of large functional proteins and protein–nucleic acid constructs via localized electroporation *Nano Lett.* **23** 3653–60

Patino C A, Pathak N, Mukherjee P, Park S H, Bao G and Espinosa H D 2022 Multiplexed high-throughput localized electroporation workflow with deep learning–based analysis for cell engineering *Sci. Adv.* **8** eabn7637

Paul C D, Bishop K, Devine A, Paine E L, Staunton J R, Thomas S M *et al* 2019 Tissue architectural cues drive organ targeting of tumor cells in zebrafish *Cell Syst.* **9** 187–206.e16

Pecci A, Ma X, Savoia A and Adelstein R S 2018 MYH9: structure, functions and role of non-muscle myosin IIA in human disease *Gene* **664** 152–67

Pegoraro A F, Janmey P and Weitz D A 2017 Mechanical properties of the cytoskeleton and cells *Cold Spring Harb. Perspect. Biol.* **9** a022038

Peinado H, Zhang H, Matei I R, Costa-Silva B, Hoshino A, Rodrigues G *et al* 2017 Pre-metastatic niches: organ-specific homes for metastases *Nat. Rev. Cancer* **17** 302–17

Pelling A E and Horton M A 2008 An historical perspective on cell mechanics *Pflugers Arch—Eur. J. Physiol.* **456** 3–12

Peng G E, Wilson S R and Weiner O D 2011 A pharmacological cocktail for arresting actin dynamics in living cells *Mol. Biol. Cell* **22** 3986–94

Peng Y, Chen Z, He Y, Li P, Chen Y, Chen X *et al* 2022 Non-muscle myosin II isoforms orchestrate substrate stiffness sensing to promote cancer cell contractility and migration *Cancer Lett.* **524** 245–58

Pereira E R, Kedrin D, Seano G, Gautier O, Meijer E F J, Jones D *et al* 2018 Lymph node metastases can invade local blood vessels, exit the node, and colonize distant organs in mice *Science* **359** 1403–7

Peters L L, White R A, Birkenmeier C S, Bloom M L, Lux S E and Barker J E 1992 Changing patterns in cytoskeletal mRNA expression and protein synthesis during murine erythropoiesis *in vivo Proc. Natl Acad. Sci. USA* **89** 5749–53

Phillip J M, Aifuwa I, Walston J and Wirtz D 2015 The mechanobiology of aging *Annu. Rev. Biomed. Eng.* **17** 113–41

Phillip J M, Wu P-H, Gilkes D M, Williams W, McGovern S, Daya J *et al* 2017 Biophysical and biomolecular determination of cellular age in humans *Nat. Biomed. Eng.* **1** 0093

Picariello H S, Kenchappa R S, Rai V, Crish J F, Dovas A, Pogoda K *et al* 2019 Myosin IIA suppresses glioblastoma development in a mechanically sensitive manner *Proc. Natl Acad. Sci. USA* **116** 15550–9

Pickup M W, Mouw J K and Weaver V M 2014 The extracellular matrix modulates the hallmarks of cancer *EMBO Rep.* **15** 1243–53

Piotrowski-Daspit A S, Tien J and Nelson C M 2016 Interstitial fluid pressure regulates collective invasion in engineered human breast tumors via Snail, vimentin, and E-cadherin *Integr. Biol.* **8** 319–31

Pittman M, Iu E, Li K, Wang M, Chen J, Taneja N *et al* 2022 Membrane ruffling is a mechanosensor of extracellular fluid viscosity *Nat. Phys.* **18** 1112–21

Plodinec M, Loparic M, Monnier C A, Obermann E C, Zanetti-Dallenbach R, Oertle P *et al* 2012 The nanomechanical signature of breast cancer *Nat. Nanotechnol.* **7** 757–65

Podhorecka M, Ibanez B and Dmoszyńska A 2017 Metformin—its potential anti-cancer and anti-aging effects *Postep. Hig.Med. Dosw.* **71** 170–5

Pogoda K, Pięta E, Roman M, Piergies N, Liberda D, Wróbel T P *et al* 2021 In search of the correlation between nanomechanical and biomolecular properties of prostate cancer cells with different metastatic potential *Arch. Biochem. Biophys.* **697** 108718

Pollmann M-A, Shao Q, Laird D W and Sandig M 2005 Connexin 43 mediated gap junctional communication enhances breast tumor cell diapedesis in culture *Breast Cancer Res.* **7** R522

Pontes B, Ayala Y, Fonseca A C C, Romão L F, Amaral R F, Salgado L T *et al* 2013 Membrane elastic properties and cell function *PLoS One* **8** e67708

Provenzano P P, Eliceiri K W, Campbell J M, Inman D R, White J G and Keely P J 2006 Collagen reorganization at the tumor-stromal interface facilitates local invasion *BMC Med.* **4** 38

Provenzano P P, Inman D R, Eliceiri K W, Beggs H E and Keely P J 2008 Mammary epithelial-specific disruption of focal adhesion kinase retards tumor formation and metastasis in a transgenic mouse model of human breast cancer *Am. J. Pathol.* **173** 1551–65

Provenzano P P and Keely P J 2011 Mechanical signaling through the cytoskeleton regulates cell proliferation by coordinated focal adhesion and Rho GTPase signaling *J. Cell Sci.* **124** 1195–205

Radmacher M, Fritz M, Kacher C M, Cleveland J P and Hansma P K 1996 Measuring the viscoelastic properties of human platelets with the atomic force microscope *Biophys. J.* **70** 556–67

Ramachandran R and Schmid S L 2018 The dynamin superfamily *Curr. Biol.* **28** R411–6

Raudenska M, Kratochvilova M, Vicar T, Gumulec J, Balvan J, Polanska H *et al* 2019 Cisplatin enhances cell stiffness and decreases invasiveness rate in prostate cancer cells by actin accumulation *Sci. Rep.* **9** 1660

Rebelo L M, de Sousa J S, Mendes Filho J and Radmacher M 2013 Comparison of the viscoelastic properties of cells from different kidney cancer phenotypes measured with atomic force microscopy *Nanotechnology* **24** 055102

Regmi S, Fu A and Luo K Q 2017 High shear stresses under exercise condition destroy circulating tumor cells in a microfluidic system *Sci. Rep.* **7** 39975

Remmerbach T W, Wottawah F, Dietrich J, Lincoln B, Wittekind C and Guck J 2009 Oral cancer diagnosis by mechanical phenotyping *Cancer Res.* **69** 1728–32

Ren Y, West-Foyle H, Surcel A, Miller C and Robinson D N 2014 Genetic suppression of a phosphomimic myosin II identifies system-level factors that promote myosin II cleavage furrow accumulation *Mol. Biol. Cell* **25** 4150–65

Rheinlaender J, Dimitracopoulos A, Wallmeyer B, Kronenberg N M, Chalut K J, Gather M C *et al* 2020 Cortical cell stiffness is independent of substrate mechanics *Nat. Mater.* **19** 1019–25

Rianna C and Radmacher M 2017a Comparison of viscoelastic properties of cancer and normal thyroid cells on different stiffness substrates *Eur. Biophys. J.* **46** 309–24

Rianna C and Radmacher M 2017b Influence of microenvironment topography and stiffness on the mechanics and motility of normal and cancer renal cells *Nanoscale* **9** 11222–30

Rianna C, Radmacher M and Kumar S 2020 Direct evidence that tumor cells soften when navigating confined spaces *Mol. Biol. Cell* **31** 1726–34

Rosendahl P, Plak K, Jacobi A, Kraeter M, Toepfner N, Otto O *et al* 2018 Real-time fluorescence and deformability cytometry *Nat. Methods* **15** 355–8

Rother J, Nöding H, Mey I and Janshoff A 2014 Atomic force microscopy-based microrheology reveals significant differences in the viscoelastic response between malign and benign cell lines *Open Biol* **4** 140046

Rotsch C, Jacobson K and Radmacher M 1999 Dimensional and mechanical dynamics of active and stable edges in motile fibroblasts investigated by using atomic force microscopy *Proc. Natl Acad. Sci.* **96** 921–6

Routhier A, Astuccio M, Lahey D, Monfredo N, Johnson A, Callahan W *et al* 2010 Pharmacological inhibition of Rho-kinase signaling with *Y*-27632 blocks melanoma tumor growth *Oncol. Rep.* **23** 861–7

Safaei S, Sajed R, Shariftabrizi A, Dorafshan S, Saeednejad Zanjani L, Dehghan Manshadi M *et al* 2023 Tumor matrix stiffness provides fertile soil for cancer stem cells *Cancer Cell Int.* **23** 143

Sancho A, Vandersmissen I, Craps S, Luttun A and Groll J 2017 A new strategy to measure intercellular adhesion forces in mature cell–cell contacts *Sci. Rep.* **7** 46152

Schiffhauer E S, Luo T, Mohan K, Srivastava V, Qian X, Griffis E R *et al* 2016 Mechanoaccumulative elements of the mammalian actin cytoskeleton *Curr. Biol.* **26** 1473–9

Scholz N 2018 Cancer cell mechanics: adhesion G protein-coupled receptors in action? *Front. Oncol.* **8** 59

Schramek D, Sendoel A, Segal J P, Beronja S, Heller E, Oristian D *et al* 2014 Direct *in vivo* RNAi screen unveils myosin IIa as a tumor suppressor of squamous cell carcinomas *Science* **343** 309–13

Seltmann K, Fritsch A W, Käs J A and Magin T M 2013 Keratins significantly contribute to cell stiffness and impact invasive behavior *Proc. Natl Acad. Sci. USA* **110** 18507–12

Sen S and Kumar S 2009 Cell–matrix de-adhesion dynamics reflect contractile mechanics *Cel. Mol. Bioeng.* **2** 218–30

Sherr C J and DePinho R A 2000 Cellular senescence *Cell* **102** 407–10

Shin D, Lee W, Lee J H and Bang D 2019 Multiplexed single-cell RNA-seq via transient barcoding for simultaneous expression profiling of various drug perturbations *Sci. Adv.* **5** eaav2249

Shintaku H, Hakamada K, Fujimoto H, Nagata T, Miyake J and Kawano S 2014 Measurement of local electric field in microdevices for low-voltage electroporation of adherent cells *Microsyst. Technol.* **20** 303–13

Shiomi A, Kaneko T, Nishikawa K, Tsuchida A, Isoshima T, Sato M *et al* 2024 High-throughput mechanical phenotyping and transcriptomics of single cells *Nat. Commun.* **15** 3812

Shiomi A, Nagao K, Kasai H, Hara Y and Umeda M 2020 Changes in the physicochemical properties of fish cell membranes during cellular senescence *Biosci. Biotechnol. Biochem.* **84** 583–93

Shiomi A, Nagao K, Yokota N, Tsuchiya M, Kato U, Juni N *et al* 2021 Extreme deformability of insect cell membranes is governed by phospholipid scrambling *Cell Rep.* **35** 109219

Shivashankar G V 2011 Mechanosignaling to the cell nucleus and gene regulation *Annu. Rev. Biophys.* **40** 361–78

Singh A and Settleman J 2009 Oncogenic K-ras 'addiction' and synthetic lethality *Cell Cycle* **8** 2676–8

Sinha B, Köster D, Ruez R, Gonnord P, Bastiani M, Abankwa D *et al* 2011 Cells respond to mechanical stress by rapid disassembly of caveolae *Cell* **144** 402–13

Sleep J, Wilson D, Simmons R and Gratzer W 1999 Elasticity of the red cell membrane and its relation to hemolytic disorders: an optical Tweezers study *Biophys. J.* **77** 3085–95

Smith A S, Nowak R B, Zhou S, Giannetto M, Gokhin D S, Papoin J *et al* 2018 Myosin IIA interacts with the spectrin-actin membrane skeleton to control red blood cell membrane curvature and deformability *Proc. Natl Acad. Sci. USA* **115** E4377–85

Solon J, Levental I, Sengupta K, Georges P C and Janmey P A 2007 Fibroblast adaptation and stiffness matching to soft elastic substrates *Biophys. J.* **93** 4453–61

Song Z, Wei B, Lu C, Huang X, Li P and Chen L 2017 Metformin suppresses the expression of Sonic hedgehog in gastric cancer cells *Mol. Med. Rep.* **15** 1909–15

Soteriou D, Kubánková M, Schweitzer C, López-Posadas R, Pradhan R, Thoma O-M *et al* 2023 Rapid single-cell physical phenotyping of mechanically dissociated tissue biopsies *Nat. Biomed. Eng.* **7** 1392–403

Stacker S A, Williams S P, Karnezis T, Shayan R, Fox S B and Achen M G 2014 Lymphangiogenesis and lymphatic vessel remodelling in cancer *Nat. Rev. Cancer* **14** 159–72

Starr D A and Fridolfsson H N 2010 Interactions between nuclei and the cytoskeleton are mediated by SUN-KASH nuclear-envelope bridges *Annu. Rev. Cell Dev. Biol.* **26** 421–44

Steeg P S 2016 Targeting metastasis *Nat. Rev. Cancer* **16** 201–18

Stevenson R P, Veltman D and Machesky L M 2012 Actin-bundling proteins in cancer progression at a glance *J. Cell Sci.* **125** 1073–9

Stoeckius M, Hafemeister C, Stephenson W, Houck-Loomis B, Chattopadhyay P K, Swerdlow H *et al* 2017 Simultaneous epitope and transcriptome measurement in single cells *Nat. Methods* **14** 865–8

Stokke B T, Mikkelsen A and Elgsaeter A 1986 Spectrin, human erythrocyte shapes, and mechanochemical properties *Biophys. J.* **49** 319–27

Stossel T P, Condeelis J, Cooley L, Hartwig J H, Noegel A, Schleicher M *et al* 2001 Filamins as integrators of cell mechanics and signalling *Nat. Rev. Mol. Cell Biol.* **2** 138–45

Strell C and Entschladen F 2008 Extravasation of leukocytes in comparison to tumor cells *Cell Commun. Signal.* **6** 10

Strilic B and Offermanns S 2017 Intravascular survival and extravasation of tumor cells *Cancer Cell* **32** 282–93

Stuelten C H, Parent C A and Montell D J 2018 Cell motility in cancer invasion and metastasis: insights from simple model organisms *Nat. Rev. Cancer* **18** 296–312

Stylianopoulos T, Martin J D, Chauhan V P, Jain S R, Diop-Frimpong B, Bardeesy N *et al* 2012 Causes, consequences, and remedies for growth-induced solid stress in murine and human tumors *Proc. Natl Acad. Sci. USA* **109** 15101–8

Suissa S and Azoulay L 2014 Metformin and cancer: mounting evidence against an association *Diabetes Care* **37** 1786–8

Sullivan W J, Mullen P J, Schmid E W, Flores A, Momcilovic M, Sharpley M S *et al* 2018 Extracellular matrix remodeling regulates glucose metabolism through TXNIP destabilization *Cell* **175** 117–132.e21

Sun S, Wong S, Mak A and Cho M 2014 Impact of oxidative stress on cellular biomechanics and rho signaling in C2C12 myoblasts *J. Biomech.* **47** 3650–6

Sunyer R, Trepat X, Fredberg J J, Farré R and Navajas D 2009 The temperature dependence of cell mechanics measured by atomic force microscopy *Phys. Biol.* **6** 025009

Surcel A, Schiffhauer E S, Thomas D G, Zhu Q, DiNapoli K T, Herbig M *et al* 2019 Targeting mechanoresponsive proteins in pancreatic cancer: 4hHydroxyacetophenone blocks dissemination and invasion by activating MYH14 *Cancer Res.* **79** 4665–78

Suresh S 2007 Biomechanics and biophysics of cancer cells *Acta Biomater.* **3** 413–38

Suresh S, Spatz J, Mills J P, Micoulet A, Dao M, Lim C T *et al* 2005 Connections between single-cell biomechanics and human disease states: gastrointestinal cancer and malaria *Acta Biomater.* **1** 15–30

Suzuki M, Shigematsu H, Shames D S, Sunaga N, Takahashi T, Shivapurkar N *et al* 2007 Methylation and gene silencing of the ras-related GTPase gene in lung and breast cancers *Ann. Surg. Oncol.* **14** 1397–404

Swaminathan V, Mythreye K, O'Brien E T, Berchuck A, Blobe G C and Superfine R 2011 Mechanical stiffness grades metastatic potential in patient tumor cells and in cancer cell lines *Cancer Res.* **71** 5075–80

Swift J and Discher D E 2014 The nuclear lamina is mechano-responsive to ECM elasticity in mature tissue *J. Cell Sci.* **127** 3005–15

Tang K, Xin Y, Li K, Chen X and Tan Y 2021 Cell cytoskeleton and stiffness are mechanical indicators of organotropism in breast cancer *Biology* **10** 259

Taubenberger A V, Girardo S, Träber N, Fischer-Friedrich E, Kräter M, Wagner K *et al* 2019 3D microenvironment stiffness regulates tumor spheroid growth and mechanics via p21 and ROCK *Adv. Biosys.* **3** 1900128

Tavares S, Vieira A F, Taubenberger A V, Araújo M, Martins N P, Brás-Pereira C *et al* 2017 Actin stress fiber organization promotes cell stiffening and proliferation of pre-invasive breast cancer cells *Nat. Commun.* **8** 15237

Physical Sciences - Oncology Centers Network T 2013 A physical sciences network characterization of non-tumorigenic and metastatic cells *Sci. Rep.* **3** 1449

Thiam H-R, Vargas P, Carpi N, Crespo C L, Raab M, Terriac E *et al* 2016 Perinuclear Arp2/3-driven actin polymerization enables nuclear deformation to facilitate cell migration through complex environments *Nat. Commun.* **7** 10997

Thomas G, Burnham N A, Camesano T A and Wen Q 2013 Measuring the mechanical properties of living cells using atomic force microscopy *JoVE* **76** e50497

Thomas J A 2015 Optical imaging probes for biomolecules: an introductory perspective *Chem. Soc. Rev.* **44** 4494–500

Thoumine O and Ott A 1997 Time scale dependent viscoelastic and contractile regimes in fibroblasts probed by microplate manipulation *J. Cell Sci.* **110** 2109–16

Tian M, Yang X, Li Y and Guo S 2020 The expression of dynamin 1, 2, and 3 in human hepatocellular carcinoma and patient prognosis *Med. Sci. Monit.* **26** e923359

Tietze S, Kräter M, Jacobi A, Taubenberger A, Herbig M, Wehner R *et al* 2019 Spheroid culture of mesenchymal stromal cells results in morphorheological properties appropriate for improved microcirculation *Adv. Sci.* **6** 1802104

Tinevez J-Y, Schulze U, Salbreux G, Roensch J, Joanny J-F and Paluch E 2009 Role of cortical tension in bleb growth *Proc. Natl Acad. Sci. USA* **106** 18581–6

Tishchenko A, Azorín D D, Vidal-Brime L, Muñoz M J, Arenas P J, Pearce C *et al* 2020 Cx43 and associated cell signaling pathways regulate tunneling nanotubes in breast cancer cells *Cancers* **12** 2798

Titushkin I and Cho M 2006 Distinct membrane mechanical properties of human mesenchymal stem cells determined using laser optical tweezers *Biophys. J.* **90** 2582–91

Toepfner N, Herold C, Otto O, Rosendahl P, Jacobi A, Kräter M *et al* 2018 Detection of human disease conditions by single-cell morpho-rheological phenotyping of blood *eLife* **7** e29213

Tse H T K, Gossett D R, Moon Y S, Masaeli M, Sohsman M, Ying Y *et al* 2013 Quantitative diagnosis of malignant pleural effusions by single-cell mechanophenotyping *Sci. Transl. Med.* **5** 212ra163

Tse J R and Engler A J 2010 Preparation of hydrogel substrates with tunable mechanical properties *Curr. Protoc. Cell Biol.* **47** https://doi.org/10.1002/0471143030.cb1016s47

Tsujita K, Satow R, Asada S, Nakamura Y, Arnes L, Sako K *et al* 2021 Homeostatic membrane tension constrains cancer cell dissemination by counteracting BAR protein assembly *Nat. Commun.* **12** 5930

Tymchenko N, Nilebäck E, Voinova M V, Gold J, Kasemo B and Svedhem S 2012 Reversible changes in cell morphology due to cytoskeletal rearrangements measured in real-time by QCM-D *Biointerphases* **7** 43

Udono M, Fujii K, Harada G, Tsuzuki Y, Kadooka K, Zhang P *et al* 2015 Impaired ATP6V0A2 expression contributes to Golgi dispersion and glycosylation changes in senescent cells *Sci. Rep.* **5** 17342

Urbanska M, Muñoz H E, Shaw Bagnall J, Otto O, Manalis S R, Di Carlo D *et al* 2020 A comparison of microfluidic methods for high-throughput cell deformability measurements *Nat. Methods* **17** 587–93

Van Vliet K J, Bao G and Suresh S 2003 The biomechanics toolbox: experimental approaches for living cells and biomolecules *Acta Mater.* **51** 5881–905

Varol R, Karavelioglu Z, Omeroglu S, Aydemir G, Karadag A, Meco H E *et al* 2022 Acousto-holographic reconstruction of whole-cell stiffness maps *Nat. Commun.* **13** 7351

Vassilopoulos A, Xiao C, Chisholm C, Chen W, Xu X, Lahusen T J *et al* 2014 Synergistic therapeutic effect of cisplatin and phosphatidylinositol 3-kinase (PI3K) inhibitors in cancer growth and metastasis of Brca1 mutant tumors *J. Biol. Chem.* **289** 24202–14

Venkatesh S K, Yin M, Glockner J F, Takahashi N, Araoz P A, Talwalkar J A *et al* 2008 MR elastography of liver tumors: preliminary results *Am. J. Roentgenol.* **190** 1534–40

Vennin C, Chin V T, Warren S C, Lucas M C, Herrmann D, Magenau A *et al* 2017 Transient tissue priming via ROCK inhibition uncouples pancreatic cancer progression, sensitivity to chemotherapy, and metastasis *Sci. Transl. Med.* **9** eaai8504

Vicente-Dueñas C, Hauer J, Cobaleda C, Borkhardt A and Sánchez-García I 2018 Epigenetic priming in cancer initiation *Trends Cancer* **4** 408–17

Vicente-Manzanares M, Ma X, Adelstein R S and Horwitz A R 2009 Non-muscle myosin II takes centre stage in cell adhesion and migration *Nat. Rev. Mol. Cell Biol.* **10** 778–90

Vichare S, Sen S and Inamdar M M 2014 Cellular mechanoadaptation to substrate mechanical properties: contributions of substrate stiffness and thickness to cell stiffness measurements using AFM *Soft Matter* **10** 1174

Von Beek C, Alriksson L, Palle J, Gustafson A-M, Grujic M, Melo F R *et al* 2021 Dynamin inhibition causes context-dependent cell death of leukemia and lymphoma cells *PLoS One* **16** e0256708

Wakatsuki T, Schwab B, Thompson N C and Elson E L 2001 Effects of cytochalasin D and latrunculin B on mechanical properties of cells *J. Cell Sci.* **114** 1025–36

Wang K and Singer S J 1977 Interaction of filamin with f-actin in solution *Proc. Natl Acad. Sci. USA* **74** 2021–5

Wang N, Butler J P and Ingber D E 1993 Mechanotransduction across the cell surface and through the cytoskeleton *Science* **260** 1124–7

Wang N, Tolić-Nørrelykke I M, Chen J, Mijailovich S M, Butler J P, Fredberg J J *et al* 2002 Cell prestress. I. Stiffness and prestress are closely associated in adherent contractile cells *Am. J. Physiol. Cell Physiol.* **282** C606–16

Wang Y, Liu S, Zhang Y and Yang J 2019 Myosin heavy chain 9: oncogene or tumor suppressor gene? *Med. Sci. Monit.* **25** 888–92

Wang Z, Zhu Z, Li C, Zhang Y, Li Z and Sun S 2021 NMIIA promotes tumorigenesis and prevents chemosensitivity in colorectal cancer by activating AMPK/mTOR pathway *Exp. Cell. Res.* **398** 112387

Ward K A, Li W-I, Zimmer S and Davis T 1991 Viscoelastic properties of transformed cells: role in tumor cell progression and metastasis formation *BIR* **28** 301–13

Watnick R S 2012 The role of the tumor microenvironment in regulating angiogenesis *Cold Spring Harb. Perspect.Med.* **2** a006676–a6

Watson A W, Grant A D, Parker S S, Hill S, Whalen M B, Chakrabarti J *et al* 2021 Breast tumor stiffness instructs bone metastasis via maintenance of mechanical conditioning *Cell Rep.* **35** 109293

Weber G F, Bjerke M A and DeSimone D W 2011 Integrins and cadherins join forces to form adhesive networks *J. Cell Sci.* **124** 1183–93

Wei S and Siegal G P 2017 Metastatic organotropism: an intrinsic property of breast cancer molecular subtypes *Adv. Anat. Pathol.* **24** 78–81

Wei Z, Guo H, Qin J, Lu S, Liu Q, Zhang X *et al* 2019 Pan-senescence transcriptome analysis identified RRAD as a marker and negative regulator of cellular senescence *Free Radic. Biol. Med.* **130** 267–77

Weiss L 1990 Metastatic inefficiency *Advances in Cancer Research* (Amsterdam: Elsevier) pp 159–211
Werb Z and Lu P 2015 The role of stroma in tumor development *Cancer J.* **21** 250–3
Wheaton W W, Weinberg S E, Hamanaka R B, Soberanes S, Sullivan L B, Anso E *et al* 2014 Metformin inhibits mitochondrial complex I of cancer cells to reduce tumorigenesis *eLife* **3** e02242
Whitburn J, Edwards C M and Sooriakumaran P 2017 Metformin and prostate cancer: a new role for an old drug *Curr. Urol. Rep.* **18** 46
Whiteside T L 2006 The role of immune cells in the tumor microenvironment *The Link Between Inflammation and Cancer* ed A G Dalgleish and B Haefner (Boston, MA: Springer) pp 103–24
Wirtz D, Konstantopoulos K and Searson P C 2011 The physics of cancer: the role of physical interactions and mechanical forces in metastasis *Nat. Rev. Cancer* **11** 512–22
Wolf K, te Lindert M, Krause M, Alexander S, te Riet J, Willis A L *et al* 2013 Physical limits of cell migration: control by ECM space and nuclear deformation and tuning by proteolysis and traction force *J. Cell Biol.* **201** 1069–84
Wottawah F, Schinkinger S, Lincoln B, Ananthakrishnan R, Romeyke M, Guck J *et al* 2005 Optical rheology of biological cells *Phys. Rev. Lett.* **94** 098103
Wu K, Fukuda K, Xing F, Zhang Y, Sharma S, Liu Y *et al* 2015 Roles of the cyclooxygenase 2 matrix metalloproteinase 1 pathway in brain metastasis of breast cancer *J. Biol. Chem.* **290** 9842–54
Wu P-H, Aroush D R-B, Asnacios A, Chen W-C, Dokukin M E, Doss B L *et al* 2018 A comparison of methods to assess cell mechanical properties *Nat. Methods* **15** 491–8
Wu P-H, Gambhir S S, Hale C M, Chen W-C, Wirtz D and Smith B R 2020 Particle tracking microrheology of cancer cells in living subjects *Mater. Today* **39** 98–109
Wullkopf L, West A-K V, Leijnse N, Cox T R, Madsen C D, Oddershede L B *et al* 2018 Cancer cells' ability to mechanically adjust to extracellular matrix stiffness correlates with their invasive potential *Mol. Biol. Cell* **29** 2378–85
Xia Y, Cai X-Y, Fan J-Q, Zhang L-L, Ren J-H, Chen J *et al* 2015 Rho kinase inhibitor fasudil suppresses the vasculogenic mimicry of B16 mouse melanoma cells both *in vitro* and *in vivo* *Mol. Cancer Ther.* **14** 1582–90
Xie K, Yang Y and Jiang H 2018 Controlling cellular volume via mechanical and physical properties of substrate *Biophys. J.* **114** 675–87
Xin Y, Chen X, Tang X, Li K, Yang M, Tai W C-S *et al* 2019 Mechanics and actomyosin-dependent survival/chemoresistance of suspended tumor cells in shear flow *Biophys. J.* **116** 1803–14
Xu W, Mezencev R, Kim B, Wang L, McDonald J and Sulchek T 2012 Cell stiffness is a biomarker of the metastatic potential of vvarian cancer cells *PLoS One* **7** e46609
Yamaguchi H and Condeelis J 2007 Regulation of the actin cytoskeleton in cancer cell migration and invasion *Biochim. Biophys.Acta (BBA)—Mol. Cell Res.* **1773** 642–52
Yates L R, Knappskog S, Wedge D, Farmery J H R, Gonzalez S, Martincorena I *et al* 2017 Genomic evolution of breast cancer metastasis and relapse *Cancer Cell* **32** 169–84.e7
Yeşilkaya F 2019 Examination of the expression levels of MACC1, filamin A and FBXW7 genes in colorectal cancer patients *North Clin. Istanbul* **7** 1–5
Yu H W, Chen Y, Huang C, Liu C, Chiou A, Wang Y *et al* 2015 β- PIX controls intracellular viscoelasticity to regulate lung cancer cell migration *J. Cell. Mol. Med.* **19** 934–47
Yu Q, Zhang B, Zhang Y-M, Liu Y-H and Liu Y 2020 Actin cytoskeleton-disrupting and magnetic field-responsive multivalent supramolecular assemblies for efficient cancer therapy *ACS Appl. Mater. Interfaces* **12** 13709–17

Zahalak G I, McConnaughey W B and Elson E L 1990 Determination of cellular mechanical properties by cell poking, with an application to leukocytes *J. Biomech. Eng.* **112** 283–94

Zanotelli M R, Bordeleau F and Reinhart-King C A 2017 Subcellular regulation of cancer cell mechanics *Curr. Opin.Biomed. Eng.* **1** 8–14

Zeng J, Zhang Y, Xu R, Chen H, Tang X, Zhang S *et al* 2023 Nanomechanical-based classification of prostate tumor using atomic force microscopy *Prostate* **83** 1591–601

Zhang C, Liu J, Wu R, Liang Y, Lin M, Liu J *et al* 2014 Tumor suppressor p53 negatively regulates glycolysis stimulated by hypoxia through its target RRAD *Oncotarget* **5** 5535–46

Zhang C, Zhang X, Huang L, Guan Y, Huang X, Tian X *et al* 2021 ATF3 drives senescence by reconstructing accessible chromatin profiles *Aging Cell* **20** e13315

Zhang G, Long M, Wu Z-Z and Yu W-Q 2002 Mechanical properties of hepatocellular carcinoma cells *WJG* **8** 243

Zhang R, Lee D M, Jimah J R, Gerassimov N, Yang C, Kim S *et al* 2020 Dynamin regulates the dynamics and mechanical strength of the actin cytoskeleton as a multifilament actin-bundling protein *Nat. Cell Biol.* **22** 674–88

Zhang W, Kai K, Choi D S, Iwamoto T, Nguyen Y H, Wong H *et al* 2012 Microfluidics separation reveals the stem-cell–like deformability of tumor-initiating cells *Proc. Natl Acad. Sci. USA* **109** 18707–12

Zhang Y, Li Y, Thompson K N, Stoletov K, Yuan Q, Bera K *et al* 2022 Polarized NHE1 and SWELL1 regulate migration direction, efficiency and metastasis *Nat. Commun.* **13** 6128

Zhang Y, Wolf G W, Bhat K, Jin A, Allio T, Burkhart W A *et al* 2003 Ribosomal protein L11 negatively regulates oncoprotein MDM2 and mediates a p53-dependent ribosomal-stress checkpoint pathway *Mol. Cell. Biol.* **23** 8902–12

Zhou M, Zheng M, Zhou X, Tian S, Yang X, Ning Y *et al* 2023 The roles of connexins and gap junctions in the progression of cancer *Cell Commun. Signal.* **21** 8

Zhu X, Suo Y, Fu Y, Zhang F, Ding N, Pang K *et al* 2021 *In vivo* flow cytometry reveals a circadian rhythm of circulating tumor cells *Light: Sci. Appl.* **10** 110

Zhuang X, Zhang H, Li X, Li X, Cong M, Peng F *et al* 2017 Differential effects on lung and bone metastasis of breast cancer by WNT signalling inhibitor DKK1 *Nat. Cell Biol.* **19** 1274–85

Zibara K, Awada Z, Dib L, El-Saghir J, Al-Ghadban S, Ibrik A *et al* 2015 Anti-angiogenesis therapy and gap junction inhibition reduce MDA-MB-231 breast cancer cell invasion and metastasis *in vitro* and *in vivo Sci. Rep.* **5** 12598

www.ingramcontent.com/pod-product-compliance
Ingram Content Group UK Ltd.
Pitfield, Milton Keynes, MK11 3LW, UK
UKHW051911100926
8556IPUK00005B/844

9 780750 340052